Editorial Board Members' Collection Series: Photonics Sensors

Editorial Board Members' Collection Series: Photonics Sensors

Guest Editors

Carlos Marques
Flavio Esposito

MDPI

Basel • Beijing • Wuhan • Barcelona • Belgrade • Novi Sad • Cluj • Manchester

Guest Editors
Carlos Marques
CICECO-Aveiro Institute of Materials
Physics Department
University of Aveiro
Aveiro
Portugal

Flavio Esposito
Department of Engineering
University of Naples "Parthenope"
Napoli
Italy

Editorial Office
MDPI AG
Grosspeteranlage 5
4052 Basel, Switzerland

This is a reprint of the Special Issue, published open access by the journal *Photonics* (ISSN 2304-6732), freely accessible at: https://www.mdpi.com/journal/photonics/special_issues/968DX56E4G.

For citation purposes, cite each article independently as indicated on the article page online and as indicated below:

Lastname, A.A.; Lastname, B.B. Article Title. *Journal Name* **Year**, *Volume Number*, Page Range.

ISBN 978-3-7258-4105-9 (Hbk)
ISBN 978-3-7258-4106-6 (PDF)
https://doi.org/10.3390/books978-3-7258-4106-6

Contents

About the Editors

Carlos Marques

Carlos Marques received a PhD degree in Physics Engineering from the UA, Portugal, in 2013. He was a Marie Curie (MC) Fellow at the Aston Institute of Photonics Technologies, Aston University, UK, until March 2016. After that, he was a research scientist (Assistant and Principal Researcher (CEEC)) in the Physics Department & I3N, University of Aveiro, Aveiro, Portugal. Currently, he is a Professor in the Physics Department, University of Aveiro, Portugal, and a full member of CICECO—Aveiro Institute of Materials, University of Aveiro, Portugal. Also, he is a researcher on the REFRESH project at the VSB Technical University of Ostrava within the Department of Physics. His work is focused on sensors and devices. His main interests include innovative optical devices and structures for sensing, integrated optics, low-cost interrogation systems, eHealth, aviation, data acquisition, predictive analytics, and optical transmission data. He has authored or co-authored four books, seven book chapters, 250+ peer-reviewed journal papers, and more than 100 conference papers with an emphasis on optical communications, data processing, and sensing. According to Google Scholar, his journal/conference articles have received 11000+ citations (h-index: 72). He was a member of the IMOC2019 conference organization team, he is a member of the organization team for AOP2024, and he will be the chair of EWOFS2027 in Aveiro. He has received several awards. Also, he is the Editor in Chief of the IEEE's *Sensors Reviews*, and he is the Associate Editor of prestigious scientific journals like the OSA's *Optics Express*, *Opto-Electronic Advances*, and *Optical Fiber Technology* (Elsevier) and also acts as a referee for several high-level international journals, receiving awards for his review quality.

Flavio Esposito

Flavio Esposito received an MSc in Electronic Engineering from the University of Naples "Federico II" and a PhD in Information Engineering from the University of Naples "Parthenope", Naples, Italy. He is currently an Assistant Professor (tenure track) of Electronics in the Department of Engineering of the University of Naples "Parthenope". His research activity is focused on the design, simulation, fabrication, and characterization of novel fiber optic devices. His work currently involves the development and application of sensors for bio-chemical measurements, radiation-exposed environments, environmental monitoring, and structural health monitoring. He has received outstanding reviewer awards from *Sensors* (2019), IEEE *Sensors Journal* (2020), and IEEE *Photonics Technology Letters* (2023). He is an Editorial Board Member of *Sensors*, *Photonics*, and *Biosensors*; he serves as an Associate Editor for IEEE *Sensors Letters* and IEEE *Sensors Reviews*.

photonics

MDPI

Article

Silica-Titania Integrated Photonics Platform-Based 1 × 2 Demultiplexer Utilizing Two Serially Cascaded Racetrack Microrings for 1310 nm and 1550 nm Telecommunication Wavelengths

Muhammad A. Butt *, Muhammad Shahbaz, Łukasz Kozłowski, Andrzej Kaźmierczak and Ryszard Piramidowicz

Institute of Microelectronics and Optoelectronics, Warsaw University of Technology, ul. Koszykowa 75, 00-662 Warszawa, Poland

* Correspondence: ali.butt@pw.edu.pl

Abstract: Herein, a numerical analysis of a 1 × 2 demultiplexer based on a silica-titania integrated photonics platform is conducted via the finite element method. The structure is composed of two coupled racetrack microrings (RTMRs) and a subwavelength grating (SWG) structure for the demultiplexing of 1310 nm and 1550 nm telecommunication wavelengths. The material platform selected for this design is highly attractive due to its refined optical, physical, and chemical properties. Moreover, silica-titania sol-gel thin-films can be deposited on glass substrates with the dip-coating method. The proposed device has a small footprint of 84 × 125 μm^2 and offers crosstalk as low as ~−6.6 dB and ~−9.04 dB for 1550 nm and 1310 nm, respectively. We are convinced that this study promotes the use of the silica-titania platform for the development of low-cost on-chip optical communication devices for signal multiplexing and demultiplexing.

Keywords: demultiplexer; silica-titania platform; racetrack microrings; subwavelength grating structure; telecommunication wavelength; wavelength division multiplexing

Citation: Butt, M.A.; Shahbaz, M.; Kozłowski, Ł.; Kaźmierczak, A.; Piramidowicz, R. Silica-Titania Integrated Photonics Platform-Based 1 × 2 Demultiplexer Utilizing Two Serially Cascaded Racetrack Microrings for 1310 nm and 1550 nm Telecommunication Wavelengths. *Photonics* **2023**, *10*, 208. https://doi.org/10.3390/photonics10020208

Received: 1 February 2023
Revised: 10 February 2023
Accepted: 13 February 2023
Published: 14 February 2023

1. Introduction

In optical communication networks, the concept of using light waves to convey data and information packets provides great benefits including high speed, high bandwidth, electro-magnetic immunity, etc. [1]. The optical fibers act as the transmission channel for optical waves in optical communication networks [2]. The effective capacity of optical fibers has subsequently been maximized using wavelength division multiplexing (WDM) technology [3,4] such that several users can share a single (physical) optical fiber and various (data) optical channels can be sent over the same single optical fiber. To distinguish these channels from one another and send them to their appropriate users, a proper photonic device is required at the user end of the network. Mach-Zehnder interferometers (MZI) [5,6], multimode interference (MMI) couplers [7–9], Y-branch devices [10,11], ring resonators [12–15], and other solutions can all be used to create optical demultiplexers based on different optical platforms, which are a crucial components of communication networks [16–19]. Moreover, small size and low crosstalk demultiplexer devices may have advantages in quantum information processing and quantum communication [20–22].

Recently, several novel demultiplexer system designs have been proposed on semiconductor platforms. For instance, a novel design of an 8-channel MMI demultiplexer utilizing slot waveguide (WG) structures that operate at 1530 nm, 1535 nm, 1540 nm, 1545 nm, 1550 nm, 1555 nm, 1560 nm, and 1565 nm was proposed [23]. The device is composed of Gallium nitride (GaN) with silicon (Si) surrounding it and constructed using seven 1 × 2 MMI couplers, 14 S-bands, and 1 input taper. The geometrical parameters were analyzed using a full vectorial-beam propagation method (FV-BPM), and simulation results show that the device has low crosstalk (CT) (−19.97–−13.77 dB) and bandwidth (1.8–3.6 nm) and can transmit 8-channels across the entire C-band (1530–1565 nm). Another work on MMI suggested a design for an optical demultiplexer utilizing slot-WG technology that can operate at wavelengths of 547 nm, 559 nm, 566 nm, and 584 nm [24]. The use of GaN and silicon oxide in the slot-WG structure was found to be beneficial. Simulation results indicate the device has low transmission loss (0.983–1.423 dB), minimal CT (13.8–18.3 dB), and wide bandwidth (1.8–3.2 nm), making it suitable for use in WDM visible light networking systems.

A 5-channel DWDM demultiplexer design using photonic crystal (PhC) hexagonal ring resonators (RRs) of airholes on a silicon slab as channel drop filters is presented [25], along with an analysis of its optimal design parameters and device performance. Wavelength control is achieved through fine-tuning coupling holes outside each ring resonator, resulting in resolved wavelengths in the range of 1535 to 1539 nm, suitable for use with erbium doped fiber amplifiers. The average coupling efficiency and CT were found to be 72.4% and −18.31 dB, respectively. In [26], a five-channel demultiplexer utilizing a simple ring resonator is proposed for the third communication window 1550–1567 nm range. The resonant wavelengths in the output ports were adjusted by altering the dielectric constant of the ring core. According to the study, this demultiplexer structure is suitable for use in WDM optical communication systems. The CT between channels was calculated, with a maximum value of −13 dB and a minimum of −24 dB. The average transmission efficiency is above 92%, with channel spacing of less than 4.2 nm and an average bandwidth of less than 1.51 nm. The quality factor for the channels ranges from 1034 to 1044. The overall size of the demultiplexer is roughly 689.61 μm^2, which according to the authors makes it practical for use in optoelectronic and photonic integrated circuits.

An ultra-compact wavelength demultiplexer operating at O-band (~1310 nm) and C-band (~1550 nm) is proposed using a subwavelength grating (SWG) [27]. The proposed scheme utilizes the SOI platform, utilizing silicon WGs (with a refractive index of approximately 3.478) on a SiO_2 substrate (with a refractive index of approximately 1.455) with a thickness of 2 μm. The device is fully embedded in a silicon oxide cladding. The device also has a wide 1-dB bandwidth of 140 nm across the O-band and 125 nm across the C-band, with low insertion losses of 0.27 dB and 0.08 dB for the central wavelength of 1310 nm and 1550 nm, respectively, and an extinction ratio of better than 15 dB for both central wavelengths. The device is more sensitive to fabrication errors in the gap between the strip WGs of a DC, as it has a greater impact on the effective refractive index of the first TE modes.

Furthermore, a compact and broadband WDM utilizing an SWG-assisted directional coupler was developed and successfully tested for separating wavelengths of 1310 nm and 1550 nm on an SOI platform [28]. The inclusion of an SWG allows for a greatly reduced size, achieved through precise phase matching at 1550 nm and a large phase mismatch at 1310 nm. The final device offers high extinction ratios (ER) of 23 dB and 19 dB at 1310 and 1550 nm, respectively, with 3-dB bandwidths of 85 nm for the O-band (ER >15 dB) and 140 nm for the C-band (ER > 10 dB). A grating coupler for fiber-chip connections based on SOI that can split wavelengths for both O and C communication bands without being affected by polarization is proposed in [29]. This was achieved by using subwavelength segmentation of silicon gratings and a special design that allows dual wavelengths to travel in opposite directions on the chip. The coupler's efficiency for TE and TM polarizations was around −4.5 dB at both 1310 nm and 1550 nm, with an average 1-dB bandwidth of ~45 nm and ~60 nm, respectively. An AWG demultiplexer using Si photonic WGs was demonstrated in [30]. The researchers designed and fabricated a 110 × 93 mm^2 AWG on an SOI platform, with a demultiplexing function observed in the wavelength range of 1500–1570 nm, a channel spacing of 6 nm, and a free spectral range of over 90 nm. The study also suggested that a channel spacing of 1 nm could be achieved in an area of 500 mm^2 through the optimization of AWGs and slab WGs. An optical demultiplexer with a 1 × 4 channel configuration was proposed using an ultra-low-loss Si_3N_4 WG with a propagation loss of approximately 3.1 dB/m [31]. It includes three 2 × 2 MZIs, each containing two 2 × 2 multimode interference couplers based on general interference. The couplers have power division ratios of −3.3 dB to −3.7 dB across a 50 nm wavelength range of 1530 nm to 1580 nm. The demultiplexer is thermally tunable, with chrome-based heaters on the delay arms of the MZI filters to control the optical phase shift and achieve a moderately low CT of 14.5 dB between adjacent channels. The insertion loss per channel is between 1.5 dB and 2.2 dB in the 1550 nm to 1565 nm wavelength range.

Inspired by the previously demonstrated demultiplexer configurations, herein, a 1 × 2 demultiplexer based on two serially cascaded racetrack microrings is proposed. The selection of the silica-titania platform is based on our ongoing project on the development of low-cost photonic devices [32]. The initial, very promising results related to the deposition of high-quality WG films via the dip-coating method are obtained. In this work, we reported the numerical modeling study of the demultiplexer device based on the silica-titania platform and fabrication methods that will be employed in the final implementation of this device.

2. Silica-Titania Platform

The choice of an appropriate material for the previously mentioned structures is one of the most important aspects of the initial stage of the fabrication process. There are a number of options to choose from; however, we would like to emphasize the use of silica-titania (SiO_2:TiO_2) as the WG material. SiO_2:TiO_2 is a material that has many possibilities in the field of photonic integrated circuits, because of its advantages [33] over other more conventional materials, for example, SiN or InP. The most convincing benefits of using SiO_2:TiO_2 are listed [34] in Figure 1.

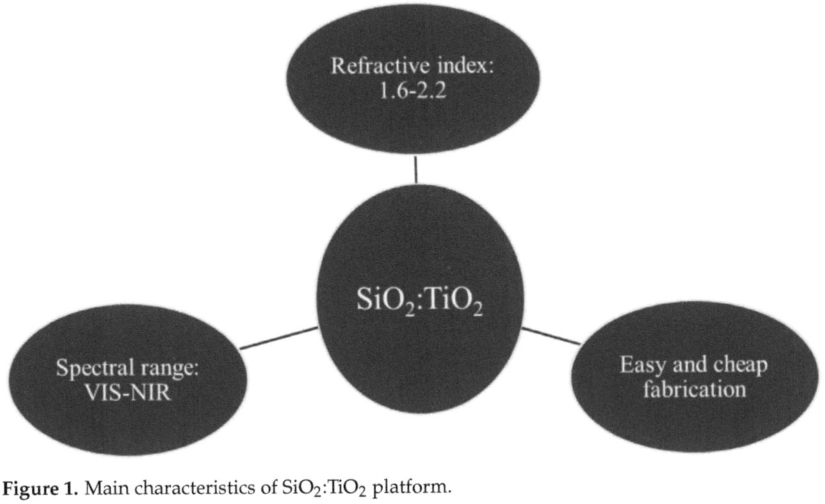

Figure 1. Main characteristics of SiO_2:TiO_2 platform.

This platform can be relatively easily manufactured by applying a thin-film of SiO_2:TiO_2 on a simple glass substrate. The preferred and simple method for SiO_2:TiO_2 film deposition is sol-gel dip-coating. The method itself is very easy to conduct and does not require any advanced equipment such as chemical vapor deposition (CVD) devices. It can be considered a great method for laboratory applications; however, it is not yet automatized, and it needs to be more developed to use for commercial purposes. Sol-gel dip-coating is conducted by dipping the substrate in the sol to be covered by the material; afterward, the substrate is withdrawn from the sol at a selected withdrawal speed and after the solvent evaporation, the thermal annealing process is conducted [35]. The process order is described in Figure 2.

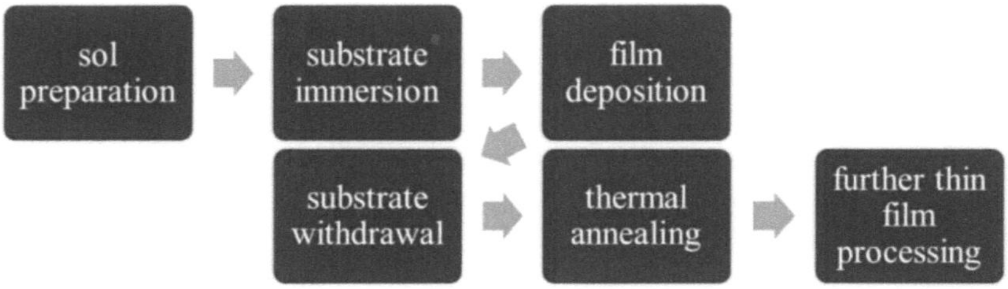

Figure 2. Graphical illustration of the sol-gel dip-coating method.

During the sol preparations, the used precursors are tetraethyl orthosilane $Si(OC_2H_5)_4$ (TEOS) and tetraethoxytitanate $Ti(OC_2H_5)_4$ (TET) for Si and Ti, respectively. First, partial hydrolysis of the solutions occurs, then they are mixed, and a homogenizing agent (ethyl) and catalyst (HCl) are added [36]. The obtained thin film is further processed to create waveguiding structures. In some special cases, the guiding structures are previously prepared in the substrate, for example, the reverse rib structures.

3. Fabrication Processes Under-Development

The next step after thin-film deposition is transferring the layout of previously designed structures to the prepared material sample. There are several technological solutions used to fabricate waveguiding structures in SiO_2:TiO_2 currently being considered by our consortium. We propose to take a deeper look at the following:

- Electron-beam (E-beam)/deep UV (DUV) lithography, combined with inductively coupled plasma—reactive ion etching (ICP-RIE)
- Optical lithography combined with wet chemical etching
- Nano-imprint lithography (NIL)

All of the three above-mentioned technological approaches differ from each other significantly in terms of fabrication cost and effort, as well as expected photonic circuit compactness. This is therefore a big advantage of using SiO_2:TiO_2, that all of them can be applied. It allows taking different approaches regarding fabrication. The comparison of these techniques is shown in Figure 3. ICP-RIE is a well-known and developed fabrication method for integrated photonics. It finds application with other materials such as SOI and InP platforms. However, it is expensive, difficult to carry, and requires advanced equipment and facilities. Nevertheless, using ICP-RIE gives nearly complete certainty of a successful fabrication [37].

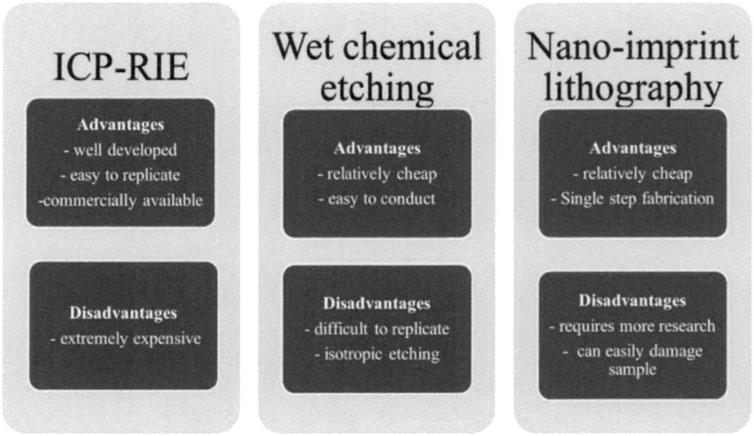

Figure 3. Pros and cons of the proposed fabrication method on the SiO_2:TiO_2 platform.

However, less developed methods such as wet chemical etching and NIL also deserve to be considered as potential fabrication techniques. Both chemical etching and NIL are relatively cheap and do not require advanced equipment or facilities. Those technologies combined with sol-gel dip-coating could result in a novel cheap approach to photonic integrated circuits fabrication. Wet chemical etching involves covering the substrate with a mask by photolithography, chemical etching with an acid solution (e.g., HF), and mask removal [38].

NIL is an etch-less fabrication method. It requires the previous fabrication of a "master stamp" and imprinting it over the deposited non-hardened WG film before the thermal annealing to form the structures [7]. However, both wet chemical etching and NIL are still under development for the SiO_2:TiO_2 platform and give a less probable successful fabrication effect than ICP-RIE.

4. Device Design and Numerical Model

The proposed 1 × 2 demultiplexer is based on double racetrack microring (RTMR) architecture, whereby two RTMRs are mutually coupled. Consequently, as each RTMR reverses the dropping direction concerning the optical signal input direction, the output of a double RTMR is acquired in the same direction as the input. The input signal, inserted into the input port is coupled to the 1st RTMR, which circulates along the racetrack circumference. When the wavelength of the inserted signal meets the resonance condition, the light is fully decoupled to the racetrack from the input WG and is decoupled to the 2nd racetrack. Following loops around the first ring couple, the two microrings together pass the light into the second microring once it reaches the resonance state of the second ring. Eventually, the light transfers into the output WG using the same technique as shown in Figure 4.

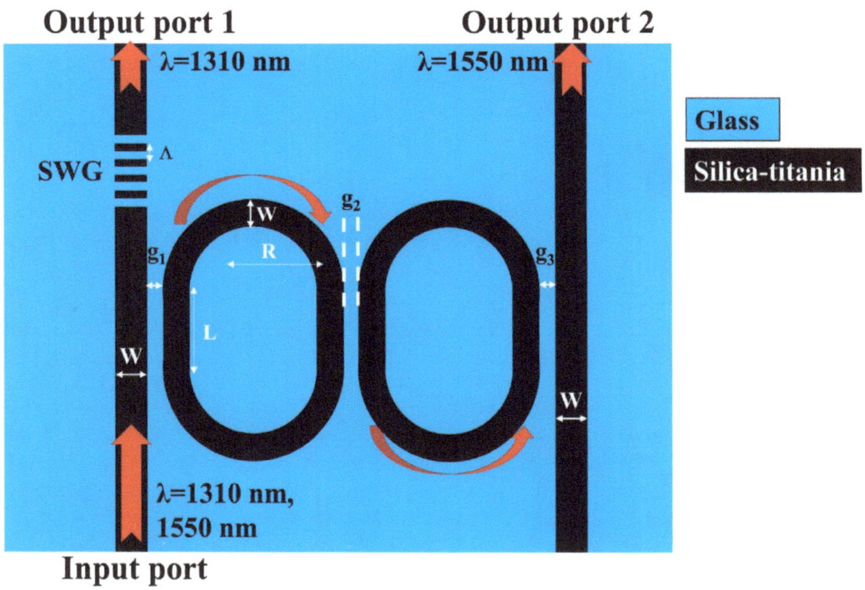

Figure 4. Schematic representation of 1 × 2 demultiplexer based on silica-titania RTMR structures and SWG segments for 1310 nm and 1550 nm telecommunication wavelength.

The radius of the RTMRs is devised in such a way that an operational wavelength of 1550 nm is coupled to a first RTMR and then transfers the power to the second RTMR and eventually, the output is collected at output port 2. The wavelength of 1310 nm can pass through the output port 1 without meeting the resonance condition of RTMR and thus not being de-coupled to the ring. To obtain an enhanced working device, the geometric parameters such as the radius of the ring and coupling gaps should be properly optimized. The radius of the RTMR is denoted as R, which is identical for both rings; and L is the coupling length, which is fixed at 10 μm. The width of the bus WG and the ring WG is denoted as W. The geometry of the SiO_2:TiO_2 ridge WG is optimized for the single mode operation at the operational wavelength of 1310 nm and 1550 nm. The height of the WG is fixed at 400 nm, which can be obtained by performing the dip-coating method twice [34]. The W of the WG is varied between 1000 nm and 2500 nm, as shown in Figure 5. For λ = 1310 nm, the WG supports a single mode for W = 1000 nm to 1800 nm. For W > 1800 nm, multimode starts to appear. Whereas W < 1800 nm is a cut-off region for λ = 1550 nm. And the WG supports single mode till W = 2500 nm. Keeping in mind the single mode operation of both λ = 1310 nm and λ = 1550 nm, W = 1800 nm is the viable dimension for designing the optical WGs for the demultiplexer. The effective refractive indices of 1.5194 and 1.4742 are employed in the design for the wavelength of 1310 nm and 1550 nm, respectively.

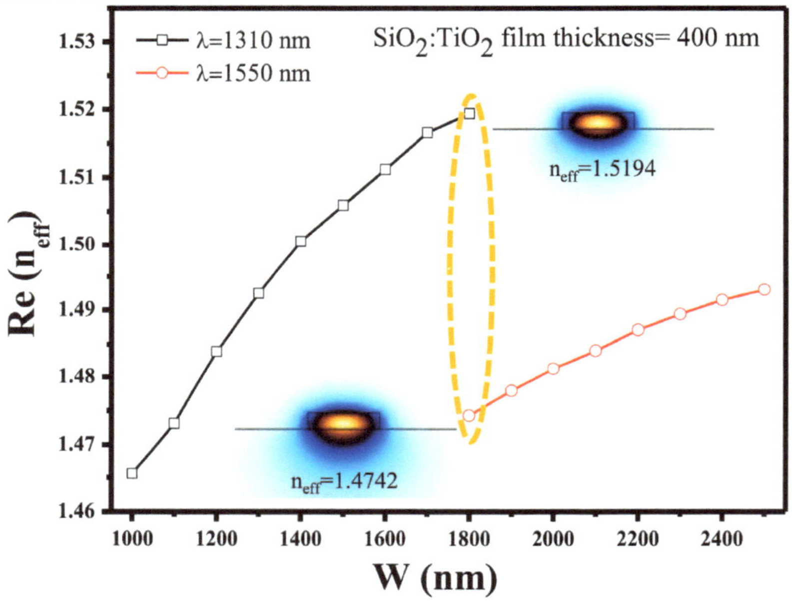

Figure 5. Real part of the effective refractive index of SiO_2:TiO_2 WG with varying W for the operational wavelength of 1310 nm and 1550 nm. The core height is fixed at 400 nm.

The coupling gap between the bus WG and the first RTMR, coupling gap between the first RTMR and second RTMR, and coupling gap between the second RTMR and drop port is denoted as g_1, g_2, and g_3, respectively. COMSOL Multiphysics software is being used to carry out the numerical analysis utilizing the finite element method (FEM). As a physics interface, the electromagnetic wave frequency domain (EWFD) has been used. In the whole demultiplexer design, the sub-domains in the WG cross-section are divided into triangular mesh components with a mesh grid size of $\lambda/100$. To simulate an open geometry, scattering boundary conditions (SBCs) are placed at the FEM simulation window's outside edges. To minimize the CT (dB) between the output ports, subwavelength grating (SWG) structures are introduced in the structure. The detailed geometric parameters of the device are shown in Table 1.

Table 1. Geometric parameters of the 1 × 2 demultiplex based on the silica-titania platform for 1310 nm and 1550 nm telecommunication wavelengths.

Parameter	Values
W	1800 nm (fixed)
R	10–22 μm (range)
L	10 μm (fixed)
g_1	80–180 nm (range)
g_2	80–370 nm (range)
g_3	60–120 nm (range)
Operational wavelength	1310 and 1550 nm (fixed)
Footprint	84 μm × 125 μm
Λ= a + b, where b = 250 nm	350–850 nm (range)
N	15

With numerous practical devices based on subwavelength structures, recent advancements in lithographic techniques on the semiconductor-on-insulator substrate with sub-10 nm patterning precision have gained recognition [39]. The three methodologies for the fabrication processes include lithography-based approaches, mechanics-enabled approaches, and post-trimming approaches. Despite being still widely used, lithography-based methods have low resolution and are expensive. With the advantages of better resolution or lower cost, mechanics-enabled and post-trimming technologies provide innovative alternatives that complement lithography-based approaches, particularly for applications in basic research and non-CMOS device manufacturing [39]. The essence of an SWG structure is a periodic arrangement of two distinct materials with a period significantly smaller than the wavelength of light [40]. The capability of SWG structures to do refractive index engineering is of ever-increasing interest. Light traveling in an SWG WG excites a mixture of the periodic structure's Bloch modes. Bloch waves are the natural modes of periodic media, just as plane waves are the inherent modes of free space. The Bloch mode has a wave vector of $k = 2\pi/\lambda$ and a temporal frequency of ω, and it propagates in the direction perpendicular to the periodic segments of SWG WG. The wavelength range that satisfies the equation $\Lambda = \lambda/2$ is regarded as a photonic bandgap (PBG) where there is no light transmission. The SWG WG segment is designed for the fill factor (FF = a/Λ) of 0.3 to 0.7, which provides a PBG for 1550 nm; and Bloch mode at 1310 nm can propagate as shown in Figure 6a. The E-field distribution in the SWG for FF = 0.61 at 1550 nm and 1310 nm is also presented in Figure 6b,c, respectively.

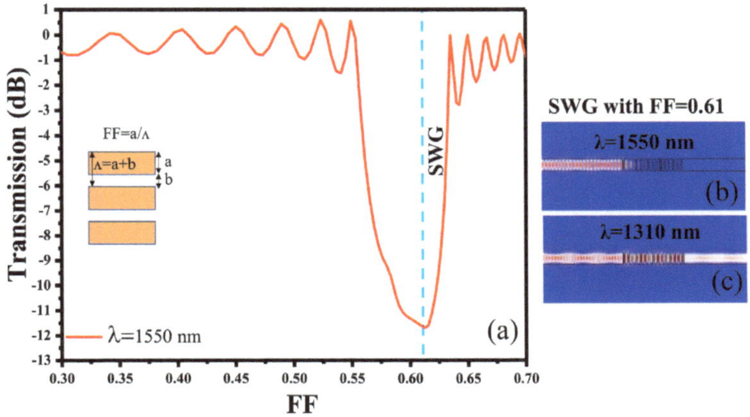

Figure 6. (**a**) Transmittance of SWG WG at varying FF for the operational wavelength of 1550 nm, (**b**) E-field distribution at λ = 1550 nm for SWG with FF = 0.61, (**c**) E-field distribution at λ = 1310 nm for SWG with FF = 0.61.

The 1 × 2 demultiplexer design is optimized step-by-step to obtain the maximum power coupling and lower CT (dB). In the first step, the optical confinement in the ring of different radiuses (R) at λ = 1310 nm and 1550 nm is analyzed. R is varied in the range of 10 μm and 22 μm whereas g_1 is fixed at 100 nm (as a starting point). From Figure 7a it can be seen that the maximum optical confinement in the ring for λ = 1310 nm and λ = 1550 nm is obtained at R = 16.75 μm and R = 16 μm, respectively. This suggests that ΔR = 0.75 μm, which is quite big; therefore, the ring designed for λ = 1550 nm will not resonate at λ = 1310 nm. However, according to the equation $m\lambda = 2\pi R n_{eff}$, lower order resonant modes (m) can provide some mode power that can be blocked by inserting SWG structures in the thru port [40,41].

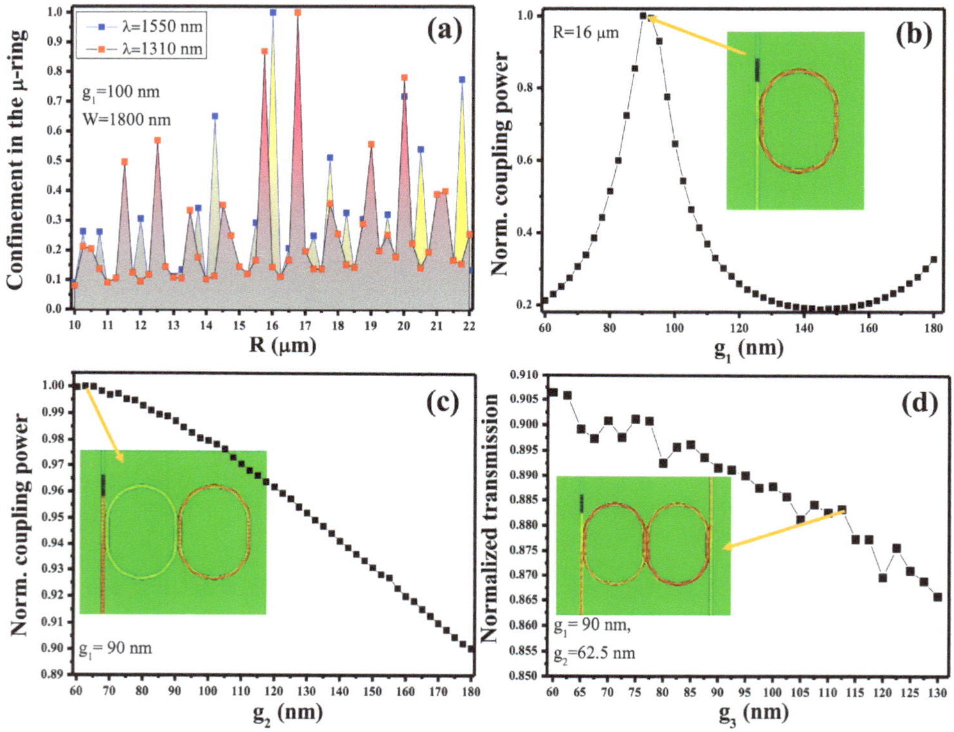

Figure 7. Optimization of the demultiplexer, (**a**) R versus E-field confinement for λ = 1310 nm and 1550 nm, (**b**) g_1 versus optimum coupling of the E-field at λ = 1550 nm in the first RTMR, (**c**) g_2 versus optimum coupling of the E-field at λ = 1550 nm in the second RTMR, (**d**) g_3 versus optimum transmission of λ = 1550 nm at output port 2.

The distance, coupling length, and refractive indices between the subsequent WGs and the microrings are the three factors that influence optical coupling. Typically, the distance between the microring and WG is reduced to maximize coupling. The coupling is also impacted by the coupling length. The coupling length denotes the length of the ring's effective curve required for the coupling phenomena to occur with the WG. Additionally, the optical coupling is influenced by the refractive indices of the materials used to make the WG, ring resonator, and medium used in between. The medium material is typically the most significant factor since it greatly influences how well light waves transmit. From Figure 7b, it can be seen that g_1 is varied between 60 nm and 180 nm, and the maximum coupling between the bus WG and a ring is obtained at g_1=90 nm. The E-field distribution in the ring at g_1 = 90 nm is indicated in the inset. The remaining geometric parameters such as R and W are fixed at 16 μm and 1800 nm, respectively.

In the next step, two rings are serially connected to the bus WG, and we aim to optimize the maximum power coupling to the second RTMR. This optimization is crucial so that most of the power can be transferred to the output WG. From Figure 7c, it can be seen that the maximum coupling in the second ring is obtained at g_2 = 62.5 nm. The E-field distribution in the structure at g_2 = 62.5 nm is shown in the inset. Now in the last step, the output WG (drop port) is side coupled to the second ring, where the coupling gap (g_3) should be optimized to obtain the maximum transmission for λ = 1550 nm. The normalized transmission of ~88% is obtained at output port 2 when g_3 = 112.5 nm and the E-distribution is revealed in the inset as shown in Figure 7d. At port 1, the normalized transmission is ~93% for the operational wavelength of 1310 nm. From these analyses, the optimized geometric parameters of the device are obtained, which are as follows: R = 16 μm, g_1 = 90 nm, g_2 = 62.5 nm, and g_3 = 112.5 nm. The CT at output port 1 and output port 2 is calculated by applying (1) and (2), respectively.

$$CT\ (dB)\ at\ output\ port\ 1 = 10 \times \log\left(\frac{P_{1550\ \mathrm{nm}}}{P_{1310\ \mathrm{nm}}}\right); \tag{1}$$

$$CT\ (dB)\ at\ output\ port\ 2 = 10 \times \log\left(\frac{P_{1310\ \mathrm{nm}}}{P_{1550\ \mathrm{nm}}}\right); \tag{2}$$

The CT of the device before the insertion of the SWG structure at output port 1 is around −1.6 dB and −9.23 dB at port 1 and port 2, respectively. The CT at port 1 is lowered to −6.6 dB by adding an SWG segment, which blocks the residual power of the 1550 nm wavelength flowing through port 1. The values of the CT with and without the SWG segment are presented in Table 2. Moreover, for the transmission bandwidth of 2 nm, the CT at Port 1 and Port 2 varies, as shown in Figures 8a and 8b, respectively. At Port 1, the CT is determined for the wavelength range of 1310 nm to 1312 nm with a step size of 0.1 nm, whereas at Port 2, the CT is also determined for the wavelength range of 1550 nm to 1552 nm with a step size of 0.1 nm.

Table 2. CT (dB) of the 1x2 demultiplexer device with and without SWG structure.

Design	Cross Talk at Output Port 1 (dB)	Cross Talk at Output Port 2 (dB)
Demultiplexer without SWG	−1.6	−9.23
Demultiplexer with SWG	−6.6	−9.04

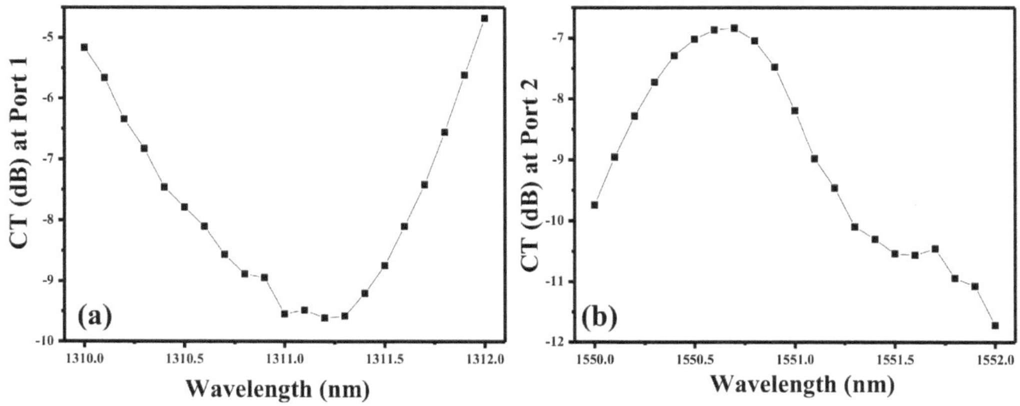

Figure 8. CT at Port 1 and Port 2 for the operational wavelength of: (**a**) 1310 nm to 1312 nm, (**b**) 1550 nm to 1552 nm.

The normalized E-field distribution for λ = 1310 nm and λ = 1550 nm is plotted in Figure 9a,b, respectively. It can be seen that when λ = 1310 nm is launched at the input port, it travels through the bus WG without coupling to the ring as it does not satisfy the resonance condition. Whereas λ = 1550 nm is launched at the input port coupled to the first ring as it satisfies the coupling condition, and then the mode power is transferred at output port 2 after passing through the second RTMR. For comparison, several demultiplexer designs established on different platforms and their performance are presented in Table 3.

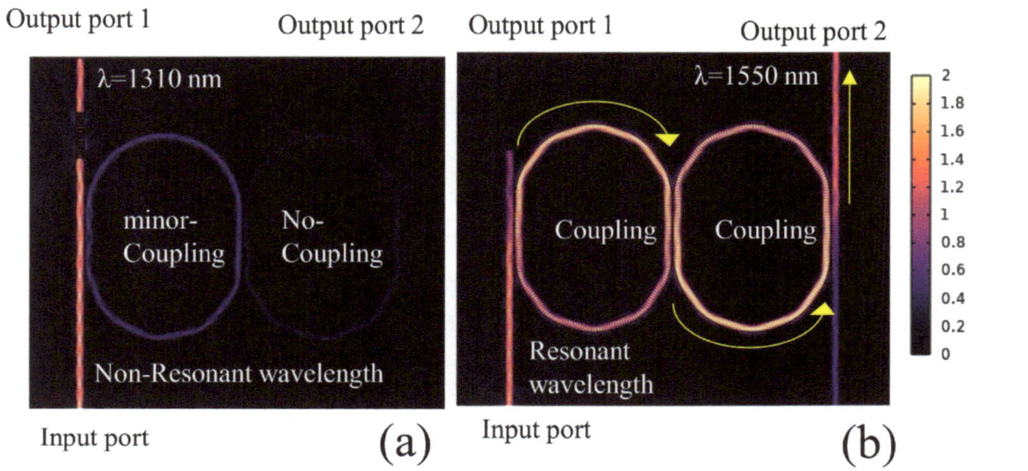

Figure 9. Normalized E-field distribution in the 1 × 2 demultiplexer design at the operational wavelength of: (**a**) 1310 nm, and (**b**) 1550 nm.

Table 3. Comparison of the proposed demultiplexer design with the previously published demultiplexers.

Ref.	Material	Design	Wavelength (nm)	Numerical/ Experimental	CT (dB)	Footprint
[23]	GaN, Si	MMI	1530–1565	Numerical	−19.97 to −13.77	-
[24]	GaN, Si	MMI	547–584	Numerical	13.8–18.3 dB	-
[25]	Si	PhC, RR	1535–1539	Numerical	−18.31	-
[26]	Si	RR	1550–1567	Numerical	−13–−24	689.61 (μm^2)
[27]	SOI	SWG	1310, 1550	Numerical	-	-
[28]	SOI	SWG	1310, 1550	Experimental	-	-
[29]	SOI	SWG	1310, 1550	Numerical	-	-
[30]	SOI	AWG	1500–1570	Experimental	-	110 × 93 (mm^2)
[31]	$Si3N_4$	MZI	1530–1580	Experimental	14.5	-
This work	SiO_2:TiO_2	RR	1550, 1310	Numerical	−6.6 and −9.04	84 μm × 125 μm

5. Conclusions

Herein, a 1 × 2 demultiplexer for telecommunication wavelengths 1310 nm and 1550 nm is proposed on a silica-titania platform. The choice of this platform is due to its highly attractive optical, physical, and chemical characteristics. The numerical analysis of the device is conducted via a finite element method. Two serially cascaded racetrack microrings (RTMRs) are side coupled to a bus WG to launch the input light. The wavelength division multiplexing (WDM) is achieved in such a configuration and the separated signals are collected via the thru port and drop port. Additionally, a subwavelength grating (SWG) structure is incorporated in the thru port to reduce the CT between the output ports. The minimum CT of −6.6 dB and −9.04 dB is obtained for 1550 nm and 1310 nm, respectively. This study can further open the possibilities of designing 1 × N demultiplexing devices, which can enhance the number of channels in communication systems.

Author Contributions: Conceptualization, M.A.B.; methodology, M.A.B.; software, M.A.B.; validation, M.A.B. and A.K.; formal analysis, M.A.B.; investigation, M.A.B., M.S. and Ł.K.; resources, R.P.; data curation, M.A.B.; writing—original draft preparation, M.A.B., M.S., Ł.K. and A.K.; writing—review and editing, M.A.B., M.S., Ł.K., A.K. and R.P.; visualization, M.A.B.; supervision, M.A.B.; project administration, M.A.B.; funding acquisition, M.A.B. All authors have read and agreed to the published version of the manuscript.

Funding: The research is co-financed by the Foundation for Polish Science from the European Regional Development Fund within the project POIR.04.04.00-00-14D6/18 "Hybrid sensor platforms for integrated photonic systems based on ceramic and polymer materials (HYPHa)" (TEAM-NET program).

Institutional Review Board Statement: Not applicable.

Informed Consent Statement: Not applicable.

Data Availability Statement: Not applicable.

Acknowledgments: We acknowledge the continuous support of the Warsaw University of Technology for the completion of this work.

Conflicts of Interest: The authors declare no conflict of interest.

References

1. Sheng, Y.-B.; Zhou, L.; Long, G.-L. One-step quantum secure direct communication. *Sci. Bull.* **2022**, *67*, 367–374. [CrossRef] [PubMed]
2. Khonina, S.; Kazanskiy, N.; Butt, M.; Karpeev, S. Optical multiplexing techniques and their marriage for on-chip and optical fiber communication: A review. *Opto-Electron. Adv.* **2022**, *5*, 210127. [CrossRef]
3. Park, J.; Joo, J.; Kwack, M.-J.; Kim, G.; Han, S.-P.; Kim, S. Three-dimensional wavelength-division multiplexing interconnects based on a low-loss SixNy arrayed waveguide grating. *Opt. Express* **2021**, *29*, 35261–35270. [CrossRef]
4. Pathak, S.; Thourout, D.; Bogaerts, W. Comparison of AWGs and Echelle gratings for wavelength division multiplexing on silicon-on-insulator. *IEEE Photonics J.* **2014**, *6*, 4900109. [CrossRef]
5. Lin, Y.-J.; Lee, S.-L.; Yao, C.-L. Four-channel coarse-wavelength division multiplexing demultiplexer with a modified Mach-Zehnder interferometer configuration on a silicon-on-insulator waveguide. *Appl. Opt.* **2003**, *42*, 2689–2694. [CrossRef] [PubMed]
6. Ahmmed, K.; Chan, H.; Li, B. Three-mode multiplexer and demultiplexer based on the Mach-Zehnder interferometer. *OSA Contin.* **2021**, *4*, 1519–1532. [CrossRef]
7. Menahem, J.; Malka, D. 1 × 4 wavelength demultiplexer C-band using cascaded multimode interference on SiN buried waveguide structure. *Materials* **2022**, *15*, 5067. [CrossRef]
8. Rouifed, M.-S.; Littlejohns, C.; Tina, G.; Qiu, H.; Penades, J.; Nedeljkovic, M.; Zhang, Z.; Liu, C.; Thomson, D.; Mashanovich, G.; et al. Ultra-compact MMI-based beam splitter demultiplexer for the NIR/MIR wavelengths of 1.55 um and 2 um. *Opt. Express* **2017**, *25*, 10893–10900. [CrossRef]
9. Chen, J.; Zhang, Y.; Shi, Y. An on-chip triplexer based on silicon Bragg grating-assisted multimode interference couplers. *IEEE Photon. Technol. Lett.* **2017**, *29*, 63–65. [CrossRef]
10. Xu, Y.; Zhou, W.; Chen, K.; Huang, X. Y-branch wavelength demultiplexer based on topological valley photonic crystals. *Opt. Laser Technol.* **2022**, *155*, 108422. [CrossRef]
11. Prajzler, V.; Mastera, R. Wavelength division multiplexing module with large core optical polymer planar splitter and multilayered dielectric filters. *Opt. Quantum Electron.* **2017**, *49*, 133. [CrossRef]
12. Mehdizadeh, F.; Soroosh, M.; Alipour-Banaei, H. An optical demultiplexer based on photonic crystal ring resonators. *Optik* **2016**, *127*, 8706–8709. [CrossRef]
13. Masilamani, S.; Punniakodi, S. Photonic crystal ring resonator based optical MUX/DEMUX design structures: A survey and comparison study. *J. Opt.* **2020**, *49*, 168–177. [CrossRef]
14. Rakshit, J.; Roy, J. All-optical ultrafast switching in a silicon microring resonator and its application to design multiplexer/demultiplexer, adder/substrator and comparator circuit. *Opt. Appl.* **2016**, *XLVI*, 517.
15. Bagheri, A.; Nazari, F.; Moravvej-Farshi, M. Tunable optical demultiplexer for dense wavelength division multiplexing systems using graphene-silicon microring resonators. *J. Electron. Mater.* **2020**, *49*, 7410–7419. [CrossRef]
16. Chang, H.; Kuo, Y.; Jones, R.; Barkai, A.; Bowers, J. Integrated hybrid silicon triplexer. *Opt. Express* **2010**, *18*, 23891–23899. [CrossRef]
17. Xu, H.; Shi, Y. On-chip silicon triplexer based on asymmetrical directional couplers. *IEEE Photon. Technol. Lett.* **2017**, *29*, 1265–1268. [CrossRef]
18. Zhu, N.; Song, J.; Wosinski, L.; He, S.; Thylen, L. Experimental demonstration of a cross-order echelle grating triplexer based on an amorphous silicon nanowire platform. *Opt. Lett.* **2009**, *34*, 383–385. [CrossRef]

19. Butt, M.; Kazanskiy, N.; Khonina, S. Miniaturized design of a 1 × 2 plasmonic demultiplexer based on metal-insulator-metal waveguide for telecommunication wavelengths. *Plasmonics* **2023**. [CrossRef]
20. Xie, Y.-M.; Lu, Y.-S.; Weng, C.-X.; Cao, X.-Y.; Jia, Z.-Y.; Bao, Y.; Wang, Y.; Yin, H.-L.; Chen, Z.-B. Breaking the rate-loss bound of quantum key distribution with asynchronous two-photon interference. *PRX Quantum* **2022**, *3*, 020315. [CrossRef]
21. Gu, J.; Cao, X.-Y.; Fu, Y.; He, Z.-W.; Yin, Z.-J.; Yin, H.-L.; Chen, Z.-B. Experimental measurement-device-independent type quantum key distribution with flawed and correlated sources. *Sci. Bull.* **2022**, *67*, 2167–2175. [CrossRef] [PubMed]
22. Wu, J.; Long, G.-L.; Hayashi, M. Quantum Secure Direct Communication with Private Dense Coding Using a General Preshared Quantum State. *Phys. Rev. Appl.* **2022**, *17*, 064011. [CrossRef]
23. Zaken, B.; Baruch, B.; Zanzury, T.; Malka, D. An 8-channel wavelength MMI demultiplexer in slot waveguide structures. *Materials* **2016**, *9*, 881. [CrossRef] [PubMed]
24. Tamir, S.; Karanov, N.; Malka, D. 1 × 4 MMI visible light wavelength demultiplexer based on a GaN slot-waveguide structure. *Photonics Nanostruct.-Fundam. Appl.* **2018**, *30*, 45–49.
25. Dhandrapati, L.; Tupakula, S. A novel 8-channel DWDM demultiplexer on silicon photonic crystal slab: Design and analysis. *Optik* **2022**, *256*, 168734. [CrossRef]
26. Naghizade, S.; Sattari-Esfahlan, S. An optical five channel demultiplexer-based simple photonic crystal ring resonator for WDM applications. *J. Opt. Commun.* **2020**, *41*, 37–43. [CrossRef]
27. Chen, J. A broadband wavelength demultiplexer assisted by SWG-based directional couplers. *Optik* **2020**, *202*, 163602. [CrossRef]
28. Wang, F.; Xu, X.; Zhang, C.; Sun, C.; Zhao, J. Design and demonstration of compact and broadband wavelength demultiplexer based on subwavelength grating (SWG). *IEEE Photonics J.* **2022**, *14*, 1–6. [CrossRef]
29. Hao, T.; Sánchez-Postigo, A.; Cheben, P.; Ortega-Moñux, A.; Winnie, N.Y. Dual-band polarization-independent subwavelength grating couplers for wavelength demultiplexing. *IEEE Photonics Technol. Lett.* **2020**, *32*, 1163–1166. [CrossRef]
30. Fukazawa, T.; Fumiaki, O.; Toshihiko, B. Very compact arrayed-waveguide-grating demultiplexer using Si photonic wire waveguides. *Jpn. J. Appl. Phys.* **2004**, *43*, L673. [CrossRef]
31. Hai, M.S.; Leinse, A.; Veenstra, T.; Liboiron-Ladouceur, O. A thermally tunable 1 × 4 channel wavelength demultiplexer designed on a low-loss Si3N4 waveguide platform. *Photonics* **2015**, *2*, 1065–1080. [CrossRef]
32. Butt, M.A.; Kaźmierczak, A.; Tyszkiewicz, C.; Karasiński, P.; Środa, E.; Olszewski, J.; Pala, P.; Martynkien, T.; Hlushchenko, D.; Baraniecki, T.; et al. HYPHa project: A low-cost alternative for integrated photonics. *Photonics Lett. Pol.* **2022**, *14*, 25–27. [CrossRef]
33. Kozlowski, L.; Shahbaz, M.; Butt, M.; Tyszkiewicz, C.; Karasiński, P.; Kazmierczak, A.; Piramidowicz, R. Low-cost Integrated Photonic Platform Developed via a Sol-gel Dip-coating Method: A Brief Review. *Sens. Transducers* **2022**, *259*, 82–92.
34. Butt, M.; Tyszkiewicz, C.; Wojtasik, K.; Karasinski, P.; Kazmierczak, A.; Piramidowicz, R. Subwavelength grating waveguide structures proposed on the low-cost silica-titania platform for optical filtering and refractive index sensing applications. *Int. J. Mol. Sci.* **2022**, *23*, 6614. [CrossRef]
35. Karasinski, P. Sol-gel derived optical waveguide films for planar sensors with phase modulation. *Opt. Appl.* **2004**, *34*, 467–475.
36. Karasinski, P.; Jaglarz, J.; Mazur, J. Low loss silica-titania waveguide films. *Photonics Lett. Pol.* **2010**, *2*, 37–39.
37. Racka-Szmidt, K.; Stonio, B.; Zelazko, J.; Filipiak, M.; Sochacki, M. A review: Inductively coupled plasma reactive ion etching of silicon carbide. *Materials* **2022**, *15*, 123. [CrossRef] [PubMed]
38. Park, H.; Cho, J.; Jung, J.; Duy, P.P.; Le, A.; Yi, J. A review of wet chemical etching of glasses in hydrofluoric acid based solution for thin film silicon solar cell application. *Curr. Photovolt. Res.* **2017**, *5*, 75–82.
39. Chen, Y.; Shu, Z.; Zhang, S.; Zeng, P.; Liang, H.; Zheng, M.; Duan, H. Sub-10 nm fabrication: Methods and applications. *Int. J. Extrem. Manuf.* **2021**, *3*, 032002. [CrossRef]
40. Kazanskiy, N.L.; Butt, M.; Khonina, S.N. Silicon photonic devices realized on refractive index engineered subwavelength grating waveguides-A review. *Opt. Laser Technol.* **2021**, *138*, 106863. [CrossRef]
41. Khonina, S.N.; Kazanskiy, N.L.; Butt, M.A. Spectral characteristics of broad band-rejection filter based on Bragg grating, one-dimensional photonic crystal, and subwavelength grating waveguide. *Phys. Scr.* **2021**, *96*, 055505. [CrossRef]

photonics

Article

Remote Operation of an Open-Path, Laser-Based Instrument for Atmospheric CO_2 and CH_4 Monitoring

Panagiotis Siozos *, Giannis Psyllakis and Michalis Velegrakis *

Institute of Electronic Structure and Laser, Foundation for Research and Technology-Hellas (IESL-FORTH), N. Plastira 100, Vassilika Vouton, 70013 Heraklion, Greece; gpsyllakis@iesl.forth.gr
* Correspondence: psiozos@iesl.forth.gr (P.S.); vele@iesl.forth.gr (M.V.)

Abstract: The technical specifications and the evaluation of the remote operation of the open-path, tunable diode laser absorption spectroscopic (TDLAS) instrument are presented. The instrument is equipped with two low optical power diode lasers in the near-infrared spectral range for the atmospheric detection of carbon dioxide, methane, and water vapors (CO_2, CH_4, and H_2O). Additionally, the instrument eliminates the requirement of retroreflectors since it detects the back reflection of the laser beam from any topographic target. The instrument was operated remotely by measuring background concentrations of CO_2 and CH_4 in the atmosphere from 24 November 2022 to 4 January 2023. The accuracy of CO_2 and CH_4 measurement retrievals on a 200 m laser path was estimated at 20 ppm (4.8%) and 60 ppb (3.1%), respectively. The CH_4 accuracy is comparable, but the CO_2 accuracy is noticeably lower than the accuracy achieved in local operation. The accuracy issues raised are studied and discussed in terms of the laser driver's cooling performance.

Keywords: greenhouse gases; remote operation; TDLAS; DIAL; open path

Citation: Siozos, P.; Psyllakis, G.; Velegrakis, M. Remote Operation of an Open-Path, Laser-Based Instrument for Atmospheric CO_2 and CH_4 Monitoring. *Photonics* **2023**, *10*, 386. https://doi.org/10.3390/photonics10040386

Received: 7 March 2023
Revised: 24 March 2023
Accepted: 29 March 2023
Published: 31 March 2023

1. Introduction

Detailed greenhouse gas (GHG) emission estimation requires extensive atmospheric concentration measurements of regions with high GHG emissions (cities, airports, power plants, oil and natural gas production and distribution facilities, landfills, cultivated fields, etc.). Calculation of the emissions from these specific locations has large uncertainties due to insufficient emission monitoring [1–3]. The installation and operation of ground-based stations are considered an optimal solution for detecting and monitoring GHG concentrations [4,5] from such locations. Therefore, robust, reliable, easy-to-maintain, remotely-operated instruments detecting GHG in the atmosphere are essential to survey GHG concentrations in the atmosphere from places with high emissions. These instruments must have low detection limits and high accuracies to measure global variations of GHG concentrations. In particular, their accuracy must be in the range of sub-ppm for carbon dioxide (CO_2) detection and a few ppb for methane (CH_4) detection.

Optical spectroscopy techniques possess high sensitivity, good selectivity, and continuous, real-time detection [6–8]. Open-path optical spectroscopy techniques such as Fourier-transform infrared (FTIR) spectroscopy and differential optical absorption spectroscopy (DOAS) have been employed to measure trace gas column concentrations. Several research groups have developed FTIR and DOAS systems for measuring column concentrations of various trace gases simultaneously, using sunlight (passive), lamps, or LEDs (active) as the light source [9–12]. However, the broadband light source has a low spectral resolution, which limits gas detection sensitivity.

The Total Carbon Column Observing Network (TCCON) operates high-sensitivity, stationary FTIR spectrometers [13] that measure total columns of CO_2, CO, CH_4, N_2O, H_2O, HF, and other atmospheric gases. In addition, the COllaborative Carbon Column Observing Network (COCCON) is a greenhouse-gas-observing network that uses common FTIR instrumental standards and data analysis procedures [14]. The main objective of the COCCON is to increase global greenhouse gas observations due to the simplicity of the FTIR spectrometers used on the network. Both networks provide measurements of column-averaged abundances of GHGs by analyzing the solar radiation using Fourier-transform infrared (FTIR) spectrometers. These two networks should be considered the standard in optical spectroscopy for GHG detection due to the instrumentation and methodology employed.

Tunable diode laser absorption spectroscopy (TDLAS) has the potential to achieve high sensitivities because it utilizes robust and reliable distributed feedback (DFB) laser diodes that emit light in the near-infrared spectral range, exhibit excellent wavelength stability, and offer narrow spectral resolutions over a wide range of temperatures and current conditions, making them suitable for use in harsh environments. Open-path TDLAS is a flexible, cost-effective sensing technology for measuring selected target gases in complex mixtures associated with evolving modern industrial applications [15,16].

Several research groups have reported the construction and deployment of open-path TDLAS sensors [17]. Zimmerman et al. reported an open-path TDLAS sensor used to measure CO_2 concentrations along a fixed 100 m path [18]. The sensor proved effective as an alarm-type system intended for personnel safety. Bailey et al. developed an open-path TDLAS instrument for CO_2 absorption measurements over a 200 m long horizontal path to a retroreflector [19]. Xia et al. presented measurements of both CO_2 and CH_4 over a 2.6 km long path using an open-path TDLAS instrument [20]. Therefore, measurement of CO_2 column concentration in a large number of regions based on open-path TDLAS is valuable for improving climate warming.

All these works and developed instruments require the establishment of reflectors; otherwise, short-range operation of the open-path instrument is a requirement [21,22]. However, the installation of retroreflectors used for efficient collection of back-reflected light is a limitation. Retroreflectors reduce the suitable number of sites and require precise alignment. Due to this, the instrument's installation becomes complex and the remote operation and maintenance of the system is difficult, reducing the robustness of the system.

In this paper, we present in detail the design and specifications, and the results of the remote operation, of an open-path TDLAS instrument to monitor atmospheric CO_2 and CH_4 concentrations. The device is designed to operate at long laser path ranges (from 100 m up to 1 km) and is capable of detecting back reflection of topographic targets, eliminating the establishment of reflectors for efficient detection. Furthermore, we present the evaluation of the performance of the device during remote operation for more than a month.

2. Materials and Methods

2.1. DIAL Methodology

The range-resolved gas assessments are calculated from the recorded signal using the differential absorption lidar (DIAL) technique [23,24]. DIAL uses the retrieval of the atmospheric reflected light from a laser that is transmitted into the atmosphere. The wavelength is alternated successively between two wavelengths, one coinciding with an absorption line of the gas of interest (λ_{on}), the other in proximity but out of the absorption line (λ_{off}). The averaged gas imprint over a range of z is calculated from the ratio of the atmospheric reflected light at the two wavelengths. It is presumed that the atmosphere and the detection system have the same properties at the two wavelengths apart from the absorption due to the gas of interest:

$$N = \frac{1}{2\left[\sigma(\lambda_{on}) - \sigma\left(\lambda_{off}\right)\right]z} ln\left[\frac{P\left(\lambda_{off}\right)}{P(\lambda_{on})}\right] \quad (1)$$

where, z is the propagation range of the laser beam from the device to the reflection surface, P (λ_{on}) and P (λ_{off}) are the recorded signals at λ_{on} and λ_{off}, N is the averaged gas concentration in the atmospheric line of the laser beam, and σ (λ) is the gas absorption cross-section which can be found in the literature [25,26].

Even though the DIAL method lies on the recorded signal intensities at two different laser wavelengths (the λ_{on} and λ_{off}), the proposed approach uses the peak amplitude obtained from the full absorption spectrum. A fitting process of the data is necessary to improve the accuracy of the measurement since the recorded signal has a low signal-to-noise ratio (SNR). The instrument operates at a low elevation (<500 m), where the atmospheric pressure has a dominant effect on broadening the absorption peaks. Therefore, a Lorentz function is selected to fit the absorption peaks since this function predicts the broadening due to pressure [27]. The ln (P (λ_{on})/P (λ_{off})) value that is required in the DIAL equation to calculate the concentration of the GHGs is equal to the difference in the maximum from the baseline of the fitted function.

The accuracy of the calculation of the atmospheric GHG concentrations is impacted by two major sources of error; the uncertainty in the distance z value between the device position and the reflective surface and the SNR which affects the accuracy of the fitting function. Thus, the error of the instrument is derived from the equation:

$$s(N) = N\sqrt{\left(-\frac{1}{z}s(z)\right)^2 + \left(\frac{1}{A}s(A)\right)^2} \tag{2}$$

where s (N) is the error of the calculated averaged gas concentration, z is the range of the laser beam, s (z) is the error of the range z, A is equal to the ln (P (λ_{on})/P (λ_{off})) that is calculated from the fitted function, s (A) is the error of A, and σ (λ) is the gas absorption cross-sections derived from the HITRAN database [25,26]. The s (A) is the standard error produced from the fitting procedure. The s (A) value is calculated from the square root of the diagonal of the parameters' covariance matrix.

The CO_2 and CH_4 content calculation from DIAL Equation (1) requires the measurement of the laser range z. The range z was calculated using the method demonstrated in our previous work [28] based on the absorption lines of water vapors recorded in the CH_4 spectrum at 6047.77 cm^{-1}. The atmospheric parameters (vapor pressure, temperature, atmospheric pressure) were recorded simultaneously with the laser back-reflection measurements using a local weather station installed in the vicinity of the device. The water vapor concentration in the atmosphere is calculated from the specific atmospheric parameters. The distance z from the laser sensor to the object is calculated using once more the DIAL method and the water vapor concentration:

$$z = \frac{1}{2\left[\sigma_{H_2O}(\lambda_{on}) - \sigma_{H_2O}\left(\lambda_{off}\right)\right]N_{H_2O}} ln\left[\frac{P_{H_2O}\left(\lambda_{off}\right)}{P_{H_2O}(\lambda_{on})}\right] \tag{3}$$

where N_{H_2O} is the water vapor concertation in the atmosphere calculated from the water station measurements, P_{H_2O} (λ_{on}) and P_{H_2O} (λ_{off}) are the recorded signals at λ_{on} and λ_{off}, and σ_{H_2O} (λ) are the gas absorption cross-sections which can be found in the literature [25,26].

2.2. Design and Manufacturing

The system (Figure 1) and the methodology for GHG concentration measurements have been presented in detail in previous work [29]. The system is equipped with two cw DFB tunable diode lasers in a 14-pin butterfly package. The first one has a central wavelength of 1.57 μm (6369 cm^{-1}) with a linewidth of 1 MHz and maximum output power of 60 mW (Toptica, #LD-1550-0060-DFB-1) and is used for the detection of atmospheric CO_2. The second one emits around 1.652 μm (6053 cm^{-1}) with a linewidth lower than 2 MHz and maximum output power of 15 mW (ld-pd, PL-DFB-1650-A-A81-PA) for the detection of CH_4. Two current and temperature controllers (Maiman, SF8075-ZIF14) regulate each laser. The emission wavelengths of the CO_2 and CH_4 detection lasers were adjusted from 6365 to 6375 cm^{-1} and 6040 to 6050 cm^{-1}, respectively.

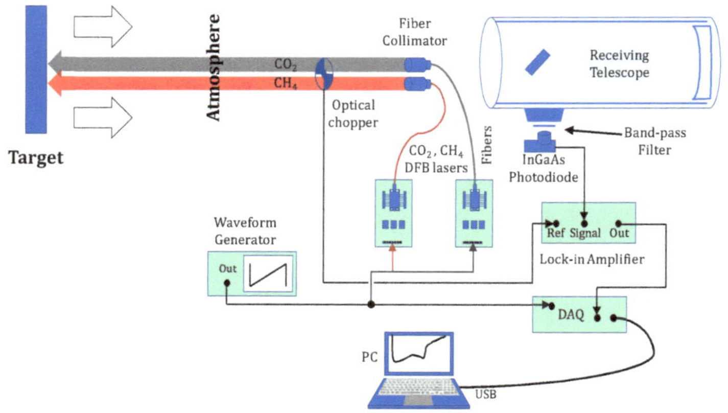

Figure 1. Diagram of the instrument's setup.

The laser beam propagates through single-mode fibers and collimates using fiber collimators with a focal length of 19 mm (Thorlabs, F280APC-1550). Finally, an optical chopper modulates the laser beam (Scitec, 360C-OEM) at a low frequency (400 Hz) before being transmitted into the atmosphere. The square pulse signal generated from the rotation of the chopper is used as a reference signal to the lock-in amplifier.

The back-reflected light is collected by an f/5 Newtonian telescope (SkyWatcher, 200/1000 PDS) with a 1 m focal length mirror and detected by a homemade amplified photodetector equipped with a 3 mm diameter InGaAs diode (Hamamatsu, G12180-130A). A dual-phase digital lock-in amplifier (Femto, LIA-200MVD-H) amplifies the photodetector's signal by using the optical chopper signal as a reference to achieve maximum sensitivity.

A full profile of the absorption peak is recorded by wavelength scanning of the laser beam instead of the two-wavelength (on-line and off-line) approach that is usually employed.

The instrument's mechanical design must be adapted to overcome various challenges that arise from its remote operation. These challenges are imposed by the constraints from the installation location and the selection of the measurement orientation, as well as the need to protect the instrument from environmental conditions. The device has an envelope of 1.45 m × 0.5 m × 0.45 m, an approximate weight of 70 kg, and was fully modeled in 3D (Figure 2).

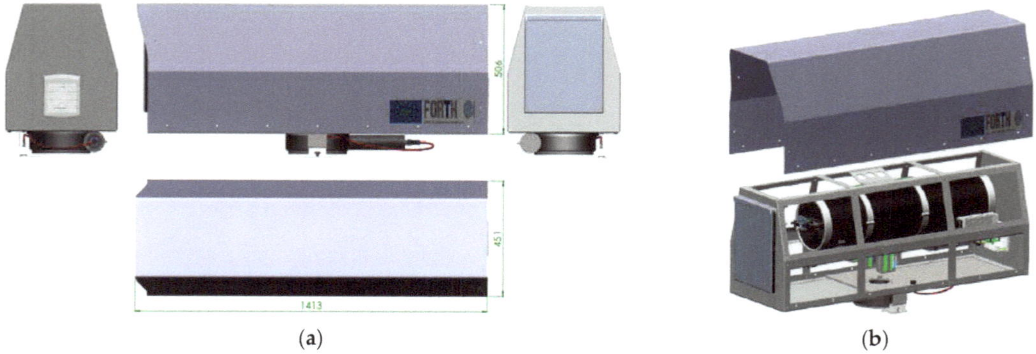

Figure 2. CAD images of the instrument's (**a**) external and (**b**) internal configuration (dimensions in mm).

To ensure that the laser beams propagate above the designated canopy, a supporting pole for the device was designed. Simulation of the device and the poll was applied using finite element methods to ensure durability at the installation height. Moreover, a slewing drive was installed to rotate the device at the desired orientation. The slewing drive was equipped with a reduction and DC motor with integrated Hall sensors for position control. In addition, an inductive sensor was installed to initialize the drive system.

The welded frame and the cover were constructed using stainless steel galvanized sheets. High temperatures during summer can affect the device's performance amid remote operation in the Mediterranean region. To address this issue, the device and the cover were externally painted white to reflect sunlight during summer and internally black on the inside to reduce unwanted light reflection. In addition, the device was equipped with a ventilation fan that generated an air stream towards the hatch to reduce contamination by dust particles during the dry summer season.

The telescope and electronics were mounted on the frame, while the laser fiber optics were attached to the telescope (Figure 3a). Moreover, a hatch was designed using two stepper motors with threaded rods. The hatch mechanism was based on an extruded aluminum V-slot frame and was driven by two synchronized stepper motors via a T8 trapezoidal axis. The hatch was sealed from aluminum profiles with integrated rubber. Terminal switches are activated in the open and closed positions and are used to control the hatch.

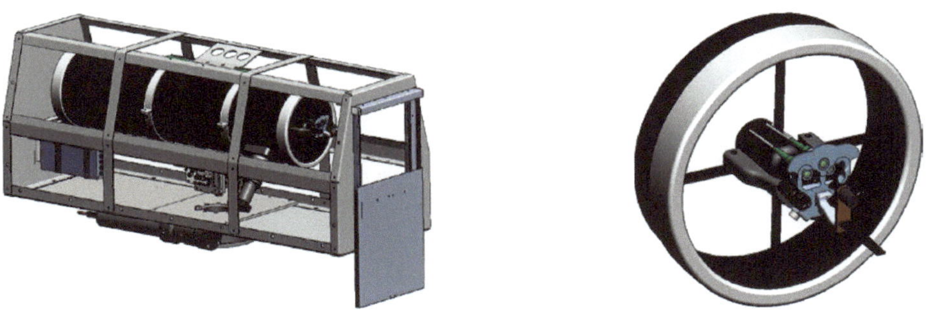

(**a**) (**b**)

Figure 3. CAD image of the (**a**) instruments internal configuration and (**b**) laser diode collimator mount.

The alignment requirement of the laser diodes is crucial for the smooth operation of the instrument. Therefore, a miniature stage that can support up to three collimators with independent movement, and one optical chopper for all lasers, was designed and manufactured (Figure 3b). The specific configuration was attached to the support cross of the telescope auxiliary mirror.

For quality assurance, all devices were subjected to a 20-day period of continuous operational indoor tests. These tests included power on/off cycle, hatch operation, sensor checking, and continuous full measurement cycles.

2.3. Hardware

A single-board computer (Raspberry Pi 3) was used for handling, remote control, communications, data handling, and transmission. The laser drivers are connected via USB connection and controlled by UART protocol. Motor functions, power distribution (supply) to the subsystems, sensor reading, and partly the data collecting is handled by a microcontroller board (Arduino DUE) with a custom shield and firmware. The microcontroller is connected to the Raspberry Pi 3 via a serial connection, which allows it to receive commands and send status updates. The 12-bit ADC of the microcontroller is used to record the value of the lock-in amplifier during the measurement (Figure 4).

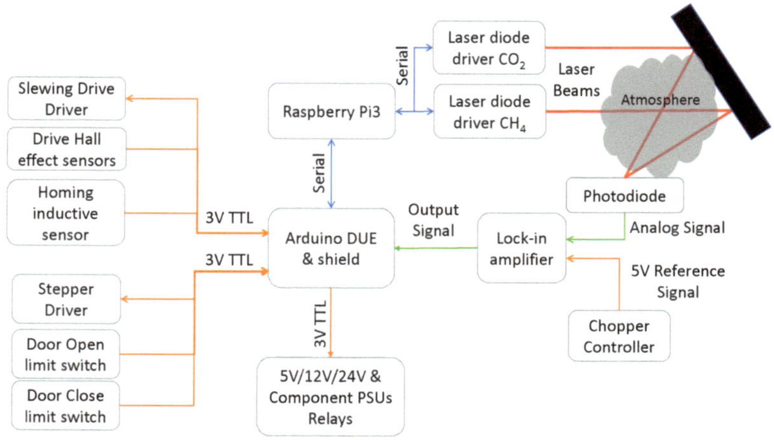

Figure 4. Device control diagram.

The system is powered by 220 V AC and turns on when connected to power as the Raspberry Pi has its own power supply directly connected to the power line. A 12 V/24 V, 100 W power supply provides power to the motor drivers, the ventilation, and the optical chopper driver. Each laser driver has an exclusive 5 V, 35 W power supply. Additionally, each of the optical components (lock-in amplifier, photodiode, and optical chopper) are powered by separate power supplies with specifications recommended by the manufacturers. The power supplies are activated by individual relays on the shield depending on the operating stage. To control the hatch stepper motors, a TB6600 guide was used. A Single DC motor driver (DRI00042) was chosen to drive the DC rotation motor. The Arduino shield also receives signals from the door terminal switches, the zero-position induction sensor, and the integrated hall sensors in the rotation motor for position control.

2.4. *Software*

Python was used to implement the control software for the instruments, allowing for seamless communication with the subsystems such as laser drivers and an Arduino shield, as well as external connectivity. After initialization at startup, the microcontroller enters a serial communication standby mode. The software performs various functions including hatch operation, azimuth movement, relay control, status reporting, and lock-in amplifier data acquisition. The laser drivers are controlled using serial hex commands, which are sent through the Raspberry Pi.

The device can operate in two modes depending on the tasks performed: field mode or calibration/debug mode. During remote operation of the instrument for GHG monitoring, the field mode is selected that enables autonomous operation without external input from a user. Minimum external commands allow the instrument to operate more reliably, providing consistent measurements of GHG concentrations. However, during the installation of the device, calibration/debug mode is activated. In this mode, a user provides the commands required to set up and calibrate the system. Furthermore, calibration/debug mode is selected to confirm the proper operation of the components and to certify the instrument's performance for remote operation. In addition, this mode is used for remote software upgrades and debugging.

On system startup, field operation is initialized by default and internet connectivity is established by an Azure server. When a measurement request is issued, the device begins the measurement cycle (Figure 5). Moreover, a graphical user interface (GUI) was developed for calibration, testing, and debugging.

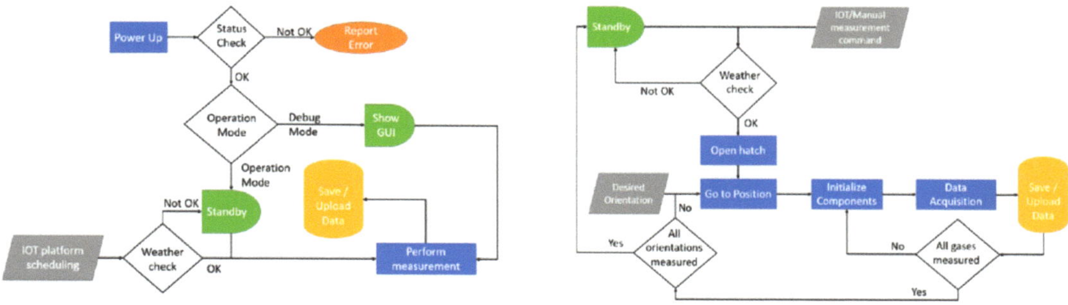

(a) (b)

Figure 5. Flowcharts (**a**) for the device operation (**b**) for the measurement procedure.

The lock-in amplifier output, the laser driver current, and the temperature are recorded. Information such as the device's date and time, position angle, door status, internal temperature, and selected laser driver status is also controlled and monitored.

2.5. Installation and Testing Location

The instrument was installed in the Kato Valsamonero village, Crete, Greece, a rural area dominated by olive trees (Figures 6 and 7). A weather station was also installed to simultaneously measure the weather conditions in the area (relative humidity, air temperature, wind speed, and direction).

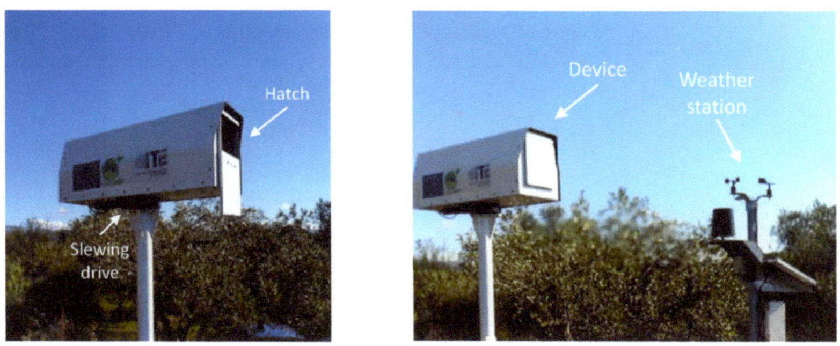

(a) (b)

Figure 6. Image (**a**) of the device with (**b**) the weather station installed in Kato Valsamonero, Crete, Greece.

Figure 7. Map of the instrument's installation location (image from the Hellenic Land Registry [30]).

Any available topographic target that provides sufficient reflection for the laser beam is suitable for the system's operation such as a nearby mountain, hill, or building. The methodology followed is unaffected by the absorption characteristics of a surface since the wavelength scan range of the lasers is 1 nm, approximately. In such a short spectral range, the variation in the absorption coefficient is expected to be negligible. The instrument is directed towards a surface at a distance of 200 ± 5 m. The distance was calculated applying the DIAL method on the water vapor peak at 6047.77 cm^{-1} recorded in the CH_4 spectra (Figure 8b). A single measurement of each GHG had a 15 min duration.

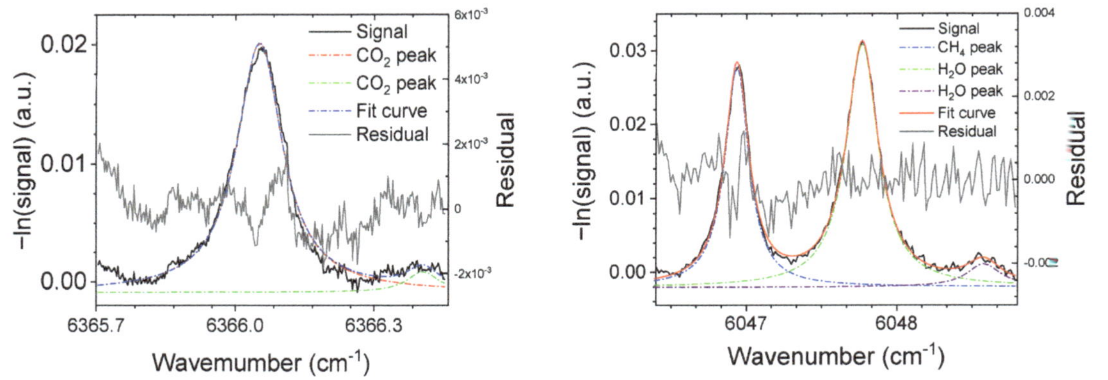

(**a**) (**b**)

Figure 8. Typical (**a**) CO_2 signal, fitted function, and the residual between the fitted profile and the CO_2 signal; and (**b**) CH_4 and H_2O signal, fitted functions, and the residual between the fitted profile and the CH_4 signal.

3. Results

Remote Operation and Measurements

The recorded signal of the CO_2 absorption line and the fitted function of a typical measurement is presented in Figure 8a. Moreover, the recorded peaks of the CH_4 absorption line with the water vapor absorption lines and their fitted functions are presented in Figure 8b.

The instrument performance and the accuracy of the methodology to measure GHG concentrations in the atmosphere were evaluated during remote operation of the device for a period higher than a month from 24 November 2022 until 4 January 2023. The recorded concentrations of the CO_2 and the CH_4 in the atmosphere are presented in Figure 9a,b, respectively. The mean reduced chi-square ($\tilde{\chi}^2$) value for CO_2 fitting is 1.015 and for CH_4 fitting is 1.070. Therefore, the fitting process produces statistically reliable ln (P (λ_{on})/P (λ_{off})) values since the $\tilde{\chi}^2$ is sufficiently small.

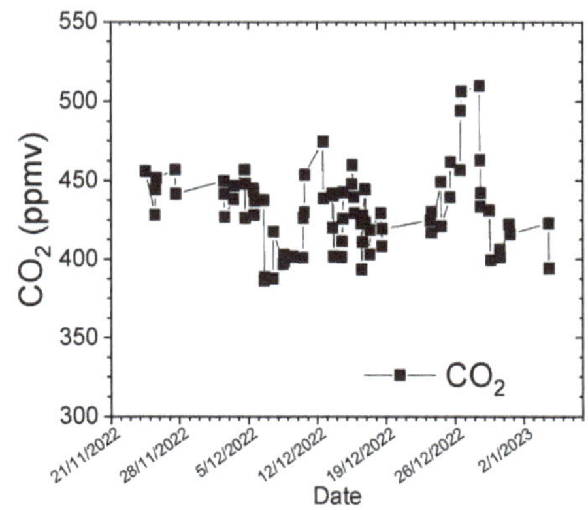

(a)

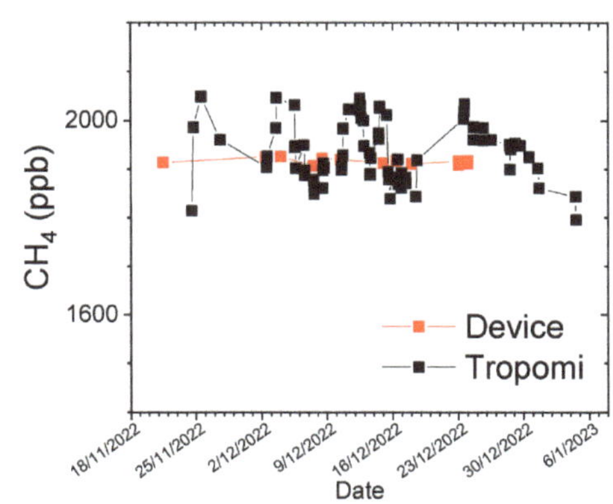

(b)

Figure 9. Concentration variation in (**a**) CO_2 and (**b**) CH_4 in the atmosphere from 24 November 2022 to 4 January 2023.

The mean CO_2 concentration value calculated during the instrument's remote operation was 430 ± 20 ppm which is 2.6% higher than the global value provided by NOAA (418.95 ppb, December 2022) [31]. Furthermore, the measurement deviation of the monitoring area compared with the reference values is smaller than the measurement error (4.6%).

However, the measured atmospheric CO_2 content variation is noticeably higher than that reported in our recent work (Figure 9a), which is attributed to the lower atmospheric temperatures during the remote operation of the device compared with the summer season measurements. Low atmospheric temperature affects the performance of the thermoelectric cooling (TEC) driver of the CO_2 laser. In particular, the TEC is ineffective in stabilizing the temperature of the laser (Figure 10). The laser temperature instability changes the wavelength of the laser and affects the recorded signal, which causes a broadening of the CO_2 absorption peak. Even though the CO_2 laser suffers from TEC stability in lower air temperatures, the CH_4 laser temperature remains stable at all temperatures due to the lower electric current required for CH_4 laser operation than the CO_2 laser. The lower current produces a smaller thermal load, which is easily compensated by the TEC module. In any case, a future upgrade of the device is necessary to enhance the laser driver's thermal insulation, enabling it to operate effectively at lower temperatures without compromising the performance of the TEC driver.

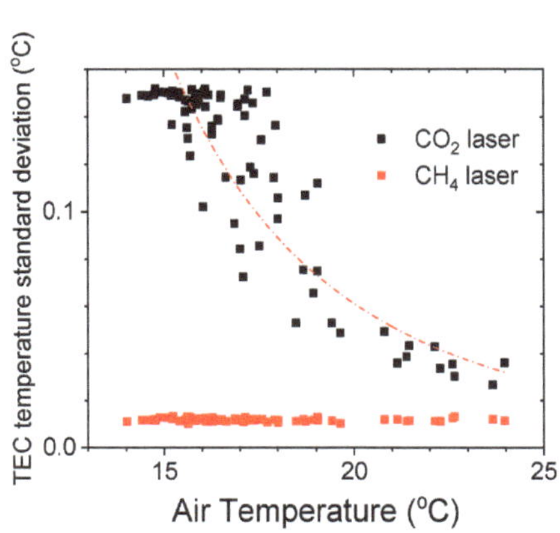

Figure 10. Standard deviation of the CO_2 and CH_4 laser temperature during a CO_2 and CH_4 atmospheric measurement at various air temperatures.

The CH_4 concentration presented no significant variation during the remote operation of the device (Figure 9b). The mean CH_4 concentration calculated during the instrument's remote operation was 1940 ± 60 ppb. This value is 1.3% higher than the global values provided by NOAA (1915.86 ppb, September 2022) [32] and 1.2% higher than the value of the Kato Valsamonero area provided by the TROPOMI instrument (1917 ± 6 ppm) of the Sentinel5P observatory [33,34] (Figure 9b). Furthermore, the measurement deviation of the monitoring area compared with the reference values is smaller than the measurement error (3.1%) and is similar to those reported by our recent work. Therefore, the device accuracy is probably even better than those derived by the error calculations.

4. Conclusions

Developing robust, reliable, and simple-to-operate and maintain instruments is crucial for effective GHG emission monitoring. These instruments must ensure accurate and precise measurements in various environments and conditions, including harsh weather, extreme temperatures, and remote locations.

In this study, we present the technical specifications and the evaluation of the remote operation of the open-path, tunable diode laser absorption spectroscopic (TDLAS) instrument. The instrument is equipped with two near-infrared lasers at 1.57 μm and 1.65 μm to measure the atmospheric concentrations of CO_2 and CH_4, respectively.

Any topographic target that provides detectable reflection to the device can be used to measure GHG, thereby eliminating the requirement of retroreflector installation. This design feature substantially simplifies the instrument's installation complexity as well as the operational and maintenance costs.

The device was installed in a rural area in Crete, Greece, and operated remotely from 24 November 2022 to 4 January 2023. During the remote operation of the instrument, any manual intervention from a user was unnecessary, enabling autonomous operation. Such a minimum external interference allows the instrument to operate more reliably, providing consistent measurements of GHG concentrations.

The DIAL method was applied to calculate the mean concentration of CO_2 and CH_4 gases in the atmosphere along the propagation path of the laser beam. The path length of the laser beam from the device to the back-reflection target is 200 ± 5 m, calculated based on the water vapor concentration in the atmosphere and the water peak height recorded in the CH_4 spectrum. Over the remote operation period of the instrument, the average atmospheric background concentrations of CO_2 and CH_4 were measured to be 430 ± 20 ppm and 1940 ± 60 ppb, respectively.

Low performance of TEC at lower air temperatures leads to lower CO2 measurement accuracy. Therefore, improving the insulation of the electronics of the laser driver will enhance the accuracy of the CO_2 measurement.

The successful remote operation of the instrument demonstrates its potential as an effective tool for measuring GHG atmospheric concentrations. The ability to remotely operate the device is essential for monitoring emissions in hard-to-reach or remote locations where continuous monitoring has previously been difficult or impossible to perform.

Author Contributions: Conceptualization, P.S. and M.V.; methodology, P.S.; data curation, G.P.; software, G.P.; writing—original draft preparation, P.S.; writing—review and editing, G.P. and M.V.; supervision, M.V. All authors have read and agreed to the published version of the manuscript.

Funding: This study was co-funded by the European Commission in the framework of the LIFE17 CCM/GR/000087 project "Innovative technologies for climate change mitigation by Mediterranean agricultural sector (LIFE ClimaMed)".

Data Availability Statement: Unprocessed data are available from the http://iot-telemetry.green-projects.gr/ (accessed on 28 March 2023) website upon request. For related information, please visit the LIFE ClimaMed project site at https://life-climamed.eu/ (accessed on 28 March 2023).

Acknowledgments: We would like to express our gratitude to Nikolaos Tsotsolas and GreenProject SA for their installation of the weather station and their operation of the internet gate, which enabled us to remotely access the instrument.

Conflicts of Interest: The authors declare no conflict of interest.

References

1. Solazzo, E.; Crippa, M.; Guizzardi, D.; Muntean, M.; Choulga, M.; Janssens-Maenhout, G. Uncertainties in the Emissions Database for Global Atmospheric Research (EDGAR) Emission Inventory of Greenhouse Gases. *Atmos. Chem. Phys.* **2021**, *21*, 5655–5683. [CrossRef]
2. European Commission; Joint Research Centre; Institute for Environment and Sustainability. *Uncertainty Estimates and Guidance for Road Transport Emission Calculations*; Publications Office of the European Union: Luxembourg, 2010; ISBN 978-92-79-15307-5.
3. Marland, G. Uncertainties in Accounting for CO_2 From Fossil Fuels. *J. Ind. Ecol.* **2008**, *12*, 136–139. [CrossRef]
4. Lloyd, C.R.; Rebelo, L.-M.; Max Finlayson, C. Providing Low-Budget Estimations of Carbon Sequestration and Greenhouse Gas Emissions in Agricultural Wetlands. *Environ. Res. Lett.* **2013**, *8*, 015010. [CrossRef]
5. *Verifying Greenhouse Gas Emissions: Methods to Support International Climate Agreements*; National Research Council (U.S.); National Research Council (U.S.); National Research Council (U.S.) (Eds.) National Academies Press: Washington, DC, USA, 2010; ISBN 978-0-309-15211-2.
6. Wang, F.; Cheng, Y.; Xue, Q.; Wang, Q.; Liang, R.; Wu, J.; Sun, J.; Zhu, C.; Li, Q. Techniques to Enhance the Photoacoustic Signal for Trace Gas Sensing: A Review. *Sens. Actuators A Phys.* **2022**, *345*, 113807. [CrossRef]
7. Picqué, N.; Hänsch, T.W. Frequency Comb Spectroscopy. *Nat. Photon* **2019**, *13*, 146–157. [CrossRef]
8. Queißer, M.; Burton, M.; Fiorani, L. Differential Absorption Lidar for Volcanic CO_2 Sensing Tested in an Unstable Atmosphere. *Opt. Express* **2015**, *23*, 6634. [CrossRef]
9. Kern, C.; Trick, S.; Rippel, B.; Platt, U. Applicability of Light-Emitting Diodes as Light Sources for Active Differential Optical Absorption Spectroscopy Measurements. *Appl. Opt.* **2006**, *45*, 2077. [CrossRef]
10. Ohyama, H.; Kawakami, S.; Tanaka, T.; Morino, I.; Uchino, O.; Inoue, M.; Sakai, T.; Nagai, T.; Yamazaki, A.; Uchiyama, A.; et al. Observations of XCO2 and XCH4 with Ground-Based High-Resolution FTS at Saga, Japan, and Comparisons with GOSAT Products. *Atmos. Meas. Tech.* **2015**, *8*, 5263–5276. [CrossRef]
11. Griffith, D.W.T.; Pöhler, D.; Schmitt, S.; Hammer, S.; Vardag, S.N.; Platt, U. Long Open-Path Measurements of Greenhouse Gases in Air Using near-Infrared Fourier Transform Spectroscopy. *Atmos. Meas. Tech.* **2018**, *11*, 1549–1563. [CrossRef]
12. Butz, A.; Guerlet, S.; Hasekamp, O.; Schepers, D.; Galli, A.; Aben, I.; Frankenberg, C.; Hartmann, J.-M.; Tran, H.; Kuze, A.; et al. Toward Accurate CO_2 and CH_4 Observations from GOSAT: GOSAT CO $_2$ AND CH $_4$ VALIDATION. *Geophys. Res. Lett.* **2011**, *38*, L14812. [CrossRef]
13. Wunch, D.; Toon, G.C.; Blavier, J.-F.L.; Washenfelder, R.A.; Notholt, J.; Connor, B.J.; Griffith, D.W.T.; Sherlock, V.; Wennberg, P.O. The Total Carbon Column Observing Network. *Phil. Trans. R. Soc. A* **2011**, *369*, 2087–2112. [CrossRef]
14. Frey, M.; Sha, M.K.; Hase, F.; Kiel, M.; Blumenstock, T.; Harig, R.; Surawicz, G.; Deutscher, N.M.; Shiomi, K.; Franklin, J.E.; et al. Building the COllaborative Carbon Column Observing Network (COCCON): Long-Term Stability and Ensemble Performance of the EM27/SUN Fourier Transform Spectrometer. *Atmos. Meas. Tech.* **2019**, *12*, 1513–1530. [CrossRef]
15. Helmut, H.T.; Ureña, A.G.; Robert, J. Donovan Environmental and Other Analytical Applications. In *Laser Chemistry: Spectroscopy, Dynamics and Applications*; John Wiley & Sons: Chichester, UK; Hoboken, NJ, USA, 2007; pp. 395–408. ISBN 978-0-471-48570-4.
16. Lackner, M. Tunable Diode Laser Absorption Spectroscopy (TDLAS) in the Process Industries—A Review. *Rev. Chem. Eng.* **2007**, *23*, 65. [CrossRef]
17. Xin, F.; Li, J.; Guo, J.; Yang, D.; Wang, Y.; Tang, Q.; Liu, Z. Measurement of Atmospheric CO2 Column Concentrations Based on Open-Path TDLAS. *Sensors* **2021**, *21*, 1722. [CrossRef] [PubMed]
18. Zimmerman, J.W.; Ii, R.A.L.; Blakley, C.S.; Frish, M.B.; Laderer, M.C.; Wainner, R.T. Tunable Diode Lase Absorption Spectrometers for CO2 Wellhead and Pipeline Leakage Monitoring: Experiences from Prototype Testing at the Illinois Basin—Decatur Project, USA. *Energy Procedia* **2014**, *63*, 4083–4094. [CrossRef]
19. Bailey, D.M.; Adkins, E.M.; Miller, J.H. An Open-Path Tunable Diode Laser Absorption Spectrometer for Detection of Carbon Dioxide at the Bonanza Creek Long-Term Ecological Research Site near Fairbanks, Alaska. *Appl. Phys. B* **2017**, *123*, 245. [CrossRef]
20. Xia, J.; Zhu, F.; Zhang, S.; Kolomenskii, A.; Dong, J.; Okada, K.; Strohaber, J.; Schuessler, H.A. Probing Greenhouse Gases in Turbulent Atmosphere by Long-Range Open-Path Wavelength Modulation Spectroscopy. *Opt. Lasers Eng.* **2019**, *117*, 21–28. [CrossRef]
21. Liang, W.; Wei, G.; He, A.; Shen, H. A Novel Wavelength Modulation Spectroscopy in TDLAS. *Infrared Phys. Technol.* **2021**, *114*, 103661. [CrossRef]
22. Li, J.; Yang, X.; Li, L.; Wang, Z.; He, L.; Wu, Z.; Du, Z. Simultaneous Standoff Sensing for Methane and Hydrogen Sulfide Using Wavelength-Modulated Laser Absorption Spectroscopy with Non-Cooperative Target. *Sens. Actuators B Chem.* **2023**, *374*, 132825. [CrossRef]
23. Ismail, S.; Browell, E.V. LIDAR | Differential Absorption Lidar. In *Encyclopedia of Atmospheric Sciences*; North, G.R., Pyle, J., Zhang, F., Eds.; Elsevier: Amsterdam, The Netherlands, 2015; pp. 277–288. ISBN 978-0-12-382225-3.

24. Differential Absorption Lidar Technique (DIAL). In *Elastic Lidar: Theory, Practice, and Analysis Methods*; Kovalev, V.A.; Eichinger, W.E. (Eds.) John Wiley: Hoboken, NJ, USA, 2004; ISBN 978-0-471-20171-7.
25. Gordon, I.E.; Rothman, L.S.; Hargreaves, R.J.; Hashemi, R.; Karlovets, E.V.; Skinner, F.M.; Conway, E.K.; Hill, C.; Kochanov, R.V.; Tan, Y.; et al. The HITRAN2020 Molecular Spectroscopic Database. *J. Quant. Spectrosc. Radiat. Transf.* **2022**, *277*, 107949. [CrossRef]
26. HITRAN on the Web. Available online: https://hitran.iao.ru/home (accessed on 1 February 2023).
27. Liou, K.-N. *An Introduction to Atmospheric Radiation*, 2nd ed.; International Geophysics Series; Academic Press: Amsterdam, The Netherlands, 2002; ISBN 978-0-12-451451-5.
28. Siozos, P.; Psyllakis, G.; Velegrakis, M. A Continuous-wave, Lidar Sensor Based on Water Vapour Absorption Lines at 1.52 Mm. *Remote Sens. Lett.* **2022**, *13*, 1164–1172. [CrossRef]
29. Siozos, P.; Psyllakis, G.; Samartzis, P.C.; Velegrakis, M. Autonomous Differential Absorption Laser Device for Remote Sensing of Atmospheric Greenhouse Gases. *Remote Sens.* **2022**, *14*, 460. [CrossRef]
30. Hellenic Land Registry. Available online: http://gis.ktimanet.gr/wms/ktbasemap/default.aspx (accessed on 2 February 2023).
31. Lan, X.P.; Tans, K.W. Thoning Trends in Globally-Averaged CO_2 Determined from NOAA Global Monitoring Laboratory Measurements. Version 2023-02. Available online: http://gml.noaa.gov/ccgg/trends/ (accessed on 4 March 2023).
32. Lan, X.; Thoning, K.W.; Dlugokencky, E.J. Trends in Globally-Averaged CH_4, N_2O, and SF6 Determined from NOAA Global Monitoring Laboratory Measurements. Version 2023-01. Available online: https://doi.org/10.15138/P8XG-AA10 (accessed on 4 March 2023).
33. Veefkind, J.P.; Aben, I.; McMullan, K.; Förster, H.; de Vries, J.; Otter, G.; Claas, J.; Eskes, H.J.; de Haan, J.F.; Kleipool, Q.; et al. TROPOMI on the ESA Sentinel-5 Precursor: A GMES Mission for Global Observations of the Atmospheric Composition for Climate, Air Quality and Ozone Layer Applications. *Remote Sens. Environ.* **2012**, *120*, 70–83. [CrossRef]
34. Copernicus Data Space Ecosystem. Available online: https://dataspace.copernicus.eu/ (accessed on 1 February 2023).

hv photonics

MDPI

Article

Through-The-Coating Fabrication of Fiber Bragg Grating Relative Humidity Sensors Using Femtosecond Pulse Duration Infrared Lasers and a Phase Mask

Stephen J. Mihailov *, Huimin Ding, Cyril Hnatovsky, Robert B. Walker, Ping Lu and Manny De Silva

Security and Disruptive Technologies Research Center, National Research Council of Canada, 100 Sussex Drive, Ottawa, ON K1A 0R6, Canada; huimin.ding@nrc-cnrc.gc.ca (H.D.); cyril.hnatovsky@nrc-cnrc.gc.ca (C.H.); robert.walker2@nrc-cnrc.gc.ca (R.B.W.); ping.lu@nrc-cnrc.gc.ca (P.L.); manjula.desilva@nrc-cnrc.gc.ca (M.D.S.)

* Correspondence: stephen.mihailov@nrc-cnrc.gc.ca

Abstract: Fiber Bragg grating (FBG) relative humidity (RH) sensors are fabricated in commercially available polyimide (PI)-coated optical fibers with diameters of 50 and 125 μm. Infrared (800 nm) femtosecond pulse duration laser pulses and a phase mask are used to inscribe Type-I and Type-II FBGs directly through the protective polyimide coatings of both 50 and 125 μm diameter fibers without typical fiber processing such as hydrogen loading, cryogenic storage, stripping, recoating or annealing. The devices are then evaluated for their performance as humidity sensors. At telecom wavelengths, the 50 μm diameter fiber devices with a 10 μm thick PI coating had a wavelength shift of the Bragg resonance at a constant temperature of 2.7 pm/%RH, whereas the 125 μm diameter fiber devices with a 17 μm thick PI coating had a wavelength shift of 1.8 pm/%RH. The humidity sensors in the 50 μm diameter fiber demonstrated a more rapid response time to small changes in humidity and a weaker hysteresis when compared to the 125 μm diameter fiber devices. No modification to the PI coatings was observed during fabrication. No difference in RH sensitivity was observed for Type-I devices when compared with Type-II devices with the same fiber. The applicability of this approach for fabricating distributed RH sensing arrays with hundreds of sensing elements on a single fiber is discussed.

Keywords: optical fiber sensor; relative humidity sensors; fiber Bragg grating; nonlinear optics

Citation: Mihailov, S.J.; Ding, H.; Hnatovsky, C.; Walker, R.B.; Lu, P.; De Silva, M. Through-The-Coating Fabrication of Fiber Bragg Grating Relative Humidity Sensors Using Femtosecond Pulse Duration Infrared Lasers and a Phase Mask. *Photonics* **2023**, *10*, 625. https://doi.org/10.3390/photonics10060625

Received: 21 April 2023
Revised: 19 May 2023
Accepted: 26 May 2023
Published: 29 May 2023

1. Introduction

The measurement and monitoring of humidity and moisture have widespread importance in many industries such as food processing, packaging and storage, agriculture, pharmaceuticals and healthcare, as well as in commercial and domestic sectors where heating, ventilation, air conditioning and refrigeration (HVACR) are important. Relative humidity (RH) is defined as the ratio of the amount of water vapor present in the atmosphere to the maximum amount that the atmosphere can hold at the existing temperature. It is the most widely used parameter to quantify the amount of water vapor in the environment. Optical fiber as a sensing platform for RH has certain advantages over more conventional electronic hygrometers, namely, their small size, immunity to electromagnetic interference and chemical inertness. Fiber sensors are typically more sensitive and offer a broader range of capabilities tailored for different applications (e.g., colorimetric, point or distributed). Fiber interferometers based on polymer-filled microcavities [1] or using exotic hydrophilic polymer/silica nanoparticle fiber coatings [2] promise to improve sensor response and sensitivity. Very high RH sensitivities have also been reported for fiber Bragg gratings (FBGs) inscribed in microstructured PMMA polymer fibers [3]. A thorough review of optical fiber RH sensors that are based on optical absorption, grating structures of both Bragg and long-period, Fabry–Perot interferometers, modal interferometers, lossy mode resonances, etc., is presented in [4]. This particular work will focus on RH sensors based on fiber Bragg gratings within polyimide (PI)-coated silica fibers.

FBG-based sensors are optical filters that are photoinduced in the core of the optical fiber using high-powered ultraviolet [5] or infrared femtosecond (fs) pulse duration lasers [6]. They have been shown to be effective optical fiber sensors for direct measurement of temperature and strain [5]. In order to measure other parameters, a transduction layer often needs to be applied to the FBG in order to convert the measurand of interest into a variation of temperature or strain that can be detected by the Bragg grating. In the case of humidity measurements, polyimide (PI) fiber optic coatings were demonstrated by Kronenberg et al. to be effective moisture-sensitive transduction layers when applied to the FBG [7]. The RH detection function is caused by the swelling of the PI layer when exposed to moisture, which induces a strain on the FBG that results in a detectable wavelength shift of the light reflected by the grating. The sensitivity of the devices to RH was dependent on the thickness of the PI coating applied to the FBG. More recently, David et al. were able to enhance the sensitivity of the RH measurement by recoating FBGs that had been chemically etched to reduce their fiber diameters [8], because the force sensitivity of the optical fiber scales inversely with the square of the fiber diameter [9]. In these instances, the fabrication of the sensors is labor-intensive as fiber stripping, etching and recoating are required. These processes also reduce the fiber reliability, making the fiber sensors extremely fragile.

It has been demonstrated that FBGs can be directly inscribed through a PI protective coating on silica fibers using infrared femtosecond irradiation and a phase mask [10–12] or by the point-by-point method [13]. The resulting through-the-coating (TTC) gratings proved to have higher mechanical strength since no fiber stripping or recoating was required. Recently, an FBG was written point-by-point through the PI coating and used as an RH sensor to monitor concrete corrosion in wastewater pipes [14].

In 2020, the femtosecond laser/phase mask approach was optimized to allow for the TTC inscription of FBGs in ultra-thin, 50 μm diameter PI-coated fibers [15] without fiber photosensitization processes such as deuterium loading, which were previously required in [12]. Namely, the spherical aberration caused by the mask substrate was balanced with conical diffraction produced by the mask at a specific distance from the mask, and the chromatic aberration of the acylindrical (i.e., free from spherical aberration) focusing lens in the exposure setup was cancelled out by the chromatic dispersion of the mask. This resulted in a significant sharpening of the line-shaped focus and permitted the direct inscription of an FBG through the polyimide coating of non-sensitized 50 μm diameter fibers.

In this work, the optimized exposure techniques presented in [15] were used to fabricate TTC FBGs in PI-coated 50 and 125 μm diameter optical fibers acquired from Fibercore. By varying the laser pulse energy and number of incident pulses, low scattering loss TTC FBGs, both thermally unstable (Type-I) and thermally stable (Type-II), were created [16] and then evaluated for their responses to RH using a damp heat environmental chamber. These low-loss gratings are ideally suited for distributed sensing arrays where potentially hundreds of grating elements can be concatenated within a single length of fiber.

2. Materials and Methods

The classical optical setup for the phase mask inscription of FBGs was used [15]. The source was a Ti-sapphire regenerative amplifier laser system operating at a central wavelength of 800 nm with a Fourier-transform-limited pulse duration of 80 fs (see Figure 1). The linearly polarized fs-beam was expanded approximately 3.5 times along the optical fiber axis and focused through a zeroth-order-nulled holographic phase mask (M) (Ibsen Photonics) with a 1.07 μm pitch using a plano-convex acylindrical lens (CL) with a focal length of 15 mm. The length of the line-focused beam along the x-axis was defined by the aperture (A) of the CL's holder as 15 mm. The linear polarization of the fs-beam was parallel to the mask grooves as the phase mask was optimized to maximize the zero-order suppression for this polarization. The front surfaces of the polyimide-coated fibers from

Fibercore were placed along the line-shaped focus and aligned parallel to the CL. Each fiber was placed at a distance $d \approx 300$ μm away from the phase mask (see Figure 1), where the confocal parameter of the fs-laser focus was the smallest, and thus the peak intensity in the focus was the highest [15]. The position of the focus in the fiber core was aligned by utilizing the techniques of nonlinear photoluminescence microscopy and dark-field microscopy [17].

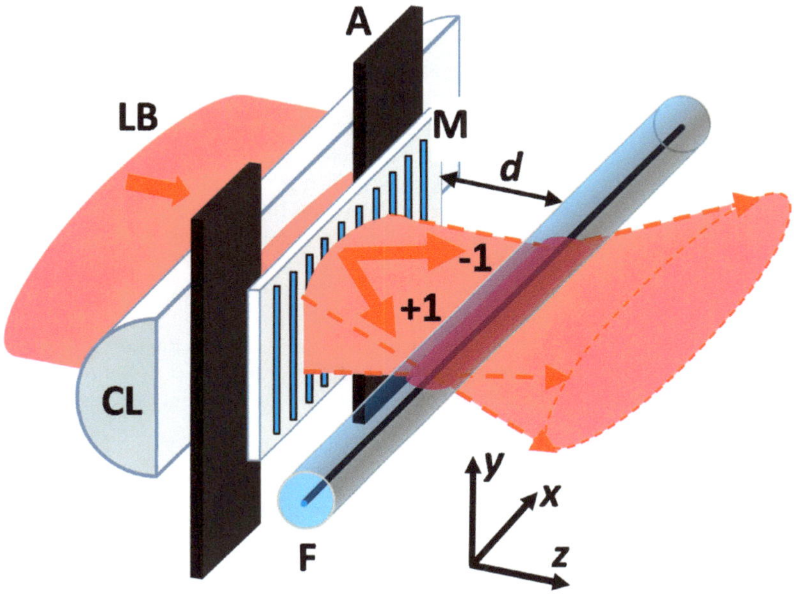

Figure 1. Schematic of the laser-writing configuration based on the phase mask technique used to photo-induce changes in the fiber core. The output laser beam (≈7 mm in diameter at the $1/e^2$-intensity level) is expanded along the x-axis. F is the fiber. LB is the femtosecond laser beam; M is the holographic phase mask, which generates only −1 and +1 diffraction orders at 800 nm; CL is the focusing acylindrical lens; A is the 15 mm wide aperture (along the x-axis) defined by the CL's holder; F is the optical fiber; and *d* denotes the mask-to-fiber distance (i.e., ≈300 μm).

Using this approach, TTC FBGs could be written in both the room temperature stable Type-I regime or the high-temperature stable Type-II regime, depending on the exposure conditions [16]. The specifications of the Ge-doped fibers used and the exposure parameters for the Type-I or Type-II modifications are given in Table 1. The fiber coating thicknesses were measured by comparison of coated and stripped fiber diameters using an optical microscope.

Table 1. Fiber parameters and exposure conditions for Type-I and Type-II index changes.

Fibercore Fiber	Diameter (μm)	PI Coating Thickness (μm)	Type-I Modification Conditions	Type-II Modification Conditions
SM1500(7.8/125)P	125	17	Beam scanning, 1 kHz pulse repetition rate, 195 μJ/pulse	1 pulse; 850 μJ/pulse
SM1500(4.2/50)P	50	10	No beam scanning, 60 s exposure μat 1 kHz, 95 μJ/pulse	3–5 pulses; 650 μJ/pulse

For the 125 μm fiber with a 7.8 μm core diameter and Type-I modification, the acylindrical lens was translated perpendicular to the fiber axis using a piezo-actuated stage in order to scan the beam across the fiber core in a single 20 μm vertical sweep with a duration of 30 s. In all other cases (i.e., Type-I Bragg gratings in the 50 μm diameter fiber and Type-II Bragg gratings in the 50 and 125 μm diameter fibers), the beam focus was centralized within the fiber cores without scanning. The bend-insensitive 50 μm diameter fiber has a smaller core size of 4.2 μm that corresponds to a higher Ge content, making the fibers more photosensitive to femtosecond pulse duration infrared radiation [10]. The pulse energies displayed in Table 1 were measured in front of the phase mask. The fabricated devices were inspected with an optical microscope to determine if there were any visible modifications to the coating surface. None were observed. A total of 12 devices were fabricated: 1 Type-I and 5 Type-II in the PI-coated 125 μm diameter fiber and 4 Type-I and 2 Type-II in the PI-coated 50 μm diameter fiber.

The devices were then tested for their sensitivity to RH using a temperature/humidity test chamber Model MCB(H)-1.2-.33-H/AC from Cincinnati Sub-Zero. The fibers were placed loosely within the chamber with no applied strain. RH levels were varied from 20 to 90%RH at a constant temperature of 40 °C and from 10 to 90%RH at a constant temperature of 60 °C in steps of 10% RH. In both cases, the chamber was programmed to change the RH gradually over 15 min and then to maintain the RH at a stable level for 30 min. RH levels were verified using internal sensors to the unit as well as an Omega HH314A Humidity Temperature meter, which possessed a 0.1%RH resolution and accuracy ± 2.5% RH. The spectral responses from the FBG sensors were monitored using a Micron Optics (Luna) Hyperion FBG interrogator.

3. Results

Examples of the transmission and reflection spectra of FBGs written in each of the Fibercore fibers are given in Figure 2. In Figure 2a,b, the transmission and reflection spectra of Type-I TTC gratings fabricated in the 125 μm and 50 μm diameter fibers are denoted by the green and blue traces, respectively. The holographic phase mask with a 1.07 μm pitch typically produces a nominal Bragg resonance wavelength of 1550 nm in standard telecommunication fibers such as SMF-28. In the case of the 125 μm diameter Fibercore fiber, the Bragg resonance wavelength was approximately 1551.4 nm. When written with the same phase mask, the Bragg wavelength of the gratings written in the 50 μm diameter fiber was longer than the Bragg wavelength of the gratings in the 125 μm diameter fiber. Indeed, the core diameter reduction is achieved by increasing the Ge doping levels of the fiber core, which, in turn, increases its effective index and hence creates a Bragg resonance at a longer wavelength for a given grating pitch. The higher Ge doping level associated with core reduction also increases the photosensitivity of the fiber to ultrafast infrared radiation [10]. Figure 2c,d display the transmission and reflection spectra, respectively, of the Type-II TTC FBGs fabricated in the 125 μm and 50 μm diameter fibers, also denoted by the green and blue traces, respectively. It is interesting to note that the appearance of shorter wavelength cladding modes in the transmission spectra, often associated with Type-II ultrafast infrared laser-induced grating writing, are increasingly separated in wavelength from the Bragg resonance as the fiber core size is reduced. This is consistent with what has been observed for FBGs written in high numerical aperture fibers [18].

Both the Type-I and Type-II FBGs fabricated in each of the fiber types were then tested for their responses to changes in RH. The spectral responses of the devices were monitored in reflection using the Hyperion interrogator, which reports a wavelength accuracy of the measurement of 1 pm. As an example, Figure 3 shows the wavelength variation of Type-I TTC FBGs written in the 50 μm diameter fiber and the 125 μm diameter fiber. For comparison, the RH levels as measured by the external Omega meter are also included.

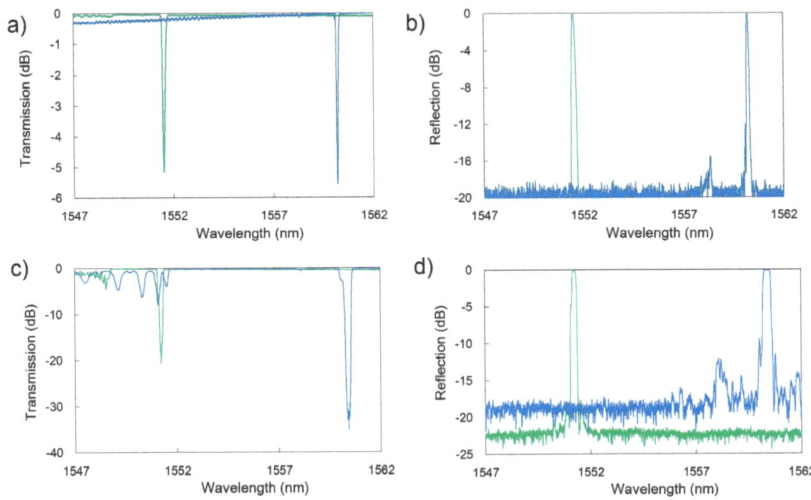

Figure 2. Example spectra of TTC FBGs written in PI-coated Fibercore fiber with a 125 μm diameter (green trace) and 50 μm diameter (blue trace). Type-I FBG transmission and reflection spectra are shown in (**a**) and (**b**), respectively, while Type-II FBG transmission and reflection spectra are shown in (**c**) and (**d**), respectively.

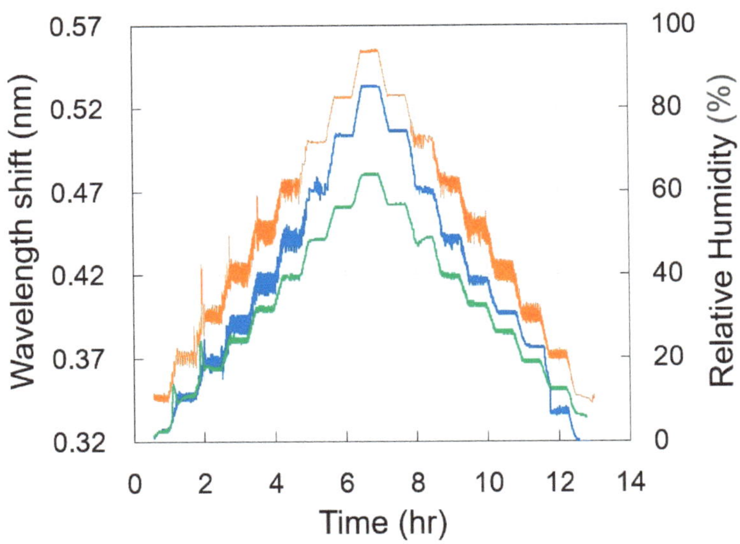

Figure 3. Example measurement of TTC FBG wavelength shift and RH as a function of time: the wavelength shift of example FBGs in PI-coated 50 μm and 125 μm diameter fibers are denoted by the blue trace and green trace, respectively. RH measurements were obtained simultaneously with the Omega HH314A Humidity Temperature meter (orange trace). The humidity chamber was maintained at 60 °C in this example. The wavelength shift was normalized to the starting time of the measurement.

Figure 3 shows that below 70% RH, the humidity within the chamber at a given 'fixed' humidity level appears to be noisy with a variation of ± 2.5%RH. A closer examination of the ramp-up to 50%RH shows this oscillation (see Figure 4). There is a direct correlation between the wavelength shift by the FBG RH sensors and the humidity measurement from the Omega HH314A humidity meter, even on the expanded time scales shown in Figure 4. According to the manufacturer, the response time of the meter is 75 s in slowly moving air. The staircase response of the wavelength shifts of the FBGs is due to the resolution of the FBG interrogator. Because of the reduced thickness of both the coating and the fiber diameter, the 50 μm fiber RH is able to better track variations in RH with time as compared to the sensor fabricated in the 125 μm diameter fiber.

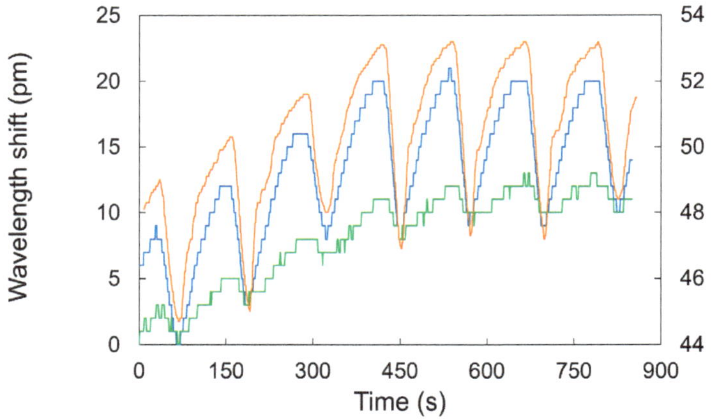

Figure 4. Comparison of response times of the 50 μm diameter (blue trace) and 125 μm diameter (green trace) TTC FBGs with changes in relative humidity as measured with the Omega HH314A humidity meter (orange trace).

By re-expressing the kind of data presented in Figure 3, where the wavelength shift for a given constant RH is averaged over 10 min, the variations in wavelength shift versus RH as a function of fiber type and refractive index change type (i.e., Type-I or Type-II) can be obtained. When measured at room temperature, the RH within the humidity chamber was not stable. The wavelength-shift measurements as a function of RH were performed at constant temperatures of 40 °C and 60 °C, as the maximum temperature for the Omega RH meter was limited to 60 °C. An example sensitivity plot of the wavelength shift versus RH of four Type-II devices, two in 50 μm and two in 125 μm diameter fibers, taken at 60 °C is presented in Figure 5. In this instance, the humidity ramp-up and ramp-down were included in the generation of the linear regression trace for each device. As some hysteresis is observable in both Figures 3 and 5, an explicit evaluation of the device hysteresis is presented in Figure 6, where the linear regression traces of data taken during humidity ramp-up and ramp-down are calculated separately. The devices that displayed the largest hysteresis were selected. From the figure, it can be seen that the total hysteresis for the 50 μm and 125 μm diameter devices is denoted by a 13 pm wavelength shift difference when the humidity returns to 10%RH. There was also a slight increase in the device sensitivity on the ramp-down, as denoted by the increase in the linear regression slope.

Figure 7 presents the wavelength shift versus humidity at 60 °C for four 50 μm diameter devices, two written in the Type-I regime and two in the Type-II regime. It can be seen that there is little difference in the response to RH between the Type-I and Type-II devices. There is also little hysteresis in the response of the devices to the ramp-up and ramp-down of the RH. The sensitivity of the devices to RH is defined by the slope of the linear regression of wavelength-versus-RH traces and is presented in Table 2a,b presents the average of the sensitivity values for all devices made with 50 μm diameter fiber (both Type-I and Type-II) and 125 μm diameter fiber at either 40 or 60 °C. The quoted error is the standard deviation of the sensitivity values.

Table 2. (**a**). Sensor sensitivities in pm/%RH as a function of temperature and Type-I and Type-II refractive index change and humidity chamber temperature. (**b**). Standard deviation of the device sensitivity response as a function of temperature.

(a)

Device		FBG Type	RH Sensitivity (pm/%RH)		Regression Slope Linearity R^2 (%)	
			40 °C	60 °C	40 °C	60 °C
50 μm	A	I	2.855	2.600	99.939	98.634
	B	I	2.731	2.548	99.800	99.169
	C	I	2.814	2.631	99.937	99.054
	D	I	2.684	2.500	99.865	99.461
	E	II	2.700	2.793	97.728	99.240
	F	II	2.845	2.690	98.870	99.099
125 μm	A	I	1.842	1.855	99.770	99.839
	B	II	1.871	1.977	99.827	99.393
	C	II	1.827	1.767	99.515	99.765
	D	II	1.828	1.779	99.593	98.763
	E	II	1.783	1.776	99.897	99.187
	F	II	1.858	1.776	99.645	99.286

(b)

Device	Temperature	Sensitivity
50 μm	40 °C	2.77 ± 0.08 pm/%RH
	60 °C	2.63 ± 0.10 pm/%RH
125 μm	40 °C	1.83 ± 0.03 pm/%RH
	60 °C	1.82 ± 0.08 pm/%RH

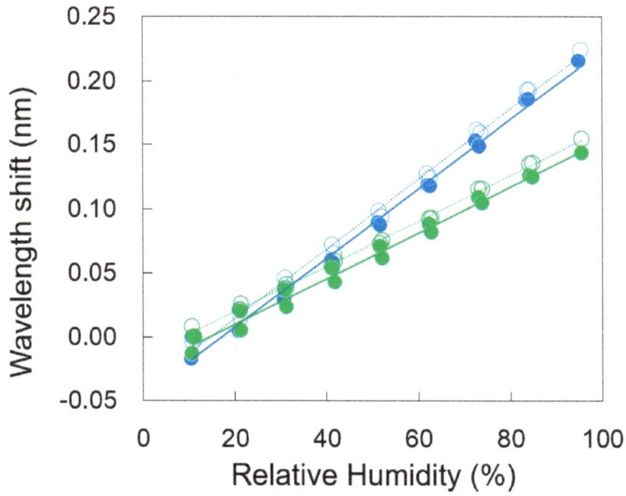

Figure 5. Measurements of Type-II TTC FBG wavelength shift as a function of RH at 60 °C for two 125 μm diameter fibers (green traces) and two devices in 50 μm diameter fibers (blue traces). Separate devices with a given fiber diameter are denoted by either open or filled bullet points. Linear regression lines are either solid or dashed for the filled or open bullet points, respectively. Slopes and errors of the regression lines for all of the devices are summarized in Table 2. RH measurements were obtained simultaneously with the Omega HH314A Humidity Temperature meter. Wavelength shifts were normalized to the starting RH value of the measurement, 10% RH.

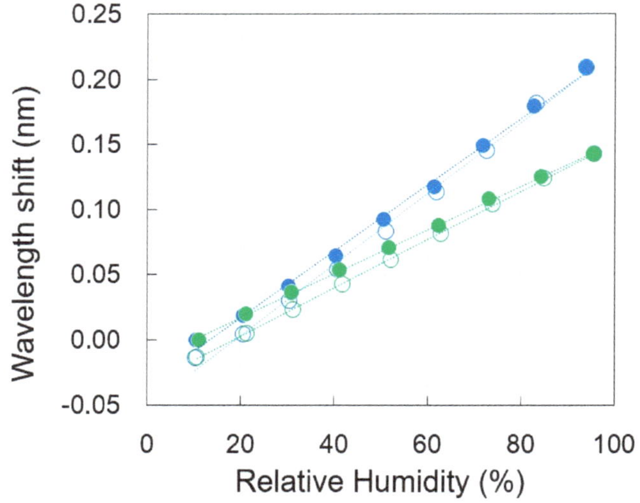

Figure 6. Hysteresis measurements at 60 °C for 125 μm (green trace) and 50 μm (blue trace) diameter devices. Solid bullets represent the wavelength shift during the humidity ramp-up, while the open bullets denote the wavelength shift during the humidity ramp-down. The dotted lines represent the linear regression of each ramp-up or ramp-down response.

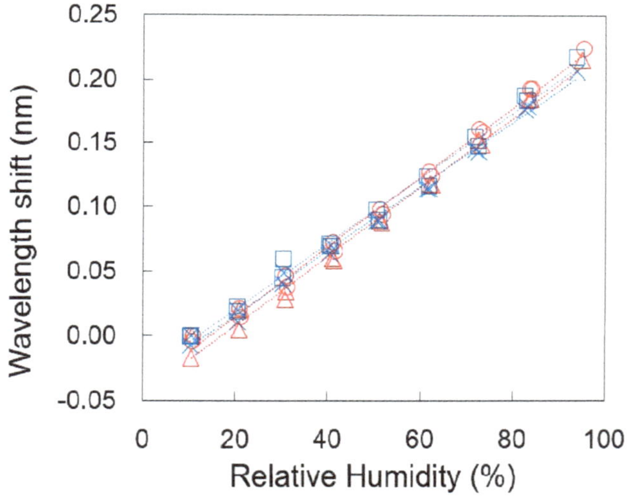

Figure 7. 50 μm fiber diameter device sensitivity at 60 °C as a function of index change type. Type-I devices are denoted by blue symbols (cross and square) and Type-II by red symbols (circle and triangle). Both humidity ramp-up and ramp-down points were included in the generation of the regression traces for each device trace.

Measurements on the Type-II devices in 50 μm diameter fiber were repeated on four different occasions at 60 °C, resulting in an average sensitivity of 2.7 ± 0.1 pm/%RH. The resulting sensitivity and error values of the repeated measurements were consistent with the overall average measurement value and error presented in Table 2b. Hysteresis of the devices improved after the fourth cycle from the worst-case wavelength difference of 13 pm shown in Figure 6 to 6 pm or ~2.5% RH, i.e., within error of the reference Omega RH probe.

4. Discussion

The response to RH of the TTC Bragg gratings in the PI-coated 125 μm diameter fiber from Fibercore listed in Table 2b is consistent with other standard PI-coated FBGs in the literature, where, for example, PI-coated FBGs from Fiberlogix and Avensys (with PI thicknesses of 14 μm and 18 μm, respectively) had wavelength sensitivities to RH of 1.3 and 1.5 pm/%RH, respectively [14,19]. Similarly, the response of the TTC Bragg gratings in the 50 μm diameter fiber is comparable to that observed for FBG RH sensors in tapered fibers. In that case, the FBGs were written in a 125 μm fiber, after which the fiber was etched to a 54 μm diameter and then coated with 20 μm of PI. The reported sensitivity of the FBG RH sensors in tapered fibers was 3.2 pm/%RH [19]. A comparison of the results from this work with some of those found in the literature is given in Table 3.

Table 3. Comparison of state-of-the-art optical fiber-based humidity sensors. (SM: single-mode; MM: multimode; PMMA: Poly(methyl methacrylate); fs PbP: femtosecond laser and the point-by-point technique).

OFS Structure	Waveguide Diameter (μm)	PI Coating Thickness (μm)	RH Sensitivity	Ref.
PI-coated FBG in SM silica fiber	125	29.3	2.58 ± 0.12 pm/%RH	[7]
PI-coated FBG in SM silica fiber and tapers	125 54	18 19	1.5 pm/%RH 3.17 pm/%RH	[19]
Fs PbP FBG in SM silica fiber	125	14	1.3 pm/%RH	[14]
Fs phase mask FBG in SM silica fiber	125 50	17 10	1.82 ± 0.08 pm/%RH 2.70 ± 0.10 pm/%RH	This work
Polymer filled F-P cavity	125	-	17.1 ± 1.7 pm/%RH	[1]
MM fibre tip with hydrophillic polymer/silica nanoparticle coating	125	-	Transmittance variation 0.43%/RH%	[2]
Microstructured PMMA polymer fiber with FBG	125	-	35 ± 3 pm/%RH	[3]

The important result to note in this work is that the sensors were produced by processing fibers as received from the manufacturer. Inscription of TTC gratings in PI-coated 125, 80 and 50 μm fibers using femtosecond lasers and a phase mask was reported by Habel et al. [12]. In that work, only Type-I FBGs were created in the PI-coated 50 μm fiber and only after the fiber had undergone the deuterium-loading fiber photosensitization process [20]. In our case, no fiber processing such as fiber photosensitization, cryogenic storage, stripping, etching, recoating or annealing was required to produce Type-I or Type-II gratings in the 50 μm diameter fiber in order to make effective RH sensors. The higher intensity irradiation required for Type-II grating formation could potentially damage the coating, causing delamination that would affect the sensitivity or performance of the device. No noticeable modification of the PI coating was observed for the FBG sensors written in either the Type-I or Type-II regimes. From Figure 7, it is clear that very similar sensitivities to RH were observed for both Type-I and Type-II sensors in the same fiber. Type-II sensors would be appropriate for measurement of moisture content in flue or exhaust gases at temperatures up to 300 °C, the rated maximum temperature for PI.

TTC inscription of FBGs lends itself to automation of the fabrication of distributed sensing arrays because extremely low scattering loss gratings can be produced in both Type-I and Type-II regimes using the phase mask method [21]. While Type-II devices do exhibit cladding-mode coupling that is potentially quite high if the gratings have high reflectivity (5 dB in the case of Figure 2c), if interrogated correctly, the lowest wavelength device first, array spacing and intensity loss can be minimized. Type-I devices do not suffer from this constraint because they have very little cladding-mode coupling. As these gratings have extremely low scattering loss, even Type-II, many can be concatenated together. To demonstrate this principle, 160 low-reflectivity Type-II gratings were inscribed through the PI coating of 125 μm telecom fibers using a phase mask array with discrete mask elements to produce sensor elements resonant in the telecommunication S-, C- and L-bands. See Figure 8. These devices were not tested for their response to humidity at the time of publication.

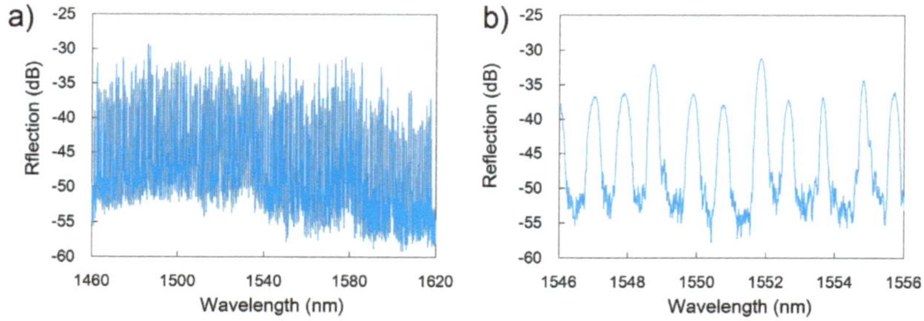

Figure 8. Reflection spectra of a 160 element TTC FBG array inscribed in PI-coated 125 μm telecom fibers: (**a**) presents the entire device spectrum over the S-, C- and L-bands, while (**b**) presents a 10 nm wavelength span centered at 1551 nm. Spectral responses were measured with a Hyperion FBG interrogator.

With proper interrogation, either using wide bandwidth FBG interrogator sources or an interrogator that combines wavelength division multiplexing and time division multiplexing (WDM/TDM) in its measurement methodology, effective distributed RH sensing systems could be created. Such a distributed sensor system is of interest to the oil and gas sector, where monitoring of RH and water vapor condensation is important for corrosion detection. With its potential low cost per unit length, simple preparation, easy operation and good sensitivity, a distributed RH sensor system based on TTC FBGs would be a good candidate for a distributed corrosion sensing system for natural gas transmission pipelines [22].

5. Conclusions

In this paper, we have shown that sensitive relative humidity sensors can be created in commercially available polyimide-coated 50 μm diameter silica optical fibers by using through-the-coating fiber Bragg grating inscription techniques based on high-powered near-infrared femtosecond pulse duration laser exposure through a phase mask. Devices stable at room temperature (Type-I) or high temperatures (Type-II) showed no difference in RH sensitivity for a given fiber geometry. The devices perform similarly to those published in the literature but are much more easily manufactured as no fiber processing techniques such as hydrogen loading, cryogenic storage, stripping, recoating or annealing are required. With the removal of these several labor-intensive processing steps that degrade fiber reliability, the described technique can be used to fabricate many robust sensors in a single fiber, potentially producing a distributed RH sensor.

Author Contributions: Conceptualization, H.D., P.L. and S.J.M.; methodology, C.H and R.B.W.; software, R.B.W.; validation, H.D.; formal analysis, H.D. and S.J.M.; investigation, H.D.; resources, M.D.S. and R.B.W.; data curation, H.D.; writing—original draft preparation, H.D. and S.J.M.; writing—review and editing, S.J.M.; visualization, S.J.M.; supervision, S.J.M.; project administration, H.D. All authors have read and agreed to the published version of the manuscript.

Funding: This research received no external funding.

Data Availability Statement: The data presented in this study are available on request from the corresponding author.

Conflicts of Interest: The authors declare no conflict of interest.

References

1. Alberto, N.; Tavares, C.; Domingues, M.F.; Correia, S.F.H.; Marques, C.; Antunes, P.; Pinto, L.L.; Ferreira, R.A.S.; André, P.S. Relative humidity sensing using micro-cavities produced by the catastrphic fuse effect. *Opt. Quant. Electron.* **2016**, *48*, 216. [CrossRef]
2. He, C.; Korposh, S.; Correia, R.; Liu, L.; Hayes-Gill, B.R.; Morgan, S.P. Optical fibre sensor for simultaneous temperature and relative humidity measurement: Towards absolute humidity evaluation. *Sens. Actuators B Chem.* **2021**, *344*, 130154. [CrossRef]
3. Woyessa, G.; Nielsen, K.; Stefani, A.; Markos, C.; Bang, O. Temperature insensitive hysteresis free highly sensitive polymer optical fiber Bragg grating humidity sensor. *Opt. Express* **2016**, *24*, 1206–1213. [CrossRef] [PubMed]
4. Ascorbe, J.; Corres, J.M.; Arregui, F.J.; Matias, I.R. Recent Developments in Fiber Optics Humidity Sensors. *Sensors* **2017**, *17*, 893. [CrossRef] [PubMed]
5. Kersey, A.D.; Davis, M.A.; Patrick, H.J.; Le Blanc, M.; Koo, K.P.; Askins, C.G.; Putnam, M.A.; Friebele, E.J. Fiber grating sensors. *J. Light. Technol.* **1997**, *15*, 1442–1463. [CrossRef]
6. Mihailov, S.J. Fiber Bragg grating sensors for harsh environments. *Sensors* **2012**, *12*, 1898–1918. [CrossRef] [PubMed]
7. Kronenberg, P.; Rastogi, P.K.; Giaccari, P.; Limberger, H.G. Relative humidity sensor with optical fiber Bragg gratings. *Opt. Lett.* **2002**, *27*, 1385–1387. [CrossRef] [PubMed]
8. David, N.A.; Wild, P.M.; Djilali, N. Parametric study of a polymer-coated fibre-optic humidity sensor Meas. *Sci. Technol.* **2012**, *23*, 035103.
9. Kou, J.-L.; Ding, M.; Feng, J.; Lu, Y.-Q.; Xu, F.; Brambilla, G. Microfiber-based Bragg gratings for sensing applications: A review. *Sensors* **2012**, *12*, 8861–8876. [CrossRef] [PubMed]
10. Mihailov, S.J.; Grobnic, D.; Walker, R.B.; Smelser, C.W.; Cuglietta, G.; Graver, T.; Mendez, A. Bragg grating writing through the polyimide coating of high NA optical fibres with femtosecond IR radiation. *Opt. Commun.* **2008**, *281*, 5344–5348. [CrossRef]
11. Bernier, M.; Trépanier, F.; Carrier, J.; Vallée, R. High mechanical strength fiber Bragg gratings made with infrared femtosecond pulses and a phase mask. *Opt. Lett.* **2014**, *39*, 3646–3649. [CrossRef] [PubMed]
12. Habel, J.; Boilard, T.; Frenière, J.-S.; Trépanier, F.; Bernier, M. Femtosecond FBG written through the coating of sensing application. *Sensors* **2017**, *17*, 2519. [CrossRef] [PubMed]
13. Dostovalov, A.V.; Wolf, A.A.; Parygin, A.V.; Zyubin, V.E.; Babin, S.A. Femtosecond point-by-point inscription of Bragg gratings by drawing a coated fiber through a ferrule. *Opt. Express* **2016**, *24*, 16232–16237. [CrossRef] [PubMed]
14. Ams, M.; Ha, P.; Taheri, S.; Clark, S.; Withford, M.J.; Bustamante, H.; Gonzalez, J.; Vorreiter, L. Fibre optic temperature and humidity sensors for harsh wastewater environments. In Proceedings of the International Conference on Sensing Technology, ICST, Sydney, Australia, 4–6 December 2017; pp. 1–3.
15. Abdukerim, N.; Grobnic, D.; Lausten, R.; Hnatovsky, C.; Mihailov, S.J. Complex diffraction and dispersion effects in femtosecond laser writing of fiber Bragg gratings using the phase mask technique. *Opt. Express* **2019**, *27*, 32536–32555. [CrossRef]
16. Smelser, C.W.; Mihailov, S.J.; Grobnic, D. Formation of Type I-IR and Type II-IR gratings with an ultrafast IR laser and a phase mask. *Opt. Express* **2005**, *13*, 5377–5386. [CrossRef] [PubMed]
17. Hnatovsky, C.; Grobnic, D.; Mihailov, S.J. Nonlinear photoluminescence imaging applied to femtosecond laser manufacturing of fiber Bragg Gratings. *Opt. Express* **2017**, *25*, 14247–14259. [CrossRef]
18. Komukai, T.; Masataka, N. Efficient Fiber Gratings Formed on High NA Dispersion-Shifted Fiber and Dispersion-Flattened Fiber. *Jpn. J. Appl. Phys.* **1995**, *34*, L1286–L1287. [CrossRef]
19. Swanson, A.J.; Raymond, S.G.; Janssens, S.; Breukers, R.D.; Bhuiyan, M.D.H.; Lovell-Smith, J.W.; Waterland, M.R. Investigation of polyimide coated fibre Bragg grating for relative humidity sensing. *Meas. Sci. Technol.* **2015**, *26*, 125101. [CrossRef]
20. Smelser, C.W.; Mihailov, S.J.; Grobnic, D. Hydrogen loading for fiber grating writing with a femtosecond laser and a phase mask. *Opt. Lett.* **2004**, *29*, 2127–2129. [CrossRef] [PubMed]
21. Hnatovsky, C.; De Silva, K.; Abdukerim, N.; Walker, R.B.; Ding, H.; Mihailov, S.J. Nanoscale morphology and thermal properties of low insertion loss fiber Bragg gratings produced using the phase mask technique and a single femtosecond laser pulse. *Opt. Express* **2022**, *30*, 47361–47374. [CrossRef] [PubMed]
22. Wright, R.F.; Badar, M.; Egbu, J.C.; Lu, P.; Buric, M.; Ohodnicki, P.R., Jr. Fully distributed optical fiber sensor for water and humidity monitoring. In *Fiber Optic Sensors and Applications XVI*; SPIE: Bellingham, WA, USA, 2019; p. 1100007.

photonics

MDPI

Communication

Optical Sensor Methodology for Measuring Shift, Thickness, Refractive Index and Tilt Angle of Thin Films

Anton Nalimov [1,2,*], Sergey Stafeev [1,2], Victor Kotlyar [1,2] and Elena Kozlova [1,2]

1 IPSI RAS—Branch of the FSRC "Crystallography and Photonics" RAS, Molodogvardeyskaya 151, Samara 443001, Russia; stafeev@ipsiras.ru (S.S.); kotlyar@ipsiras.ru (V.K.); kozlova.elena.s@gmail.com (E.K.)

2 Technical Cybernetics Department, Samara National Research University, Moskovskoye Shosse 34, Samara 443086, Russia

* Correspondence: anton@ipsiras.ru; Tel.: +7-8463325787

Abstract: We propose a simple optical method and device design for the non-contact determination of small shift, thickness, refractive index, and tilt angle of thin films. The proposed sensor consists of a laser light source, a third- or two-order spiral amplitude zone plate with a high numerical aperture, and a CCD camera connected to a computer. It is shown that the third-order zone plate transforms the incident Gaussian beam into a three-petal rotating beam. By measuring the rotation angle of the three-petal intensity distribution, one can measure the following: a minimum shift along the optical axis of about 7 nm (the wavelength is 532 nm), a change in the plate thickness by 3 nm, a change in the tilt angle of the plate by 0.1 degrees, and a change in the refractive index by 0.01.

Keywords: spiral zone plate; thickness sensor; displacement sensor

1. Introduction

Rotating laser beams became known after works [1–4]. It was shown in [1] that rotating laser beams can be formed as an axial superposition of Laguerre–Gauss beams with certain numbers. In [2,3], rotating laser beams were experimentally formed as a superposition of Laguerre–Gauss beams using a Wolaston prism in a laser cavity [3] and a diffractive optical element (DOE) [2]. In [4], rotating laser beams were also experimentally formed using DOEs, but these beams were a superposition of diffraction-free Bessel beams. In [5], a spiral conical beam was studied; it was formed using a DOE with transmission $\exp(-i\alpha r\varphi)$, where (r, φ) are the polar coordinates in the beam's cross-section. In [6], a two-lobe rotating beam consisting of two Laguerre–Gauss beams with different wavelengths was observed. Methods for increasing the formation efficiency of two-lobe rotating beams were considered in [7,8]. The formation efficiency of 57% was achieved in [8]. In [9], the orbital angular momentum of rotating beams, which are a superposition of Bessel–Gaussian beams, was analyzed, and the rotation speed of such beams was measured in [10]. It was shown in [11] that rotating laser beams can have zero orbital angular momentum. In [12], laser beams with four rotating lobes in their cross-section were formed. Rotating laser beams are used to increase the longitudinal resolution of optical microscopes. For example, in [13], a three-dimensional localization of an individual fluorescent molecule with a resolution of several nanometers was obtained using a rotating two-petal laser beam. In [14], a second-order spiral zone plate was utilized to determine the diffusion parameters of fluorescent microspheres that are 100 nm in diameter. In [15], a two-lobe rotating beam was formed by a meta surface with an efficiency of 70%. A two-lobe rotating beam was also been employed in [16] for the three-dimensional localization of quantum dots with a resolution of about 10 nm. Additionally, a two-lobe rotating beam was formed in [17] using a second-order spiral axicon. In all the works listed above, the angular velocity of the rotation of the beams in the cross-section did not exceed 50 deg/μm.

Citation: Nalimov, A.; Stafeev, S.; Kotlyar, V.; Kozlova, E. Optical Sensor Methodology for Measuring Shift, Thickness, Refractive Index and Tilt Angle of Thin Films. *Photonics* **2023**, *10*, 690. https://doi.org/10.3390/photonics10060690

Received: 3 April 2023
Revised: 13 June 2023
Accepted: 14 June 2023
Published: 15 June 2023

In this work, a three-petal rotating laser beam formed using a third-order spiral zone plate (ZP) is considered. Three local intensity maxima are formed in the beam that passed the ZP. If the ZP is displaced along the optical axis, then these three local intensity maxima will rotate around the optical axis. If we measure the rotation angle of the three-petal intensity distribution using a CCD camera, then we will be able to determine the shift of the ZP. By placing between the CCD camera and the ZP a thin film, it is possible to determine its thickness (or the change in thickness) by measuring the rotation angle of the three-petal beam. By shifting the film across the optical axis, one can measure the microrelief (thickness change) of the film over the cross-section by measuring the rotation angle of the three-petal beam. If a thin cuvette is placed between the ZP and the CCD camera, where a liquid flows, then utilizing the rotation angle of the three-petal beam, the refractive index of the liquid (or rather, the change in the refractive index), can be measured. In addition, if a thin plate placed between the ZP and the CCD camera is tilted by a small angle around an axis perpendicular to the optical axis, this tilt angle can be evaluated utilizing the angle of rotation of the three-petal beam. Note that the angular velocity of an intensity distribution rotation is shown in this work to be approximately 136 deg/μm, which is almost two times higher than in similar works on determining the 3D coordinates of nanoobjects [13,16]. For example, in [13], the angular rotational velocity of the two-lobe intensity pattern was 45 deg/μm.

2. Theoretical Background

The theory of rotating optical vortices makes it possible to determine the rotation angle and relate it to the distance propagated by the laser beam [11]. Taking into account that the distance propagated by light per time is also related to the parameters of the medium, particularly the refractive index, it becomes clear that the rotation angle can be related to the characteristics of the medium. Below, we will demonstrate how the angle of rotation of the generated intensity distribution can be used to measure characteristics such as a small shift, the thickness, the refractive index, and the tilt angle of thin films. It should also be mentioned that the theory can be applied not only in the classical case of superposition of optical vortices with different topological charges but also in the case of plane wave diffraction on a spiral axicon [17]. However, the petal laser beam can also be formed using a spiral ZP [18].

Let us compare the rotation velocity of the intensity lobes after the spiral axicon [17] and the spiral ZP.

The transmission function of the phase spiral ZP has the following form:

$$T_{ZP}(r,\varphi) = \exp\left(im\varphi - \frac{ikr^2}{2f}\right), \tag{1}$$

where (r, φ) are polar coordinates in the plate plane, m is an order, k is a wave number of the light with the wavelength λ, and f is the focal distance of the parabolic lens. We can derive the transmission function of the binary amplitude spiral ZP from Equation (1):

$$T_{BZP}(r,\varphi) = \frac{1}{2}\left(1 + \text{sgn}\left(\cos\left(m\varphi - \frac{kr^2}{2f}\right)\right)\right), \tag{2}$$

where $\text{sgn}(x) = \{1,\ x > 0;\ -1,\ x < 0\}$ is a sign function.

For a binary spiral axicon [17], the transmission function is as follows:

$$T_{BA}(r,\varphi) = \text{sgn}[\cos(k\alpha r + m\varphi)] \tag{3}$$

where α determines the numerical aperture of axicon (NA = sin α).

The rotation velocity of an m-lobe optical vortex after an axicon (3) with a period T was shown in [17] to be as follows:

$$\frac{\Delta\varphi}{\Delta z} \approx \frac{\pi\lambda}{mT^2}, \tag{4}$$

where $\Delta\varphi$ is the angle of rotation of the intensity pattern when changing the distance Δz along optical axis. The rotation velocity of the intensity lobes after the focus of the spiral ZP in (1) can be rewritten as follows:

$$\frac{\Delta\varphi}{\Delta z} \approx \frac{\pi\sqrt{N}}{mf}, \tag{5}$$

where N is the number of the N-th ring of the ZP. In the comparison of (4) and (5), we see that in both cases, the rotation velocity is constant (the tilt angle changes linearly with the distance) and depends inversely on the spiral number m. Therefore, the rotation velocity in both cases is less for the spiral element with a higher m. The minimum axicon period in (4) can be equal to the wavelength $T = \lambda$. In our investigation, we proposed that the focal length in (5) equals to the wavelength $f = \lambda$. In this case, the velocity (5) will be higher than the velocity (4) in $\sqrt{N}$ times. The use of the ZP instead of the axicon reduces the distance along the optical axis at which the beam rotates by the same angle. That is, a high numerical aperture and a short focal length of the ZP enable an increase in the angular velocity of the intensity distribution rotation by more than two times, relative to a spiral axicon [17]. With a higher rotation speed, the reference beam formed by a spiral ZP, in contrast to the beam formed by a spiral axicon [17], will make it possible to determine the characteristics of thin films more accurately. In view of this, in our work, we proposed an optical sensor technology based on a spiral ZP.

3. Optical Scheme of Measurements

In our measurement method, we propose a spiral binary amplitude Fresnel zone plate (2) with a focal distance of $f = \lambda$, where $\lambda = 532$ nm is a wavelength. The binary relief of the element is proposed to be manufactured in a layer of aluminum with a thickness of $h = 60$ nm and deposited on a glass substrate. The numerical aperture of this ZP is close to 1. The fabrication of an amplitude ZP using electron lithography technology is simpler than that of a phase ZP [19]. However, the rotation of intensity maxima and its velocity are the same for both amplitude and phase zone plates.

The optical shift sensor works as follows (Figure 1a): A linearly polarized Gaussian beam emitted by the laser is expanded by a micro objective and collected by a spherical lens onto the surface of the spiral ZP. In the transmitted beam, three local intensity maxima are formed at some distance along the optical axis behind the ZP (Figure 1b). If the CCD camera is moved along the optical axis, then the intensity distribution in the beam's cross-section, in the form of several petals, will rotate around the optical axis. By measuring the rotation angle of the petals, the shift of the ZP or CCD camera can be obtained. Light from circles of different radii propagates to different output transverse planes, causing the intensity petals to rotate. Additionally, because of the helicity of the ZP, branches of the spiral in different places intersect circles of different radii. Thus, the produced intensity maxima are visible and rotate until the ZP begins to work as a whole, that is, when light from each ZP point comes to each point of the beam.

As will be shown below, the use of a short-focus ZP instead of an axicon [17] makes it possible to increase the sensitivity of the displacement measurement method by several times. In addition, as it will be shown below, using a third-order spiral, in contrast to the second-order axicon used in [17], makes it possible to increase the accuracy of measuring the rotation angle by averaging over three lobes. At the same time, the magnitude of the order of the ZP is not so significant. You can use ZP or the second or third order.

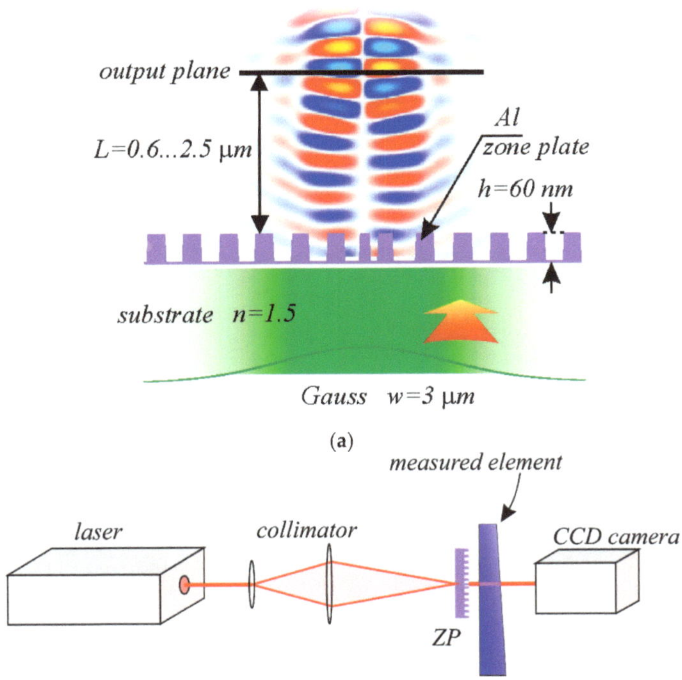

Figure 1. (**a**) Transmittance of a Gaussian beam through the spiral aluminum ZP. (**b**) Optical scheme of a sensor for measurement of thickness, displacement, refractive index, and tilt angle of a thin film.

4. Estimation of the Rotation Velocity Using Numerical Simulation

In this section, using numerical simulation, we find the dependence of the rotation angle on the shift distance of the output plane. We consider the ZP with $m = 2$, $m = 3$, and $m = 4$. The size of the ZP in the simulation is chosen to be 8×8 μm. The relief pattern of the spiral ZP for $m = 3$ is shown in Figure 2. The simulation is carried out using the FDTD method implemented in the FullWave (RSoft) software package 2018.03, 64-bit; the simulation grid in all three coordinates is $\lambda/30$. The incident wave is a Gaussian beam with a waist radius of $w = 3$ μm, limited by an aperture with a radius equal to 4 μm.

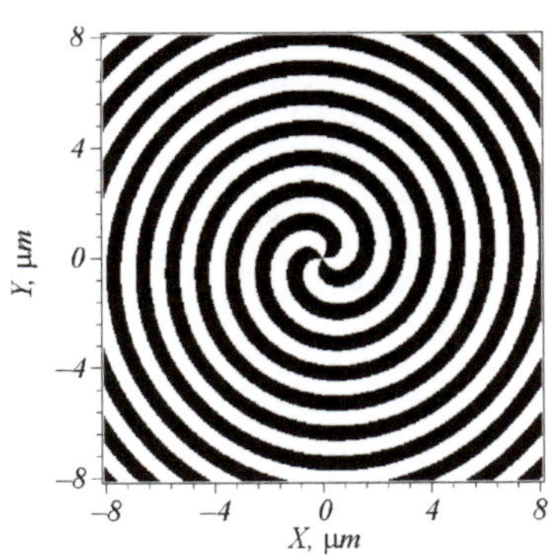

Figure 2. A spiral zone plate with $m = 3$ (black color—transmittance is 0; white—transmittance is 1).

Figure 3a demonstrates dependences of the tilt angle φ of the intensity maxima on the observation distance from the relief along the optical axis L (μm) in free space calculated for different orders m.

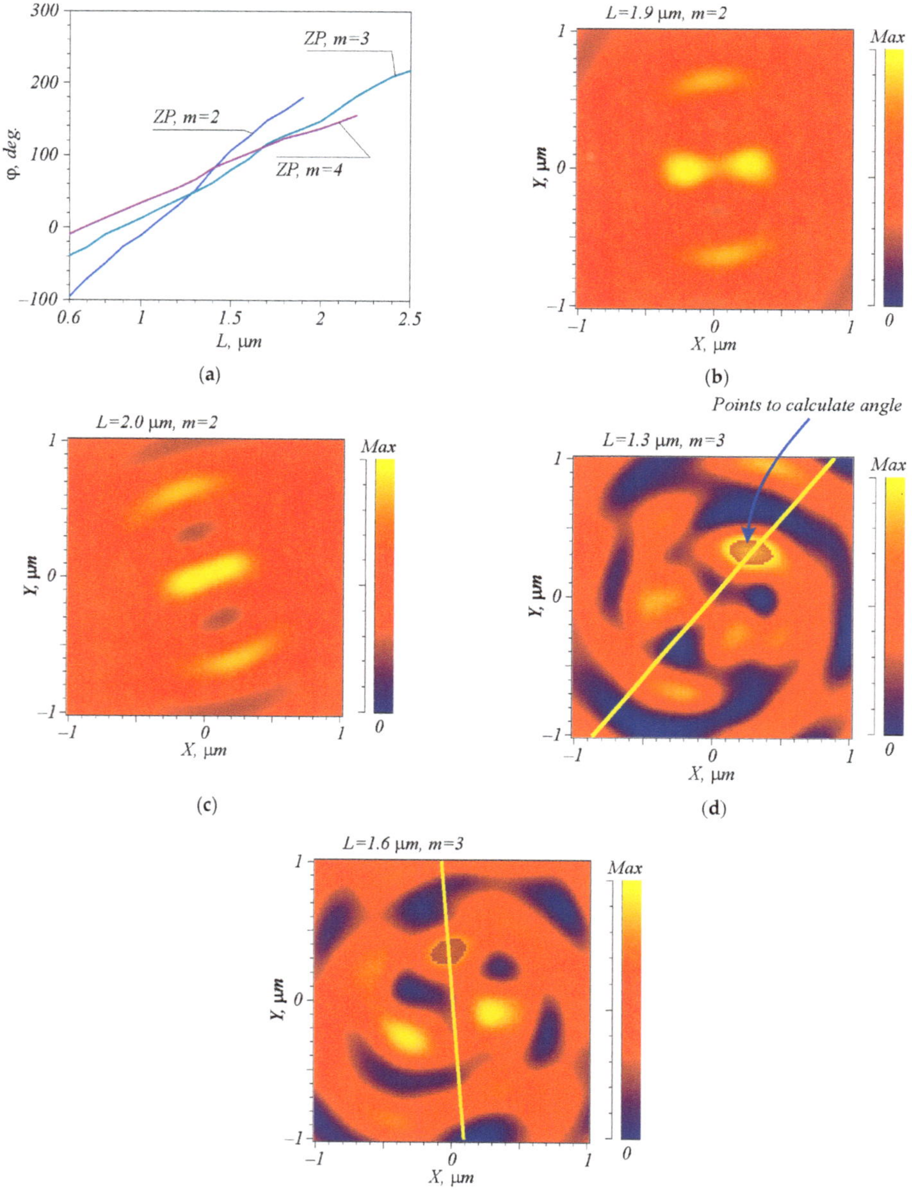

Figure 3. Dependence of the rotation angle φ of a single intensity maximum on the observation distance L from the upper edge of the ZP relief for m = 2, 3, 4 (**a**); the intensity distribution at L = 1.9 μm (**b**) and L = 2.0 μm (**c**) for m = 2; and the intensity distribution at L = 1.3 μm (**d**) and L = 1.6 μm (**e**) for m = 3. Marked points around the intensity maximum are points that were utilized to calculate rotation angle.

The plot of the tilt angle φ is seen from Figure 3a to have the longest length for the case $m = 3$, up to the value $L = 2.5$ µm. It is seen that the higher is the ZP order m, the lower is the rotation rate of the intensity maxima. The angular velocity is the highest for the order of the ZP $m = 2$, but starting from the distance $L = 2$ µm, the intensity maxima merge, which makes it difficult to recognize the angle φ (Figure 3c). The simulation for $m = 3$ shows that before the focus, three intensity lobes had not yet formed. Only after the focus ($L = f = \lambda = 0.532$ µm) do they appear and rotate up to a distance of approximately 2 µm (Figure 3a). After this distance, the intensity pattern more and more resembles an inhomogeneous light ring, and three local maxima can no longer be distinguished. At the same time, the contrast of images for values $m > 3$ or $m < 3$ is worse, as can be seen from the example in Figure 3b,c. For $m > 3$ (for example, $m = 4$), an intensity peak appears on the optical axis, and all the intensity maxima rotating around the optical axis become closer to each other. Since that, we simulate a ZP with the order $m = 3$.

The rotation angle of a single intensity maximum (it is indicated in Figure 3d by a blue arrow) is determined as follows: First, three points with the maximum intensity in the beam's cross-section are found. Then, a line connecting the selected intensity maximum with the middle of the segment connecting the centers of the other two maxima is found. This line is shown in Figure 3 in yellow. Next, the angle of this yellow line relative to the horizontal x-axis is found. In Figure 3d, points (indicated by the blue arrow) are used to obtain the average coordinates of the intensity maximum, so points greater than 0.85 of the local intensity maximum are utilized. The averaging helps to obtain more accurate positions of peaks.

At a distance of about 2 µm behind the focal plane ($f = 0.6$ µm), the rotation angle changes from $-50°$ to $210°$ (the rotation angle range is about $260°$), as shown in Figure 3a. The "velocity" of the intensity distribution rotation is 136 deg/µm, which means it takes 2.63 µm to make a full turn. For example, for the case of $m = 2$, it is 211 deg/µm, and for $m = 4$, the velocity is 102 deg/µm. Examples of evaluations of the rotation angle of the intensity maximum at $L = 1.3$ µm and $L = 1.6$ µm are shown in Figure 3d,e (ZP with $m = 3$). The rotation angle relative to the x axis is $49.3°$ for $L = 1.3$ µm and $95°$ for $L = 1.6$ µm. The calculation is performed in an automatic mode with a shift of the observation plane by 0.1 µm, and the intensity is taken into account greater than 0.8 of the maximum. Note that, in this case, the rotation velocity of the beam produced by ZP with $m = 2$ is almost 2.5 times higher than in [17], where a spiral axicon with $m = 2$ is proposed. The simulation shows that for the chosen parameters, the rotation can be distinguished with a resolution of $1°$. This means that the minimum shift possible to measure this sensor is $1000/136 \approx 7$ nm. The rotation of the intensity pattern in Figure 3 is counterclockwise. This is because the order of the spiral in Figure 2 is positive ($m > 0$). If the spiral order is negative ($m < 0$), then the intensity lobes would rotate clockwise with increasing L.

The contribution to the intensity at a distance of up to 2 µm from the ZP is seen from Figure 3b,c to be made by the first 1–2 rings of the ZP. If we assume that $N = 1.5$, then the square root will be equal to $\sqrt{N} = 1.22$ and the velocity (5) will be equal for our case (Figure 3): $\Delta\varphi/\Delta z \approx 1.22\pi/(3 \times 0.532) = 2.4$ rad/µm $= 137.5$ deg/µm. This value is slightly higher than the simulation result (136 deg/µm). It is seen from Figure 3a that with an increase in the helix order m, the velocity decreases according to Formula (5). Moreover, the ratio of the velocities obtained from Figure 3a is exactly equal to the ratio of the spiral orders: $211/136 = 3/2$ and $136/102 = 4/3$.

5. Simulation Results of the Thin Transparent Film's Thickness Measurement

Utilizing the rotation angle, not only the shift of the ZP or CCD camera can be evaluated. In this section, we describe the accuracy of measuring the thickness of thin films achievable by this sensor. We assume that a thin dielectric film is placed between the spiral ZP (Figure 2) and the observation plane (CCD camera in Figure 1b). Then, the rotation angle of the three-petal intensity distribution will be proportional to the film's thickness.

Figure 4 shows the simulation diagram (a) and the rotation angle φ as a function of the film's thickness T (b).

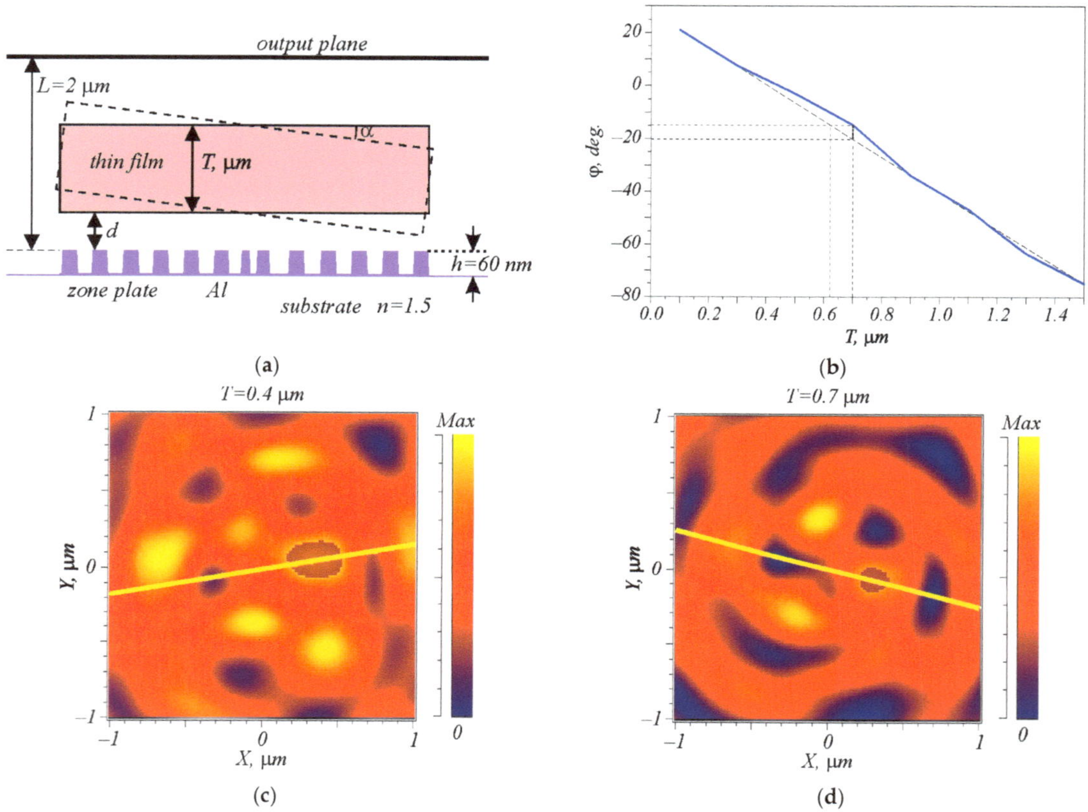

(a) (b) (c) (d)

Figure 4. Simulation scheme (**a**); the rotation angle φ as a function of the film's thickness T, located between the ZP and the observation plane (CCD camera) (**b**); intensity in the output plane at film thicknesses T = 0.4 μm (**c**) and T = 0.7 μm (**d**).

Figure 4c,d shows the intensity distributions of the three-petal beam passed through the ZP (Figure 2) for the thicknesses of the measured films (Figure 4a) T = 0.4 μm (Figure 4c) and T = 0.7 μm (Figure 4d). The refractive index of the film is n = 1.5; the distance between the film and the ZP is d = 0.2 μm. The film's thickness varies in the range from 0.1 to 1.5 μm with a step of 0.2 μm, and the film's tilt angle α = 0. The output plane (CCD camera) is placed at a distance of L = 2 μm from the upper edge of the ZP relief.

The simulation shows that an increase in the film's thickness by 1 μm (Figure 4b) causes the rotation of the intensity maxima by −68.5°, which is approximately two times less than in the case shown in Figure 3a. This is explained by the chosen refractive index of the film, which leads the eikonal (phase delay) to slowly increase twice, relative to the thickness of the film. The rotation direction in this case is the opposite (Figure 4b). This occurs because when rays fall to a medium with a higher refractive index, the angle of rays to the optical axis becomes smaller, and the focal length f of the ZP increases by the value T, which results in the intensity maxima rotation becoming slower (5). Because the rotation velocity in the medium is lower, the beam rotation angle at a distance of L = 2 μm (Figure 4a) is also smaller. Additionally, the thicker the film, the lower the rotation angle of the beam will be at the same distance. There is an effect of rotations in the opposite direction (clockwise). If the film's thickness is fixed and shifted along the z axis between the ZP and the CCD camera, then no rotation of the intensity maxima occurs.

There are two criteria for evaluating the quality of the measurement—the sensitivity of the method and the accuracy of the measurement. We obtain the rotation angle, utilizing averaging of points coordinates in a petal with intensity higher than a threshold (it equals 0.8 max); this method has very high sensitivity. This is demonstrated in Table 1. In this case, the resolution in the output plane along the x and y axes coincides with the simulation grid and is equal to $\lambda/30 = 0.0177$ μm. Table 1 shows the rotation angles φ obtained for different film thicknesses T.

Table 1. Rotation angles measured to three decimal places for indicated film thicknesses.

Thickness T, μm	φ, Deg.
0.740	91.7415
0.720	94.5576
0.710	96.0270
0.705	97.3740
0.702	98.2742
0.701	98.2742

It can be seen from Table 1 that for the given simulation parameters, the change in thickness by 1 nm cannot be distinguished. This is because the simulation grid step along the z axis has also been 0.0177 μm. That is, changes in the film's thickness may not lead to a change in the simulation scheme, since the modeling grid is coarser. However, it has been shown that a change in the thickness by $\Delta T = 3$ nm can be detected by this method. To evaluate the accuracy of measuring the film's thickness, the standard deviation of the difference between the resulting profile and the straight line closest to this graph (Figure 4b, dotted line) has been calculated. The obtained standard deviation is $\sigma = 2.39°$, i.e., the root-mean-square error in evaluating the thickness is expected to be ± 35 nm.

6. Simulation Results of the Thin Transparent Film's Tilt Angle Measurement

In this section, we show that the optical sensor with a spiral ZP can evaluate the tilt angle α of a thin transparent film placed between the ZP and the CCD camera. The film turns around an axis perpendicular to the optical axis (Figure 4a). Figure 5 demonstrates the resulting rotation angle φ of the three intensity maxima as a function of the tilt angle α of the film. The simulation parameters are as follows: the film's thickness T is 0.5 μm, the refractive index n is 1.5, and the film tilt angle α varies from 0 to 7 degrees. The distance between the film and the ZP is increased to $d = 0.6$ μm.

It follows from Figure 5 that the rotation of the film, on the one hand, affects the accuracy of determining its thickness. A change in the angle α in the range from 0 to 7° leads to a rotation of the intensity maxima by 3°, which corresponds to increases in the maximum error in measuring the thickness of ± 44 nm. However, on the other hand, Figure 5 demonstrates that the optical sensor can measure the tilt angles of the film too. If the thickness of the film is known and is necessary to evaluate the tilt angle of the film, then according to the plot in Figure 5, this method can measure the tilt angle in the range from 0° to 2° with an accuracy of 0.1°, or in the range from 3° to 7° with the same accuracy.

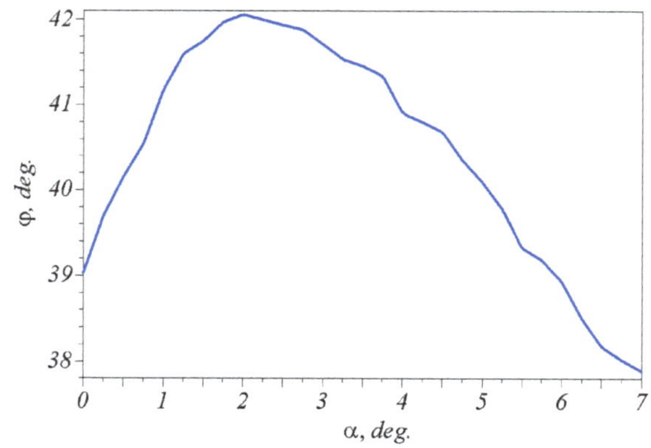

Figure 5. Dependence of the rotation angle of the three-petal intensity distribution φ on the tilt angle of the film α. Both angles are measured in degrees.

7. Simulation Results of the Thin Transparent Film's Refractive Index Measurement

It is clear that a change in the refractive index of an object between the ZP and the CCD camera also introduces a difference in the eikonal of the light wave, and the output three-petal beam rotates. Therefore, this optical sensor can evaluate the refractive index of a thin film or liquid in a cuvette if its thickness is known. The dependence of the rotation angle of the intensity maxima φ on the refractive index n in the range from 1 to 1.5 for the film's thickness T = 0.5 μm is shown in Figure 6.

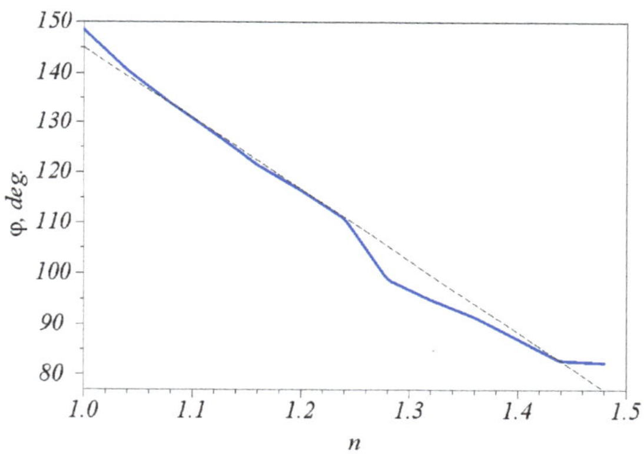

Figure 6. Dependence of the beam rotation angle φ on the refractive index of the film n.

It follows from Figure 6 that increasing the refractive index by one results in decreasing the angle φ by 141.7°. It also follows from Figure 6 that if the refractive index increases by one or two tenths, then the focal length of the ZP almost does not change and the rotation velocity is approximately equal to the rotation velocity without the film (Figure 3). However, since the speed of light inside the film is lower, the rotation angle of the beam in the output plane is smaller relative to the case without the film. Therefore, the rotation becomes clockwise (Figure 6), in contrast to the case without the film (Figure 3), where the rotation is anticlockwise. With a further increase in the refractive index of the film, the focal length of the ZP increases, and the rotation velocity decreases. Therefore, we do not show the plot for $n > 1.5$ in Figure 6. The dependence shown in Figure 6 is not a straight line. The standard deviation between the obtained plot and the straight line (dashed line in Figure 6) is 2.63°, which corresponds to the accuracy of determining the refractive index $n \pm 0.019$.

However, the sensitivity of this method is better, and the minimal resolution in measuring the refractive index is $\Delta n = 1/141 = 0.007$.

8. Experimental Measurement of the Rotation Velocity

Previously, we made a second-order ZP. Therefore, we added this section to the comparison of the simulation and the experiment for a second-order ZP. We used electron beam direct recording technology and subsequent plasma etching a second-order helical ZP ($m = 2$) with a diameter of 12 μm and a focal distance equal to the wavelength ($f = \lambda = 532$ nm), which was fabricated in a thin film (60 nm) of aluminum deposited on a silicon substrate. The experimental setup (Figure 7) included the following: a laser, a polarizer P_1, a micro-objective MO_1, focusing light on the ZP, and a micro-objective MO_2, depicting the focal plane on the CCD camera sensor. The light from the MGL-F-532-700 laser ($\lambda = 532$ nm) passed through the P_1 linear polarizer and was focused 4× by the Olympus RMS4X (MO_1) objective onto the investigated ZP (sample). Images of the transmitted beam were detected using a 100× Nikon objective (NA = 0.95) on a UCMOS10000KPA CCD camera. The camera was moved along the optical axis to register the rotation of the intensity lobes.

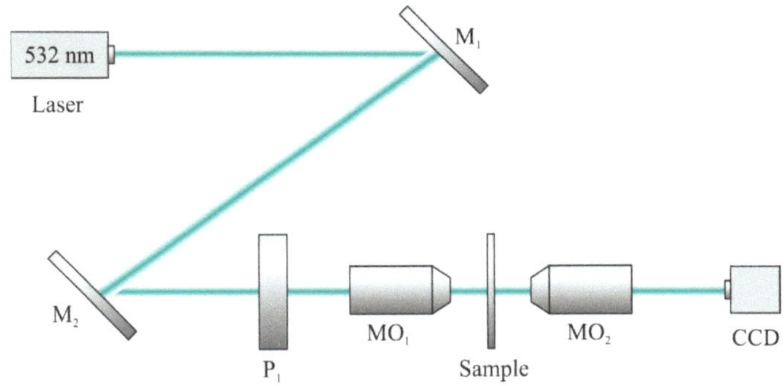

Figure 7. Experimental setup. Laser—MGL-F-532-700 ($\lambda = 532$ nm); M_1, M_2,—mirrors; P_1—linear polarizer; MO_1—a 4× objective lens (Olympus RMS4X); MO_2—a 100× objective lens (Nikon 100×/0.95); CCD—CCD camera (UCMOS10000KPA).

Images of a mask in a thin chromium film (45 nm) (a) and of the fabricated ZP in the thin aluminum film (b) derived by an electron microscope are shown in Figure 8.

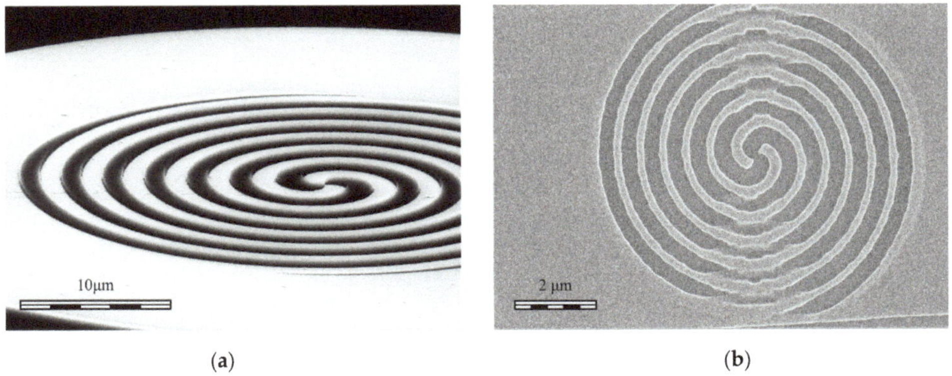

Figure 8. SEM images of a chromium mask (**a**) and the fabricated ZP of the second order in the aluminum film (**b**).

Figure 9 shows two intensity distributions recorded by the CCD camera at distances $L = 1.3$ μm (a) and $L = 1.7$ μm (b) after the ZP. Two intensity lobes are seen in Figure 9 to rotate by almost 91° after 0.4 μm. It follows from these measurements that the rotation velocity of the intensity lobes is approximately 227.5 deg/μm. This is more than the velocity obtained by the simulation (Figure 3a): 211 deg/μm. The error (about 7%) appears because of the accuracy of measuring the displacement of the CCD camera; this accuracy is about 0.1–0.2 μm.

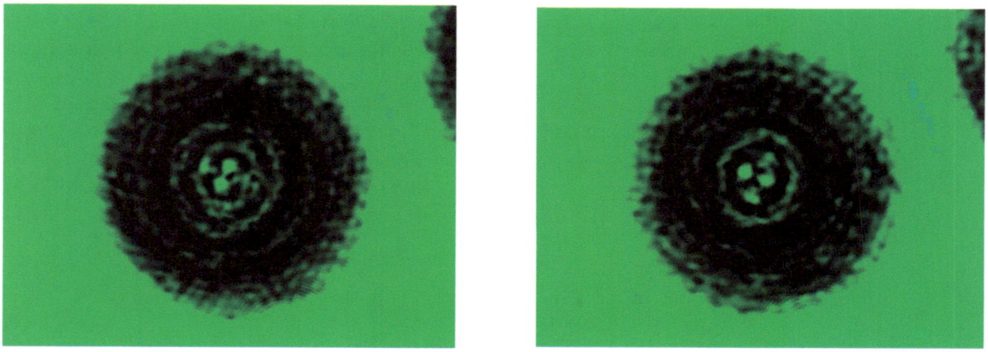

(**a**) (**b**)

Figure 9. Measured intensity patterns of light passed the ZP of the second order at distances: $L = 1.3$ μm (tilt angle $\varphi = 66.5°$) (**a**), $L = 1.7$ μm ($\varphi = 157.5°$) (**b**).

9. Conclusions

Based on the FDTD method, a simple optical sensor methodology is developed. It can evaluate several parameters at once: a shift, a thickness or relief profile, a tilt angle, and a refractive index. Rotating laser beams containing a two-petal intensity distribution have been used for a long time to increase the longitudinal resolution of optical microscopes. The two-petal point spread function makes it possible to determine the three-dimensional position of an individual molecule or quantum dot with an accuracy of 10 nm–20 nm [13,16]. In this work, a rotating beam was generated using a light modulator. In order to miniaturize the optical sensor, in this paper, we propose using a microelement in the form of a third-order spiral binary amplitude zone plate with a high numerical aperture and a size of only 8 × 8 μm, made in a thin aluminum film 60 nm thick. It is shown that shifting the output plane by about 2 μm from the ZP surface results in an almost linear rotation of the three-petal beam by about 260° (Figure 3a). The achieved angular rotation velocity of 130 deg/μm is more than two times higher than that achieved in similar works [8,13]. This paper also shows that if a thin film with a thickness ranging from 200 nm to 1400 nm is placed between the ZP and the output plane (CCD camera), then the change in the film's thickness can be evaluated utilizing the rotation angle of the three-petal intensity pattern (Figure 4b) with an accuracy of 35 nm. The minimum detectable change in the film's thickness is 3 nm. This optical sensor (Figure 1) is shown to make it possible to evaluate the refractive index of thin films if their thickness is known. In this case, the accuracy of determining the refractive index is 0.019. It can also measure the tilt angle α by several degrees with a thin transparent film. In this case, the accuracy is about 0.1°. Although this sensor methodology cannot achieve outstanding accuracy in measuring these values, it is based on a spiral zone plate, which is simple to fabricate. It is easy and quite multifunctional.

Author Contributions: Conceptualization, V.K. and A.N.; methodology, V.K., A.N., S.S. and E.K.; software, A.N.; validation, V.K.; formal analysis, V.K.; investigation, V.K., A.N., S.S. and E.K.; resources, A.N. and E.K.; data curation, A.N. and S.S.; writing—original draft preparation, A.N.; writing—review and editing, V.K. and E.K.; visualization, A.N., S.S. and E.K.; supervision, V.K.; project administration, V.K.; funding acquisition, V.K. All authors have read and agreed to the published version of the manuscript.

Funding: This research was funded by the Russian Science Foundation, grant number #23-12-00263 (in parts "Theoretical background", "Optical scheme of measurements", "Estimation of the rotation velocity by numerical simulation", "Simulation results of the thin transparent film thickness measurement", "Simulation results of the thin transparent film tilt angle measurement", and "Experimental measurement of the rotation velocity") and by the State's assignment of Federal Scientific Research Center "Crystallography and Photonics" of Russian Academy of Sciences (in part "Simulation results of the thin transparent film refractive index measurement").

Institutional Review Board Statement: Not applicable.

Informed Consent Statement: Not applicable.

Data Availability Statement: Not available.

Conflicts of Interest: The authors declare no conflict of interest.

References

1. Abramochkin, E.; Volostnikov, V. Spiral-type beams. *Opt. Commun.* **1993**, *102*, 336–350. [CrossRef]
2. Abramochkin, E.; Losevsky, N.; Volostnikov, V. Generation of spiral-type laser beams. *Opt. Commun.* **1997**, *141*, 59–64. [CrossRef]
3. Schechner, Y.Y.; Piestun, R.; Shamir, J. Wave propagation with rotating intensity distributions. *Phys. Rev. E* **1996**, *54*, R50. [CrossRef] [PubMed]
4. Paakkonen, P.; Lautanen, J.; Honkanen, M.; Kuittinen, M.; Turunen, J.; Khonina, S.N.; Kotlyar, V.V.; Soifer, V.A.; Friberg, A.T. Rotating optical fields: Experimental demonstration with diffractive optics. *J. Mod. Opt.* **1998**, *45*, 2355. [CrossRef]
5. Daria, V.R.; Palima, D.Z.; Gluckstad, J. Optical twists in phase and amplitude. *Opt. Express* **2011**, *19*, 476–481. [CrossRef] [PubMed]
6. Bekshaeva, A.Y.; Soskinb, M.S.; Vasnetsovb, M.V. Angular momentum of a rotating light beam. *Opt. Commun.* **2005**, *249*, 367–378. [CrossRef]
7. Greengard, A.; Schechner, Y.Y.; Piestun, R. Depth from diffracted rotation. *Opt. Lett.* **2006**, *31*, 181–183. [CrossRef] [PubMed]
8. Pavani, S.R.P.; Piestun, R. High-efficiency rotating point spread functions. *Opt. Express* **2008**, *16*, 3484–3489. [CrossRef] [PubMed]
9. Litvin, I.A.; Dudley, A.; Forbes, A. Poynting vector and orbital angular momentum density of superpositions of Bessel beams. *Opt. Express* **2011**, *19*, 16760–16771. [CrossRef] [PubMed]
10. Rop, R.; Dudley, A.; Lopez-Mariscal, C.; Forbes, A. Measuring the rotation rates of superpositions of higher-order Bessel beams. *J. Mod. Opt.* **2012**, *59*, 259–267. [CrossRef]
11. Kotlyar, V.V.; Khonina, S.N.; Skidanov, R.V.; Soifer, V.A. Rotation of laser beams with zero of the orbital angular momentum. *Opt. Commun.* **2007**, *274*, 8–14. [CrossRef]
12. Dorrah, A.H.; Zamboni-Rached, M.; Mojahedi, M. Controlling the topological charge of twisted light beams with propagation. *Phys. Rev. A* **2016**, *93*, 063864. [CrossRef]
13. Backlund, M.P.; Lew, M.D.; Backer, A.S.; Sahl, S.J.; Grover, G.; Agrawal, A.; Piestun, R.; Moerner, W.E. The Double-Helix Point Spread Function Enables Precise and Accurate measurement of 3D Single-Molecule Localization and Orientation. In Proceedings of the Single Molecule Spectroscopy and Superresolution Imaging VI, SPIE BiOS, San Francisco, CA, USA, 2–7 February 2013; Volume 8590, p. 85900L. [CrossRef]
14. Li, H.; Yun, X.; Zhang, Y.; Wang, F.; Huang, W. Optimization of Fresnel-zones-based Double Helix Point Spread Function and measurement of particle diffusion coefficient. *Opt. Commun.* **2022**, *502*, 127411. [CrossRef]
15. Jin, C.; Zhang, J.; Guo, C. Metasurface integrated with double-helix point spread function and metalens for three-dimensional imaging. *Nanophotonics* **2019**, *8*, 451–458. [CrossRef]
16. Volostnikov, V.G.; Vorontsov, E.N.; Kotova, S.P.; Losevsky, N.N.; Prokopova, D.V.; Razueva, E.V.; Samagin, S.A. Generation of two-lobe light fields with a rotating intensity distribution under propagation for single emitter spectroscopy. *EPJ Web Conf.* **2016**, *132*, 02012. [CrossRef]
17. Kotlyar, V.V.; Stafeev, S.S.; Nalimov, A.G.; Schulz, S.; O'Faolain, L. Two-petal laser beam near a binary spiral axicon with topological charge 2. *Opt. Laser Technol.* **2019**, *119*, 105649. [CrossRef]

18. Savelyeva, A.A.; Kozlova, E.S. Comparison of the shape of focal spots in terms of intensity and energy flux for a high-aperture zone plate and a spiral zone plate. *Comput. Opt.* **2022**, *46*, 531–536. [CrossRef]
19. Kozlova, E.; Stafeev, S.; Podlipnov, V.; Fomchenkov, S.; Kotlyar, V. Theoretical and Experimental Study of Spiral Zone Plates in Aluminum Thin Film. In Proceedings of the IEEE Explore: 2021 International Conference on Information Technology and Nanotechnology (ITNT), Samara, Russia, 20–24 September 2021; pp. 1–4. [CrossRef]

photonics

Communication

Immobilized Sorption-Colorimetric Microprobes for Chemical Analysis

Sergey Kovalev, Tatyana Tikhomirova and Vladimir Apyari *

Department of Chemistry, Lomonosov Moscow State University, 119991 Moscow, Russia; rabota.mincz@mail.ru (S.K.); tikhomirova-tatyana@yandex.ru (T.T.)
* Correspondence: apyari@mail.ru; Tel.: +7-495-939-46-08

Abstract: Here, we propose a concept of immobilized sorption-colorimetric microprobes for preconcentration and sensing of colored analytical forms. Individual particles of sorbents distributed over a small area of 1 mm × 1 mm and attached to an easy-to-made strip with an adhesive layer were considered as the probes producing colorimetric responses through preconcentration of colored analytical forms. These responses were then directly recorded with a scanner at 1200 dpi, separated from a background, and processed to obtain information about the amounts of analytes. The food dyes Fast Green FCF, and Ponceau-4R were used as the proof-of-concept colored analytes. The microprobes based on silica modified with quaternary ammonium bases and on alumina were studied. Some features of the analytes' adsorption by the probes and their scanometric sensing were found. It was shown that the proposed method is applicable for the determination of 1–7 mg L^{-1} of the dyes.

Keywords: chemical sensors; lab on chip; optical sensors; solid-phase extraction; colorimetry; digital image analysis; food dyes

Citation: Kovalev, S.; Tikhomirova, T.; Apyari, V. Immobilized Sorption-Colorimetric Microprobes for Chemical Analysis. *Photonics* **2023**, *10*, 787. https://doi.org/10.3390/photonics10070787

Received: 7 June 2023
Revised: 2 July 2023
Accepted: 4 July 2023
Published: 7 July 2023

1. Introduction

Miniaturization of devices for sorption preconcentration is one of the steadily developing areas of analytical chemistry. Minicolumns for extraction of analytes from aqueous and gaseous phases have been developed and successfully used. The use of such columns made it possible to reduce the volume of analyzed samples and to automate the analysis by injecting the entire volume of eluate directly into the instrument. Thus, flow atomic absorption methods for the determination of metals [1] or sorption-chromatographic methods for the determination of organic compounds [2,3] are known. These methods involve the quantitative extraction of analytes when the process is carried out in a dynamic mode [4].

Another method of the miniaturized solid-phase extraction involves implementation in a static mode (microsorption), when only 1–4% of the amounts of analytes is extracted by a sorbent, such as a fiber fixed in a glass tube, followed by thermal desorption directly in a gas chromatograph. All of these methods require sophisticated equipment, and many of them are time-consuming.

Methods of sorption preconcentration that exclude a desorption stage have been developed when the determination of the microcomponent is carried out directly on the surface of a sorbent. In this case, spectral methods of analysis are mainly used for detection. Recently, miniature devices for carrying out solid-phase extraction of analytes with the aim of their subsequent determination on the sorbent surface have attracted the attention of researchers. A procedure has been developed for the determination of antibiotics in natural waters and milk, based on sorption preconcentration of antibiotics on a small amount (~0.3 mg) of europium hydroxide precipitate due to the complex formation between Eu(III) and an antibiotic. The determination was carried out using digital colorimetry (smartphone) by sensitized fluorescence of europium on its hydroxide [5]. In a similar context, it should also be mentioned such a method of preconcentration as dispersive liquid–liquid microextraction, which uses a minimum amount of an extractant (up to several tens of microliters). Using this approach, a method has been developed for the determination of penicillins in the form of associates with methylene blue, based on measuring the colorimetric characteristics of the extracts using a smartphone [6].

The studies of sorption processes, even on a single particle of sorbent, have been reported [7–10]. For example, a microprobe based on a single crystal formed by Tb^{3+} ions and 1,3,5-benzenetribenzoic acid ligands was used for the selective determination of phosphate ions [7]. The use of such microprobes will make it possible, for example, to reduce the time of analysis and extract analytes from a small sample volume, which is especially important for the analysis of biological objects. When extracting colored compounds, it will be possible to use colorimetry with an ordinary laboratory scanner as a recording device.

Modern digital imaging devices (scanners and digital photo- and video cameras) allow one to easily and quickly register the colorimetric characteristics of one or more objects simultaneously. They have been applied in various fields of science, including analytical chemistry [11–14]. In most cases, these devices allow for capturing excellent-quality images with high resolution, which corresponds to a good locality of the colorimetric signal measurement. This opens up the possibility of a simple and affordable study of small samples, which is in line with the sharp today's trend toward the miniaturization of chemical analysis and its transition from the bench to the hand [15]. Examples of the use of household color-recording devices have been described, e.g., for the determination of drugs [16], ascorbic acid using a smartphone [17], and copper in alcohol using a digital camera [18].

In this paper, a concept of immobilized sorption-colorimetric microprobes for the preconcentration and sensing of colored analytical forms is proposed. Individual sorbent particles or their arrays fixed on an adhesive substrate were considered as the probes (see Section 2.1). The contact of the probes with the analyzed solution results in the extraction of colored analytical forms and the appearance of a colorimetric response, which can be measured directly on the surface of the solid phase (see Section 2.2). The food dyes Fast Green FCF and Ponceau 4R were chosen as the analytes.

2. Materials and Methods

2.1. Reagents and Instruments

Solutions of Ponceau 4R (Sigma-ALDRICH, Burlington, MA, USA, ≥75%), Fast Green FCF (Sigma-ALDRICH, ≥85%), $c = 0.1\ g\ L^{-1}$, and $1\ mol\ L^{-1}$ hydrochloric acid were used in work. Sorbents: silica chemically modified with quaternary ammonium bases (CMS-QAB, specific surface area $500\ m^2\ g^{-1}$, particle size 50 μm, pore diameter 6 nm), and γ-alumina (specific surface area $140\ m^2\ g^{-1}$, particle size 100–200 μm, pore diameter 8 nm) were used as the sorption-colorimetric microprobes. An Epson Perfection 4180 Photo office scanner was used to take images.

To operate microprobes, we used a simple homemade device holder, as shown in Figure 1. The device consisted of a transparent polymer plate with an adhesive zone prepared by gluing a transparent double-sided tape GPT020F (3M Company, St. Paul, MN, USA).

A sorbent was applied to the center of the adhesive zone using a pattern with a cut-out square hole of 1 mm × 1 mm or 2 mm × 2 mm with gentle pressure on it to attach it to the adhesive tape. After that, the excess sorbent was removed, and the pattern was removed from the adhesive zone. As a result, an area remained on the adhesive layer containing the separate immobilized sorption-colorimetric microprobes or their dense array, depending on the applied amount of the sorbent.

2.2. Methods

A typical analytical procedure using the immobilized sorption colorimetry microprobes is schematically shown in Figure 2. An aliquot of the dye solution and 0.5 mL of $1\ mol\ L^{-1}$ hydrochloric acid solution were added into 15 mL-test tubes and diluted with distilled water up to 10 mL. The solution was mixed and transferred into a beaker. The holder with microprobes was immersed in the solution and kept for 30 min. The device was then removed from the beaker, washed with water, and dried in air. In the case of a small sample volume analysis, the solution was dropped onto a horizontally located holder with the immobilized microprobes and held for 30 min. Next, the analyzed solution was washed off with distilled water, and the device was dried in air. Blank microprobes were prepared using the same procedure without the addition of the dyes.

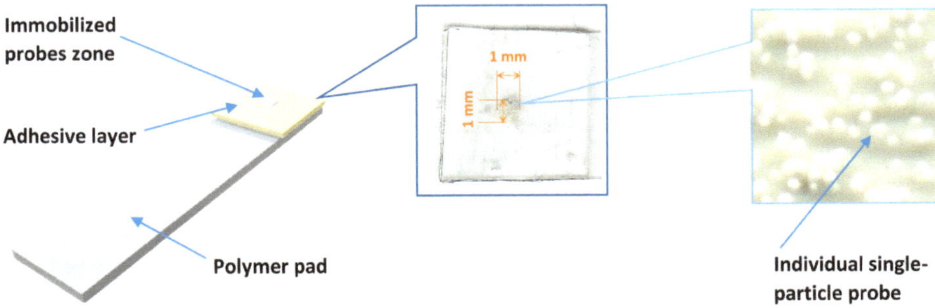

Figure 1. Design of a simple device carrying immobilized sorption-colorimetric microprobes.

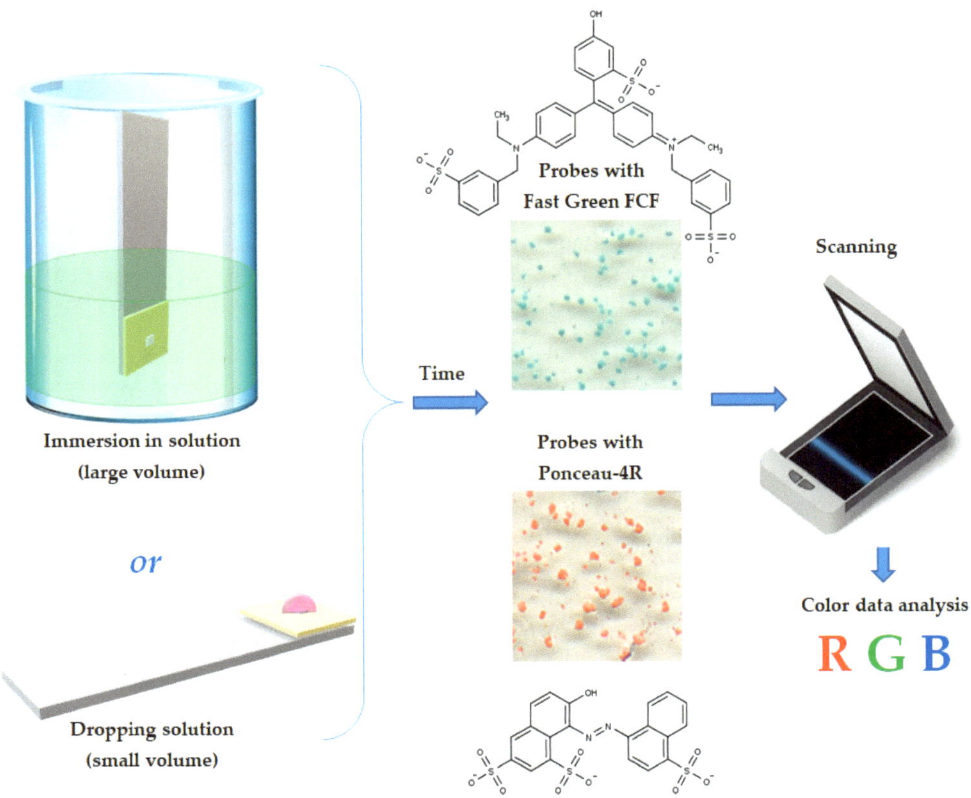

Figure 2. Scheme of immobilized sorption-colorimetric microprobes application (by the example of synthetic food dyes Fast Green FCF and Ponceau-4R). This figure has been designed using assets from Freepik.com.

The holders with microprobes were scanned at a resolution of 1200 dpi; digital image files in TIF format were saved to a computer and analyzed using the standard Histogram procedure using Adobe Photoshop software in the RGB color space. A square area of 1 mm × 1 mm or 2 mm × 2 mm was selected as a region of interest (ROI). The channel intensities were found from the histograms of their distribution, visually dividing the maximum in the histogram in half and considering the position of the secant as a desired value.

3. Results and Discussion

The concept of immobilized sorption-colorimetric microprobes proposed in this work implies the use of individual sorption elements (for example, individual particles of sorbent) or their small arrays for the extraction of colored analytical forms with subsequent registration of the colorimetric response directly on the surface of the microprobes. For the convenience of handling microprobes, an easy-to-made device was proposed (Figure 1), which uses an adhesive layer of double-sided adhesive tape to fix them on the plate. The number of applied microprobes can be controlled by using patterns with holes of different sizes and different sorbent filling densities. It was established that the microprobes are firmly held on the surface of the adhesive layer, and they are not washed off during their use, washing, and drying. The devices can be easily placed on the scanner glass and scanned at high resolution in both reflection and transmission modes.

Various sorbents can be used to create the microprobes. The main requirements for them are high efficiency and sorption rate in relation to the studied analytical form, providing sensitivity and rapid determination, as well as the absence of intrinsic color, which reduces background noise during the colorimetric measurement. In this work, silica chemically modified with quaternary ammonium bases (CMS-QAB), and γ-alumina were chosen for the preparation of the microprobes. These sorbents have a rigid, wide-pore matrix structure, which provides a high mass transfer rate. The presence of anion-exchange centers guarantees the efficiency of extraction of anionic dyes, which were the model analytes in this study. The sorption microprobes were used to determine the anionic food dyes Fast Green FCF (FG) and Ponceau-4R (P-4R). Previously [19], we determined the distribution coefficients in the Henry region for the sorption isotherms on γ-alumina (5×10^2 cm^3 g^{-1} for FG and 1×10^3 cm^3 g^{-1} for P-4R). When using CMS-QAB, these values were determined as 3×10^3 cm^3 g^{-1}, which guarantees high extraction efficiency for the dyes.

To quantify color, the RGB color system was used, formed by three primary colors—red (R), green (G), and blue (B) [20]. The sensitivity of one or another channel to a change in the content of the analytical form depends on its color. For the green dye FG, the most sensitive channel is R; for P-4R, it is B.

3.1. *Peculiarities of Recording the Colorimetric Signal of Microprobes*

To optimize the conditions for recording the signal of microprobes, it is necessary to set the scanner resolution, at which the channel intensity values do not depend on it. The chosen resolution should also ensure the measurement of individual sorbent granules but should not lead to a strong slowdown in scanning and a significant increase in the file size. Figure 3 shows the change in channel intensities with increasing the scanning resolution. At low resolutions, individual microprobes cannot be recognized in the scanned image. Instead, a number of pixels possessing average intensity values between the microprobes and the background can be observed. Increasing resolution results in appearing images of the individual microprobes. Simultaneously their signals can be more effectively separated from the background. It can also be seen that the almost constant values of channel intensities are observed starting from 1200 dpi. At lower resolutions, the errors of the response measurement increase remarkably because of the small number of pixels in the image. At higher resolutions, a significant increase in the file size is noted without a remarkable increase in the image quality. Therefore, in further experiments, the scanning was carried out at the resolution of 1200 dpi.

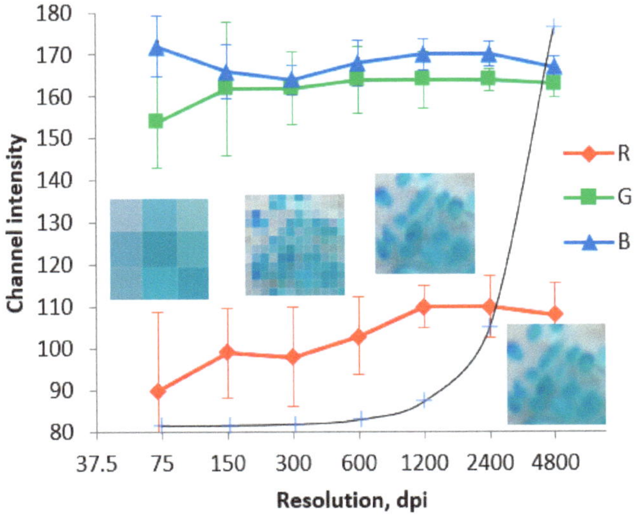

Figure 3. Dependence of RGB channel intensities on the scanning resolution for FG on CMS-QAB (the thin black line illustrates the increase in image file size; the inserts represent the same small array of microprobes scanned at 72, 300, 1200, and 4800 dpi, respectively).

The effect of the scanning mode (transmission or reflection) on the quality of microprobes images and their response in the presence of an analyte was studied. The results are represented in Figure 4. In this figure, the left part of the histogram represents R, G, and B channel intensities for the blank microprobes recorded in the transmission and reflection mode. It can be seen that the reflection mode provides higher intensity values (brighter image). After the adsorption of FG (the right part of the histogram), all channel intensities decrease, and the R channel provides the highest change, i.e., the highest analytical response. Obviously, the reflection mode ensures the greatest difference between the blank and the probe in the R channel.

It can also be concluded from the figure that the image sharpness is much higher when scanning is performed in the transmission mode. Therefore, this mode can be recommended for studying the faceting and structure of sorbent particles. However, it can be seen from the presented histograms that the channel intensities and their change during dye sorption are smaller in the transmission mode than those in the reflection one.

From an analytical point of view, it is important that a sensory system provides the maximum response during dye sorption. Therefore, in further studies, the reflection mode was considered the optimal one.

Changing the size of the microprobe zone from 1 mm × 1 mm to 2 mm × 2 mm has practically no effect on the value of the analytical response. The 2 mm × 2 mm zone was measured at various points on the image using a 1 mm × 1 mm region of interest (ROI). It was shown that the value of relative standard deviation (RSD) for five measurements at different points of the sample did not exceed 7%. In order to miniaturize the device, we used the size of the microprobe zone of 1 mm × 1 mm in all further experiments.

When measuring macroscopic amounts of sorbents, the colorimetric characteristics of the substrate have practically no effect on the recorded signal. In contrast to this variant, the layer of microprobes immobilized on an adhesive substrate is characterized by a small thickness and greater inhomogeneity. The immobilized sorption-colorimetric microprobes can be distributed unevenly and incompactly on the adhesive substrate. This makes it necessary to separate their signal from the signal associated with the background. For this purpose, it was proposed to carry out a colorimetric analysis using the histograms of the channel intensity distribution provided by the standard application of Adobe Photoshop.

This possibility is illustrated in Figure 5a. In this figure, a square area marked with a dotted line is a region of interest (ROI), i.e., an area selected on the image that is used for calculating the average intensities of R, G, and B channels. Adobe Photoshop provides a histogram of intensity distribution for each of the three color channels. Examples of such distributions for R channels built for different ROI are shown in the top parts of each image in Figure 5.

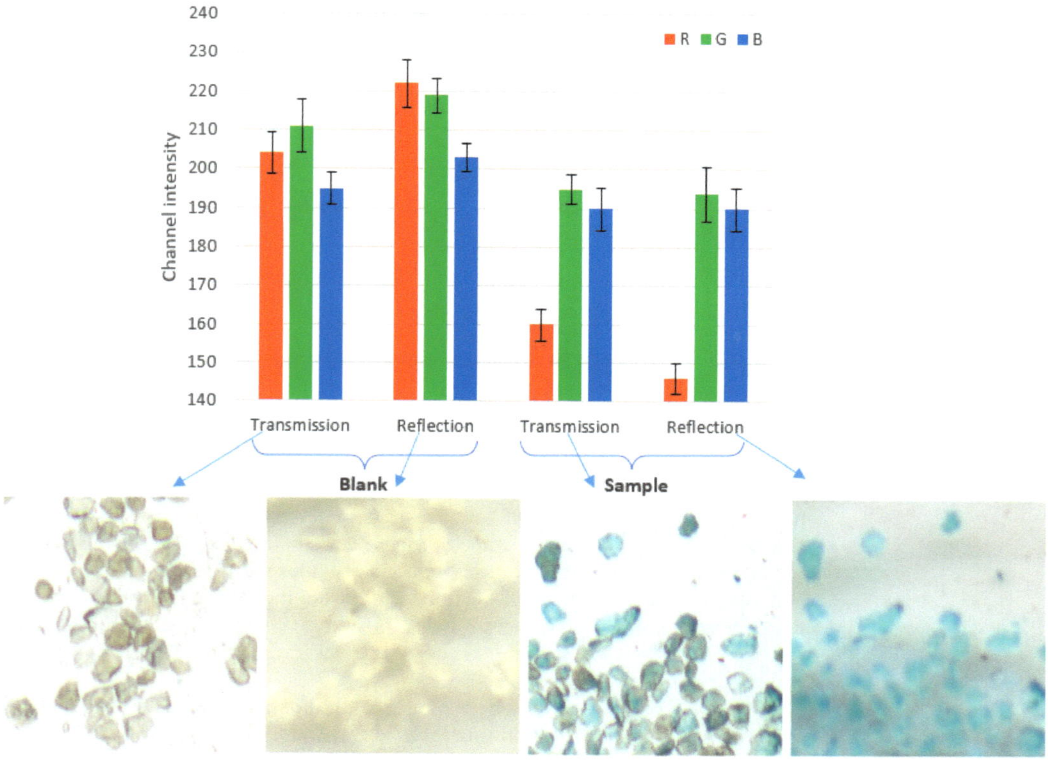

Figure 4. Channel intensities and images of the CMS-QAB-based immobilized sorption-colorimetric microprobes recorded in the transmission and reflection scanning modes. Blank is pure CMS-QAB adsorbent, sample is CMS-QAB treated with solution of FG.

The choice of an area completely filled with the sorbent as an ROI gives a histogram with a maximum R channel intensity distribution at 145. A gradual decrease in the proportion of the sorbent and an increase in the proportion of the background within the ROI leads to two peaks appearing on the histogram. The left peak is still associated with the "useful" signal from the microprobes, while the right one is due to the background contribution. It can be seen from the presented images that the microprobe signal can be extracted with sufficient accuracy even when the microprobe proportion in the ROI is only about 10%. The standard deviation of the recorded "useful" signal is 1.5, which is 2% of the difference between the background and the microprobe signals. In this case, the background contribution can be effectively separated and does not interfere with the "useful" signal.

An important issue is also the possibility of reducing the size of the microprobes array down to single particles of the sorbent. This capability was demonstrated by reducing the size of ROI (Figure 5b). It can be seen from the figure that a decrease in ROI size from 1 mm × 1 mm to 85 μm × 85 μm leads to a depletion of the histogram and its narrowing.

However, the microprobe signal can still be determined even at an ROI size of about a single sorbent particle, and its standard deviation is only two units.

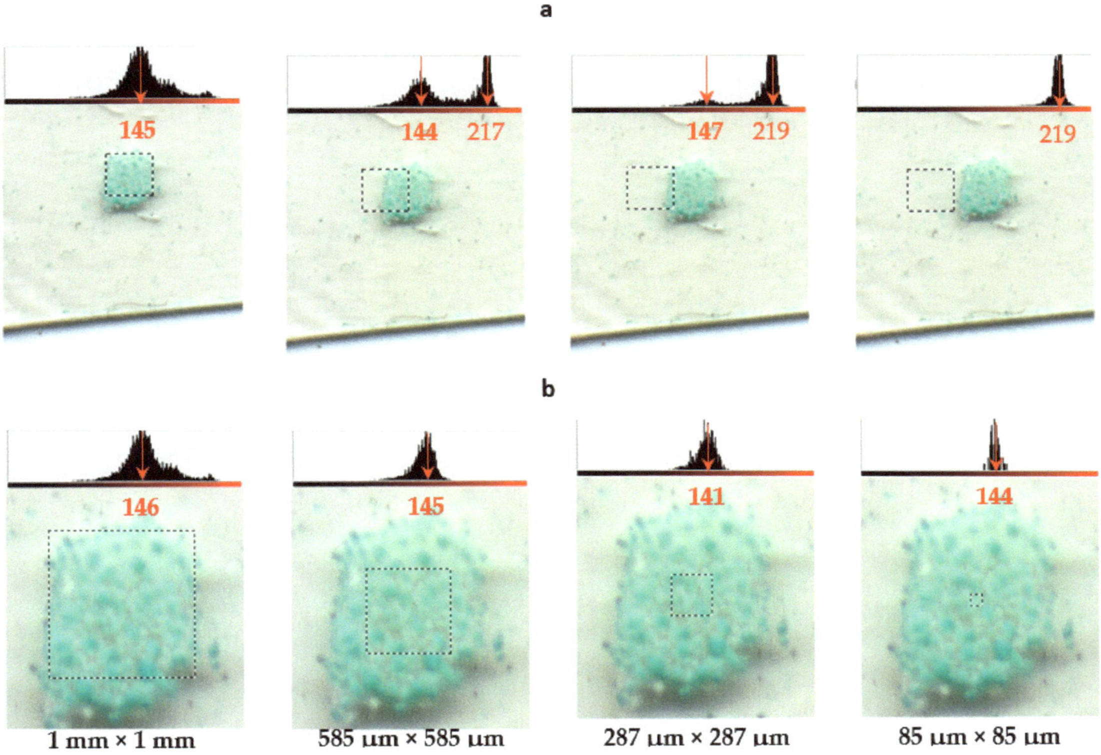

Figure 5. Histogram of R channel intensity distribution for FG on alumina depending on (**a**) position and (**b**) size of ROI (the channel intensities associated with the dye and the background are indicated with a bold and a plain red font, respectively; (**b**): the number of pixels within ROI of the indicated size is 8836, 3025, 729, and 64, respectively).

3.2. Prospects for the Analytical Application of Microprobes

For the practical use of microprobes in chemical analysis, it is necessary to establish: the time to reach sorption equilibrium and the dependence of the analytical response on the concentration of an analyte in solution. The study of recovery depending on the volume of an analyzed sample is of particular interest to assess the possibility of using microprobes for small sample volumes.

It was found that 30 min is sufficient to reach equilibrium. Calibration plots of channel intensities of the microprobes based on CMS-QAB and alumina on the concentration of dyes FG and P-4R were plotted (Figure 6). It was established that for CMS-QAB in the range of 0–0.01 mg mL^{-1} FG, the graph is well described (R^2 = 0.996) by an exponent $R = y_0 + Ae^{-c/t}$, where R is the corresponding channel intensity, c is the concentration of FG, y_0, A and t are parameters of the exponential fitting.

The sensitivity coefficient (slope) in the initial section, which can be calculated as the ratio of the parameters A/t, is 2.0×10^4. The plot for P-4R is also well described by the exponent for channel B (R^2 = 0.995) in the range of 0–0.01 mg mL^{-1} with a sensitivity coefficient of 1.9×10^4.

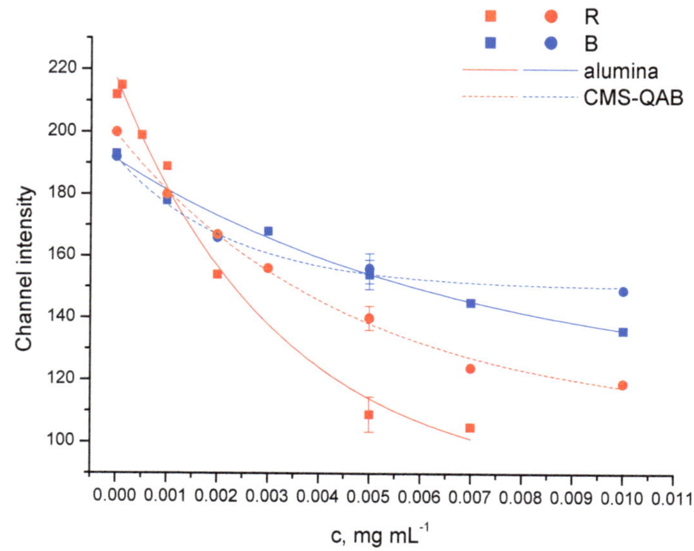

Figure 6. Dependence of the optimal channel intensity of the immobilized sorption-colorimetric microprobes on the concentration of FG (channel R) and P-4R (channel B) (extraction on CMS-QAB was carried out at pH 6.0; on alumina—from 0.05 mol L^{-1} HCl).

When alumina was used as a sorbent, the dyes were extracted from 0.05 mol L^{-1} HCl [21]. The calibration curve for P-4R on alumina does not differ much from the one on CMS-QAB. On the contrary, the R channel intensity sharply decreases with increasing the concentration of FG, which limits the determination range to 0.007 mg mL^{-1}. The sensitivity coefficient at the initial section of the calibration curve (4.2×10^4) is twice as high as compared to CMS-QAB. A similar effect was noted earlier when using diffuse reflectance spectroscopy [22]. It can be associated with a change in the conformation of FG molecules on the surface of the studied sorbents.

The estimated limits of detection lay below 0.001 mg mL^{-1}, which indicates good prospects in sensing low concentrations of the dyes. Also, it can be supposed that the microprobes will exhibit good sensitivity in detecting other colored analytical forms that are firmly adsorbed on the chosen solid phase.

In spite of the good sensitivity of the proposed method, it is worth noting that it possesses some selectivity limitations, which are inherent to molecular spectroscopy and colorimetry. They are associated with interferences from other colored compounds and analytical forms. These problems can be normally solved by choosing proper sample preparation, analytical reagents, or mathematical treatment of the colorimetric responses. As an example, it can be supposed that P-4R will not interfere with the determination of FG within the calibration range since the intensity of the R channel, which is an analytical response for FG, is almost unchanged for different concentrations of P-4R. While it is difficult to determine P-4R in the presence of FG since the intensity of the B channel changes with changing concentration of FG. In this case, it is possible to determine P-4R by the difference, taking into account the contribution to the intensity of channel B from FG.

To assess the possibility of using microprobes to determine small-volume samples, the dependence of the analytical response on the sample volume was studied (Figure 7). To perform this, the solution of FG was applied to the microprobe by dropping. It was established that the colorimetric response, which is a decrease in the R channel intensity, is observed when applying even as low as 10 μL of the analyzed solution, and it remains constant with further increase in the volume. It can be explained by the small number of microprobes within the detection zone, resulting in a small amount of the analyte adsorbed.

The quantity of adsorbed analyte is normally proportional to its equilibrium concentration in solution. The small amount of adsorbed dye means small changes in its concentration in the solution. Therefore, the colorimetric response does not depend on the volume of the solution.

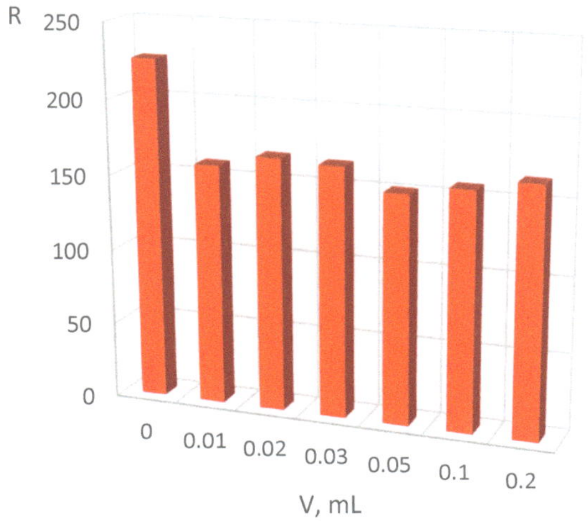

Figure 7. R channel intensity of the alumina-based sorption-colorimetric microprobes depending on the volume of analyzed FG solution (the column at V = 0 mL is a response of the microprobe zone treated with the blank solution).

These results indicate the possibility of using the sorption-colorimetric microprobes to analyze samples in a wide range of volumes, up to several tens of microliters. It is promising from the point of view of minimizing the reagent consumption, as well as solving problems when the available sample volume is limited.

4. Conclusions

A concept of immobilized sorption-colorimetric microprobes for the determination of colored analytical forms has been proposed and tested by the example of the food dyes Fast Green FCF and Ponceau-4R. A design of an easy-to-made microprobe holder carrying a 1 mm × 1 mm sensing zone has been developed. Alumina and silica, chemically modified with quaternary ammonium bases, have shown good prospects as microprobe sorbents. The conditions for recording the colorimetric analytical response using a scanner have been optimized: channels—R and B for FG and P-4R, respectively, resolution—1200 dpi, and registration mode—reflection.

It has been established that the calibration curves for both dyes on CMS-QAB and for P-4R on alumina in the range of 0–0.01 mg mL^{-1} can be fitted with an exponential equation $y = y_0 + Ae^{-c/t}$. In the case of FG on alumina, the R channel intensity drops sharply with increasing concentration, which limits the determination range by 0.007 mg mL^{-1}. The sensitivity coefficient at the initial section (4.2×10^4) is twice as high compared to CMS-QAB. It has been shown that for samples of small volume (0.01–0.2 mL), the analytical response does not depend on the sample volume.

The proposed immobilized sorption-colorimetric microprobes are easy to prepare, convenient to use, cheap, and suitable for the analysis of small sample volumes, which opens broad prospects for their practical application in the determination of colored analytical forms.

Author Contributions: Conceptualization, T.T. and V.A.; methodology, T.T. and V.A.; validation, S.K., T.T. and V.A.; formal analysis, S.K. and V.A.; investigation, S.K. and T.T.; resources, T.T.; data curation, V.A.; writing—original draft preparation, T.T. and V.A.; writing—review and editing, S.K.;

visualization, V.A.; supervision, T.T. All authors have read and agreed to the published version of the manuscript.

Funding: This research received no external funding.

Data Availability Statement: Data will be made available on request.

Acknowledgments: This work was carried out within the State Contract number AAAA-A21-121011990021-7 and according to the Development program of the Interdisciplinary Scientific and Educational School of Lomonosov Moscow State University «The future of the planet and global environmental change».

Conflicts of Interest: The authors declare no conflict of interest.

References

1. Bol'shov, M.A.; Karandashev, V.K.; Tsisin, G.I.; Zolotov, Y.A. Flow methods for the determination of elements in solutions based on sorption preconcentration and inductively coupled plasma mass spectrometry. *J. Anal. Chem.* **2011**, *66*, 548–564. [CrossRef]
2. Borisova, D.R.; Statkus, M.A.; Tsysin, G.I.; Zolotov, Y.A. On-line coupling of solid-phase extraction of phenols on porous graphitic carbon and LC separation on C18 silica gel column via subcritical water desorption. *Sep. Sci. Technol.* **2016**, *51*, 1979–1985. [CrossRef]
3. Tsizin, G.I.; Statkus, M.A.; Zolotov, Y.A. Preconcentration of organic substances on low-polar adsorbents in the flow systems of analysis. *J. Anal. Chem.* **2018**, *73*, 1030–1042. [CrossRef]
4. Tsizin, G.; Statkus, M. *Sorption Preconcentration of Microcomponents under Dynamic Conditions*; LENAND: Moscow, Russia, 2016; 480p. (In Russian)
5. Amelin, V.G.; Shogaha, Z.A.C.; Bol'shakov, D.S. Using a smartphone for determining tetracyclines in water and milk by the sensitized solid state fluorescence of europium on its hydroxide. *J. Anal. Chem.* **2021**, *76*, 1211–1216. [CrossRef]
6. Amelin, V.G.; Shogah, Z.A.C.; Bolshakov, D.S. Microextraction-colometric determination and identification of penicillin antibiotics in medicines using a smartphone and chemometric analysis. *Mosc. Univ. Chem. Bull.* **2022**, *77*, 163–170. [CrossRef]
7. Asha, K.S.; Bhattacharjee, R.; Mandal, S. Complete transmetalation in a metal–organic framework by metal ion metathesis in a single crystal for selective sensing of phosphate ions in aqueous media. *Angew. Chem. Int.* **2016**, *55*, 11528–11532. [CrossRef] [PubMed]
8. Huang, W.; Li, H.; Ling, Y.; Lin, Y.; Lei, Y.; Jin, L.; Yu, H.; He, Y. Imaging adsorption of iodide on single Cu_2O microparticles reveals the acid activation mechanism. *J. Hazard. Mater.* **2021**, *420*, 126539. [CrossRef]
9. Lei, Y.; Zhang, G.; Zhang, Q.; Yu, L.; Li, H.; Yu, H.; He, Y. Visualization of gaseous iodine adsorption on single zeolitic imidazolate framework-90 particles. *Nat. Commun.* **2021**, *12*, 4483. [CrossRef] [PubMed]
10. Rademann, J.; Barth, M.; Brock, R.; Egelhaaf, H.-J.; Jung, G. Spatially resolved single bead analysis: Homogeneity, diffusion, and adsorption in cross-linked polystyrene. *Chem. Eur. J.* **2001**, *7*, 3884–3889. [CrossRef]
11. Monogarova, O.V.; Oskolok, K.V.; Apyari, V.V. Colorimetry in chemical analysis. *J. Anal. Chem.* **2018**, *73*, 1076–1084. [CrossRef]
12. Apyari, V.V.; Gorbunova, M.V.; Isachenko, A.I.; Dmitrienko, S.G.; Zolotov, Y.A. Use of household color-recording devices in quantitative chemical analysis. *J. Anal. Chem.* **2017**, *72*, 1127–1137. [CrossRef]
13. Duangdeewong, C.; Sitanurak, J.; Wilairat, P.; Nacapricha, D.; Teerasong, S. Microfluidic paper-based analytical device for convenient use in measurement of iodate in table salt and irrigation water. *Microchem. J.* **2020**, *152*, 104447. [CrossRef]
14. Akrivi, E.A.; Vlessidis, A.G.; Giokas, D.L.; Kourkoumelis, N. Gold-modified micellar composites as colorimetric probes for the determination of low molecular weight thiols in biological fluids using consumer electronic devices. *Appl. Sci.* **2021**, *11*, 2705. [CrossRef]
15. Giokas, D.L.; Prodromidis, M.I. From the bench to the hand: Point of need/point of care technologies and analytical assays for field testing and medical diagnostics. *Microchim. Acta* **2022**, *189*, 280. [CrossRef]
16. Monogarova, O.V.; Chaplenko, A.A.; Oskolok, K.V. Multisensory colorimetric analysis of drugs dydrogesterone, troxerutin and ademetionine using barcodes. *J. Pharm. Pharmacol.* **2021**, *9*, 64–72. [CrossRef]
17. Santos, V.B.; da Silva, E.K.N.; de Oliveira, L.M.A.; Suarez, W.T. Low cost in situ digital image method, based on spot testing and smartphone images, for determination of ascorbic acid in Brazilian Amazon native and exotic fruits. *Food Chem.* **2019**, *285*, 340–346. [CrossRef]
18. Pessoa, K.D.; Suarez, W.T.; dos Reis, M.F.; Franco, M.O.K.; Moreira, R.P.L.; dos Santos, V.B. A digital image method of spot tests for determination of copper in sugar cane spirits. *Specrochim. Acta A* **2017**, *185*, 310–316. [CrossRef]
19. Tikhomirova, T.I.; Ramazanova, G.R.; Apyari, V.V. Effect of nature and structure of synthetic anionic food dyes on their sorption onto different sorbents: Peculiarities and prospects. *Microchem. J.* **2018**, *143*, 305–311. [CrossRef]
20. Schults, E.V.; Monogarova, O.V.; Oskolok, K.V. Digital colorimetry: Analytical possibilities and prospects of use. *Mosc. Univ. Chem. Bull.* **2019**, *74*, 55–62. [CrossRef]

21. Tikhomirova, T.I.; Ramazanova, G.R.; Apyari, V.V. Sorption of Ponceau 4R anionic dye from aqueous solutions on aluminum oxide and polyurethane foam. *Russ. J. Phys. Chem.* **2014**, *88*, 2192–2196. [CrossRef]
22. Ramazanova, G.R.; Tikhomirova, T.I.; Apyari, V.V. Sorption of food dyes on polyurethane foam and aluminum oxide. *Mosc. Univ. Chem. Bull.* **2013**, *68*, 175–180. [CrossRef]

hv *photonics*

MDPI

Communication

Study on Cortisol Sensing Principle Based on Fluorophore and Aptamer Competitive Assay on Polymer Optical Fiber

Vivek Semwal [1,*], Jonas Højgaard [2], Emil Møller [2], Ole Bang [1] and Jakob Janting [1,*]

[1] DTU Electro, Department of Electrical and Photonics Engineering, Technical University of Denmark, 2800 Kongens Lyngby, Denmark; oban@dtu.dk

[2] Pisco Group, Rønnegade 1, 3. th., 2100 Copenhagen, Denmark

* Correspondence: vivse@dtu.dk (V.S.); jajant@dtu.dk (J.J.)

Abstract: In this study, we present a polymer optical fiber fluorophore/aptamer competitive assay-based cortisol sensing principle. We developed a low-cost, two-fiber perpendicular design for fluorophore-based sensing with less input light interference and high output signal intensity. The design is suitable for narrow stokes shift fluorophores. We have demonstrated the cortisol sensing principle based on the competition between tagged and normal cortisol. To date, the sensing design has exhibited a slow response, and we identified possible modifications for improvement. Our estimation shows that with miniaturization and a modified sensor assay compartment design, a less than one-hour response time can be achieved. The reported sensing principle and low-cost new design will be helpful for the future development of fluorophore-based fiber optic aptasensors that can potentially be used in a wet environment for online sensing.

Keywords: fluorophore; aptamer; optical fiber; competitive assay; sensing

Citation: Semwal, V.; Højgaard, J.; Møller, E.; Bang, O.; Janting, J. Study on Cortisol Sensing Principle Based on Fluorophore and Aptamer Competitive Assay on Polymer Optical Fiber. *Photonics* **2023**, *10*, 840. https://doi.org/10.3390/photonics10070840

Received: 15 June 2023
Revised: 11 July 2023
Accepted: 19 July 2023
Published: 20 July 2023

1. Introduction

Cortisol is a stress hormone that plays a crucial role in humans and animals. In humans, it is an important biomarker for various diseases such as cardiovascular, immune, renal, skeletal, and endocrine system-based diseases [1,2]. Abnormal cortisol levels can cause irritation, depression, obesity, fatigue, bone fragility, and increased amino acid levels in the blood [2]. Cortisol can also affect, e.g., fish welfare, growth rate, and production of meat because stressed fish consume most of their energy for stress-related activities.

In the literature, the best limits of detection (LOD) for cortisol are 36 fg/mL [3] and 25.9 fg/mL [4]. The LOD for chromatographic techniques [2] and ordinary immunoassays like enzyme-linked immunosorbent assays (ELISA) [5] are approximately 1 pg/mL and 50 pg/mL, respectively. The concentration range of cortisol in humans and aquaculture is μg/mL and pg/mL, respectively, and therefore a broad range of LODs are of interest [1,2]. Currently used measurement techniques require time-consuming sampling and laboratory analysis. Further, available methods for cortisol sensing are expensive and cannot be used for online monitoring. Optical fiber sensors are attractive because they have several advantages, such as the feasibility of online measurements in water over long distances [1]. The glass optical fiber cortisol sensor developed by Usha et al. [4] is of high performance, but at the expense of being bulky and complicated to fabricate, requiring operation in transmission mode and feeding/removal of an analyte. In our polymer optical fiber (POF) immunosensor design, we have used, for the first time, a simpler design comprising perpendicular POFs, a hydrogel, and a luminescent competitive assay.

Aptamers are single-stranded DNA and are widely used for small molecule detection in biosensors [6]. Aptamers are developed by the systematic evolution of ligands by exponential enrichment (SELEX). In this process, a multi-round in vitro process is used to isolate aptamers from pools of single-stranded oligonucleotides with randomized sequences [7,8]. In recent years, aptamers have become the first choice for the recognition of specific molecules and sensing applications because they have several advantages over antibodies, such as low cost, easy synthesis, stability over a wide range of temperatures and pH levels, easy modification, etc. [6,7].

Fluorescence is one of the most dominant signal transductions for aptamer-based assays. This is due to its non-destructive, highly sensitive nature, good signal intensity, and ease of modification [9,10]. In recent years, various fluorescence techniques have revolutionized the field of receptor–ligand interactions and sensing. For the detection of molecules, different fluorescence-based strategies have been developed using the combination of aptamer and fluorophore, such as FRET [11,12], aptamer-based ELISA, fluorescence polarization [13,14] and quenching [15,16], competition assays, etc. In this study, we have shown the aptamer and fluorescence-based competitive binding assay principles for the sensing of cortisol. We have developed a new perpendicular fiber design to measure the signal with less input signal interference. In this study, the AF488 hydrazide fluorophore was used because it has several properties, such as high fluorescence quantum yield, low self-quenching, pH-insensitivity, high photostability, etc., which make this fluorophore an ideal candidate for robust sensing. The excitation and emission wavelengths of AF488H are 493 and 519 nm, respectively. Hence, the Stokes shift of AF488H is narrow. LED-based excitation systems have very broad spectra; therefore, expensive filters and complex instrumentation are normally required. The proposed new design provides an easy and low-cost replacement for the complex and expensive instrumentation for narrow Stokes shift fluorophores. This new design also enables the possibilities of online monitoring and remote sensing in a wet environment, which is challenging for an electronics-based sensor design. There are several challenges associated with the use of electronics-based sensors in wet environments, such as ineffective sensing in humid environments, the possibility of circuit damage, and corrosion [17,18]. Therefore, these sensors require very robust packaging, which increases their cost and complexity. Here, we successfully demonstrate a cortisol-sensing principle using a fluorophore and aptamer competitive binding assay. To the best of our knowledge, this is the first study on a fluorophore/aptamer competitive assay using a two-fiber perpendicular design. We also describe reasons for the current slow response of the sensor. To improve the response time, we discuss possible modifications in the sensing design.

2. Sensing Principle

The sensing mechanism is based on a competitive binding assay consisting of fluorophore-tagged cortisol molecules and an aptamer for cortisol recognition (see Figure 1). A thin layer of polyacrylamide gel in which the aptamer is physically immobilized is placed inside a compartment. Both tagged and normal cortisol can diffuse in and out through the gel, and depending on the amount of normal cortisol in the surrounding liquid, the concentration of the recognized/captured cortisol or tagged cortisol will vary inside the gel just in front of the fibers [1]. This will change the fluorescence intensity. Two POFs are connected to the compartment in a perpendicular direction, one to send the excitation light and the other to receive the output fluorescence light. This perpendicular design is good when the separation between the excitation and emission wavelengths of the tagged fluorophore is small (~20 nm). The POFs are connected perpendicularly to reduce the interference of the input signal in the emission signal of the fluorescence output spectra, something that would otherwise require filtration and, therefore, intensity loss.

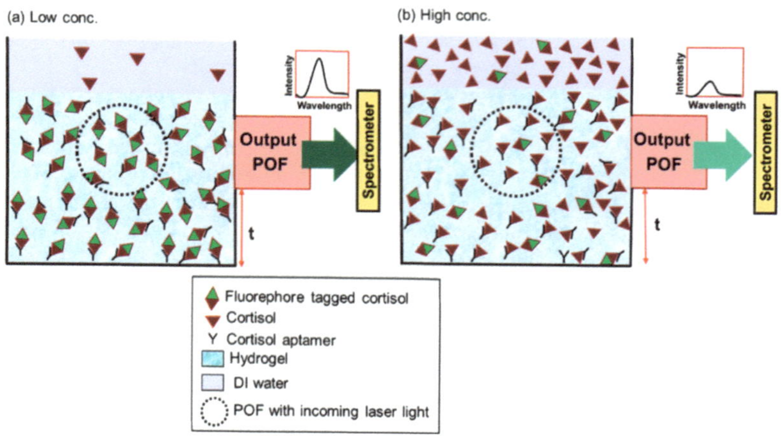

Figure 1. Schematic diagram of the sensing principle: (**a**) low cortisol concentration; (**b**) high cortisol concentration. (The dotted circle indicates the POF with incoming light; t is the distance between the bottom of the assay compartment and the fibers.)

3. Fabrication

3.1. Materials

Cortisol 3-(O-carboxymethyl)oxime (C3CMO), hydrocortisone, N-(3-dimethylaminopropyl)-N′-ethylcarbodiimide hydrochloride (EDC), anhydrous dimethylsulfoxide (DMSO), 1-hydroxy-7-azabenzotriazole (HOAt) (0.6 M in dimethylformamide (DMF)), isopropyl alcohol (IPA), acrylamide, N,N′-methylenebis(acrylamide), phosphate-buffered saline (PBS), TLC plates, TLC saturation pads, and preparative TLC (PTLC) plates (20 cm × 20 cm × 0.2 cm silica gel on glass plates) were procured from Sigma-Aldrich, Denmark. Alexa Fluor 488 hydrazide (AF488H), 2-(N-morpholino)ethanesulfonic acid buffer (MES), and (BupH™ MES buffered saline: 0.1 M MES, 0.9% sodium chloride, pH 4.7) were purchased from Fisher Scientific, Denmark. N,N,N′,N′-Tetramethyl ethylenediamine (TEMED) and ammonium persulfate (APS) were purchased from Merck. The LED (490 nm) light source was procured from Thorlabs. A 488 nm cyan blue laser diode was procured from eBay. SMA 905 connectors and a crimp tool were purchased from FiberFin, USA. Syringe filters (0.2 micron) were purchased from GE Healthcare Life Sciences. Step- index polymer optical fiber with a PMMA core diameter of approx. 980 µm and fluorinated cladding thickness of approx. 10 µm with a black polyethylene jacket, making the total outer diameter approx. 2.2 mm, was purchased from Edmund Optics. The cortisol aptamer reported by Dalirirad et al., (ATGGGCAATGCGGGGTGGAGAATGGTTGCCGCACTTCGGC) was purchased from biomers.net GmbH [19].

3.2. Labeling of Cortisol

The sensing mechanism is based on fluorescence; therefore, first, we labeled the cortisol with the AF488 hydrazide (AF488H) fluorophore. Further, for the conjugation, we chose the cortisol derivative cortisol 3-(O-carboxymethyl)oxime (C3CMO). The reason for choosing these two molecules is that EDC cross-linking can be used between the carboxyl group of C3CMO and the hydrazide primary amine on AF488H. The C3CMO and AF488H conjugation and chemical identification procedures are provided in detail in our previous work [1]. Since the chemical separation was modified compared to our earlier work, it is described in detail here. To separate the reactants and products, we used PTLC plates. First, we drew a horizontal line using a pencil from one vertical edge to the other 2 cm from both vertical bottom edges of the PTLC plate. Then, 200 µL of crude liquid were dispensed over the 16 cm long line, which is shown in Figure 2a,d. In Figure 2a–c, the images were captured in normal ambient light, while in Figure 2d–f they were captured with UV lamp (365 nm) illumination. The lines of liquid were dispensed as small drops on the PTLC plate using a 100 μL pipette. Since DMSO was used as a solvent in the labeling process of cortisol, it was evaporated in a vacuum oven at 45 °C and 0.02 mbar for 30 min. The same process was repeated until all of the product solution was dispensed. After that, any remaining DMSO was evaporated in a vacuum oven at 45 °C and 0.02 mbar for at least 3 h. For separation of the AF488H-tagged cortisol from the mixture, we mixed an eluent solvent consisting of 160 mL IPA and 40 mL MES buffer and poured it onto a TLC saturation pad inside a TLC separation chamber and waited for 5 min to equilibrate the solvent inside the chamber. We placed the PTLC plate inside the chamber immediately after taking it from the vacuum oven, and the chamber was closed with a glass plate and covered with aluminum foil. The separation step took 5 h for the eluent to reach the top of the PTLC plate. After that, the plate was removed. Next, the plate was kept inside the fume hood to dry overnight. In this process, the separation between the AF488H and AF488H-tagged cortisol was small and not very clear, which is shown in Figure 2b,e. Due to the limited separation, we again kept the PTLC plate inside the vacuum oven overnight, and the next day it was again kept inside the eluent chamber using the same protocol. After repeating a process like that, we observed a nice separation between the AF488H- and AF488H-linked cortisol, as shown in Figure 2c,f. Figure 2c,f shows the pictures of the PTLC plate with nicely separated lines of AF488H-tagged cortisol and AF488H. The AF488H was identified by dispensing AF488H separately on a PTLC plate placed inside the eluent chamber and measuring and comparing the height of the line, and the AF488H-tagged cortisol was identified using liquid chromatography–mass spectrometry (LC–MS) [1]. The silica gel containing AF488H-tagged cortisol was scraped off the glass plates and washed with stirring in 20 mL Milli-Q water for at least an hour before centrifugation in 2 mL vials at 13.4 krpm for 5 min. The water containing AF488H-tagged cortisol was collected from the vials. A 0.5 mL volume of water was added to the vials; they were vortexed until the silica gel was loose again and then placed on a rocking table overnight. Then the vials were centrifuged again, and the AF488H-tagged cortisol containing water was again collected. The washing step was repeated 3 times, and finally, it was filtered through the 0.2 μM syringe filter.

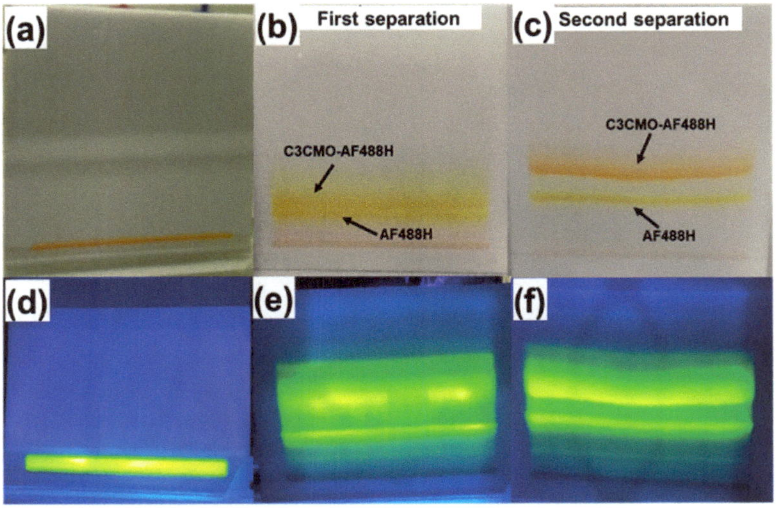

Figure 2. PTLC plates: (**a**) before being placed inside the eluent chamber; (**b**) first separation of tagged cortisol from reactants and by-products; (**c**) second separation of tagged cortisol from reactants and by-products; (**d–f**) are the images of (**a–c**) in 365 nm illumination.

3.3. Synthesis of Hydrogel Containing Aptamer and Tagged Cortisol and Porosity Test of Hydrogel

We used polyacrylamide hydrogel [20] for sensor aptamer immobilization. The porosity and transparency of the hydrogel can be changed by varying the monomer to crosslinker ratio. Thus, it was optimized by testing ratios of 19:1, 29:1, and 37.5:1. A good porous and transparent hydrogel capable of immobilizing the aptamer while allowing in/out transport of tagged/untagged cortisol was achieved with the 37.5:1 ratio between the monomer and crosslinker. For the synthesis of polyacrylamide hydrogel containing the sensor assay, 35 µL of AF488H-tagged cortisol and 20 µL (10 µmol) of cortisol aptamer were first mixed in 80 µL of phosphate-buffered saline (PBS). The cortisol aptamer and AF488H-tagged cortisol solution was then mixed with 140 µL of 30 wt.% acrylamide/bisacrylamide (37.5:1) using a magnetic stirrer in the presence of N2 gas. The nitrogen gas was purged to remove the dissolved oxygen from the solution because it may deteriorate the polymerization reaction. The solution was then mixed with 14 µL of APS (10 wt.% in DI water) and 2 µL of TEMED. After mixing with the TEMED, the gelation started, and the gel was formed within 1 min. Porosity was tested by placing small drops of AF488H- and AF488H-tagged cortisol molecules on top of the hydrogel discs shown in Figure 3a,c, respectively. Using UV lamp (365 nm) illumination, Figure 3b,d shows the penetration of the AF488H- and AF488H-tagged cortisol molecules after 24 h inside the hydrogel, which means the synthesized polyacrylamide hydrogel was porous enough that AF488H- and AF488H-tagged cortisol molecules could diffuse inside it. As shown in Figure 3d, the intensity of the green color of AF488H-tagged cortisol inside the hydrogel was very low due to the low concentration of AF488H-tagged cortisol in that specific solution.

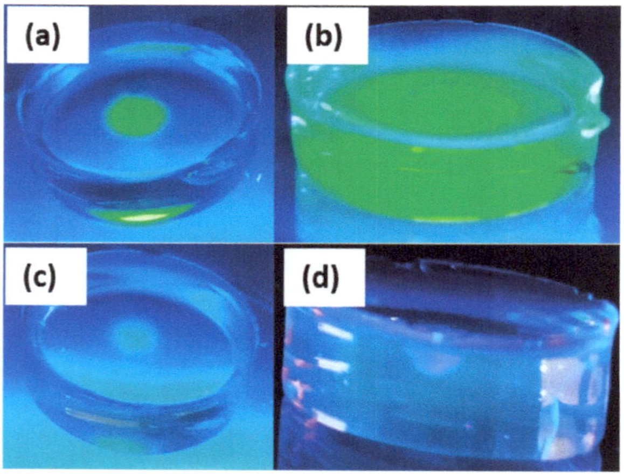

Figure 3. Images of hydrogel disc under the illumination of a 365 nm UV lamp: (**a**) right after placing a drop of AF488H solution onto it; (**b**) after 24 h with AF488H; (**c**) right after placing a drop of AF488H- tagged cortisol onto it; and (**d**) after 24 h with AF488H-tagged cortisol.

3.4. Components of Sensing Setup

The absorption and emission wavelengths of AF488H are 493 and 519 nm, respectively. Therefore, to excite this fluorophore, we tested a 490 nm LED and a 488 nm laser diode. The optical spectra of the LED and laser diode are shown in Figure 4a. The bandwidths (FWHM) of the LED and laser diode are 40 nm and 10 nm, respectively. We recorded the emission spectra of the hydrogel containing AF488H-tagged cortisol in our perpendicular fiber sensing setup, using the LED and laser diode as excitation sources. The output spectra are shown in Figure 4b. The Stokes shift between the absorption and emission wavelengths of AF488H is 26 nm, which means that the 40 nm LED spectrum covers both the absorption and emission wavelengths of AF488H. This results in a strong overlapping of the excitation and emission wavelengths when the LED is used, as clearly seen in Figure 4b. In contrast, for the laser diode, there is a clear separation between excitation and emission peaks, which means that they can easily be separated with a filter. Therefore, we used the 488 nm laser diode as an excitation source.

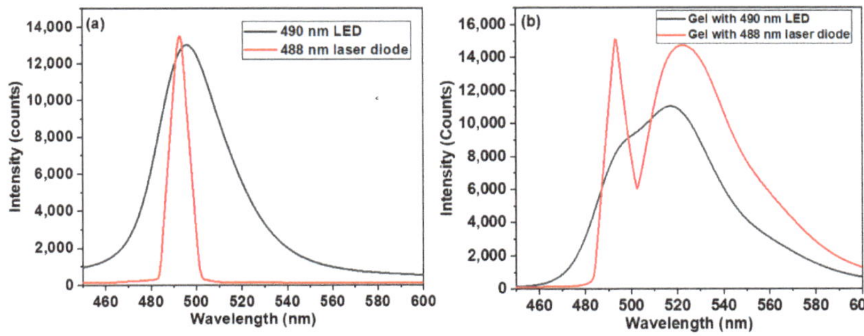

Figure 4. (**a**) The output spectra of the 490 nm LED and 488 nm laser diode. (**b**) The spectra of the 490 nm LED and 488 nm laser diode together with AF488H fluorescence after passing through the C3CMO-AF488H-containing hydrogel.

We note that our earlier proposed sensor design had the sensor assay compartment on the tip of the POF [1] and used 490 nm LED excitation light. The fluorescence signal was thus picked up with the same fiber using a fiber splitter and two filters (both of LED excitation and of fluorophore emission), which increased the size, complexity, and cost of the sensor. The excitation and emission filters also decreased the intensity of the output signal. Therefore, we used a new design (two perpendicular fibers) to solve all issues related to low intensity, cost, and size. The new perpendicular design was inspired by the fact that fluorophores emit light in all directions; therefore, to reduce the interference of incoming light, fluorescence can be collected from the perpendicular direction.

We fabricated a 3D-printed sensing compartment with two 1 mm cylindrical slots for POF attachment. The actual photo of the sensing assay compartment is shown in Figure 5. Two POFs are connected to the assay compartment using UV glue. The diameter and height of the cylindrical assay compartment are 3 mm. The tip of the POF was polished with P1200 and P4000 sandpaper in deionized water (DI) and then cleaned in DI water to obtain a better input and output signal. Both input and output fibers were connected to SMA 905 connectors using a crimp/swage hand tool. The hydrogel was prepared according to the aforementioned protocol in Section 3.3. The hydrogel was placed inside the cavity, and the sensing compartment was stored at room temperature for 30 min for complete polymerization of the hydrogel. After 30 min, the sensing compartment was used for measurements.

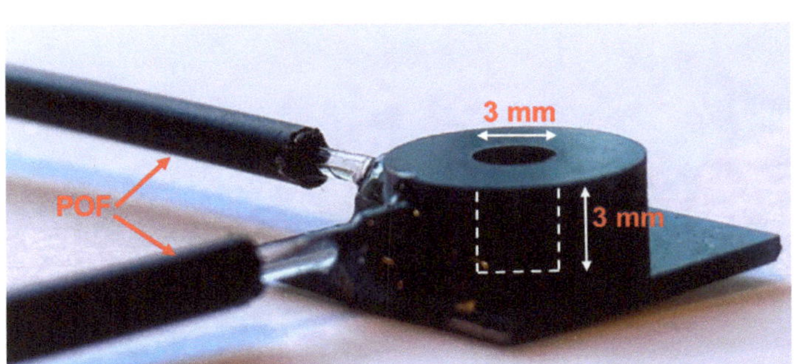

Figure 5. Image of the fabricated sensing assay compartment. The dashed line indicates the compartment.

4. Results and Discussion

First, we fabricated two 3D-printed compartments connected with POF and containing gel, including aptamer and tagged cortisol. Figure 6a shows the fluorescence spectra at different times when the sensing compartment was immersed in the cortisol solution (50 μg/mL). Similarly, Figure 6b shows fluorescence spectra in the presence of pure DI water. The fluorescence signal intensity decreased with time in the cortisol solution (50 μg/mL) and DI water.

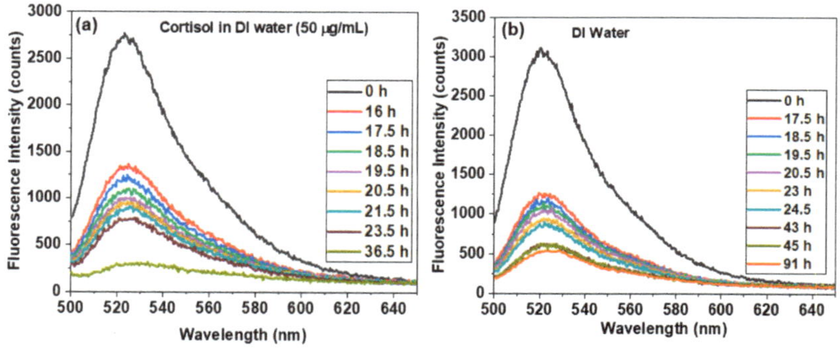

Figure 6. Fluorescence intensity vs wavelength with time: (**a**) in cortisol (50 μg/mL) solution; (**b**) in DI water.

The signal intensities at different times were extracted at the emission wavelength of the AF488H (520 nm) from Figure 6 and plotted. Figure 7 shows this plot of the fluorescence intensity variation of the sensor in DI water and cortisol solution with time. Initially, in Figure 7, the rate of change of fluorescence intensity appeared similar for cortisol solution and DI water until 20 h of exposure. After 20 h, the rate of change of fluorescence intensity in the presence of cortisol solution was faster compared to that of the DI water sample. The inset of Figure 7 shows that the variation of the fluorescence intensity in the presence of cortisol solution was almost linear, while in the presence of DI water, the fluorescence variation was almost saturated after 40 h. There are two possible reasons for this kind of fluorescence variation. For the initial 20 h, the rate of change of fluorescence in the presence of DI water and cortisol solution showed a similar trend due to the leaching of the unbound AF488H-tagged cortisol molecules from the hydrogel.

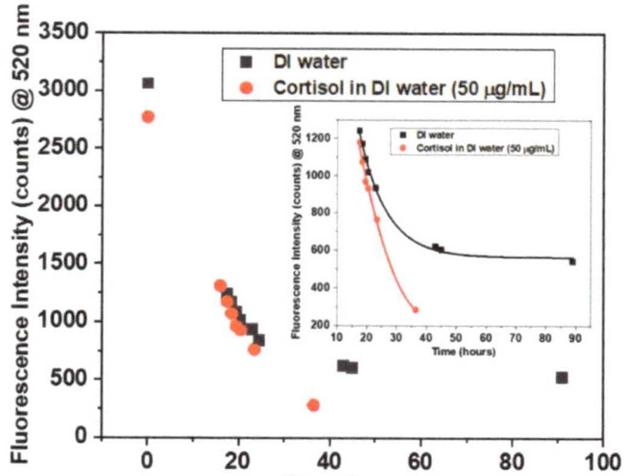

Figure 7. Fluorescence intensity vs. time with exposure to cortisol (50 μg/mL) solution. The inset shows the fluorescence intensity vs. time from 20 h to 90 h.

These excess AF488H-tagged cortisol molecules were present in the hydrogel and were not captured by aptamers. In the presence of an aqueous environment, these unbound/free AF488H-tagged cortisol molecules could escape from the hydrogel all the way out of the compartment, but the AF488H-tagged cortisol molecules, which were captured by the aptamer, could not as easily escape from the hydrogel. These aptamer-captured AF488H-tagged cortisol molecules escaped more easily when competing cortisol was present. After 20 h, when the unbound molecules were almost leached out of the hydrogel, the aptamer-captured tagged cortisol molecules started contributing to the difference in the sensing response in DI water and cortisol solution. In other words, the aptamer-captured tagged cortisol molecules were not displaced from the hydrogel in the presence of DI water; therefore, the fluorescence intensity remained almost unchanged and saturated at a relatively high level. In the presence of the cortisol solution, the fluorescence intensity dropped quickly because the AF488H-tagged cortisol molecules were displaced by the cortisol molecules. This displacement was due to the competitive nature of this assay. These measurements show the competition between the AF488H-tagged cortisol and cortisol molecules. The reason for this competition is the affinity of the cortisol molecules to the aptamer. The affinity of the cortisol molecules towards the aptamer was higher compared to that of the AF488H-tagged cortisol molecule because the aptamer was designed specifically for the cortisol molecule. The affinity of the AF488H-tagged cortisol molecules should be lower because, after tagging with the AF488H fluorescence molecules on the cortisol, the overall molecular structure of the AF488H-tagged cortisol was changed compared to that of cortisol; therefore, due to this molecular structural change, the affinity should be lower. The reason for the slow sensor response is the following: firstly, the excess amount of AF488H-tagged cortisol in the hydrogel and, secondly, the thickness of the hydrogel. The excess amount of AF488H-tagged cortisol in the hydrogel took around 20 h to leach, and then real sensing started. That means the actual sensing started after 20 h. Due to the thick hydrogel, the cortisol molecules took a long time to enter the hydrogel and, likewise, for the displaced tagged cortisol, to exit. In Figure 3, we can see the AF488H- and AF488H-tagged cortisol took around 24 h to completely enter the 8 mm thick hydrogel disc. In our sensor, the thickness of the hydrogel was approximately 2.6 mm. That means the cortisol molecules were completely inside the hydrogel in 8–9 h. A possible way to improve the sensing response is through quantification of the AF488H-tagged cortisol and thickness optimization of the hydrogel. We needed to use the AF488H-tagged cortisol in such a way that the amount of unbound AF488H-tagged cortisol remained very small in the hydrogel, and that is possible after performing the quantification of the AF488H-tagged cortisol. Thickness optimization of the hydrogel was also needed to improve the sensing response. The thickness of the hydrogel should be a few hundred microns to reduce the response time to 1 h. If the hydrogel thickness is reduced, the fluorescence signal will decrease, which can influence the sensitivity of the sensor. Therefore, there is a trade-off between the thickness of the hydrogel and the fluorescence signal; hence, optimized thickness is required. In Figure 1, we can see that in our fabricated sensing compartment, the fibers were placed at height t above the bottom, and, therefore, we had an extra t height of hydrogel in our sensing, which was below the fiber. We needed to modify the sensing assay compartment and remove this extra hydrogel layer of thickness t from the assay compartment volume. A thin layer of hydrogel was possible with a new assay compartment and small diameter fibers. Using these suggestions, a fast response time could be achieved. In general, fluorophore/aptamer competitive assay-based sensing is performed in liquid, and narrow Stokes-shift fluorophores require complex instrumentation. We have demonstrated fluorophore/aptamer competitive assays inside the hydrogel with a new optical fiber design that enables online sensing in a wet environment, and no complex instrumentation is needed to separate the excitation and emission signals of the fluorophore. In the past, several techniques have been explored for the detection of cortisol. A comparison of different cortisol-sensing techniques is presented in Table 1.

Table 1. Comparison of different cortisol sensing techniques.

Technique	Sensitivity	Response Time	Merits	Limitations	Ref.
Lateral Flow assay	ng/L	15–20 min	Rapid screening, disposable, good for biofluids	Not feasible for online measurements, wet environment and non-reusable.	[21]
Electrochemical	ng/L	~hours	Good selectivity	Not feasible for wet environment, complex instrumentations.	[22]
Impedance spectroscopy	ng/mL	Not mentioned	Real-time, continuous monitoring, reusable	Not feasible for wet environment, complex instrumentations.	[23]
Colorimetric	ng/L	20–30 min	Simple, low cost	Not feasible for online measurements, wet environment and non-reusable. Need a lot of optimizations	[24]
Fluorophore/ aptamer-based competitive assay	μg/mL	~hours	Simple, low cost, robust, feasible for online measurements, remote sensing and wet environment	Need gel thickness optimization, non-reusable.	Present work

5. Conclusions

In this study, we developed a simple and low-cost design for fluorophore- and aptamer-based optical fiber sensing. Using the laser diode and perpendicular design compartment, one can easily obtain the fluorescence signal without using any filters, splitters, or complex instruments for optical fiber sensing. The hydrogel entrapment method was used to immobilize the aptamer inside the hydrogel just in front of the fiber for excitation and emission pick-up. We have shown single-time cortisol sensing using the competition between the tagged and normal cortisol. Using the same strategies, one can label according to the analyte of interest and follow the same technique for sensing. The reported technique can also be utilized for the affinity test of the aptamers. The reported sensing design needs a few optimizations to improve the response time. We hope that the reported sensing principle and the new design will be helpful for the researcher to develop cost-effective fluorophore- and aptamer-based fiber optic sensors.

Author Contributions: Conceptualization, V.S. and J.J.; methodology, V.S. and J.J.; validation, V.S. and J.J.; formal analysis, V.S., O.B. and J.J.; investigation, V.S., O.B. and J.J.; resources, V.S., J.H., E.M., O.B. and J.J.; data curation, V.S.; writing—original draft preparation, V.S. and J.J.; writing—review and editing, V.S., O.B. and J.J.; visualization V.S. and J.J.; supervision, O.B. and J.J.; project administration, J.J.; funding acquisition, J.J. All authors have read and agreed to the published version of the manuscript.

Funding: This project was funded by the Green Development and Demonstration Programme, GUDP, Denmark. (Grant number: 34009-20-1769).

Institutional Review Board Statement: Not applicable.

Informed Consent Statement: Not applicable.

Data Availability Statement: Data underlying the results presented in this paper are not publicly available at this time but may be obtained from the authors upon reasonable request.

Conflicts of Interest: The authors declare no conflict of interest.

References

1. Janting, J.; Inglev, R. Towards an online polymer optical fiber cortisol sensor for aquaculture. In Proceedings of the 27th International Conference on Optical Fiber Sensors, The Westin Alexandria, VA, USA, 8–12 July 2020; pp. 1–4. [CrossRef]
2. Kaushik, A.; Vasudev, A.; Arya, S.K.; Pasha, S.K.; Bhansali, S. Recent advances in cortisol sensing technologies for point-of-care application. *Biosens. Bioelectron.* **2014**, *53*, 499–512. [CrossRef]
3. Mota, V.C.; Martins, C.I.M.; Eding, E.H.; Canário, A.V.M.; Verreth, J.A.J. Water cortisol and testosterone in Nile tilapia (*Oreochromis niloticus*) recirculating aquaculture systems. *Aquaculture* **2017**, *468*, 255–261. [CrossRef]
4. Usha, S.P.; Shrivastav, A.M.; Gupta, B.D. A contemporary approach for design and characterization of fiber-optic-cortisol sensor tailoring LMR and ZnO/PPY molecularly imprinted film. *Biosens. Bioelectron.* **2017**, *87*, 178–186. [CrossRef]
5. Fanouraki, E.; Papandroulakis, N.; Ellis, T.; Mylonas, C.C.; Scott, A.P.; Pavlidis, M. Water Cortisol Is a Reliable Indicator of Stress in European Sea Bass, Dicentrarchus labrax. *Behaviour. Bioacti.* **2017**, *145*, 1267–1281.
6. Song, S.; Wang, L.; Li, J.; Fan, C.; Zhao, J. Aptamer-based biosensors. *TrAC—Trends Anal. Chem.* **2008**, *27*, 108–117. [CrossRef]
7. Ali, M.H.; Elsherbiny, M.E.; Emara, M. Updates on aptamer research. *Int. J. Mol. Sci.* **2019**, *20*, 1–23. [CrossRef]
8. Tuerk, C.; Gold, L. Systematic evolution of ligands by exponential enrichment: RNA ligands to bacteriophage T4 DNA polymerase. *Science* **1990**, *249*, 505–510. [CrossRef]
9. Pickup, J.C.; Hussain, F.; Evans, N.D.; Rolinski, O.J.; Birch, D.J.S. Fluorescence-based glucose sensors. *Biosens. Bioelectron.* **2005**, *20*, 2555–2565. [CrossRef]
10. Zhong, W. Nanomaterials in fluorescence-based biosensing. *Anal. Bioanal. Chem.* **2009**, *394*, 47–59. [CrossRef] [PubMed]
11. Jepsen, M.D.; Sparvath, S.M.; Nielsen, T.B.; Langvad, A.H.; Grossi, G.; Gothelf, K.V.; Andersen, E.S. Development of a genetically encodable FRET system using fluorescent RNA aptamers. *Nat. Commun.* **2018**, *9*, 18. [CrossRef]
12. Pehlivan, Z.S.; Torabfam, M.; Kurt, H.; Ow-Yang, C.; Hildebrandt, N.; Yüce, M. Aptamer and nanomaterial based FRET biosensors: A review on recent advances (2014–2019). *Microchim. Acta* **2019**, *186*, 563. [CrossRef]
13. Hendrickson, O.D.; Taranova, N.A.; Zherdev, A.V.; Dzantiev, B.B.; Eremin, S.A. Fluorescence polarization-based bioassays: New horizons. *Sensors* **2020**, *20*, 7132. [CrossRef] [PubMed]
14. Cruz-Aguado, J.A.; Penner, G. Fluorescence polarization based displacement assay for the determination of small molecules with aptamers. *Anal. Chem.* **2008**, *80*, 8853–8855. [CrossRef] [PubMed]
15. Trettnak, W. Optical sensors based on fluorescence quenching. In *Fluorescence Spectroscopy: New Methods and Applications*; Springer: Berlin/Heidelberg, Germany, 1993; pp. 79–89. [CrossRef]
16. Geddes, C.D. Optical halide sensing using fluorescence quenching: Theory, simulations and applications—A review. *Meas. Sci. Technol.* **2001**, *12*, R53–R88. [CrossRef]
17. Baylakoğlu, I.; Fortier, A.; Kyeong, S.; Ambat, R.; Conseil-Gudla, H.; Azarian, M.H.; Pecht, M.G. The detrimental effects of water on electronic devices. *Adv. Electr. Electron. Eng.* **2021**, *1*, 100016. [CrossRef]
18. Zhang, H.; Xia, C.; Feng, G.; Fang, J. Hospitals and laboratories on paper-based sensors: A Mini Review. *Sensors* **2021**, *21*, 5998. [CrossRef]
19. Dalirirad, S.; Steckl, A.J. Aptamer-based lateral flow assay for point of care cortisol detection in sweat. *Sens. Actuators B Chem.* **2019**, *283*, 79–86. [CrossRef]
20. Semwal, V.; Shrivastav, A.M.; Verma, R.; Gupta, B.D. Surface plasmon resonance based fiber optic ethanol sensor using layers of silver/silicon/hydrogel entrapped with ADH/NAD. *Sens. Actuators B Chem.* **2016**, *230*, 485–492. [CrossRef]
21. Shima, D.; Daewoo, H.; Andrew, S.J. Aptamer-Based Lateral Flow Biosensor for Rapid Detection of Salivary Cortisol. *ACS Omega* **2020**, *5*, 32890–32898.
22. Roushani, M.; Hosseini, H.; Hajinia, Z.; Rahmati, Z. Rationally designed of hollow nitrogen doped carbon nanotubes double shelled with hierarchical nickel hydroxide nanosheet as a high performance surface substrate for cortisol aptasensing. *Electrochim. Acta* **2021**, *388*, 138608. [CrossRef]
23. Pali, M.; Jagannath, B.; Lin, K.C.; Upasham, S.; Sankhalab, S.; Upashama, S.; Muthukumar, S.; Prasad, S. CATCH (Cortisol Apta WATCH): 'Bio-mimic alarm' to track Anxiety, Stress, Immunity in human sweat. *Electrochim. Acta.* **2021**, *390*, 138834. [CrossRef]
24. Niu, C.; Ding, Y.; Zhang, C.; Liu, J. Comparing two cortisol aptamers for label-free fluorescent and colorimetric biosensors. *Sens. Diagn.* **2022**, *1*, 541–549. [CrossRef]

photonics

Article

Distributed Strain Measurements Based on Rayleigh Scattering in the Presence of Fiber Bragg Gratings in an Optical Fiber

Valerii Matveenko and Grigorii Serovaev *

Institute of Continuous Media Mechanics, Ural Branch, Russian Academy of Sciences, 614018 Perm, Russia; mvp@icmm.ru
* Correspondence: serovaev@icmm.ru

Abstract: This paper addresses the challenge of strain measurement using distributed fiber-optic sensors based on Rayleigh scattering in the presence of fiber Bragg gratings (FBGs) with a reflectivity level of 70% within the optical fiber. The reflectivity of such FBGs complicates distributed strain measurements that rely on the cross-correlation algorithm. The cases where the scanning ranges of a backscatter reflectometer include the resonant wavelengths of the FBGs and those beyond their limits, resulting in either a complete absence of a useful signal or the emergence of insensitive zones near the FBGs, are considered. An approach is proposed that employs a windowed Fourier transform with Hann window function for signal processing. This method effectively eliminates insensitive zones in distributed strain measurements based on Rayleigh scattering.

Keywords: point fiber-optic sensors; fiber Bragg grating; distributed fiber-optic sensors; strain registration; strain gradient; Rayleigh scattering; optical frequency domain reflectometry

Citation: Matveenko, V.; Serovaev, G. Distributed Strain Measurements Based on Rayleigh Scattering in the Presence of Fiber Bragg Gratings in an Optical Fiber. *Photonics* **2023**, *10*, 868. https://doi.org/10.3390/photonics10080868

Received: 10 July 2023
Revised: 21 July 2023
Accepted: 25 July 2023
Published: 26 July 2023

1. Introduction

The quality of the Structural Health Monitoring (SHM) system depends on the effectiveness of the used sensitive elements that register crucial parameters of the monitored object. Fiber-optic sensors (FOSs) demonstrate high sensitivity, the ability to operate in a wide range of temperatures and environmental conditions, and independence from electromagnetic interference, which make them highly relevant in the field of structural integrity assessment. The small dimensions of the optical fiber make it possible to attach FOSs both on the surface of the monitored object and to embed them into the structure's material, opening up possibilities for registering the mechanical state at the manufacturing stage. FOSs find applications in aerospace [1–3], geotechnical [4–6], civil engineering [7–9], medical [10–12], and many other [13] fields.

FOSs are sensitive to strain and/or temperature changes as well as derivatives of these parameters (pressure, force, displacement, etc.) in a certain section of the optical fiber or throughout the entire fiber under test (FUT). Depending on the size of the sensitive zone, FOSs are usually divided into categories of single point and distributed sensors [14].

Typically, in the fabrication of the point FOSs, the optical fiber undergoes preliminary treatment in the area where the sensitive element should be located. For instance, to record an FBG, which is the most common point of FOS for measuring strain, a periodic change in the refractive index is induced via laser radiation in the section of the fiber-optic core [15].

On the other hand, for distributed measurements, it is possible to use an optical fiber that is not subjected to special processing. Distributed FOSs are based on the measurement of optical scattering resulting from the interaction of light with the material in an optical fiber. There are three main types of scattering used in distributed FOSs: Rayleigh (elastic), Brillouin, and Raman scattering [16].

Among the main methods of optical reflectometry that are utilized for distributed measurements are Optical Time Domain Reflectometry (OTDR) and Optical Frequency Domain Reflectometry (OFDR). The principle of operation for OTDR involves introducing optical pulses into the optical fiber and detecting the intensity of reflected and scattered light based on the delay time between the injected and reflected signals. In contrast, OFDR uses a continuous wave (CW) tunable laser source (TLS), the optical signal from which, after passing through the reference and measuring arms of the interferometer, is mixed and detected using a photodetector. By applying the Fast Fourier Transform (FFT) to the received optical signal, it is possible to obtain the intensity distribution of the reflected and scattered light along the length of the optical fiber [17].

Among distributed FOSs, the method based on measuring the spectral shift in Rayleigh backscattering via OFDR with the addition of polarization measurements has the best spatial resolution, which makes it possible to carry out distributed measurements of strain and temperature using a standard single-mode optical fiber [18].

Measurements using point FBG sensors are based on the analysis of the resonant wavelength of the reflected spectrum. The most common method of simultaneous interrogation of multiple FBGs on a single optical fiber is Wavelength Division Multiplexing (WDM), where each FBG corresponds to a specific section of the interrogator's operating wavelength range. This approach limits the maximum possible number of simultaneously interrogated FBGs to the width of the wavelength range of the interrogator channel [19]. There are studies related to the use of OTDR and OFDR methods for interrogating a large number of weakly reflecting FBGs, which can increase the number of interrogated FBGs by up to several thousand [20–22]. When using these methods, addressing the spectral shadowing effect becomes important [23,24]. The use of optical fibers with a large number of weak reflection FBGs instead of standard telecommunication optical fibers, in which Rayleigh scattering is measured, has a number of advantages. The larger reflection amplitude in this case improves the registration of strain at a higher sampling rate. However, it significantly increases the cost of optical fiber production [25].

Despite the evident benefits of distributed strain measurement systems using the OFDR method based on Rayleigh scattering, these systems significantly fall behind traditional FBG interrogators employing WDM in terms of sensor sampling rate (hundreds of Hz versus several kHz) [26]. The combination of both methods for strain measurement on a single optical fiber may allow to use the distinctive advantages of each method: a high sampling rate for point sensors located in critical sections of the monitored structure using WDM of FBGs, and a high spatial resolution for distributed FOSs based on Rayleigh scattering by OFDR.

The problem of correspondence between the readings of point and distributed FOSs is significant for their practical application. In [27], on the example of full-scale fatigue and crack growth tests, the authors demonstrated that strain measurements conducted using distributed FOSs based on Rayleigh scattering are in good agreement with the readings of point FBG sensors and classical foil strain gages in areas with moderate strain gradients. However, measuring strain in regions with high strain gradients using distributed FOSs becomes challenging and may result in information loss due to the complexity of cross-correlation analysis. In [28], various parameters affecting the reliability of strain measurements using distributed FOSs based on Rayleigh scattering in cases of gradient strain distribution along an optical fiber are considered. A close correspondence between point and distributed FOSs is demonstrated when selecting the optimal sensor gage length. In [29], an approach is proposed for calibrating distributed FOS readings based on a finite number of point FBG sensor measurements.

Typically, comparisons of point and distributed FOSs are conducted on different optical fibers located near each other, but not in the same location, which may lead to differences in the stress–strain state in the area where the sensors are placed. A more reliable comparison would involve using two measuring systems on a single optical fiber, ensuring full compliance with the stress–strain state in the region where the FOSs are located.

In practical applications of strain measurement via point FOSs, it becomes particularly necessary to estimate the strain distribution in the vicinity of FBG locations, especially when measuring gradient strain fields. Distributed fiber-optic sensors based on Rayleigh scattering can address this issue by providing high spatial resolution in strain measurement.

This paper addresses the challenges of strain measurement using distributed FOSs based on Rayleigh scattering in the presence of fiber Bragg gratings with a reflection coefficient of 70% in the measured optical fiber. This situation can lead to significant noise in the measuring signal or the emergence of insensitive zones where strain is not determined. To overcome these issues, an approach is proposed that involves applying the Hann window function to the primary signal recorded with the detector of an optical backscatter reflectometer and selecting a scanning range that excludes resonant FBG wavelengths.

2. Materials and Methods

An FBG recorded in a single-mode optical fiber enables the measurement of strain and temperature at the grating location based on an analysis of the shift in the resonant wavelength of the optical spectrum reflected by the grating. Multiple FBGs with different resonant wavelengths can be recorded on a single optical fiber, allowing for the simultaneous interrogation of all gratings using the WDM method. This approach enables strain measurement at various points of a structure when the optical fiber is attached to its surface or embedded within the material.

The maximum number of measurement points is limited by the spectral wavelength range of the FBG sensor interrogation system (interrogator). Increasing the number of FBGs in the optical fiber results in a reduced strain measurement range for each sensor. The most optimal configuration includes 10–15 FBGs per interrogator channel, which can significantly limit the ability to assess gradient strain field distributions. This number of FBGs in the optical fiber provides a strain measurement range of ±(2500–4100) microstrain for each sensor. In this study, 5 mm long FBGs that have a reflection coefficient of 70% and feature a main peak width of 0.25 nm in the reflected spectrum written using the phase mask method are considered.

In contrast to point FOSs based on FBGs, where only a small region of the optical fiber containing the written grating is sensitive to changes in strain and temperature, distributed FOS utilize the entire length of the optical fiber as a sensitive element. Among the fiber-optic methods for distributed strain measurement, the approach based on Rayleigh scattering measurement using OFDR offers the best spatial resolution.

Rayleigh scattering along the optical fiber arises as a result of the interaction of the injected light with inhomogeneities in the optical fiber, the size of which is smaller than the wavelength of the input radiation. The scattering distribution, unique for each fiber under test, remains constant from measurement to measurement, provided there is no external influence.

To conduct measurements using a backscatter reflectometer, a specific section of the fiber is selected on the reflectogram of the fiber under test (FUT). This selected area is divided into sub-regions of a predefined length (gage length) with a certain spacing (distance between sensors), which can be considered as individual sensors. In this case, the sub-regions may overlap, and the center of each sub-region represents one of the points on the distributed measurement diagram for strain or temperature.

When an external influence (strain, temperature) affects the optical fiber, the signal shifts in the frequency domain relative to the reference signal for each of the sub-regions. To calculate strain or temperature, the spectral shift within a specific window is estimated relative to the reference signal via cross-correlation analysis. The spectrum shift ($\Delta\nu$) of the measured sub-region of the optical fiber is similar to the spectrum shift or the resonant wavelength shift ($\Delta\lambda$) of the Bragg grating and is related to the change in strain and temperature as follows:

$$\frac{\Delta\lambda}{\lambda} = -\frac{\Delta\nu}{\nu} = K_{\varepsilon} \cdot \varepsilon + K_T \cdot \Delta T, \tag{1}$$

where K_T and K_{ε} are the temperature and strain sensitivity coefficients. For most germanium-doped silica glass optical fibers $K_T = 6.45 \cdot 10^{-6}\text{C}^{-1}$, $K_{\varepsilon} = 0.78$.

With a constant strain in the measured section of the optical fiber, the temperature change can be expressed as:

$$\Delta T = -\frac{\overline{\lambda}}{cK_T}\Delta\nu, \quad (2)$$

where $\overline{\lambda}$ is the central scanning wavelength, which is the midpoint wavelength of the applied wavelength scanning range of the tunable laser source and c is the speed of light in a vacuum.

Similarly, the change in strain, considering temperature compensation, can be expressed as:

$$\varepsilon = \frac{\overline{\lambda}}{cK_\varepsilon}(\Delta\nu^T - \Delta\nu), \quad (3)$$

where $\Delta\nu^T$ is the spectral shift measured in the section of the optical fiber that does not experience a change in mechanical strain.

Utilizing an optical fiber with recorded FBGs for strain measurement enables the combination of the advantages of two types of fiber-optic sensors: point FOSs based on FBGs and distributed FOSs. Point FOSs, due to their higher sampling rate, allow for the strain response measurement at FBG locations under dynamic loading on the structure. Distributed FOSs, on the other hand, provide the measurement of strain distribution under the static loading of the structure with high spatial resolution.

Performing distributed measurements in the presence of considered FBGs in an optical fiber can be challenging due to the high reflectivity of FBGs at wavelengths within the scanning range of the TLS of a backscatter reflectometer. The presence of such FBGs in the measuring optical fiber is common for many operational objects in which the measurement system is already installed and cannot be modified.

The following section presents the results of experiments on distributed strain measurement under the uniaxial tension of an optical fiber with varying numbers of FBGs recorded in it. Distributed strain measurements based on Rayleigh scattering were conducted using an OBR4600 reflectometer from Luna Innovations.

3. Results and Discussion

Under the uniaxial tension of an optical fiber not attached to the material, a uniform strain distribution is formed along its axis. Figure 1 displays a strain distribution plot obtained via OBR4600 in the presence of a 5 mm long FBG in the measurement area. The resonant wavelength of the FBG is 1520 nm, and the scanning range of the optical backscatter reflectometer was selected outside the main reflective peak of the FBG, specifically in the range 1589–1611 nm.

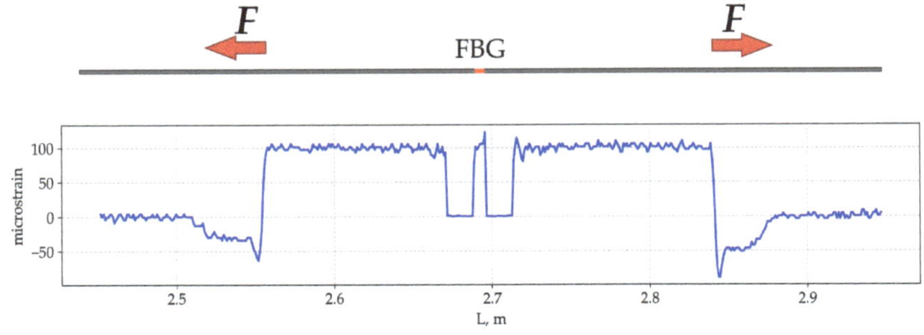

Figure 1. Strain distribution measured using an optical fiber with an FBG under tensile loading at a scanning wavelength range of OBR system selected outside the main reflective peak of the FBG.

The obtained results showed that using a reflectometer scanning wavelength range outside the FBG resonant wavelength led to the emergence of insensitive areas of 20 mm length in the measured strain distribution along the optical fiber in the vicinity of the FBG location. However, strain readings are preserved within the FBG zone.

The case of distributed strain measurement using an optical fiber with multiple FBGs holds particular practical significance, since it enables the supplementation of strain data obtained using point sensors with strain distribution information in the vicinity of the FBGs, as well as the estimation of the strain distribution in other sections of the optical fiber.

The experiment was carried out by applying tension load to three sections of an optical fiber, two of which are outside the FBG location zone, according to the scheme in Figure 2. One section includes five FBGs, each 5 mm long, with a distance of 10 mm between centers and resonant wavelengths of 1520.4, 1535.2, 1549.9, 1564.7, and 1579.9 nm, respectively.

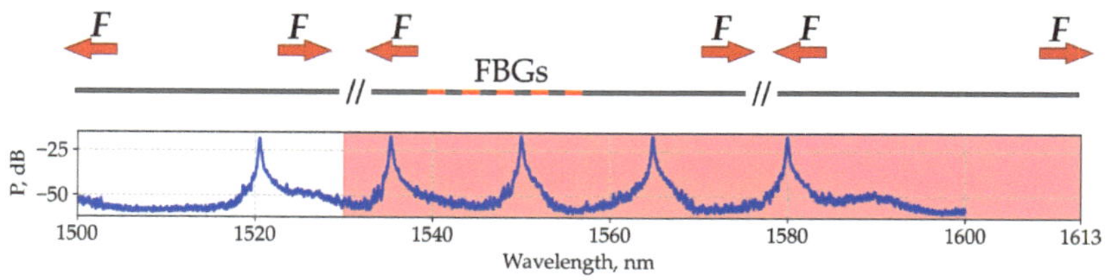

Figure 2. The scheme of loaded sections of an optical fiber and scanning range of the backscatter reflectometer.

It is important to note that the choice of the scanning range affects the spatial resolution of the measured reflectograms Δz and the maximum strain measurement range ε_{range} in accordance with the following relations [30,31]:

$$\Delta z = \frac{c}{2n\Delta\nu} = \frac{\lambda_{start}\lambda_{end}}{2n(\lambda_{end} - \lambda_{start})}, \quad (4)$$

$$\varepsilon_{range} = \frac{\lambda_{end} - \lambda_{start}}{\overline{\lambda}}, \quad (5)$$

where λ_{start} is the initial scanning wavelength and λ_{end} is the final scanning wavelength of the TLS. Thus, a larger scan range allows for better spatial resolution and the measurement of higher strain levels.

In this regard, when stretching an optical fiber with five FBGs, the maximum possible range of reflectometer scanning wavelengths of 1530–1613 nm was selected (red area in Figure 2). The result of the distributed strain measurement for this experiment is illustrated in Figure 3.

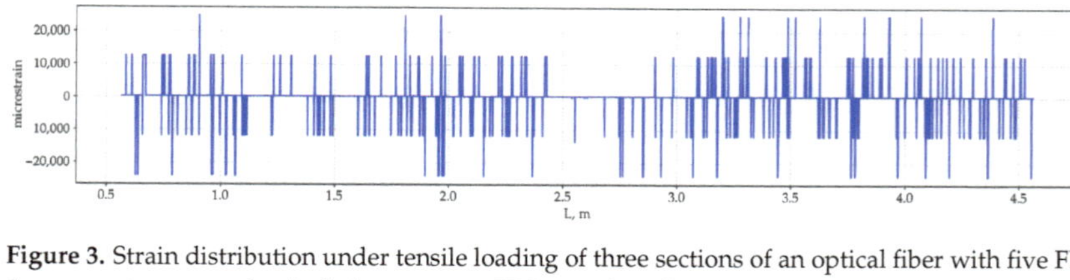

Figure 3. Strain distribution under tensile loading of three sections of an optical fiber with five FBGs for a scanning range that includes resonant FBG wavelengths.

The inclusion of resonant wavelengths in the scanning range of the TLS results in a noise level within the strain distribution not allowing us to isolate the useful signal.

Selecting a scanning range of 1591–1613 nm, which does not include any of the resonant FBG wavelengths, as shown in Figure 4, enables the acquisition of the tensile strain distribution of the three sections of the optical fiber, as depicted in Figure 5. It is important to note that a reduction in the scanning range results in a decrease in the strain measurement range.

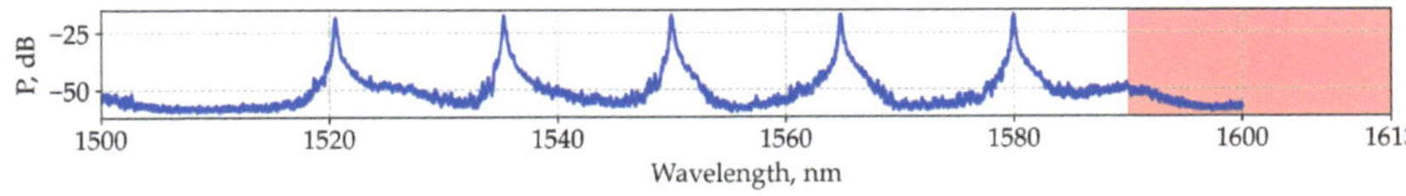

Figure 4. Scanning range (red area) of the backscatter reflectometer outside the resonant wavelengths of the FBGs.

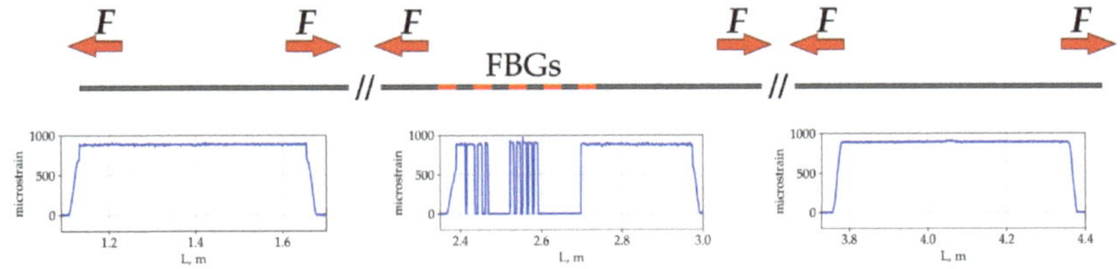

Figure 5. Loading scheme of an optical fiber with five FBGs and the distribution of strain in loaded sections at the scanning range of the backscatter reflectometer outside the resonant wavelengths of the FBGs.

The strain distribution in the loaded sections of an optical fiber that does not contain FBGs exhibits the expected uniform character. Similar to the case with one FBG considered earlier, the loaded section with five FBGs contains insensitive zones in the vicinity of the FBGs. The strain distribution in this section is shown in Figure 6, where the zones of FBG location along the length of the optical fiber are marked in red.

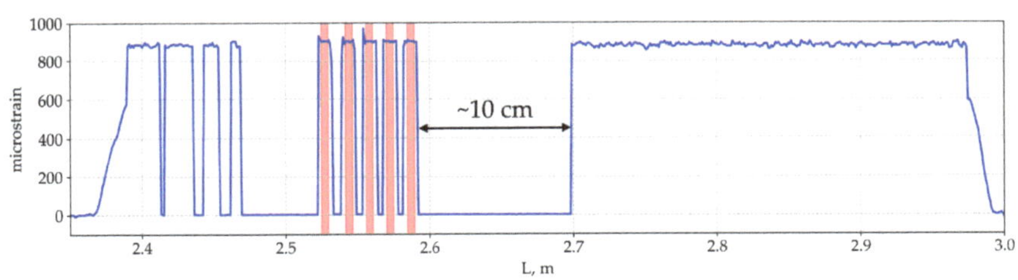

Figure 6. Strain distribution of the stretched section of an optical fiber with five FBGs (red areas).

The maximum length of the insensitive zone was 100 mm, which largely limits the application of the optical fiber with recorded FBGs with the considered optical characteristics for measuring non-uniform strain fields.

The presence of insensitive zones may be attributed to signal smearing resulted from the FFT procedure. In this case, the reflectogram displays a considerable amplitude difference between the sections of Fresnel reflection in the FBG zones and the Rayleigh scattering, due to the high reflectivity of the FBGs used (Figure 7).

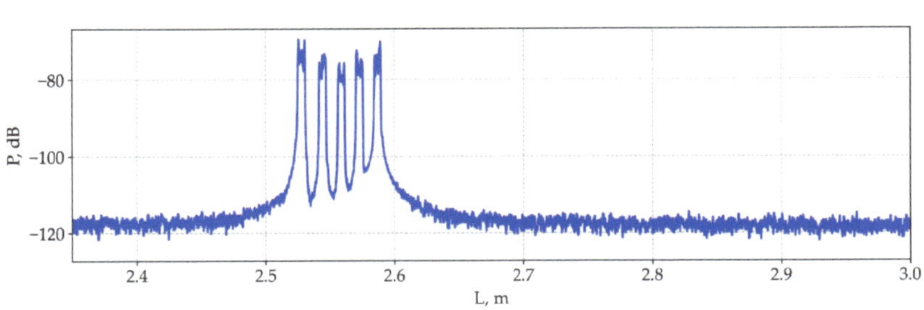

Figure 7. Reflectogram of the stretched section of an optical fiber with five FBGs.

It is well known that using a Discrete Fourier Transform (DFT) for a signal of finite length results in spectral leakage, which is expressed as smearing (broadening) of the spectrum in the frequency domain. This phenomenon occurs because some signal frequencies do not align with the basis functions of the DFT over a given measurement interval, leading to contributions across the entire spectral range. Due to spectral leakage and smearing under the FFT, the Rayleigh scattering regions near FBG locations are suppressed and may exhibit an unstable spectral composition, which is essential for calculating strain through the cross-correlation of the initial and current signals.

To mitigate the spectral leakage associated with a finite measurement interval, the original signal within this interval is multiplied by a window function before applying the DFT. This window function ensures signal continuity at the boundaries of the measurement interval. One of the most widely used window functions is the Hann window, which is defined by the following relation [32]:

$$w(n) = 0.5 - 0.5\cos(\frac{2\pi n}{N-1}), \quad n = 0,1,2,\ldots,N-1, \tag{6}$$

where n is the sample number and N is the number of data points.

The windowed discrete Fourier transform of the signal $x(n)$ can be written as:

$$X(k) = \sum_{n=0}^{N-1} x(n)\cdot w(n)e^{-j\frac{2\pi}{N}kn}, \quad 0 \le k \le N-1, \tag{7}$$

where $X(k)$ is the complex transformed signal at frequency index k, consisting of amplitudes and phases of the basis functions.

Figure 8 demonstrates the effect of the application of windowed Fourier transform on the frequency spectrum of a signal composed of the sum of two harmonic oscillations of different amplitudes but close frequencies.

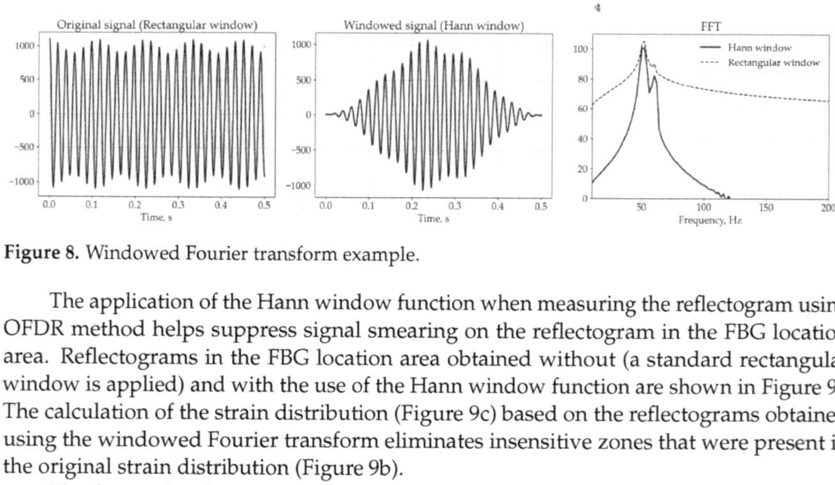

Figure 8. Windowed Fourier transform example.

The application of the Hann window function when measuring the reflectogram using OFDR method helps suppress signal smearing on the reflectogram in the FBG location area. Reflectograms in the FBG location area obtained without (a standard rectangular window is applied) and with the use of the Hann window function are shown in Figure 9a. The calculation of the strain distribution (Figure 9c) based on the reflectograms obtained using the windowed Fourier transform eliminates insensitive zones that were present in the original strain distribution (Figure 9b).

The downside of using a windowed Fourier transform is that, due to the suppression of the original signal in the initial and final sections of the measurement interval, the window function narrows the scanning range. This limitation negatively impacts the maximum strain measurement range ε_{range}.

The approach of applying the windowed Fourier transform is demonstrated using an example of a specimen with an embedded optical fiber containing three FBGs in the region of cutouts, as shown in Figure 10. The presence of cutouts in the specimen results in a non-uniform strain distribution along its length under uniaxial tension.

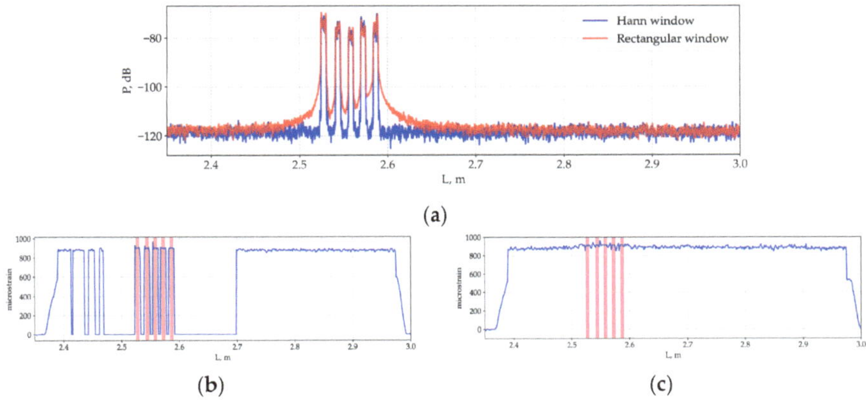

Figure 9. Reflectograms of a section of an optical fiber with five FBGs calculated with and without the Hann window function applied (**a**). Strain distribution obtained based on reflectograms without (**b**) and with (**c**) the Hann window function applied.

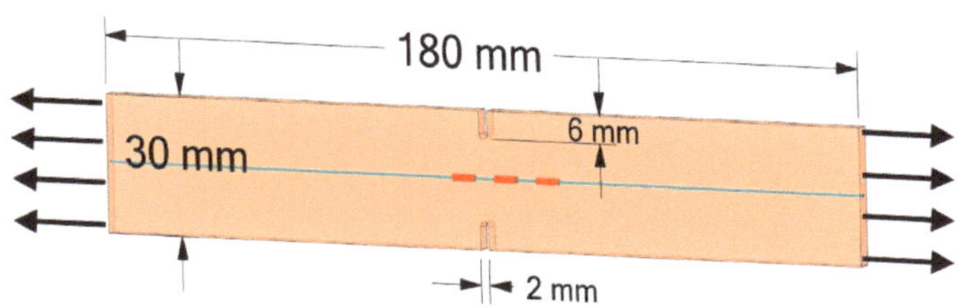

Figure 10. Specimen scheme with an embedded optical fiber with three FBGs.

Figure 11a displays the strain distributions along the length of the specimen, measured using distributed FOS based on Rayleigh scattering at three levels of tensile force. These measurements were conducted using a standard signal processing algorithm without applying the Hann window function and with a scanning range of the TLS that does not include resonant FBG wavelengths. These distributions provide limited information due to the presence of insensitive zones, which do not allow for the assessment of non-uniform changes in strain along the length of the specimen.

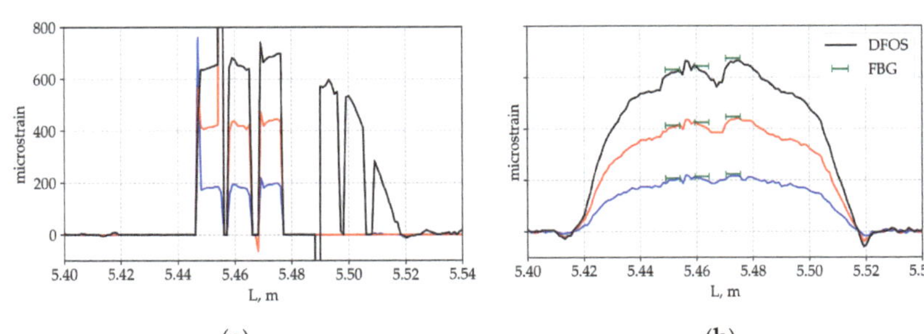

Figure 11. Strain distributions along the specimen under three load levels, obtained using a distributed FOS (DFOS) with three FBGs (**a**) without using the Hann window function; and (**b**) using the Hann window function and corresponding FBG sensor measurements.

Figure 11b presents similar strain distributions obtained using the Hann window function in signal processing.

When using the windowed Fourier transform, it is possible to evaluate the gradient nature of strain distribution along the entire length of the specimen under study. In addition to distributed strain measurements, at each loading stage, strain measurements were conducted using point FBG sensors based on recording the shift of the FBG resonant wavelength with the Hyperion si255 interrogator. The readings of point FOSs are shown in Figure 11b as green segments, the length of which corresponds to the length of the FBGs used, and the location is determined based on the reflectogram obtained using a backscatter reflectometer. The readings of the two measuring systems are in close agreement with each other.

Combined measurements with point and distributed FOSs on one optical fiber in this study were conducted sequentially by switching the FUT between two interrogation systems (OBR4600 and Hyperion si255) at each loading step. However, the simultaneous interrogation of FBG sensors and distributed FOSs based on Rayleigh scattering is possible. Such a system should employ distinct operational wavelength ranges for point and distributed measurements to prevent interference between them. In the present study, the OBR4600 and the FBG interrogator had overlapping operational wavelength ranges (1530–1613 nm for OBR4600 and 1500–1600 nm for Hyperion si255 FBG interrogator). To separate the distributed interrogation system from the resonant wavelengths of FBGs, the scanning wavelength range of the OBR4600 was narrowed and shifted to longer wavelengths, which consequently reduced the maximum measurement strain range. It is possible to develop an interrogation optical system that utilizes different operational wavelength ranges for distributed and point FOS, the input signals from which would be combined in the fiber under test and the reflected signals would be separated for subsequent analysis.

Despite the advantages of distributed FOSs, FBG interrogation systems still offer superior capabilities for dynamic strain measurements. Additionally, FBG interrogators are more accessible in terms of price and market availability. Therefore, potential applications where both point and distributed sensors are necessary include cases where it is crucial to combine the benefits of both systems: measuring strain with a high acquisition rate at critical points of the structure, which FBG sensors can provide, and simultaneously requiring high spatial resolution to assess the strain distribution in areas with potential strain gradients where damage may occur. This can be essential in aircraft SHM systems, where the structure is subjected to both dynamic and static loads, as well as in SHM systems for civil engineering structures. Another possible application is related to the research and development stage of various products, where distributed FOSs can offer valuable information on the mechanical state of an object under different loads during laboratory tests. However, due to the high cost, it may be more appropriate to install point FBG sensors, calibrated with distributed FOSs, in the final product.

4. Conclusions

This paper addresses the problem of strain measurement using distributed FOSs based on Rayleigh scattering in the presence of one or more high reflectivity FBGs in the measuring optical fiber. This situation complicates measurements based on the cross-correlation algorithm of the current and reference spectral signals. It is demonstrated that, when using the scanning range of the backscatter reflectometer, which includes the resonant wavelengths of the FBGs, a high noise level is observed, making it difficult to isolate the useful signal. Using scanning wavelengths outside the resonant wavelengths of FBGs eliminates this noise but leads to the presence of insensitive zones in the vicinity of FBGs, which make it difficult to analyze the strain distribution.

The proposed approach of using the windowed Fourier transform with the Hann window function suppresses signal spreading in the vicinity of the FBGs and eliminates the presence of insensitive zones. The application of this approach is demonstrated in the example of a specimen with an embedded optical fiber and recorded FBGs, in which a gradient strain distribution is realized under tension. This approach can be useful for increasing the reliability of comparing the results of measuring for point and distributed

FOSs, as well as for obtaining more detailed information on the strain state of objects on which point FOSs based on the high reflectivity FBGs are installed.

Author Contributions: Conceptualization, V.M. and G.S.; methodology, V.M. and G.S.; validation, V.M.; investigation, G.S.; resources, V.M. and G.S.; data curation, G.S.; writing—original draft preparation, G.S.; writing—review and editing, V.M.; visualization, G.S.; supervision, V.M.; project administration, V.M.; funding acquisition, V.M. All authors have read and agreed to the published version of the manuscript.

Funding: This paper was prepared in the framework of the program for the creation and development of the world-class scientific center for 2020–2025 with the financial support of the Ministry of Science and Higher Education of the Russian Federation (Agreement No. 075-15-2022-329 of 21 April 2022).

Institutional Review Board Statement: Not applicable.

Informed Consent Statement: Not applicable.

Data Availability Statement: Not available.

Conflicts of Interest: The authors declare no conflict of interest.

References

1. Rocha, H.; Semprimoschnig, C.; Nunes, J.P. Sensors for Process and Structural Health Monitoring of Aerospace Composites: A Review. *Eng. Struct.* **2021**, *237*, 112231. [CrossRef]
2. Iele, A.; Leone, M.; Consales, M.; Persiano, G.V.; Brindisi, A.; Ameduri, S.; Concilio, A.; Ciminello, M.; Apicella, A.; Bocchetto, F.; et al. Load Monitoring of Aircraft Landing Gears Using Fiber Optic Sensors. *Sens. Actuators A Phys.* **2018**, *281*, 31–41. [CrossRef]
3. Aniskovich, V.A.; Budadin, O.N.; Kozelskaya, S.O.; Kutyurin, Y.u.G.; Rykov, A.N.; Skleznev, A.A.; Gnusin, P.I.; Yuranev, O.A. Integration of FOS into a Composite Cylindrical Body Made of Cfrp by Continuous Winding. *Test. Diagn.* **2022**, *25*, 16–23. [CrossRef]
4. Schenato, L. A Review of Distributed Fibre Optic Sensors for Geo-Hydrological Applications. *Appl. Sci.* **2017**, *7*, 896. [CrossRef]
5. Zheng, Y.; Zhu, Z.-W.; Xiao, W.; Deng, Q.-X. Review of Fiber Optic Sensors in Geotechnical Health Monitoring. *Opt. Fiber Technol.* **2020**, *54*, 102127. [CrossRef]
6. Li, Y.; Wang, H.; Cai, W.; Li, S.; Zhang, Q. Stability Monitoring of Surrounding Rock Mass on a Forked Tunnel Using Both Strain Gauges and FBG Sensors. *Measurement* **2020**, *153*, 107449. [CrossRef]
7. Barrias, A.; Casas, J.; Villalba, S. A Review of Distributed Optical Fiber Sensors for Civil Engineering Applications. *Sensors* **2016**, *16*, 748. [CrossRef]
8. Bado, M.F.; Casas, J.R. A Review of Recent Distributed Optical Fiber Sensors Applications for Civil Engineering Structural Health Monitoring. *Sensors* **2021**, *21*, 1818. [CrossRef]
9. Kashaganova, G.; Kozbakova, A.; Kartbayev, T.; Balbayev, G.; Togzhanova, K.; Alimseitova, Z.; Orazaliyeva, S. Research of a Fiber Sensor Based on Fiber Bragg Grating for Road Surface Monitoring. *Electronics* **2023**, *12*, 2491. [CrossRef]
10. Presti, D.L.; Santucci, F.; Massaroni, C.; Formica, D.; Setola, R.; Schena, E. A Multi-point Heart Rate Monitoring Using a Soft Wearable System Based on Fiber Optic Technology. *Sci. Rep.* **2021**, *11*, 21162. [CrossRef]
11. De Tommasi, F.; Lo Presti, D.; Virgili, F.; Massaroni, C.; Schena, E.; Carassiti, M. Soft System Based on Fiber Bragg Grating Sensor for Loss of Resistance Detection during Epidural Procedures: In Silico and In Vivo Assessment. *Sensors* **2021**, *21*, 5329. [CrossRef] [PubMed]
12. Nedoma, J.; Fajkus, M.; Martinek, R.; Nazeran, H. Vital Sign Monitoring and Cardiac Triggering at 1.5 Tesla: A Practical Solution by an MR-Ballistocardiography Fiber-Optic Sensor. *Sensors* **2019**, *19*, 470. [CrossRef] [PubMed]
13. Gorshkov, B.G.; Yüksel, K.; Fotiadi, A.A.; Wuilpart, M.; Korobko, D.A.; Zhirnov, A.A.; Stepanov, K.V.; Turov, A.T.; Konstantinov, Y.A.; Lobach, I.A. Scientific Applications of Distributed Acoustic Sensing: State-of-the-Art Review and Perspective. *Sensors* **2022**, *22*, 1033. [CrossRef]
14. Joe, H.-E.; Yun, H.; Jo, S.-H.; Jun, M.B.G.; Min, B.-K. A Review on Optical Fiber Sensors for Environmental Monitoring. *Int. J. Precis. Eng. Manuf.-Green Technol.* **2018**, *5*, 173–191. [CrossRef]
15. Hill, K.O.; Meltz, G. Fiber Bragg Grating Technology Fundamentals and Overview. *J. Light. Technol.* **1997**, *15*, 1263–1276. [CrossRef]
16. Pendão, C.; Silva, I. Optical Fiber Sensors and Sensing Networks: Overview of the Main Principles and Applications. *Sensors* **2022**, *22*, 7554. [CrossRef]
17. Lu, P.; Lalam, N.; Badar, M.; Liu, B.; Chorpening, B.T.; Buric, M.P.; Ohodnicki, P.R. Distributed Optical Fiber Sensing: Review and Perspective. *Appl. Phys. Rev.* **2019**, *6*, 041302. [CrossRef]
18. Soller, B.J.; Gifford, D.K.; Wolfe, M.S.; Froggatt, M.E. High Resolution Optical Frequency Domain Reflectometry for Characterization of Components and Assemblies. *Opt. Express* **2005**, *13*, 666. [CrossRef]

19. Sahota, J.K.; Gupta, N.; Dhawan, D. Fiber Bragg Grating Sensors for Monitoring of Physical Parameters: A Comprehensive Review. *Opt. Eng.* **2020**, *59*, 1. [CrossRef]
20. Dey, K.; Roy, S.; Kishore, P.; Shankar, M.S.; Buddu, R.K.; Ranjan, R. Analysis and Performance of Edge Filtering Interrogation Scheme for FBG Sensor Using SMS Fiber and OTDR. *Results Opt.* **2021**, *2*, 100039. [CrossRef]
21. Zhu, M.; Murayama, H. Fast Demodulation of OFDR Based Long Length FBG Sensing System for Noisy Signals. *Opt. Express* **2018**, *26*, 19804. [CrossRef] [PubMed]
22. Childers, B.A.; Froggatt, M.E.; Allison, S.G.; Moore, T.C., Sr.; Hare, D.A.; Batten, C.F.; Jegley, D.C. Use of 3000 Bragg Grating Strain Sensors Distributed on Four 8-m Optical Fibers during Static Load Tests of a Composite Structure. *Proc. SPIE* **2001**, *4332*, 133–142. [CrossRef]
23. Yang, N.; Zhang, J.; Yu, L.; Zheng, P.; Zhao, B.; Sun, H.; Liu, Q. A Flexible Multiplexing Method of FBG Sensors Based on OFDR. *Optik* **2022**, *267*, 169764. [CrossRef]
24. Yuksel, K.; Moeyaert, V.; Megret, P.; Wuilpart, M. Complete Analysis of Multireflection and Spectral-Shadowing Crosstalks in a Quasi-Distributed Fiber Sensor Interrogated by OFDR. *IEEE Sens. J.* **2012**, *12*, 988–995. [CrossRef]
25. Kreger, S.T.; Rahim, N.A.A.; Garg, N.; Klute, S.M.; Metrey, D.R.; Beaty, N.; Jeans, J.W.; Gamber, R. Optical Frequency Domain Reflectometry: Principles and Applications in Fiber Optic Sensing. *Proc. SPIE* **2016**, *9852*, 98520T. [CrossRef]
26. Di Sante, R. Fibre Optic Sensors for Structural Health Monitoring of Aircraft Composite Structures: Recent Advances and Applications. *Sensors* **2015**, *15*, 18666–18713. [CrossRef] [PubMed]
27. Davis, C.; Knowles, M.; Rajic, N.; Swanton, G. Evaluation of a Distributed Fibre Optic Strain Sensing System for Full-Scale Fatigue Testing. *Procedia Struct. Integr.* **2016**, *2*, 3784–3791. [CrossRef]
28. Matveenko, V.; Kosheleva, N.; Serovaev, G.; Fedorov, A. Measurement of Gradient Strain Fields with Fiber-Optic Sensors. *Sensors* **2022**, *23*, 410. [CrossRef]
29. Cheng, L.; Cigada, A.; Lang, Z.-Q.; Zappa, E. Calibrating Static Measurement Data from Distributed Fiber Optics by the Integration of Limited FBG Sensors Based on the Extended Kernel Regression Method. *Meas. Sci. Technol.* **2019**, *30*, 125102. [CrossRef]
30. Kreger, S.T.; Sang, A.K.; Gifford, D.K.; Froggatt, M.E. Distributed Strain and Temperature Sensing in Plastic Optical Fiber Using Rayleigh Scatter. *Proc. SPIE* **2009**, *7316*, 73160A. [CrossRef]
31. Froggatt, M.; Moore, J. High-Spatial-Resolution Distributed Strain Measurement in Optical Fiber with Rayleigh Scatter. *Appl. Opt.* **1998**, *37*, 1735. [CrossRef] [PubMed]
32. Harris, F.J. On the Use of Windows for Harmonic Analysis with the Discrete Fourier Transform. *Proc. IEEE* **1978**, *66*, 51–83. [CrossRef]

photonics

Article

FP Interferometric Optic Fiber Humidity Sensor Based on Acrylate AB Adhesive Film

Qiying Wang [1], Fangfang Zhang [1], Mengfan Zhang [1], Kunpu Zhang [2], Yaqian Zhang [1,*], Guanjun Wang [1,*], Zhuhua Hu [1] and Qian Deng [1]

1 School of Information and Communication Engineering, Hainan University, Haikou 570228, China
2 School of Materials and Chemical Engineering, Beijing University of Chemical Technology, Beijing 100029, China
* Correspondence: 993888@hainanu.edu.cn (Y.Z.); wangguanjun@hainanu.edu.cn (G.W.)

Abstract: In this paper, an optical fiber humidity sensor based on acrylate AB adhesive film is studied, and two methods—a bubbling method and a dual pressure assisted method—for preparing thin films are proposed. The forms will make the thin film lighter and make the sensor more sensitive. Using a glass tube to connect the acrylate AB adhesive film to the single mode optical fiber, the humidity sensor is obtained. Through humidity response experiments, the sensor's sensitivity reaches 167.5 pm/% RH, and the response time reaches 4.8 s/% RH. At the same time, the experiments show that the sensor has good repeatability and stability. Finally, the influence of temperature on the working process is analyzed, and we provide a method for improvement through FBG.

Keywords: optical fiber sensor; FP interference; acrylate AB adhesive; humidity sensor

Citation: Wang, Q.; Zhang, F.; Zhang, M.; Zhang, K.; Zhang, Y.; Wang, G.; Hu, Z.; Deng, Q. FP Interferometric Optic Fiber Humidity Sensor Based on Acrylate AB Adhesive Film. *Photonics* **2023**, *10*, 873. https://doi.org/10.3390/photonics10080873

Received: 1 June 2023
Revised: 19 July 2023
Accepted: 20 July 2023
Published: 27 July 2023

1. Introduction

Relative humidity is a physical quantity that cannot be ignored in numerous areas of daily life and industrial production, such as human health [1], environmental monitoring [2], industrial production [3], structural health testing [4], electronic device manufacturing [5], and other fields. For a comfortable life and efficient production, it is necessary to effectively monitor relative humidity through sensing devices. Compared with traditional electronic humidity sensors, optical fiber humidity sensors have the advantages of compact structure, high sensitivity, anti-electromagnetic interference, and being able to work in harsh environments such as high temperature and flammability [6]. Currently, there has been a lot of research on optical fiber humidity sensors, including optical fiber interferometer [7], Fiber Bragg Grating [8], photonic crystal fiber [9], Resonators [10], refractive index sensitive fiber [11], etc. Among them, the interferometric fiber optic humidity sensors have attracted more attention due to their flexible design and wide measurement range. In particular, the Fabry–Pérot (FP) interferometric humidity sensors have a simple structure and are easy to implement, and they also show good performance in humidity detection. Changpeng Lang [12] et al. used the tensile curing method to fabricate an optic fiber FPI (Fabry–Pérot interferometer) of polymer microrods based on the end face of the fiber core. This humidity sensor has the advantages of simple fabrication process, low cost, high sensitivity, and fast response. Bo Wang [13] et al. proposed a highly sensitive humidity optical fiber sensor based on full agar FPI. The sensor is made by coating a thin layer of humidity sensitive agar on a single mode optical fiber using an end immersion method. The sensitivity is 4.20 nm/% RH, and the response time is shorter than 340 ms, with excellent characteristics.

In addition, with the development of optical fiber sensing technology, optical fibers can be combined with biochemical sensitive materials with different characteristics, such as agar [14], chitosan [15], SnO_2 [16], graphene [17], and so on, to achieve the detection of more physical quantities, and this aspect has been widely reported. Kai Ni [18] et al. proposed a new type of humidity sensor based on the MZI (Mach–Zehnder interferometer) principle. The sensor is formed by fusing the waist expansion between two SMFs and is modified by depositing chitosan on the surface of the single mode optical fiber, possessing a sensitivity of 119.6 pm/% RH. Rang Chu [19] et al. spliced a section of dual core optical fiber with a standard single mode optical fiber to make the splice point smaller and produced a highly sensitive graphene oxide humidity sensor based on a side polished dual core optical fiber Michelson interferometer. Although the measurement range is small, it has ultra-high humidity sensitivity. Ruijie Tong [20] et al. designed a new type of MZI based on photonic crystal fibers coated with graphene quantum dots and polyvinyl alcohol film for relative humidity detection. As the relative humidity changes, the refractive index and volume of the film change, and films with different thicknesses have different sensitivity and spectral movement directions.

Acrylate AB adhesive has the advantages of low cost, simple operation, and good stability and is widely used in daily life as an adhesive [21]. Its molecules have hydrophilic groups [22]. The refractive index of the adhesive film changes after adsorbing water molecules, giving it a more sensitive humidity response. Aiming at the problems of complex fabrication and high process requirements in traditional optical fiber humidity sensors, this paper focuses on the principle of FP interference, combining optical fiber with acrylate AB film and fabricating an optical fiber humidity sensor. The proposed sensor has a simple manufacturing process, stable performance, and good performance in response time. As a humidity sensor, it has good application prospects and practical value.

2. Structure and Principle

Considering the efficient measurement of optical fiber FPI and the adsorption characteristics of acrylate AB adhesive, a new type of optical fiber biochemical sensor can be developed by combining their advantages, and the connection method can be determined based on their physical characteristics. A FP interference optical fiber humidity sensor based on acrylate AB adhesive film is proposed, with its structural diagram shown in Figure 1.

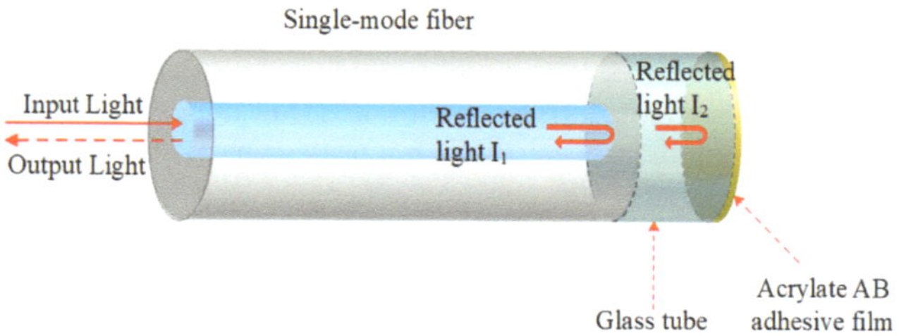

Figure 1. Structural diagram of FP interference optical fiber humidity sensor based on acrylate AB adhesive film.

A section of glass tube is welded at the end of a single mode fiber (SMF, Coring G.652. D), and an FP cavity is formed by covering the end of the glass tube with an acrylic AB adhesive film, forming the humidity sensor. Because the acrylate AB adhesive film is very thin, the three-beam interference can be approximated as two-beam interference. SMF/air is the first reflective surface, and air/thin film is the second reflective surface. The refractive index of the light transmission medium changes: n_0, n_1, n_2 are the refractive indices of the SMF fiber core, air and adhesive film, respectively. According to the Fresnel reflection effect, their reflectivity is [23]:

$$R_1 = \left(\frac{n_1 - n_0}{n_1 + n_0}\right)^2, R_2 = \left(\frac{n_2 - n_1}{n_2 + n_1}\right)^2, \quad (1)$$

Part of the light emitted by the light source is reflected at M_1 to form a reflected light I_1, and the light passing through M_1 reaches M_2 to form a reflected light I_2. There is a phase difference between the two reflected lights, so the intensity of the interference light I generated by mutual interference is [24]:

$$I = I_1 + I_2 + 2\sqrt{I_1 I_2}\cos\left(\frac{4\pi nd}{\lambda} + \varphi_0\right), \tag{2}$$

where n is the refractive index of the FP cavity, d is the distance between two reflective surfaces, λ is the wavelength of the incident light, and φ_0 is the initial phase difference of the interference beam.

When the phase difference between I_1 and I_2 meets the phase matching condition, a cancellation interference spectrum is generated, and its peak is given by the following formula:

$$\frac{4\pi nd}{\lambda_m} + \varphi_0 = 2\pi m, m = 0, 1, 2, \ldots, \tag{3}$$

where λ_m is the wavelength of the mth-order peak. According to Equation (3), it can be obtained that:

$$nd = \lambda_m \frac{2\pi m - \varphi_0}{4\pi}, m = 0, 1, 2, \ldots, \tag{4}$$

As can be seen from Equation (4), λ_m is only affected by the refractive index of the sensor FP cavity n and the distance between the two reflective surfaces d. When the acrylate AB adhesive film absorbs water molecules, the thickness and refractive index of the film change, resulting in a shift in the interference spectrum.

3. Fabrication of Sensors

The sensor structure is fabricated from the platform shown in Figure 2.

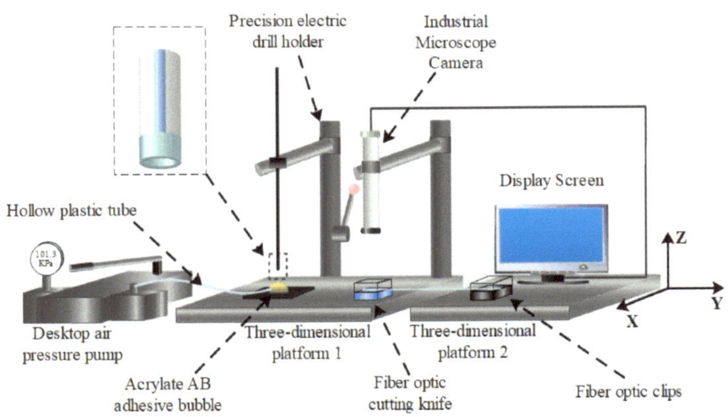

Figure 2. Sensor Manufacturing Platform.

The main body of the platform are two three-dimensional platforms that can be moved in the XYZ axis direction. The left of three-dimensional platform 1 is fixed with a precision electric drill bracket (AOBEN AB8101, Jinhua, China), and the right is fixed with an industrial microscope camera and a fiber optic cutting knife (Fujikura CT50, Shanghai, China).

The upper end of three-dimensional platform 2 is fixed with a display screen connected to an industrial microscope camera and an optical fiber fixing clip, which are adjusted by knobs on the two three-dimensional platforms to align the grooves of the optical fiber cutting knife with the grooves of the optical fiber fixing clip. The camera of the industrial microscope camera is adjusted to the position corresponding to the optical fiber cutter, and the length of the cutting optical fiber and glass tube can be observed on the display screen.

A desktop air pressure pump (ConST 162, Beijing, China) is placed on the left side of three-dimensional platform 1, a section of hollow plastic pipe with a length of about 40 cm is selected, one end of the pipe is connected to the air outlet of the pressure pump through solidified acrylic AB adhesive, and the handle of the desktop air pressure pump is pressed to generate air flow in the hollow plastic pipe.

Through this platform, it is easy to achieve the cutting of glass tubes and the preparation of acrylic AB adhesive films required for experiments, thus improving the production efficiency and success rate of sensors.

The fabrication of an optical fiber humidity sensor based on acrylate AB adhesive film can be divided into the following 5 steps, as shown in Figure 3.

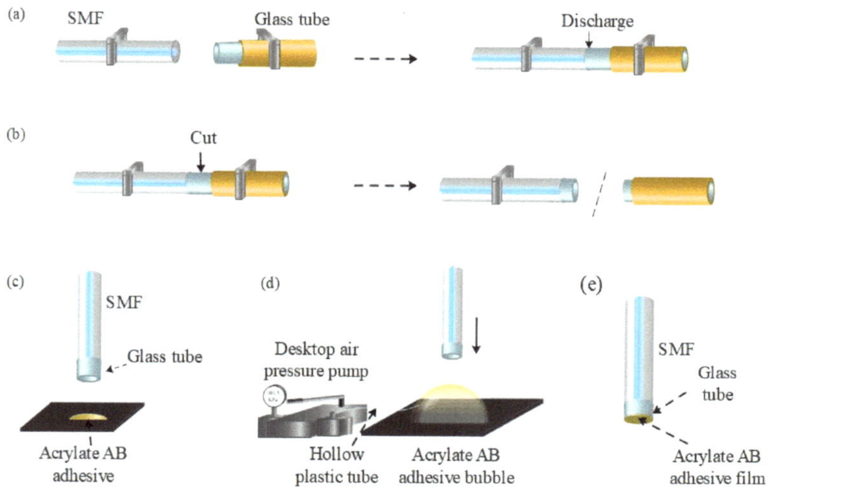

Figure 3. Manufacturing steps of an optical fiber humidity sensor based on acrylate AB adhesive film. (**a**) Fuse SMF and glass tube. (**b**) Cut glass tube. (**c**) Place optical fibers and acrylic AB adhesive. (**d**) Bubble blow and film transfer. (**e**) Sensor structure.

Step 1: Select a suitable length of hollow glass tube (outer diameter and inner diameter are 125 μm and 75 μm, respectively) After burning off about 1 cm of the yellow coating, cut the end flat and place it on the right side of the optical fiber fusion splicer. Cut the end surface of the single mode optical fiber that has been stripped of the protective layer and wiped clean, and place it on the left side of the optical fiber fusion splicer. Select appropriate discharge parameters to fuse the glass tube and the single mode optical fiber together, as shown in Figure 3a.

Step 2: Place one end of the glass tube on the optical fiber cutting knife of the production platform, and place one end of the single mode optical fiber on the optical fiber fixing clip, and fix it. Adjust the length of the cut glass tube by adjusting the Y-axis of three-dimensional platform 2, as shown in Figure 3b.

Step 3: Fix the single mode optical fiber onto the precision electric drill bracket on three-dimensional platform 1. The end of the glass tube is vertically downward, about 5 cm from the platform surface, which can prevent accidental contact and facilitate timely contact during bubble blowing. Next, evenly mix adhesive A and adhesive B of the acrylate adhesive in a 1:1 ratio and place them directly below the single-mode optical fiber fused with a glass tube, as shown in Figure 3c.

Step 4: Insert a hollow plastic pipe into the acrylic adhesive and start a desktop air pressure pump to pressurize, and an acrylic AB adhesive bubble will be generated at the end of the plastic pipe. The generation process is shown in Figure 3d.

Step 5: Observe the generated acrylic AB glue bubbles. When the bubbles are large and relatively stable, press down on the handle of the precision electric drill bracket to drive the single mode optical fiber fused to the glass tube to tilt downward, so that the glass tube probe gently touches the acrylic AB glue bubbles, which can transfer the thinnest film on top of the bubble to the end of the glass tube. Due to intermolecular forces, a thin film will form on the surface of the glass tube probe, as shown in Figure 3e.

The acrylic AB adhesive bubbles generated by the bubbling method are shown in Figure 4. When the glue of a bubble comes into contact with the gas, due to intermolecular interactions, surface tension can be generated on the surface layer. The viscosity of the film controls the surface tension, which effectively slows down the breaking speed of the bubbles and helps ensure sufficient time for subsequent experimental operations.

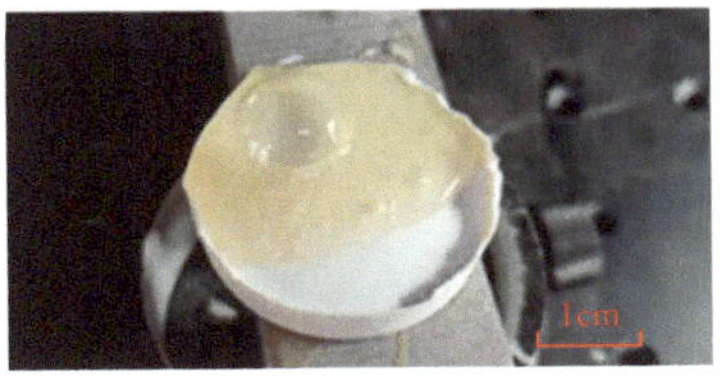

Figure 4. Physical image of acrylic AB adhesive bubble.

The physical object of the fabricated optical fiber sensor structure based on acrylate AB adhesive film is shown in Figure 5.

Figure 5. SMF connected to glass tube and covered with acrylic AB adhesive film.

In addition, a dual pressure assisted method can be used to more accurately control the production of acrylate AB adhesive films, with some steps different from the above.

First, the outer diameter and inner diameter of the glass tube selected by the dual pressure assistance method are 200 μm and 126 μm, respectively. The processed glass tube in the first step is not fused to a single mode optical fiber, but rather is directly fixed to the precision electric drill bracket. Then, a thin film is formed at the end of the glass tube through the bubbling method in the fourth and fifth steps, as shown in Figure 6a.

Secondly, an internal pressure system is composed of a needle tube and a spiral micrometer. As shown in Figure 6b, a secondary pressure is applied to the acrylic AB adhesive film at the end from the inside of the glass tube, allowing the film to expand and become thinner, andwait for 30 min until the film is initially cured.

Finally, after the adhesive film has been cured, a processed SMF is inserted into the glass tube from the end without a film, and the connection interface between the SMF and the glass tube is fixed with UV adhesive to make the sensor probe structure stable and compact, as shown in Figure 6c.

The fabrication process of the dual pressure assisted method is more complex, but it can help quantify the pressurization process. The film thickness produced by this method can reach 4.44 μm so that the fabricated sensor structure has better response characteristics. In addition, the methods of sensor fabrication through the bubble blowing and dual pressure assisted methods are not limited to acrylic AB adhesive. They can also be applied to other sensing materials, allowing easier and faster methods to produce sensors with better performance.

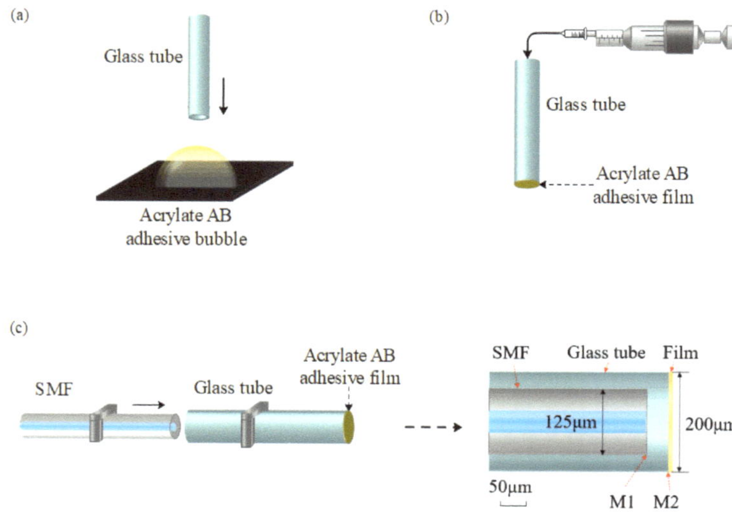

Figure 6. Schematic diagram of dual pressure assistance method. (**a**) Transfer adhesive film to glass tube. (**b**) Apply pressure through the needle tube and spiral micrometer. (**c**) Connect SMF to glass tube.

4. Experiments and Discussions

Measure the humidity performance indicators of the sensor proposed in this article, such as sensitivity, response time, repeatability, and stability, and analyze the impact of temperature on the humidity sensor.

4.1. Humidity Response Experiment

4.1.1. Experimental Device

In order to obtain reliable data and a stable humidity environment, a corresponding humidity detection system has been designed based on experimental requirements and detection content, as shown in Figure 7. The system includes two major parts: a modulation and demodulation system and a humidity control system.

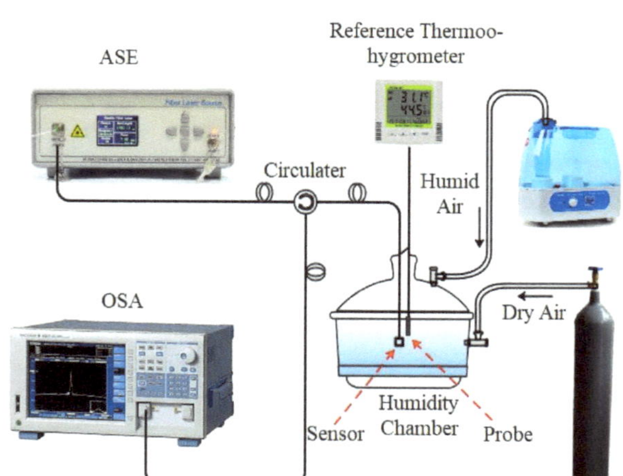

Figure 7. Structure Diagram of Humidity Detection System.

Modulation and demodulation systems include an ASE broadband light source (KG-ASE-CL-17-SP, Beijing, China) with a wavelength range of 1528–1603 nm, an OSA (Yokogawa AQ6370D, Shanghai, China), whose resolution is maintained at 0.1 nm, an optical fiber circulator, and a humidity sensor based on the acrylate AB film. During the experiment, the light emitted from the ASE broadband light source passes through the circulator, with one output SMF connected to the sensor and the other output SMF connected to the OSA.

Humidity control systems include a hygrometer, sealed adjustable devices (humidity boxes), humidifiers, and nitrogen cylinders. The hygrometer has a humidity resolution of 0.1%RH with an accuracy of ±1.5% RH and a temperature resolution of 0.1 °C with an accuracy of ±0.2 °C. It is used to monitor the humidity and temperature levels in the experimental device. The nitrogen cylinders and humidifiers are connected to the humidity box through a conduit to provide dry and wet air for the experiment. Adjusting the switch at the connection can change the flow rate of dry and wet air, thereby controlling the humidity in the humidity box. Through multiple measurements, it can be determined that the humidity in the device can be adjusted to a minimum of 3% RH and a maximum of 98% RH.

4.1.2. Sensitivity

Sensitivity is one of the main indicators to measure the performance of a sensor, which refers to the ratio of the change value of the output quantity Δy to the change value of the input quantity Δx under the steady-state operation of the sensor:

$$Sensitivity = \frac{\Delta y}{\Delta x}, \tag{5}$$

Figure 8 shows the sensor interference spectrum when the relative humidity rises from 10% RH to 90% RH in steps of 5% RH at room temperature (27 °C) for the sensor based on acrylate AB adhesive film. As the relative humidity increases, the interference spectrum moves towards a longer wavelength direction.

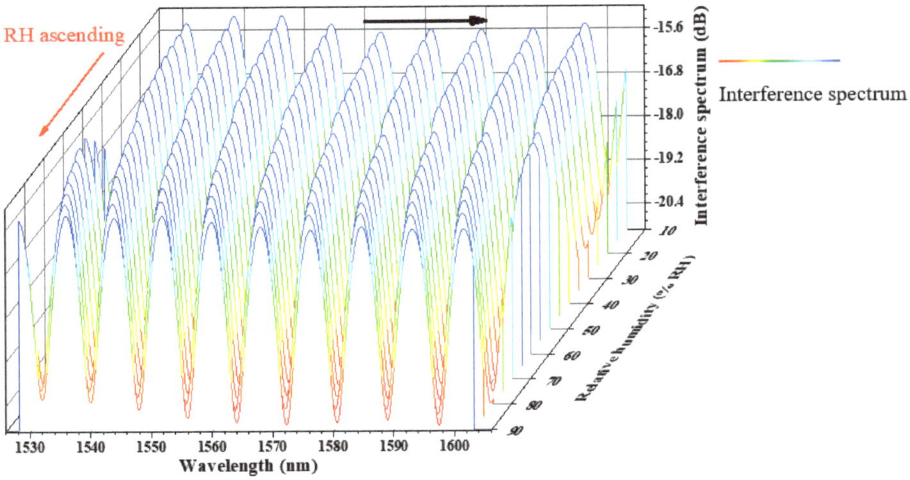

Figure 8. When the humidity increases, the interference spectrum moves towards a longer wavelength direction.

Figure 9 shows the scatter plot after sampling using the peak value of a wave peak in Figure 8 and the curve obtained through exponential fitting. From the scatter diagram, it can be seen that the interference spectrum of the sensor moves slightly with the change in humidity when the humidity is low, and the interference spectrum movement distance of the sensor gradually increases as the humidity gradually increases. In addition, the calculation results of the fitting curve have supported that during the humidity rise process, there is an exponential correlation between the peak point of the interference spectrum and the RH gradient rise; the variance is about 0.9994, and the RH ascending sensitivity is 167.5 pm/% RH.

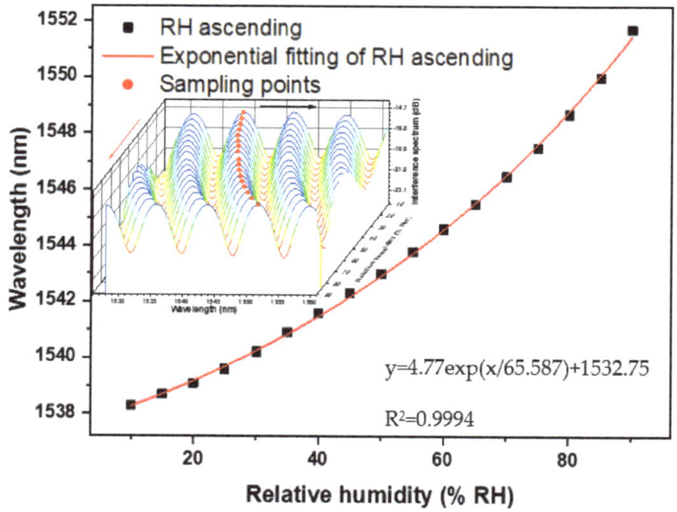

Figure 9. Scatter plots and exponential fitting plots of sampled values at RH ascending.

Similarly, when the interference spectrum of the sensor is in an environment with a relative humidity of 90% RH, the relative humidity of the environment is reduced from 90% RH to 10% RH in steps of 5% RH, and the change in the reflected interference spectrum is recorded. As shown in Figure 10, when the relative humidity decreases, and the interference spectrum moves towards a shorter wavelength direction.

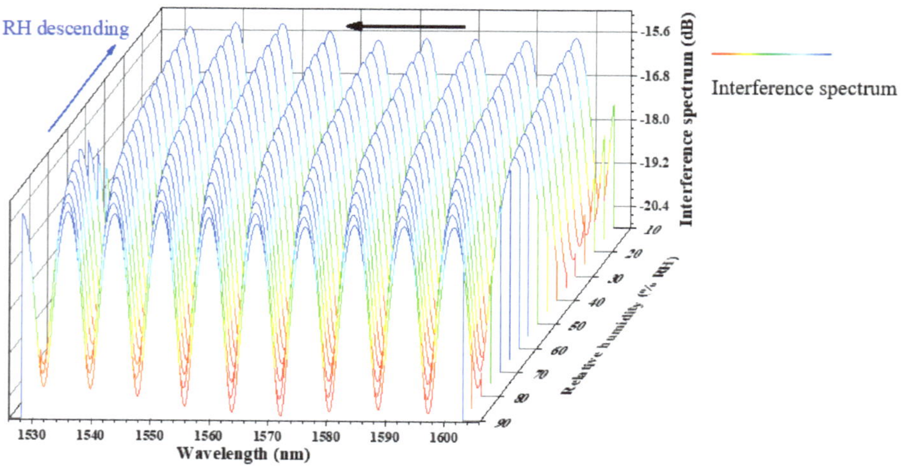

Figure 10. When the humidity decreases, the interference spectrum moves towards a shorter wavelength direction.

Figure 11 is an exponential fitting curve of the peak value of a wave peak. According to the fitting calculation results, there is an exponential correlation between the peak point of the interference spectrum and the RH gradient decrease during the humidity decrease process, with a variance of 0.99984 and a decrease sensitivity of 161.3 pm/% RH.

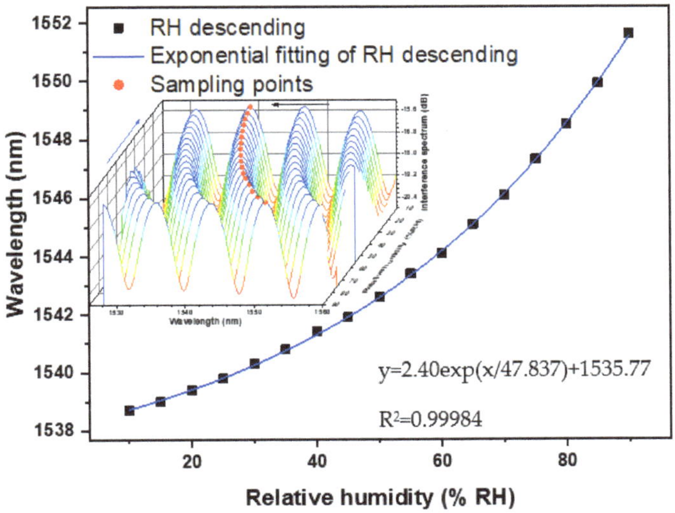

Figure 11. Scatter plots and exponential fitting plots of sampled values at RH descending.

We produced 33 sensors during the experiment, and they all exhibited the same behavior, indicating that they have stable and good performance.

4.1.3. Response Time

In sensor response time measurement experiments, sudden changes in the humidity environment are needed to reduce the impact on response time measurement. Therefore, based on the above humidity detection system, a modified conical flask filled with different saturated salt solutions was selected for the experiment. The improved response time experimental detection device is shown in Figure 12.

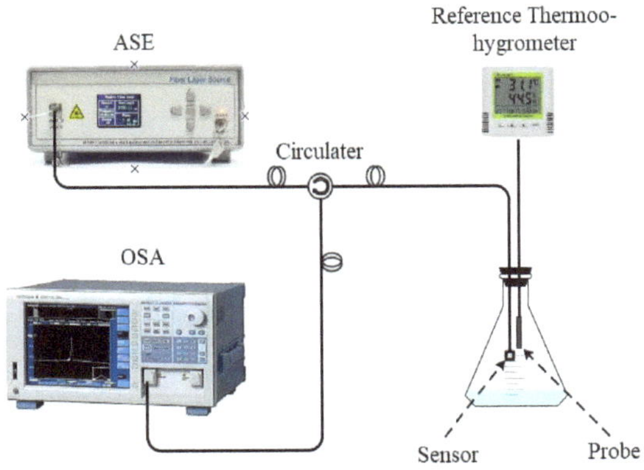

Figure 12. Humidity Response Time Detection Device for Sensors.

Conical flasks filled with $MgCl_2$ saturated solution, $Mg(NO_3)_2$ saturated solution, and NaCl saturated solution, respectively (regarded as device 1, 2, 3), are selected to provide three different relative humidity environments, 39.0% RH, 55.5% RH, and 76.9% RH (measured before the experiment). In this experiment, the response time is the time required to rise from 5% of the wavelength difference to 95%, based on the spectral wavelength corresponding to stabilization in two humidity environments.

$$Time_{Response} = \frac{Time_{95\%} - Time_{5\%}}{\Delta RH}, \quad (6)$$

At room temperature (27 °C), the sensor is placed in the air in device 1. After the waveform stabilizes, the recording of the interference spectrum of the sensor begins. By quickly transferring the sensor to device 2, it is possible to measure the response time of the sensor from one relative humidity environment to another. After the transfer, the sensor is placed in the environment for 5 min to basically reach a stable state, and then the sensor is transferred to device 3 again to constantly observe the changes in the interference spectrum. After it stabilizes, the previous operation is repeated in the order of humidity from high to low, until it is completely stable, and the recording ends.

Figure 13 shows the humidity response time curve of the sensor. When the relative humidity around the sensor suddenly changes from 39.0% RH to 55.5% RH, it takes about 96 s to move from the interference spectrum to stabilization. When it is stabilized in device 2, due to the impact of indoor humidity on device 2 during the transfer, the humidity is measured and becomes 55.9% RH.

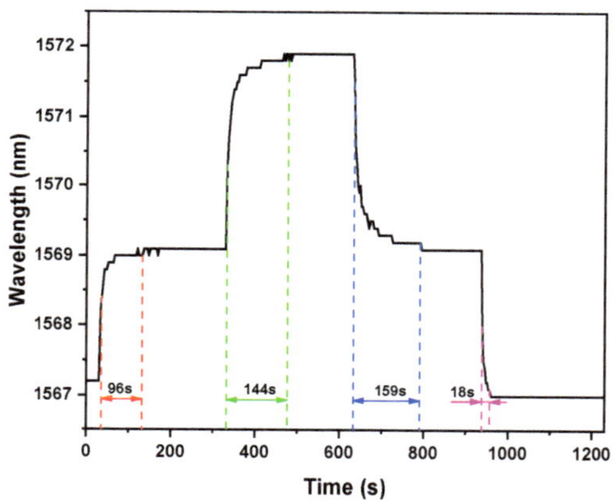

Figure 13. The time curve that sensor achieves stable performance after being transferred to different conical bottle devices.

Next, the sensor is transferred to conical flask 3, and the time for the interference spectrum to move from the beginning to stabilization is approximately 144 s. Between spectral stabilization and the transfer sensor, there may be a shift in the interference light due to changes in the external environment affecting the interior of the device as the experiment progresses. After stabilization, the humidity in device 3 becomes 76.0% RH.

Then, using the same method, the sensor is transferred from device 3 to device 2, and finally it is transferred to device 1 until it is stable. The time required for the sensor to transfer to a stable state twice is approximately 159 s and 18 s. Based on the above data, it can be obtained that the response time for humidity rises is 6.5 s/% RH, and the response time for humidity decreases is 4.8 s/% RH.

The reason why the last response time is significantly different from the previous one may be that the relative humidity in device 1 is the lowest, and it takes the shortest time to recover from exposure to air. However, device 2 and device 3 require a certain amount of time to recover to their original humidity environment after exposure to air. Because the interference spectrum of the sensor varies with the humidity in the device, this process has a certain impact on the measurement of response time.

4.1.4. Repetitiveness

In order to evaluate the repeatability of the sensor, it is necessary to conduct a second humidity sensitivity experiment based on the sensor sensitivity experiment in Section 4.1.2. In this work, after completing the first sensitivity experiment, a second humidity sensitivity detection experiment was conducted on the sensor immediately. Figures 14 and 15, respectively, show the experimental results of two humidity increase/decrease experiments by the sensor.

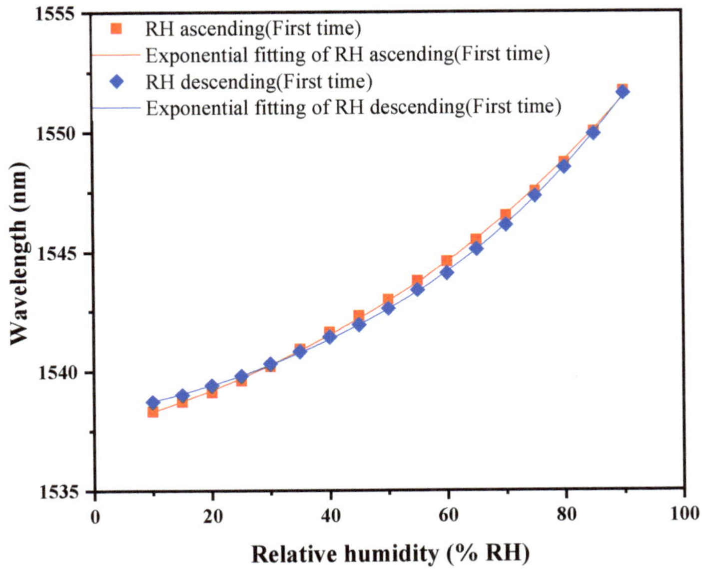

Figure 14. Scatter plots and exponential fitting results of the first sensitivity experiment.

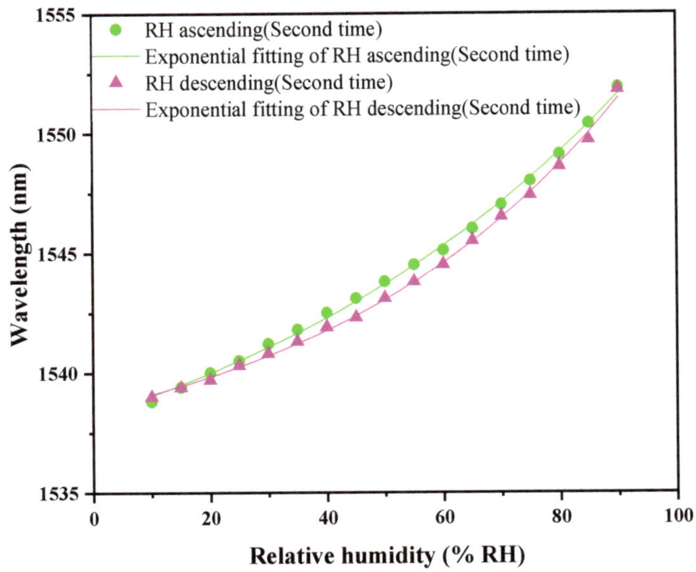

Figure 15. Scatter plots and exponential fitting results of the second sensitivity experiment.

From the results of two humidity rise and fall experiments, we can see that the sensor always maintains a good exponential response to the relative humidity of the external environment. Comparing the two experimental results, this brings up the important fact that the proposed sensor has a relatively good repeatability in humidity detection, which further verifies the reversibility of the sensor.

4.1.5. Stability

An experiment was developed to detect the stability of the sensor using the system in Figure 12. Under normal temperature (27 °C) and pressure, in the above device 1, 2, 3 the stability of the proposed sensor was tested in a humidity environment with relative humidity of 39.1% RH, 57.1% RH, and 76.9% RH (measured during the experiment), respectively. After placing the sensor in the current humidity environment for 5 min to ensure that the humidity environment within the device reaches a stable level, the recording begins. the interference spectrum is recorded every 10 min, with a total recording time of 210 min and a total of 22 records. The experimental results of the stability experiment are shown in Figure 16.

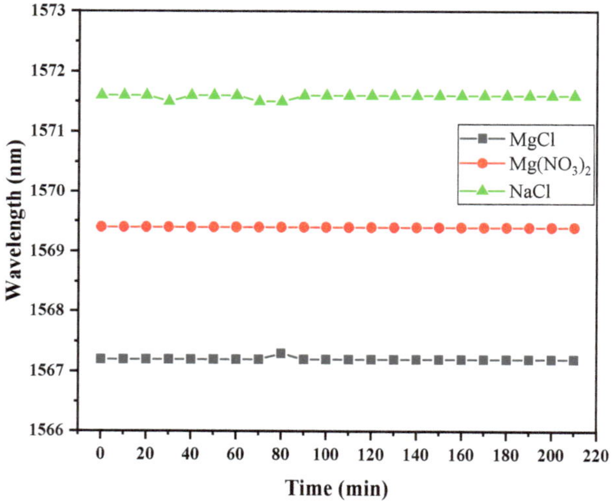

Figure 16. Stability experiment of sensor.

According to the experimental results, when the humidity environment is almost constant, the interference wavelength remains within a certain range. In different humidity environments, the maximum fluctuation of the humidity sensor within 60 min is 0.1 nm, which exhibits good stability.

Due to the limitations of experimental conditions, it is impossible to completely avoid small disturbances caused by changes in environmental conditions during the experimental process, such as changes in room temperature, and changes in three humidity environments experienced during sensor transfer. However, through the analysis of sensor sensitivity, response time, repeatability, reversibility, and stability, it can still be seen that the sensor has excellent performance in humidity sensing.

4.2. Temperature Impact Analysis

In humidity response experiments, it can be observed that changes in ambient temperature have an impact on the sensor. Therefore, through the temperature detection system shown in Figure 17, we further explored the impact of temperature on the sensing process.

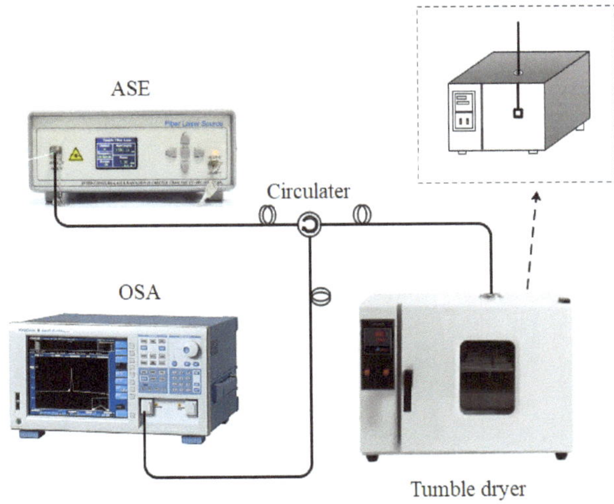

Figure 17. Temperature Detection System.

In the temperature response experiment of the sensor, another sensor with the same structure is used for temperature detection experiment. The humidity is maintained at 30% RH, and the temperature increases from 25 °C to 45 °C. The interference spectrum changes with the ambient temperature as shown in Figure 18.

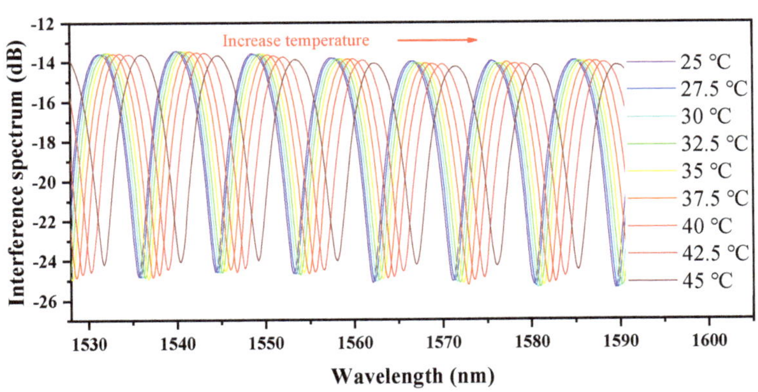

Figure 18. Temperature response curve of the sensor.

This picture sets forth the important fact that as the temperature increases, the interference spectrum moves towards a longer wavelength direction. Because we conducted the experiment in an environment with low humidity, we can basically eliminate this interference. The increase in temperature causes a change in the refractive index of the material, resulting in a shift in the interference spectrum. the peak point of the leftmost wave peak is taken, and a plot of the interference spectrum change when the temperature rises from 25 °C to 45 °C is obtained, as shown in Figure 19. From the curve, it shows that the moving range of the interference spectrum is 1531.2 dB to 1536.1 dB, and the temperature sensitivity of the sensor is 245 pm/°C.

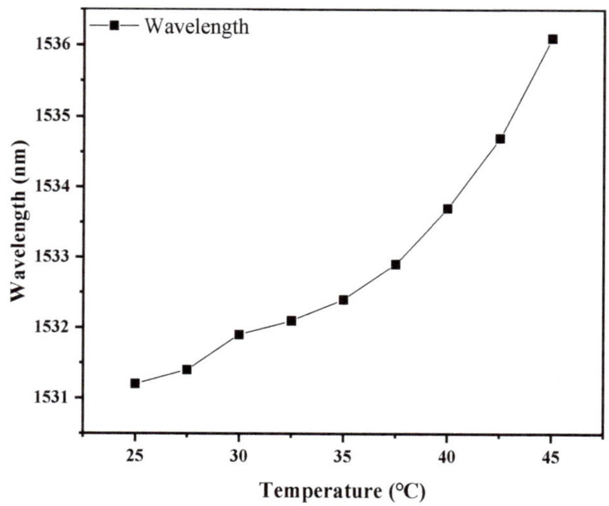

Figure 19. The sensor interference spectrum moves towards a longer wavelength as the temperature increases.

Temperature calibration is achieved by splicing a section of fiber Bragg gratings (FBG) onto the fiber of the sensor [25]. The sensitivity matrix is as follows:

$$\begin{bmatrix} \Delta\lambda_{FBG} \\ \Delta\lambda_{FP} \end{bmatrix} = \begin{bmatrix} S_{RH\_FBG} & S_{T\_FBG} \\ S_{RH\_FP} & S_{T\_FP} \end{bmatrix} \begin{bmatrix} \Delta RH \\ \Delta T \end{bmatrix}, \quad (7)$$

where S_{RH_FBG} and S_{RH_FP} are the relative humidity coefficients of the fiber Bragg grating and the FP cavity; S_{T_FBG} and S_{T_FP} are the temperature coefficients of the fiber Bragg grating and the FP cavity; ΔRH and ΔT are changes in relative humidity and temperature; $\Delta\lambda_{FBG}$ and $\Delta\lambda_{FP}$ are the wavelength shifts of the fiber Bragg grating and the FP cavity, respectively. The variation of the interference spectrum of the improved sensor at different ambient temperature is shown in Figure 20.

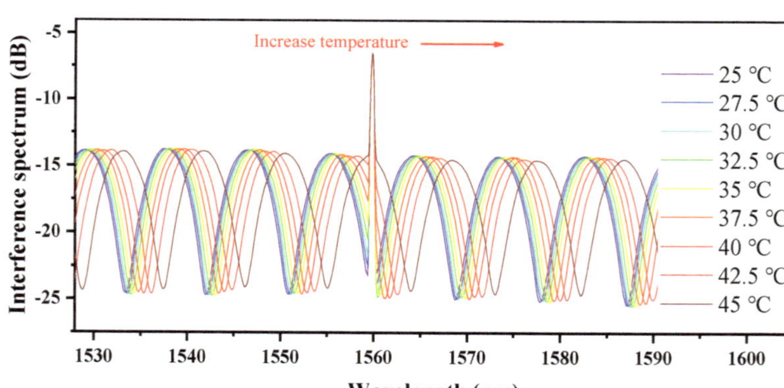

Figure 20. After improving the sensor structure, the interference spectrum shift decreases as the temperature increases.

Similarly, as the temperature increases, the interference spectrum moves towards a longer wavelength direction. The variation curve of the interference spectrum with temperature rise obtained by taking the peak point of the leftmost wave peak is shown in Figure 21. is the curve supports the fact that the interference spectrum of the sensor with FBG has a moving range of 1537.6 dB to 1540.4 dB, and its temperature sensitivity is 140 pm/°C.

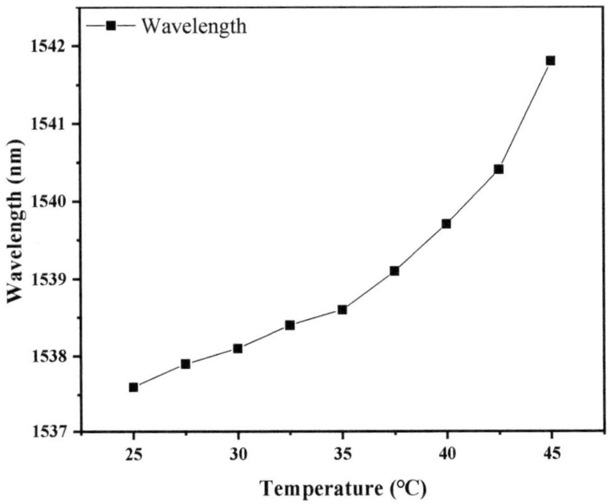

Figure 21. Temperature response curve of sensor with FBG.

The improved structure of the sensor greatly reduces its sensitivity to temperature, making it more suitable for use in daily environments.

This work is the first time that acrylate AB adhesive has been used for humidity sensor research. From the results, it can be seen that the sensor has good performance in sensitivity, stability, and response time, indicating that acrylate AB adhesive has great potential in humidity sensing.

4.3. Performance Comparison

In Table 1, we compare the performance of some fiber optic humidity sensors based on FP interference in recent years. Though they have different humidity-sensitive materials, the sensors mentioned in this article still perform outstandingly.

Table 1. Performance Comparison of Fiber Optic Humidity Sensors Based on FP.

The Structure of Sensors	Material	Range	Sensitivity	Time
FPI	Agar film	25–95% RH	4.2 nm/% RH	2018 [13]
FPI	Agarose	43–63% RH	22.5 pm/% RH	2016 [25]
FPI	agarose gel	16–85% RH	22.5 pm/% RH	2017 [26]
FPI	Chitosan	35–95% RH	280 pm/% RH	2016 [27]
FPI	polyimide	20–90% RH	22.1 pm/% RH	2020 [28]
FPI	PI film	20–90% RH	22.07 pm/% RH	2018 [29]
FPI	PVA film	46–75% RH	248.9 pm/% RH	2021 [30]
FPI	GQDs-PVA	13.47–81.34% RH	117.25 pm/% RH	2019 [31]
FPI	CAB film	8.8–88.1% RH	307 pm/% RH	2013 [32]
FPI	Acrylate AB glue	10–90% RH	172.5 pm/% RH	our work

In addition, Acrylic AB adhesive is cheap and has good stability, its adhesive performance can optimize the production process, and it has potential in humidity sensing.

5. Conclusions

A method of fabricating an FP cavity through acrylate AB adhesive film using a bubbling method and a dual pressure assisted method is proposed. These methods can quickly and efficiently fabricate ultra-thin film structures. In addition, the dual pressure assist method not only pressurizes the acrylate AB adhesive to generate bubbles, but also inflates and pressurizes the film, which can make the film thinner and make the structure more sensitive. Then, through humidity response experiments, various characteristics of the sensor are verified, and the sensitivity of humidity rise and fall is 167.5 pm/% RH and 161.3 pm/% RH, respectively. The response time for humidity rise is 6.5 s/% RH, and the response time for humidity drop is 4.8 s/% RH. In addition, as a humidity sensitive material for FP cavities, acrylate AB film exhibits good repeatability and stability. Finally, the temperature response detection is completed, and the structure is improved.

Author Contributions: Conceptualization, Q.W., F.Z., M.Z. and G.W.; methodology, Q.W., M.Z. and G.W.; software, Q.W.; validation, Q.W., F.Z. and M.Z.; formal analysis, Q.W., F.Z. and K.Z.; investigation, Q.W., F.Z. and M.Z.; resources, G.W.; data curation, Q.W. and K.Z.; writing—original draft preparation, Q.W.; writing—review and editing, Y.Z., G.W. and Q.D.; visualization, Q.W. and K.Z.; supervision, G.W., Z.H. and Q.D.; project administration, Y.Z. and G.W.; funding acquisition, Y.Z. and G.W. All authors have read and agreed to the published version of the manuscript.

Funding: This research was funded by National Natural Science Foundation of China (61762033) and Natural Science Foundation of Hainan Province (617079).

Institutional Review Board Statement: Not applicable.

Informed Consent Statement: Not applicable.

Data Availability Statement: Not applicable.

Acknowledgments: The authors thanks for the support of National Natural Science Foundation of China (61762033) and Natural Science Foundation of Hainan Province (617079). And thanks to G.W. for providing access to instrumentation.

Conflicts of Interest: The authors declare no conflict of interest.

References

1. Niu, H.; Yue, W.; Li, Y.; Yin, F.; Gao, S.; Zhang, C.; Kan, H.; Yao, Z.; Jiang, C.; Wang, C. Ultrafast-response/recovery capacitive humidity sensor based on arc-shaped hollow structure with nanocone arrays for human physiological signals monitoring. *Sens. Actuators B Chem.* **2021**, *334*, 129637. [CrossRef]
2. Zhang, Y.; Wu, L.; Wang, X.; Yu, J.; Bin Ding, B. Super hygroscopic nanofibrous membrane-based moisture pump for solar-driven indoor dehumidification. *Nat. Commun.* **2020**, *11*, 3302. [CrossRef] [PubMed]
3. Farahani, H.; Wagiran, R.; Hamidon, M.N. Humidity sensors principle, mechanism, and fabrication technologies: A comprehensive review. *Sensors* **2014**, *14*, 7881–7939. [PubMed]
4. Li, H.-N.; Li, D.-S.; Song, G.-B. Recent applications of fiber optic sensors to health monitoring in civil engineering. *Eng. Struct.* **2004**, *26*, 1647–1657. [CrossRef]
5. Kolpakov, S.A.; Gordon, N.T.; Mou, C.; Zhou, K. Toward a new generation of photonic humidity sensors. *Sensors* **2014**, *14*, 3986–4013. [CrossRef]
6. Ascorbe, J.; Corres, J.M.; Arregui, F.J.; Matias, I.R. Recent developments in fiber optics humidity sensors. *Sensors* **2017**, *17*, 893. [CrossRef]
7. Sun, L.-P.; Li, J.; Jin, L.; Ran, Y.; Guan, B.-O. High-birefringence microfiber Sagnac interferometer based humidity sensor. *Sens. Actuators B Chem.* **2016**, *231*, 696–700. [CrossRef]
8. Shivananju, B.N.; Yamdagni, S.; Fazuldeen, R.; Kumar, A.K.S.; Nithin, S.P.; Varma, M.M.; Asokan, S. Highly sensitive carbon nanotubes coated etched fiber bragg grating sensor for humidity sensing. *IEEE Sens. J.* **2014**, *14*, 2615–2619. [CrossRef]
9. Noor, M.Y.M.; Kassim, N.M.; Supaat, A.S.M.; Ibrahim, M.H.; Azmi, A.I.; Abdullah, A.S.; Peng, G.D. Temperature-insensitive photonic crystal fiber interferometer for relative humidity sensing without hygroscopic coating. *Meas. Sci. Technol.* **2013**, *24*, 105205. [CrossRef]
10. Zheng, Y.; Dong, X.; Zhao, C.; Li, Y.; Shao, L.; Jin, S. Relative humidity sensor based on microfiber loop resonator. *Adv. Mater. Sci. Eng.* **2013**, *2013*, 815930. [CrossRef]
11. Tan, Y.; Sun, L.-P.; Jin, L.; Li, J.; Guan, B.-O. Temperature-insensitive humidity sensor based on a silica fiber taper interferometer. *IEEE Photonics Technol. Lett.* **2013**, *25*, 2201–2204. [CrossRef]

12. Lang, C.; Liu, Y.; Cao, K.; Li, Y.; Qu, S. Ultra-compact, fast-responsive and highly-sensitive humidity sensor based on a polymer micro-rod on the end-face of fiber core. *Sens. Actuators B Chem.* **2019**, *290*, 23–27. [CrossRef]
13. Wang, B.; Tian, J.; Hu, L.; Yao, Y. High sensitivity humidity fiber-optic sensor based on all-agar Fabry–Perot interferometer. *IEEE Sens. J.* **2018**, *18*, 4879–4885. [CrossRef]
14. Massaroni, C.; Caponero, M.A.; D'amato, R.; Lo Presti, D.; Schena, E. Fiber Bragg grating measuring system for simultaneous monitoring of temperature and humidity in mechanical ventilation. *Sensors* **2017**, *17*, 749. [CrossRef]
15. Hu, P.; Dong, X.; Ni, K.; Chen, L.H.; Wong, W.C.; Chan, C.C. Sensitivity-enhanced Michelson interferometric humidity sensor with waist-enlarged fiber bitaper. *Sens. Actuators B Chem.* **2014**, *194*, 180–184. [CrossRef]
16. Aldaba, A.L.; Lopez-Torres, D.; Elosua, C.; Auguste, J.-L.; Jamier, R.; Roy, P.; Arregui, F.; Lopez-Amo, M. SnO_2-MOF-Fabry-Perot optical sensor for relative humidity measurements. *Sens. Actuators B Chem.* **2018**, *257*, 189–199. [CrossRef]
17. Wang, Y.; Shen, C.; Lou, W.; Shentu, F.; Zhong, C.; Dong, X.; Tong, L. Fiber optic relative humidity sensor based on the tilted fiber Bragg grating coated with graphene oxide. *Appl. Phys. Lett.* **2016**, *109*, 031107. [CrossRef]
18. Ni, K.; Chan, C.C.; Chen, L.; Dong, X.; Huang, R.; Ma, Q. A chitosan-coated humidity sensor based on Mach-Zehnder interferometer with waist-enlarged fusion bitapers. *Opt. Fiber Technol.* **2017**, *33*, 56–59. [CrossRef]
19. Chu, R.; Guan, C.; Bo, Y.; Shi, J.; Zhu, Z.; Li, P.; Yang, J.; Yuan, L. All-optical graphene-oxide humidity sensor based on a side-polished symmetrical twin-core fiber Michelson interferometer. *Sens. Actuators B Chem.* **2019**, *284*, 623–627. [CrossRef]
20. Tong, R.-J.; Zhao, Y.; Chen, M.-Q.; Peng, Y. Relative humidity sensor based on small up-tapered photonic crystal fiber Mach–Zehnder interferometer. *Sens. Actuators A Phys.* **2018**, *280*, 24–30. [CrossRef]
21. Ben Ayed, E.; Magnin, A.; Putaux, J.-L.; Boufi, S. Vinyltriethoxysilane-functionalized starch nanocrystals as Pickering stabilizer in emulsion polymerization of acrylic monomers. Application in nanocomposites and pressure-sensitive adhesives. *J. Colloid Interface Sci.* **2020**, *578*, 533–546. [CrossRef] [PubMed]
22. Chung, Y.C.; Park, J.E.; Choi, J.W.; Chun, B.C. Synthesis and characterizations of antifungal polyurethanes with enhanced tensile and shape recovery performances. *Adv. Polym. Technol.* **2018**, *37*, 3392–3400. [CrossRef]
23. Hu, J.S. Foundation of optical coating. *Opt. Eng.* **2005**, *2*, 123–125.
24. Zhu, C.; Perman, J.A.; Gerald, R.E.; Ma, S.; Huang, J. Chemical detection using a metal–organic framework single crystal coupled to an optical fiber. *ACS Appl. Mater. Interfaces* **2019**, *11*, 4393–4398. [CrossRef] [PubMed]
25. Wang, C.; Zhou, B.; Jiang, H.; He, S. Agarose filled Fabry–Perot cavity for temperature self-calibration humidity sensing. *IEEE Photonics Technol. Lett.* **2016**, *28*, 2027–2030. [CrossRef]
26. Nazari, M.; Amini, A.; Hill, M.R.; Cheng, C.; Samali, B. Physical and chemical reaction sensing in a mixed aqueous solution via metal-organic framework thin-film coated optical fiber. *Microw. Opt. Technol. Lett.* **2020**, *62*, 72–77. [CrossRef]
27. Zhao, Y.; Yuan, Y.; Gan, W.; Yang, M. Optical fiber Fabry–Perot humidity sensor based on polyimide membrane: Sensitivity and adsorption kinetics. *Sens. Actuators A Phys.* **2018**, *281*, 48–54.
28. Zhu, H.; He, J.-J.; Shao, L.; Li, M. Ultra-high sensitivity optical sensors based on cascaded two Fabry-Perot interferometers. *Sens. Actuators B Chem.* **2018**, *277*, 152–156. [CrossRef]
29. Wang, Y.; Huang, Q.; Zhu, W.; Yang, M. Simultaneous measurement of temperature and relative humidity based on FBG and FP interferometer. *IEEE Photonics Technol. Lett.* **2018**, *30*, 833–836. [CrossRef]
30. Chen, M.-Q.; Zhao, Y.; Wei, H.-M.; Zhu, C.-L.; Krishnaswamy, S. 3D printed castle style Fabry-Perot microcavity on optical fiber tip as a highly sensitive humidity sensor. *Sens. Actuators B Chem.* **2021**, *328*, 128981. [CrossRef]
31. Zhao, Y.; Tong, R.-J.; Chen, M.-Q.; Xia, F. Relative humidity sensor based on hollow core fiber filled with GQDs-PVA. *Sens. Actuators B Chem.* **2019**, *284*, 96–102. [CrossRef]
32. Xu, W.; Huang, W.-B.; Huang, X.-G.; Yu, C.-Y. A simple fiber-optic humidity sensor based on extrinsic Fabry–Perot cavity constructed by cellulose acetate butyrate film. *Opt. Fiber Technol.* **2013**, *19*, 583–586. [CrossRef]

photonics

Article

Nonlocal Hydrodynamic Model with Viscosive Damping and Generalized Drude–Lorentz Term

Milan Burda and Ivan Richter *

Faculty of Nuclear Sciences and Physical Engineering, Czech Technical University in Prague, Břehová 7, 115 19 Prague, Czech Republic; milan.burda@fjfi.cvut.cz
* Correspondence: ivan.richter@fjfi.cvut.cz

Abstract: The response of plasmonic metal particles to an electromagnetic wave produces significant features at the nanoscale level. Different properties of the internal composition of a metal, such as its ionic background and the free electron gas, begin to manifest more prominently. As the dimensions of the nanostructures decrease, the classical local theory gradually becomes inadequate. Therefore, Maxwell's equations need to be supplemented with a relationship determining the dynamics of current density which is the essence of nonlocal plasmonic models. In this field of physics, the standard (linearized) hydrodynamic model (HDM) has been widely adopted with great success, serving as the basis for a variety of simulation methods. However, ongoing efforts are also being made to expand and refine it. Recently, the GNOR (general nonlocal optical response) modification of the HDM has been used, with the intention of incorporating the influence of electron gas diffusion. Clearly, from the classical description of fluid dynamics, a close relationship between viscosive damping and diffusion arises. This offers a relevant motivation for introducing the GNOR modification in an alternative manner. The standard HDM and its existing GNOR modification also do not include the influence of interband electron transitions in the conduction band and other phenomena that are part of many refining modifications of the Drude–Lorentz and other models of metal permittivity. In this article, we present a modified version of GNOR-HDM that incorporates the viscosive damping of the electron gas and a generalized Drude–Lorentz term. In the selected simulations, we also introduce Landau damping, which corrects the magnitude of the standard damping constant of the electron gas based on the size of the nanoparticle. We have chosen a spherical particle as a suitable object for testing and comparing HD models and their modifications because it allows the calculation of precise analytical solutions for the interactions and, simultaneously, it is a relatively easily fabricated nanostructure in practice. Our contribution also includes our own analytical method for solving the HDM interaction of a plane wave with a spherical particle. This method forms the core of calculations of the characteristic quantities, such as the extinction cross-sections and the corresponding components of electric fields and current densities.

Keywords: hydrodynamic model; spherical metal nanoparticle; nonlocal response; general nonlocal optical response; viscosive damping; Drude–Lorentz term

Citation: Burda, M.; Richter, I. Nonlocal Hydrodynamic Model with Viscosive Damping and Generalized Drude–Lorentz Term. *Photonics* **2023**, *10*, 913. https://doi.org/10.3390/photonics10080913

Received: 16 July 2023
Revised: 6 August 2023
Accepted: 7 August 2023
Published: 9 August 2023

1. Introduction

Metallic nanoparticles are currently being intensively studied both in terms of describing their interaction with electromagnetic radiation and for utilizing their unique properties. Their application as plasmonic nanostructured materials can find use in various fields, such as sensors utilizing the effect of phosphorescence [1], extraordinary optical transmission (EOT) [2], photothermal applications [3], Raman spectroscopy (SERS) [4,5], designs of highly sensitive gas sensors [6–9], unique anti-reflective coatings to enhance the efficiency of solar cells [10], integrated optical or quantum signal processing [11], battery research [12,13], biomedical applications [4,14,15], and even in designing metamaterials with unique properties such as negative refractive index [16], among others.

The existing research and development of new applications in the aforementioned directions, necessarily relies on adequately precise calculations. At the current level of knowledge, approximate solutions in the form of quasistatic models appear insufficient [17,18], even though in the early days of plasmonics, these theories provided many satisfactory explanations [19]. A fundamental breakthrough for the entire field of plasmonics was represented by Mie's theory [20], which accurately determined the position and value of resonant extinction maxima and associated quantities characterizing the behavior of spherical particles interacting with incident electromagnetic waves.

The demand for accuracy in description, and thus the relevance of simulations derived from them, increases in parallel with the growing technological capabilities of preparations of nanostructures and nanostructured surfaces [21–23]. As an example, sensor applications based on SERS technology can be mentioned, where the sensitivity of the sensor relies on the amplification of the field at the specifically designed site where the binding of the detected molecule is intended to occur.

Just as the quasistatic theory has proven to be inadequate, the classical solutions of Maxwell's equations now also appear unsatisfactory, particularly when a high precision of results is required to simulate the interaction of plasmonic nanostructures with electromagnetic waves. Based on numerous experiments, it is becoming evident that standard simulations, in some cases, significantly overestimate the intensity of the electromagnetic field in the vicinity of sharp edges and interfaces of nanostructures [24]. In the case of nanostructures with characteristic dimensions on the order of a few nanometers, as mentioned above, the classical theory also inaccurately determines the position of resonant maxima for characteristic quantities.

These shortcomings can become a significant problem in practice. A clear example can be found in sensor applications, where in addition to accurately determining the resonant frequency, the intensity of the electric field in specific detection locations of the sensor also plays a significant role. The mentioned requirements for high calculation accuracy have given rise to the need to incorporate nonlocal response.

In contrast to the classical description (Maxwell's equations and Ampere's law), the nonlocal theory assumes a more complex relationship between current density and electric field evolution than a simple proportional relationship. The most widely used nonlocal model in the field of plasmonics has become the so-called standard hydrodynamic model (HDM) [25]. However, there are also alternative nonlocal models [26,27].

Using the HDM, Mie theory has already been generalized [28,29], and several other analytical solutions have been found for metallic interfaces [30], such as the generalized Fresnel equations [31,32]. For the structure of an infinitely long cylinder, the need for finding an accurate solution has been demonstrated, as the so-called curl-free approximation exhibits significant numerical drawbacks in the form of spurious resonances below the plasma frequency [33].

While the standard HDM has become a successful mathematical tool for predicting various phenomena, such as the blue shift of the main extinction maximum in gold and silver spherical nanoparticles with diameters smaller than 10 nm, theoretical research in this field continues with the aim of achieving the most accurate description. Recently, a modification of HDM called GNOR (general nonlocal optical response) has been proposed [34,35] which incorporates the diffusion of the electron gas. The influence of Landau (Kreibig) damping, which is inversely proportional to the nanoparticle size [36–38], is also being discussed. Simultaneously, efforts are being made to find solutions for a more accurate nonlinearized form of HDM [39,40], and the question of viscosive damping of the electron gas is gaining prominence [41–44]. The hydrodynamic model itself is based on the concept of a jellium model which can be interpreted as the electron fluid moving with respect to a positively charged background of metal ions. From this perspective, the question directly arises of how much of the overall material response of a metal belongs to the electron fluid itself and how much to the ion background. The standard HDM, for example, does not consider the influence of energy transitions of electrons within the conduction band, and thus this response is implicitly attributed from a mathematical perspective to the ion background of a metal. From the description of fluid dynamics, there is also a well-known connection between diffusion and viscosive damping, which suggests incorporating the influence of diffusion in an alternative way, different from the GNOR modification. These aforementioned insights are, in our opinion, a relevant stimulus for considering further possible modifications of the hydrodynamic model, which is also the main goal of our article.

The contributions of our article are thus numerous. Among other things, we provide a relatively detailed mathematical procedure for solving HDM in the case of the interaction of a metallic spherical particle with a planar wave. We also compare selected calculations between the classical (Mie) and nonlocal HD models. However, the main contribution lies in presenting a possible approach for further generalizing the standard HDM through the modification of the existing GNOR-HDM by incorporating viscosive damping and considering the general form of the Drude–Lorentz term. Furthermore, in analogy with the previous approach, we compare the computations obtained by solving each variant of the HDM.

The remaining sections of the article are organized as follows: in the second section of the article, we briefly recapitulate the procedure of solving the problem of the interaction between a plane wave and a spherical metal nanoparticle within the framework of the HD model, and subsequently present important new results in the form of an explicit expression for the field expansion coefficients; in the third section, the effective extinction cross sections of gold and silver particles determined according to the classical Mie theory and the standard HD model are compared. Additionally, the influence of Kreibig damping on the behavior of the mentioned quantities is briefly analyzed; the fourth section is dedicated to the generalization of the GNOR-HDM, so that the new model incorporates viscosive damping of the electron gas and the general Drude–Lorentz term. Furthermore, the procedure for the possible solution of our modified GNOR-HDM is discussed; the final fifth section of the article is focused on comparing the results of individual variants of HD models, with the emphasis on comparing the calculations of selected quantities using the standard HD model and our generalized GNOR-HDM, also referred to as the HD model with viscosive damping; finally, the paper is concluded with the conclusions together with possible future activities.

2. Review of the Standard Hydrodynamic Model and Calculation Procedure

In this section, we review the standard hydrodynamic model calculation procedure [29], and subsequently also show our implementation of the technique (Section 3), which will be later used for the generalized hydrodynamic model (Section 4). From a mathematic point of view, the HD model and its modifications represents a system of two linear differential equations for the vector functions of the electric field $\boldsymbol{E}$ and the induced current density $\boldsymbol{J}$. In this article, we will use simplified notation for vector functions $\boldsymbol{E} \equiv \boldsymbol{E}(r, \theta, \varphi, \omega), \boldsymbol{J} \equiv \boldsymbol{J}(r, \theta, \varphi, \omega)$, where ω is the angular frequency and r, θ, and φ are the spherical coordinates (see Figure 1).

In the traditional HDM, the electrical field and the current density are given by the solutions of the wave equation and the hydrodynamic equation-of-motion [38]. We will concentrate here on the specific spherical geometry of a nanoparticle, although some parts of the calculation are more general (primarily the determination of the nonlocal wave number). Clearly, derived expansion coefficients of the fields are generally dependent on the particle shape and analytically can be determined only for a spherical geometry considered here.

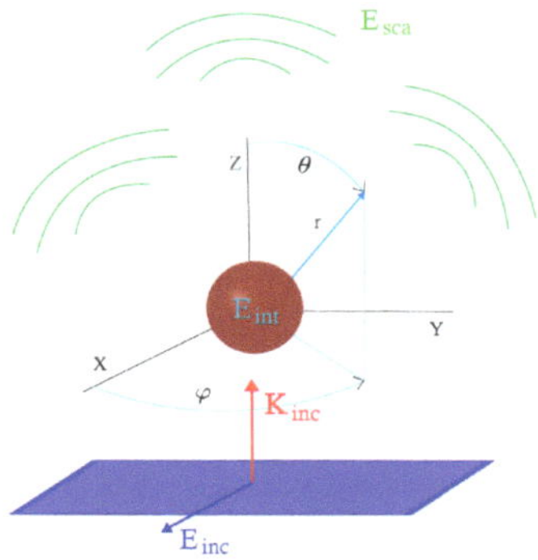

Figure 1. Schematic representation of a plane wave interacting with a spherical metallic particle. A plane wave with the electric intensity $\boldsymbol{D}_{inc}$ and wave vector $\boldsymbol{k}_{inc}$ is incident along the Z-axis onto a spherical particle. The electric field $\boldsymbol{E}_{int}$ is induced within the particle and simultaneously the scattered electric field $\boldsymbol{E}_{sca}$ is formed in the vicinity of a particle.

The basic HD model can be formulated with Equations (1) and (2) as follows:

$$\beta^2 \nabla(\nabla . \boldsymbol{J}) + \omega(\omega + i\gamma)\boldsymbol{J} = i\omega\omega_p^2 \varepsilon_0 \boldsymbol{E}, \quad (1)$$

$$\nabla \times (\nabla \times \boldsymbol{E}) - \frac{\omega^2}{c^2}(\varepsilon_t - \varepsilon_{eg})\boldsymbol{E} = i\omega\mu_0 \boldsymbol{J}. \quad (2)$$

Here, the constants ε_0, μ_0 and c represent vacuum permittivity, vacuum permeability, and the speed of light, respectively. Next, quantities of the HD model ε_{eg} and ε_t indicate the permittivity of electron gas and the total permittivity of a metal, respectively.

The next assumption is concerned with the expression for the permittivity of an electron gas. The HDM modification unfortunately provides no clue which particular form should represent the electron gas permittivity. Clearly, one possibility is to use the expression often formulated in the literature; for example in [45,46]. This expression for the electron gas permittivity is, in fact, the Drude–Lorentz relation for the permittivity, and can be stated as follows:

$$\varepsilon_{eg} = \varepsilon_\infty - \frac{\omega_p^2}{\omega(\omega + i\gamma)}. \quad (3)$$

Here, the difference $(\varepsilon_t - \varepsilon_{eg})$ has the meaning of permittivity of the ionic background of the metal. Other parameters of the HD model are the plasma angular frequency ω_p, the attenuation constant γ, and the nonlocal constant $\beta^2 \equiv 3/5v_F^2$, where v_F is the Fermi velocity. At this point, it is appropriate to provide specific values for the parameters. According to [38], for gold, the values are $\omega_p = 1.3673{\cdot}10^{16}$ Hz, $\gamma = \omega_p/127$ and $v_F = 1.39{\cdot}10^6$ m·s^{-1}, in the case of silver, the values are $\omega_p = 1.3627{\cdot}10^{16}$ Hz, $\gamma = \omega_p/360$ and $v_F = 1.39{\cdot}10^6$ m·s^{-1}. The mentioned values of parameters are used in the analysis of selected quantities in the third and fifth part of this article.

For solving this system of equations, at first it is necessary to determine both longitudinal and transversal wave numbers, as applied, e.g., in [29]. We note that while the longitudinal wave number belongs to the electric field with the vector of electric intensity oscillating in the direction of the wave vector of the field, the transversal wave number belongs to the field which oscillates perpendicularly with respect to the direction of propagation. Following [29], the transversal and longitudinal field components can be described as $\boldsymbol{E}_t = \nabla \times \boldsymbol{\Psi}$ and $\boldsymbol{E}_l = \nabla\Phi$, respectively, where $\boldsymbol{E}_t$ is the transversal electric field, $\boldsymbol{\Psi}$ is some vector function, $\boldsymbol{E}_l$ is the longitudinal electric field and Φ some scalar function. Although these assumptions are not absolutely general, they are typically used (as in Equation (1)) and allow us to obtain the analytical results, as required. It should be also noted that there exists an alternative procedure based on the mathematical theory of generalized functions [2], but this approach brings another difficulty connected with the fulfillment of the boundary conditions and thus is not very practical. A reader can find a classification of HD models from the perspective of such a generalized function (including also the derivation of Green functions) in [47]. Returning to our approach, for the spherical geometry considered, $\boldsymbol{E}_t$ is therefore necessarily a spherical vector function and the Φ function satisfies the scalar wave equation. The fields $\boldsymbol{E}_t$ and $\boldsymbol{E}_l$, hence, satisfy the following Equations (4) and (5), respectively; this follows from the definition of vector spherical harmonics.

$$\Delta \boldsymbol{E}_l = -k_l^2 \boldsymbol{E}_l, \tag{4}$$

$$\Delta \boldsymbol{J}_l = -k_l^2 \boldsymbol{J}_l. \tag{5}$$

Further, these relations (4) and (5) will be substituted into Equations (1) and (2), to obtain the following equations

$$-\beta^2 k_l^2 \boldsymbol{J}_l + \omega(\omega + i\gamma)(\boldsymbol{J}_l + \boldsymbol{J}_t) = i\omega\omega_p^2 \varepsilon_0 (\boldsymbol{E}_l + \boldsymbol{E}_t), \tag{6}$$

$$k_t^2 \boldsymbol{E}_t - \frac{\omega^2}{c^2}(\varepsilon_t - \varepsilon_{eg})(\boldsymbol{E}_l + \boldsymbol{E}_t) = i\omega\mu(\boldsymbol{J}_l + \boldsymbol{J}_t). \tag{7}$$

These two equations can be further converted to describe the longitudinal and transversal fields separately. This will enable assuming the fields can be described by scalar functions and hence the mathematical operators of rotation and gradient can be applied to them, providing for the longitudinal case

$$-\beta^2 k_l^2 \boldsymbol{J}_l + \omega(\omega + i\gamma)\boldsymbol{J}_l = i\omega\omega_p^2 \varepsilon_0 \boldsymbol{E}_l, \tag{8}$$

$$-\frac{\omega^2}{c^2}(\varepsilon_t - \varepsilon_{eg})\boldsymbol{E}_l = i\omega\mu \boldsymbol{J}_l, \tag{9}$$

and similarly, for the transversal field and transversal wave number

$$\omega(\omega + i\gamma)\boldsymbol{J}_t = i\omega\omega_p^2 \varepsilon_0 \boldsymbol{E}_t, \tag{10}$$

$$k_t^2 \boldsymbol{E}_t - \frac{\omega^2}{c^2}(\varepsilon_t - \varepsilon_{eg})\boldsymbol{E}_t = i\omega\mu \boldsymbol{J}_t. \tag{11}$$

This will further allow expressing the longitudinal wave number from Equations (8) and (9), and correspondingly, the transversal wave number, from Equations (10) and (11), as

$$k_t^2 = \frac{\omega^2}{c^2}\left(\varepsilon_t - \varepsilon_{eg} - \frac{\omega_p^2}{\omega(\omega + i\gamma)}\right), \tag{12}$$

$$k_l^2 = \frac{1}{\beta^2}\left(\omega(\omega + i\gamma) - \frac{\omega_p^2}{(\varepsilon_t - \varepsilon_{eg})}.\right) \tag{13}$$

Unfortunately, if we assume that ε_{eg} is zero, as it follows from standard Maxwell equations, then nonphysical results appear, specifically, the extinction cross section will follow wrong functional dependence and the extinction maxima will not follow correct spectral position for bigger particles, in accordance with the Mie theory. Here, we will track the calculation method, introduced in [29], to obtain the correct results. The idea of this technique is based on expressing the unknown permittivity of an electron gas ε_{eg} as $-\omega_p^2/\omega(\omega + i\gamma)$ (see Equations (3) and (12)). Applying this, the transversal wave number follows the predictions for the electric and magnetic fields from the Mie theory [48]. The HD model, however, as compared to the standard Mie theory, also considers the longitudinal wave number (which often has a significantly larger value). Clearly, from the wave number formulas, one can further proceed with the calculations of the scattered field around a particle and the induced field in a particle (i.e., the absorbed field). First, we must specify the proper boundary conditions. Although Maxwell boundary conditions allow discontinuity of the tangential magnetic field component, the Mie theory assumes their continuity. The same boundary conditions are used in the case of solving HDM in [29]. Second, the tangential electric field is naturally continuous on the boundary of a particle. Finally, the last boundary condition determines the zero normal component of the electric current density on the particle boundary. Overall, following the expansion of the fields in the spherical coordinates (θ, φ), we obtain five boundary conditions on the particle boundary

$$\boldsymbol{E}_{\theta}^{inc}(a) + \boldsymbol{E}_{\theta}^{s}(a) = \boldsymbol{E}_{\theta}^{p}(a), \quad \boldsymbol{E}_{\varphi}^{inc}(a) + \boldsymbol{E}_{\varphi}^{s}(a) = \boldsymbol{E}_{\varphi}^{p}(a), \tag{14}$$

$$\boldsymbol{H}_{\theta}^{inc}(a) + \boldsymbol{H}_{\theta}^{s}(a) = \boldsymbol{H}_{\theta}^{p}(a), \quad \boldsymbol{H}_{\varphi}^{inc}(a) + \boldsymbol{H}_{\varphi}^{s}(a) = \boldsymbol{H}_{\varphi}^{p}(a), \tag{15}$$

$$\boldsymbol{J}_{r}(a) = 0. \tag{16}$$

Here, $\boldsymbol{E}_{\varphi}^{p}(a)$, $\boldsymbol{E}_{\theta}^{p}(a)$ and $\boldsymbol{H}_{\varphi}^{p}(a)$, $\boldsymbol{H}_{\theta}^{p}(a)$ are the components of the total electric and magnetic field inside a particle, in the directions of spherical coordinates φ and θ, respectively. Similarly, $\boldsymbol{E}_{\varphi}^{s}(a)$, $\boldsymbol{E}_{\theta}^{s}(a)$ and $\boldsymbol{H}_{\varphi}^{s}(a)$, $\boldsymbol{H}_{\theta}^{s}(a)$ are the components of the scattered fields, and $\boldsymbol{E}_{\varphi}^{inc}(a)$, $\boldsymbol{E}_{\theta}^{inc}(a)$ and $\boldsymbol{H}_{\varphi}^{inc}(a)$, $\boldsymbol{H}_{\theta}^{inc}(a)$ represent the incident fields. All fields are considered on the surface of a particle. Before starting the determination of expansion coefficients of the fields, let us make some remarks. We will follow the procedure of decomposing the transversal fields into series of vector spherical wave functions $\boldsymbol{M}$ and $\boldsymbol{N}$ of both even (index e) and odd (index o) components which we will denote as $\boldsymbol{M}_{nme}^{h}$, $\boldsymbol{M}_{nme}$, $\boldsymbol{N}_{nme}^{h}$, $\boldsymbol{N}_{nme}$, $\boldsymbol{M}_{nmo}^{h}$, $\boldsymbol{M}_{nmo}$, $\boldsymbol{N}_{nmo}^{h}$, $\boldsymbol{N}_{nmo}$. Here, the index n belongs to the radial part whereas the index m represents the azimuthal part of the wave function, respectively.

In order to describe the longitudinal fields, it is necessary to find the solution of Equation (5) in spherical coordinates. We will thus define the scalar even and odd functions Φ_{nme} and Φ_{nmo}. It is also convenient to distinguish the mentioned functions according to the wave number in the argument of their radial functions, therefore we add the given wave number as the superscript of the respective functions. The index n takes values from $1, 2 \ldots \infty$, the index m should be restricted only to the values -1 or 1 because we consider the case of a single spherical particle and an incident plane wave. Clearly, in the case of a Gaussian beam (or an evanescent wave), this symmetry is broken. The vector functions with no upper index are expressed with radial spherical Bessel functions $j_n(r)$, vector functions with index h include a spherical Hankel function of the first type $h_n(r)$. Index o denotes the odd function $\sin(m\varphi)$ and index e denotes the even function $\cos(m\varphi)$.

The mathematical form of the vector spherical harmonics in the spherical coordinates r, θ, and φ are well known, and can be found, for example in [20] or [49]. The explicit expression of the gradient of the $\Phi_{nme}^{k_l}$ and $\Phi_{nmo}^{k_l}$ functions necessary for the decomposition of the longitudinal fields is thus as follows:

$$\nabla \Phi_{nme}^{k_l} = \begin{pmatrix} k_l \cos(m\varphi) P_{n,m}(\cos\theta) \frac{d\, j_n(k_l r)}{d\,(k_l r)} \\ \cos(m\varphi) \frac{d}{d\theta}(P_{n,m}(\cos\theta)) r^{-1} j_n(k_l r) \\ -\sin(m\varphi) \frac{m\, P_{n,m}(\cos\theta)}{\sin\theta} r^{-1} j_n(k_l r) \end{pmatrix}, \tag{17}$$

$$\nabla \Phi_{nmo}^{k_l} = \begin{pmatrix} k_l \sin(m\varphi) P_{n,m}(\cos\theta) \frac{d\, j_n(k_l r)}{d\,(k_l r)} \\ \sin(m\varphi) \frac{d}{d\theta}(P_{n,m}(\cos\theta)) r^{-1} j_n(k_l r) \\ \cos(m\varphi) \frac{m\, P_{n,m}(\cos\theta)}{\sin\theta} r^{-1} j_n(k_l r) \end{pmatrix}. \tag{18}$$

Using the vector harmonics M and N and the gradients of the scalar functions mentioned above, it is possible to decompose the fields. For the selected index m and n, the following relations can be found:

$$E^{inc}_{n,m} = \left[M^{k_0}_{nmo} - iN^{k_0}_{nme}\right], \tag{19}$$

$$H^{inc}_{n,m} = \left[M^{k_0}_{nme} + iN^{k_0}_{nmo}\right], \tag{20}$$

$$E^{p}_{n,m} = \left[B_e M^{k_t}_{mne} + B_o M^{k_t}_{mno} - iA_e M^{k_t}_{mne} - iA_o M^{k_t}_{mno} + C_e \nabla\Phi^{k_l}_{mne} + C_o \nabla\Phi^{k_l}_{mno}\right], \tag{21}$$

$$H^{p}_{n,m} = \frac{k_t}{\omega\mu_0}\left[A_e M^{k_t}_{nme} + A_o M^{k_t}_{nmo} + iB_e N^{k_t}_{nme} + iB_o N^{k_t}_{nmo}\right], \tag{22}$$

$$J_{m,n} = -T\left[iA_e M^{k_t}_{mne} + iA_o M^{k_t}_{mno} - iB_e N^{k_t}_{nme} - iB_o N^{k_t}_{nmo}\right] + S\left[C_e \nabla\Phi^{k_l}_{mne} + C_o \nabla\Phi^{k_l}_{mno}\right], \tag{23}$$

$$E^{s}_{n,m} = \left[-\beta_e M^{h,k_0}_{nme} - \beta_o M^{h,k_0}_{nmo} + i\alpha_e N^{h,k_0}_{nme} + i\alpha_o N^{h,k_0}_{nmo}\right], \tag{24}$$

$$H^{s}_{n,m} = -\frac{k_0}{\omega\mu_0}\left[i\beta_e M^{h,k_0}_{nme} + i\beta_o M^{h,k_0}_{nmo} + i\alpha_e N^{h,k_0}_{nme} + i\alpha_o N^{h,k_0}_{nmo}\right]. \tag{25}$$

Relations (19) and (20) hold for the incident plane wave, see Figure 1, and its electric and magnetic components, respectively. The electric and magnetic fields and the induced current density inside the particle can be determined by Equations (21)–(23). Similar to the incident field around the particle, relations for the scattered field can be established in the form of Equation (24) for the electric field and (25) for the magnetic field.

The individual wave numbers are marked as follows: k_0 is the (vacuum) wave number of the incident and scattered field, k_t and k_l are the wave numbers of the transversal and longitudinal fields inside the nanoparticle. Parameters S and T take the following form $S = i\omega\varepsilon_0(\varepsilon_t + \varepsilon_{eg})$, $T = i\omega\varepsilon_0(\varepsilon_t + \varepsilon_{eg}) - \frac{ik_t^2}{\mu\omega}$. To obtain the total fields E, H and current density J, it is necessary to sum the components in Equations (19)–(25), over all n and m. Before summation, one must determine ten unknown coefficients, $A_e, A_o, B_e, B_o, \alpha_e, \alpha_o, \beta_e, \beta_o, C_e, C_o$ by substituting the components from Equations (19)–(25) into Equations (14)–(16). In our situation of the nonlocal case, the gradients in Equations (17) and (18) will be applicable. Next, we can proceed with solving five boundary conditions using the finding that the expansion coefficients $A_o, B_e, \alpha_o, \beta_e, C_o$ are, in fact, zero. That is implied by the fact that we have only five linear equations for the ten unknowns. It is thus possible to determine the expansion coefficients in the form shown below. Here, it is convenient to define the following effective labelling: $j^{k_t}_n = j_n(k_t a)$, $h^{k_0}_n = h_n(k_0 a)$, $Z^{j,k_0}_n = \frac{d(k_0 r j_n(k_0 r))}{d(k_0 r)}|_a$, $dj^{k_l}_n = \frac{dj_n(k_l r)}{d(k_l r)}|_a$, and similarly for other indices. The expansion coefficients A^n_e, α^n_e, D^n_e, B^n_o, β^n_o, after some algebraic manipulation, take the following form:

$$A^n_e = \frac{k_0 k_t \left(Z^{h,k_0}_n j^{k_0}_n - Z^{j,k_0}_n h^{k_0}_n\right)}{k_0^2\left(\frac{k_t^2}{k_0^2} Z^{h,k_0}_n j^{k_t}_n - Z^{j,k_t}_n h^{k_0}_n + h^{k_0}_n j^{k_l}_n j^{k_t}_n \frac{T\, n\,(n+1)}{a\, S\, k_l\, dj^{k_l}_n}\right)}, \tag{26}$$

$$\alpha^n_e = \frac{j^{k_0}_n}{h^{k_0}_n} - \frac{k_t j^{k_t}_n}{k_0 h^{k_0}_n} A^n_e, \tag{27}$$

$$D^n_e = \frac{i\, T\, n(n+1) j^{k_t}_n}{k_t\, a\, S\, k_l\, dj^{k_l}_n} A^n_e, \tag{28}$$

$$B_o^n = \frac{Z_n^{j,k_0} h_n^{k_0} - Z_n^{h,k_0} j_n^{k_0}}{Z_n^{j,k_t} h_n^{k_0} - Z_n^{h,k_0} j_n^{k_t}}, \tag{29}$$

$$\beta_o^n = \frac{Z_n^{j,k_t} j_n^{k_0} - Z_n^{j,k_0} j_n^{k_t}}{Z_n^{j,k_t} h_n^{k_0} - Z_n^{h,k_0} j_n^{k_t}}. \tag{30}$$

In order to facilitate the reader's possible work when deriving these expansion coefficients for the fields, we present here their explicit expressions. Now, from the knowledge of the α_e^n and β_o^n coefficients, it is clearly possible to evaluate the effective cross-section of extinction:

$$\sigma_{ext} = \frac{2\pi}{k_0^2} \sum_{n=1}^{\infty} (2n+1) Re(\alpha_e^n + \beta_o^n). \tag{31}$$

The expansion coefficients A_e^n, α_e^n, D_e^n, B_o^n, β_o^n are also necessary for calculation of induced current density in a particle, electric and magnetic fields in the surroundings and within the volume of a particle. The coefficients have a general form for the spherical particle, therefore parametric modifications of the HD model would change only the formulas for the wave numbers and electron gas permittivities.

One possible modification of the HD model is the so-called general nonlocal optical response (GNOR) theory [34,38]. This theory should unify quantum-pressure convection effects and induced charge-diffusion kinetics. GNOR also describes size-dependent damping and corresponding frequency shifts. By considering the diffusion effect within the GNOR model, the nonlocal constant is modified as follows: $\beta^2 \to \beta^2 + D(\omega + i\gamma)$, where D is the diffusion constant that can be estimated using the relation $D \cong v_F^2/\gamma$. The diffusion effect modifies the nonlocal constant which manifests as a change in the longitudinal wave number. This effect could be easily added to our model, too. Another possible modification concerns the correction of the damping constant value with respect to the particle size. Experimental measurements of extinction spectra of small plasmonic particles using EELS (electron energy loss spectroscopy) have revealed broadening and shifts of the resonance peaks which can be explained by the size-dependent dependence of the damping constant. Phenomenologically, as proposed by Kreibig, this phenomenon can be described by a correction to the damping constant, where $\gamma \to \gamma + Av_F/r$. From the theoretical explanation of this phenomenon found by Landau [37], it follows that the enhanced damping is a result of the direct excitation of electron-hole pairs in the metal within the region of highly confined surface plasmon field. For a spherical metal particle, one can deduce [34] an estimate $A \sim 1$, this estimation was used in our model. The effect of this attenuation on the spectral behavior of the extinction will be shown in the next section.

3. Model Implementation—Comparison of Local and Standard Nonlocal HD Model

In this section, we will briefly show the results of our implementation of the standard nonlocal HD model, presented in the previous section, in comparison with the local method, to demonstrate the differences and thus the correct implementation of the HDM. For such a demonstration, we will use the quantity of the extinction cross section which allows identifying positions of the resonance maxima and revealing the extent to which the spherical particle interacts with incident electromagnetic radiation. Let us note that the absorption spectra play a dominant role, especially for small particles (below 10 nm in radius). Figures 2 and 3 display a comparison of the extinction cross section between Mie's and HDM results, for a spherical gold (Figure 2) and silver (Figure 3) particle in a vacuum and water environment.

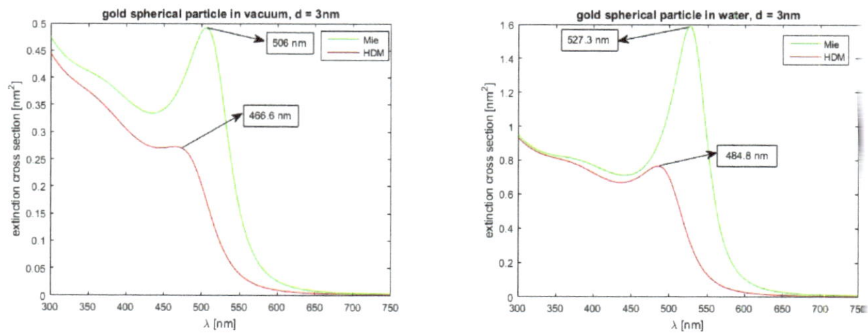

(**a**) (**b**)

Figure 2. Comparison of the extinction cross sections of a gold spherical particle with a diameter of 3 nm according to the classical Mie theory and the nonlocal HD model, (**a**) for a vacuum environment and (**b**) for a water environment.

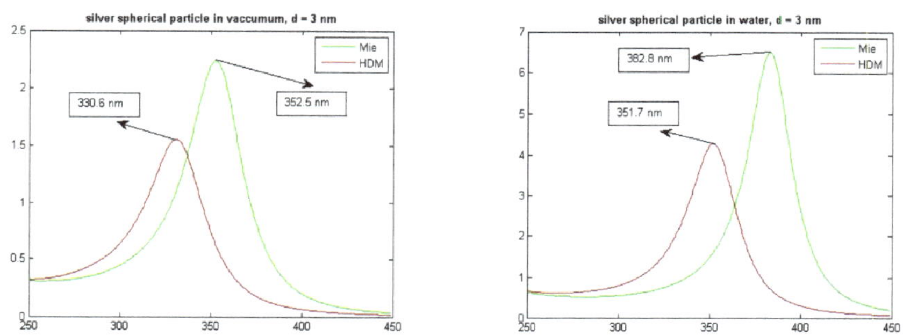

(**a**) (**b**)

Figure 3. Comparison of the extinction cross-sections of a silver spherical particle with a diameter of 3 nm according to the classical Mie theory and the nonlocal HD model, (**a**) for a vacuum environment and (**b**) for a water environment.

As is seen and is well known, HD theory predicts a significant blue shift of the position of extinction maxima, in comparison to the Mie theory. As one can see, these differences are more significant in a water environment. As expected, in a water environment, both models predict circa three times larger extinction maxima and also maximum spectral shifts towards longer wavelengths. Concerning the nonlocality, a blue shift of the nonlocal curve is slightly more significant in a water environment. Further, in Figure 4, again for the case of a vacuum and water environment, we can follow the effect of Landau's damping, incorporated into our standard HD model. Evidently, it modifies the shape of the extinction spectra curve, as compared to the standard HDM. Additionally, Landau's damping changes the local maximum extinction values and values of the extinction for shorter wavelengths (both decrease) whereas for longer wavelengths, a more relaxed declination to zero value is visible, in comparison with the standard HD model without Landau's damping.

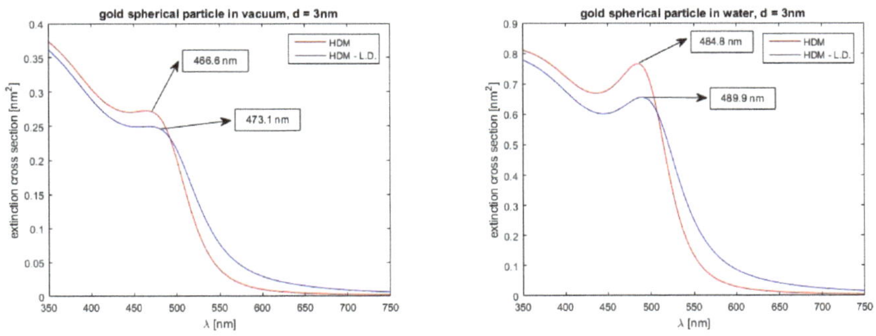

(**a**) (**b**)

Figure 4. Comparison of the extinction cross-sections of a silver spherical particle with a diameter of 3 nm according to the standard HDM and the HDM–L.D. (HDM with Landau damping) model, (**a**) for a vacuum environment and (**b**) for a water environment.

Finally, in Figures 5 and 6, we can see main differences in distribution of the X-component of the electric field calculated by Mie's theory and standard HD model, for two selected wavelengths ($\lambda = 527.3$ nm—maximum prediction for the Mie model, and $\lambda = 486$ nm for the HD model). It is apparent that, in accordance with the results of others [24], the HD model predicts smaller values of the electric field inside a particle (and also in its vicinity). Let us also note an interesting fact, that the electric field is weaker around the particle surface for the HD model. It should be noted, because of this HDM, in agreement with the Mie scattering (and even quasistatic approximation) that the hot spots, represented with the scattered field, are positioned along the Y direction, as expected.

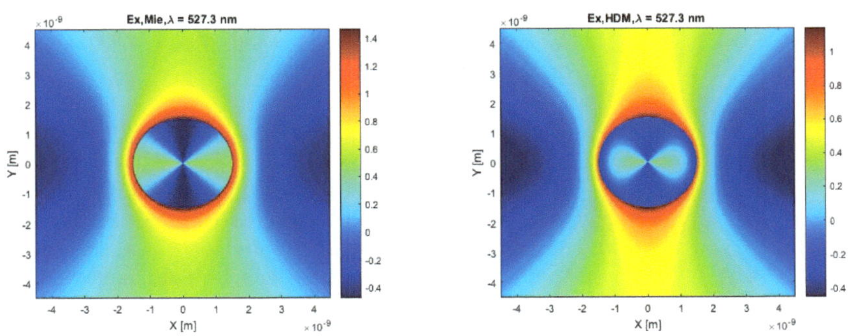

(**a**) (**b**)

Figure 5. The X-component of the electric field calculated for the wavelength of λ = 527.3 nm and the case of a gold spherical particle with a diameter of 3 nm in a water environment, according to (**a**) Mie theory and (**b**) HDM.

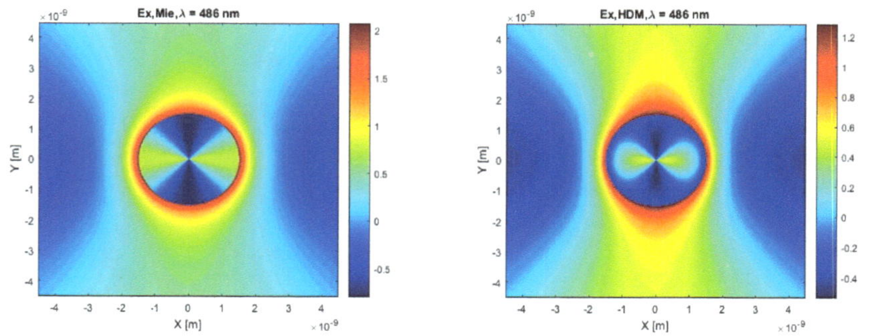

(a) (b)

Figure 6. The X-component of the electric field calculated for a wavelength of λ = 486 nm and the case of a gold spherical particle with a diameter of 3 nm in a water environment, according to (**a**) Mie theory and (**b**) HDM.

4. Nonlocal Hydrodynamic Model with Viscosive Damping and Generalized Drude–Lorentz Term

The standard HDM, analyzed in the previous section, is based on the perception of an electron gas as a charged continuum which can be described, in a certain approximation with a model, relying on the equations of motion of a fluid flow. Recently, an extension of the HDM has been proposed that includes the diffusion of an electron gas; a phenomenon occurring regularly in liquids and gases. Within the physics of fluids, the Stokes–Navier equations play a crucial role, which, among other issues, establishes a close connection between diffusion and viscosive damping. Therefore, it is natural to generalize the standard HDM by introducing viscosive damping of the electron fluid. Starting from the derivation of HDM in [25] and the Stokes–Navier equations for the case of convective form [50], it is straightforward to arrive at the initial Equation (32), which can be further modified to find the generalized variant of HDM.

$$\rho_m(-i\omega + \boldsymbol{v}\cdot\nabla)\boldsymbol{v} = -e\rho_e(\boldsymbol{E} + \boldsymbol{v}\times\boldsymbol{B}) - \rho_m\gamma\boldsymbol{v} - \frac{\rho_m}{i\omega}\beta^2\nabla\nabla\cdot\boldsymbol{v} + \mu_v \triangle \boldsymbol{v} + \frac{1}{3}\mu_v\nabla\nabla\cdot\boldsymbol{v}. \quad (32)$$

The left-hand side of Equation (32) represents the total derivative of the velocity vector v of the electron fluid with respect to time. The first term on the right-hand side represents the Lorentz force acting on an individual volume element of the electron fluid. The second term is the standard damping term, the third term is the standard nonlocal expression, and the last two terms of the equation involve viscosive damping. The individual parameters of the model are as follows: B is the magnetic field, ρ_m is the mass density of the electron fluid, e is the charge of the electron, ρ_e is the charge density, β is the same nonlocal constant as in the standard HDM, and terms μ_v and $\frac{1}{3}\mu_v$ represent the relevant damping terms of viscosive attenuation in the convective form of the Navier–Stokes equation.

To proceed further, it is first convenient to assume the behavior of the electron gas density in the form: $n(r,t) \approx n_0 + n_1(r,t)$, where n_0 is a constant equilibrium density and $n_1(r,t)$ is a linear deviation. Subsequently, it is possible to linearize the Equation (32), which leads to neglecting the influence of the magnetic field and introducing an approximate expression for the current density, charge density and mass density as: $\boldsymbol{J} \approx en_0\boldsymbol{v}$, $\rho_e \approx en_0$, and $\rho_m \approx m_e n_0$, where m_e is the electron mass. If we multiply Equation (32) by the term $-e/m_e$ and perform the mentioned linearization, we obtain:

$$\left(\omega^2 + i\omega\gamma\right)\boldsymbol{J} = in_0\frac{\omega e^2}{m_e}\boldsymbol{E} - \beta^2\nabla\nabla\cdot\boldsymbol{J} + \frac{i\omega}{m_e n_0}\mu_v\Delta\boldsymbol{J} + \frac{i\omega}{3m_e n_0}\mu_v\nabla\nabla\cdot\boldsymbol{J}. \quad (33)$$

If we introduce a new damping parameter, $\gamma_v = \frac{\mu_v}{m_e n_0}$, which is analogous to kinematic viscosity in the classical approach, and further, according to the common definition, the plasma frequency $\omega_p = \sqrt{\frac{e^2 n_0}{m_e \varepsilon_0}}$, then by straightforward adjustments of Equation (33), we obtain a modified novel version of the first equation of the standard HDM as follows:

$$\left(\beta^2 - \frac{1}{3} i\omega\gamma_v\right)\nabla(\nabla\cdot\boldsymbol{J}) - i\omega\gamma_v\Delta\boldsymbol{J} + \omega(\omega + i\gamma)\boldsymbol{J} = i\omega\omega_p^2\varepsilon_0\boldsymbol{E}. \tag{34}$$

Now the question arises whether it is possible to further generalize Equation (34). Since the HDM can be considered as a more sophisticated version of the Drude model for the electron dispersion in a metal, it is worth considering a similar modification approach as in the case of the actual Drude model. As it is well known, the original Drude model of the permittivity of a metal can be modified by the additional Lorentzian terms, resulting in the Drude–Lorentz model [51,52], to account for intraband (within the conduction band) and interband (between bands) electron energy transitions. By a simple analysis of the first equation of the HDM, i.e., the current Equation (34), by substituting the transverse fields, it can be determined that precisely the term $\omega(\omega + i\gamma)\boldsymbol{J}$ is crucial for the Drude-type response of the permittivity ε_{eg} of a free electron gas. Let us assume a generalization: $\omega(\omega + i\gamma) \rightarrow \omega(\omega + i\gamma) + \xi(\omega)$, where $\xi(\omega)$ is an as-yet undetermined function. For better clarity, let us denote $U = \omega(\omega + i\gamma) + \xi(\omega)$ and introduce an HDM with viscosive damping and a generalized Drude–Lorentz term in the following form:

$$\left(\beta^2 - i\frac{1}{3}\omega\gamma_v\right)\nabla(\nabla\cdot\boldsymbol{J}) - i\omega\gamma_v\Delta\boldsymbol{J} + U\boldsymbol{J} = i\omega\omega_p^2\varepsilon_0\boldsymbol{E}, \tag{35}$$

$$\nabla \times (\nabla \times \boldsymbol{E}) - k_0^2(\varepsilon_t - \varepsilon_{eg,v})\boldsymbol{E} = i\omega\mu_0\boldsymbol{J}. \tag{36}$$

The generalized HDM defined by Equations (35) and (36) includes the three unknown parameters γ_v, U, $\varepsilon_{eg,v}$ for which it will be necessary to establish the three conditions. The term $\varepsilon_{eg,v}$ has a similar meaning as ε_{eg} in the case of classical HDM, representing the permittivity of the electron gas, but it is generally different from ε_{eg}. The total permittivity ε_t of the metal is the same as in the classical HDM and is determined by tabulated values [53]. All these three parameters are clearly frequency dependent, however, for the sake of clarity in the mathematical notation, it will not be shown here.

Finding a solution for the generalized HDM is possible in a similar way to the classical HDM. From Equation (36), it is possible to express the relationship for current density J and substitute it into Equation (35). In this way, an equation for the total electric field E only is obtained, which can be solved separately for its transverse and longitudinal components. In the case of a transverse field, it is advantageous to use the relationships between vector spherical harmonics $\nabla \times \boldsymbol{M} = k_t\boldsymbol{N}$, $\nabla \times \boldsymbol{N} = k_t\boldsymbol{M}$ and operator identities: $\nabla \times \nabla \times \boldsymbol{\Psi} = \nabla\nabla\cdot\boldsymbol{\Psi} - \Delta\boldsymbol{\Psi}$, $\nabla\cdot\nabla \times \boldsymbol{\Psi} = 0$, valid for any vector function $\boldsymbol{\Psi}$. In the case of a longitudinal field, it is beneficial to use the identities $\nabla\nabla\cdot\nabla\,\varphi = \nabla\Delta\varphi$ and $\nabla \times \nabla\varphi = 0$. Based on the mentioned procedure, the relationships for the transversal and longitudinal wave numbers can be obtained in the following form:

$$k_t^4 - \left(\frac{iU}{\omega\gamma_v} + k_0^2(\varepsilon_t - \varepsilon_{eg,v})\right)k_t^2 - \frac{ik_0^2}{\omega\gamma_v}\left(\omega_p^2 - U(\varepsilon_t - \varepsilon_{eg,v})\right) = 0, \tag{37}$$

$$k_{l,v}^2 = \frac{1}{\beta^2 + \frac{4}{3}i\omega\gamma_v}\left(U - \frac{\omega_p^2}{(\varepsilon_t - \varepsilon_{eg,v})}\right). \tag{38}$$

Equations (37) and (38) serve as a starting point for finding the additional relationships for the parameters γ_v, U, and $\varepsilon_{eg,v}$. The classical HDM already includes a closely unspecified parameter ε_{eg}. However, to ensure that the HDM solution is not too far from the experimentally measured values, a condition $k_t = k_0\sqrt{\varepsilon_t}$ has been introduced for the transverse wave number [29] which is widely utilized, and which will be also applied here. Considering that the standard solution of Equation (37) as a quadratic equation would provide only one condition, it seems reasonable to proceed by evaluating the brackets in (37), which allows us to obtain a pair of additional expressions without the need for the introduction of ad hoc parameters. One possibility is to set the second term in Equation (37) equal to k_t and the third term equal to zero. However, it can be easily verified that such a solution implies a zero value for the longitudinal wave number and therefore the absence of the nonlocal response. One option is to set the second term in Equation (37) equal to k_t^2 and the third term equal to zero. However, such a solution implies a zero value for the longitudinal wave number, and therefore the absence of a nonlocal response. Therefore, as a solution, it is suggested to set the first bracket in Equation (37) to zero and assign the value of k_t^4 to the last term in Equation (37). From Equation (37), we can easily obtain two conditions in the following form:

$$i\omega\gamma_v k_0^2(\varepsilon_t - \varepsilon_{eg,v}) - U = 0, \tag{39}$$

$$U(\varepsilon_t - \varepsilon_{eg,v}) - \omega_p^2 = i\omega\gamma_v k_0^2 \varepsilon_t^2. \tag{40}$$

To establish the final condition, it is possible to start from the following consideration. Let us assume that the diffusion phenomenon is well phenomenologically described by the GNOR modification [34] of the HDM, although the magnitude of the diffusion constant D may still be the subject of investigation [38]. Thanks to the fact that the fourth term in the right-hand side of Equation (32) is responsible for the diffusion according to the classical description of fluid dynamics, this diffusion is indeed included in Equation (32). From the comparison of the GNOR model with the classical HD, it follows that diffusion manifests only in the change in the nonlocal parameter as: $\beta^2 \to \beta^2 + D(\gamma - i\omega)$. Under the assumption that diffusion is the main accompanying phenomenon caused by viscosive damping, it can be presumed that such viscosive damping will manifest in a similar way, primarily through a certain change in the nonlocal constant β. Based on the mentioned considerations and Equation (38), the third condition can be written in the following form:

$$\gamma_v = -\frac{3}{4} iD\omega^{-1}(\gamma - i\omega). \tag{41}$$

From the given conditions (39)–(41), the relationships for the parameters U and $\varepsilon_{eg,v}$ can be derived. Specifically, from Equations (39) and (41), it is straightforward to obtain the following relationship for $\varepsilon_{eg,v}$ as:

$$\varepsilon_{eg,v} = \varepsilon_t - \frac{4}{3}\frac{U}{D(\gamma - i\omega)k_0^2}. \tag{42}$$

Subsequently, from Equations (40)–(42), the relationship for U can be derived in the following form:

$$U = k_0\left(\frac{3}{4}D(\gamma - i\omega)\left(\omega_p^2 + i\omega k_0^2 \varepsilon_t^2\right)\right)^{\frac{1}{2}}. \tag{43}$$

From the above Equations (41)–(43), it is finally possible to determine the longitudinal wave number defined by (38). As mentioned before, for the transverse wave number, the relationship $k_t = k_0\sqrt{\varepsilon_t}$ holds. The value of the diffusion constant can be approximated as $D \sim v_F^2/\gamma$ according to [38,53] which we have also used in our calculations. It remains to be mentioned that the expansion coefficients of the fields, including the substitution relations denoted by S and T, given in the second section, can be still used in the same form, but with a logical replacement of k_l, ε_{eg}, and $\omega(\omega + i\gamma)$ by $k_{l,v}$, $\varepsilon_{eg,v}$, and U. For the calculation of extinction, the same approach as in the case of the classical HDM, is clearly now applicable as well. The Formulas (26) to (30), together with Equation (31), can be used, with the difference that the longitudinal wave number is now determined with the new corresponding relation, i.e., according to Equation (38).

5. Results of HDM with Generalized Drude–Lorentz Term and Viscosive Damping

In this section, we turn to the results of the generalized HD model. From the outcome of our simulations, it is possible to come up with the following findings. First, in Figure 7, our viscosive damping generalization of the HDM model in comparison with the standard HDM, for a gold spherical nanoparticle, provides the extinction cross-section (standard HDM and our HDM–Vis.D. model) for a gold spherical particle (3 nm) in a vacuum (Figure 7a) and in water (Figure 7b). Additionally, the HDM–Vis.D. model has caused a small red correction to the blue shift of extinction local maxima (about 4 to 8 nm). Additionally, local maxima for gold spherical particles, for the smallest particles, have disappeared, in comparison with the results from the standard HDM.

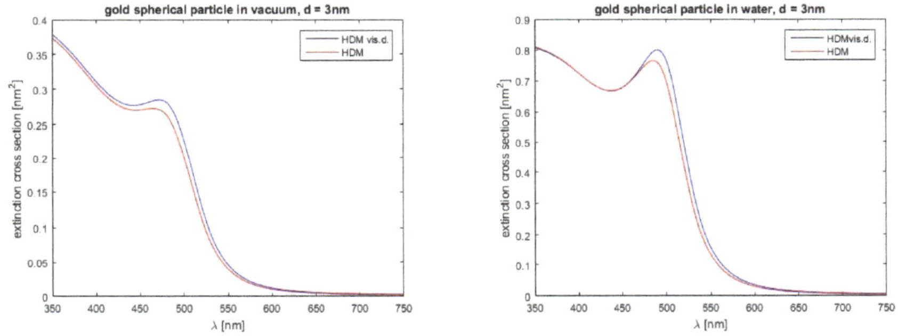

(a) (b)

Figure 7. Comparison of the extinction cross-sections of a gold spherical particle with a diameter of 3 nm according to the HDM and the HDM–Vis.D model: (**a**) for a vacuum environment and (**b**) for a water environment.

This is visible in Figure 7 where a blue curve, which represents the hydrodynamical model with viscosive damping, reaches in both cases (vacuum or water) higher values. Our model predicts a red correction to the blue shift of about 3 nm in both cases (vacuum or water), the extinction cross-section in the main resonance peak, as predicted from our model, has about a 5% higher value. Both HDM models also show that the resonance maximum is larger and narrower in the case of water surroundings. It is also seen in Figure 7 that both models converge to the same values in the case of a water environment for shorter wavelengths (350–450 nm).

Next, Figure 8 shows a similar comparison of both models, now for a silver spherical nanoparticle in a vacuum and a water environment. The permittivity spectral dependence is for silver obtained by the Drude–Lorentz dispersion model. In Figure 8, we can notice again weak differences between calculated values of the two models, as expected. Similar to gold, our model again predicts a red correction to the blue shift, somewhat higher values of extinction at the resonance peak (more noticeable for a vacuum) and larger half-width values (also more obvious for a vacuum).

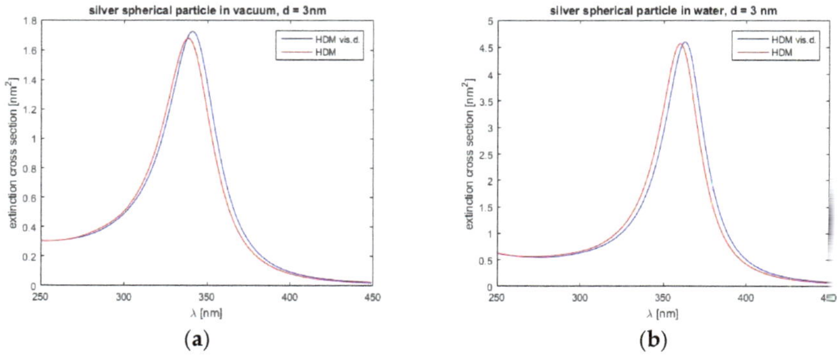

Figure 8. Comparison of the extinction cross-sections of a silver spherical particle with a diameter of 3 nm according to the HDM and the and HDM–Vis.D. model: (**a**) for a vacuum environment and (**b**) for a water environment.

Further, Figure 9 shows additional modification of calculations of the extinction spectra which are represented in Figure 7. This modification, the Landau damping, denotes additional size-dependent damping of localized plasmons on the interface of a metal structure. Figure 9 shows that the extinction cross-section of a gold spherical particle in a water environment has almost the same value at shorter wavelengths (between 350–400 nm), similar to Figure 7, whereas a particle in a vacuum exhibits little difference between the two model predictions, at these short wavelengths.

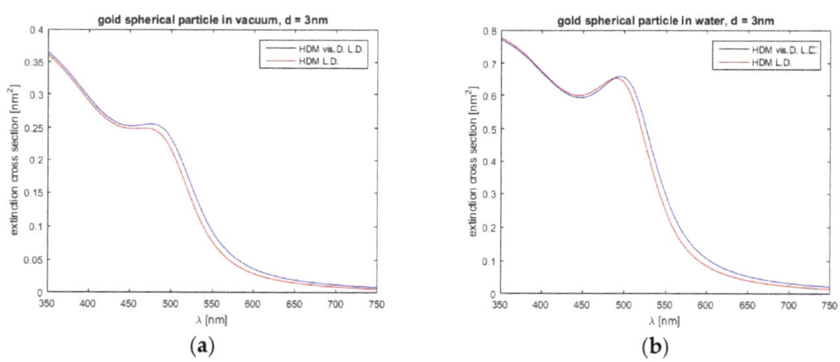

Figure 9. Comparison of the extinction cross-sections of a gold spherical particle with a diameter of 3 nm according to the HDM–L.D and the HDM–Vis.D. L.D. (both HDM and HDM–Vis.D. with Landau damping): (**a**) for a vacuum environment and (**b**) for a water environment.

The HDM model with both viscosive and Landau damping, again predicts corrections to the blue shift for both cases of surrounding environment and it also shows slightly larger values in the resonance maxima, as compared to Figure 7. It is clearly visible that additional inclusion of Landau damping into the model causes a lowering of resonance maxima and slower convergence to zero values of the extinction cross-section for long wavelengths (up to 650 nm).

As we can see in Figure 10, again our model with both viscosive and Landau damping, compared to the standard HDM, predicts corrections of blue shift and a slightly larger resonance maximum value. Clearly, silver particles exhibit a stronger influence of Landau damping, thus the resonance maxima peak is significantly larger and narrower (compared to Figure 7). Descent of the extinction of the resonance wavelength up to 400 nm is surprisingly faster. Our model with viscosive and Landau damping predicts new, although very weak, resonance on shorter wavelengths. This effect may be caused by the Drude–Lorentz model for metal permittivity data.

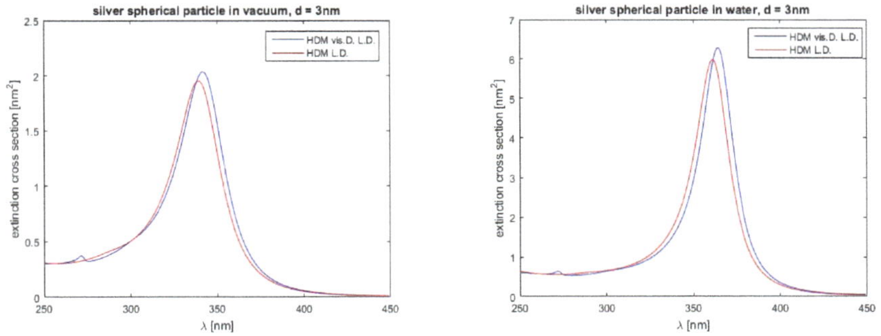

(a) (b)

Figure 10. Comparison of the extinction cross-sections of a silver spherical particle with a diameter of 3 nm according to the HDM–L.D and the HDM–Vis.D. L.D. (both HDM and HDM–Vis.D. with Landau damping): (**a**) for a vacuum environment and (**b**) for a water environment.

Next, Figure 11 shows the calculated X-component of the electric field for our HDM with viscosive damping and classical HDM at the frequency of the resonant maximum of the HDM. At the first glance of the figures, a difference in the negative field values is noticeable. The calculated profile of the electric field inside a particle using HDM can help us evoke the idea of the direction of the incident plane wave, but this deduction using the field distribution in the particle calculated by HDM with viscosive damping is not so apparent.

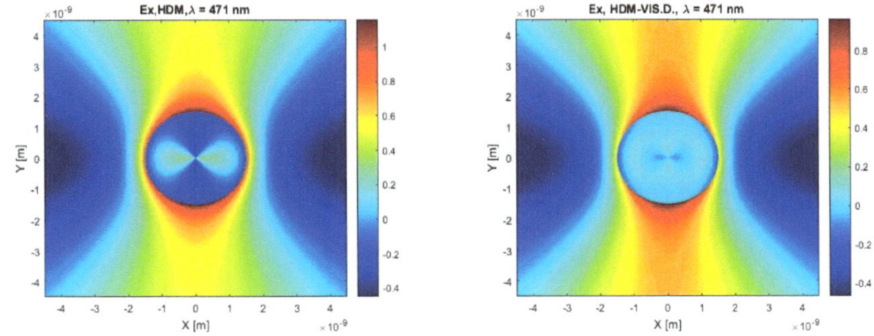

(a) (b)

Figure 11. The X-component of the electric field calculated for the wavelength of λ = 471 nm and the case of a gold spherical particle with a diameter of 3 nm in a vacuum environment, according to (**a**) HDM and (**b**) HDM–Vis.D.

Figure 12 represents the electric current distribution in the plane cut through a spherical particle in the X and Y axes. Both graphs display a symmetric distribution of the current density in the X and Y direction. At first sight, these current profiles are very similar, but more detailed examination reveals that our HDM with viscosive damping predicts higher absolute values of the current density, and the area around the maxima value is situated away from the center of the particle (in the middle of each hemisphere) while classical HDM provides the maxima values near the center of the particle. It can be also mentioned, that classical HDM predicts a slower descent to zero values of the current density near the boundary of a particle.

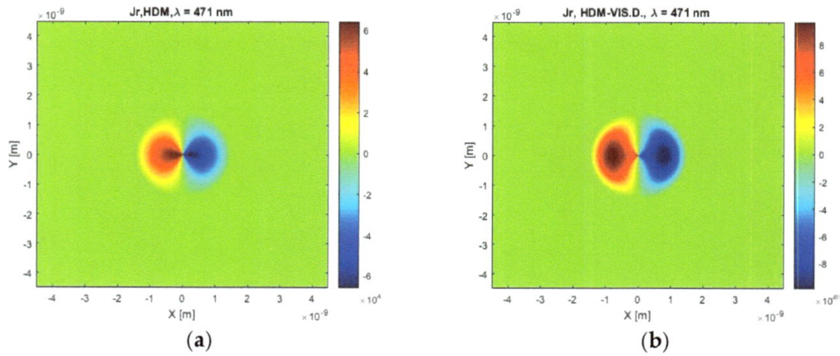

Figure 12. The radial component of the current density calculated for the wavelength of λ = 471 nm and the case of a gold spherical particle with a diameter of 3 nm in a vacuum environment, according to (**a**) HDM and (**b**) HDM–Vis.D.

In Figure 13, when comparing the X-component electric fields calculated by the HDM and our HDM with viscosive damping, we can notice three major differences. The first distinction is that our model predicts only a twice as strong positive maximum value than the maximum negative value, whereas the classical HDM predicts this ratio to be about three. Next, the majority of the area inside a particle exhibits an inverse value of the field, i.e., where the HDM calculations display negative values, our model shows positive values and the converse. The last difference is related to the close vicinity of a particle, in the case of the classical HDM, the particle is surrounded by a region with maximum positive values of the field, whereas positive values of the field obtained by our model are located mainly in the direction of the Y axes.

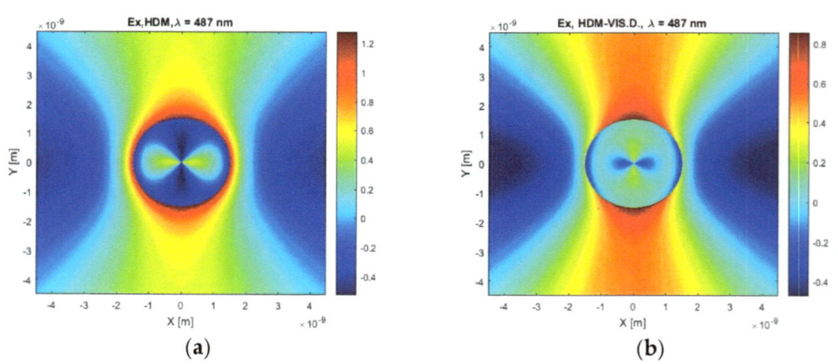

Figure 13. The X-component of the electric field calculated for the wavelength of λ = 487 nm and the case of a gold spherical particle with a diameter of 3 nm in a water environment, according to (**a**) HDM and (**b**) HDM–Vis.D.

Figure 14 shows the density of the energy displayed in a plane cut that is defined by the X and Y axes. Now the reader can see, that the classical HDM predicts a larger range of the energy density values than the previous case with a vacuum environment. If we compare maximum and minimum values of the energy density within a particle, we find that the classical HDM predicts an unusually large difference in these values, moreover, there are visible areas inside the particle of a near zero density. Our model, however, predicts a different shape of distribution density of energy inside a particle and a significantly larger minimal value.

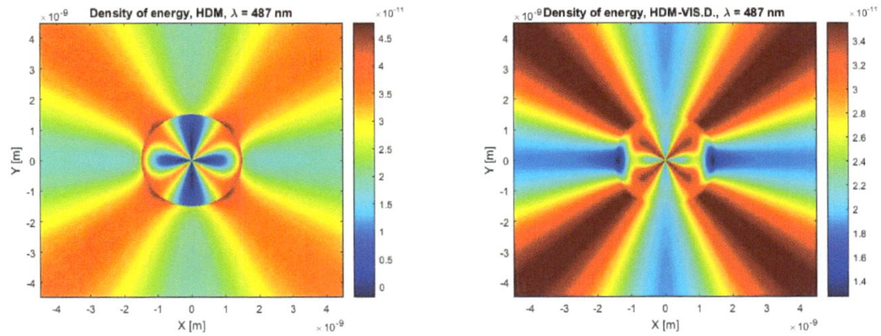

(a) **(b)**

Figure 14. The density of electromagnetic field energy calculated for the wavelength of λ = 487 nm and the case of a gold spherical particle with a diameter of 3 nm in a water environment, according to (**a**) HDM and (**b**) HDM–Vis.D.

Additionally, Figure 15 shows a very similar situation to Figure 13, however our model predicts larger attenuation (positive maxima values) in comparison with the classical HDM. Figure 15b, corresponding to the HDM with viscosive dumping, shows that the lowest values of the X component of the electric field are located outside the particle, whereas the HDM predicts that these values are located inside the particle and their absolute values are larger as compared to the HDM with viscosive dumping.

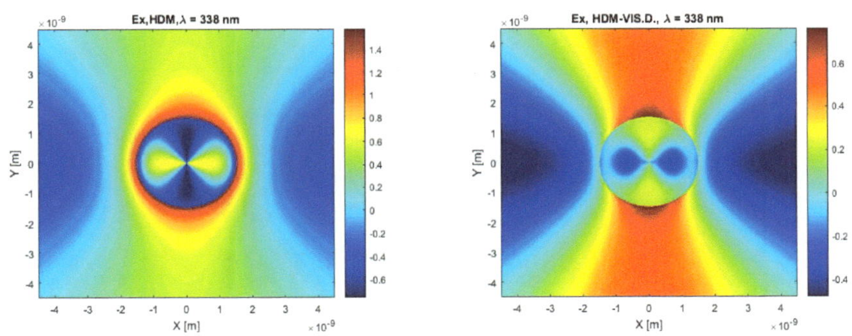

(a) **(b)**

Figure 15. The X-component of the electric field calculated for the wavelength of λ = 338 nm and the case of a silver spherical particle with a diameter of 3 nm in a vacuum environment, according to (**a**) HDM and (**b**) HDM–Vis.D.

Next, Figure 16 shows us the distribution of the energy density in the plane cut defined by the X and Y axes. Differences between the HDM and our model are clearly visible. HDM predicts that the areas with maximal energy density are located mainly inside a particle and its vicinity. It can be further read from the figures that our model predicts higher values of the lowest energy density. The lowest energy is in both cases of HD models, located inside the particle, but the area of the highest energy density calculated by our model is oriented diagonally and mainly outside of the particle.

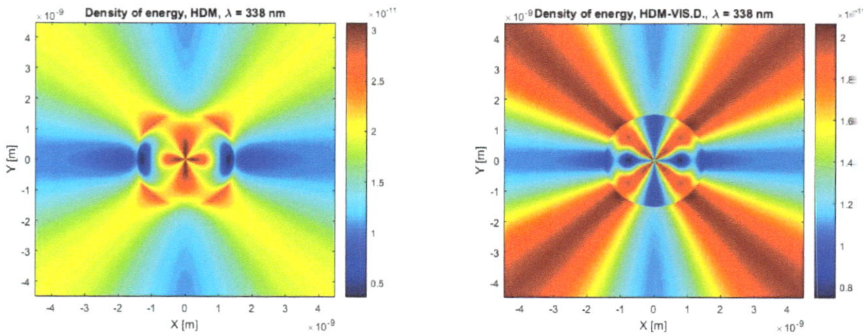

(**a**) (**b**)

Figure 16. The density of electromagnetic field energy calculated for the wavelength of λ = 338 nm and the case of a silver spherical particle with a diameter of 3 nm in a vacuum environment, according to (**a**) HDM and (**b**) HDM–Vis.D.

As the reader can see from Figure 17, both models predict different distributions of the current density, but every calculated pattern is symmetric to the X axis and antisymmetric to the Y axis. A further finding from these calculations is that our model shows a two times larger current density, which is situated near the edge of the particle, whereas the area with the largest current density calculated by HDM is located near the center of the particle. Another interesting point is that our model also shows that inside a particle there is a large area where the current density is nearly zero.

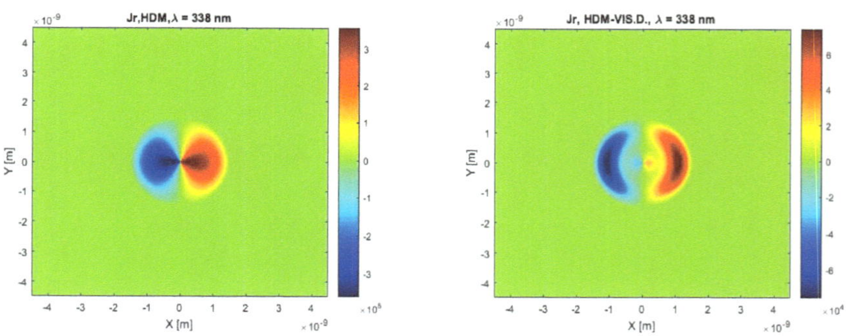

(**a**) (**b**)

Figure 17. The radial component of the current density calculated for the wavelength of λ = 338 nm and the case of a silver spherical particle with a diameter of 3 nm in a vacuum environment, according to (**a**) HDM and (**b**) HDM–Vis.D.

Additionally, Figure 18 displays, similarly, to Figure 16, the X-component of the electric field but for a water environment and at a different wavelength. Although these two cases are not directly comparable, we can see that the shape of the field distribution of the electric field is similar to that predicted with the HDM. In contrast to the HDM calculation, however, of our model shows the opposite distribution of positive and negative values of the electric field in a particle. Very apparent, is also a different distribution of the electric field outside of a particle, as it is determined by our model in comparison with the HDM. Again, our model predicts significantly smaller negative and positive values of the electric field.

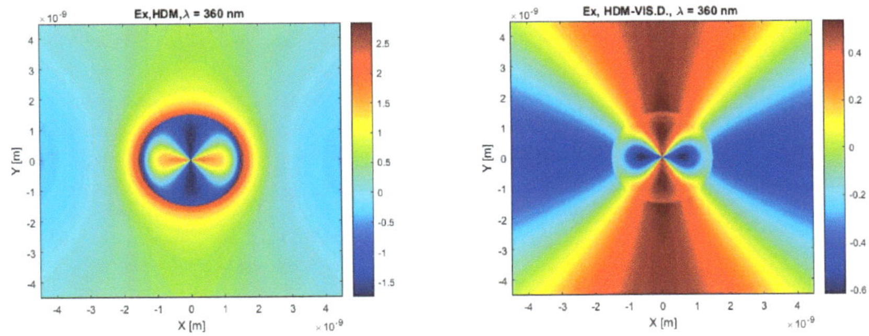

(**a**) (**b**)

Figure 18. The X-component of the electric field calculated for the wavelength of λ = 360 nm and the case of a silver spherical particle with a diameter of 3 nm in a water environment, according to (**a**) HDM and (**b**) HDM–Vis.D.

Finally, Figure 19 shows quite important data, demonstrating another disadvantage of the classic HDM. It should be noted that the same formula was used for these calculations of energy density as for the calculations shown in Figure 16. A quite strange finding is that standard HDM shows negative values of energy density in some places inside a particle volume; these non-physical values are not found when applying the same calculations to a gold particle. The standard HDM, unlike our model, displays large energy densities on the edge of a particle which are about three times larger than the maximum density calculated by our model. With these presented results we clearly demonstrated the applicability of our HD model with viscosive damping and generalized Drude–Lorentz term.

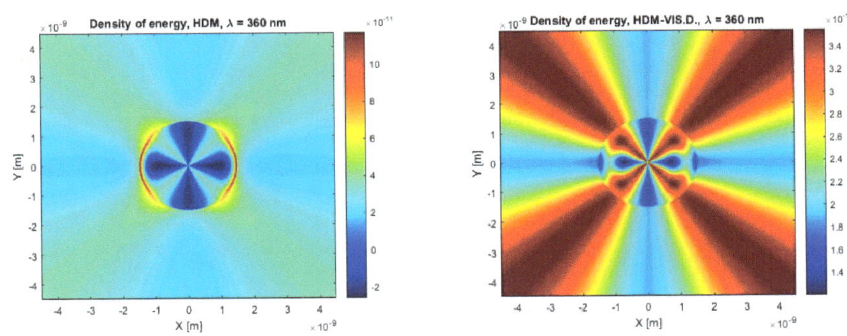

(**a**) (**b**)

Figure 19. The density of electromagnetic field energy calculated for the wavelength of λ = 360 nm in the case of a silver spherical particle with a diameter of 3 nm in a water environment, according to (**a**) HDM and (**b**) HDM–Vis.D.

6. Conclusions

In this article, we have studied theoretically and numerically the linearized nonlocal plasmonic interaction of light with a simple spherical metallic nanoparticle. We have concentrated on understanding the interaction and developing a simple model capable of predicting the longitudinal nonlocal response based on the linearized hydrodynamic model, generalizing the standard nonlocal HD model. Using this model, we have studied the electric field interaction with a nanoparticle immersed in a dielectric surrounding media (such as air or water). Our generalization conveys the inclusion of the generalized Drude–Lorentz term and viscosive current damping in connection to Landau damping. We have demonstrated the applicability of our extended model by comparing the extinction cross-section predictions of both gold and silver spherical nanoparticles. The results confirm the generally assumed conclusion, that metallic particles with a diameter of a few nanometers possess a blue resonance shift. It was also found that the HD model predicts a lower electric field intensity inside and around the particle. The modification in the form of Landau's attenuation, corrects the blue shift and affects the shape of the spectral course of the extinction.

Author Contributions: Methodology, M.B. and I.R.; Software, M.B.; Validation, M.B.; Formal analysis, M.B.; Investigation, M.B. and I.R.; Writing—original draft, M.B. and I.R.; Supervision, I.R.; Project administration, I.R. All authors have read and agreed to the published version of the manuscript.

Funding: The research has been performed with the support of the Czech Science Foundation, grant number GA21-05259S, and the MEYS of the Czech Republic grant number CZ.02.1.01/0.0/0.0/16_019/0000778 of the European Structural and Investments Funds.

Institutional Review Board Statement: Not applicable.

Informed Consent Statement: Not applicable.

Data Availability Statement: Data available from the authors upon personal request.

Conflicts of Interest: The authors declare no conflict of interest.

References

1. Ostrowski, J.C.; Mikhailovsky, A.; Bussian, D.A.; Summers, M.A.; Buratto, S.K.; Bazan, G.C. Enhancement of phosphorescence by surface-plasmon resonances in colloidal metal nanoparticles: The role of aggregates. *Adv. Funct. Mater.* **2006**, *16*, 1221–1227. [CrossRef]
2. Landström, L.; Brodoceanu, D.; Piglmayer, K.; Bäuerle, D. Extraordinary optical transmission through metal-coated colloidal monolayers. *Appl. Phys. A* **2006**, *84*, 373–377. [CrossRef]
3. Indhu, A.R.; Keerthana, L.; Dharmalingam, G. Plasmonic nanotechnology for photothermal applications—An evaluation. *Beilstein J. Nanotechnol.* **2023**, *14*, 380–419. [CrossRef]
4. Tian, F.; Bonnier, F.; Casey, A.; Shanahan, A.E.; Byrne, H.J. Surface enhanced Raman scattering with gold nanoparticles: Effect of particle shape. *Anal. Methods* **2014**, *6*, 9116–9123. [CrossRef]
5. Balčytis, A.; Nishijima, Y.; Krishnamoorthy, S.; Kuchmizhak, A.; Stoddart, P.R.; Petruškevičius, R.; Juodkazis, S. From fundamental toward applied SERS: Shared principles and divergent approaches. *Adv. Opt. Mater.* **2018**, *6*, 1800292. [CrossRef]
6. Proença, M.; Rodrigues, M.S.; Borges, J.; Vaz, F. Gas sensing with nanoplasmonic thin films composed of nanoparticles (Au, Ag) dispersed in a CuO matrix. *Coatings* **2019**, *9*, 337. [CrossRef]
7. Rodrigues, M.S.; Borges, J.; Lopes, C.; Pereira, R.M.S.; Vasilevskiy, M.I.; Vaz, F. Gas sensors based on localized surface plasmon resonances: Synthesis of oxide films with embedded metal nanoparticles: Theory, simulation, and sensitivity enhancement strategies. *Appl. Sci.* **2021**, *11*, 5388. [CrossRef]
8. Long, N.V.; Teranishi, T.; Yang, Y.; Thi, C.M.; Cao, Y.; Nogami, M. Iron oxide nanoparticles for next generation gas sensors. *Int. J. Metall. Mater. Eng.* **2015**, *1*, 119. [CrossRef]
9. Tittl, A.; Giessen, H.; Liu, N. Plasmonic gas and chemical sensing. *Nanophotonics* **2014**, *3*, 157–180. [CrossRef]
10. Temple, T.L.; Bagnall, D.M. Optical properties of gold and aluminium nanoparticles for silicon solar cell applications. *J. Appl. Phys.* **2011**, *109*, 084343. [CrossRef]
11. Chang, D.E.; Sørensen, A.S.; Hemmer, P.R.; Lukin, M.D. Quantum optics with surface plasmons. *Phys. Rev. Lett.* **2006**, *97*, 053002. [CrossRef]
12. Zhu, Z.; Ni, Y.; Lv, Q.; Geng, J.; Xie, W.; Li, F.; Chen, J. Surface plasmon mediates the visible light–responsive lithium–oxygen battery with Au nanoparticles on defective carbon nitride. *Proc. Natl. Acad. Sci. USA* **2021**, *118*, e2024619118. [CrossRef]
13. Marques Mota, F.; Yu, S.; Chae, K.; Che Mohamad, N.A.R.; Kim, D.H. An analysis of the promise of Li–O_2 and Li–S batteries incorporating plasmonic metal nanostructures. *Mater. Today Energy* **2022**, *27*, 101033. [CrossRef]
14. Wu, T.; Shao, Y.; Wang, Y.; Cao, S.; Cao, W.; Zhang, F.; Liao, C.; He, J.; Huang, Y.; Hou, M.; et al. Surface plasmon resonance biosensor based on gold-coated side-polished hexagonal structure photonic crystal fiber. *Opt. Express* **2017**, *25*, 20313. [CrossRef]
15. Chen, J.; Gong, M.; Fan, Y.; Feng, J.; Han, L.; Xin, H.L.; Cao, M.; Zhang, Q.; Zhang, D.; Lei, D.; et al. Collective plasmon coupling in gold nanoparticle clusters for highly efficient photothermal therapy. *ACS Nano* **2022**, *16*, 910–920. [CrossRef]
16. Parsons, J.; Hendry, E.; Sambles, J.R.; Barnes, W.L. Localized surface-plasmon resonances and negative refractive index in nanostructured electromagnetic metamaterials. *Phys. Rev. B* **2009**, *80*, 245117. [CrossRef]
17. Fan, X.; Zheng, W.; Singh, D.J. Light scattering and surface plasmons on small spherical particles. *Light Sci. Appl.* **2014**, *3*, e179. [CrossRef]
18. Dutta, A.; Tiainen, V.; Toppari, J.J. Numerical study on the limit of quasi-static approximation for plasmonic nanosphere. *AIP Conf. Proc.* **2020**, *2220*, 050012.

19. Maier, S.A. *Plasmonics: Fundamentals and Applications*; Springer: New York, NY, USA, 2007; ISBN 978-0-387-33150-8.
20. Quinten, M. *Optical Properties of Nanoparticle Systems: Mie and Beyond*; Wiley-VCH: Weinheim, Germany, 2011; ISBN 978-3-527-41043-9.
21. Baig, N.; Kammakakam, I.; Falath, W. Nanomaterials: A review of synthesis methods, properties, recent progress, and challenges. *Mater. Adv.* **2021**, *2*, 1821–1871. [CrossRef]
22. Qiao, J.; Qi, L. Recent progress in plant-gold nanoparticles fabrication methods and bio-applications. *Talanta* **2021**, *223*, 121396. [CrossRef]
23. Liu, Y.; Plate, P.; Hinrichs, V.; Köhler, T.; Song, M.; Manley, P.; Schmid, M.; Bartsch, P.; Fiechter, S.; Lux-Steiner, M.C.; et al. Size- and density-controlled deposition of Ag nanoparticle films by a novel low-temperature spray chemical vapour deposition method—Research into mechanism, particle growth and optical simulation. *J. Nanopart. Res.* **2017**, *19*, 141. [CrossRef]
24. Toscano, G.; Raza, S.; Yan, W.; Jeppesen, C.; Xiao, S.; Wubs, M.; Jauho, A.-P.; Bozhevolnyi, S.I.; Mortensen, N.A. Nonlocal response in plasmonic waveguiding with extreme light confinement. *Nanophotonics* **2013**, *2*, 161–166. [CrossRef]
25. Hiremath, K.R.; Zschiedrich, L.; Schmidt, F. Numerical solution of nonlocal hydrodynamic Drude model for arbitrary shaped nano-plasmonic structures using Nédélec finite elements. *J. Comput. Phys.* **2012**, *231*, 5890–5896. [CrossRef]
26. McMahon, J.M.; Gray, S.K.; Schatz, G.C. Calculating nonlocal optical properties of structures with arbitrary shape. *Phys. Rev. B* **2010**, *82*, 035423. [CrossRef]
27. Ginzburg, P.; Zayats, A.V. Localized surface plasmon resonances in spatially dispersive nano-objects: Phenomenological treatise. *ACS Nano* **2013**, *7*, 4334–4342. [CrossRef]
28. García De Abajo, F.J. Nonlocal effects in the plasmons of strongly interacting nanoparticles, dimers, and waveguides. *J. Phys. Chem. C* **2008**, *112*, 17983–17987. [CrossRef]
29. David, C.; García De Abajo, F.J. Spatial nonlocality in the optical response of metal nanoparticles. *J. Phys. Chem. C* **2011**, *115*, 19470–19475. [CrossRef]
30. Benedicto, J.; Pollès, R.; Ciracì, C.; Centeno, E.; Smith, D.R.; Moreau, A. Numerical tool to take nonlocal effects into account in metallo-dielectric multilayers. *J. Opt. Soc. Am. A* **2015**, *32*, 1581. [CrossRef]
31. David, C.; Mortensen, N.A.; Christensen, J. Perfect Imaging, Epsilon-near zero phenomena and waveguiding in the scope of nonlocal effects. *Sci. Rep.* **2013**, *3*, 2526. [CrossRef]
32. Pitelet, A.; Mallet, É.; Centeno, E.; Moreau, A. Fresnel coefficients and Fabry-Perot formula for spatially dispersive metallic layers. *Phys. Rev. B* **2017**, *96*, 041406. [CrossRef]
33. Raza, S.; Toscano, G.; Jauho, A.-P.; Wubs, M.; Mortensen, N.A. Unusual resonances in nanoplasmonic structures due to nonlocal response. *Phys. Rev. B* **2011**, *84*, 121412. [CrossRef]
34. Mortensen, N.A.; Raza, S.; Wubs, M.; Søndergaard, T.; Bozhevolnyi, S.I. A generalized non-Local optical response theory for plasmonic nanostructures. *Nat. Commun.* **2014**, *5*, 3809. [CrossRef]
35. Svendsen, M.K.; Wolff, C.; Jauho, A.-P.; Mortensen, N.A.; Tserkezis, C. Role of diffusive surface scattering in nonlocal plasmonics. *J. Phys. Condens. Matter* **2020**, *32*, 395702. [CrossRef]
36. Karimi, S.; Moshaii, A.; Abbasian, S.; Nikkhah, M. Surface plasmon resonance in small gold nanoparticles: Introducing a size-dependent plasma frequency for nanoparticles in quantum regime. *Plasmonics* **2019**, *14*, 851–860. [CrossRef]
37. Khurgin, J.B.; Sun, G. Landau Damping—The ultimate limit of field confinement and enhancement in plasmonic structures. In *Quantum Plasmonics*; Bozhevolnyi, S.I., Martin-Moreno, L., Garcia-Vidal, F., Eds.; Springer Series in Solid-State Sciences; Springer International Publishing: Cham, Switzerland, 2017; Volume 185, pp. 303–322, ISBN 978-3-319-45819-9.
38. Raza, S.; Bozhevolnyi, S.I.; Wubs, M.; Asger Mortensen, N. Nonlocal optical response in metallic nanostructures. *J. Phys. Condens. Matter* **2015**, *27*, 183204. [CrossRef]
39. Huynh, D.-N.; Moeferdt, M.; Matyssek, C.; Wolff, C.; Busch, K. Ultrafast three-wave-mixing in plasmonic nanostructures. *Appl. Phys. B* **2016**, *122*, 139. [CrossRef]
40. Moeferdt, M.; Kiel, T.; Sproll, T.; Intravaia, F.; Busch, K. Plasmonic modes in nanowire dimers: A study based on the hydrodynamic Drude model including nonlocal and nonlinear effects. *Phys. Rev. B* **2018**, *97*, 075431. [CrossRef]
41. Petrov, A.S.; Svintsov, D. Viscosity-limited drift instabilities in two-dimensional electron systems. *Phys. Rev. Appl.* **2022**, *17*, 054026. [CrossRef]
42. Diaw, A.; Murillo, M.S. A Viscous quantum hydrodynamics model based on dynamic density functional theory. *Sci. Rep.* **2017**, *7*, 15352. [CrossRef]
43. Tokatly, I.; Pankratov, O. Hydrodynamic theory of an electron gas. *Phys. Rev. B* **1999**, *60*, 15550–15553. [CrossRef]
44. Scholl, J.A.; Koh, A.L.; Dionne, J.A. Quantum plasmon resonances of individual metallic nanoparticles. *Nature* **2012**, *483*, 421–427. [CrossRef] [PubMed]
45. Alabastri, A.; Tuccio, S.; Giugni, A.; Toma, A.; Liberale, C.; Das, G.; Angelis, F.; Fabrizio, E.; Zaccaria, R. Molding of plasmonic resonances in metallic nanostructures: Dependence of the non-linear electric permittivity on system size and temperature. *Materials* **2013**, *6*, 4879–4910. [CrossRef] [PubMed]
46. Wubs, M. Classification of scalar and dyadic nonlocal optical response models. *Opt. Express* **2015**, *23*, 31296–31312. [CrossRef] [PubMed]
47. Eremin, Y.A. Scattering/Scattering Theory. In *Encyclopedia of Modern Optics*; Elsevier: Amsterdam, The Netherlands, 2005; pp. 326–330, ISBN 978-0-12-369395-2.

48. Mishchenko, M.I.; Travis, L.D.; Lacis, A.A. *Scattering, Absorption, and Emission of Light by Small Particles*; Cambridge University Press: Cambridge, UK, 2002; ISBN 978-0-521-78252-4.
49. Batchelor, G.K. *An Introduction to Fluid Dynamics*, 1st ed.; Cambridge University Press: Cambridge, UK, 2000; ISBN 978-0-521-66396-0.
50. West, P.R.; Ishii, S.; Naik, G.V.; Emani, N.K.; Shalaev, V.M.; Boltasseva, A. Searching for better plasmonic materials. *Laser Photonics Rev.* **2010**, *4*, 795–808. [CrossRef]
51. Rakić, A.D.; Djurišić, A.B.; Elazar, J.M.; Majewski, M.L. Optical properties of metallic films for vertical-cavity optoelectronic devices. *Appl. Opt.* **1998**, *37*, 5271. [CrossRef]
52. Palik, E.D. (Ed.) *Handbook of Optical Constants of Solids II*; Academic Press: Cambridge, MA, USA, 1991; ISBN 978-0-12-544422-4.
53. Kreibig, U.; Fragstein, C.V. The limitation of electron mean free path in small silver particles. *Z. Physik.* **1969**, *224*, 307–323. [CrossRef]

photonics

MDPI

Article

Influence of Adhesive Bonding on the Dynamic and Static Strain Transfers of Fibre Optic Sensors

Chloé Landreau [1,2], Adriana Morana [2], Nicolas Ponthus [1], Thomas Le Gall [1], Jacques Charvin [1], Sylvain Girard [2] and Emmanuel Marin [2,*]

[1] Avnir Energy, F-01150 Saint-Vulbas, France; chloe.landreau@univ-st-etienne.fr (C.L.); n.ponthus@avnir.fr (N.P.); t.legall@avnir.fr (T.L.G.); j.charvin@avnir.fr (J.C.)

[2] Laboratoire Hubert Curien UMR 5516, Institut d'Optique Graduate School, CNRS, Université Jean Monnet Saint-Etienne, F-42000 Saint-Etienne, France; adriana.morana@univ-st-etienne.fr (A.M.); sylvain.girard@univ-st-etienne.fr (S.G.)

* Correspondence: emmanuel.marin@univ-st-etienne.fr

Abstract: The influence of the bonding procedure (the adhesive type, application procedure, etc.) on the static and dynamic strain transfers of bonded optical fibre sensors is studied theoretically and experimentally at room temperature. The achievable performances with four different types of adhesives (three urethane and one epoxy adhesive), and with different fibre types, are evaluated: acrylate-coated, polyimide-coated, and bare single-mode optical fibres. Static strain measurements, ranging from 20 to 200 μ strain, are performed using both fibre Bragg gratings (FBGs) and optical frequency domain reflectometry (OFDR), and are compared to reference strain-gauge measurements, and to the proposed analytical model, which is developed on the basis of stress equilibrium relations. This model is valid for bonding to all types of linear and elastic materials, as long as there is no sliding between the host material, the adhesive, and the optical fibre. The results agree between the analytical model and the experiments. Regarding the dynamic sinusoidal strain measurements, the studied dynamic range is from 10 to 100 Hz, and only the FBGs are tested. The results demonstrate that the sensitivities of strain sensors based on bonded uncoated fibres or bonded polyimide-coated fibres are comparable to those of strain gauges, and that it is possible to use bonded FBGs for precise dynamic strain measurements.

Keywords: fibre Bragg gratings; optical frequency domain reflectometry; adhesives; static strain tests; dynamic strain tests; analytical model; strain transfer efficiency

Citation: Landreau, C.; Morana, A.; Ponthus, N.; Le Gall, T.; Charvin, J.; Girard, S.; Marin, E. Influence of Adhesive Bonding on the Dynamic and Static Strain Transfers of Fibre Optic Sensors. *Photonics* **2023**, *10*, 996. https://doi.org/10.3390/photonics10090996

Received: 17 July 2023
Revised: 25 August 2023
Accepted: 29 August 2023
Published: 31 August 2023

1. Introduction

Silica-based optical fibres are widely used as the sensitive elements of point or distributed strain sensors for structural health monitoring. Due to their numerous advantages, they can actually serve as a real alternative to strain gauges: they are small and lightweight, are insensitive to most electromagnetic perturbations, and can operate over a wide temperature range with appropriate coatings. Different interrogation schemes have been developed to allow wavelength- or/and time-multiplexed measurements. Furthermore, this technology offers the possibility of remote measurements far from the acquisition system [1,2], a crucial advantage for operation in severe environments in which the interrogators could not survive. A wide range of optical fibre strain sensors exist, from distributed sensors based on Rayleigh and Brillouin scattering [1,2], to point sensors, such as Fabry–Pérot cavities [3], long-period gratings (LPGs) [4,5], or fibre Bragg gratings (FBGs) [6–8]. This article focuses on static and dynamic strain measurements based on Rayleigh scattering and FBGs.

The architectures of fibre sensors for strain measurements are very diverse. For example, Matveenko et al. [9] and Luyckx et al. [10] have investigated the embedding of FBGs in composites to perform local internal strain measurements. Barrias et al. [11] have also studied embedded fibres, but in concrete beams, and using the distributed OFDR (optical frequency domain reflectometry) technique. While these embedded optical fibre sensors can give access to unique information inside a structure, they cannot be repaired or replaced if needed. In another way, Guo et al. [12] and Cheng et al. [13] proposed packaged structure FBG sensors, suitable for local surface strain measurements. Although these packaged sensors are reusable, and exhibit enhanced strain sensitivities compared to bare FBGs, their complex design and size—10 cm long for the one developed by Guo et al. [12]—are not suitable for all applications. Here, we focus on another integration method, wherein the optical fibres are directly bonded on the surface of the object to be monitored. This method, which is less costly and easier to implement, is actually the subject of many studies [14–21]. In this context, the literature suggests various approaches to setting up analytical models that describe the static strain transfer mechanism between the material of interest and the bonded optical fibre sensor. The model remains the same for FBG and OFDR measurements. In [14,15,20], the authors proposed analytical models based on the calculation of the stress equilibrium relation [22], while in [16,17], the analytical studies are based on the shear lag model [23]. Contrary to the others, Liang et al. [15] have included the viscoelastic properties of adhesives, considering a standard linear solid model [24]. Nevertheless, we observed from these studies that favouring the viscoelasticity of adhesive in the model does not have a decisive impact on the consistency between the analytical results and experiments. Moreover, the finite elements method (FEM) has been exploited in order to study the static strain transfer phenomenon between a host material and the fibre core [17–19,21], and has shown a good agreement with analytical models.

Overall, analytical models, FEM, and experiments agree on the importance of the coating properties and the adhesive length and thickness on the fibre-sensing performance. In this paper, a revisited analytical model of a strain transfer mechanism based on the previous work by Liang et al. [15] is described. Additionally, static strain experiments at room temperature using FBG and OFDR techniques are performed, studying optical fibres with different coatings, bonded on aluminium beams, using different adhesives. Then, the results are compared with the analytical model, and with strain-gauge measurements. Finally, the strain transfer efficiency for the dynamic strain is experimentally studied using only FBGs, again at room temperature, to ensure that the bonded fibres are adapted to dynamic measurements.

Our results show that strain sensors based on FBGs and OFDR can perform as well as a strain gauge, and that the removal of the fibre coating allows for an increase in the sensitivity of the sensor. These results are validated via the analytical model, too. Bonded FBGs can also be used to monitor dynamic processes, as their responses do not change as a function of the frequency.

2. Analytical Model

The perfect strain transfer between the material being monitored and the sensitive part of the optical fibre strain sensor is essential to obtaining reliable measurements describing the material response. In this way, setting up an analytical model can provide a better understanding of the strain transfer phenomenon, and could allow for the maximisation of the sensor sensitivity.

2.1. Implementation of the Analytical Model

The analytical model described in this article is based on [15]. In our work, the model does not consider the viscoelasticity of the adhesive, but was adapted for an uncoated fibre, a single-coated fibre, and a double-coated fibre. The mechanical model is based on a layered stacking of the host material, the adhesive, the coating (which can be single, double, or absent), the cladding, and the core.

The following assumptions have also been made in this model:

1. All materials are linear and elastic;
2. There is no sliding at the interfaces, so the layers are perfectly bonded together;

3. The host material, which in our case will be aluminium, experiences a uniform longitudinal strain;
4. The core and cladding are considered as a single unit, as they share the same mechanical properties.

The model is, therefore, valid regardless of the host material, as long as the first three assumptions can be applied.

As described in Figure 1, considering a single-coated fibre, when the host material is subjected to longitudinal strain, it deforms by U_{mat}, whereas the fibre core and cladding deform by only U_{fibre}. The huge difference between U_{mat} and U_{fibre} is explained by the high elasticity of the adhesive and coating layers, which partly absorb the strain energy of the host material. This difference can be calculated using the following formula:

$$\begin{aligned} U_{mat}(x) - U_{fibre}(x) &= \Delta U_{coat} + \Delta U_{adh} \\ &= \int_{H-h-r_{coat}+r_{fibre}}^{H-h} \gamma_{coat}(x,y)dy + \int_{H-h}^{H} \gamma_{adh}(x,y)dy \end{aligned} \tag{1}$$

where r_{coat} is the coating radius, r_{fibre} is the radius of the fibre (i.e., {core + cladding}), γ_{coat} is the coating shear strain, γ_{adh} is the adhesive shear strain, h is the adhesive bondline thickness between the host material and the optical fibre, and H is the total adhesive bondline thickness. The two shear strain values are obtained via the application of stress equilibrium relations [22] to a small segment of both layers, and considering them as linear. We obtain:

$$\gamma_{coat}(x,y) = -\frac{E_{fibre}}{2G_{coat}} \cdot [h + r_{coat} - H + y] \cdot \frac{d\varepsilon_{fibre}(x)}{dx} \tag{2}$$

and,

$$\gamma_{adh}(x,y) = \frac{E_{fibre}}{G_{fibre}} \cdot \frac{r_{coat}}{2Dh} \cdot [D(y-H) + 2\pi r_{coat}(H-h-y)] \cdot \frac{d\varepsilon_{fibre}(x)}{dx} \tag{3}$$

where E_{fibre} is the fibre Young's modulus, G_{fibre} and G_{coat} are, respectively, the fibre and the coating shear moduli, ε_{fibre} is the fibre strain, and D is the bonding width. Then, by substituting Equations (2) and (3) into Equation (1), we obtain, for the single coating fibre:

$$U_{mat}(x) - U_{fibre}(x) = -\frac{1}{\xi^2} \frac{d\varepsilon_{fibre}(x)}{dx} \tag{4}$$

where:

$$\frac{1}{\xi^2} = \left[\frac{r_{coat}^2 - r_{fibre}^2}{4\,G_{coat}} + \frac{h\,r_{coat}}{4\,G_{adh}} \cdot \left(1 + \frac{2\,\pi\,r_{coat}}{D}\right) \right] \cdot E_{fibre} \tag{5}$$

Applying the same reasoning to a double-coated fibre, and an uncoated fibre, this last value is equal to:

$$\begin{aligned} \frac{1}{\xi^2} = \Bigg[& \frac{r_{soft\_coat}^2 - r_{fibre}^2}{4\,G_{soft\_coat}} + \frac{r_{hard\_coat}^2 - r_{soft\_coat}^2}{4\,G_{hard\_coat}} \\ & + \frac{h\,r_{hard\_coat}}{4\,G_{adh}} \cdot \left(1 + \frac{2\,\pi\,r_{hard\_coat}}{D}\right) \Bigg] \cdot E_{fiber} \end{aligned} \tag{6}$$

for a double-acrylate-coated fibre, where r_{soft_coat} and r_{hard_coat} are the soft- and hard-coating radius, and G_{soft_coat} and G_{hard_coat} are the soft- and hard-coating shear moduli, and:

$$\frac{1}{\xi^2} = \left[\frac{h\,r_{fibre}}{4\,G_{adh}} \cdot \left(1 + \frac{2\,\pi\,r_{fibre}}{D}\right) \right] \cdot E_{fiber} \tag{7}$$

for an uncoated fibre.

Then, by differentiating Equation (4) with respect to x, we obtain the following differential equation:

$$\frac{d^2\varepsilon_{fibre}(x)}{dx^2} - \xi^2 \cdot \varepsilon_{fibre}(x) = -\xi^2 \cdot \varepsilon_{mat} \tag{8}$$

To solve this equation, the boundary conditions are such that the strain transfer is zero on the unbonded parts of the fibre. Thus, we have:

$$\varepsilon_{fibre}(x) = \varepsilon_{mat} \cdot \left[1 - \frac{cosh(\xi \cdot x)}{cosh\left(\xi \cdot \frac{L}{2}\right)}\right] \tag{9}$$

where L is the bonding length, and ε_{mat} is the strain experienced in the host material.

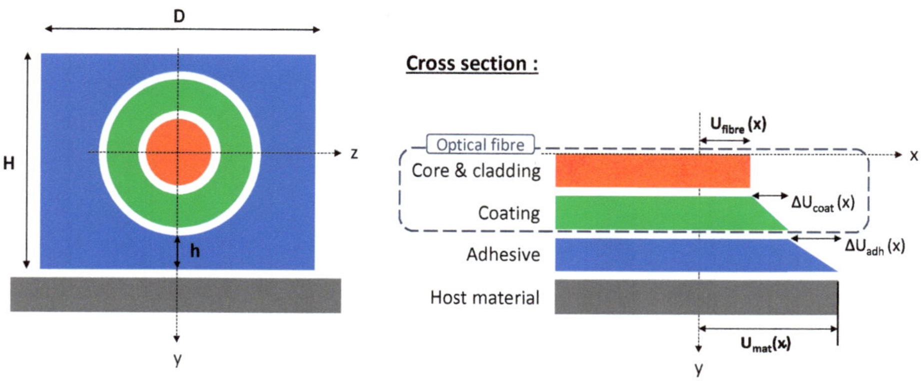

Figure 1. Mechanical model for strain transfer analysis.

In this article, the investigated optical fibres were either double-acrylate-coated fibres, mono-layer polyimide-coated, or uncoated. All fibres have a pure silica core and a fluorine-doped cladding, and they were all manufactured by EXAIL [25], allowing their use in radiation-rich environments [26]. Moreover, the choice of these fibres was guided by knowledge of their optical and mechanical characteristics. These values were used as input parameters for all the modelling procedures, and are listed in Table 1.

Table 1. Optical fibres mechanical and geometrical properties (from [17,27,28]).

	Uncoated Fibre	Polyimide Fibre	Double-Acrylate Fibre
		Core + cladding	
Radius (μm)	62.6	62.35	62.6
Young's modulus (MPa)	72,000	72,000	72,000
Poisson ratio	0.17	0.17	0.17
		First Coating	
Radius (μm)	-	79.4	95
Young's modulus (MPa)	-	2500	4 (Ref. [17]) 1 (Ref. [27])
Poisson ratio		0.17	0.498
		Second Coating	
Radius (μm)	-	-	122.5
Young's modulus (MPa)	-	-	1000 (Ref. [17]) 1150 (Ref. [27])
Poisson ratio			0.368

Figure 2 illustrates the simulated strain transfer efficiency profile along the core of a bonded double-acrylate-coated fibre, a polyimide-coated fibre, and an uncoated fibre, when the host material is deformed by ε_{mat}. It shows the strong influence that the bonding length L, and the coating, have on the sensor sensitivity.

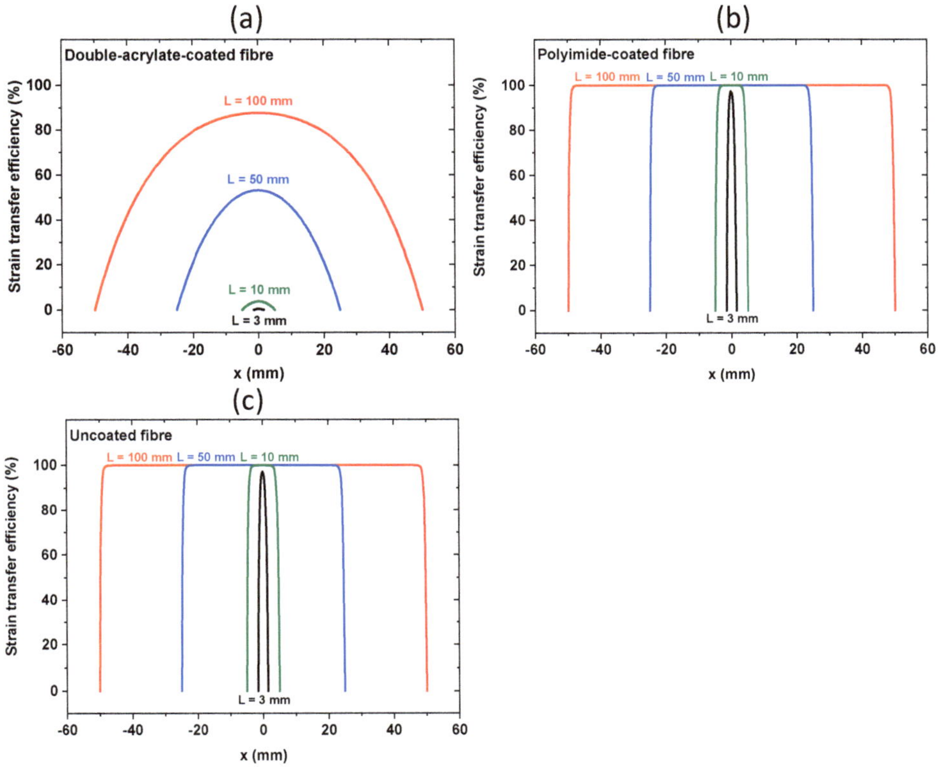

Figure 2. The simulated strain transfer efficiencies along a bonded fibre for different bonding lengths, from 3 to 100 mm. The results were obtained considering (**a**) a double-acrylate-coated fibre, (**b**) a polyimide-coated fibre, and (**c**) an uncoated fibre. For these simulations, Young's moduli of the first soft coating and the second hard coating of the double-acrylate fibre are, respectively, equal to 1 MPa and 1150 MPa, the Young's modulus of the adhesive is 2800 MPa, and the width and thickness of the adhesive are 4 mm and 0.1 mm, respectively.

Finally, as the experimental strain measurements shown in Sections 4.1 and 4.2 were averaged on the bonding length, we also averaged the strain field expressed in (9):

$$\bar{\varepsilon}_{fibre} = \varepsilon_{mat} \cdot \left[1 - \frac{2}{\xi \cdot L} \, tanh\left(\frac{\xi \cdot L}{2}\right)\right] \quad (10)$$

In conclusion, the length L, the width D, and the thickness h of the bond, as well as the shear modulus of the coating and the adhesive, appear to be the parameters that need to be adjusted to improve the efficiency of the strain transfer. However, as we will see below, not all of these parameters have the same impact on the strain transfer efficiency.

2.2. Influences of Parameters on the Strain Transfer

Figure 3a,b show the impact of the geometrical parameters on the strain transfer efficiency, and Figure 3c,d show the impact of the mechanical parameters.

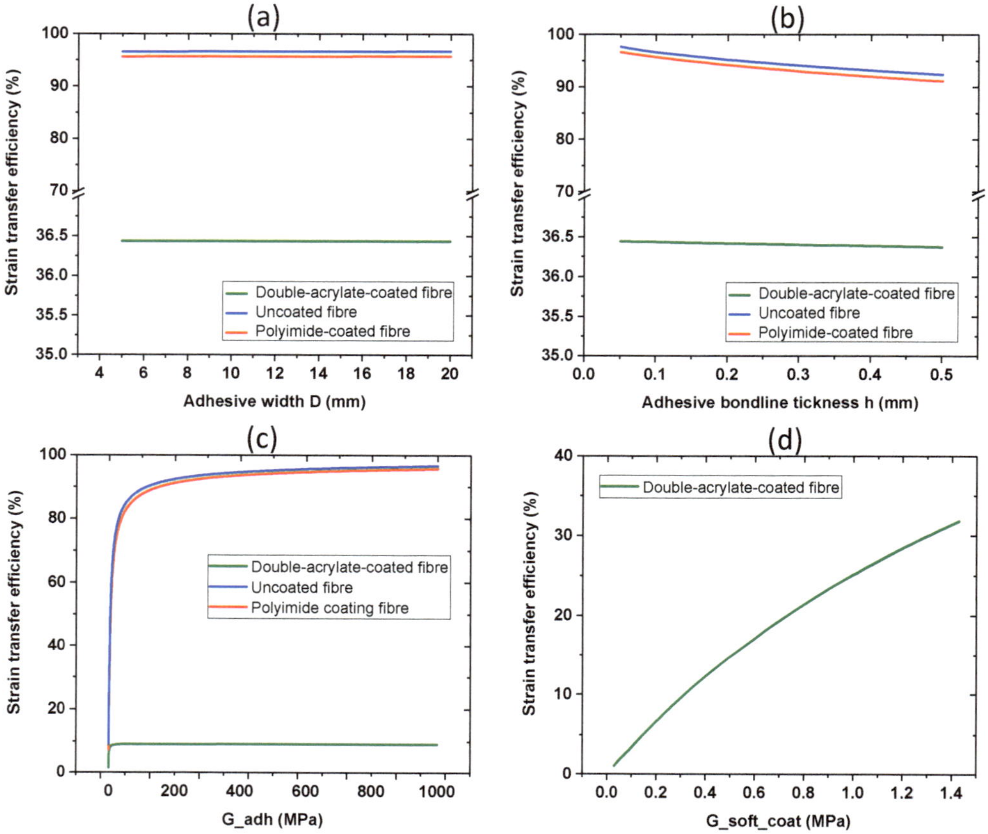

Figure 3. Modelling of the strain transfer efficiency as a function of the geometrical parameters: adhesive width D (**a**), and adhesive bondline thickness h (**b**), and the mechanical parameters: adhesive shear modulus G_{adh} (**c**), and soft acrylate coating shear modulus G_{soft_coat} (**d**), for a bonded double-acrylate-coated fibre, a polyimide-coated fibre, and an uncoated fibre. The parameters used for the simulation were: L = 20 mm; h = 0.1 mm; D = 10 mm; G_{soft_coat} = 0.3 MPa; G_{adh} = 1000 MPa.

From these simulations, we can conclude that, compared to the bonding length, the adhesive bondline thickness h between the host material and the fibre has very little influence on the strain transfer, and that the bonding width D has no impact. Regarding the mechanical parameters, for a double-acrylate-coated fibre, the impact of the shear modulus of the soft acrylate coating on the strain transfer efficiency is very high. However, we cannot act directly on its value without a direct collaboration with the fibre manufacturer. On the other hand, we could more easily act on the shear modulus of the adhesive, in order to increase the strain transfer efficiency. However, given the very low shear modulus of the soft acrylate coating, the efficiency of the transfer as a function of the shear modulus of the adhesive converges very quickly towards its maximum. With a polyimide-coated fibre and an uncoated fibre, this convergence occurs less quickly, so the choice of the right adhesive can be crucial. Finally, as the impacts of the hard coating shear modulus of a double-acrylate-coated fibre, and the coating shear modulus of a polyimide-coated fibre do not influence significantly the strain transfer efficiency, they have not been represented in Figure 3.

3. Materials and Method

3.1. Optical Fibres and Fibre Bragg Gratings (FBGs)

We used 3 mm long Type II FBGs for our experiments. They were written through the fibre coatings via the point-by-point technique, using a Pharos IR-femtosecond laser and the harmonic generator HIRO @515 nm from Light Conversion. The inscription parameters included a pulse width of 190 fs, a laser energy of 70 μJ, and a ×40 objective with a numerical aperture of 0.75. They were inscribed either through the double-acrylate or polyimide coating, as described in Table 1. In addition, all FBGs were thermally treated for 15 h at 80 °C, so that unstable photoinduced refractive index contribution was erased.

3.2. Optical Fibres Bonding

3.2.1. The Adhesives Used and the Bonding Protocol

The strain transfer has been experimentally studied using four different adhesives, differing in terms of their chemical nature and Young's modulus. As this latter parameter is of interest for our study, the adhesives were chosen to cover a wide range of Young's moduli. Details of each of these adhesives can be found in Table 2.

Table 2. Adhesives properties.

Adhesive	Intended for	Chemical Nature	Young's Modulus (MPa)
X280 [29]	Gauge bonding	Epoxy	2800
NOA 63 [30]	Optic	Urethane	1654
NOA 81 [30]	Optic	Urethane	1378
NOA 60 [30]	Optic	Urethane	931

Optical fibres were bonded to aluminium beams that had been carefully prepared in order to ensure a perfect bond between the two entities, and guarantee the validity of the second hypothesis, on which the analytical model is based (Section 2.1). The cleaning process incorporates:

1. A coarse clean with a degreasing soap;
2. Cleaning with an ethanol solution to remove ink marks and any other dirt;
3. A large-area mild etching, using a phosphoric-acid solution;
4. A local-area sanding step, to promote glue adherence;
5. A large, and then localised and intensive cleaning, using the phosphoric-acid solution to remove metal residues caused by the sanding;
6. A large, and then localised and intensive neutralisation, using an ammonia-based solution.

3.2.2. Bonding for FBG Measurements

FBGs were bonded, as described in the previous section, to aluminium beams using the four different adhesives, over a length of 11.4 ± 0.8 mm. According to Figure 2, a bonding length of 10 mm ensures a 100% strain transfer efficiency, at least for the polyimide-coated and the uncoated fibres. Table 3 shows the combination of beam, adhesive, and optical fibre used in each experiment.

Finally, strain gauges (CEA-06-032UW-120 from Micro-Measurement, Malvern, PA, USA) were glued on each sample using the X280 adhesive, so that the strains measured with FBGs could be compared to the strain-gauge measurements.

Table 3. Details of the combination of beams, adhesives, and fibres.

Aluminium Beam		Optical Fibre Coating	Adhesive
Item	Dimension		
1	180 × 23.75 × 1.2 mm^3	Double-acrylate Double-acrylate	NOA 81 X280
2	180 × 23.75 × 1.2 mm^3	Double-acrylate Double-acrylate	NOA 60 NOA 63
3	180 × 23.75 × 1.2 mm^3	Polyimide Polyimide	NOA 81 X280
4	180 × 23.75 × 1.2 mm^3	Polyimide Polyimide	NOA 60 NOA 63
5	180 × 22.50 × 1.2 mm^3	Uncoated Uncoated	NOA 81 X280
6	180 × 22.50 × 1.2 mm^3	Uncoated Uncoated	NOA 60 NOA 63

3.2.3. Bonding for OFDR Measurements

OFDR measurements were only performed using polyimide-coated fibres. Indeed, as we will see below, the FBG results showed that acrylate-coated fibres exhibit a poor performance, and that uncoated fibres are very fragile.

A single aluminium beam (280 × 25 × 2 mm^3) was prepared, as described in paragraph 3.2.1, and a polyimide-coated optical fibre was bonded using the four adhesives, each over a length of 5 cm. The chosen length allowed us to study a longer length than that used for bonding FBGs, and enabled us to obtain a deformation distribution profile along the bonded fibre, exploiting the high spatial resolution of the OFDR. Two fibres were bonded on one side of the beam, and two others on the other side. In addition, a strain gauge was bonded to one side of the beam using the X280 adhesive.

The uncertainties regarding the bonding, the aluminium beam, and the position and orientation of the fibres were estimated via preparing other samples, but using the same adhesive on each. These measured uncertainties were added to the experimental results.

3.3. *Static Strain Experiment*

Static strain experiments were carried out using a tensile testing machine (AGX-V from Shimadzu, Kyoto, Japan), and applying static forces from 100 to 350 N to each sample (Figure 4). In this tensile configuration, the strain field is uniform along the aluminium samples.

During testing, the Bragg wavelength evolutions were recorded through an FBG interrogator, the Gator from PhotonFirst (Alkmaar, The Netherlands), at a sampling frequency of 19 kHz, in the spectral range between 1520 and 1540 nm. The Bragg wavelength evolutions were converted into strain variations, using the strain sensitivity coefficient of 1.2 pm/με [6,8]. Spectral shift measurements were recorded through an optical backscatter reflectometer, OBR 4600 from Luna, working at 1550 nm, with a 10 mm gauge length and a 1 mm sensor spacing. For the conversion into strain variations, the strain sensitivity used was 0.77 ppm/με [31]. As was shown for the OFDR, the strain sensitivity does not change significantly (less than 1%) for the uncoated or coated fibres. Finally, the strain-gauge static measurements were obtained using an extensometer quarter bridge, the P-350AF from Vishay.

The tests were repeated three times, and all results were averaged, to improve the graph's readability. Thus, all the graphs presented in this article represent averaged experimental values, and their standard deviations as uncertainties.

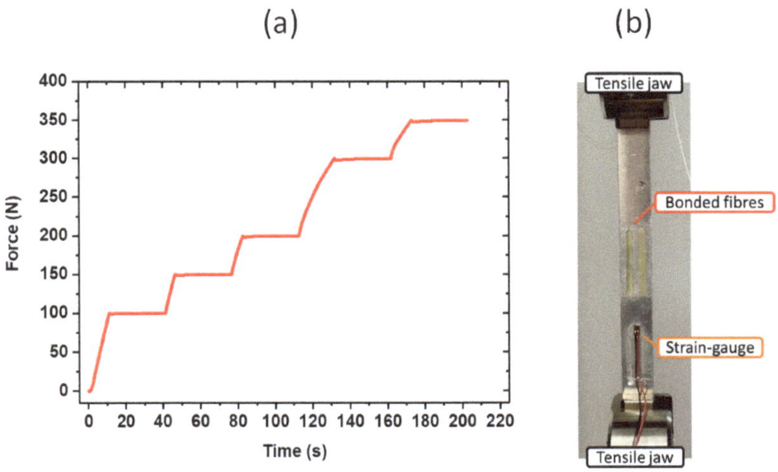

Figure 4. (**a**) The force levels applied with the tensile machine to (**b**) the aluminium beam, fitted with bonded fibres and a strain gauge.

3.4. Dynamic STRAIN Experiments

In the dynamic experiments, aluminium beams were embedded on both sides: one side was screwed to a vibration exciter (Type 4809 from Brüel & Kjær, Naerum, Denmark), and the other side was screwed to a mechanical piece, linked with an optical table (Figure 5). In this way, the natural frequency of the beam was sufficiently far away (>100 Hz) to be able to only study the bonded FBG behaviour, as the beam response remained flat in the studied [10:100] Hz frequency range. The FBGs were glued close to an embedded side, as it is in these areas that the deformations in the beam are greater. Finally, because of the too-slow acquisition rate of the OFDR (lower than 0.2 Hz), it was not possible to study the dynamic strain transfer using this method.

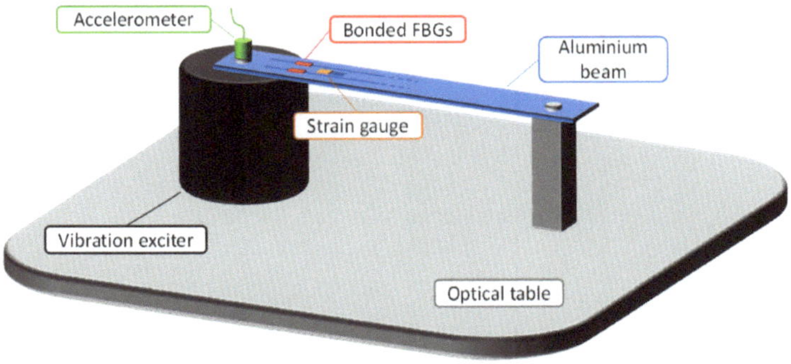

Figure 5. Schema of the experimental setup.

An accelerometer (352C33 from PCB Piezotronics, Depew, NY, USA) fixed in the axis of the vibration exciter allowed us to record the applied acceleration to the aluminium beam, through the data acquisition system LAN-XI from Brüel & Kjær, and the associated software BK Connect. Additionally, as for the static experiments, the FBG data were recorded through the Gator interrogator system, at a sampling frequency of 19 kHz.

Experimentally, the dynamic tests consisted of applying a sinusoidal displacement of constant amplitude to the beam, at frequencies from 10 to 100 Hz. Four experiments were carried out, with four different constant displacements: 0.028 mm, 0.055 mm, 0.085 mm, and 0.14 mm. These displacement values were chosen in accordance with the performance of the vibration exciter, which had a 4 mm displacement limit and a 44.5 N force limit (Figure 6). These performances led to a maximum achievable acceleration of:

$$a_{max} = \frac{F_{limit}}{m_e + m_{sample}} = \frac{44.5}{0.06 + 0.015} \approx 594 \text{ m/s}^2 \tag{11}$$

where F_{limit} is the maximum force the exciter can provide, m_e is the mass of the exciter moving element, and m_{sample} is the mass of an aluminium sample. Using this maximal acceleration, a limit frequency f_{limit}, above which the vibration exciter will no longer be able to provide sufficient force to maintain a displacement of 4 mm, can be defined. This limit is equal to:

$$f_{limit} = \frac{1}{2\pi}\sqrt{\frac{a_{max}}{d_{max}}} = \frac{1}{2\pi}\sqrt{\frac{594}{0.004}} \approx 61 \text{ Hz} \tag{12}$$

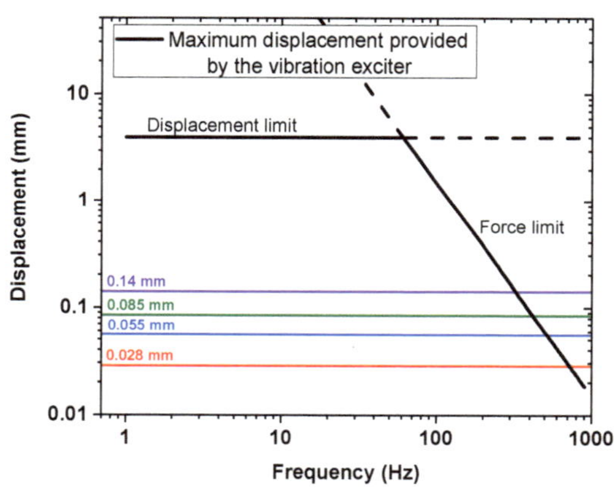

Figure 6. The maximum displacement that the vibration exciter can provide, according to its displacement and force performance limits as a function of the frequency. The experimentally applied displacements are also represented.

After this frequency limit, the maximum possible displacement d_{max} decreases with increasing frequency, following the function:

$$d_{max} = \frac{a_{max}}{(2\pi \cdot f)^2} = \frac{594}{(2\pi \cdot f)^2}, \quad with\ f > f_{limit} \tag{13}$$

According to Figure 6, a larger displacement could have been applied to the sample but was not, as we had no prior knowledge of the behavior of the bonded fibres.

4. Results and Discussion

This section is divided into three parts. The first and second parts focus on the experimental results of the static tests, and their comparison with the analytical model established in Section 2. The first part pertains to the measurements obtained from the FBGs, while the second part pertains to the measurements obtained from the OFDR. Finally, the last part discusses the dynamic measurements performed on the bonded FBGs.

4.1. Static Strain Study on the Bonded FBGs

4.1.1. Experimental Results

The experimental results for the bonded FBGs are shown in Figure 7. This figure reviews the strain measured for the polyimide-coated, uncoated, and double-acrylate-coated FBGs, bonded with each of the adhesives, as a function of the strain gauge measurement. The results were obtained through repeating each experiment three times, and calculating the means and standard deviations. Several conclusions can be drawn from these results:

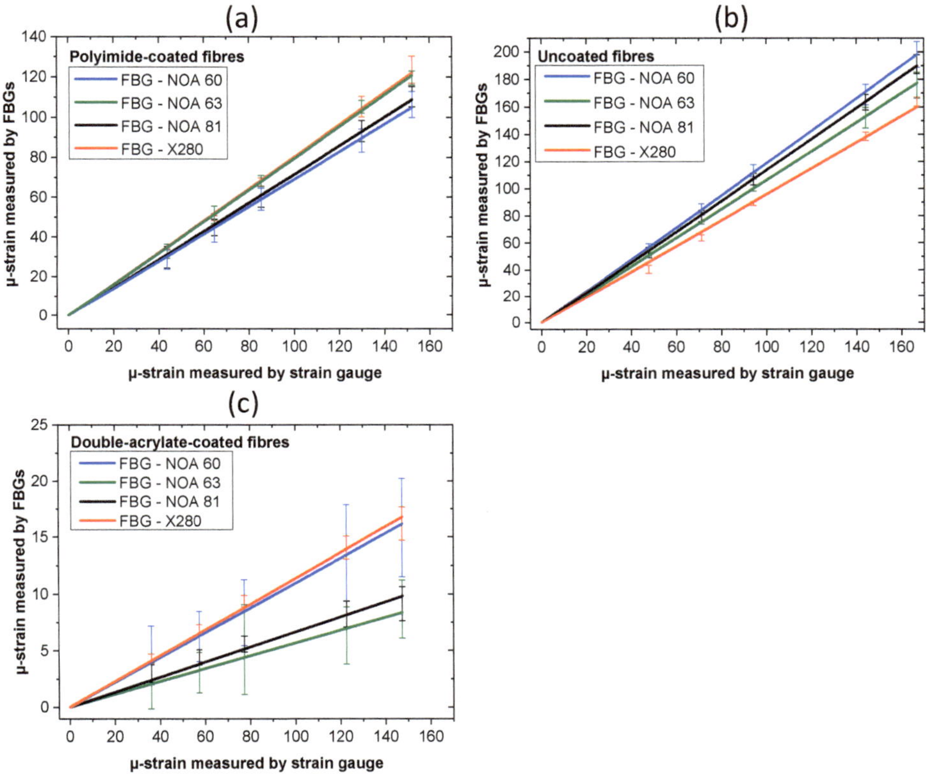

Figure 7. Strain measured via the bonded FBGs written in (**a**) a polyimide-coated fibre, (**b**) an uncoated fibre, and (**c**) a double-acrylate-coated fibre, as a function of the strain measured via the strain gauges.

- Firstly, the various adhesives transfer the deformation differently from the tensile specimen to the FBGs. As the bonding length has a huge influence on the strain transfer (Figure 2), and as this length is experimentally challenging to control, due to the various adhesive viscosities, the relative efficiency of the adhesive changes from one fibre to the other. This hypothesis will be verified via comparison with the analytical model in the next section.
- Secondly, we can see that the gratings written in the double acrylate-coated fibres are poorly sensitive to deformation in the aluminium sample. This is attributed to the high elasticity of these coating materials.
- Finally, we can state that bonded Bragg gratings inscribed on uncoated or polyimide-coated fibres are very sensitive to strain, and that uncoated gratings give similar or even better results than the strain gauges. For example, in Figure 7b, the FBG written on an uncoated fibre bonded with the NOA63 adhesive measured 176.5 ± 9.4 μ strain,

as the strain gauge measured 167 μ strain, and the theoretical value obtained via Hooke's law is 178 μ strain.

As mentioned at the beginning of Section 3.3, in pure tensile configuration, the strain field is uniform along the specimen, except close to the embedment. It can be seen that the strain gauges, bonded to different aluminium beams and in different positions, provide similar strain measurements for Figure 7a,c. The strain measurements shown in Figure 7b are higher, but this can be explained by the narrower width of the aluminium beam (Table 3). However, by multiplying the strains measured via the gauges by the width of their host aluminium beam, we can compare all the gauges together. Figure 8 presents these results, showing that the gauges, when not co-positioned, measure similar strains. Thus, the tensile specimens do, indeed, experience uniform strain fields, and the co-positioning of FBGs and strain gauges is not essential in our study.

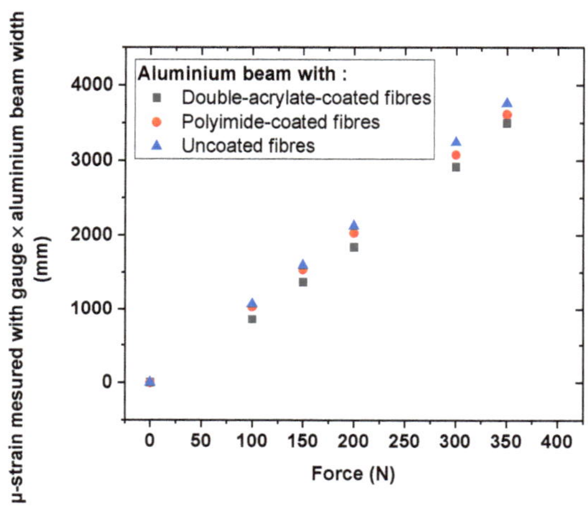

Figure 8. The experimental strains measured via strain gauges multiplied by the width of their host aluminium beam, as a function of the tensile force applied to the beam.

4.1.2. Comparison with the Analytical Model

Figure 9 compares the experimental results to the analytical model outputs. The latter were obtained using the physical quantities given in Tables 1 and 2, and considering the exact experimental bonding lengths and widths. The adhesive bondline thickness h is not possible to control or measure without the destruction of the samples. However, as the adhesive bondline thickness has little influence on the efficiency of strain transfer, according to the analytical model (Figure 3b), we therefore assumed that it was equal to 0.1 mm.

The analytical model corresponds quite well to the experiments, even though we can see some discrepancies. These can indicate inaccuracies between the geometrical and mechanical parameters used as inputs into the model, and the real parameters. These discrepancies are particularly noticeable on the double-acrylate fibre. This is explained by the uncertainty in the value of the Young's modulus (and, therefore, the shear modulus) of the soft acrylate coating, which, according to the literature, can vary between 1 MPa [27] and 4 MPa [17]. The results from the analytical model are extremely sensitive to this uncertainty, as shown in Figure 3b, with the efficiency of the strain transfer increasing from 10.4% to 30.4% as Young's modulus increases from 1 to 4 MPa (or the shear modulus increases from about 0.3 MPa to 1.3 MPa). The results of the model were calculated using each of these two extreme values from the literature, and we can see that the experimental results range between the two analytical results, demonstrating the consistency of the model.

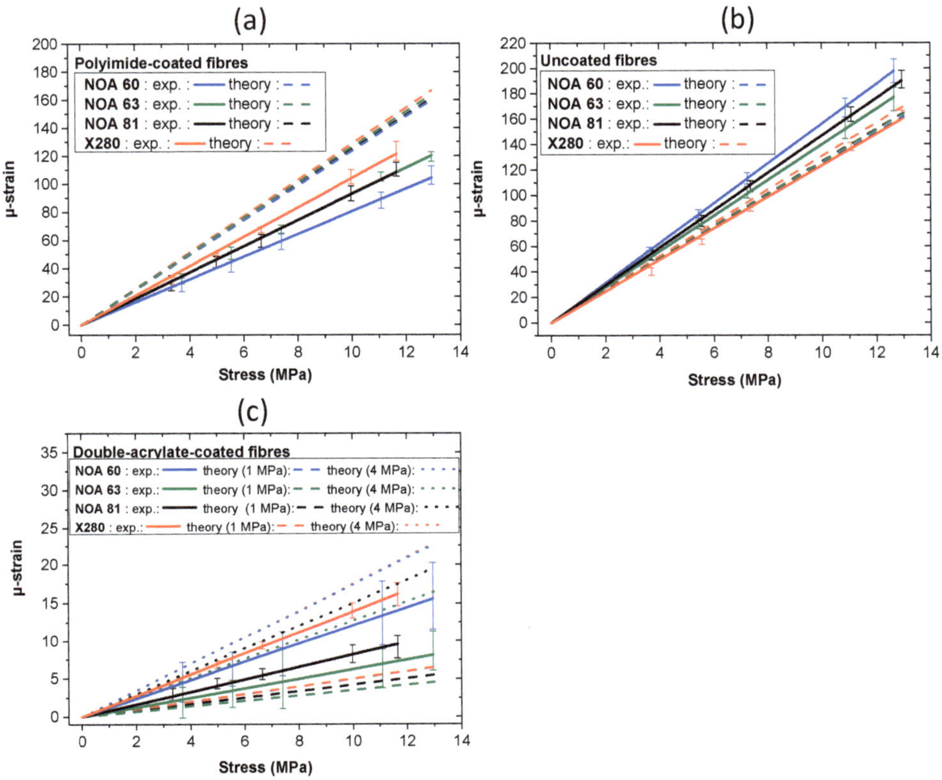

Figure 9. The experimental and analytical strains of the bonded FBGs written in (**a**) polyimide-coated fibres, (**b**) uncoated fibres, and (**c**) double-acrylate-coated fibres, as a function of the strain applied to the host material (some analytical data overlap). For the double-acrylate-coated fibres, the analytical solutions were calculated twice: the first time with E_{soft_coat} = 1 MPa, and the second time with E_{soft_coat} = 4 MPa.

4.2. OFDR Static Strain Study

Experimental Results

As briefly discuss in Section 3.2.3, the uncertainties regarding the bonding, the aluminium beam, and the position and the orientation of the fibres were experimentally estimated through bonding 5 cm of polyimide fibre at four points on either side of the beam, using the same adhesive, and testing the sample with the tensile machine. One of these measurements is represented in Figure 10, where the uncertainties were estimated using the NOA 60 adhesive. The results from the four bonded areas were averaged, and a dispersion was calculated. In this way, we find that the bonding, the aluminium beam, and the position and the orientation of the fibres induce 15% of the error in the OFDR strain measurements. This error, which exceeds the experimental repeatability error, is represented in the results shown in Figure 12.

Figure 11 shows the spectral shift profile observed along a polyimide-coated fibre. Using the strain sensitivity coefficient of a fibre interrogated via OFDR, as defined in Section 3.3, we obtained a strain profile, of the opposite sign, similar to the spectral shift profile, itself comparable to the theoretical profiles illustrated in Figure 2, with a gradual increase and decrease in the strain sensitivity of the fibre. The average strains were then calculated for each deformation level, to obtain the results shown in Figure 12.

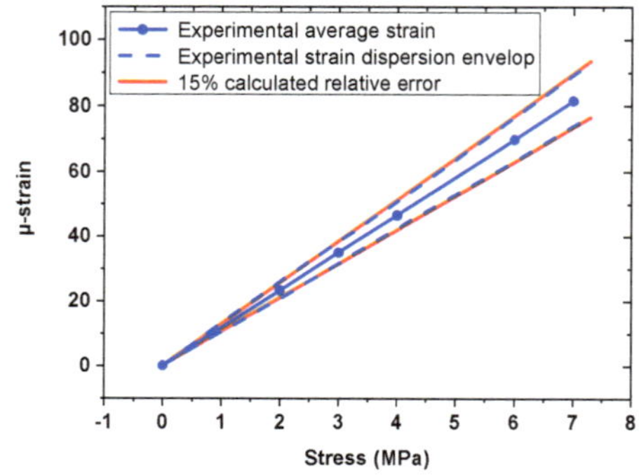

Figure 10. The experimental results of the determination of errors due to the position and orientation of the fibre, bonding, and beam, when a polyimide-coated fibre is bonded on an aluminium beam over 5 cm, using the NOA 60 adhesive. The dispersion of experimental measurements can be perfectly approximated with a relative error of 15%.

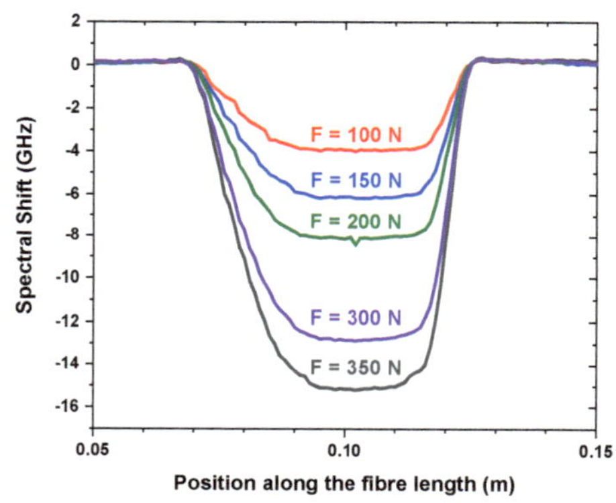

Figure 11. The experimental spectral shift profile observed through OFDR measurements along a polyimide-coated fibre bonded over 5 cm to an aluminium beam using the NOA 81 adhesive.

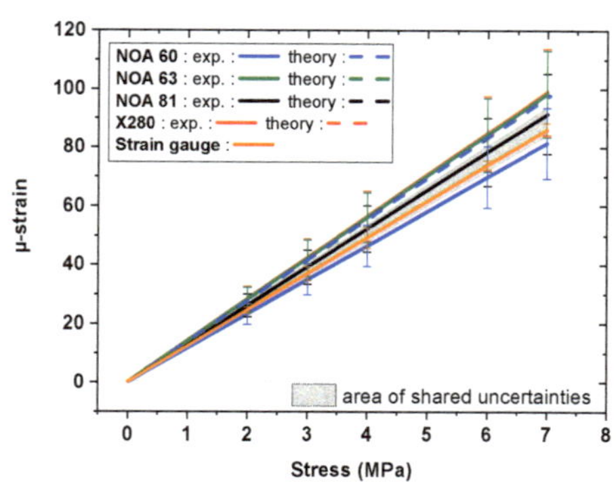

Figure 12. Strain measured via the four bonded polyimide-coated fibres, and via the strain gauge as a function of the stress applied to the tensile specimen; the results from the analytical model are also shown, in dotted lines (all overlapping).

Figure 12 shows the strains measured via the bonded polyimide-coated fibre as a function of the stress applied to the tensile test specimen. To each of the results, the 15% measured uncertainty has been added, in the form of an error bar. In these results, we were able to highlight a zone in which all the adhesives behaved in the same way. We can observe that the strain gauge measurements correspond well with this highlighted area.

The results from the analytical model are also shown in Figure 12. It appears that the analytical model gives us almost-identical deformation responses for all adhesives, to within 0.12 μ strain. This result agrees with the experimental results, and can be explained by the fact that, after a certain bonding length, the Young's modulus of the adhesives no longer has an impact, as the transfer is almost perfect (Figure 13).

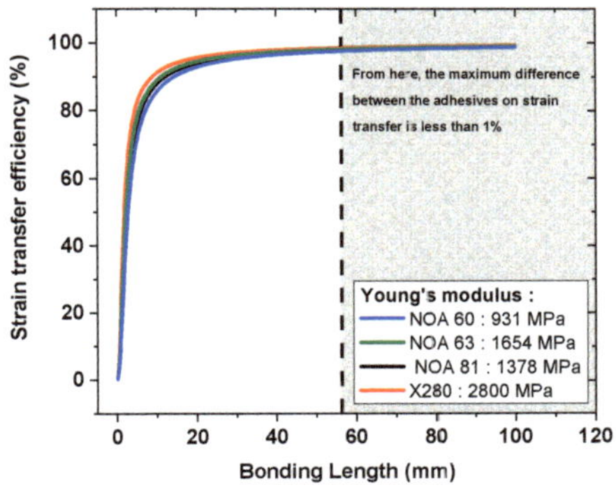

Figure 13. The analytical results showing the strain transfer efficiencies as a function of the bonding length, for the different Young's moduli of the adhesives, for a polyimide-coated fibre.

4.3. Dynamic Strain Experiences of the Bonded FBGs

Figure 14 reports the frequency response function (FRF), defined as the ratio between the strain measured via the FBGs, and the acceleration applied to the aluminium specimen, as a function of the excitation frequency, obtained for each fibre coating and each adhesive. Each of these curves is the average of the ratio between the measured strain and the applied acceleration of the four experiments performed at different displacements, between 0.028 mm and 0.14 mm. As constant displacements were applied, i.e., accelerations increasing with frequency, all the FRFs decreased linearly with frequency. We do not observe any irregular behaviour in these results, and we can, therefore, conclude that, up to 100 Hz, the strain transfer from the specimen to the bonded FBG remains unchanged. Finally, we observed a difference in the sensitivity of the measurements between the different optical fibre coatings and the different adhesives, as demonstrated by the static tests and the analytical model.

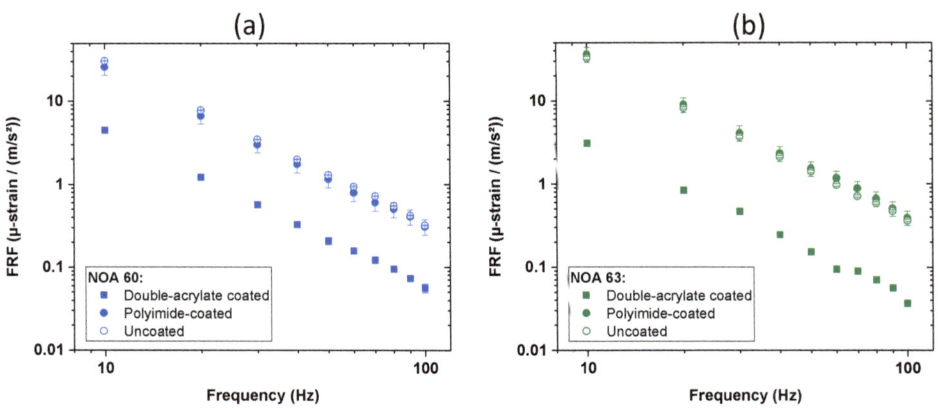

Figure 14. *Cont.*

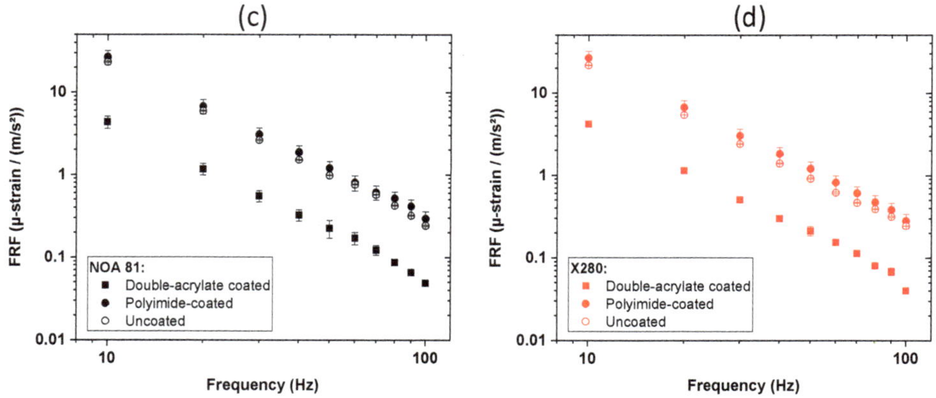

Figure 14. The frequency response function (FRF) of the double-acrylate-coated, polyimide-coated and uncoated FBGs bonded with the adhesive (**a**) NOA60, (**b**) NOA63, (**c**) NOA81 and (**d**) X280, when subjected to dynamic excitations, with constant displacements, from 10 to 100 Hz.

5. Conclusions

Through this combined experimental and theoretical study, it has been shown that the use of polyimide-coated fibres and uncoated fibres should be prioritised to ensure reliable strain measurements through the FBG or OFDR techniques. Nevertheless, uncoated fibres turned out to be very fragile and, for long-term structural health monitoring, the polyimide-coated fibre appears to be the best choice, as it offers almost the same sensitivity. Concerning FBGs written in acrylate-coated fibres, they display a very weak strain sensitivity. Yet, we could exploit them to measure large strains, which polyimide-coated fibres could not resist. However, this would entail the calibration of the bonded acrylate-coated FBG, and the determination of the limit of acrylate adhesion. The experiments and analytical model have also proven that the bonding length is essential to ensuring the most perfect possible strain transfer, and that, below a ~50 mm bonding length, the Young's modulus of the adhesive needs to be maximised for polyimide-coated and uncoated fibres.

Then, it was demonstrated that the bonded optical fibre sensors were functional over a dynamic range from 0 to 100 Hz. The precautions concerning bonding and coatings should be the same as those taken for static applications.

Further research is required, to study the impact of ageing on the strain transfer efficiency, particularly in relation to the other constraints (temperature and/or radiation) of the harsh environments for use in which sensors made using fluorine-doped optical fibres are very attractive.

Author Contributions: Conceptualisation, E.M., A.M. and C.L.; methodology, E.M., A.M., N.P., T.L.G. and C.L.; formal analysis, C.L.; writing—original draft preparation, C.L.; writing—review and editing, E.M., A.M. and S.G.; supervision, E.M., A.M., S.G., T.L.G., N.P. and J.C. All authors have read and agreed to the published version of the manuscript.

Funding: This research received no external funding.

Institutional Review Board Statement: Not applicable.

Informed Consent Statement: Not applicable.

Data Availability Statement: The data supporting the findings of this study are available from the corresponding authors upon reasonable request.

Conflicts of Interest: The authors declare no conflict of interest.

References

1. Bao, X.; Chen, L. Recent Progress in Distributed Fiber Optic Sensors. *Sensors* **2012**, *12*, 8601–8639. [CrossRef] [PubMed]
2. Hartog, A.H. *An Introduction to Distributed Optical Fibre Sensors*; Series in Fiber Optic Sensors; CRC Press: Boca Raton, FL, USA; Taylor & Francis Group: Abingdon, UK, 2017; ISBN 978-1-4822-5957-5.
3. Rao, Y.-J. Recent Progress in Fiber-Optic Extrinsic Fabry–Perot Interferometric Sensors. *Opt. Fiber Technol.* **2006**, *12*, 227–237. [CrossRef]
4. Vengsarkar, A.M.; Lemaire, P.J.; Judkins, J.B.; Bhatia, V.; Erdogan, T.; Sipe, J.E. Long-Period Fiber Gratings as Band-Rejection Filters. *J. Light. Technol.* **1996**, *14*, 58–65. [CrossRef]
5. James, S.W.; Tatam, R.P. Optical Fibre Long-Period Grating Sensors: Characteristics and Application. *Meas. Sci. Technol.* **2003**, *14*, R49–R61. [CrossRef]
6. Ferdinand, P. Capteurs à fibres optiques à réseaux de Bragg. *Mes. Mécaniques Dimens.* **1999**, *6735*, 1–24. [CrossRef]
7. Hill, K.O.; Meltz, G. Fiber Bragg Grating Technology Fundamentals and Overview. *J. Light. Technol.* **1997**, *15*, 1263–1276. [CrossRef]
8. Kashyap, R. *Fiber Bragg Gratings*; Optics and Photonics; Academic Press: San Diego, CA, USA, 1999; ISBN 978-0-12-400560-0.
9. Matveenko, V.P.; Shardakov, I.N.; Voronkov, A.A.; Kosheleva, N.A.; Lobanov, D.S.; Serovaev, G.S.; Spaskova, E.M.; Shipunov, G.S. Measurement of Strains by Optical Fiber Bragg Grating Sensors Embedded into Polymer Composite Material. *Struct. Control Health Monit.* **2018**, *25*, e2118. [CrossRef]
10. Luyckx, G.; Voet, E.; Lammens, N.; Degrieck, J. Strain Measurements of Composite Laminates with Embedded Fibre Bragg Gratings: Criticism and Opportunities for Research. *Sensors* **2010**, *11*, 384–408. [CrossRef]
11. Barrias, A.; Casas, J.R.; Villalba, S. Application Study of Embedded Rayleigh Based Distributed Optical Fiber Sensors in Concrete Beams. *Procedia Eng.* **2017**, *199*, 2014–2019. [CrossRef]
12. Guo, Y.; Kong, J.; Liu, H.; Hu, D.; Qin, L. Design and Investigation of a Reusable Surface-Mounted Optical Fiber Bragg Grating Strain Sensor. *IEEE Sens. J.* **2016**, *16*, 8456–8462. [CrossRef]
13. Cheng, Y.; Lu, X.; Gong, Y.; Wu, Y.; Rao, Y. Derrick Safety Monitoring System Based on Fiber Bragg Grating Strain Sensors. *Photonic Sens.* **2013**, *3*, 237–240. [CrossRef]
14. Ansari, F.; Libo, Y. Mechanics of Bond and Interface Shear Transfer in Optical Fiber Sensors. *J. Eng. Mech.* **1998**, *124*, 385–394. [CrossRef]
15. Liang, M.; Chen, N.; Fang, X.; Wu, G. Strain Transferring Mechanism Analysis of the Surface-Bonded FBG Sensor. *Appl. Opt.* **2018**, *57*, 5837. [CrossRef] [PubMed]
16. Her, S.-C.; Huang, C.-Y. The Effects of Adhesive and Bonding Length on the Strain Transfer of Optical Fiber Sensors. *Appl. Sci.* **2016**, *6*, 13. [CrossRef]
17. Wan, K.T.; Leung, C.K.Y.; Olson, N.G. Investigation of the Strain Transfer for Surface-Attached Optical Fiber Strain Sensors. *Smart Mater. Struct.* **2008**, *17*, 035037. [CrossRef]
18. Floris, I.; Sangiorgio, V.; Adam, J.M.; Uva, G.; Rapido, M.; Calderón, P.A.; Madrigal, J. Effects of Bonding on the Performance of Optical Fiber Strain Sensors. *Struct. Control Health Monit.* **2021**, *28*, e2782. [CrossRef]
19. Duck, G.; Renaud, G.; Measures, R. The Mechanical Load Transfer into a Distributed Optical Fiber Sensor Due to a Linear Strain Gradient: Embedded and Surface Bonded Cases. *Smart Mater. Struct.* **1999**, *8*, 175–181. [CrossRef]
20. Shen, W.; Wang, X.; Xu, L.; Zhao, Y. Strain Transferring Mechanism Analysis of the Substrate-Bonded FBG Sensor. *Optik* **2018**, *154*, 441–452. [CrossRef]
21. Li, W.Y.; Cheng, C.C.; Lo, Y.L. Investigation of Strain Transmission of Surface-Bonded FBGs Used as Strain Sensors. *Sens. Actuators Phys.* **2009**, *149*, 201–207. [CrossRef]
22. Hejazi, F.; Chun, T.K. *Advanced Solid Mechanics: Simplified Theory: Simplified Theory*, 1st ed.; CRC Press: Boca Raton, FL, USA, 2021; ISBN 978-1-00-314682-7.
23. Dillard, D.A. Fundamentals of Stress Transfer in Bonded Systems. In *Adhesion Science and Engineering*; Elsevier: Amsterdam, The Netherlands, 2002; pp. 1–44, ISBN 978-0-444-51140-9.
24. Meidav, T. Viscoelastic Properties of the Standard Linear Solid. *Geophys. Prospect.* **1964**, *12*, 80–99. [CrossRef]
25. Available online: https://www.exail.com/ (accessed on 11 July 2023).
26. Girard, S.; Kuhnhenn, J.; Gusarov, A.; Brichard, B.; Van Uffelen, M.; Ouerdane, Y.; Boukenter, A.; Marcandella, C. Radiation Effects on Silica-Based Optical Fibers: Recent Advances and Future Challenges. *IEEE Trans. Nucl. Sci.* **2013**, *60*, 2015–2036. [CrossRef]
27. Pillon, J.; Louf, F.; Boiron, H.; Rattier, M.; Peter, E.; Boucard, P.-A.; Lefèvre, H.C. Thermomechanical Analysis of the Effects of Homogeneous Thermal Field Induced in the Sensing Coil of a Fiber-Optic Gyroscope. *Finite Elem. Anal. Des.* **2022**, *212*, 103826. [CrossRef]
28. Ismardi, A.; Sugandi, G.; Lizarna, M.A. Design and Simulation of Double Planar Spring Based on Polyimide for Electrodynamic Vibration Energy Harvesting. *J. Phys. Conf. Ser.* **2019**, *1170*, 012061. [CrossRef]
29. HBM. Available online: https://www.hbm.com/fr/ (accessed on 11 July 2023).

30. NOA. Available online: https://www.norlandprod.com/adhesiveindex2.html (accessed on 11 July 2023).
31. Boiron, H.; Pillon, J.; Peter, E.; Robin, T.; Villedieu, T.; Morana, A.; Girard, S.; Boukenter, A.; Marin, E.; Lefèvre, H. Optical Fiber Strain and Temperature Coefficients Determination Based on Rayleigh-OFDR. In *Optical Fiber Sensors Conference 2020 Special Edition*; Optica Publishing Group: Washington, DC, USA, 2021; p. T3.42.

photonics

Article

UV Absorption Spectrum for Dissolved Oxygen Monitoring: A Low-Cost Proposal for Water Quality Monitoring

Aika Miura, Lorena Parra *, Jaime Lloret and Mónica Catalá-Icardo

Instituto de Investigación para la Gestión Integrada de Zonas Costeras, Universitat Politècnica de València, Gandía C/Paranimf, 1, 46730 Grao de Gandia, Spain; asilmiu@doctor.upv.es (A.M.); jlloret@dcom.upv.es (J.L.); mocaic@qim.upv.es (M.C.-I.)

* Correspondence: loparbo@doctor.upv.es

Abstract: One of the key indicators of water quality is dissolved oxygen. Even though oxygen is important in environmental monitoring, the sensors for dissolved oxygen are expensive and require periodic maintenance due to the use of membranes. In this paper, we propose using ultraviolet light absorption to estimate dissolved oxygen saturation in water samples. The absorption spectrum of dissolved oxygen in the ultraviolet range is investigated over a water matrix with different levels of complexity. First, the difference between different water matrixes is studied. The results indicate similar variations between river water and tap water matrices for comparative purposes. Both samples present much higher absorbance signals than distilled water. Thus, the rest of the tests were performed with only three water matrixes (ultrapure, distilled, and river water). By aerating, water samples were completely saturated. Then, nitrogen gas was used to remove dissolved oxygen from samples to obtain saturations of 75, 50, 25, and 3%. The absorption was measured from 190 to 380 nm, using LLG-uniSPEC 2. The obtained data were used to generate regression models for selected wavelengths (190, 210, 240, and 250 nm). The differences beyond 260 nm for the studied dissolved oxygen saturations were null. The generated models had correlation coefficients from 0.99 to 0.97 for ultrapure water, 0.98 to 0.95 for distilled water, and 0.90 to 0.83 for river water. The maximum differences were found between samples with 75 and 100% of saturation.

Keywords: optical sensor; oxygen saturation; water quality; absorbance; nitrogen gas; UV-C; spectral signature

Citation: Miura, A.; Parra, L.; Lloret, J.; Catalá-Icardo, M. UV Absorption Spectrum for Dissolved Oxygen Monitoring: A Low-Cost Proposal for Water Quality Monitoring. *Photonics* **2023**, *10*, 1336. https://doi.org/10.3390/photonics10121336

Received: 12 October 2023
Revised: 12 November 2023
Accepted: 28 November 2023
Published: 1 December 2023

1. Introduction

Most aquatic life depends on dissolved oxygen (DO) availability in the water that diffuses across the air–water interface. Even in the current context of global warming, which imposes the study of the gases that cause it, special attention should be paid to oxygen transfer. The amount of DO in water bodies is a primary indicator of water quality [1]. Oxygen loss is one of the most important changes occurring in an ocean, increasingly modified by human activities that have raised temperatures, CO_2 levels, and nutrient inputs and have altered the abundances and distributions of marine species [2]. It is important to know the DO variation in the water column to identify the factors that influence it. In estuaries and other coastal systems strongly influenced by their watershed, oxygen declines have been caused by increased loadings of nutrients (nitrogen and phosphorus) and organic matter. Those inputs are primarily from agriculture, sewage, and the combustion of fossil fuels [2]. Globally, the number and size of anoxic and hypoxic areas, regions with low or null concentrations of DO, have grown dramatically in recent years [3]. Although very high nutrient inputs and intense eutrophication are required to drive a weakly stratified ecosystem to hypoxia or anoxia, even moderate increases in nutrient inputs and eutrophication can lead to these conditions in a strongly stratified water body [4].

The presence of DO in water can be measured in two different ways. It can be expressed as the amount of oxygen in a given volume of water or as the solubility of oxygen. The solubility of oxygen ranges from 14.6 mg/L at 0 °C to approximately 7 mg/L at 35 °C under a pressure of 760 mmHg in fresh water [5]. In other words, oxygen is susceptible to variations in temperature, atmospheric pressure, and the content of salts or other substances in the water. Some authors [6] have proposed an indicator to assess the state of a river in terms of its deviations in the DO. Concentrations below 1.0–1.5 mg/L for a few hours can kill warm-water animals. Cold-water species require 2–3 mg/L more DO than warm-water species [7]. Loss of oxygen from the ocean (deoxygenation) can thus trigger major unforeseen changes in the structure and function of the midwater ecosystem [8].

The lack of marine environmental data compared with onshore environments is mainly caused by the high cost of probes, instruments, and installation. Without the data, there is no control over the impact of measures on the conservation and regeneration of native fishery resources or the delimitation of comprehensive protection areas. Therefore, defining the authorized activities in these reserve networks is impossible. The proposal of low-cost optical sensors for analyzing indicators such as DO in water can contribute to the environmental monitoring and evaluation of coastal and marine areas. The use of physical sensors is not only a low-cost option but also a non-destructive methodology that can operate if the environment does not damage the system. This is a risk shared by all detection methods in the aquatic environment. In addition, a low-maintenance system can be used in remote areas that are difficult to access.

The spectral signature is a concept mainly used in remote sensing but with a high application in spectroscopy. The spectral signature is the unique pattern or combination of peaks and troughs in a spectrum produced by a substance when it interacts with different radiation. On the one hand, spectral signatures are used in remote sensing to differentiate surfaces with different characteristics, such as urban areas, industrial areas, crops, and natural areas. Nevertheless, some studies are focused on identifying more specific compounds. Thus, it has been possible to identify different chemical substances from satellite images, such as chlorophylls [9] and hazardous chemicals [10]. Based on the current development of hyperspectral cameras, this topic will become more important in the following years. On the other hand, data obtained from Raman spectroscopy and spectroradiometer are being collected in databases to characterize different compounds, such as Paracetamol [11], or different polymers [12]. Nonetheless, as far as we are concerned, no study has focused on the application of spectra analyses to quantify the presence of DO in water samples. Thus, we propose to analyze the potential of spectral signatures for the quantification of DO in different types of water samples.

In this paper, we will perform DO detection using the spectrophotometric technique for different DO saturation levels and for different water matrixes. As a dissolved gas present in the water, DO is colorless to the naked eye, except using a chemical agent. We decided not to use visible wavelengths (between 380 and 750 nm) for the study because the dissolved oxygen molecule is a gas and is not visible within the visible light spectrum. Furthermore, according to previous studies and bibliographies already cited here, we excluded the infrared wavelengths (750 to 1100 nm) and chose to study in the ultraviolet range (190 to 380 nm). The main novelty of this paper is the measurement of the DO absorption spectrum in the UV range. The use of the UV spectrum to determine DO concentrations eliminates the main limiting factors that other methodologies have: interference in the visible spectrum due to visible compounds present in water, as it depends on the properties of light; and the need for physical contact of the sensor with the samples in cases of methodologies with chemical principles. Most of the absorption spectrum presented in the biography is for atoms instead of molecules. The existing commercial sensors for DO monitoring are based on membranes and phosphorescence quenching, using visible light [13]. In addition, studies like this are fundamental for the development of new technologies for mobile networks for monitoring coastal and oceanic waters.

2. Related Work

In this section, we detail the state-of-the-art in the different areas covered in the introduction. First, we summarize the current proposals for DO monitoring systems. Finally, the ongoing uses of spectrophotometry are outlined.

Previous studies have proposed different methodologies to analyze oxygen levels in the water. In [14], they proposed a system consisting of a submersible ceramic vessel containing a soil-based anodic chamber. Based on the study, the calibration model obtained with a two-factorial design of temperature experiments and DO can be used to successfully predict the DO value of an unknown water sample. The authors in [15] developed a microbial biosensor for the in situ, continuous, and online monitoring of DO. They monitored concentrations along several depths of lake water. Meanwhile, in [16], a submersible microbial fuel cell proved to be effective with an external resistance of 1000 Ω to determine DO-level monitoring in various environmental waters. Another option is the use of remote sensing, as proposed in [17], to monitor DO levels, using numerical simulation algorithms of DO concentrations, using the TELEMAC-WAQTEL-O2 model. In addition to these works mentioned, in [18], they proposed a cost-effective, accessible, and sustainable system for measuring the DO levels present in bodies of water based on a sensor for oxygen saturation in human blood.

The literature shows the determination of DO levels using combined methods. For example, the authors in [19] concluded that DO sensor technology will become the focus of future research and mainstream development. The smart DO sensor can intelligently process signals by using analogue and digital signal processing technologies and performing real-time dynamic compensation and correction for the temperature, pressure, salinity, and other interference factors, thereby greatly reducing the manual operation and management cost and directly or indirectly reducing human errors [19]. Nevertheless, there are currently not many studies on the spectral signature of DO in water due to several factors that influence oxygen availability, such as temperature, salinity, pH, and organic matter. DO concentrations were summarized and analyzed as the crucial drivers of hypoxia in Northwestern and Southern Hong Kong and Mirs Bay over the past three decades [20].

Previous studies have already developed methodologies with spectrophotometry to determine the DO in water. For example, in [21], the authors used ultraviolet spectrophotometry to determine the DO in marine and estuarine water by the Winkler method. They reached a p-value < 0.05 as a result, as well as a true correlation between the A and B variables, with $r > 0.98$. Furthermore, the authors of [22] used this same spectrophotometric variant of the Winkler method in fresh water, using the ultraviolet range. The results presented demonstrated the ease of applying ultraviolet spectrophotometry to analyze DO via the Winkler method in natural waters within a wide salt range [21]. A previous work [23] presented the groundwork for the implementation of a monitoring system of DO levels that uses a Light-Emitting Diode (LED) and Light-Dependent Resistance (LDR). To analyze different levels of DO, they used sodium sulfite as a cheap option with fast oxidation [23]. Furthermore, in [24], a low-cost system was studied for data gathering in coastal waters, using Wireless System Networks to monitor water exchange, which is very useful for quality monitoring.

Using spectrophotometry, the authors of [25] concluded that their method is simple, rapid, sensitive, precise, and useful for the determination of DO in RW samples. The water sample was fixed for DO via Winkler's procedure and was directly used for the spectrophotometric measurements by the proposed method. The wavelength used in this case was 644 nm because the liberated iodine bleaches the violet color of azure B (which is measured at 644 nm). Thus, it was concluded that the amount of I2 released is equivalent to the oxygen dissolved in the water. In other studies, different optical techniques are used to determine DO, such as a polydimethylsiloxane membrane coated with a platinum octaethylporphyrin and a multilayer optical–fluidic sensor for in situ measurement [26,27]. In addition, several authors have developed mechanisms to monitor DO in water as an indicator of the quality and health of aquatic ecosystems [28–30].

As mentioned above, no papers are focused on the analysis of the spectral signatures for the determination of the DO concentration in water samples. The existing sensors are mainly based on technologies with high requirements for maintenance of the use of reagents.

3. Materials and Methods

This section describes the samples, the equipment used to reach the results, and how the samples were prepared to find saturation levels of DO.

3.1. Samples Collection

For the tests, nitrogen gas was used, and three types of water were prepared for analysis in the spectrophotometer: UW, DW, and RW. We use different samples to go from an ultrapure sample to real samples in which other compounds can act as interferents and affect the measurements. The use of different water matrices in the study is important to ensure that the DO measurement is reliable; in this case, the absorbance signals derive mainly from the DO concentrations and not from possible compounds present in the samples. We selected a wide variety of samples. The included samples include pure water samples and water samples with a modified matrix, such as tap water with a content of residual-free chlorine: 1 mg/L and no presence of pollutants and a natural water sample.

The UW samples have the highest degree of purity due to the Barnstead Smart2Pure [31] purification system for laboratory applications (see Figure 1A3). They have the highest degree of reliability regarding interference by other organic or inorganic substances and undergo the deionization process. The necessary volume of UW water is generated every time that a sample is needed. The volume of each sample is 3 mL. In addition to it, with a slightly lower degree of purity but with a filtration and distillation process, we used distilled water for DW samples (Figure 1A2). The total volume of DW necessary to conduct the experiments was collected at the beginning of the experiment and stored in a 500 mL plastic squeeze bottle dispenser.

Figure 1. Sample preparation schematic. (**A1**) River in which RW samples are collected. (**A2**) Distilled water for DW samples. (**A3**) Barnstead Smart2Pure purification system for UW samples. (**B**) DO850 Portable Optical DO sensor used to verify DO saturations. (**C**) The nitrogen gas system with pressure gauge. (**D**) The method used for nitrogen introduction into sample cuvette measurements.

Finally, as an approach to real samples of natural water, such as river, coastal, and oceanic waters, we used a sample of water from a river close to the worksite (see Figure 1A1). The sample was collected at the moment of being used in an opaque laboratory glass bottle of 200 mL.

3.2. Samples Preparation

Before the preparation of the samples, around 150 pretests were carried out to find the ideal saturation levels to build a DO gradient. The tests analyzed time and gas flow to estimate different DO concentrations. In the end, the five ideal concentration levels were established based on the time and flow of nitrogen gas inserted into the sample.

The accuracy test of the methodology was carried out using the DO8500 Portable Optical DO instrument [32] on the 5 replicates of each pre-sample. In each replica, we applied the same time and the same amount of gas to verify whether the DO concentration values were equal. Table 1 establishes the samples, along with the concentrations.

Table 1. Details of nitrogen introduction in sample preparation.

Range Saturation (%)	Time (s)	Flux (Bars)
Blank (3)	180	3.
25	30	3.5
50	15	3.5
75	5	3.5
100	0	3.5

In the preparation of the samples, we chose to create a gradient of oxygen levels to analyze the behavior between the wavelengths. All samples were aerated, mixing for 1 min to raise the saturation to 100%. The samples were brought to saturation at five levels of DO. For the DO saturations, the DO8500 Portable Optical DO [32] was used to verify that the samples achieved all five levels (3% (blank), 25%, 50%, 75%, and 100%; see Figure 1B).

Nitrogen gas in samples to remove oxygen is possible because nitrogen physically displaces oxygen from the sample by bubbling. In other words, it is the reverse methodology to bubble oxygen, using the aerator to obtain 100% oxygen saturation in the samples. Before measurements, DW was used to verify the efficacy of bubbling nitrogen to eliminate the DO present in water. A total of 49 measurements were taken, and the DO and temperature levels were quantified at the beginning and the end of each nitrogen gas introduction. The time and amount of nitrogen chosen were variable until stable and reproducible levels were established for posterior testing.

The amounts of nitrogen introduced into the samples were controlled by the inlet flow (bars) of the gas (Figure 1C) and by its residence time (seconds) inside the sample (Figure 1D). The cuvette, which was used in the spectrophotometers containing 3 mL of water, was used as a container for the sample. A capillary tube with constant outlet pressure was used. The time was measured with a timer.

3.3. Measuring Equipment

The spectrophotometer (LLG-uniSPEC 2 [33]) was used for measurements. The spectrophotometer used can perform readings in ultraviolet, visible light, and infrared wavelengths (from 190 to 1100 nm). Among some features, it has a single beam system grid with 1200 lines/mm, silicon photodiode detector, 4 position cell holder (10 mm), deuterium and tungsten lamps, and automatic wavelength adjustment. Each measurement was taken in real time and saved in an Excel sheet. Three values were taken from each sample to avoid variation from reading errors. The measuring range chosen goes from 190 nm to 380 nm, using the basic operation mode of the spectrophotometer.

3.4. Performed Analysis

To perform all the measures, we chose a range that goes from 190 nm to 380 nm, taking measurements 5 nm by 5 nm for the UW, DW, and RW samples. Figure 2 shows the data collection scheme of samples analyzed with the spectrophotometer. From each type of water (UW, DW, and RW) we took 5 samples and analyzed them three times for each DO saturation level. That means that we analyzed 15 samples, considering 3 blanks for each water matrix. In total, we studied the samples 471 times, covering all wavelengths. All samples were prepared according to the properties and wavelengths shown in Table 2. Changes in temperature and atmospheric pressure influence the DO concentration. In fresh water, DO reaches 14.6 mg/L at 0 °C and approximately 9.1, 8.3, and 7.0 mg/L at 20, 25, and 35 °C, respectively, and 1 atm pressure. At temperatures of 20 and 30 °C, the level of saturated DO is 9.0–7.0 mg/L [34]. The samples in this work were collected and handled with the aim of avoiding variation in DO due to external factors. The analyses were carried out in a closed environment, without temperature variation (at 22 °C), and in the pretests, the temperature of the sample was measured before and after aeration and remained constant (approximately 25 °C). After that, the reading was then read on the spectrophotometer, without shaking and with the cuvette sealed. Special attention was paid during the measuring procedure to avoid the introduction of oxygen in the samples, using parafilm to seal the cuvettes.

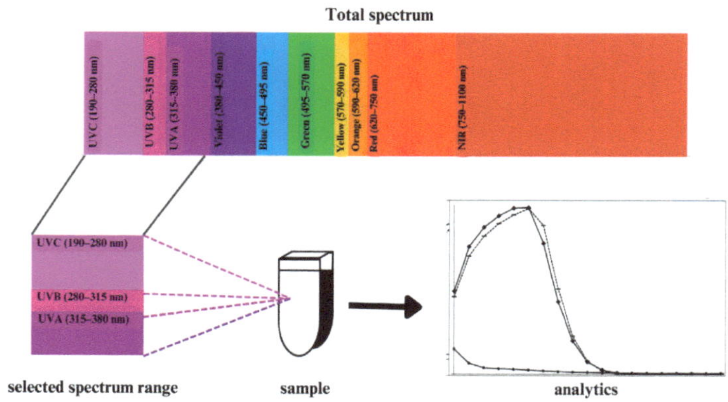

Figure 2. Detection scheme, using a spectrophotometer and UC wavelength range.

In addition, we used a sample of tap water (TW) to perform comparisons between the matrices used to identify differences in absorption without DO. For this, a sample of UW was used for the blank (with 3% of DO), and two other samples of DW and RW (with 3% of DO) were used. Thus, it is possible to see the absorption contrasts between the samples since, in this part of the spectrum, there are many compounds present in the water that absorb light. This informs us if it is possible to have a single calibrator (valid for any type of water matrix) or if we need a calibrator for each type of sample.

In the sampling structure, comparisons between matrices are made first. Then, each of them is measured in five different concentrations. With the results, we make a calibration for each sample, from the simplest UW to the most complex RW, so that we find out if it is possible to measure the DO concentration at a single wavelength.

4. Results

In this section, we present the obtained results from the laboratory analyses. First, we show the results for sample matrices, and then we show the results for UW, DW, and RW in the second, third, and fourth subsections, respectively.

4.1. Results of Water Matrix Samples

We can identify in Figure 3 that the TW and RW samples present an expected behavior of near saturation of optical absorbance between the wavelengths 195 and 220 nm. Thus, to keep the experimental test bench simple, only RW is further used. This fact is due to the presence of volatile compounds and other substances present, such as organic matter, heavy metals, gases, etc. The sample used for the blank was the UW sample that does not contain substances causing optical interference. The absorbance levels of the DW sample are very low compared to those of RW and TW, with a maximum at the wavelength of 190 nm, with 0.40 AU. This result may have occurred because the distilled water came from a laboratory container with stagnant water without circulation, unlike the river and tap samples. The AU decay for the DW samples may just be a clipping of the DO concentration decay as the wavelength increases, as happens for the TW and RW samples after 220 nm. Therefore, with low DO concentrations in the sample, it is more difficult to detect the DO concentration through the absorbance signal.

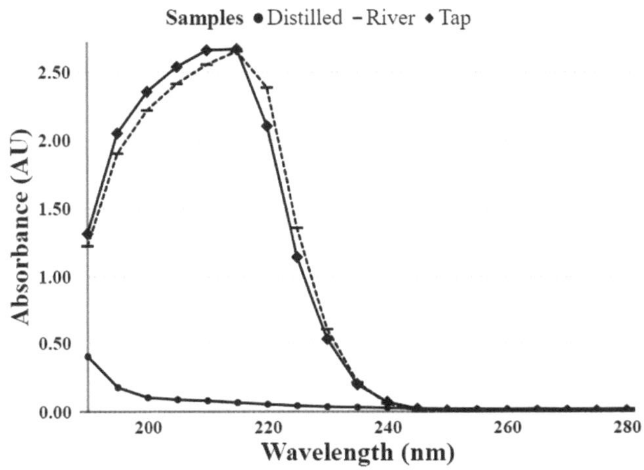

Figure 3. Plot of the absorbance signal with respect to the wavelength for DW, TW, and RW samples.

The highest absorbance values are at the wavelength of 215 nm (2.67 AU for the TW sample and 2.65 AU for the RW). Note that saturations with high absorbance values show a peak up to 215 nm. From there, there is a notable decay reaching almost 0 AU at the wavelength of 245 nm (0.02 AU for TW and RW samples and 0.01 AU for DW). In addition, the differences between the matrices used are significantly large, so blanks of each analyzed matrix were used. In other words, for blanks (considering 3% of DO), we must use ultrapure for UW, distilled for DW, and river for RW. Considering that water turbidity has greater absorption at wavelengths of the visible spectrum, for our study (at wavelengths below 280 nm), we did not observe that turbidity did not cause interference in the samples, so much so that, as you can see in Figure 3, the river water samples show a similar signature to tap samples below 280 nm. Nevertheless, the correction of the turbidity effect using NIR light will be implemented in the future.

4.2. Results of UW

We show, in Figure 4, the absorbance for the UW for the wavelengths from 190 to 280 nm. In addition, after verifying that there are no considerable variations after the 280 by 380 nm wavelength, we chose to present the graphic analysis up to 280 nm for better visualization. We can mainly differentiate two regions with different trends. Error bars were added based on the UW sample standard deviation. Considering 100% DO saturation, the standard deviation is 0.3637; for 75% saturation, the value is 0.1250; for 50% saturation, it is 0.0070; and for 25% DO saturation, the result is 0.0059.

There are notable differences in absorbance values between the analyzed samples in the first region, from 190 to 220 nm. As the percentage of DO saturation increases, the absorbance signal value increases. It is important to note that the values of 25 and 50% are overlapped (lighter color of blue). Moreover, in the second region, from 230 to 280 nm, the absorbance of the samples does not increase further as the wavelength increases, with the absorbance values tending toward 0 AU. Thus, just the first region can be used to estimate the samples' DO saturation. From 200 to 220 nm, it can be used for this purpose and is safely detectable at wavelengths up to approximately 230 nm.

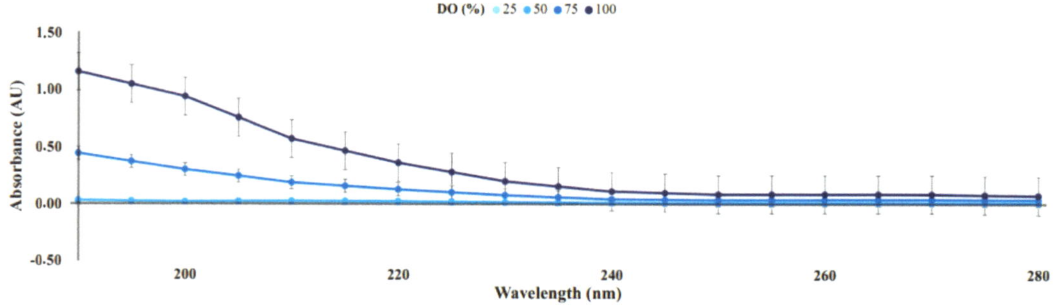

Figure 4. Absorbance signals for UW samples in the UVC range.

4.3. Results of DW

Figure 5 displays the absorbance for the wavelengths from 190 to 280 nm (UVC range) because, after that, until 380, all values become stable, without differences between all four saturations. As explained in Section 3.4, a 3% DW sample was used for the blank for spectrophotometer calibration to maintain the most realistic comparisons.

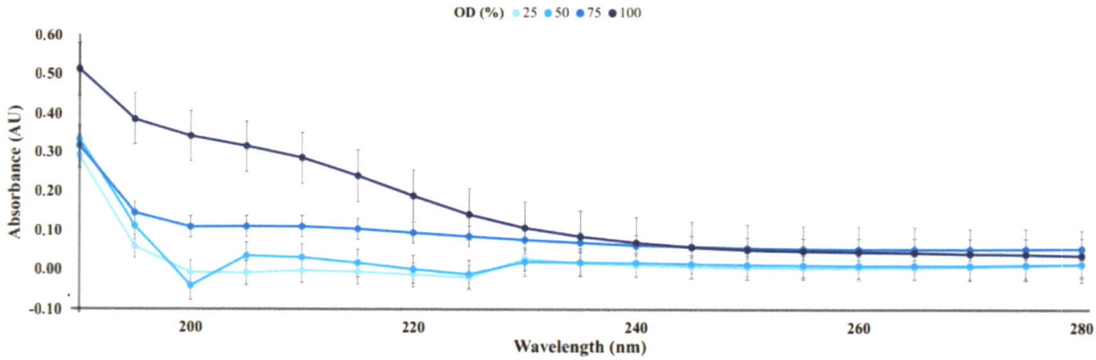

Figure 5. Absorbance signals for DW samples in the UVC range.

Furthermore, between 190 to 220 nm, we can see that the samples follow a clear trend and absorbance increases as the saturation increases. Nonetheless, some values had negative values, a minimum of −0.04 AU (50% of DO). In general terms, the absorbances decrease with the wavelength increase for all ranges and samples. There are some small regions where the differences between the saturations are more significant, which is between 195 and 215 nm (25% sample with −0.04 AU and 100% sample with 0.34 AU). In these regions, the possibility to estimate the DO saturation in the samples with light absorption is promising, and some wavelengths can be used for that purpose. About the standard deviation, for 100% DO samples, the value is 0.1450; for 75% saturation, the value is 0.0612; for 50% saturation, it is 0.0779; and for 25% DO saturation, the result is 0.0672. In a further subsection, we analyze which wavelength has better accuracy in regard to predicting the DO saturation.

4.4. Results of RW

Subsequently, Figure 6 represents the absorbance of the samples for the river in the UVC region of the spectrum. Despite the noise due to the presence of biomaterials present in the river water, it is possible to perceive higher values of the absorbance signal depending on the increase in DO saturation in the water. Just like the other matrices, the standard deviation was calculated. For 100% samples, the value is 0.0091; for 75%, the value is 0.0055; for 50% saturation, it is 0.0031; and for 25% samples, the result is 0.0058.

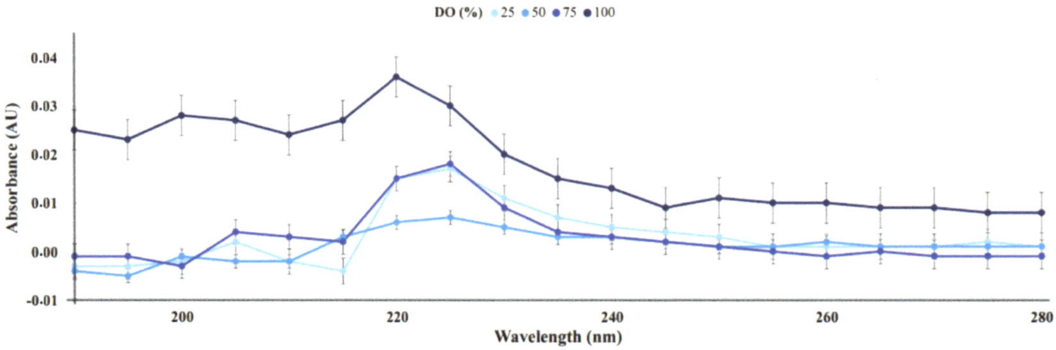

Figure 6. Absorbance signals for RW samples in the UVC range.

In general terms, the sample with the highest saturation (100% of DO) has higher absorbance, and samples with 25% and 50% have lower absorbances. Nonetheless, possibly due to compounds present in the sample, we cannot see that the samples follow a clear trend. We can see a detection peak for all samples, between 215 nm and 230 nm, before a noticeable decrease in the absorbance levels. The maximum value for 100% of the DO sample is 0,036 AU, and the minimum for 50% of the DO sample is 0.007 AU. Nonetheless, the absorbance of the sample with 25% of DO saturation appears above the sample with 50% of DO saturation. Even so, estimating the DO differences present in the water is considered a good range.

5. Discussion

In this section, we discuss the observed differences in the absorption of tested samples. First, we analyze the impact of these differences and the potential to be used as an optical method for sensing the DO in water. Then, the comparison between the proposed approach and existing methods for DO monitoring is presented. Finally, the main limitations of the proposed method are summarized.

5.1. Potential Use of DO Absorption in UV Spectrum as an Optical Approach for DO Monitoring

After analyzing the results for the absorbance values of the chosen saturations, it was necessary to determine in which wavelength the DO in water is more likely to be identified. Moreover, it is necessary to evaluate the level of correlation between the wavelength and the DO concentrations for all of the samples.

At this point, the objective of obtaining and analyzing the absorbances for all wavelength ranges was to find the most appropriate wavelength to perform a calibration. This calibration can be used in the future to determine the saturation of DO in natural samples. The DO results obtained for the three water matrixes show a maximum range of detection between 190 and 250 nm. For each studied wavelength, the *p*-value (ρ) and the correlation coefficient (CC) were evaluated for the three water matrixes. Table 2 lists the values of CC and ρ for selected wavelengths and indicates the type of generated model. These wavelengths were selected to maximize the differences between samples. Models were obtained with Statgraphics Centurion XVII [35], using the Simple Regression Tool. The type of model was selected according to the comparison of alternative models.

Table 2. Correlation coefficients and ρ-values for UW, DW, and RW samples between 190 and 280 nm.

Wavelength	UW Samples			UW Samples			RW Samples		
	Model	CC	ρ	Model	CC	ρ	Model	CC	ρ
190	Squared Root-Y Squared-X	0.985	0.002	Squared Root-X	0.952	0.013	Squared-X	0.840	0.075
210	Squared Root-Y Squared-X	0.990	0.001	Squared-X	0.981	0.003	Squared-X	0.903	0.035
240	Squared-X	0.975	0.004	Squared Root-Y	0.973	0.005	Double Squared Root	0.856	0.065
250	Squared Root-Y Squared-X	0.980	0.003	Squared Root-Y	0.960	0.001	Double Squared	0.834	0.079

At 210 nm, we see the highest values for CC for all three samples (between 0.9032 and 0.9902) and two different models that adapt better, the Squared Root-Y Squared-X (UW sample) and Squared-X (DW and RW samples). To reduce the complexity of the system, we also represented the model Squared-X for UW. Thus, selecting the same model for all samples is more reliable.

Figures 7 and 8 depict the models and their equations for UW. In all figures, the blue line indicates the calculated model, and the dotted lines represent the confidence (green) and prediction intervals (grey). We can see that the loss of accuracy, according to the regression and confidence intervals, from changing the models is too high to be accepted. Thus, the system uses the Squared Root-Y Squared-X. Then, Figures 9 and 10 portray the calibration graphics at 210 nm for DW and RW. We can see that, as the complexity of the matrix increases, the regression and confidence intervals increase as a response to the decrease in CC. This is mainly caused by the presence of interferents in real samples, which almost saturate the absorption of UV-C even in the blank.

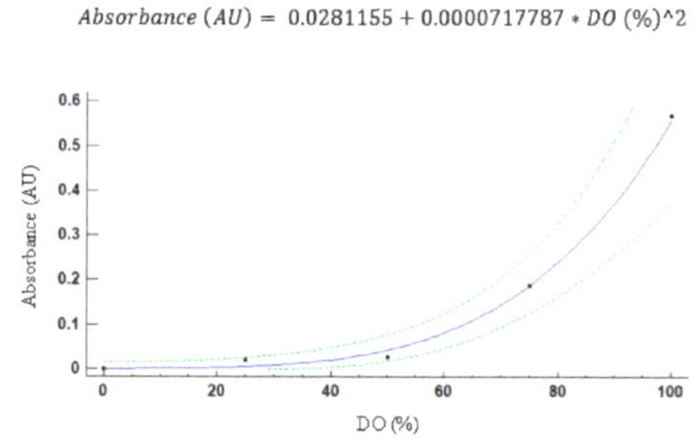

Figure 7. Plot of absorbance signals and DO saturation with UW results. Calibration at 210 nm (Squared Root-Y Squared-X Model).

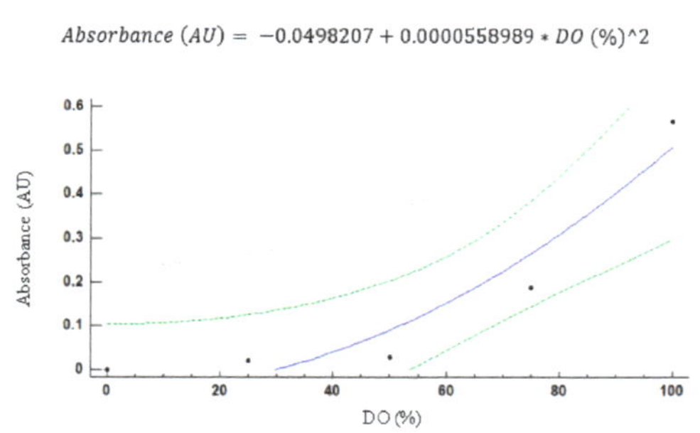

Figure 8. Plot of absorbance signals and DO saturation with UW results. Calibration at 210 nm (Squared-X Model).

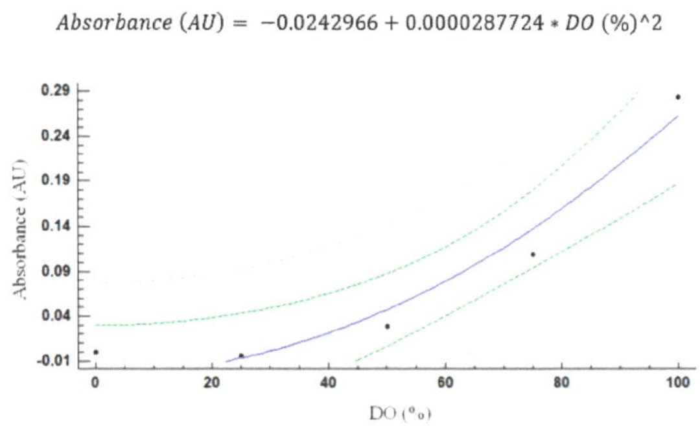

Figure 9. Plot of absorbance signals and DO saturation with DW results. Model calibration at 210 nm (Squared-X Model).

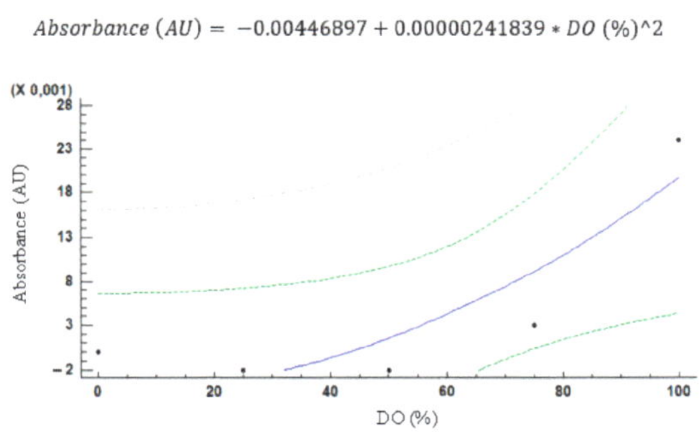

Figure 10. Plot of absorbance signals and DO saturation with RW results. Model calibration at 210 nm (Squared-X Model).

Additional analysis includes the Analysis of Variance (ANOVA) with the different saturation levels for each sample, including the entire measured range with the mean values for each wavelength. These results can be seen in Table 3. The results of the UW showed the largest differences among DO values. Nevertheless, the results of the multiple ranges test, using the least significant difference for the DW, are characterized by the clearest differences among the levels of DO. In the RW, the differences are only significant between the saturation of 100% and the rest of the saturations.

Table 3. Results of ANOVA tests for the different saturation levels.

Saturation	Absorbance for Each Type of Sample		
	UW	DW	RW
25%	0.015 ^{a}	0.020 ^{a}	0.001 ^{a}
50%	0.017 ^{a}	0.031 ^{ab}	0.003 ^{a}
75%	0.126 ^{a}	0.089 ^{b}	0.003 ^{a}
100%	0.352 ^{b}	0.158 ^{c}	0.018 ^{b}
p-value	0.0001	0.0001	0.0000

5.2. Comparison with Existing Proposals

Existing proposals which measured DO in water, explored in the Introduction, offered limited information about the accuracy of the proposed approaches. Given the wide variety of metrics, such as the determination coefficient, CC, square root of mean error (RMSE), ρ, and standard deviation (σ), fairly comparing the results has become complex. Moreover, the DO can be measured in different ways, such as saturation and mg/L; without knowing the temperature at which other authors have performed their experiments, converting the units is impossible.

The authors of [13–15] determined the correlation coefficient values of the sensors and models to be between 0.99 and 0.88. In our case, the models have correlation coefficients from 0.99 to 0.97 for UW, 0.98 to 0.95 for DW, and 0.90 to 0.83 for RW. The authors of [20] indicated the accuracy of their sensor based on the ρ ($\rho < 0.05$). In our case, for UW and DW, all ρ values were accomplished with $\rho < 0.05$. Nonetheless, for the RW results, only data at 210 nm accomplish the $\rho < 0.05$. The rest of the ρ is between 0.06 and 0.08. The standard deviation and RMSE metrics cannot be compared since the units are different.

5.3. Limitations of Presented Tests

The main limitation of the proposed approach is that, in complex environments, many compounds might absorb the UV spectrum, which can be considered interferences. Nonetheless, the absorption peak of different compounds changes. Our aim is to include at least two-to-four LED emitters at different regions of the UV spectrum, as shown in Table 2, to reduce the impact of those interferences. Most of those interferences absorb light at other wavelengths beyond the UV wavelength. However, since it is necessary to use another methodology for this improvement, researchers should consider carrying out these new tests with a new data framework and on more sample matrices. This new methodology includes creating and calibrating a new sensor and integrating it with the ones developed in [23,24], not simply using the current data. Thus, the combination of this UV-absorption sensor with a sensor with multiple LEDs emitting at different regions of the visible and IR spectrum is proposed. By combining the absorption at the UV spectrum with the existing visible and IR spectrum with the use of Artificial Neural Networks (ANNs), as proposed in [36], it could be possible to detect a considerable range of interferences and estimate the DO concentration.

Another problem encountered with the obtained data is the loss of accuracy when the DO saturation is below 50%. Since the objective for these sensors is to be used in natural environments, such as rivers or oceans, the most common concentrations are around 100%. When the DO saturation drops below 60%, most fish experience breathing problems. Therefore, the interesting range for our sensor is from 60 to 100% of DO saturation, and the data accomplish this requirement. Although each sample needs a white itself, the calibration of the future sensor can prevail by detecting unique-to-unknown real samples based on further studies of UV spectrum detection.

6. Conclusions

In this paper, the absorbance spectrum of DO in the UV range was studied in water samples of different matrixes, UW, DW, and RW. The aim was to establish if it is possible to use the absorbance spectrum to determine the DO saturation.

The data collected in the laboratory indicate that the different water matrixes with a DO % near 0% have different absorption spectra. This is caused by the presence of different compounds in each water matrix which are absorbed in this region, such as pesticides or organic compounds. We determined that the region from 190 to 250 nm is the spectrum in which DO has the highest absorption. A total of four wavelengths were used to generate a series of models that estimate the saturation of DO in different water samples. Even though the models have problems detecting the DO saturation at low values, the critical DO values are between 100 and 60% for the environmental application. Through standard deviation analyses, we can see that, despite some overlapping points between saturations (as in the UW graph, between 25% and 50% of OD), the error bars show us that, for measurements carried out at higher saturations, the answer is consistent within the trend of the line.

In future work, we will combine the absorption spectrum of UV with visible and IR ranges. Instead of regression models to estimate the DO saturation, ANN and other AI-based mechanisms will be used to detect the presence of interferences in seawater samples, in which interferents will generally be at low concentrations. Furthermore, real samples with a bigger sample size should be conducted on actual water samples, including TW, seawater, fresh water, and reclaimed water, among others, to enhance the applicability of the study. Then, the sensor will be applied in real scenarios to be part of a marine observatory as a low-cost option to monitor water quality.

Author Contributions: Conceptualization, J.L. and M.C.-I.; methodology, J.L. and M.C.-I.; software, A.M. and L.P.; validation, A.M. and L.P.; formal analysis, A.M. and L.P.; investigation, A.M. and L.P.; resources, J.L. and M.C.-I.; data curation, A.M. and L.P.; writing—original draft preparation, A.M. and L.P.; writing—review and editing, L.P. and J.L.; visualization, J.L.; supervision, J.L. and M.C.-I.; project administration, J.L.; funding acquisition, J.L. All authors have read and agreed to the published version of the manuscript.

Funding: This study forms part of the ThinkInAzul program and was supported by MCIN with funding from the European Union's NextGenerationEU (PRTR-C17.I1) and by Generalitat Valenciana (THINKINAZUL/2021/002).

Institutional Review Board Statement: Not applicable.

Informed Consent Statement: Not applicable.

Data Availability Statement: The data used to support the findings of this study are available from the corresponding author upon request.

Conflicts of Interest: Authors declare that there is no conflict of interest regarding the publication of this paper or between the authors.

References

1. Janzen, J.G.; Schulz, H.E.; Lamon, A.W. Measurements of dissolved oxygen concentration at water surface. *Eng. San. E Amb.* **2008**, *13*, 278–283.
2. Breitburg, D.; Levin, L.A.; Oschlies, A.; Grégoire, M.; Chavez, F.P.; Conley, D.J.; Garçon, V.; Gilbert, D.; Gutiérrez, D.; Isensee, K.; et al. Declining oxygen in the global ocean and coastal waters. *Science* **2018**, *359*, eaam7240. [CrossRef] [PubMed]
3. Köster, M.; Wardenga, R.; Blume, M. Microscale investigations of microbial communities in coastal surficial sediments. *Mar. Ecol.* **2008**, *29*, 89–105. [CrossRef]
4. Kemp, W.M.; Sampou, P.A.; Garber, J.; Tuttle, J.; Boynton, W.R. Seasonal depletion of oxygen from bottom waters of Chesapeake Bay: Relative roles of benthic and planktonic respiration and physical exchange processes. *Mar. Ecol. Prog. Ser.* **1992**, *85*, 137–152. [CrossRef]
5. Kamykowski, D.Z.; Zentara, S.J. Hypoxia in the world ocean as recorded in the historical data set. *Deep Sea Res. Part A Ocean. Res. Papers* **1990**, *37*, 1861–1874. [CrossRef]
6. Posada, E.; Mojica, D.; Pino, N.; Bustamante, C.; Monzon Pineda, A. Establishment of Environmental Quality Indices of Rivers According to the Behavior of Dissolved Oxygen and Temperature. Applied to the Medellín River, in the Valley of Aburra in Colombia. *Dyna* **2013**, *80*, 192–200.
7. Global Seafood Alliance. Available online: https://www.aquaculturealliance.org/advocate/dinamica-del-oxigeno-disuelto/ (accessed on 13 August 2022).
8. Wishner, K.F.; Seibel, B.A.; Roman, C.; Deutsch, C.; Outram, D.; Shaw, C.T.; Birk, M.A.; Mislan, K.A.S.; Adams, T.J.; Moore, D.; et al. Ocean deoxygenation and zooplankton: Very small oxygen differences matter. *Sci. Adv.* **2018**, *4*, eaau5180. [CrossRef]
9. Sawicki, A.; Willows, R.D.; Chen, M. Spectral signatures of five hydroxymethyl chlorophyll derivatives chemically derived from chlorophyll b or chlorophyll f. *Photosynth. Res.* **2019**, *140*, 115–127. [CrossRef]
10. Gwon, Y.; Kim, D.; You, H.; Nam, S.H.; Kim, Y.D. A Standardised Procedure to Build a Spectral Library for Hazardous Chemicals Mixed in River Flow Using Hyperspectral Image. *Remote Sens.* **2023**, *15*, 477. [CrossRef]
11. Gupta, R.S.; Deshmukh, R.R.; Kshirsagar, A.V. Spectral Database of Pharmaceutical Common Excipients and Paracetamol API Using ASD Field Spec 4 Spectroradiometer. *Med.-Leg. Update* **2021**, *21*, 1012–1018.
12. Ramanna, S.; Morozovskii, D.; Swanson, S.; Bruneau, J. Machine Learning of polymer types from the spectral signature of Raman spectroscopy microplastics data. *arXiv* **2022**, arXiv:2201.05445. [CrossRef]
13. Lakowicz, J.R.; Weber, G. Quenching of fluorescence by oxygen. Probe for structural fluctuations in macromolecules. *Biochemistry* **1973**, *12*, 4161–4170. [CrossRef] [PubMed]
14. Olias, L.G.; Otero, A.R.; Cameron, P.J.; Di Lorenzo, M. A soil microbial fuel cell-based biosensor for dissolved oxygen monitoring in water. *Electroch. Acta* **2020**, *362*, 137108. [CrossRef]
15. Song, N.; Yan, Z.; Xu, H.; Yao, Z.; Wang, C.; Chen, M.; Zhao, Z.; Peng, Z.; Wang, C.; Jiang, H. Development of a sediment microbial fuel cell-based biosensor for simultaneous online monitoring of dissolved oxygen concentrations along various depths in lake water. *Sci. Tot. Environ.* **2019**, *673*, 272–280. [CrossRef] [PubMed]

16. Zhang, Y.; Angelidaki, I. A simple and rapid method for monitoring dissolved oxygen in water with a submersible microbial fuel cell (SBMFC). *Biosens. Bioelectron.* **2012**, *38*, 189–194. [CrossRef] [PubMed]
17. Abouelsaad, O.; Matta, E.; Omar, M.; Hinkelmann, R. Numerical simulation of Dissolved Oxygen as a water quality indicator in artificial lagoons–Case study El Gouna, Egypt. *Region. Stud. Marin. Sci.* **2022**, *56*, 102697. [CrossRef]
18. Shaghaghi, N.; Nguyen, T.; Patel, J.; Soriano, A.; Mayer, J. Monitoring, DOxy: Dissolved Oxygen Monitoring. In Proceedings of the 2020 IEEE Global Humanitarian Technology Conference (GHTC), Seattle, WA, USA, 29 October–1 November 2020; pp. 1–4. [CrossRef]
19. Wei, Y.; Jiao, Y.; An, D.; Li, D.; Li, W.; Wei, Q. Review of Dissolved Oxygen Detection Technology: From Laboratory Analysis to Online Intelligent Detection. *Sensors* **2019**, *19*, 3995. [CrossRef] [PubMed]
20. Qian, W.; Zhang, S.; Tong, C.; Sardans, J.; Peñuelas, J.; Li, X. Long-term patterns of dissolved oxygen dynamics in the Pearl River Estuary. *J. Geo. Res. Biog.* **2022**, *127*, 1–14. [CrossRef]
21. Ramírez, T.G. Aplicación De La Espectrofotometria Ultravioleta Para Determinar El Oxigeno Disuelto En Agua Por El étodo Winkler. *Bol. Investig. Mar. Cost.* **1992**, *21*, 95–100.
22. Duval, W.S.; Brockington, P.J.; von Melville, M.S.; Geen, G.H. Spectrophotometric determination of dissolved oxygen concentration in water. *J. Fish. Res. Board Canada* **1974**, *31*, 1529–1530. [CrossRef]
23. Miura, A.S.; Parra, M.; Lloret, J.; Rodilla, M. LED optical sensor prototype to determine dissolved oxygen saturation in water. In Proceedings of the 2021 Global Congress on Electrical Engineering, Valencia, Spain, 10–12 December 2021; pp. 115–120. [CrossRef]
24. Parra, L.; Viciano-Tudela, S.; Carrasco, D.; Sendra, S.; Lloret, J. Low-cost microcontroller-based multiparametric probe for coastal area monitoring. *Sensors* **2023**, *23*, 1871. [CrossRef] [PubMed]
25. Narayana, B.; Mathew, M.; Cherian, T.; Pasha, C. Spectrophotometric determination of dissolved oxygen in water. *Indian J. Chem. Technol.* **2005**, *12*, 593–595.
26. Penso, C.M.; Rocha, J.L.; Martins, M.S.; Sousa, P.J.; Pinto, V.C.; Minas, G.; Silva, M.M.; Goncalves, L.M. PtOEP–PDMS-based optical oxygen sensor. *Sensors* **2021**, *21*, 5645. [CrossRef] [PubMed]
27. Mahoney, E.J.; Hsu, H.H.L.; Du, F.; Xiong, B.; Selvaganapathy, P.R.; Fang, Q. Optofluidic Dissolved Oxygen Sensing with Sensitivity Enhancement Through Multiple Reflections. *IEEE Sens. J.* **2019**, *19*, 10452–10460. [CrossRef]
28. Pittaluga, F.; Aleffi, I.F.; Bettoso, N.; Blasutto, O.; Celio, M.; Codarin, A.; Cumani, F.; Faresi, L.; Guiatti, D.; Orlandi, C.; et al. The SHAPE Project: An Innovative Approach to Understanding Seasonal and Diel Dissolved Oxygen Dynamics in the Marano and Grado Lagoon (Adriatic Sea) under the WFD/2000/60/CE. *J. Mar. Sci. Eng.* **2022**, *10*, 208. [CrossRef]
29. Tholen, C.; Parnum, I.; Rofallski, R.; Nolle, L.; Zielinski, O. Investigation of the Spatio-Temporal Behaviour of Submarine Groundwater Discharge Using a Low-Cost Multi-Sensor-Platform. *J. Mar. Sci. Eng.* **2021**, *9*, 802. [CrossRef]
30. Oliveira, V.H.; Sousa, M.C.; Morgado, F.; Dias, J.M. Modeling the Impact of Extreme River Discharge on the Nutrient Dynamics and Dissolved Oxygen in Two Adjacent Estuaries (Portugal). *J. Mar. Sci. Eng.* **2019**, *7*, 412. [CrossRef]
31. Barnstead Smart2Pure. Available online: https://es.vwr.com/store/product/13527973/ultrapure-water-systems-barnsteadtm-smart2puretm (accessed on 22 June 2022).
32. Apera Instruments. Available online: https://aperainst.com/do8500-portable-optical-dissolved-oxygen-meter-kit (accessed on 22 June 2022).
33. LLG Labware. Available online: https://shop.llg-labware.com/info1297-spektralphotometer-llg-unispec-2-und_lang_DE.htm?UID=5500633accf1000000000000 (accessed on 3 October 2022).
34. Patel, H.; Vashi, R.T. Chapter 2—Characterization of Textile Wastewater. In *Characterization and Treatment of Textile Wastewater*; Elsevier: Amsterdam, The Netherlands, 2015; pp. 21–71. [CrossRef]
35. Statgraphics Centurion XVIII. Available online: https://www.statgraphics.com/download (accessed on 15 November 2022).
36. Rocher, J.; Parra, L.; Jimenez, J.M.; Lloret, J.; Basterrechea, D.A. Development of a Low-Cost Optical Sensor to Detect Eutrophication in Irrigation Reservoirs. *Sensors* **2021**, *21*, 7637. [CrossRef]

photonics

Article

The Determination of the Sensitivity of Refractive Index Sensors

Ilya M. Efimov *, Nikolay A. Vanyushkin and Ashot H. Gevorgyan

Institute of High Technologies and Advanced Materials, Far Eastern Federal University, 10 Ajax Bay, Russky Island, 690922 Vladivostok, Russia; vanyushkin.na@dvfu.ru (N.A.V.); gevorgyan.aa@dvfu.ru (A.H.G.)
* Correspondence: efimov.im@dvfu.ru; Tel.: +7-914-666-82-00

Abstract: A new approach to determining the sensitivity of refractive index sensors is proposed. It has been shown that relative and absolute sensitivity show different results, and also, for the first time, it is demonstrated that relative sensitivity has advantages over absolute sensitivity. In addition, the influence of the relative width of the photonic band gap and the difference in the refractive indices of the layers on the sensitivity are examined and the corresponding dependences of these parameters are obtained. We propose these parameters as a convenient tool for optimizing the sensitivity of sensors based on defective photonic crystals. Finally, results are obtained regarding the behavior of the defect mode at the center of the photonic band gap of one-dimensional photonic crystals.

Keywords: sensitivity; refractive index sensors; photonic crystals

1. Introduction

With the advent of optical methods of substance research, humanity has entered a new era of optics and photonics, and optical sensors are under active development in science and industry. Optical sensors are instruments that use the optical properties of an object to measure various parameters. In various fields such as manufacturing, robotics, medicine, and even home appliances, optical sensors are already widely used. They may be active, in which case they generate their own light and measure their reflection, transmission, or absorption, or passive, in which case they detect only light from other sources [1–4]. Research in the field of optical sensors is of great importance because the use of optical sensors has a number of advantages over other methods of analyzing matter; in particular, they are a non-contact method and have high accuracy [5]. Of course, there are also drawbacks, including certain difficulties associated with the unambiguous interpretation of signals [5].

This is the reason for a great number of papers written in the field of optical sensors, in particular refractive index sensors [6–37], aiming to find solutions to the following issues: extending their operating range, improving stability and reliability, decreasing costs, and increasing the availability, resolution, and sensitivity of the sensors. In particular, works [11–37] study the dependence of the sensitivity of refractive index sensors on various parameters. Works [11–13] use different operating ranges to increase sensitivity and also carry out comparative analyses of identical structures in different ranges. Works [14–26] focus on the effect of different types of sensor structures on the sensitivity of optical devices. The sensitivity of optical devices depends on the properties of the object under investigation; thus, the sensitivity of the same sensor can be different when used for different objects. The influence of such parameters is considered in works [11,13,24–33]. Works [16,22,31–37] investigate the influence of incidence angle on sensitivity.

One can see that sensitivity is widely used in the comparative analysis given in these articles and is often presented as a final parameter in order to demonstrate the advantage of a new sensor with respect to sensors with lower sensitivity. The present paper introduces an alternative approach to determining sensitivity, taking into account the resolution of

Citation: Efimov, I.M.; Vanyushkin, N.A.; Gevorgyan, A.H. The Determination of the Sensitivity of Refractive Index Sensors. *Photonics* **2024**, *11*, 56. https://doi.org/10.3390/photonics11010056

Received: 16 November 2023
Revised: 20 December 2023
Accepted: 2 January 2024
Published: 4 January 2024

the sensor. The question of which parameters should be considered when optimizing sensitivity is also considered.

Sensitivity is used to characterize refractive index sensors in most works [6–37]. In our work, we will consider this parameter using a photonic sensor based on a 1D photonic crystal (PC) with a defect layer (DL) as an example. A PC [38–41] is a periodic structure of layers with different refractive indices. PCs have a unique property of having a certain range of frequencies, called a photonic bandgap (PBG), in which electromagnetic waves cannot propagate through the PC. If a DL is added to the periodic structure of the PC, the periodicity of the structure is violated, which leads to changes in the transmission and reflection spectra in the entire region. In particular, this is represented by the appearance of a narrow band within the PBG, called the defect mode (DM), which is highly sensitive to changes in the DL parameters and is used for sensing.

2. Theory

2.1. Model of the Refractive Index Sensor

Our model of a refractive index sensor consists of two ideal PCs with a DL between them. Each PC consists of N periodic unit cells, where each cell is a pair of layers with thickness $h_{1,2}$ and refractive index $n_{1,2}$.

Between the two PCs, a DL is located, and in our model, it is the refractive index of the DL that changes under external influence. The DL in the PC violates its periodic structure and changes the light transmission properties, and DMs appear. The position of the DM depends on the many physical properties of the DL, such as the refractive index and the thickness of the DL.

In our case, the ideal PCs are mirrored with respect to the DL, with the DL bordering the layer that has the higher refractive index. It was shown in [25,26] that this configuration provides the highest sensitivity of the sensor and field localization in the DL.

Figure 1 shows a schematic diagram of our structure.

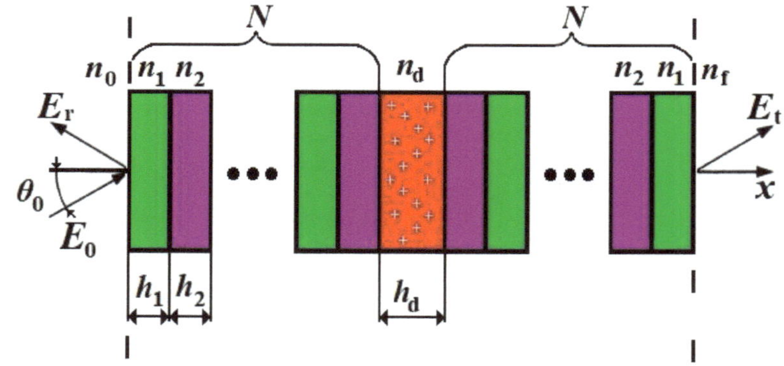

Figure 1. Schematic diagram of the structure under investigation. N is the number of unit cells, θ_0 is the angle of incidence, E_r is the reflected wave, E_0 is the incident wave, E_t is the transmitted wave, and n_0 and n_f are refractive indices of the structure surroundings.

In this work, we used the transfer matrix method [42–45] to calculate the transmission and reflection spectra of the investigated structure. The transfer matrix for the j-th layer in the structure can be written as:

$$M_j = \begin{pmatrix} \cos k_j\, h_j & \frac{-i}{p_j}\sin k_j\, h_j \\ -ip_j \sin k_j\, h_j & \cos k_j\, h_j \end{pmatrix}, \quad (1)$$

where $k_j = \frac{2\pi}{\lambda} n_j \cos\theta_j$, θ_j is the angle of refractive in the j-th layer, which is determined from Snell's law as: $\theta_j = \cos^{-1}\sqrt{1 - \frac{n_0^2 \sin^2\theta_0}{n_j^2}}$, $p_j = n_j \cos\theta_j$, n_0 is the refractive index of the external medium from, where the wave is incident, and θ_0 is angle of incidence.

Then, the matrix M of a unit cell in the periodic part of the structure is obtained by successive multiplication of the matrices M_j $(j = 1, 2)$ of the layers contained in the cell, and finally, the transfer matrix of the whole structure has the form:

$$m = (M_1 M_2)^N M_d (M_2 M_1)^N = \begin{pmatrix} m_{11} & m_{12} \\ m_{21} & m_{22} \end{pmatrix}, \tag{2}$$

where M_d is the transfer matrix of the DL.

The transmission coefficient has the following form:

$$t = \frac{2p_0}{(m_{11} + p_f m_{12}) p_0 + (m_{21} + p_f m_{22})}, \tag{3}$$

and the energy transmittance has the form:

$$T = \frac{p_f}{p_0} |t|^2, \tag{4}$$

where the indices 0, f denote the corresponding parameters of the medium bordering the PC on the left and right, respectively.

2.2. Principal Operating Mechanism of the Sensor

The refractive index of the DL can be obtained from the model as follows. To complete this, let us look at the transmission spectra in Figure 2. The black arrow shows the defect mode shift by $d\lambda$ when the refractive index of the defect layer changes, while all other parameters are the same. This shift can be used to determine the change in the DL refractive index n_d. This is the basis of sensor operation. In particular, the main idea of sensitivity is to show how much the DM shifts at the same change of the refractive index.

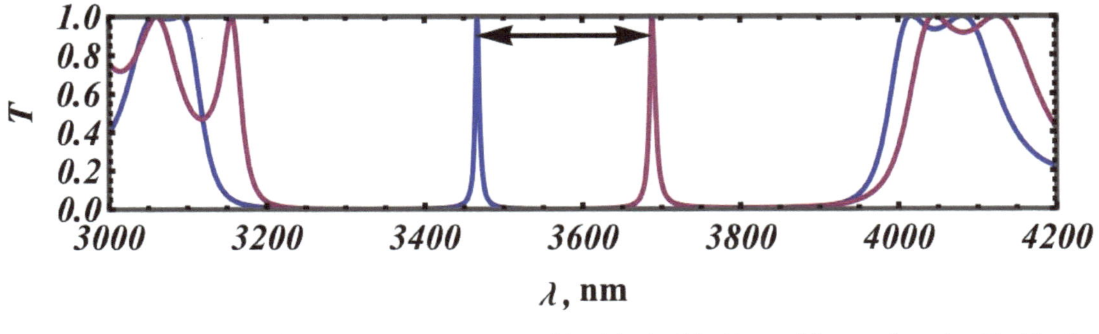

Figure 2. Transmission spectra of the defective PC with two different values of n_d. The blue line is the spectrum for a pure defective layer, and the red line is for a defective layer with inclusions.

2.3. Absolute and Relative Sensitivities

As we have already mentioned, in most of the works [9–37] devoted to refractive index sensors, a parameter such as sensitivity is used to characterize them. In the present work, we will call such sensitivity the absolute sensitivity, and in this case, the absolute sensitivity will be defined as follows:

$$S_a = \frac{d\lambda}{dn}, \tag{5}$$

Here, dn is the change in the refractive index of the medium and dλ is the spectrum or mode shift that occurs due to the change in the refractive index.

The absolute sensitivity S_a of refractive index sensors is an important characteristic that determines how well the sensor can detect a change in the refractive index of interest. However, absolute sensitivity does not take into account the operating wavelength range and has a length dimension such as nm.

There is, however, a less commonly used analogue of absolute sensitivity. This is relative (or dimensionless) sensitivity S_r (see, in particular, [11,12,33]):

$$S_r = \frac{1}{\lambda_0}\frac{d\lambda}{dn}, \tag{5}$$

where λ_0 is the working wavelength.

The relative sensitivity S_r, as well as the absolute sensitivity S_a, represents the shift in the DM position with the change in the refractive index of the DL. In addition, the relative sensitivity takes into account the working range of the wavelengths and is a dimensionless quantity.

For the frequency ν range, the same rules apply:

$$S_a = \frac{d\nu}{dn}, \; S_r = \frac{1}{\nu_0}\frac{d\nu}{dn}, \tag{7}$$

From the works [11–37], it can be seen that the relative and absolute sensitivities do not agree with each other. In some cases, when comparing two sensors, a higher absolute sensitivity can correspond to a lower relative sensitivity and vice versa. In particular, Table 1 shows the comparative characteristics of the works [11–16].

Table 1. Comparative table of sensitivity at different wavelengths [11–16].

S_a, $\frac{nm}{RIU}$	λ_0, nm	S_r, RIU^{-1}	Article
1020	5293	0.1927	[11]
80	5295	0.0151	
347	1004	0.3456	
710	2026	0.3504	
260	1560	0.1667	[14]
197	1600	0.1231	
198	1620	0.1222	
173	1640	0.1055	
80	1800	0.0444	
5018	7299	0.6875	[15]
5092	7299	0.6977	
5031	7293	0.6899	
5013	7335	0.6834	
500	523	0.9560	[27]
496	533	0.9316	
490	541	0.9058	
487	547	0.8897	
475	557	0.8520	
454	577	0.7875	
405	611	0.6639	
145	764	0.1891	[28]
144	766	0.1881	
144	769	0.1879	
144	773	0.1863	
1300	1530	0.8497	[29]
515	1550	0.3322	

When looking at Table 1, it can be seen that within a single article, there is usually an agreement between the absolute and relative sensitivity, which is explained by the fact that within a single article, the authors try to stay within approximately the same wavelength range, but when considering different operating wavelengths, the situation changes. Thus, in most papers [9–37], it can be seen that the sensitivities of different structures at different operating wavelengths are compared with those from other articles. According to these comparative characteristics, conclusions are drawn about the performance of the structures, which are not always correct.

There is also another factor that should be taken into account, which is not present in the use of absolute sensitivity. We can conclude that in practice, not every shift $\mathrm{d}\lambda$ and therefore not every change $\mathrm{d}n$ in the refractive index can be detected when solving the inverse problem of finding the unknown refractive index from a measured spectrum. We need to consider the resolving power of the optical instrument to determine the minimum detectable $\mathrm{d}\lambda$ shift.

2.4. Optical Sensor Resolution

The resolution R of an optical sensor determines its ability to discriminate between small changes in the optical signal. Let us look at the resolving power of an optical spectrometer. The resolving power of a spectrometer determines its ability to separate closely spaced spectral lines. In other words, to distinguish details in the spectrum with high accuracy. For example, we have two peak wavelengths λ and $\lambda + \mathrm{d}\lambda$, and they can be separated (or resolved) with a resolution of at least $R = \lambda/\mathrm{d}\lambda$; otherwise, they merge [46–50].

The resolution is a dimensionless quantity [46–50] and in this paper, we will consider the following situation. Consider two working wavelengths of 1000 and 2000 nm. Let the same resolving power for the two wavelengths be given and let be $R = 1000$. This means that it is possible to resolve peaks $\mathrm{d}\lambda = 1$ and 2 nm, correspondingly, and distinguish peaks with the same wavelength difference $\mathrm{d}\lambda$ at different working wavelengths λ_0.

3. Results and Discussion

3.1. Comparing Absolute to Relative Sensitivity

The main idea of this work is to show that the absolute sensitivity S_a is not a reliable parameter when evaluating sensor performance. This sensitivity depends proportionally on the operating wavelength and does not correctly reflect the sensor parameters.

To illustrate this fact, let us start with a numerical experiment. Let us consider the following 1D PC shown in Figure 1. The parameters of the model are as follows: $n_1 = 1.5$, $n_2 = 2.0$, $n_0 = n_\mathrm{f} = 1$, $N = 10$, $\theta_0 = 0$, h_1 and h_2 take the values $h_1 = h_2 = 500\,k$ nm, and $h_d = 1050\,k$ nm, where k is the proportionality factor of the structure. It should be noted that the dispersion of refractive indices is not considered in this work because it affects sensitivity and complicates the analysis of the results. Since the working wavelengths of the spectra considered are different, dispersion can make a significant difference in the PC refractive indices' values and thus the sensitivity of such a sensor, so this factor has been excluded for the clarity of the results. Otherwise, for arbitrary realistic refractive indices for PC, the conclusions of the model are valid. With these parameters, the following spectra can be obtained, as shown in the series of plots in Figure 3.

Such a large difference in refractive indices was chosen to clearly illustrate the DM shift. In the above series of spectra, the first PBG is shown whose center wavelength λ_B can be obtained from the Bragg condition:

$$\lambda_\mathrm{B} = 2\Lambda n_\mathrm{m}, \tag{8}$$

where $\Lambda = h_1 + h_2$ is the unit cell period of the PC and $n_\mathrm{m} = \frac{n_1 h_1 + n_2 h_2}{h_1 + h_2}$ is the average refractive index of the unit cell.

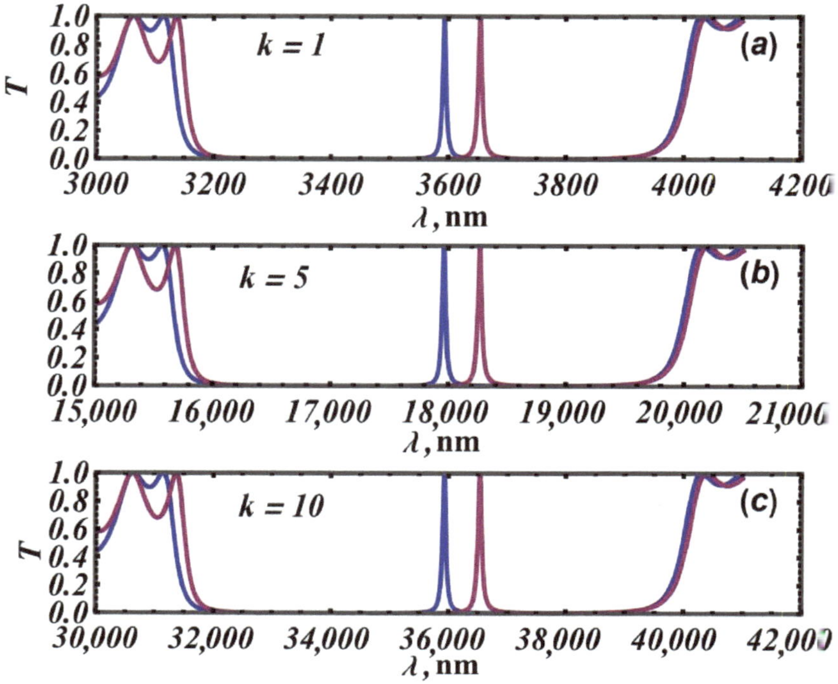

Figure 3. Transmission spectra at $n_d = 1.75$ and 1.90 (blue and red lines). The structure parameters are $n_1 = 1.5$, $n_2 = 2.0$, $N = 10$, $h_1 = h_2 = 500\ k$ nm, and $h_d = 1050\ k$ nm, (**a**) $k = 1$, (**b**) $k = 5$, (**c**) $k = 10$.

From the series of spectra in Figure 3 and Formula (8), it can be seen that the position of the first PBG is proportional to k. Also, it can be seen that the PBG widens; so, for $k = 1$, the width of the PBG is about 800 nm, and for $k = 10$, it is already 8000 nm. The shift in the DM also increases with increasing k for the same change in the refractive index of the DL. This directly affects the sensitivity S_a. Let us take a closer look at this influence in Table 2.

Table 2. Comparative values of different structures.

k	λ_B, nm	S_a, $\frac{\text{nm}}{\text{RIU}}$	S_r, RIU^{-1}
1	3500	440.33	
2	7000	880.65	
3	10,500	1320.98	0.1258074
4	14,000	1761.30	
5	17,500	2201.63	
10	35,000	4403.26	

It can be seen that the sensitivity S_a increases proportionally as k increases, and this is logical because as k increases, the shift in DM dλ increases (the same change in the refractive index of the DL), while the relative sensitivity remains constant. According to the results of such a numerical experiment, it can be said that the two sensitivities produce different results. This can be seen when different operating ranges are considered in the same paper and their sensitivities are compared, or when different papers consider different operating ranges and compare their sensitivities. Furthermore, the definition of absolute sensitivity does not take into account the effect of the resolving power of the sensor. For this purpose, let us now consider the resolving power of the device.

As mentioned in Section 2.4, the higher the resolving power, the easier it is to determine the shift of the DM peak. Let us now consider a sensor with resolving power of $R = 1000$.

With this resolution it will be possible to distinguish a DM peak shift of 3.5 nm at a wavelength of 3500 nm and a DM peak shift of 7.0 nm at a wavelength of 7000 nm. Taking this into account, let us try to determine the minimum detectable change in the refractive index of the DL at constant resolving power; a comparison is given in Table 3.

Table 3. Comparative values of different structures at $R = 1000$.

k	S_a, $\frac{nm}{RIU}$	S_a, $\frac{THz}{RIU}$	S_r, RIU^{-1}	Min $\Delta\lambda$,nm	Δn_d
1	440.33	680.83	0.1258074	3.5	0.002
2	880.65	340.42		7.0	
5	2201.63	136.17		17.5	
10	4403.26	68.08		35.0	

As can be seen at a constant resolving power, structures with a different absolute sensitivity and the same relative sensitivity have the same ability to detect the change in the refractive index of the DL. We note another advantage of relative sensitivity, which is that the numerical relative sensitivity does not differ when the frequency or wavelength spectrum is considered. This result allows us to state that the relative sensitivity is a correct reflection of the structural properties, whereas the absolute sensitivity can be misleading.

3.2. *Convenient Parameters for Optimizing the Sensitivity of a Sensor Based on 1D PCs with Defects*

In this paper, it is also proposed to consider the influence of the relative and absolute PBG width on sensitivity ($\Delta\lambda/\lambda_B$ and $\Delta\lambda$). These parameters are affected by the large number of parameters of the structure. In particular, the PBG width can be varied by changing the layer thicknesses ratio h_1/h_2 with constant $n_1 = 1.5$ and $n_2 = 2.0$ and constant unit cell period $\Lambda = 1000$ nm. When this is carried out, the mean refractive index n_m of the unit cell changes and the width of the PBG changes correspondingly. The relative width of the PBG differs from the absolute one in that it is divided by the wavelength λ_B of the PBG center (see [44]). Figure 4 shows the dependence of the absolute sensitivity on the width of the PBG and that of the relative sensitivity on the relative width of the PBG for different numbers of unit cells N.

In the series in Figure 4, the green arrow shows the direction of increase in h_1 at the constant period of the unit cell of the PC. It can be seen that as the ratio h_1 to h_2 increases, the sensitivity increases and then starts to decrease. The dependence of the width of the PBG on the ratio h_1 to h_2 is considered in more detail in [51]. The red line marks the maximum PBG width, which corresponds to the quarter-wave stack $n_1h_1 = n_2h_2 = \lambda_B/4$. After the maximum point of the PBG width, a decrease in this width can be observed when the ratio h_1/h_2 is further increased. At the same time, the sensitivity decreases. It can also be observed in the series in Figure 4 that for the first and second ratio h_1/h_2, the sensitivity values are different. However, one can see that these two sensitivities approach the same value as the number of periods increases. It can be assumed that in the case of an infinite PC with a defect, the two parts of the curve merge completely and the two sensitivities become equal.

It is important to note that the sensitivity values presented in the series in Figure 4 were obtained for the first DM mode; it can be shown that for other DMs, the general dependencies were preserved but shifted up. By changing the structural parameters for the series in Figure 4, the position of the center of the PBG λ_B was changed. Therefore, the thickness of the DL h_d was adjusted according to the center of the PBG so that the position of the DM λ_{DM} remained at the center of the PBG.

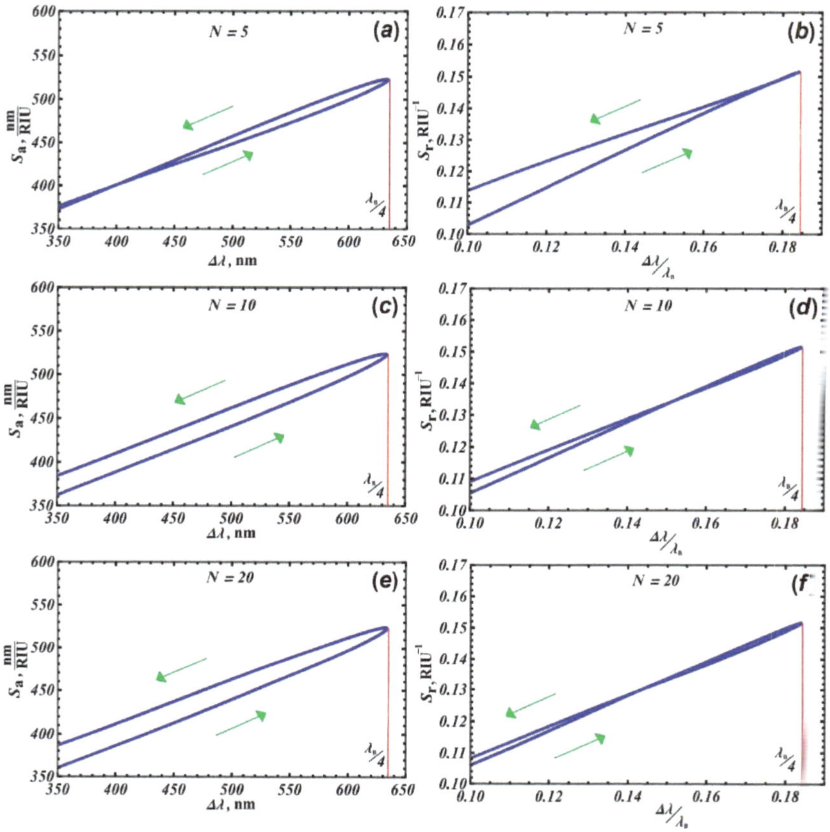

Figure 4. Dependence of absolute sensitivity on the width of the PBG (**a**,**c**,**e**) and that of the relative sensitivity on the relative width of the PBG (**b**,**d**,**f**). Here, $d\lambda$ is varied by changing h_1. The green arrows indicate the increase in h_1 over a constant period. The red line marks the fulfilment of the quarter-wave stack. λ_B is the wavelength of the PBG center. The other parameters are the same as in Figure 3a.

In this paper, we argue that the width of the PBG, in particular the relative width of the PBG, is a convenient parameter for optimizing sensors based on 1D defect PCs. One of the main conclusions of this work is the following: for the relative sensitivity, we can unambiguously state that the maximum sensitivity is reached at the maximum PBG width. In particular, using this conclusion, it can be argued that to optimize a sensor based on a defective PC, the maximum of the relative width of the PBG should be aimed for. Thus, for the first PBG, it is necessary to use the quarter-wave stack.

Next, we will consider the dependence of sensitivity on the difference between the refractive indices $\Delta n = |n_1 - n_2|$, i.e., optical contrast, with a fixed mean refractive index.

Figure 5 shows the dependence of the relative PBG width on the refractive index difference (a) and the dependence of the relative sensitivity on the refractive index difference (b) under the following conditions: $n_m = const$ and $h_1 n_1 = h_2 n_2$.

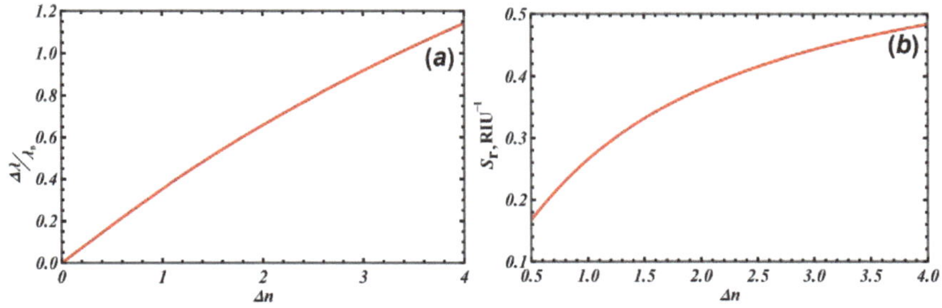

Figure 5. Dependence of the relative width of the PBG (**a**) and relative sensitivity (**b**) on the refractive index difference. The parameters are as follows: $n_m = 1.75$ and $h_1 n_1 = h_2 n_2 = \lambda_B/4$. The other parameters are the same as in Figure 3a.

Figure 5a shows that by increasing the refractive index difference of the unit cell layers, the relative bandwidth becomes larger, thus increasing the relative sensitivity. However, several interesting features should be noted. The dependence of the PBG width on Δn is not a linear function, but it is close to linear for sufficiently small values of Δn. In Figure 5b, we can see that the dependence of the relative sensitivity is correlated with the increase in the PBG width, but the width increases almost linearly, while the growth of the relative sensitivity slows down significantly and tends to a constant value (this is shown by our calculations).

When calculating the width of the PBG, the thickness of the DL is not taken into account as the calculation is for an infinite PC. However, to calculate the sensitivity, we need a certain thickness for the DL. The thickness of the DL was determined in the same way as in the series in Figure 4, i.e., the search is for such a thickness for the DL that the DM will be in the center of the PBG. In the case of the series in Figure 4, the values of the DL thickness were close to each other but still different.

When the thickness of the DL was calculated for Figure 5b, it was found that the thickness of the DL remained constant, i.e., the position of the DM depends only indirectly on h_1, n_1, h_2, n_2 the position of the DM depends primarily on the cell period Λ, the mean refractive index n_m, the parameters of the DL itself, the thickness h_d, and the refractive index of the DL n_d.

Figure 6 shows the evolution of the difference between the DM position and PBG center, with the change in the DL thickness h_d and the optical contrast Δn.

The position of the DM depends primarily on the cell period Λ, the mean refractive index n_m, the parameters of the DL itself, the thickness h_d, and the refractive index of the DL n_d. Thus, in Figure 6, it can be observed that the position of the DM does not shift when the refractive index difference Δn changes, as can be seen by the dark blue line in the center. This means that for the DM to be in the center of the PBG, the thickness of the DL h_d must take on a certain constant value, even when Δn changes.

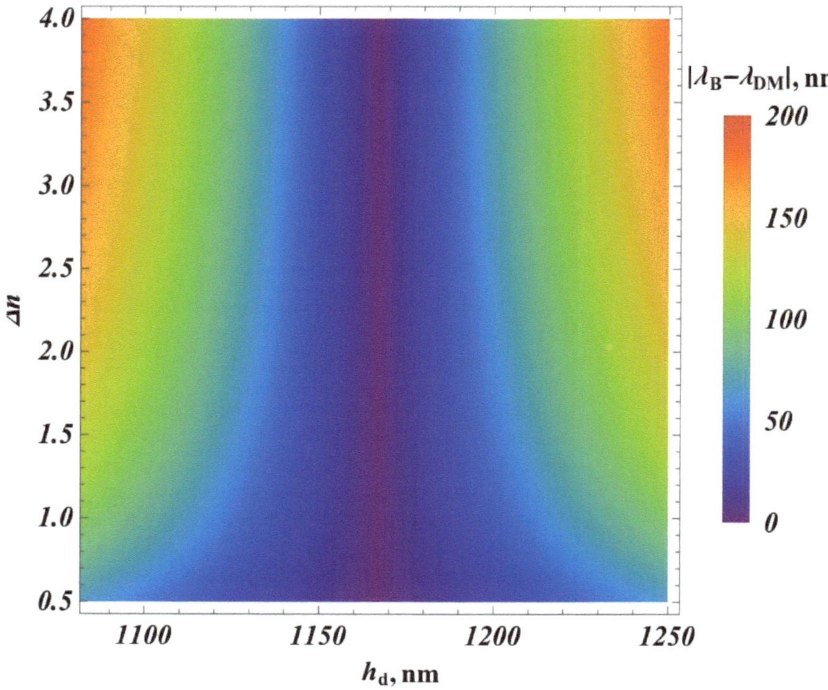

Figure 6. Dependence of the difference in the DM position λ_{DM} and the PBG center λ_B on the DL thickness h_d and the refractive index difference Δn. The parameters are the same as in Figure 5b.

4. Conclusions

In conclusion, we considered two types of sensitivity: relative and absolute. For the same structures, these two types of sensitivity were found to give different results. It was determined which parameters, such as operating wavelength and resolution, are also necessary for a correct evaluation; as a consequence, it has been shown that it is not correct to use absolute sensitivity in comparative evaluations of refractive index sensors. On the other hand, relative sensitivity has a number of advantages over its counterpart, and it has also been shown that it is relative sensitivity that is worth using in the comparative characterization of different structures at different operating wavelengths.

The effect of the width of the PBG on relative sensitivity has been shown. The maximum relative sensitivity was found to be at the maximum of the relative width of the PBG, i.e., at the quarter-wave stack. The dependence of the sensitivity on the refractive index difference was also studied. In this case, an increase in sensitivity is observed as the width of the PBG increases. However, the width increases almost linearly, whereas the increase in relative sensitivity slows down significantly and tends to a constant value.

In addition to the parameters of the DL itself, such as the thickness h_d and the refractive index of the DL n_d, the position of the DM in the center of the PBG was found to depend directly on the mean refractive index n_m and the cell period Λ.

Author Contributions: Conceptualization I.M.E., N.A.V. and A.H.G.; formal analysis N.A.V.; software I.M.E.; writing—original draft preparation I.M.E.; writing—review and editing I.M.E., N.A.V. and A.H.G.; project administration A.H.G. All authors have read and agreed to the published version of the manuscript.

Funding: This work was supported by the Foundation for the Advancement of Theoretical Physics and Mathematics "BASIS" (grant No. 21-1-1-6-1).

Institutional Review Board Statement: Not applicable.

Informed Consent Statement: Not applicable.

Data Availability Statement: Data are contained within the article.

Conflicts of Interest: The authors declare no conflicts of interest.

References

1. Santos, J.L.; Faramarz, F. *Handbook of Optical Sensors*; CRC Press: Boca Raton, FL, USA, 2014.
2. Ferreira, M.F.; Castro-Camus, E.; Ottaway, D.J.; López-Higuera, J.M.; Feng, X.; Jin, W.; Jeong, Y.; Picqué, N.; Tong, L.; Reinhard, B.M.; et al. Roadmap on optical sensors. *J. Opt.* **2017**, *19*, 083001. [CrossRef] [PubMed]
3. Tong, L. Micro/nanofibre optical sensors: Challenges and prospects. *Sensors* **2018**, *18*, 903. [CrossRef] [PubMed]
4. Kersey, A.D. A review of recent developments in fiber optic sensor technology. *Opt. Fiber Technol.* **1996**, *2*, 291–317. [CrossRef]
5. Grattan, K.T.; Sun, T. Fiber optic sensor technology: An overview. *Sens. Actuators A Phys.* **2000**, *82*, 40–61. [CrossRef]
6. Ma, S.; Xu, Y.; Pang, Y.; Zhao, X.; Li, Y.; Qin, Z.; Liu, Z.; Lu, P.; Bao, X. Optical Fiber Sensors for High-Temperature Monitoring: A Review. *Sensors* **2022**, *22*, 5722. [CrossRef] [PubMed]
7. Wang, Z.; Wang, G.; Kumar, S.; Marques, C.; Min, R.; Li, X. Recent advancements in resonant fiber optic gyro—A review. *IEEE Sens. J.* **2022**, *22*, 3195502. [CrossRef]
8. Chai, H.; Zheng, Z.; Liu, K.; Xu, J.; Wu, K.; Luo, Y.; Liao, H.; Debliquy, M.; Zhang, C. Stability of metal oxide semiconductor gas sensors: A review. *IEEE Sens. J.* **2022**, *22*, 5470–5481. [CrossRef]
9. Li, J.; Huang, X.; Tu, L.; Zhang, T.; Wang, L. A review of building detection from very high resolution optical remote sensing images. *GISci. Remote Sens.* **2022**, *59*, 1199–1225. [CrossRef]
10. Jayawickrema, U.M.N.; Herath, H.M.; Hettiarachchi, N.K.; Sooriyaarachchi, H.P.; Epaarachchi, J.A. Fibre-optic sensor and deep learning-based structural health monitoring systems for civil structures: A review. *Measurement* **2022**, *199*, 111543. [CrossRef]
11. Efimov, I.M.; Vanyushkin, N.A.; Gevorgyan, A.H.; Golik, S.S. Optical biosensor based on a photonic crystal with a defective layer designed to determine the concentration of SARS-CoV-2 in water. *Phys. Scr.* **2022**, *97*, 055506. [CrossRef]
12. Luan, E.; Shoman, H.; Ratner, D.M.; Cheung, K.C.; Chrostowski, L. Silicon Photonic Biosensors Using Label-Free Detection. *Sensors* **2018**, *18*, 3519. [CrossRef] [PubMed]
13. Dutta, H.S.; Goyal, A.K.; Srivastava, V.; Pal, S. Coupling light in photonic crystal waveguides: A review. *Photonics Nanostruct. Fundam. Appl.* **2016**, *20*, 41–58. [CrossRef]
14. Dutta, H.S.; Suchandan, P. Design of a highly sensitive photonic crystal waveguide platform for refractive index based biosensing. *Opt. Quantum Electron.* **2013**, *45*, 907–917. [CrossRef]
15. Ahmed, A.M.; Ahmed, M. Ultra-high sensitive 1D porous silicon photonic crystal sensor based on the coupling of Tamm/Fano resonances in the mid-infrared region. *Sci. Rep.* **2019**, *9*, 6973. [CrossRef] [PubMed]
16. Abdol, S.O.; Babak, A. Novel biosensors based on Weyl semimetals. *Phys. Scr.* **2022**, *97*, 125502. [CrossRef]
17. El Beheiry, M.; Liu, V.; Fan, S.; Levi, O. Sensitivity enhancement in photonic crystal slab biosensors. *Opt. Express* **2010**, *18*, 22702–22714. [CrossRef] [PubMed]
18. Wu, F.; Fan, C.; Zhu, K.; Wu, J.; Qi, X.; Sun, Y.; Xiao, S.; Jiang, H.; Chen, H. Tailoring electromagnetic responses in a coupled-grating system with combined modulation of near-field and far-field couplings. *Phys. Rev. B* **2022**, *105*, 245417. [CrossRef]
19. Bijalwan, A.; Bipin, K.S. Analysis of one-dimensional photonic crystal based sensor for detection of blood plasma and cancer cells. *Optik* **2021**, *226*, 165994. [CrossRef]
20. Dinodiya, S.; Anami, B. Biosensor Based on One-Dimensional Photonic Crystal for Poliovirus Detection. In *Advancement in Materials, Manufacturing and Energy Engineering*; Springer Nature: Berlin, Germany, 2021; Volume 1.
21. Miyan, H.; Agrahari, R.; Gowre, S.K.; Mahto, M.; Jain, P.K. Computational study of a compact and high sensitive photonic crystal for cancer cells detection. *IEEE Sens. J.* **2022**, *22*, 3298–3305. [CrossRef]
22. Oskoui, A.; Shojaei, S.S.; Abdollahipour, B. Polarization dependent light propagation in WTe 2 multilayer structure. *Sci. Rep.* **2023**, *13*, 13169. [CrossRef]
23. Suthar, B.; Anami, B. Enhanced optical sensor for waterborne bacteria-based photonic crystal using graded thickness index. *Appl. Nanosci.* **2023**, *13*, 5399–5406. [CrossRef]
24. Olyaee, S.; Samira, N. A high-quality factor and wide measurement range biosensor based on photonic crystal nanocavity resonator. *Sens. Lett.* **2013**, *11*, 483–488. [CrossRef]
25. Efimov, I.M.; Vanyushkin, N.A.; Golik, S.S.; Gevorgyan, A.H. Sensor with enhanced performance based on photonic crystal with a defect layer. *Comput. Opt.* **2023**, *47*, 572–579. [CrossRef]
26. Efimov, I.M.; Vanyushkin, N.A.; Gevorgyan, A.H. Peculiarities of the Electromagnetic Field Distribution Inside a 1D Photonic Crystal with a Defect Layer. *Bull. Russ. Acad. Sci. Phys.* **2022**, *86* (Suppl. S1), S60–S65. [CrossRef]
27. Al-Dossari, M.; Awasthi, S.K.; Mohamed, A.M.; Abd El-Gawaad, N.S.; Sabra, W.; Aly, A.H. Bio-alcohol sensor based on one-dimensional photonic crystals for detection of organic materials in wastewater. *Materials* **2022**, *15*, 4012. [CrossRef] [PubMed]
28. Edappadikkunnummal, S.; Chembra Vasudevan, R.; Dinesh, S.; Thomas, S.; Desai, N.R.; Kaniyarakkal, S. Detection of Hemoglobin Concentration Based on Defective One-Dimensional Photonic Crystals. *Photonics* **2022**, *9*, 660. [CrossRef]

29. Hu, J.; Daoxin, D. Cascaded-ring optical sensor with enhanced sensitivity by using suspended Si-nanowires. *IEEE Photonics Technol. Lett.* **2011**, *23*, 842–844.
30. Baraket, Z.; Osswa, S.; Mounir, K. Design of magnetic field direction's sensor based on a 1D tunable magneto-photonic crystal. *Opt. Quantum Electron.* **2022**, *54*, 637. [CrossRef]
31. Daher, M.G.; Jaroszewicz, Z.; Zyoud, S.H.; Panda, A.; Hasane Ahammad, S.K.; Abd-Elnaby, M.; Eid, M.M.; Rashed, A.N. Design of a novel detector based on photonic crystal nanostructure for ultra-high performance detection of cells with diabetes. *Opt. Quantum Electron.* **2022**, *54*, 701. [CrossRef]
32. Almawgani, A.H.; Daher, M.G.; Taya, S.A.; Mashagbeh, M.; Colak, I. Optical detection of fat concentration in milk using MXene-based surface plasmon resonance structure. *Biosensors* **2022**, *12*, 535. [CrossRef]
33. Yupapin, P.; Trabelsi, Y.; Vigneswaran, D.; Taya, S.A.; Daher, M.G.; Colak, I. Ultra-high-sensitive sensor based on surface plasmon resonance structure having Si and graphene layers for the detection of chikungunya virus. *Plasmonics* **2022**, *17*, 1315–1321. [CrossRef]
34. Almawgani, A.H.; Daher, M.G.; Taya, S.A.; Colak, I.; Patel, S.K.; Ramahi, O.M. Highly sensitive nano-biosensor based on a binary photonic crystal for cancer cell detection. *Opt. Quantum Electron.* **2022**, *54*, 554. [CrossRef]
35. Taya, S.A.; Daher, M.G.; Colak, I.; Ramahi, O.M. Highly sensitive nano-sensor based on a binary photonic crystal for the detection of mycobacterium tuberculosis bacteria. *J. Mater. Sci. Mater. Electron.* **2021**, *32*, 28406–28416. [CrossRef]
36. Panda, A.; Pukhrambam, P.D.; Wu, F.; Belhadj, W. Graphene-based 1D defective photonic crystal biosensor for real-time detection of cancer cells. *Eur. Phys. J. Plus* **2021**, *136*, 809. [CrossRef]
37. Khani, S.; Hayati, M. Optical biosensors using plasmonic and photonic crystal band-gap structures for the detection of basal cell cancer. *Sci. Rep.* **2022**, *12*, 5246. [CrossRef] [PubMed]
38. Winn, J.; Fink, Y.; Fan, S.; Joannopoulos, J.D. Omnidirectional reflection from a one-dimensional photonic crystal. *Opt. Lett.* **1998**, *23*, 1573–1575. [CrossRef] [PubMed]
39. Vinogradov, A.; Dorofeenko, A.V.; Erokhin, S.G.; Inoue, M.; Lisyansky, A.A.; Merzlikin, A.M.; Granovsky, A.B. Surface state peculiarities in one-dimensional photonic crystal interfaces. *Phys. Rev. B* **2006**, *74*, 045128. [CrossRef]
40. Wu, F.; Lu, G.; Guo, Z.; Jiang, H.; Xue, C.; Zheng, M.; Chen, C.; Du, G.; Chen, H. Redshift gaps in one-dimensional photonic crystals containing hyperbolic metamaterials. *Phys. Rev. Appl.* **2018**, *10*, 064022. [CrossRef]
41. Wu, F.; Liu, T.; Xiao, S. Polarization-sensitive photonic bandgaps in hybrid one-dimensional photonic crystals composed of all-dielectric elliptical metamaterials and isotropic dielectrics. *Appl. Opt.* **2023**, *62*, 706–713. [CrossRef]
42. Joannopoulos, J.D. *Photonic Crystals: Molding the Flow of Light*; Princeton University Press: Princeton, NJ, USA, 1995; 305p.
43. Yeh, P. *Optical Waves in Layered Media*; Wiley: New York, NY, USA, 1988.
44. Yariv, A.; Yeh, P. *Optical Waves in Crystals*; Wiley: New York, NY, USA, 1984.
45. Vanyushkin, N.A.; Gevorgyan, A.H.; Golik, S.S. Scattering of a plane wave by an inhomogeneous 1D dielectric layer with gradient refractive index. *Opt. Mater.* **2022**, *127*, 112306. [CrossRef]
46. Fan, X.; White, I.M.; Shopova, S.I.; Zhu, H.; Suter, J.D.; Sun, Y. Sensitive optical biosensors for unlabeled targets: A review. *Anal. Chim. Acta* **2008**, *620*, 8–26. [CrossRef]
47. Yao, J. Optoelectronic oscillators for high speed and high-resolution optical sensing. *J. Light. Technol.* **2017**, *35*, 3489–3497. [CrossRef]
48. Qin, J.; Jiang, S.; Wang, Z.; Cheng, X.; Li, B.; Shi, Y.; Tsai, D.P.; Liu, A.Q.; Huang, W.; Zhu, W. Metasurface micro/nano-optical sensors: Principles and applications. *ACS Nano* **2022**, *16*, 11598–11618. [CrossRef]
49. Van de Velde, F.; Knutsen, S.H.; Usov, A.I.; Rollema, H.S.; Cerezo, A.S. 1H and 13C high resolution NMR spectroscopy of carrageenans: Application in research and industry. *Trends Food Sci. Technol.* **2002**, *13*, 73–92. [CrossRef]
50. Wang, K.; Vincent, M. High-resolution photoelectron spectroscopy of molecules. *Annu. Rev. Phys. Chem.* **1995**, *46*, 275–304. [CrossRef]
51. Vanyushkin, N.A.; Gevorgyan, A.H.; Golik, S.S. Approximation of one-dimensional rugate photonic crystals using symmetric ternary photonic crystals. *Optik* **2021**, *242*, 167343. [CrossRef]

photonics

MDPI

Article

Investigation of Multiple High Quality-Factor Fano Resonances in Asymmetric Nanopillar Arrays for Optical Sensing

Huawei Chen [1], Xinye Fan [1,2,3,4,*], Wenjing Fang [1,3,4], Shuangshuang Cao [1], Qinghe Sun [1], Dandan Wang [1], Huijuan Niu [1,3,4], Chuanchuan Li [2], Xin Wei [2], Chenglin Bai [1,3,4] and Santosh Kumar [5,*]

1 School of Physics Science and Information Engineering, Liaocheng University, Liaocheng 252000, China
2 Institute of Semiconductors, Chinese Academy of Sciences, Beijing 100083, China
3 Shandong Provincial Key Laboratory of Optical Communication Science and Technology, Liaocheng 252000, China
4 Liaocheng Key Laboratory of Industrial-Internet Research and Application, Liaocheng 252000, China
5 Department of Electronics and Communication Engineering, K L Deemed to be University, Guntur 522302, India
* Correspondence: fanxinye@lcu.edu.cn (X.F.); santosh@kluniversity.in (S.K.)

Abstract: A novel asymmetric all-dielectric metasurface supporting multiple Fano resonances with high quality-factor through the excitation of quasi-bound states in the continuum is theoretically investigated. It is demonstrated that two resonances in the near-infrared wavelength are excited by the symmetry-protected bound state in the continuum, which can be transformed into the electric dipole and the toroidal dipole quasi-BIC resonance with high quality-factor by breaking the symmetry of metasurface. Moreover, the sensing properties based on different liquid refractive indexes are researched theoretically. The results show that the maximum quality-factor of the Fano resonance peak is 8422, and the sensitivity can reach 402 nm/RIU, with a maximum figure of merit of 2400 RIU^{-1}. This research is believed to further promote the development of optical sensing and nonlinear optics.

Keywords: all-dielectric; high Q-factor; metasurface

Citation: Chen, H.; Fan, X.; Fang, W.; Cao, S.; Sun, Q.; Wang, D.; Niu, H.; Li, C.; Wei, X.; Bai, C.; et al. Investigation of Multiple High Quality-Factor Fano Resonances in Asymmetric Nanopillar Arrays for Optical Sensing. *Photonics* **2024**, *11*, 68. https://doi.org/10.3390/photonics11010068

Received: 3 November 2023
Revised: 3 January 2024
Accepted: 5 January 2024
Published: 8 January 2024

1. Introduction

As a two-dimensional metamaterial, metasurfaces can be artificially engineered into subwavelength structures to obtain electromagnetic properties not found in natural materials [1,2]. By optimizing the structural parameters of the metasurface, high quality-factor (Q-factor) Fano resonances are achieved, which have important roles in the fields of optical refractive index sensing [3,4], optical switches [5], nonlinear optics [6,7], and slow light [8]. The high Q-factor Fano resonance not only enhances the light-matter interactions but also has a high sensitivity to tiny environmental changes [9,10].

High Q-factor Fano resonances are obtained in many ways. Mohammadi et al. presented a photonic crystal ring resonator including three different photonic crystal sensors with Q-factors up to 5365 [11]. Zhou et al. designed a guided-mode resonance sensor with a shallow subwavelength structure with a Q-factor up to 8000 [12]. All-dielectric metasurfaces offer greater freedom and design flexibility, allowing for more diversified functionality [13]. Different optical functions can be achieved by precisely designing nanostructures that allow precise modulation of several parameters of light, such as polarization, wavelength, phase, and amplitude [14]. At the same time, the all-dielectric metasurface is capable of wide-bandwidth functionality, allowing simultaneous operation of resonances in multiple frequency ranges. In addition, all-dielectric metasurface sensors have better tunability. Wavelength selectivity of the sensor is achieved by adjusting the geometry of the structure, material properties or electromagnetic excitation.

In recent years, silicon nitride has been an important member of the silicon-based family for making resonators [15,16]. Yang et al. designed a refractive index sensor based on bilayer silicon nitride with a Q-factor of 7605 [17]. Guo et al. proposed an all-dielectric metasurface based on a silicon nitride substrate, where the Q-factor of the resonance peak can reach 54757 [18]. In addition, silicon nitride is widely used in a variety of resonators (drum resonators [19], micro-ring resonators [20], waveguide-coupled resonators [21,22]), oscillators [23], and semiconductor lasers [24]). Silicon nitride-based optics can reach Q-factors of the order of 10^6 in most cases, with the highest reaching 2.6×10^8. Although silicon nitride resonators have extremely high Q-factors, all-dielectric metasurface sensors composed of silicon and silicon dioxide are still very necessary because silicon and silicon dioxide are easily manufactured using current manufacturing techniques [25]. The all-dielectric metasurface composed of silicon and silica can excite multiple Fano resonance peaks simultaneously. More resonant peaks are better suited for multi-channel sensing applications [26].

The all-dielectric metasurface is believed to be a great way to obtain high Q-factor Fano resonance due to its greater freedom, design flexibility, and compatibility with complementary metal oxide semiconductor (CMOS) processes [27]. The excitation of Fano resonance in metasurfaces has evolved from a single resonance peak to multiple resonance peaks with the development of sensing technology [28,29]. Multi-Fano resonances have potential applications in the fields of multi-wavelength sensing and multi-channel biosensing [30]. Yang et al. proposed an all-dielectric metasurfaces consisting of an array of square nanopores, which excited four Fano resonances by introducing nanopore asymmetry [31]. Wang et al. investigated an all-dielectric metasurface consisting of an asymmetric nanocylindrical dimer with five Fano resonance peaks that were excited [28].

An efficient way for generating high Q-factor Fano resonances in all-dielectric metasurfaces is through the utilization of bound states in the continuum (BIC) [32,33]. The BIC is a non-radiative state characterized by a resonance frequency embedded in the continuum spectrum of radiative modes in the surrounding space. The coupling of the resonant mode to the surrounding external channels is prohibited when symmetry is maintained, and this type of BIC is referred to as a symmetry-protected BIC [34,35], which is considered to be an ideal BIC arising from an infinitely high Q-factor and extremely narrow resonance width [36,37]. However, this ideal BIC cannot be observed because of the absence of far-field radiation and the inability to be excited by incident waves. It is necessary to create a radiation channel to the outside by breaking the symmetry of the result or by oblique incidence of the light source [38]. Through this method, the symmetry-protected BIC is switched to the quasi-BIC mode so that the Q-factor of the resonance becomes finite [39,40].

In this paper, an all-dielectric metasurface that can excite multiple Fano resonance peaks in the near-infrared spectral region is designed and simulated. The metasurface is consisting of five rectangular silicon blocks periodically placed on a silica-based substrate. Changing the length of two of the rectangular silicon blocks breaks the symmetry of the structure, and thus, four Fano resonance peaks are excited. The Q-factors of the four Fano resonance peaks all reach 10^3, and the highest can reach 8422, the sensitivity is up to 402 nm/RIU, and the figure of merit (FOM) can reach up to 2400 RIU^{-1}. The preparation process of the structure is investigated as a way to demonstrate the realizability of the structure. It is shown that the structure has potential applications in the field of refractive index sensors.

2. Structure Design

The schematics of the designed metasurface structure are shown in Figure 1. Figure 1a shows the array layout of the structure, which is periodically distributed on a substrate with the material SiO_2. The structure of a single cell is shown in Figure 1b, where the labeled P_x and P_y denote the period in the x and y directions, respectively, and h is the height of the silicon block. Figure 1c shows the top view of the structure, the length of the two rectangular blocks on the left (L_1) is 280 nm, the length of the two rectangular blocks on the right (L_2) is 170 nm, the length of the rectangular blocks in the center (L_3) is 660 nm, the width of the rectangular blocks (w) is 150 nm, and the distance between the rectangular blocks (g) is 100 nm. The structure is asymmetrical along the y-axis since $L_1 \neq L_2$. $\delta = L_1 - L_2$ as an asymmetry parameter is used to explore the effect of asymmetry on the excitation of the Fano resonance peaks.

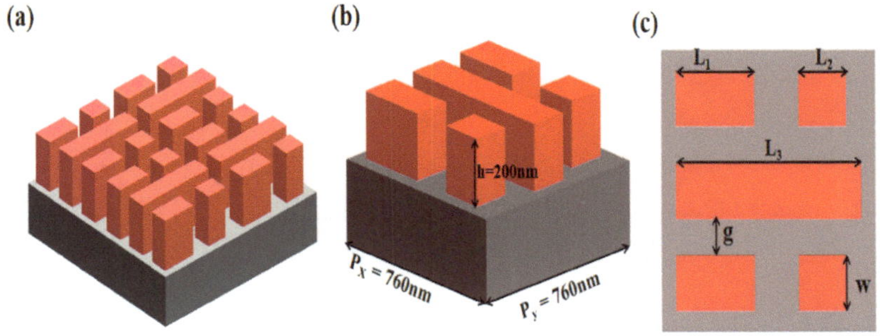

Figure 1. Schematic of the metasurface structure consisting of five rectangular columns. (**a**) Structural layout of the designed metasurface; (**b**) diagram of the structure of a single cell; (**c**) top view of a single cell.

The system of Maxwell's equations is transformed into finite difference equations by the finite difference in time domain (FDTD) method, enabling the simulations to be conducted. With this method, the distribution information of the electromagnetic field can be calculated when the boundary conditions and initial values are known, which greatly reduces the difficulty of the calculation. In the simulation, it is necessary to set the appropriate mesh size. The resolution of the mesh needs to be detailed enough to capture the details and rapidly changing features that appear in the simulation. Excessive mesh size leads to a decrease in the accuracy of the simulation results, as tiny structures and details cannot be adequately represented. If the mesh size is too small, it will lead to an increase in the computational effort, thus decreasing the computational efficiency. Therefore, the mesh was set to 8 nm × 8 nm × 8 nm during the simulation. In order to simulate the behavior of a system in three dimensions, boundary conditions need to be set. In this case, the X and Y directions are set to periodic boundary conditions, which means the simulation accurately represents the behavior of the system across multiple repetitions. On the other hand, the Z direction is set to Perfectly Matched Layer (PML). This is a type of boundary condition that effectively absorbs outgoing waves in the system, preventing reflections and ensuring accurate results. By combining periodic boundary conditions in the X and Y directions with a PML boundary condition in the Z direction, a comprehensive and reliable simulation can be achieved. The y-polarized plane wave perpendicular to the z-axis as a light source is used in the simulations. The plane wave is used in the simulation process. Plane waves propagate through space at a constant velocity without changing shape or decaying. Optical parameters of silicon and silicon dioxide used in the simulations are referenced in the Palik manual.

3. Results and Discussion

The transmission spectrum of the structure in the liquid refractive index of 1.35 state is simulated, while the background refractive index is set to n = 1.35. When $L_1 = L_2$ ($\delta = 0$ nm), the transmission spectrum is obtained as shown by the black solid line in Figure 2a. Two Fano resonance peaks marked as R1 and R2 are observed at the wavelength of 1123 nm and 1163 nm, respectively. The asymmetric metasurface transmission spectrum is shown in Figure 2b. Two new Fano resonance peaks, labeled R3 and R4, are innovatively excited at 1202 nm and 1221 nm when the length of L_2 is changed to 170 nm ($\delta = 110$ nm). As the symmetry of the metasurface is broken, radiation channels are established, allowing energy to leak outward, resulting in a symmetry-protected BIC converting to a quasi-BIC.

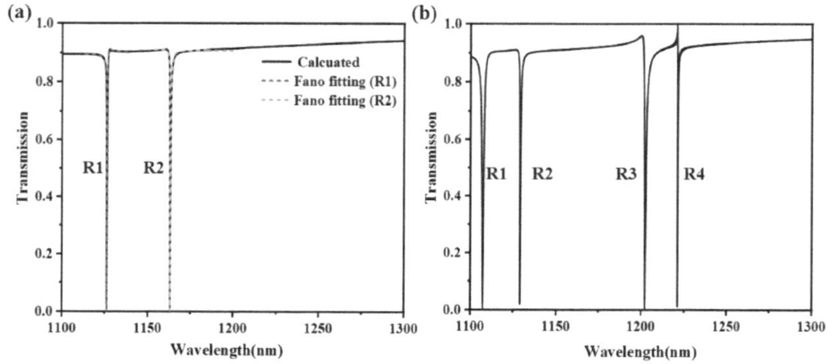

Figure 2. Simulated transmission spectra of the metasurface. (**a**) Transmission spectrum of symmetric structure and equation fitting plot; (**b**) transmission spectrum of asymmetric structure.

In addition, the excited Fano resonance peaks of the periodic all-dielectric metasurface can be analyzed and fitted by the Fano model with Equation (1) [36]:

$$T = \left| a_1 + i a_2 + \frac{b}{\omega - \omega_0 + i\gamma} \right|^2 \tag{1}$$

where a_1, a_2, and b are constant real numbers, ω_0 is the resonant frequency, and γ is the damping loss. The Fano curves fitted by Equation (1) are shown as red and green dashed lines in Figure 2a. The fitting results are in general agreement with the simulation results. Equation (2) is extracted by Equation (1), and the Q-factor of the Fano resonance can be obtained by Equation (2).

$$Q = \frac{\omega_0}{2\gamma} \tag{2}$$

The Q-factors of the Fano resonance peaks are calculated when $\delta = 110$ nm: 1054 ($\omega_0 = 1.122$ eV and $\gamma = 0.5323 \times 10^{-3}$ eV), 2377 ($\omega_0 = 1.101$ eV and $\gamma = 0.2316 \times 10^{-3}$ eV), 1336 ($\omega_0 = 1.336$ eV and $\gamma = 0.3868 \times 10^{-3}$ eV), 8422 ($\omega_0 = 1.0174$ eV and $\gamma = 0.0604 \times 10^{-3}$ eV).

The relationship between the Q-factor and the degree of asymmetry is discussed. $\alpha = \Delta S/S$ as the degree of asymmetry, where S is the area of the silicon block when symmetric and ΔS is the area of the reduced silicon block when asymmetric. As ΔS decreases, the Q-factor is larger. After observing several sets of data, the Q-factor of the excited Fano resonance is found to be related to the degree of asymmetry as $Q \propto \alpha^{-2}$, as shown in Figure 3a.

In order to further investigate the nature of the multiple resonance peaks generated by this metasurface, the electromagnetic field distributions of the resonance peaks under asymmetry are simulated, and their near-field distribution is shown in Figure 4. For the resonance peak R1, the magnetic field forms two clockwise loops in the x-z plane, and the electric field is linear in the x-y plane, which is a typical electric dipole (ED) feature. The resonance mode of R3 is also considered to be ED, but its direction is opposite to that of R1. The resonance mode of R2 can be seen in the figure as magnetic dipoles (MDs), in which the current forms three loops with the same direction in the y-z plane and the magnetic field is linearly polarized in the x-y plane. Two current loops in opposite directions can be observed in the x-y plane of R4, and the resulting magnetic field creates a large loop in the x-z plane, which is typical of toroidal dipole (TD) resonance.

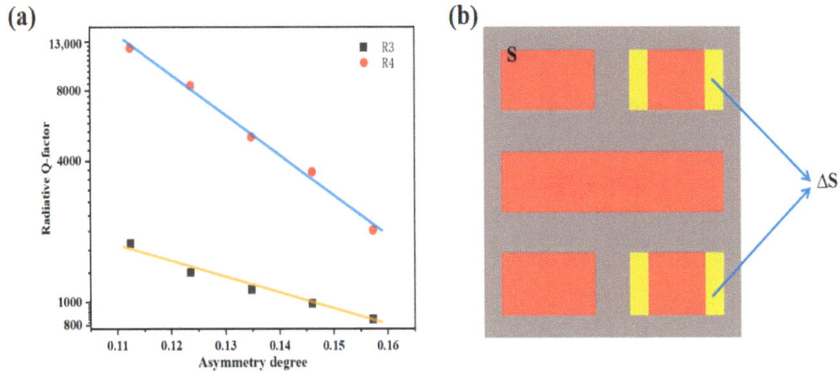

Figure 3. (**a**) Plot of the Q-factor of the excited Fano resonance versus the degree of asymmetry α; (**b**) schematic representation of ΔS and S.

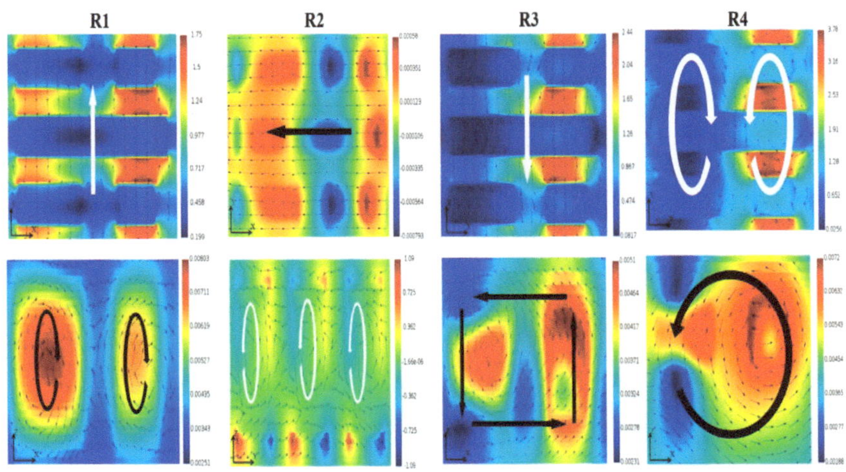

Figure 4. Near−field distribution of the electromagnetic field corresponding to the four Fano resonance peaks in the asymmetric case. The electric field vector distributions are shown as white arrows and the magnetic field vector distributions are shown as black arrows.

The characteristics of the transmission spectrum for different geometrical parameters are also discussed to obtain the optimal parameter, and the simulation results are shown in Figure 5. Figure 5a shows the transmission spectrums of the silicon block with a high between 180 nm and 220 nm, and it is clear that the resonance peaks are sensitive because the four resonance peaks red-shift as h increases, especially R1. The transmission spectrums of the silicon block with a wide w between 130 nm and 170 nm are presented in Figure 5b, where all the four resonance peaks are red-shifted with w increasing. In addition, the linewidth of R1 decreases as w increases, and the modulation depth of R4 becomes deeper as w increases. The transmission spectrum corresponding to the change in length of the intermediate silicon block (L_3) is displayed in Figure 5c. The L_3 produces little effect on the transmission spectrum of the metasurface. Figure 5d parades the transmission spectra at different periods. The Fano resonance peaks show different degrees of red-shift as the period increases. However, compared with h and w, this red-shift is slight, while the modulation depth of R4 undergoes a little change.

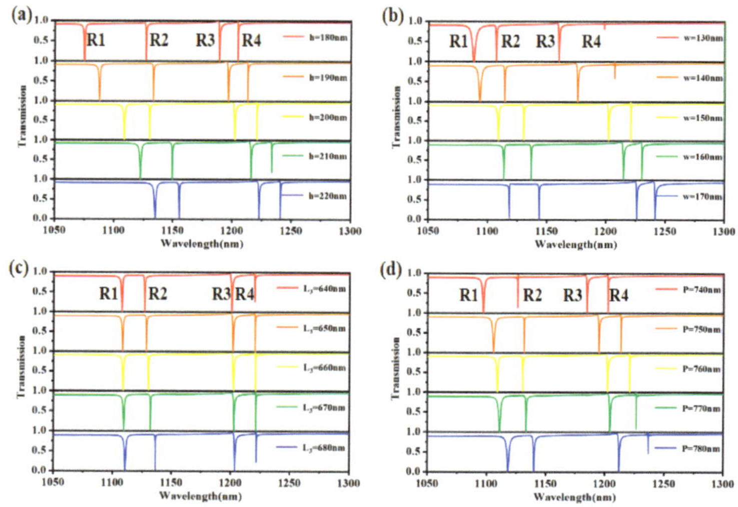

Figure 5. Transmission spectra of the metasurface with varying geometrical parameters. (**a**) Height of the silicon block (h), (**b**) width of the silicon block (w), (**c**) length of the middle silicon block (L_3), and (**d**) period of the metasurface (P).

In order to investigate the influence of light absorption and scattering caused by liquid and materials on the transmission spectrum of the structure, different extinction coefficients (k) are set for symmetric and asymmetric cases, respectively. The simulation results are shown in Figure 6. The linewidth of the resonance peak increases with increasing extinction coefficient; however, the modulation depth and Q-factor decrease. The sensitivity of the resonance peaks to the extinction coefficient varies, with R4 being the most sensitive and R1 the least sensitive. What this result presents is that the same extinction coefficient will have different effects on the Fano resonance formed by different resonant modes. These simulations demonstrated that the presence of a large intrinsic absorption in the liquid and materials used would cause the resonance to attenuate or even disappear.

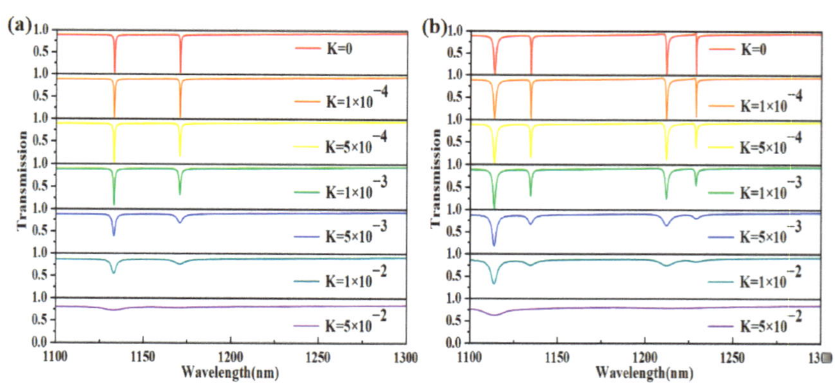

Figure 6. (**a**) Transmission spectra of symmetric structures with different extinction coefficients, (**b**)Transmission spectra of asymmetric structures with different extinction coefficients.

The proposed metastructure is shown to be investigated as a refractive index sensor due to its advantages of excitation of high Q-factor Fano resonance and local field enhancement. The potential application of metasurfaces in the field of refractive index sensors is verified by analyzing the changing patterns of transmission spectra at different ambient refractive indexes. Figure 7 show the transmission spectra of R1, R2, R3, and R4 at ambient refractive index (n) in steps of 0.01 from 1.33 to 1.37, respectively. As can be seen from Figure 7, the Fano resonance peaks are significantly red-shifted despite the little change in refractive index. The sensing performance of the metasurface is believed to be determined by two metrics, the sensitivity S and the FOM. The sensitivity is calculated by Equation (3) [41]:

$$S = \frac{\Delta\lambda(\text{nm})}{\Delta n(\text{RIU})} \tag{3}$$

in which $\Delta\lambda$ is the offset of the resonance peak and Δn is the difference in refractive index. Refractive index unit (RIU) represents the ratio of the speed of light relative to the speed of light in a vacuum as it enters the medium from a vacuum. The sensitivities of the Fano resonance peaks are 129, 333, 403, 348 by calculation. Figure 7b,d show the wavelengths corresponding to the four resonance peaks at different refractive indexes, and the slope of the curve corresponding to each resonance peak is the sensitivity of that resonance peak. The FOM is calculated by Equation (4) [42]:

$$\text{FOM} = \frac{S(\text{nm/RIU})}{\text{FWHM}(\text{nm})} \tag{4}$$

in which S is the sensitivity and FWHM is the full width at half peak of the resonance peak. The FOM corresponding to the four resonance peaks is calculated to be 123, 701, 448, and 2400, respectively.

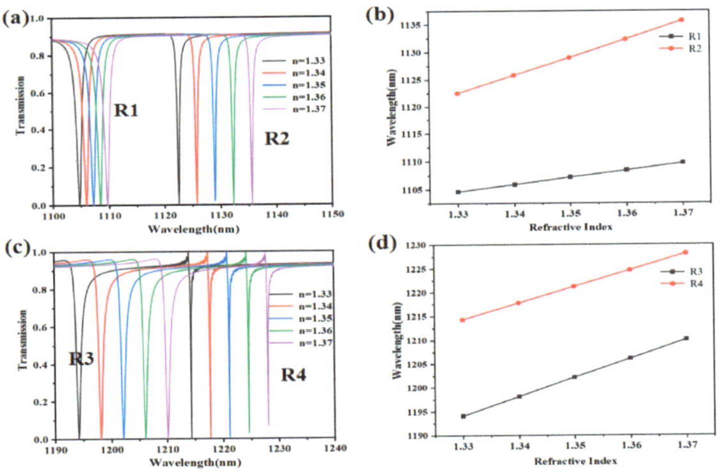

Figure 7. Transmission spectra of metasurfaces at different refractive indexes. (**a**) Transmission spectra of R1 and R2 at different refractive indexes; (**b**) wavelength plots corresponding to R1 and R2 at different refractive indexes; (**c**) transmission spectra of R3 and R4 at different refractive indexes; (**d**) wavelength plots corresponding to R3 and R4 at different refractive indexes.

In addition, the proposed metastructure is compared with the existing work for number of resonance peaks, sensitivity, and FOM, as shown in Table 1. The results show that the metastructure proposed in this paper can generate more resonance peaks with higher sensitivity and FOM, which is more suitable for multichannel sensing applications.

Table 1. The number of resonance peaks, sensitivity, and FOM of this sensor are compared with other types.

Sensor Type	Number of Resonance Peaks	S (nm/RIU)	FOM (RIU^{-1})	Ref.
Double square hollow	4	287.5	389	[29]
U-shaped silicon cylinder	1	203	29	[43]
Split-ring disk	2	282	4	[44]
Optical sensor based on a photonic crystal metasurface	2	178	445	[45]
Bilayer Silicon Nitride Photonic Crystal Sensor	1	937.64	n.r. [a]	[17]
All-dielectric metasurface based on a silicon nitride substrate	2	746	18650	[18]
Five rectangular blocks of silicon	4	403	2400	This work

[a] not reported.

The proposed sensors are mainly based on silicon on insulator (SOI) fabrication. SOI device processes are compatible with standard silicon CMOS processes, which are easy and inexpensive to fabricate, making them ideal for large-scale semiconductor chip integration. The preparation process of the proposed metastructure is shown in Figure 8. The device preparation procedure consists of the following steps: cleaning of the SOI substrate, low-pressure chemical vapor deposition (LPCVD), spin coating of the resist, electron beam lithography (EBL), development, inductively coupled plasma (ICP) etching and removal of the resist.

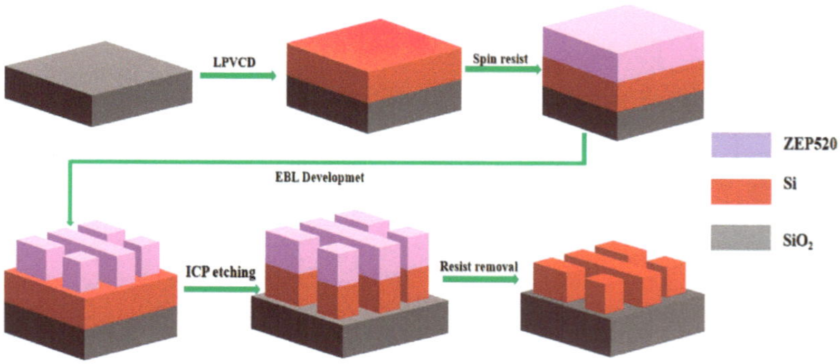

Figure 8. The fabrication process flow chart of the metastructure.

4. Conclusions

In summary, we proposed and theoretically demonstrated a structure-asymmetric all-dielectric metasurface based on asymmetric nanopillar arrays. Four sharp and high-spectral-contrast Fano resonances are excited with the best Q-factor of 8422. The results of the electromagnetic field distribution explicate that the resonances at 1107 nm and 1202 nm are governed by the ED mode, while the resonances at 1129 nm and 1221 nm are governed by the MDs mode and TD mode, respectively. In addition, the relationship between the different extinction coefficients (K) and transmission spectrum indicates that the resonances are attenuated or even disappear, occurring when there is a large intrinsic absorption in the liquid. Finally, the sensing characteristics based on different liquid refractive indices are analyzed to certify that the designed all-dielectric metasurface is employed as a refractive index sensor with the maximum sensitivity, and the FOM value is

402 nm/RIU and 2400 RIU^{-1}, respectively. This dielectric metasurface with multiple sharp resonances offers a good platform for multichannel sensing as well as optical modulation.

Author Contributions: Writing—original draft preparation, H.C.; writing—review and editing, X.F., W.F., S.C., Q.S., D.W., C.L., X.W. and S.K.; project administration, funding acquisition—X.F., W.F., H.N., C.B. and S.K. All authors have read and agreed to the published version of the manuscript.

Funding: This work is supported by the Cultivation Plan for Young Scholars in Universities of Shandong Province (2021RC085); The Natural Foundation of Shandong Province (ZR2021MF053, ZR2022MF253, ZR2021MF070, and ZR2022MF305); The Open Fund of the Key State Laboratory (BUPT, IPOC) (IPOC2021B07); The Double-Hundred Talent Plan of Shandong Province, China; Liaocheng University (318052341).

Institutional Review Board Statement: Not applicable.

Informed Consent Statement: Not applicable.

Data Availability Statement: The data supporting the findings of this study are available from the corresponding authors upon reasonable request.

Conflicts of Interest: The authors declare no conflict of interest.

References

1. Zhang, F.; Pu, M.; Gao, P.; Jin, J.; Li, X.; Guo, Y.; Ma, X.; Luo, J.; Hong, Y.; Luo, X. Simultaneous Full-Color Printing and Holography Enabled by Centimeter-Scale Plasmonic Metasurfaces. *Adv. Sci.* **2020**, *7*, 1903156. [CrossRef]
2. Alhalaby, H.; Principe, M.; Zaraket, H.; Vaiano, P.; Aliberti, A.; Quero, G.; Crescitelli, A.; Di Meo, V.; Esposito, E.; Consales, M.; et al. Design and Optimization of All-Dielectric Fluorescence Enhancing Metasurfaces: Towards Advanced Metasurface-Assisted Optrodes. *Biosensors* **2022**, *12*, 264. [CrossRef] [PubMed]
3. Zhang, Y.; Liu, W.; Li, Z.; Li, Z.; Cheng, H.; Chen, S.; Tian, J. High-quality-factor multiple Fano resonances for refractive index sensing. *Opt. Lett.* **2018**, *43*, 1842–1845. [CrossRef] [PubMed]
4. Song, S.; Yu, S.; Li, H.; Zhao, T. Ultra-high Q-factor toroidal dipole resonance and magnetic dipole quasi-bound state in the continuum in an all-dielectric hollow metasurface. *Laser Phys.* **2022**, *32*, 025403. [CrossRef]
5. Gupta, M.; Srivastava, Y.K.; Singh, R. A Toroidal Metamaterial Switch. *Adv. Mater.* **2018**, *30*, 1704845. [CrossRef]
6. Butet, J.; Martin, O.J. Fano resonances in the nonlinear optical response of coupled plasmonic nanostructures. *Opt. Express* **2014**, *22*, 29693–29707. [CrossRef]
7. Xiao, S.; Qin, M.; Duan, J.; Liu, T. Robust enhancement of high-harmonic generation from all-dielectric metasurfaces enabled by polarization-insensitive bound states in the continuum. *Opt. Express* **2022**, *30*, 32590–32599. [CrossRef]
8. Huang, L.; Li, H.; Yu, S.; Zhao, T. Analogue of electromagnetically induced transparency inspired by bound states in the continuum and toroidal dipolar response in all-dielectric metasurfaces. *Photonic Nanostruct.* **2022**, *51*, 101041. [CrossRef]
9. Modi, K.S.; Kaur, J.; Singh, S.P.; Tiwari, U.; Sinha, R.K. Extremely high figure of merit in all-dielectric split asymmetric arc metasurface for refractive index sensing. *Opt. Commun.* **2020**, *462*, 125327. [CrossRef]
10. Yang, H.; Chen, Y.; Liu, M.; Xiao, G.; Luo, Y.; Liu, H.; Li, J.; Yuan, L. High Q-Factor Hybrid Metamaterial Waveguide Multi-Fano Resonance Sensor in the Visible Wavelength Range. *Nanomaterials* **2021**, *11*, 1583. [CrossRef]
11. Mohammadi, M.; Seifouri, M. Numerical investigation of photonic crystal ring resonators coupled bus waveguide as a highly sensitive platform. *Photonic Nanostruct.* **2019**, *34*, 11–18. [CrossRef]
12. Zhou, Y.; Guo, Z.; Zhou, W.; Li, S.; Liu, Z.; Zhao, X.; Wu, X. High-Q guided mode resonance sensors based on shallow sub-wavelength grating structures. *Nanotechnology* **2020**, *31*, 325501. [CrossRef] [PubMed]
13. Li, H.; Hu, C.; Jiang, J.H.; Wu, J.; Wen, W.; Hou, B. Photonic Type-III Nodal Loop and Topological Phase Transitions at Bilayer Metasurfaces. *Front. Mate.* **2022**, *9*, 909381. [CrossRef]
14. Archetti, A.; Lin, R.J.; Restori, N.; Kiani, F.; Tsoulos, T.V.; Tagliabue, G. Thermally reconfigurable metalens. *Nanophotonics* **2022**, *11*, 3969–3980. [CrossRef] [PubMed]
15. Fong, K.Y.; Pernice, W.H.; Tang, H.X. Frequency and phase noise of ultra-high Q silicon nitride nanomechanical resonators. *Phys. Rev. B* **2012**, *85*, 4506. [CrossRef]
16. Ji, X.; Roberts, S.; Corato-Zanarella, M.; Lipson, M. Methods to achieve ultra-high quality factor silicon nitride resonators. *APL Photonics* **2021**, *6*, 071101. [CrossRef]
17. YiYang, Y.; Liu, Y.; Yang, S.; Wu, Y.; Tian, H. Double-layered silicon-nitride photonic crystal slab guided-mode-resonance high-sensitivity sensor application for refractive index sensing and nanoparticle detection. *J. Opt. Soc. Am. B* **2021**, *38*, 1927–1933. [CrossRef]
18. Guo, L.; Zhang, Z.; Xie, Q.; Li, W.; Xia, F.; Wang, M.; Feng, H.; You, C.; Yun, M. Toroidal dipole bound states in the continuum in all-dielectric metasurface for high-performance refractive index and temperature sensing. *Appl. Surf. Sci* **2023**, *615*, 156408. [CrossRef]

19. Zhou, X.; Venkatachalam, S.; Zhou, R.; Xu, H.; Pokharel, A.; Fefferman, A.; Zaknoune, M.; Collin, E. High-Q silicon nitride drum resonators strongly coupled to gates. *Nano Lett.* **2021**, *21*, 5738–5744. [CrossRef]
20. Frankis, H.C.; Kiani, K.M.; Su, D.; Mateman, R.; Leinse, A.; Bradley, J.D. High-Q tellurium-oxide-coated silicon nitride microring resonators. *Opt Lett.* **2019**, *44*, 118–121. [CrossRef]
21. Spencer, D.T.; Bauters, J.F.; Heck, M.J.; Bowers, J.E. Integrated waveguide coupled Si3N4 resonators in the ultrahigh-Q regime. *Optica* **2014**, *1*, 153–157. [CrossRef]
22. Ji, X.; Jang, J.K.; Dave, U.D.; Corato-Zanarella, M.; Joshi, C.; Gaeta, A.L.; Lipson, M. Exploiting Ultralow Loss Multimode Waveguides for Broadband Frequency Combs. *Laser Photonics Rev.* **2020**, *15*, 2000353. [CrossRef]
23. Domeneguetti, R.R.; Zhao, Y.; Ji, X.; Martinelli, M.; Lipson, M.; Gaeta, A.L.; Nussenzveig, P. Parametric sideband generation in CMOS-compatible oscillators from visible to telecom wavelengths. *Optica* **2021**, *8*, 316–322. [CrossRef]
24. Jin, W.; Yang, Q.F.; Chang, L.; Shen, B.; Wang, H.; Leal, M.A.; Wu, L.; Gao, M.; Feshali, A.; Paniccia, M.; et al. Hertz-linewidth semiconductor lasers using CMOS-ready ultra-high-Q microresonators. *Nat. Photonics* **2021**, *15*, 346–353. [CrossRef]
25. Yu, F.; Chen, J.; Huang, L.; Zhao, Z.; Wang, J.; Jin, R.; Chen, J.; Wang, J.; Miroshnichenko, A.E.; Li, T.; et al. Photonic slide rule with metasurfaces. *Light Sci. Appl.* **2022**, *11*, 77. [CrossRef]
26. Ye, Y.; Yu, S.; Li, H.; Gao, Z.; Yang, L.; Zhao, T. Triple Fano resonances metasurface and its extension for multi-channel ultra-narrow band absorber. *Results Phys.* **2022**, *42*, 106025. [CrossRef]
27. Varasteanu, P.; Radoi, A.; Tutunaru, O.; Ficai, A.; Pascu, R.; Kusko, M.; Mihalache, I. Plasmon-Enhanced Photoresponse of Self-Powered Si Nanoholes Photodetector by Metal Nanowires. *Nanomaterials* **2021**, *11*, 2460. [CrossRef]
28. Wang, W.; Zheng, L.; Xiong, L.; Qi, J.; Li, B. High Q-factor multiple Fano resonances for high-sensitivity sensing in all-dielectric metamaterials. *OSA Continuum.* **2019**, *2*, 2818. [CrossRef]
29. Li, H.; Yu, S.; Yang, L.; Zhao, T. High Q-factor multi-Fano resonances in all-dielectric double square hollow metamaterials. *Opt. Laser Technol.* **2021**, *140*, 107072. [CrossRef]
30. Biasco, S.; Beere, H.E.; Ritchie, D.A.; Li, L.; Davies, A.G.; Linfield, E.H.; Vitiello, M.S. Frequency-tunable continuous-wave random lasers at terahertz frequencies. *Light Sci. Appl.* **2019**, *8*, 43. [CrossRef]
31. Yang, L.; Yu, S.; Li, H.; Zhao, T. Multiple Fano resonances excitation on all-dielectric nanohole arrays metasurfaces. *Opt. Express* **2021**, *29*, 14905–14916. [CrossRef]
32. Fan, K.; Shadrivov, I.V.; Padilla, W.J. Dynamic bound states in the continuum. *Optica* **2019**, *6*, 169–173. [CrossRef]
33. Chen, X.; Fan, W. Tunable Bound States in the Continuum in All-Dielectric Terahertz Metasurfaces. *Nanomaterials* **2020**, *10*, 623. [CrossRef]
34. Mao, L.; Cheng, P.; Liu, K.; Lian, M.; Cao, T. Sieving nanometer enantiomers using bound states in the continuum from the metasurface. *Nanoscale Adv.* **2022**, *4*, 1617–1625. [CrossRef]
35. Cerjan, A.; Jörg, C.; Vaidya, S.; Augustine, S.; Benalcazar, W.A.; Hsu, C.W.; von Freymann, G.; Rechtsman, M.C. Observation of bound states in the continuum embedded in symmetry bandgaps. *Sci. Adv.* **2021**, *7*, eabk1117. [CrossRef]
36. Sadrieva, Z.; Frizyuk, K.; Petrov, M.; Kivshar, Y.; Bogdanov, A. Multipolar origin of bound states in the continuum. *Phy. Rev. B* **2019**, *100*, 115303. [CrossRef]
37. Koshelev, K.; Lepeshov, S.; Liu, M.; Bogdanov, A.; Kivshar, Y. Asymmetric Metasurfaces with High-Q Resonances Governed by Bound States in the Continuum. *Phys. Rev. Lett.* **2018**, *121*, 193903. [CrossRef]
38. Mi, Q.; Sang, T.; Pei, Y.; Yang, C.; Li, S.; Wang, Y.; Ma, B. High-quality-factor dual-band Fano resonances induced by dual bound states in the continuum using a planar nanohole slab. *Nanoscale Res. Lett.* **2021**, *16*, 150. [CrossRef]
39. Leitis, A.; Tittl, A.; Liu, M.; Lee, B.H.; Gu, M.B.; Kivshar, Y.S.; Altug, H. Angle-multiplexed all-dielectric metasurfaces for broadband molecular fingerprint retrieval. *Sci. Adv.* **2019**, *5*, eaaw2817. [CrossRef]
40. Reineke Matsudo, B.; Sain, B.; Carletti, L.; Zhang, X.; Gao, W.; de Angelis, C.; Huang, L.; Zentgraf, T. Efficient Frequency Conversion with Geometric Phase Control in Optical Metasurfaces. *Adv. Sci.* **2022**, *9*, e2104508. [CrossRef]
41. Abbas, M.A.; Zubair, A.; Riaz, K.; Huang, W.; Teng, J.; Mehmood, M.Q.; Zubair, M. Engineering multimodal dielectric resonance of TiO2 based nanostructures for high-performance refractive index sensing applications. *Opt. Express* **2020**, *28*, 23509–23522. [CrossRef] [PubMed]
42. Liu, G.D.; Zhai, X.; Wang, L.L.; Lin, Q.; Xia, S.X.; Luo, X.; Zhao, C.J. A High-Performance Refractive Index Sensor Based on Fano Resonance in Si Split-Ring Metasurface. *Plasmonics* **2018**, *13*, 15–19. [CrossRef]
43. Qin, M.; Pan, C.; Chen, Y.; Ma, Q.; Liu, S.; Wu, E.; Wu, B. Electromagnetically Induced Transparency in All-Dielectric U-Shaped Silicon Metamaterials. *Appl. Sci.* **2018**, *8*, 1799. [CrossRef]
44. Zhang, Q.; Wen, X.; Li, G.; Ruan, Q.; Wang, J.; Xiong, Q. Multiple magnetic mode-based Fano resonance in split-ring resonator/disk nanocavities. *ACS Nano* **2013**, *7*, 11071–11078. [CrossRef]
45. Romano, S.; Zito, G.; Torino, S.; Calafiore, G.; Penzo, E.; Coppola, G.; Cabrini, S.; Rendina, I.; Mocella, V. Label-free sensing of ultralow-weight molecules with all-dielectric metasurfaces supporting bound states in the continuum. *Photonics Res.* **2018**, *6*, 726–733. [CrossRef]

photonics

Article

Specific and Simultaneous Detection of *E. coli* O157:H7 and Shiga-like Toxins Using a Label-Free Photonic Immunosensor

Ana Fernández [1], Manuel Hernández [2], Yolanda Moreno [3,*] and Jorge García-Hernández [2]

1 Lumensia Sensors S.L., 46020 Valencia, Spain; afernandez@lumensia.com
2 Centro Avanzado de Microbiología de Alimentos, Biotechnology Department, Universitat Politècnica de València, 46022 Valencia, Spain; mhernand@btc.upv.es (M.H.); jorgarhe@btc.upv.es (J.G.-H.)
3 Instituto de Ingeniería de Agua y del Medioambiente, Universitat Politècnica de València, 46022 Valencia, Spain
* Correspondence: ymoren@upv.es

Abstract: The current study outlines the advancement of an innovative technique for the simultaneous detection of *E. coli* O157:H7 and its Shiga-like toxins in food samples by utilizing a photonic label-free biosensor coupled with a microfluidic system. This detection method relies on ring resonator transduction that is functionalized with specific bioreceptors against O157:H7 on silicon nitride surfaces capable of binding specifically to the antigen bacterium and its verotoxins. This experiment included the characterization of selected monoclonal and polyclonal antibodies employed as detection probes through ELISA immunoassays exposed to target bacterial antigens. A thorough validation of photonic immunosensor detection was conducted on inoculated minced beef samples using reference standards for *E. coli* O157:H7 and its verotoxins (VTx1 and VTx2) and compared to gold-standard quantification. The lowest limit-of-detection values of 10 CFU/mL and 1 ppm were achieved for the detection of bacteria and its verotoxins. In this study, the lowest limit of quantification (LoQ) achieved for bacterial quantification was 100 CFU/mL, and, for verotoxins, it was 2 ppm. This study confirmed the effectiveness of a new quality control and food hygiene method, demonstrating the rapid and sensitive detection of *E. coli* O157:H7 and its verotoxins. This innovative approach has the potential to be applied in food production environments.

Keywords: *E. coli* O157; food safety; photonic immunosensor; label-free method; ring resonators; photon transduction

Citation: Fernández, A.; Hernández, M.; Moreno, Y.; García-Hernández, J. Specific and Simultaneous Detection of *E. coli* O157:H7 and Shiga-like Toxins Using a Label-Free Photonic Immunosensor. *Photonics* **2024**, *11*, 374. https://doi.org/10.3390/photonics11040374

Received: 21 February 2024
Revised: 3 April 2024
Accepted: 4 April 2024
Published: 16 April 2024

1. Introduction

Escherichia coli O157:H7 (*E. coli* O157:H7) is considered one of the most hazardous types of bacteria, responsible for severe illnesses such as hemorrhagic colitis and hemoytic uremic syndrome, particularly affecting young and immunocompromised individuals [1]. Despite notable advancements in disinfection techniques within the food manufacturing and agricultural sectors, controlling *E. coli* O157:H7 remains a persistent challenge [2]. In the case of ground beef, the presence of *E. coli* O157:H7 at one colony-forming unit (CFU)/25 g or above is considered a substantial health risk, given the robust proliferative capacity of *E. coli* O157:H7 [3]. As a result, the swift and sensitive monitoring of *E. coli* O157:H7 is crucial to ensuring food and water safety and facilitating prompt and accurate disease diagnosis and treatment [4].

The scientific committees associated with the European Union have established preventative measures and applicable recommendations for avoiding possible food-borne infections caused by strains of verotoxigenic/Shiga toxin-producing/enterohemorrhagic *Escherichia coli* (VTEC/STEC/EHEC) [5]. According to the Report of the Scientific Committee of the Spanish Agency for Food Safety and Nutrition (AESAN), preventive measures for the contamination of this pathogen should be adopted, covering the entire food chain. These measures include good agricultural practices, biosecurity programs on farms with livestock, good hygiene practices and the inspection of slaughterhouses, and good practices in the processing of vegetables for fresh consumption. Additionally, according to the recommendations established by the European Union, food industry workers must adopt procedures based on the principles of Hazard Analysis and Critical Control Points (HACCP), and consumers must be taught good conservation practices and the cooking of food [6].

In the EU, regulatory microbiological criteria for outbreaks have been established for the absence of specific STEC strains with the highest potential risk of severe disease, while, in other countries, testing for specific STEC strains may be required during outbreak processing as a measure of process performance [5]. Most *E. coli* O157:H7 strains produce type 2 Shiga toxin, with its subunits binding to the surface of enterocytes. The subunit entering the cells halts protein synthesis by disrupting the function of the large ribosomal subunit [7]. The Shiga toxin can cause symptoms such as stomach pain, diarrhea, inflammation, secretion of intestinal fluids, ulcers, and, in severe cases, conditions like hemorrhagic enteritis and hemolysis, especially in infants and young children, often leading to sepsis or meningitis. Contaminated food is a significant source of infection with verocytotoxigenic *E. coli*.

The European Commission regulation regarding compliance with the microbiological criteria applicable to food products establishes the absence of *E. coli* O157:H7 required in 25 g or 25 mL [8]. Several strategies have been employed for the detection of *E. coli* O157:H7, including colony counting (considered the gold standard) as well as techniques such as enzyme-linked immunosorbent assay (ELISA), polymerase chain reaction (PCR), and quartz crystal microbalance resonators (QCM).

Traditional culture-based methods involve the ability of bacteria to grow and multiply, requiring 24 h of incubation under laboratory conditions [9]. Traditional methods are considered the "gold standards" and consist of culture techniques with specific and selective enrichment media [10], followed by biochemical tests or the counting of microorganisms indicating contamination in the food. These techniques are effective and low cost, but they can take a longer time to obtain results and have low sensitivity and specificity. These methods can be slow and time-consuming in obtaining results, which can delay the diagnosis and treatment of diseases [10]. Additionally, these methods can be expensive and require highly trained personnel to perform the tests.

Nucleic acid-based strategies, particularly those involving polymerase chain reaction (PCR), have shown promise due to their high sensitivity and rapid detection capabilities. The PCR methods used to detect trace amounts of *E. coli* O157:H7 often involve amplifying the bacteria present in a sample. These approaches are known for being labor-intensive, time-consuming, and expensive. Additionally, they necessitate highly trained personnel, leading to delays in obtaining results and hindering the timely prevention of epidemic outbreaks [11]. On the other hand, methods based on nucleic acids have emerged as powerful tools in the detection of pathogens in food. The advantages of these assays, together with their ease of use and susceptibility to automation, make them very attractive for applications in food, in order to overcome the long enrichment cultivation stage. Polymerase chain reaction (PCR), multiplex PCR, real-time PCR, hybridization, microarrays, and nucleic acid sequence-based amplification offer high sensitivity and specificity, allowing the detection of multiple pathogens in one single sample and providing quick results [12]. However, it is important to highlight that the use of the PCR technique in the detection of pathogens in food has some limitations. One of the main limitations is the need to know the target DNA sequence to design specific primers [13]. This means that a specific set of primers is needed for each pathogen, which can limit the detection capacity of the technique.

Multiplex PCR is an efficient and sensitive technique for the detection of *E. coli* O157:H7 and its verotoxins in food samples. This technique is a variant of PCR and can detect multiple pathogens in a single reaction [14], reducing sample analysis time and costs. Furthermore, multiplex PCR is capable of detecting pathogens present in low quantities in the sample, making it a highly sensitive technique. However, multiplex PCR also has some limitations in its application. The presence of inhibitors in the sample can affect the sensitivity and specificity of the technique, which can result in false-negative or false-positive results. Furthermore, multiplex PCR requires careful validation to ensure the specificity and sensitivity of the technique for each pathogen and toxin to be detected.

Immunological methods involve obtaining antibody responses, which can be assessed through various techniques, such as an enzyme-linked immunosorbent assay (ELISA). ELISA typically requires approximately 24 h for completion [15,16]. However, these methods demand considerable time, skilled personnel, and costly equipment [17,18]. The ELISA test is a highly sensitive and specific technique. But, to achieve an effective ELISA technique, it is necessary to isolate the antigens of interest and generate specific antibodies by immunizing animals.

An enzyme-linked immunosorbent assay (ELISA) is a highly sensitive and selective method used to detect specific antigens or antibodies in various biological samples [19,20]. The immobilization of the target antigen on a plastic substrate and its subsequent detection using a labeled selective antibody is a crucial feature of ELISA [21]. The numerical values obtained from ELISA measurements are crucial for quantitatively determining the concentration of specific antigens in a sample. The intensity of the signal generated by the labeled antibody (often an enzyme-linked reaction) is directly proportional to the amount of antigen present in the sample.

While ELISA provides relatively rapid results compared to traditional culture techniques, it is essential to note that the sensitivity might differ from more time-consuming methods. The current ELISA used to detect *E. coli* O157:H7 is commonly based on cELISA. In this case, anti-*E. coli* O157:H7 antibodies are immobilized on a plate to capture *E. coli* O157:H7 antigens and subsequently bind the primary antibodies and secondary antibodies linked to an enzyme or directly bind to a conjugated antibody. This method is widely used in laboratories for its speed (3–5 h). It provides a balance between speed and accuracy for routine testing, but it represents a loss of sensitivity [22].

The lateral flow immunoassay is a widely used detection method in the food industry due to its sensitivity and specificity in detecting *E. coli* O157:H7 in food samples [23]. The simplicity, rapidity, and low cost of the lateral flow immunoassay make it a valuable tool for the detection of pathogens in food, especially in settings where access to laboratories is limited [24]. However, it should be noted that, compared to other pathogen detection methods, the lateral flow immunoassay may have limitations in terms of sensitivity and specificity. Furthermore, this technique can produce false positives or false negatives, which can have serious consequences for food safety [25].

The progress of nanotechnology across diverse domains has led to the development of biosensors as precise diagnostic tools. These biosensors offer a solution to the demand for selective, rapid, and accurate procedures in identifying microorganisms within medical, food, or environmental samples [26]. The use of biosensors in the detection of pathogens and other contaminants in food offers numerous advantages compared to conventional detection methods, such as speed and sensitivity [27]. A notable aspect of biosensors is their high specificity, which means that they can detect the presence of a particular pathogen in a food sample with high precision [28]. Furthermore, biosensors can also be designed to simultaneously detect multiple pathogens or contaminants in food, reducing analysis time and increasing detection accuracy [29].

The implementation of photonic biosensors on silicon photonic-integrated circuits (PICs) indeed offers several advantages, making them a promising technology for various applications, including point-of-care diagnostics [30]. In summary, as determined by LoD and LoQ, sensitivity is a fundamental aspect of biosensor performance [31]. Achieving a high sensitivity enhances the effectiveness of biosensors across various applications where the detection and quantification of specific substances are critical. The sensitivity of waveguide sensors is intricately linked to the extent of overlap between the evanescent field and the sample being analyzed [32,33].

Compared to traditional methods, optical biosensors offer a faster and more effective response to possible food contamination. This is because optical biosensors can detect pathogens in food samples in short periods of time, in some cases within minutes [34].

These devices can detect extremely low levels of pathogens in food samples, allowing for early detection and a rapid response to potential disease outbreaks [34]. Additionally, optical biosensors are also highly specific, meaning that they can distinguish between different species of pathogens. Finally, it is worth noting that optical biosensors are also portable and easy to use, making them ideal for use in field environments [34].

Hence, it is of utmost importance to develop novel techniques for the rapid and sensitive detection of *E. coli* O157:H7, aiming to prevent potential disease outbreaks resulting from the public consumption of contaminated food. Emerging technologies, including isothermal amplification methods, biosensors, surface-enhanced Raman spectroscopy, paper-based diagnostics, and smartphone-based digital methods, represent innovative approaches in *E. coli* O157:H7 detection. In the realm of optical biosensors, two primary types are distinguished: the first relies on any potential alteration in the internal optical properties of a biomolecule due to its interaction with the target analyte, encompassing changes in absorption, emission, polarization, or luminescence reduction. The second type employs markers and optical probes. These systems, driven by reagent-mediated detection, utilize changes in the optical response of an intermediate, typically an analyte-sensitive dye molecule, to monitor the concentration of the analyte [35]. These nanobiosensors use electrochemical, fluorescence, colorimetric, and other techniques [36–38].

A biosensor falls into the category of bio-affinity sensors when the bio-identifying component is an antigen/antibody, DNAzymes, or DNA from a single-stranded DNA (ssDNA)/RNA sequence immobilized on a solid substrate through linker molecules that specifically interact with the target. Other classifications include enzyme sensors and sensor receptors, which involve ligand binding and biosensors for whole cells, depending on the type of physical probe utilized [39].

Fluorescent biosensor technologies are gaining significance in detecting *E. coli* O157:H7, offering practical advantages such as rapidity, portability, ease of use, sensitivity, and cost-effectiveness. Fluorescent dyes allow for direct measurement due to their stability compared to natural enzymes. Nevertheless, the challenge of low sensitivity persists, particularly in colorimetric detection methods [40]. The applicability of a detection method is also an important factor to consider. Some methods may be more suitable for detecting *E. coli* O157:H7 (culture, real-time PCR) or its verotoxins (ELISA, lateral flow immunoassays) or for use on certain food samples, while other methods may have broader applications. Optical and electrochemical biosensors are an example of this due to their limited applicability.

Microfluidic-based biosensors characterized by microchannels for fluidic samples are increasingly gaining relevance because they also facilitate on-chip immunoassays [40]. These systems enable the concurrent in situ execution of various laboratory processes, including detection, sampling, separation, and mixing [40,41].

Biosensors designed for *E. coli* O157 detection still rely on labor-intensive manual procedures, limiting their practical use in laboratories. While existing optical microfluidic biosensors exhibit accurate sensitivities, their high cost and complex assembly processes remain areas that require improvement [42].

This research aims to develop and optimize a photonic immunosensor for the simultaneous detection of *E. coli* O157:H7 and its verotoxins. This allows for the early detection and quantification of viable bacteria or their verotoxins, enabling the prediction of pathogen contamination in meat products, fresh vegetables, and prepared dishes. The significance of this technology in overcoming existing limitations in current pathogen detection systems in the food industry is highlighted by its potential applications in predicting and quantifying pathogen contamination.

2. Materials and Methods

2.1. Reagents and Antibodies

2.1.1. Functionalization

The functionalization reagents comprised CTES (carboxyethylsilanetriol, disodium salt 25% in MilliQ water, ABCR) at 1%, EDC (1-Ethyl-3-(3-dimethylaminopropyl) carbodiimide,

Sigma)/NHS (*N*-hydroxysuccinim) (Merck KGaA, Darmstadt, Germany), and MES (2-(*N*-morpholino) ethanesulfonic acid (ThermoFisher, Waltham, MA, USA) at 0.1 M.

The antibodies employed in this study included a polyclonal (rabbit) anti-*E. coli* O157:H7 antibody (BS-1563R, Bioss Antibodies, Woburn, MA, USA) and an anti-*E. coli* O157:H7 core monoclonal (mouse) antibody (clone E28) (CerTest Biotec, Zaragoza, Spain). The polyclonal and monoclonal antibodies were tested against target antigens derived from *E. coli* O157:H7 to evaluate the primary antibodies. The specific antigens used were the MT-25STX *E. coli* O157 VT1 recombinant protein and the MT-25VT2 *E. coli* O157 VT2 recombinant protein.

A rabbit polyclonal antibody anti-fish obtained from Eurofins Inmunolab (Reinbek, Germany) was also included as a negative control.

2.1.2. Indirect ELISA

The primary polyclonal antibody (Bioss Antibodies, Woburn, MA, USA) and the selected monoclonal anti-*E. coli* O157:H7 core monoclonal (mouse) antibody (clone E28) (CerTest Biotec, Zaragoza, Spain) were assessed against different concentrations of the inactivated *E. coli* O157 antigen (native extract) MT-28EC7U.

The indirect enzyme-linked immunosorbent assay (iELISA) reagents comprised commercial TMB substrate (Thermo Scientific, Waltham, MA, USA), H2SO4 solution (for slowing down the reaction), and 0.1 M hydrochloric acid from Scharlab (Barcelona, Spain). The secondary ELISA antibodies included the GARPO polyclonal anti-rabbit IgG (whole molecule) peroxidase antibody produced in goat and the GAMPO polyclonal anti-mouse IgG (whole molecule) peroxidase antibody produced in goat from Abcam (Cambridge, UK).

Each sample was inoculated in duplicate in 0.05 M carbonate buffer with a pH of 9.6 (Merck, Darmstadt, Germany). Furthermore, a phosphate-buffered saline (PBS) solution served as a negative control. In contrast, a standard solution of inactivated *E. coli* O157 antigen (native extract) MT-28EC7U (CerTest Biotech, Zaragoza, Spain) was employed as a positive control.

A total of seven serial dilutions from 1 mL of the stock of an inactivated *E. coli* O157 antigen (MT-28EC7U, CerTest Biotech, Spain) were prepared in 0.05 M carbonate buffer with a pH of 9.6 to assess antibody specificity: 1/10, 1/50, 1/80, 1/100, 1/1000, and 1/5000.

Serial dilutions of both MT-25STX *E. coli* O157 VT1 and MT-25VT2 *E. coli* O157 VT2 recombinant proteins (CerTest Biotech, Spain) in 0.05 M carbonate buffer with a pH of 9.6 were prepared to assess antibody specificity. The concentrations of the final dilutions of VTx1 were 48.4 ppm, 24.2 ppm, 12.1 ppm, 6.05 ppm, 3.025 ppm, and 1.1909 ppm. And the concentrations of the final dilutions of VTx2 were 38 ppm, 19 ppm, 9.5 ppm, 4.75 ppm, 2.375 ppm, and 1.1875 ppm.

The emitted absorbance was measured at 450 and 650 nm using a Varioskan Flash multimode spectral scanning plate reader (Multilabel Victor 1420 Counter). The positive control was defined as a sample with an OD450 nm value 2.1 times higher than the negative control (P/N 2.1).

2.1.3. Immunosensor Validation

To validate the detection capability of the photonic technique and the conducted detection tests against *E. coli* O157:H7, a total of five 1/10 serial dilutions from an overnight *E. coli* O157:H7 CECT 4972 (Spanish Type Culture Collection, Valencia, Spain) culture were prepared to inoculate in TSBm+n broth medium (Triptone Soya Broth + Novobiomiocin supplement) at a specified concentration range of 10 to 106 CFU/mL.

To validate the biosensor detection capability against *E. coli* O157:H7 verotoxins, serial dilutions of both MT-25STX *E. coli* O157 VT1 and MT-25VT2 *E. coli* O157 VT2 recombinant proteins (CerTest Biotech, Spain) were prepared to be inoculated in TSBm+n broth medium (Triptone Soya Broth + Novobiomiocin supplement). The concentrations of the final dilutions of VTx1 ranged from 1 to 50 ppm: 48.4 ppm, 24.2 ppm,12.1 ppm, 6.05 ppm,

3.025 ppm, and 1.1909 ppm. And the concentrations of the final dilutions of VTx2 were 38 ppm, 19 ppm, 9.5 ppm, 4.75 ppm, 2.375 ppm, and 1.1875 ppm.

Additionally, the biosensor response was further evaluated through multiple replicates of artificially contaminated minced beef meat samples obtained from the same batch (25 g of each batch was suspended in 250 mL of TSBm+n). The samples were subsequently inoculated with both different concentrations of *E. coli* O157 culture ranging from 10 to 106 CFU/mL and dilutions of VTx1 and VTx2 ranging from 1 ppm to 50 ppm. The batches of minced beef samples were collected over three months from a meat processing facility in the Valencia region. These samples had undergone a prior assessment using quantitative polymerase chain reaction (qPCR) and traditional plate counting to verify the presence of the bacteria *E. coli* O157:H7, confirming negative results. Significantly, these batches were obtained directly from their original packaging and were intended for immediate consumption. They were purchased from a local grocery store and stored at a temperature of 4 °C until employed in the assays.

Concentration tests were conducted, followed by confirmation of the bacteria presence through the separation and concentration of the microorganism using immunomagnetic particles coated with anti-*E. coli* O157:H7 antibodies (Dynabeads anti-*E. coli* O157, 71003, Applied Biosystems by Thermo Fisher Scientific Baltics UAB, Vilnius, Lithuania). For each isolated colony which was examined, one drop was dispensed to be confirmed by a latex agglutination test for the identification of the *E. coli* serogroup 0157:H7 from Oxoid (TSMX9410, RIM latex test for *E. coli* O157:H7, Oxoid, Termo Fisher-Scientific). These incubation conditions align with established recommendations and standards for bacterial culture and growth [43].

A data-reading setup according to [44] was employed for reading the signal transduction related to the resonance of 100 Photonic-Integrated Circuits (PICs) designed by Lumensia Sensors (Spain). One hundred photonic biosensors were integrated on silicon photonic-integrated circuits (PICs). A two-channel microfluidic system connected to the PICs and a peristaltic pump were connected to the data-reading setup.

2.2. Indirect ELISA

To assess the suitability of the developed biosensor for detecting *E. coli* O157:H7, it was necessary to characterize the antibodies used. This involved evaluating the binding capacity of both polyclonal and monoclonal antibodies specific to *E. coli* O157:H7. An iELISA protocol was designed for this purpose, based on established methods from the existing literature [19,20].

This new protocol was based on previous descriptions with modifications [45,46]. The primary polyclonal antibody (Bioss Antibodies, Woburn, MA, USA) and the selected monoclonal anti-*E. coli* O157:H7 core monoclonal (mouse) antibody (clone E28) (CerTest Biotec, Zaragoza, Spain) were assessed against different concentrations of the inactivated *E. coli* O157 antigen (native extract) MT-28EC7U.

A 96-well ELISA microplate was loaded with 100 µL/well of the corresponding concentration of the *E. coli* O157:H7 inactivated antigen (native extract) MT-28EC7U. Additionally, replicates from a batch of minced beef meat spiked with inactivated *E. coli* O157:H7 (native extract) MT-28EC7U (CerTest Biotech, Zaragoza, Spain), as above described in 2.1, were included to assess antibody specificity. The wells underwent three washes with PBS with 0.05% Tween-20. Subsequently, all the wells were blocked with 1% BSA in PBS by adding 100 µL to each well and were incubated for one hour at room temperature (25 °C).

Following the washes, 100 µL of a PBS solution containing 1 ppm of the selected antibody specific for *E. coli* O157 was added to all the wells previously inoculated with the antigen. The plates were then incubated at 37 °C for 1 h, with a control group lacking immobilized antibodies. Subsequently, three successive washes were conducted with PBST, and the wells were filled with 100 µL of a goat anti-rabbit–HRP conjugate, followed by another incubation at 37 °C for 1 h. After three additional washes with PBST, the detection reaction was initiated by adding 100 µL of the substrate based on 1,2-diaminobenzene

(OPD) (4 mg of 1,2-diaminobenzene and 15 L of H_2O_2 in 10 mL of citrate buffer, pH 4.5). After a 15 min incubation period, the reaction was halted by adding 50 µL of 2 M sulfuric acid. Finally, the emitted absorbance was measured at 450 and 650 nm using a Varioskan Flash multimode spectral scanning plate reader (Multilabel Victor 1420 Counter). The positive control was defined as a sample with an OD450 nm value 2.1 times higher than the negative control (P/N 2.1).

2.3. PIC Fabrication and Functionalization Process

The optical photonic-integrated circuits (PICs) developed for this study underwent fabrication in a controlled clean room environment, specifically within a class 10–100 clean room. The process involved utilizing an electron beam writing technique on a positive resist layer made of polymethylmethacrylate (PMMA) with a thickness of 100 nm [44]. The manufacturing process of the photonic-integrated circuits (PICs) began with preparing a silicon wafer, which served as the substrate for the PICs, and positive photoresist was deposited onto the wafer surface. The circuit patterns were defined on the photoresist layer using lithography techniques (electron beam process or ultraviolet light). This involved exposing the photoresist to light through a photomask containing the desired circuit pattern. After exposure, the photoresist was developed, removing the unexposed areas and leaving behind the desired circuit patterns on the wafer. A layer of chromium was deposited onto the wafer using an evaporation process. After deposition, the remaining photoresist was dissolved or lifted off, leaving the metal patterns on the wafer. The silicon nitride layer on the wafer was etched to define the waveguide structures required for photonic circuits. A layer of silicon oxide was deposited onto the wafer to cover the circuitry.

The detection of *E. coli* O157:H7 and its verotoxins (VTx1 and VTx2) was achieved through the functionalization of 100 photonic biosensors (PICs). These biosensors were obtained from various batches and wafers, and the functionalization process followed the established procedure outlined by [44].

The process began with the oxidation of the PIC surface, which was achieved by immersing it in 5 mL of a 0.1 M hydrochloric acid solution provided by Scharlab (Barcelona, Spain) for 30 min (Figure 1). The oxidation step was carried out on an orbital shaker plate at 30 rpm. Subsequently, the surface underwent a meticulous rinse with deionized water (DIW) and was dried.

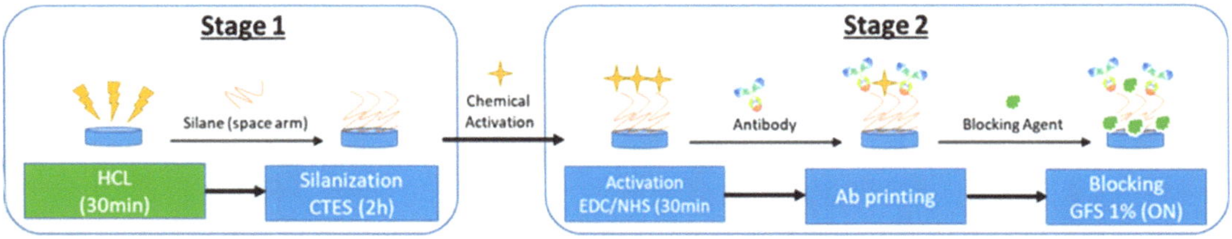

Figure 1. Graphic representation of PIC functionalization process ([46]).

After the oxidation step, the surface underwent silanization with a carboxyethylsilanetriol (CTES) solution for 2 h. Simultaneously, the activation of the carboxylic group within the CTES organosilane on the surface was accomplished by introducing a mixture of carbodiimide and *N*-hydroxysuccinimide (EDC/NHS), followed by a 30 min incubation at room temperature. The surface was then thoroughly rinsed and dried under a stream of airflow.

The covalent immobilization of anti-*E. coli* O157:H7 antibodies on the biosensor surface were conducted with the specific anti-*E. coli* O157:H7 rabbit polyclonal antibody (Bioss Antibodies, Woburn, MA, USA) and the anti-*E. coli* O157:H7 monoclonal (mouse) antibody (clone E28) (CerTest Biotec, Zaragoza, Spain). This immobilization process was carried out in an oriented manner (Figure 1).

After the covalent immobilization of anti-*E. coli* O157:H7 antibodies on the biosensor surface, the surfaces were washed with phosphate-buffered saline (PBS), dried, and then blocked. The blocking step involved incubating the surfaces overnight with a solution containing 1% cold water fish skin gelatin (GFS) in PBS. This ensured the effective blocking of the non-specific binding sites on the surface (Figure 1).

Throughout these steps, precise control and alignment were crucial to ensure the accuracy and functionality of the resulting PICs. The completed PICs could then be further processed and integrated into a two-channel microfluidic cartridge (Figure 2).

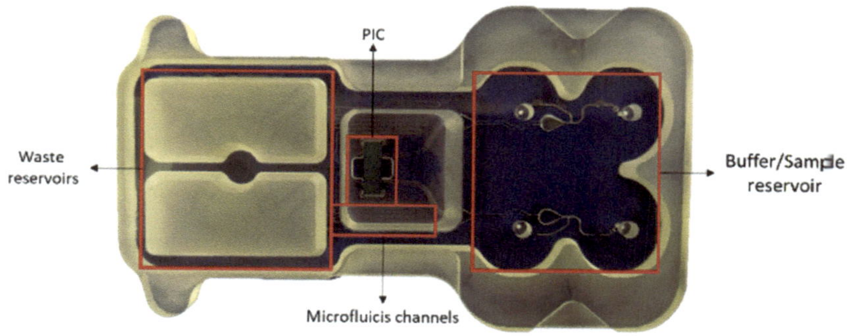

Figure 2. PIC ensembled into microfluidic cartridge.

2.4. *Photon Transduction Setup Reader*

Photon transduction is the fundamental principle underlying the detection process of this system. Resonant cavities, specifically ring resonators (RRs) fabricated with silicon nitride technology, are employed for this purpose. The analysis of the refractive indices obtained from these ring resonators (RRs) has been proposed for various applications. In these transducers, a change in the refractive index of light occurs when the analyte of interest interacts with the bioreceptor anchored on the surface of the ring. This enables the correlation of the analyte concentration attached to the ring with the observed signal. The optical transduction systems used on biosensors offer unique characteristics conferring several advantages.

Indeed, these biosensors leverage significantly smaller sample volumes by integrating microfluidic and nanophotonic systems, enabling analyses at the nanometric or micrometric scale. Furthermore, the label-free detection capability of these biosensors allows for real-time monitoring, simplifying procedures by reducing the necessary steps and reagents.

The photonic biosensors integrated on silicon photonic-integrated circuits (PICs) offer label-free performance with high detection sensitivities. Notably, they are disposable, making them particularly interesting for point-of-care diagnostics as they do not necessitate specialized personnel [30].

The current detection system is designed to capture the sensor's transduction signal during sample analysis. It involves the integration of three essential components forming a measurement setup: a photonic-integrated circuit (PIC) designed by Lumensia Sensors (Spain), a two-channel microfluidic system connected to the PIC, and a peristaltic pump. Both microfluidic channels are linked to the microfluidic system attached to the PIC, through which the samples to be detected flow (Figure 2). Four sealed reservoirs are housed per channel, in which the flow direction is carried out in the following order: a reservoir to introduce the washing buffer; another one, below, in which the test sample to be analyzed is introduced; a hole in which the adhered PIC is housed and embedded with the washing buffer first and then the problem sample; and, finally, a waste reservoir which is reached by all the fluids which have passed through the PIC housing. The system ensures that there is no cross-reactivity since both channels are sealed and separated (Figure 2).

A data-reading setup according to [44] was employed for the signal transduction related to resonance. This reading system was based on software and hardware developed by Lumensia Sensors, enabling the translation of the optical signal into resonance measured in picometers (pm).

2.5. Immunosensor Validation for E. coli O157:H7 and Shiga-like Toxins

The performance and reliability of the detection technique were evaluated by flowing different minced meat samples spiked with antigens of *E. coli* O157:H7 over previously functionalized photonic biosensors (PICs). Various experimental tests were conducted to determine this alternative detection system's optimal sensitivity and limit of detection (LoD). This critical step facilitated a comprehensive assessment of the technique's effectiveness in detecting and responding to various *E. coli* O157:H7 antigens, including a strain of the bacterium and its verotoxins VT1 and VT2 in isolation, contributing to its validation and reliability.

Different samples of minced meat from the same batch, inoculated with *E. coli* O157:H7 CECT 4972 (range of 10 to 10^6 CFU/mL, as detailed in Section 2.2), were included in the analysis to assess the sensor's detection efficiency. Moreover, the sensor's detection efficiency within a food matrix was also stablished by inoculating different dilutions of verotoxin 1 (VTx1) and verotoxin 2 (VTx2) to compare the immunosensor's response against both types of antigens. Moreover, serial dilutions of verotoxin 1 and verotoxin 2 produced by *E. coli* O157:H7, only suspended in the pre-enrichment medium, as described in Section 2.1, were used to evaluate the specific immunosensor detection.

Current biosensor trends focus on improving accessibility, robustness, and the overall efficiency of the detection process. Integrating these components creates a comprehensive biosensor system capable of detecting and analyzing analytes with high sensitivity and precision. The combination of optical transduction on a PIC, two microfluidics channels, and a controlled sample flow (peristaltic pump) enhances the performance of the biosensor. The setup is specifically designed to facilitate the controlled flow of samples over the sensor for analysis [44].

The next step in the experimental procedure involved systematically flowing the samples and subjecting them to bacterial sensing using the setup detector optimized for PICs, which was fabricated by Lumensia Sensors [44]. A critical aspect of this step was facilitating the controlled flow of samples by attaching the microfluidic adhesive layer. The coupled microfluidic system, integrated with eight ring resonators arranged within the sensor and distributed across two channels, enhanced the detection capabilities by enabling the concurrent measurement of two samples. The sensor incorporated eight ring resonators, which are optical components which respond to changes in the refractive index. These resonators were distributed across two channels, each containing four resonant rings [44]. The configuration, with multiple resonators and channels, contributed to a high sensitivity in detection. The system could detect concentrations down to the ng/mL scale [44].

Additionally, the immunosensors were securely housed within the previously functionalized PICs obtained according to the procedure outlined in Section 2.4. Tests were conducted using the same batch of sensors that had been manufactured and printed on the same date. The sensors were used to detect samples with the same levels of contamination on the same matrix on different dates, up to six months and one year after their manufacturing and functionalization.

Indeed, the setup for reading resonance data is a crucial component of the biosensor system, allowing for the interpretation of the optical signals generated during the interaction between the biosensor and the target analytes (*E. coli* O157:H7 bacterium, VTx1, and VTx2). The hardware and software components developed by Lumensia Sensors play a crucial role in translating these optical signals into measurable resonance values, typically expressed in picometers (pm). This reading system is essential for monitoring and quantifying the binding events on the biosensor surface, providing valuable information about the presence and concentration of the target analytes. The integration of this advanced reading system enhances the overall performance and reliability of the biosensor platform [44].

A peristaltic pump was utilized to control the flow rate of the sample dilutions, operating at a rate of 15 μL/min. This precise flow method ensured that the prepared samples reached the immunosensors, where the crucial reaction between the *E. coli* O157:H7 antigen and the functionalized antibodies specific to that antigen occurred in the *E. coli* O157 immunodetection assay (Figure 3). A specific flowing protocol consisting of the following steps was implemented. Initially, TSBm+n broth was run for 3 min to establish a reference signal. Following this, the bacterial sample, previously diluted in the same broth, was flowed for 15 min, allowing the immunosensors to interact with the *E. coli* O157:H7 bacterium and its verotoxins in the sample. Subsequently, a cleaning buffer was employed to prepare the system for subsequent analyses and ensure its cleanliness and readiness. For the next five minutes, a cleaning buffer (TSBm+n) was flowed. This step effectively cleared any residual materials and contaminants from the system.

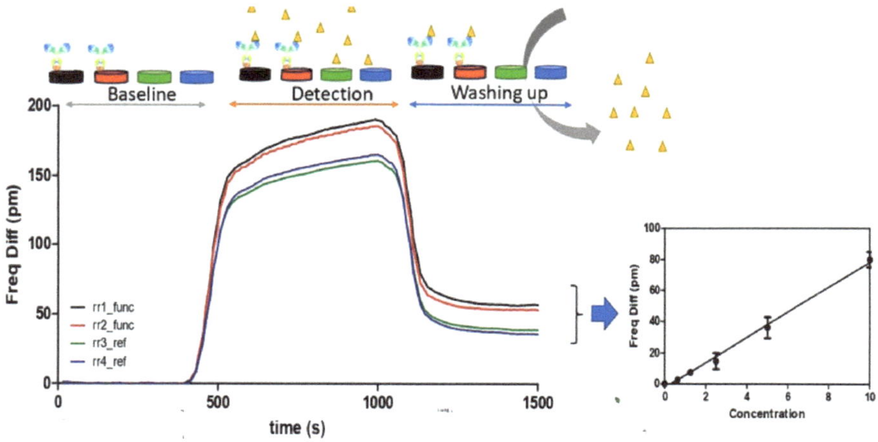

Figure 3. Biosensor sensogram detection of *E. coli* O157:H7 or its verotoxins. A schematic sensogram obtained by the immunosensor setup reader against the target antigen is included. Here are represented the resonance obtained versus the test time. The hardware and software components of the setup reader translate the optical signals obtained in the detection sensogram into measurable resonance values: freq. diff (pm) against the concentration of the target antigen (log CFU/mL).

As a graphical representation of the biosensor's response to the analyte over time, a sensogram represents the observation of the resonance signal (Figure 3). This provides insights into the biosensor's sensitivity, specificity, and overall performance in detecting and quantifying *E. coli* O157:H7. In this sense, the *X*-axis, indicating time in seconds, and the *Y*-axis, representing resonance values in picometers (pm), for monitoring and analyzing the biosensor's response. The positive difference in the resonance values between the rings functionalized with specific antibodies and the reference rings is a reliable indicator of the presence of the target analyte. This differential signal is crucial for detecting and quantifying the concentration of *E. coli* O157:H7 or its verotoxins in the sample.

The units of measurement, such as colony-forming units per milliliter (CFU/mL) for *E. coli* O157:H7 or parts per million (ppm) for verotoxins (VTx1 or VTx2), are indicative of the quantity of the analyte present in the sample. A higher concentration of the *E. coli* O157:H7 antigen results in a more pronounced increase in the resonance values on the sensogram. This tool serves for the quantitative and indirect estimation of the concentration of the target analyte.

2.6. Data Analysis

The evaluation of the method's sensitivity and specificity involved a double-blind assay, where negative minced beef meat samples were intentionally spiked with *E. coli* O157:H7 and its verotoxins 1 and 2. The obtained results were then subjected to a statistical analysis to determine their significance [47,48]. Multiple repetitions of the detection and quantification using the biosensor were conducted for each concentration of this study (strains and verotoxins), under identical conditions, utilizing similar chemicals and instrumentation elements. This approach helped ensure the reliability and consistency of the method across different repetitions and concentrations.

The statistical analysis was conducted to assess each variable's impact using an ANOVA test. Additionally, variations in the frequency of positive samples were determined through a chi-square test at a significance level of 95%. The data analysis was performed using the Systat version 9 software (SPSS Inc., Chicago, IL, USA). Statistically significant differences, as determined by a one-way analysis of variance (ANOVA), were considered when the p-values were less than or equal to 0.05.

3. Results and Discussion

3.1. iELISA and Sensitivity Studies

The evaluation of the absorbance values obtained for each antibody against different *E. coli* O157:H7 antigens (obtained and described in Section 2.3 Reagents and Antibodies) is a crucial step in characterizing the binding capacity and specificity of the antibodies. The absorbance values obtained through the iELISA method typically indicate the amount of antigen–antibody complexes formed during the assay. Higher absorbance values suggest stronger binding between the antibodies and the target antigens, demonstrating specificity and sensitivity (Figures 4 and 5).

Specificity and sensitivity results are typically generated by plotting the absorbance values against known analyte concentrations. Each analyte (commercial *E. coli* O157:H7 strain, commercial VTx, and commercial VTx2) were used with known concentrations. For *E. coli* O157:H7, the sensitivity and specificity in terms of absorbance versus concentration in the dilution factor allowed for quantifying the bacterial concentration in the samples based on the obtained absorbance values. The resultant curves established a comparison between the signal (OD 450 nm) and the actual concentration of *E. coli* O157:H7 (inactivated *E. coli* O157 antigen MT-28EC7U, MT-25STX, CerTest Biotec, Zaragoza, Spain) and its verotoxins VTx1 and VTx2 (*E. coli* O157 VT1 recombinant protein and MT-25VT2, *E. coli* O157 VT2 recombinant protein, CerTest Biotec, Zaragoza, Spain).

The observed results, in the case of the commercial *E. coli* O157:H7 strain, show that for each anti-*E. coli* O157:H7 monoclonal antibody there is a greater affinity for the entire concentration range used, since higher absorbance values are obtained for each range of concentrations. Similarly, for verotoxin 1 (VTx1) and verotoxin 2 (VTx2), the results of specificity and sensibility in terms of absorbance versus ppm (parts per million) provide a means to quantify the concentration of these toxins in the samples based on the absorbance measurements. For both verotoxins, the affinity of the polyclonal antibody is greater across the entire range of concentrations used, since higher absorbance values are obtained for each range of concentrations of each verotoxin.

The observed trend, where higher concentrations of the commercial *E. coli* O157:H7 strain and its verotoxins (VTx1 and VTx2) result in higher absorbance values, is consistent with expectations in an indirect enzyme-linked immunosorbent assay (iELISA). This behavior indicates that the monoclonal antibody and the polyclonal antibody, selected as probes for the biosensor developed, respond appropriately to varying concentrations of the target *E. coli* O157:H7 and its verotoxins, respectively. This evaluation of the binding capacity of each antibody and their specificity and sensitivity results are consistent with the previous bibliography, strengthening the reliability of the results [22,49–55].

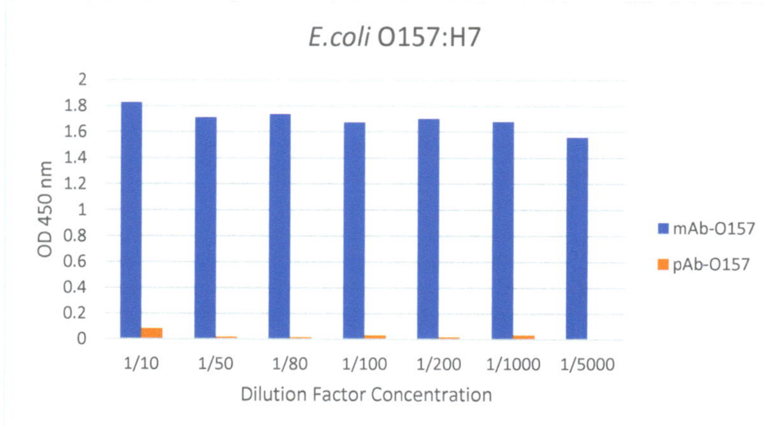

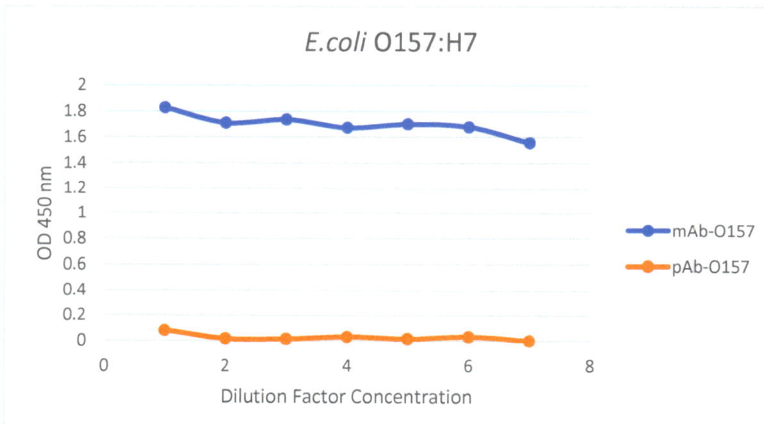

Figure 4. Specificity and sensitivity iELISA immunoassay results against *E. coli* O157:H7. Absorbance values (OD 450 nm) of monoclonal and polyclonal immunosensor antibodies versus bacterial concentration in the dilution factor of a commercial inactivated antigen of *E. coli* O157:H7.

(a)

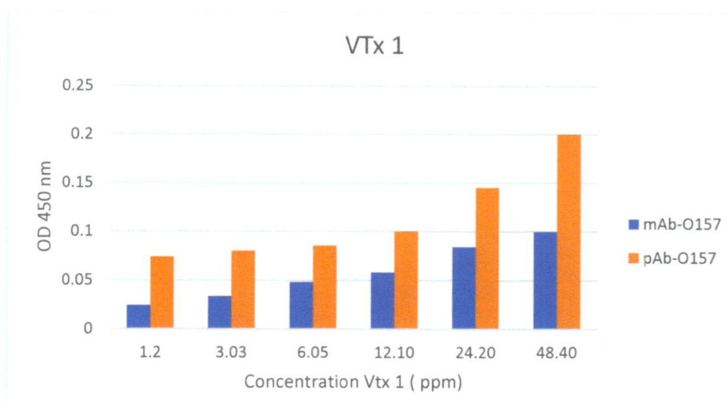

Figure 5. *Cont.*

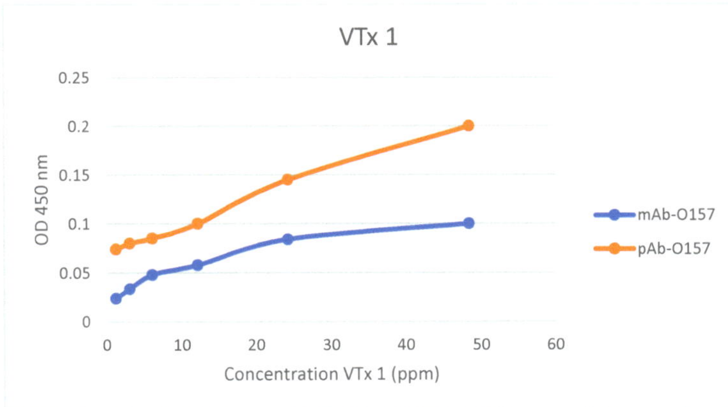

(b)

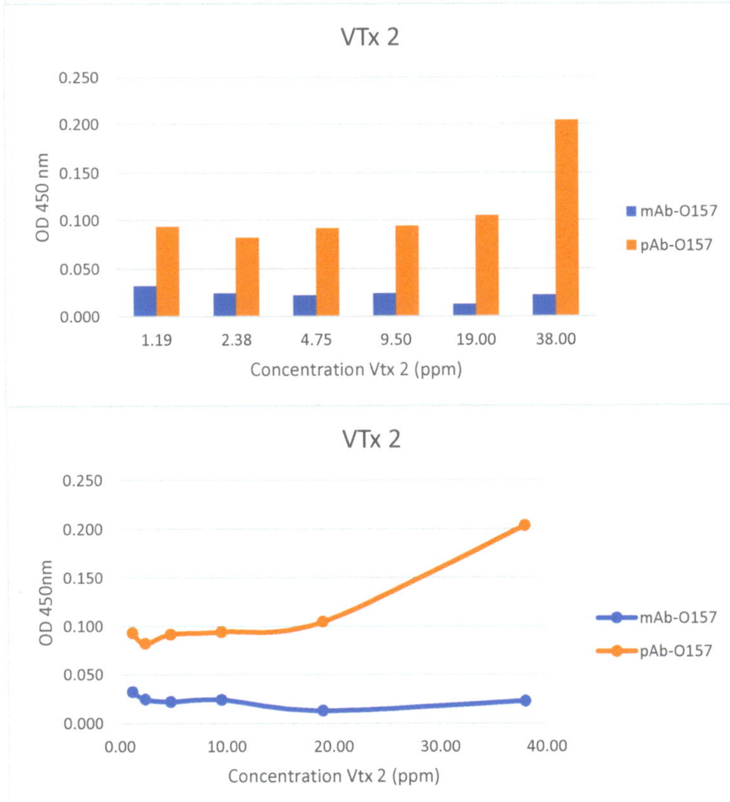

Figure 5. Specificity and sensitivity iELISA immunoassay results against *E. coli* O157:H7 verotoxins 1 and 2. Absorbance values (OD 450 nm) of monoclonal and polyclonal immunosensor antibodies versus verotoxin concentration in ppm. (**a**) iELISA antibody absorbances values increasing dilutions of commercial VTx1; (**b**) iELISA antibody absorbance values increasing dilutions of commercial VTx2.

Observing optimal results in terms of affinity at both high and low concentrations of *E. coli* O157:H7 and its verotoxins (VTx1 and VTx2) is a positive finding. This suggests that the selected polyclonal antibody is effective across various concentrations of verotoxins and does not saturate at high concentrations. The high binding efficiency at both low and high concentrations of the entire bacterium *E. coli* O157:H7 antigen is an important characteristic, indicating the versatility and sensitivity of the selected monoclonal antibody as an immunosensor probe [50].

3.2. Immunosensor Specificity and Sensitivity

The detection sensitivity of the immunosensor method for *E. coli* O157:H7 was evaluated over a concentration range from 10 to 10^6 CFU/mL. This study aimed to determine the most effective enrichment method for detecting *E. coli* O157:H7. For this purpose, the assay was conducted using both a pure culture of the bacteria and food samples inoculated with the bacteria. The validation process included several dilutions of Shiga-like toxins VTx1 and VTx2 with a range of concentrations for evaluation, and the assessment involved analyzing the presence or absence of the pathogen. The results of the comparison between the response detection of the immunosensor and quantification through a culture of *E. coli* O157:H7 (CFU/mL) (Table S1, Supplementary Materials file attached) and quantification through a chemical concentration of verotoxins (ppm) (Table S2, Supplementary Materials file attached) were compared with those obtained following gold standards.

A low *p*-value suggests that the observed results are unlikely to be due to random chance, strengthening the confidence in the effectiveness of the biosensor for detecting *E. coli* O157:H7 and its verotoxins. In this sense, the statistical significance obtained ($p = 0.0026$) further supports the reliability of the detection method (see Section 2.6 Materials and Methods Section and the results reflected in Tables S1 and S2, Supplementary Materials file attached).

The biosensor in development has achieved a 100% observed agreement or relative accuracy with the reference methods. This level of agreement is crucial for validating the biosensor as a reliable alternative for detecting *E. coli* O157:H7, especially compared to established reference methods (Figure 6).

Sensitivity is a crucial parameter in biosensor performance as it reflects the ability to correctly identify true-positive cases. This case study demonstrates the effectiveness of the biosensor method in detecting samples contaminated with the *E. coli* O157:H7 antigen while correctly identifying samples without the microorganism or its verotoxins (VTx1 and VTx2) (Tables S1 and S2, Supplementary Materials).

Specificity is another parameter that reflects the ability of the biosensor to accurately identify samples that do not contain the target microorganism or its associated toxins. The obtained predictive value (PPV) and negative predictive value (NPV) of 100% suggest that the biosensor method can distinguish between true negatives and true positives for the presence of *E. coli* O157:H7 and its verotoxins. These values confirm the accuracy and reliability of the biosensor method in detecting positive samples and correctly identifying negative samples.

The results suggest that either of the two tested methods can effectively detect *E. coli* O157:H7 and its verotoxins in minced beef samples, whether spiked with the bacterium strain or Shiga-like toxins. Both methods have comparable specificity levels and can accurately identify the presence of *E. coli* O157:H7 antigens in the sample (Tables S1 and S2, Supplementary Materials).

This method presents reproducibility based on the specific binding of *E. coli* O157:H7 antigens observed in 97.6% of the contaminated samples. This value suggests a 97.6% probability of obtaining the same detection result when analyzing identical samples under standard reproducibility conditions at different times (Tables S1 and S2, Supplementary Materials). This is a valuable characteristic, as it indicates that the methods can consistently produce accurate results across multiple analyses and under varying conditions.

The generation of calibration curves is crucial in assessing the immunosensor method's potential for providing a quantitative response. Calibration curves relate the measured parameter—in this case, the resonance in picometers (pm)—to known concentrations of *E. coli* O157:H7 and its verotoxins (VTx1 and VTx2). This allows for establishing a relationship that can be used to quantify the concentration of the target analyte in unknown samples.

This is essential for enumerating *E. coli* O157:H7 on food safety assessments to confirm the acceptability of a food product [8].

The sensitivity of the biosensor to different concentrations of *E. coli* O157:H7 and its verotoxins is reinforced by the dependent relationship between the observed concentration of the bacterium or its verotoxin versus the optical signal it generates. As observed in the data presented in Figure 6, the optical signals (resonance notch shifts) are stronger for samples with higher concentrations and weaker for more diluted samples, which is consistent with the biosensing principles. Figure 6, illustrating the micro-ring resonance notch shift in picometers (pm) during the *E. coli* O157:H7 experiment with different dilution factors and concentrations, provides valuable information about the biosensor's performance.

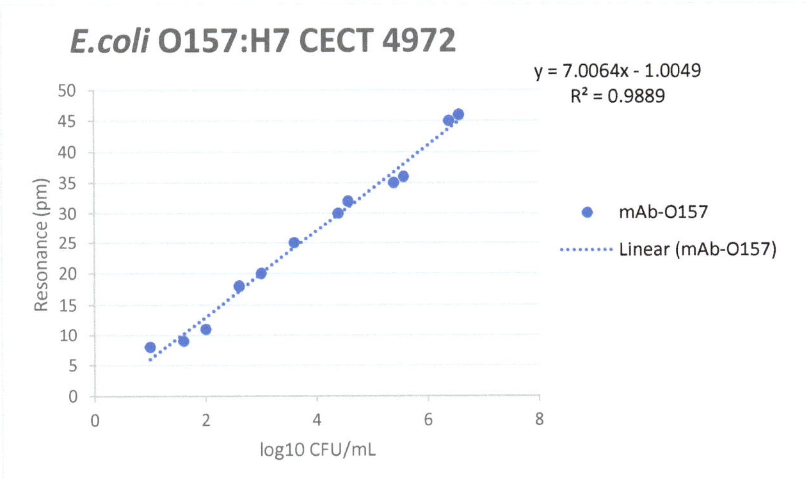

(A)

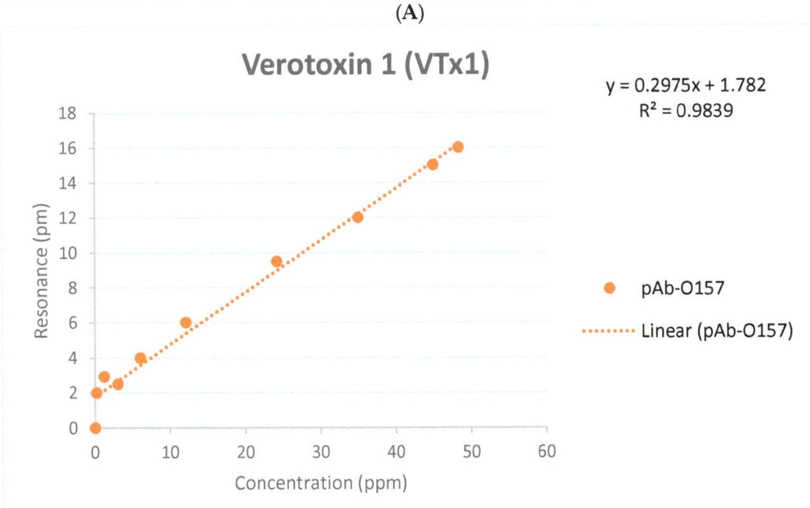

(B)

Figure 6. *Cont.*

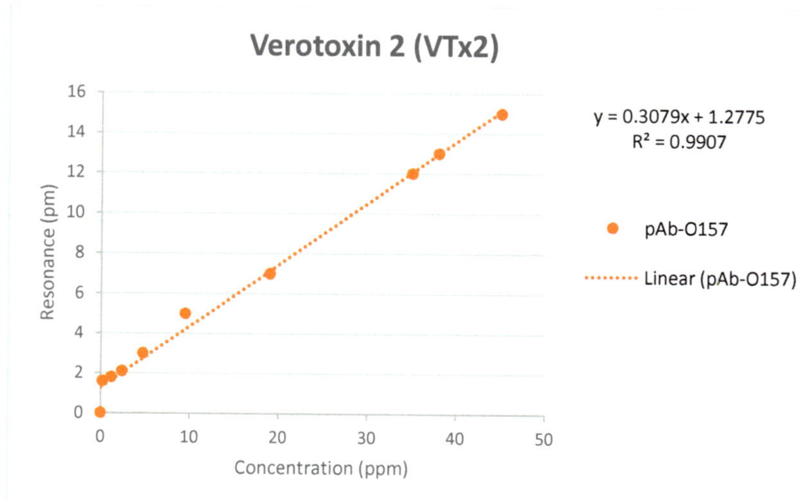

(C)

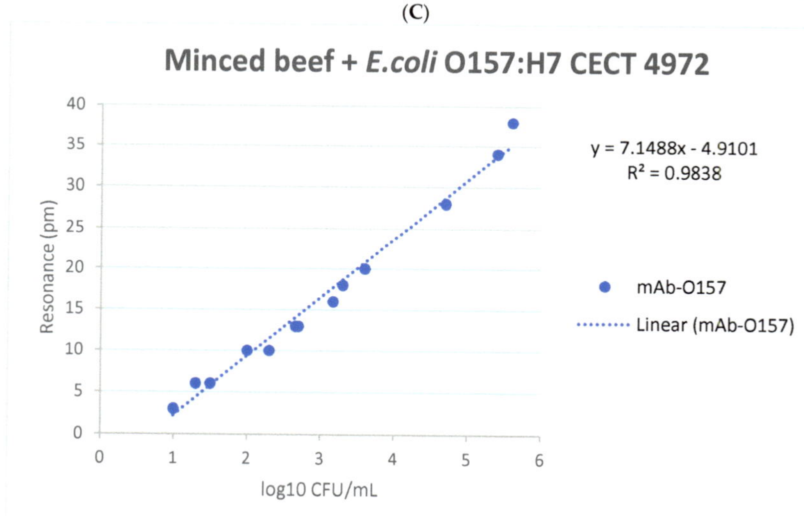

(D)

Figure 6. *Cont.*

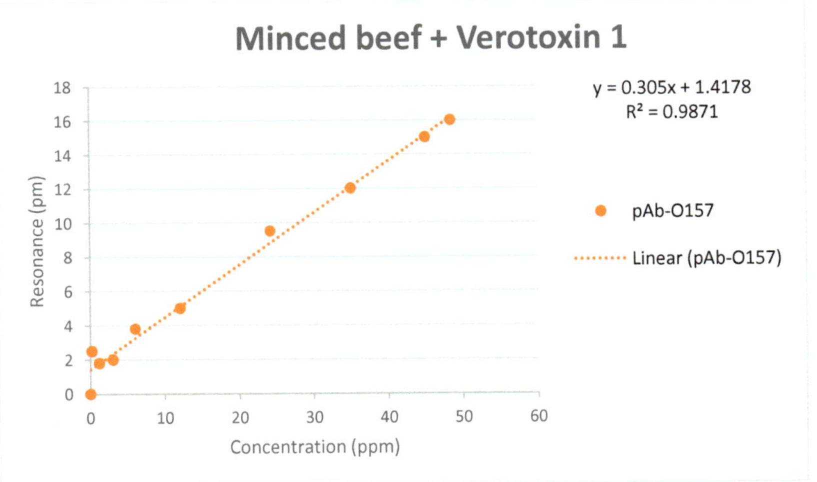

(E)

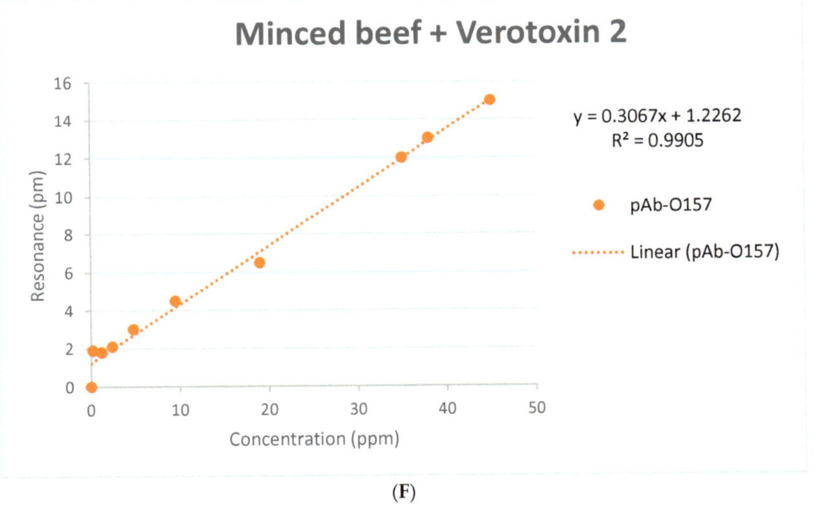

(F)

Figure 6. Validation calibration curves against antigens of *E. coli* O157:H7. (**A**) Biosensor calibration curve against *E. coli* O157:H7. (**B**) Biosensor calibration curve against verotoxin 1 (VTx1). (**C**) Biosensor calibration curve against verotoxin 2 (VTx2) (**D**) Biosensor calibration curve against *E. coli* O157:H7-spiked minced beef samples. (**E**) Biosensor calibration curve against VTx1-spiked minced beef samples. (**F**) Biosensor calibration curve against VTx2-spiked minced beef samples.

In a quantitative analysis, the working interval refers to a range of values for which the test method exhibits adequate precision, trueness, and linearity. By examining the six curves (Figure 6) presented graphically, establishing the working interval for each calibration curve facilitated the estimation of key parameters. A reliable working interval ensures that the method performs accurately across a specified concentration range of 10 to 106 CFU/mL for the entire bacterium *E. coli* O157:H7 and across a range of 1 to 50 ppm for its verotoxins.

The LoD is a critical parameter that indicates the lowest concentration of the target analyte that the biosensor can reliably detect. A notable obtained result is achieving a 100% detection rate for samples containing as few as 10 CFU/mL of *E. coli* O157:H7 or

1 ppm of VTx1 or VTx2. This indicates that the biosensor method can reliably detect low concentrations of *E. coli* O157:H7 and its verotoxins (Figure 6).

The results of the detection of spiked samples with dilution factors up to 10 CFU/mL and concentrations up to 1 ppm/mL indicate that the biosensor has an LoD of 10 CFU/mL for the detection *E. coli* O157:H7 and an LoD of 1 ppm/mL for the detection of its verotoxins (Figure 6A–C).

To calculate the Limit of Detection (LoD), the formula LoD = 3.3 * s0 was applied. To calculate the Limit of Quantification (LoQ), the formula LoQ = 10 * s0 was used. The Upper Limit of Quantification (ULoQ) was determined based on the specific requirements of the analysis. The threshold spread (s0) for the detection method was estimated by performing at least six determinations of samples at the calculated breakpoint concentration. The CFU/mL of *E. coli* O157:H7 and the ppm of VTx1 and VTx2 in the enrichment cultures were obtained from Tables S1 and S2, respectively. The unit of measurement used for resonance was picometers (pm), and these data were obtained from a laboratory setup reader after processing the spiked samples, as detailed in Section 2.

The lower limit of quantification (LoQ) represents the lowest concentration of the analyte (*E. coli* O157:H7 and verotoxins) that can be reliably quantified with acceptable precision and accuracy. In this study, the LoQ was approximately 100 CFU/mL for the bacteria and two ppm for verotoxin 1 and verotoxin 2. This indicates that the biosensor method can reliably quantify concentrations above this threshold. The limit of detection (LoD) is the lowest concentration of the analyte that can be reliably detected, although not necessarily quantified, by the method. In this study, the LoD was positioned at around 10 CFU/mL for pathogen detection and 1 ppm for verotoxins detection. This signifies the method's sensitivity in detecting even lower concentrations, albeit without precise quantification. The upper limit of quantification (ULoQ) represents the highest concentration of the analyte that can be reliably quantified without introducing significant measurement errors. The ULoQ was observed to be approximately 10^6 CFU/mL for the bacteria and 48 ppm for verotoxin 1 and verotoxin 2. This boundary indicates the upper range of concentrations within which the biosensor method can provide an accurate and reliable quantification.

This immunosensor method offers practical advantages related to the working interval, the limit of detection (LoD), and storage life that make it a promising alternative for detecting *E. coli* O157:H7 compared to other immunosensors that have already been developed (Table 1). The working intervals of technologies based on immunosensors with electrochemical impedance spectroscopy [56] or cyclic voltammetry [57] cover a wide linear range. However, the biosensor under development extends working interval further as quartz crystal microbalance [58].

Table 1. Detection immunosensor advantages and disadvantages related to other immunosensor methods against *E. coli* O157:H7.

Detection Technology	Working Interval (CFU/mL)	Limit of Detection (CFU/mL)	Storage Life	Reference
Surface plasmon resonance	10^2–10^3	6×10^2	-	[59]
Electrochemical impedance spectroscopy	10^4–10^7	10^4	-	[56]
Electrochemical impedance spectroscopy	10^3–10^5	10^3	1 week	[60]
Cyclic voltammetry	10^5–10^9	7.374×10^4	1 week	[57]
Quartz crystal microbalance	10^3–10^8	10^3	1 week	[58]
Photonic immunosensor	10^1–10^6	10^1	6 months	This work

The limit of detection (LoD) for the sensor under development compared with the LoD of other immunosensor technologies is the lowest (Table 2). Another advantage that systems based on immunosensors confer is their storage life capacity, assuming competitive advantages for this biosensor which are comparable to and improved with respect to cyclic voltammetry, electrochemical impedance spectroscopy, quartz crystal microbalance, and, indeed, surface plasmon resonance [59].

Table 2. Assessment of the immunosensor under development against the different detection methods of *E. coli* O157:H7, depending on its speed in testing, cost, applicability, sensitivity, and specificity.

Detection Technology	Speed in Testing	Cost	Applicability	Sensitivity	Specificity
Plate culture	24 h	Low	Wide	Low	Low
Lateral flow immunoassay	30–45 min	Moderate	Limited	Low	Low
Enzyme-linked Immunosorbent Assay (ELISA)	3–5 h	High	Wide	Low	Low
qPCR	6 h	High	Wide	Moderate	Moderate
Multiplex PCR	6 h	High	Wide	Moderate	Moderate
Hybridization	18–24 h	Moderate	Limited	Moderate	Moderate
Microarrays	48–72 h	High	Wide	High	High
Optical biosensors	45 min–3 h	High	Limited	High	Moderate
Electrochemical biosensors	18–24 h	Moderate	Limited	High	Moderate
Photonic immunosensor	30–45 min	Low	Wide	High	High

The results suggesting that the developed immunosensor is equally effective in detecting *E. coli* O157:H7 and its verotoxins and that it demonstrates efficiency comparable to quantitative polymerase chain reaction (qPCR) are significant findings [60–62]. This method eliminates the need for complex bacterial processing steps, such as lysis buffer and DNA purification kits. This simplification suggests that the immunosensor may streamline the analytical process, potentially reducing the time and resources required for sample preparation (Table 2). The use of specific antibodies in the immunosensor method is advantageous in terms of specificity [63]. This specificity reduces the likelihood of false positives or non-specific detections, which can challenge PCR methods.

The mention of label-free detection in the immunosensor method implies that it does not require additional labeling or modification of the target analyte for detection, which can simplify the assay and reduce potential sources of variability.

Among the novel techniques developed, the lateral flow test stands out, which seems to be a solution in relation to test time and the simultaneous detection of verotoxins and the rapid detection of *E. coli* O157. However, the sensitivity of the traditional colloidal gold immunochromatographic test strip (CG-ICTS) method is very low (105 CFU/mL), and its application is limited with respect to this photonic immunosensor [64]. Although the limit of detection (LoD) of many modified test strips has decreased slightly, this has meant that they have become more expensive due to the reagents used in their development [65,66].).

The ELISA assay is one of the most used methods for the detection of different antigens of *E. coli* O157 [67,68]. Among the practical advantages, it should be noted that the results are obtained in 3–5 h, and it does not require complex infrastructure. However, the results obtained by ELISA assume sensitivities of ng/mL, lower than those obtained by the sensor under development for *E. coli* O157:H7 verotoxins, whose LoD is around 1 ppm (Figure 6B,C).

Few portable and lab-on-a-chip electrochemical biosensors have been manufactured for the determination of *E. coli* [69]. These biosensors, most of them potentiometric, are low-cost, small, and highly sensitive and selective sensors, but they still require longer detection times than the immunosensor developed in this work (Table 2).

Many detection immunosensors have been developed to detect *E. coli* O157:H7, including GeneChip [70], genosensor [71], the surface plasmon resonance (SPR)-based assay [72,73], and the infrared spectrometer-based assay by Fourier transform (FTIR) [74]. Although these methods combine some attractive features to achieve high sensitivity, high performance, or rapid detection, all of them require more expensive instruments than the sensor under development and trained technical personnel.

A multiplex ring system in biosensor development offers the advantage of the simultaneous, comprehensive, and efficient detection of the *E. coli* O157:H7 pathogen and its verotoxins [75,76]. This capability of the developed biosensor to detect multiple antigens simultaneously, without the need for separate analyses, represents a significant advancement and a departure from traditional systems like PCR and ELISA, where each antigen of *E. coli* O157:H7 usually has to be analyzed separately [55,75].

The main advantage of the coupled microfluidic system is its ability to allow the simultaneous measurement of two samples, which is achieved by distributing eight annular resonators in two channels, each of which contains four resonant rings. The distribution of resonant rings across multiple channels allows for parallel processing, reducing the time required for analysis. The system's capability to achieve sensitivities down to a ng/mL scale with response times less than 30 min contributes to the efficiency of the detection process.

4. Conclusions

This study underscores the potential of a novel photonic biosensor as a valuable and reliable tool for detecting *E. coli* O157:H7 in food samples. Its sensitivity, specificity, and accuracy confirm it as a promising technology with applications in the food industry, contributing to enhancing food safety measures.

The biosensor's fabrication process involved silicon nitride and employed CMOS-compatible techniques. The immunosensor's surface consisted of eight ring resonators, and these resonators were functionalized with specific antibodies selected for detecting *E. coli* O157:H7 antigens in meat matrices.

In the preliminary validation of the biosensor, 100 fabricated PICs (photonic-integrated circuits) were used. The rings on these PICs were functionalized with specific antibodies, and a microfluidic layer was attached to facilitate controlled sample flow. The biosensor demonstrated the successful detection of various concentrations of bacteria and verotoxins. This initial validation highlights the promising performance, reliability, and feasibility of the innovative detection technique based on the photonic biosensor.

The selection of specific antibodies, including a monoclonal antibody with heightened specificity at lower antigen concentrations and a polyclonal antibody with a higher binding, reflects its capacity across a broader concentration range. It is important to note that no significant results were observed at very high antigen concentrations. This nuanced observation underscores both the potential and limitations of the method in different analytical contexts, providing valuable insights into the performance of the chosen antibodies under varying conditions.

This alternative approach demonstrates a rapid diagnostic capacity and provides results in four hours, allowing for prompt action in response to potential food microbiological contamination. The biosensor's ability to examine a significant number of samples makes it a valuable tool for industries involved in food production and processing.

Supplementary Materials: The following supporting information can be downloaded at https://www.mdpi.com/article/10.3390/photonics11040374/s1: Table S1: Detection inmunosensor results againts *E. coli* O157:H7; and Table S2: Detection inmunonsensor results of verotoxins 1 and verotoxin 2 of *E. coli* O157:H7.

Author Contributions: Conceptualization, A.F., J.G.-H., M.H. and Y.M.; methodology, A.F., J.G.-H., M.H. and Y.M.; validation, A.F., J.G.-H., M.H. and Y.M.; investigation, A.F., J.G.-H., M.H. and Y.M.; data curation, A.F., J.G.-H., M.H. and Y.M.; writing—original draft preparation, A.F.; writing—review and editing, J.G.-H., M.H. and Y.M.; funding acquisition, A.F., J.G.-H., M.H. and Y.M. All authors have read and agreed to the published version of the manuscript.

Funding: This research was financially supported by the AVI (Valencian Innovation Agency) and received funding from the European Union through FEDER funds. The project, identified as INNEST/2021/338, is known as the "BACTERIO project", which focuses on "Integrated Photonics for integral microbiological control in the agri-food sector".

Institutional Review Board Statement: Not applicable.

Informed Consent Statement: Not applicable.

Data Availability Statement: Data are contained within the article and Supplementary Materials.

Conflicts of Interest: Ana Fernández was employed by the company Lumensia Sensors S.L. The remaining authors declare that the research was conducted in the absence of any commercial or financial relationships that could be construed as a potential conflict of interest.

Sample Availability: Samples of the compounds are available from the authors.

References

1. Zhu, P.; Shelton, D.R.; Li, S.; Adams, D.L.; Karns, J.S.; Amstutz, P.; Tang, C.M. Detection of *E. coli* O157:H7 by immunomagnetic separation coupled with fluorescence immunoassay. *Biosens. Bioelectron.* **2011**, *30*, 337–341. [CrossRef] [PubMed]
2. Laster, B.A.; Harris, K.B.; Lucia, L.M.; Castillo, A.; Savell, J.W. Efficacy of trimming chilled beef during fabrication to control *Escherichia coli* O157:H7 surrogates on subsequent subprimals. *Meat Sci.* **2012**, *90*, 420–425. [CrossRef] [PubMed]
3. Hassan, A.-R.H.A.-A.; de la Escosura-Muñiz, A.; Merkoçi, A. Highly sensitive and rapid determination of *Escherichia coli* O157:H7 in minced beef and water using electrocatalytic gold nanoparticle tags. *Biosens. Bioelectron.* **2015**, *67*, 511–515. [CrossRef] [PubMed]
4. Chan, K.Y.; Ye, W.W.; Zhang, Y.; Xiao, L.D.; Leung, P.H.; Li, Y.; Yang, M. Ultrasensitive detection of *E. coli* O157:H7 with biofunctional magnetic bead concentration via nanoporous membrane based electrochemical immunosensor. *Biosens. Bioelectron.* **2013**, *41*, 532–537. [CrossRef] [PubMed]
5. Food and Agriculture Organization of United Nations (FAO); World Health Organization (WHO). *Shiga Toxin-Producing Escherichia coli (STEC) and Food: Attribution, Characterization, and Monitoring*; WHO: Geneva, Switzerland; FAO: Rome, Italy, 2018; ISBN 978-92-5-130682-6 (FAO). ISBN 978-92-4-151427-9 (WHO).
6. Scientific Committee of the Spanish Agency for Food Safety and Nutrition (AESAN). *Preventative Measures and Applicable Recommendations for Avoiding Possible Food-Borne Infections Caused by Strains of Verotoxigenic/Shiga Toxin-Producing/Enterohemorrhagic Escherichia coli (VTEC/STEC/EHEC)*; AESAN-2012-007; AESAN: Majadahonda, Spain, 2012.
7. Wu, W.; Zhao, S.; Mao, Y.; Fang, Z.; Lu, X.; Zeng, L. A sensitive lateral flow biosensor for *Escherichia coli* O157:H7 detection based on aptamer mediated strand displacement amplification. *Anal. Chim. Acta* **2015**, *861*, 62–68. [CrossRef] [PubMed]
8. Commission Regulation (EC). No 2073/2005 of 15 November 2005 on microbiological criteria for foodstuffs. *Off. J. Eur. Union.* **2005**, *L338*, 1–26.
9. Carvalho, F.J.G.; Mohammed Ahmed Sheikh, H.; Selvin, R. Advances in screening, detection and enumeration of *Escherichia coli* using nanotechnology-based methods: A review. *J. Biomed. Nanotechnol.* **2018**, *14*, 829–846. [CrossRef] [PubMed]
10. Umesha, S.; Manukumar, Y.H. Advanced molecular diagnostic techniques for detection of food-borne pathogens: Current applications and future challenges. *Crit. Rev. Food Sci. Nutr.* **2018**, *58*, 84–104. [CrossRef] [PubMed]
11. Luo, C.; Tang, H.; Cheng, W.; Yan, L.; Zhang, D.; Ju, H.; Ding, S. A sensitive electrochemical DNA biosensor for specific detection of Enterobacteriaceae bacteria by exonuclease III-assisted signal amplification. *Biosens. Bioelectron.* **2013**, *48*, 132–137. [CrossRef]
12. Huertas-Caro, C.; Urbano-Cáceres, E.; Torres-Caycedo, Y.M. Diagnóstico molecular una alternativa para la detección de patógenos en alimentos. *Rev. Habanera Cienc. Médicas* **2019**, *18*, 513–528.
13. Fratamico, P.M.; Bhunia, A.K.; Smith, J.A. *Foodborne Pathogens: Microbiology and Molecular Biology*; CRC Press LLC eBooks: Boca Raton, FL, USA, 2005.
14. Li, Y.; Mustapha, A. Simultaneous Detection of *Escherichia coli* O157:H7, Salmonella, and Shigella in Apple Cider and Produce by a Multiplex PCR. *J. Food Prot.* **2004**, *67*, 27–33. [CrossRef] [PubMed]
15. Daly, P.; Collier, T.; Doyle, S. PCR-ELISA detection of *Escherichia coli* in milk. *Lett. Appl. Microbiol.* **2002**, *34*, 222–226. [CrossRef] [PubMed]
16. Noble, R.T.; Weisberg, S.B. A review of technologies for rapid detection of bacteria in recreational waters. *J. Water Health* **2005**, *3*, 381–392. [CrossRef] [PubMed]
17. Pebdeni, A.B.; Mousavizadegan, M.; Hosseini, M. Sensitive detection of *S. aureus* using aptamer-and vancomycin-copper nanoclusters as dual recognition strategy. *Food Chem.* **2021**, *361*, 130137. [CrossRef] [PubMed]
18. Acharya, G.; Chang, L.C.; Savran, C. An optical biosensor for rapid and labelfree detection of cells. *J. Am. Chem. Soc.* **2006**, *128*, 3862–3863. [CrossRef] [PubMed]
19. Hornbeck, P. Double-Immunodiffusion Assay for Detecting Specific Antibodies. *Curr. Protoc. Immunol.* **2017**, *116*, 2–3. [CrossRef] [PubMed]
20. Ngo, T.T. Immunoassay. *Curr. Opin. Biotechnol.* **1991**, *2*, 102–109. [CrossRef]
21. Plested, J.S.; Coull, P.A.; Gidney, M.A.J. ELISA. *Methods Mol. Med.* **2003**, *71*, 243–261. [PubMed]
22. Zhang, X.; Li, M.; Zhang, B.; Chen, K.; He, K. Development of a Sandwich ELISA for EHEC O157:H7 Intimin γ1. *PLoS ONE* **2016**, *11*, e0162274. [CrossRef]
23. Sajid, M.; Kawde, A.; Daud, M. Designs, formats and applications of lateral flow assay: A literature review. *J. Saudi Chem. Soc.* **2015**, *19*, 689–705. [CrossRef]

24. Posthuma-Trumpie, G.A.; Korf, J.; Van Amerongen, A. Lateral flow (immuno)assay: Its strengths, weaknesses, opportunities and threats. A literature survey. *Anal. Bioanal. Chem.* **2009**, *393*, 569–582. [CrossRef] [PubMed]
25. Singh, J.P.; Sharma, S.; Nara, S. Evaluation of gold nanoparticle based lateral flow assays for diagnosis of enterobacteriaceae members in food and water. *Food Chem.* **2015**, *170*, 470–483. [CrossRef] [PubMed]
26. Watts, H.J.; Lowe, R.C.; Pollard-Knight, D.V. Optical biosensor for monitoring microbial cells. *Anal. Chem.* **1994**, *66*, 2465–2470. [CrossRef] [PubMed]
27. Riu, J.; Giussani, B. Electrochemical biosensors for the detection of pathogenic bacteria in food. *Trends Anal. Chem.* **2020**, *125*, 115863. [CrossRef]
28. Ye, Y.; Guo, H.; Sun, X. Recent progress on cell-based biosensors for analysis of food safety and quality control. *Biosens. Bioelectron.* **2019**, *126*, 389–404. [CrossRef] [PubMed]
29. Velusamy, V.; Arshak, K.; Korostynska, O.; Oliwa, K.; Adley, C.C. An overview of foodborne pathogen detection: In the perspective of biosensors. *Biotechnol. Adv.* **2010**, *28*, 232–254. [CrossRef] [PubMed]
30. González-Guerrero, A.B.; Dante, S.; Duval, D.; Osmond, J.; Lechuga, L.M. Advanced photonic biosensors for point-of-care diagnostics. *Procedia Eng.* **2011**, *25*, 71–75. [CrossRef]
31. Shrivastava, A.; Gupta, V.B. Methods for the determination of limit of detection and limit of quantitation of the analytical methods. *Chron. Young Sci.* **2011**, *2*, 21. [CrossRef]
32. Fan, X.; White, I.M.; Shopova, S.I.; Zhu, H.; Suter, J.D.; Sun, Y. Sensitive optical biosensors for unlabeled targets: A review. *Anal. Chim. Acta* **2008**, *620*, 8–26. [CrossRef]
33. De Vos, K.; Bartolozzi, I.; Schacht, E.; Bienstman, P.; Baets, R. Silicon-on-Insulator microring resonator for sensitive and label-free biosensing. *Opt. Express* **2007**, *15*, 7610–7615. [CrossRef]
34. Narsaiah, K.; Jha, S.N.; Bhardwaj, R.; Sharma, R.; Kumar, R. Optical biosensors for food quality and safety assurance—A review. *J. Food Sci. Technol.* **2012**, *49*, 383–406. [CrossRef] [PubMed]
35. Donagh, C.M.; Burke, C.S.; Crai BD, M. Optical biosensors. *Chem. Rev.* **2008**, *108*, 422–461. [CrossRef]
36. Pebdeni, A.B.; Hosseini, M.; Ganjali, M.R. Fluorescent turn-on aptasensor of Staphylococcus aureus based on the FRET between green carbon quantum dot and gold nanoparticle. *Food Anal. Methods* **2020**, *13*, 2070–2079. [CrossRef]
37. Lazcka, O.; Del Campo, F.J.; Munoz, F.X. Pathogen detection: A perspective of traditional methods and biosensors. *Biosens. Bioelectron.* **2007**, *22*, 1205–1217. [CrossRef] [PubMed]
38. Nurliyana, M.R.; Wibowo, K.M.; Muslihati, A.; Saim, H.; Ahmad, S.A.; Sari, Y.; Mansor, Z. The detection method of *Escherichia coli* in water resources: A review. *IOP Conf. Ser. J. Phys. Conf. Ser.* **2017**, *995*, 012065. [CrossRef]
39. Ionescu, R.E. Biosensor platforms for rapid detection of *E. coli* bacteria. In *Escherichia coli–Recent Advances on Physiology, Pathogenesis and Biotechnological Applications*; IntechOpen: London, UK, 2017. [CrossRef]
40. Bange, A.; Halsall, H.B.; Heineman, W.R. Microfluidic immunosensor systems. *Biosens. Bioelectron.* **2005**, *20*, 2488–2503. [CrossRef] [PubMed]
41. Luka, G.; Ahmadi, A.; Najjaran, H.; Alocilja, E.; DeRosa, M.; Wolthers, K.; Malki, A.; Aziz, H.; Althani, A.; Hoorfar, M. Microfluidics integrated biosensors: A leading technology towards lab-on-a-chip and sensing applications. *Sensors* **2015**, *15*, 30011–30031. [CrossRef] [PubMed]
42. Kant, K.; Shahbazi, M.A.; Dave, V.P.; Ngo, T.A.; Chidambara, V.A.; Than, L.Q.; Bang, D.D.; Wolff, A. Microfluidic devices for sample preparation and rapid detection of foodborne pathogens. *Biotechnol. Adv.* **2018**, *36*, 1003–1024. [CrossRef]
43. *ISO 16654:2001/Amd 1:2017*; Microbiology of food and animal feeding stuffs. Horizontal method for the detection of Escherichia coli O157. Amendment 1: Annex B: Result of interlaboratory studies. International Organization for Standardization: Geneva, Switzerland, 2017.
44. Fernández Blanco, A.; Hernández Pérez, M.; Moreno Trigos, Y.; García-Hernández, J. Development of Optical Label-Free Biosensor Method in Detection of Listeria monocytogenes from Food. *Sensors* **2023**, *23*, 5570. [CrossRef]
45. Ashkenazi, S.; Cleary, T.G. Rapid method to detect Shiga toxin and Shiga-like toxin I based on binding to globotriosyl ceramide (Gb3), their natural receptor. *J. Clin. Microbiol.* **1989**, *27*, 1145–1150. [CrossRef]
46. Hegde, N.V.; Hulet, R.; Jayarao, B.M.; Muldoon, M.; Lindpaintner, K.; Kapur, V.; DebRoy, C. Detection of the top six non-O157 Shiga toxin-producing *Escherichia coli* O groups by ELISA. *Foodborne Pathog. Dis.* **2012**, *9*, 1044–1048. [CrossRef] [PubMed]
47. Guttikonda, S.; Wang, W.W.; Suresh, M.R. Molecular zipper assays: A simple homosandwich with the sensitivity of pcr. *J. Pharm. Pharm. Sci.* **2004**, *7*, 7–16. [PubMed]
48. Vytrasova, J.; Zachová, I.; Cervenka, L.; Štepanková, J.; Pejchalová, M. Non-specific reactions during immunomagnetic sepa-ration of Listeria. *Food Technol. Biotechnol.* **2005**, *43*, 397–401.
49. Kehl, K.S.; Havens, P.; Behnke, C.E.; Acheson, D.W. Evaluation of the premier EHEC assay for detection of Shiga toxin-producing *Escherichia coli*. *J. Clin. Microbiol.* **1997**, *35*, 2051–2054. [CrossRef] [PubMed]
50. Teel, L.D.; Daly, J.A.; Jerris, R.C.; Maul, D.; Svanas, G.; O'Brien, A.D.; Park, C.H. Rapid detection of Shiga toxin-producing *Escherichia coli* by optical immunoassay. *J. Clin. Microbiol.* **2007**, *45*, 3377–3380. [CrossRef] [PubMed]
51. Skinner, C.; Patfield, S.; Stanker, L.H.; Fratamico, P.; He, X. New high-affinity 4 monoclonal antibodies against Shiga toxin 1 facilitate the detection of hybrid Stx1/Stx2 in vivo. *PLoS ONE* **2014**, *9*, e99854. [CrossRef] [PubMed]

52. He, X.; McMahon, S.; Skinner, C.; Merrill, P.; Scotcher, M.C.; Stanker, L.H. Development and characterization of monoclonal antibodies against Shiga toxin 2 and their application for toxin detection in milk. *J. Immunol. Methods* **2013**, *389*, 18–28. [CrossRef] [PubMed]
53. Ching, K.H. Lateral Flow Immunoassay. *Methods Mol. Biol.* **2015**, *1318*, 127–137.
54. Chen, M.; Yu, Z.; Liu, D.; Peng, T.; Liu, K.; Wang, S.; Xiong, Y.; Wei, H.; Xu, H.; Lai, W. Dual gold nanoparticle late flow immunoassay for sensitive detection of *Escherichia coli* O157:H7. *Anal. Chim. Acta* **2015**, *876*, 71–76. [CrossRef]
55. Ching, K.H.; He, X.; Stanker, L.H.; Lin, A.V.; McGarvey, J.A.; Hnasko, R. Detection of shiga toxins by lateral flow assay. *Toxins* **2015**, *7*, 1163–1173. [CrossRef]
56. Radke, S.M.; Alocilja, E.C. A high density microelectrode array biosensor for detection of *E. coli* O157:H7. *Biosens. Bioelectron.* **2005**, *20*, 1662–1667. [CrossRef] [PubMed]
57. Zhao, G.Y.; Xing, F.F.; Deng, S.P. A disposable amperometric enzyme immunosensor for rapid detection of Vibrio parahaemolyticus in food based on agarose/nano–Au membrane and screen-printed electrode. *Electrochem. Commun.* **2007**, *9*, 1263–1268. [CrossRef]
58. Su, X.L.; Li, Y.B. A self-assembled monolayer-based piezoelectric immunosensor for rapid detection of *Escherichia coli* O157:H7. *Biosens. Bioelectron.* **2004**, *19*, 563–574. [CrossRef] [PubMed]
59. Waswaa, J.; Irudayarajb, J.; DebRoy, C. Direct detection of *E. coli* O157:H7 in selected food systems by a surface plasmon resonance biosensor. *LWT* **2007**, *40*, 187–192. [CrossRef]
60. Li, D.J.; Feng, Y.Y.; Zhou, L.; Ye, Z.Z.; Wang, J.P.; Ying, Y.B.; Ruan, C.M.; Wang, R.H.; Li, Y.B. Label-free capacitive immunosensor based on quartz crystal Au electrode for rapid and sensitive detection of *Escherichia coli* O157:H7. *Anal. Chim. Acta* **2011**, *687*, 89–96. [CrossRef] [PubMed]
61. Li, B.; Chen, J.-Q. Real-Time PCR Methodology for Selective Detection of Viable *Escherichia coli* O157:H7 Cells by Targeting Z3276 as a Genetic Marker. *Appl. Environ. Microbiol.* **2012**, *78*, 5297–5304. [CrossRef] [PubMed]
62. Li, L.; Zhang, X.H.; He, K.W.; Jiang, P.; Zhao, P.; Ye, Q.; Luan, X.; Wen, L.; Ni, Y.; Zhou, J.; et al. Duplex PCR Procedure for the Detection of EHEC O157:H7. *Acta Agric. Boreali-Sin.* **2011**, *26*, 58–66. [CrossRef]
63. Ibekwe, A.M.; Grieve, C.M. Detection and quantification of *Escherichia coli* O157:H7 in environmental samples by real-time PCR. *J. Appl. Microbiol.* **2003**, *94*, 421–431. [CrossRef] [PubMed]
64. Xie, Q.Y.; Wu, Y.H.; Xiong, Q.R.; Xu, H.Y.; Xiong, Y.H.; Liu, K.; Jin, Y.; Lai, W.H. Advantages of fluorescent microspheres compared with colloidal gold as a label in immunochromatographic lateral flow assays. *Biosens. Bioelectron.* **2014**, *54*, 262–265. [CrossRef]
65. Shan, S.; Lai, W.; Xiong, Y.; Wei, H.; Xu, H. Novel strategies to enhance lateral flow immunoassay sensitivity for detecting foodborne pathogens. *Agric. Food Chem.* **2015**, *63*, 745–753. [CrossRef]
66. Bruno, J.G. Application of DNA aptamers and quantum dots to lateral flow test strips for detection of foodborne pathogens with improved sensitivity versus colloidal gold. *Pathogens* **2014**, *3*, 341–355. [CrossRef] [PubMed]
67. Wang, C.; Xing, K.; Zhang, G.; Yuan, M.; Xu, S.; Liu, D.; Chen, W.; Peng, J.; Hu, S.; Lai, W.-H. Novel ELISA based on fluorescent quenching of DNA-stabilized silver nanoclusters for detecting *E. coli* O157:H7. *Food Chem.* **2019**, *281*, 91–96. [CrossRef] [PubMed]
68. Pang, B.; Zhao, C.; Li, L.; Song, X.; Xu, K.; Wang, J.; Liu, Y.; Fu, K.; Bao, H.; Song, D.; et al. Development of a low-cost paper-based ELISA method for rapid *Escherichia coli* O157:H7 detection. *Anal. Biochem.* **2018**, *542*, 58–62. [CrossRef] [PubMed]
69. Perumal, V.; Hashim, U. Advances in biosensors: Principle, architecture and applications. *J. Appl. Biomed.* **2014**, *12*, 1–15. [CrossRef]
70. Kim, H.; Kane, M.D.; Kim, S.; Dominguez, W.; Applegate, B.M.; Savikhin, S. A molecular beacon DNA microarray system for rapid detection of *E. coli* O157:H7 that eliminates the risk of a false negative signal. *Biosens. Bioelectron.* **2007**, *22*, 1041–1047. [CrossRef]
71. Abdalhai, M.H.; Fernandes, A.M.; Xia, X.; Musa, A.; Ji, J.; Sun, X. Electrochemical genosensor to detect pathogenic bacteria (*Escherichia coli* O157:H7) as applied in real food samples (fresh beef) to improve food safety and quality control. *J. Agric. Food Chem.* **2015**, *63*, 5017–5025. [CrossRef] [PubMed]
72. Wang, Y.; Ye, Z.; Si, C.; Ying, Y. Monitoring of *Escherichia coli* O157:H7 in food samples using lectin based surface plasmon resonance biosensor. *Food Chem.* **2013**, *136*, 1303–1308. [CrossRef]
73. Vaisocherova-Lisalova, H.; Visova, I.; Ermini, M.L.; Springer, T.; Song, X.C.; Mrazek, J.; Lamacova, J.; Scott Lynn, N., Jr.; Sedivak, P.; Homola, J. Low-fouling surface plasmon resonance biosensor for multi-step detection of foodborne bacterial pathogens in complex food samples. *Biosens. Bioelectron.* **2016**, *80*, 84–90. [CrossRef]
74. Chen, J.; Shi, X.; Gehring, A.G.; Paoli, G.C. Automated immunomagnetic separation for the detection of *Escherichia coli* O157:H7 from spinach. *Int. J. Food Microbiol.* **2014**, *179*, 33–37. [CrossRef]
75. Rani, A.; Ravindran, V.B.; Surapaneni, A.; Shahsavari, E.; Haleyur, N.; Mantri, N.; Ball, A.S. Evaluation and comparison of recombinase polymerase amplification coupled with lateral-flow bioassay for *Escherichia coli* O157:H7 detection using different genes. *Sci. Rep.* **2021**, *11*, 1881. [CrossRef]
76. Hu, J.; Wang, Y.; Su, H.; Ding, H.; Sun, X.; Gao, H.; Geng, Y.; Wang, Z. Rapid analysis of *Escherichia coli* O157:H7 using isothermal recombinase polymerase amplification combined with triple-labeled nucleotide probes. *Mol. Cell. Probes* **2020**, *50*, 101501. [CrossRef] [PubMed]

photonics

Article

Comparison of Lifetime-Based Pressure-Sensitive Paint Measurements in a Wind Tunnel Using Model Pitch–Traverse and Pitch–Pause Modes

Christian Klein *, Daisuke Yorita and Ulrich Henne

Institute of Aerodynamics and Flow Technology, German Aerospace Center (DLR), Bunsenstrasse 10, 37073 Göttingen, Germany; daisuke.yorita@dlr.de (D.Y.); ulrich.henne@dlr.de (U.H.)
* Correspondence: christian.klein@dlr.de

Abstract: In order to improve the data productivity of a wind tunnel test, the model under investigation in the wind tunnel is moved continuously with a predetermined constant angular speed in the so-called pitch–traverse mode. Alternatively, the wind tunnel model can be moved in the so-called pitch–pause mode, in which it keeps its position for a certain (measurement) time at a fixed pitch position, after which it is moved to the next pitch position. The latter procedure is more time-consuming, so, for the same time interval, the number of measured data points taken in the pitch–pause mode is less than that for the pitch–traverse mode. Since wind tunnel test time can be quite expensive, in most wind tunnel tests where only conventional forces and pressures are recorded with conventional measuring systems, the wind tunnel model is moved in the pitch–traverse mode in order to obtain as much aerodynamic data as possible during the tunnel runtime. The application of the Pressure-Sensitive Paint (PSP) technique has been widely used in wind tunnel testing for the purpose of providing pressure data on wind tunnel models with high spatial resolution. The lifetime-based PSP method has several advantages over the intensity-based method since it often has higher accuracy. Up until now, the lifetime-based PSP technique has mainly been used for wind tunnel testing, where the test model has been moved to the pitch–pause mode. The traditional lifetime method using on-chip accumulation requires multiple (~1000) excitation light pulses to accumulate enough luminescence (fluorescence or phosphorescence) photons on the camera sensor to provide acceptable signal-to-noise ratios and, therefore, it may seem to be not compatible with a continuously moving wind tunnel model. Nevertheless, the present study verifies the application of lifetime-based PSP utilizing on-chip accumulation with a continuously moving wind tunnel model which would make the entire PSP data acquisition compatible with that of the conventional measurements (forces and pressures), as mentioned above. In this paper, the applicability of the lifetime-based PSP technique to a continuously moving wind tunnel model (in pitch–traverse mode) is investigated with the help of measurements in the transonic wind tunnel in Göttingen (TWG). For this investigation, PSP was applied on the delta-wing model DLR-F22, which is to be tested in TWG. The pressure distribution on the wind tunnel model was measured using the PSP lifetime method for both model movement modes (pitch–pause and pitch–traverse mode) so that the corresponding PSP results could be directly compared with each other. In addition, an error analysis of the PSP results was carried out and compared with the conventional pressure measurement results, hence providing an assessment of the accuracy of the PSP results; finally, a recommendation for future PSP measurements could be given.

Keywords: pressure-sensitive paint (PSP); lifetime-based measurements; pitch–traverse; pitch–pause; transonic flow; DLR-F22 wind tunnel model

Citation: Klein, C.; Yorita, D.; Henne, U. Comparison of Lifetime-Based Pressure-Sensitive Paint Measurements in a Wind Tunnel Using Model Pitch–Traverse and Pitch–Pause Modes. *Photonics* **2024**, *11*, 546. https://doi.org/10.3390/photonics11060546

Received: 8 May 2024
Revised: 3 June 2024
Accepted: 5 June 2024
Published: 7 June 2024

1. Introduction

Pressure-Sensitive Paint (PSP) is an optical pressure measurement technique widely used in wind tunnel testing [1–3]; it enables the acquisition of pressure distributions on a model surface from one set of images without the need to drill pressure tap holes and to connect tubes inside the investigated wind tunnel model. Thus, it is an excellent measurement system for investigating wind tunnel models having a number of different configurations which are to be examined successively, as is the case for the DLR-F22 delta-wing model [4] in the transonic wind tunnel in Göttingen (TWG). A great advantage of the PSP technique is that a surface pressure distribution over the whole model surface can be obtained quantitatively and with high spatial resolution. Two different measurement methods using PSP (the intensity [3] and lifetime-based [3,5] methods) have already been applied to testing in various wind tunnels. The lifetime-based method makes use of the dependence of PSP luminescent lifetime on the pressure. Two types of lifetime-based methods have so far been used; one uses pulsed excitation, whereas the other uses continuously modulated excitation [3]. In this study, we focus on a two-gate method using pulsed excitation. In the two-gate method, two luminescent images are acquired at different times relative to the excitation pulse so that images are available at different times on the luminescence excitation and decay curve. To achieve this, a pulsed excitation light source and a fast-shutter camera are required. Two different measurement strategies for the two-gate lifetime-based PSP have been reported in the literature so far: (a) if both gate images are acquired from a single excitation pulse, the method is referred to as the "single-shot lifetime technique" [6]; (b) with the "on-chip accumulation" method [7], the camera shutter is opened with a constant time delay after every excitation light pulse so that with multiple excitation pulses the resulting PSP luminescence images can be accumulated on the camera-sensor chip itself.

A major advantage of the lifetime-based method is that the relationship between the ratio of the two gated images and pressure is independent of the excitation light intensity distribution. Since both gate images are acquired only at wind-on conditions, problems arising from model deformation or displacement due to wind-on/wind-off differences do not occur. Nevertheless, in many cases of practical applications, spatial gate ratio variations (non-uniform lifetime patterns) have been observed [8]. This spatial variation causes critical errors in the final pressure calculation. Therefore, one single gate ratio obtained also under wind-off conditions is required to correct for these non-uniform lifetime patterns [9]. In recent years, groups at both universities and research establishments have made great progress in the development of different PSP lifetime measurement methods for wind tunnel testing, single-shot lifetime PSP, and on-chip accumulation lifetime PSP [9–13].

In 2016, Mébarki and Benmeddour reported [19] a successful measurement using the PSP lifetime method in the NRC transonic (blowdown) wind tunnel on a so-called GBU-38 wind tunnel model, where the model was moved continuously (in pitch–traverse mode) at 3°/s, 6°/s or 9°/s. In their work, they used pulsed UV-LED systems in combination with CMOS-camera technology, with which PSP image acquisition rates of 49 Hz could be achieved for the single-shot lifetime method. In their work, they state [19] (p. 2) that "the traditional lifetime method using on-chip accumulation ... is not compatible with a continuously moving model". The implications of this statement must be evaluated with due consideration paid to the chosen pitch–traverse rates (angular speeds). Furthermore, the model pitch–traverse rate chosen for a wind tunnel measurement also depends on the operating mode of the wind tunnel itself. For example, model traverse rates of just 0.1 °/sec are used in continuously operating transonic wind tunnels such as TWG [20] or the European Transonic Windtunnel (ETW) [21]. This low model traverse rate is one order of magnitude smaller than that mentioned by Mébarki and Benmeddour in their intermittently operating blowdown wind tunnel. At significantly smaller model traverse rates compared to those reported by Mébarki and Benmeddour, the possibility of using the traditional PSP lifetime method using on-chip accumulation for measurements with a continuously moving wind tunnel model should, therefore, be examined.

The traditional PSP lifetime method with on-chip accumulation has typically made use of CCD-sensor technology; CMOS-based cameras do not have this feature up to now. In 2015, the Sony Corporation (the top seller of CCD image sensors in 2014) announced the cessation of their CCD image sensor manufacturing by the end of March 2017, with the last shipments of their finished product being in March 2020 [22]. Therefore, CCD-sensor technology, in the long term, will no longer be available for PSP measurements. Nevertheless, many universities and research establishments still have in use many such CCD cameras with on-chip accumulation technology. From the authors' point of view, the investigation carried out here for the first time using CCD-camera technology with continuously moving wind tunnel models is still of scientific and industrial importance since this camera technology is still currently being used for wind tunnel measurements.

In this paper, the current status of the lifetime-based PSP method with on-chip accumulation using CCD-sensor technology on a continuously moving wind tunnel model is reported. The main topics of this work are (i) the optimization of the camera and LED excitation source settings for high-quality PSP data generation whilst the wind tunnel model is moving in pitch–traverse or pitch–pause mode; (ii) the automation of data acquisition for high data productivity; and (iii) the achievement of accuracy and precision of the lifetime-based PSP method for the pitch–traverse and pitch–pause modes.

This current work reviews the first wind tunnel test campaign of the generic triple-delta wing model DLR-F22 in TWG and reveals the possibilities of using the lifetime-based PSP method for a continuously moving wind tunnel model in pitch–traverse mode.

2. Lifetime-Based PSP-Technique

The principle behind the pressure measurement by the lifetime-based PSP method is based on a photo-physical phenomenon of luminophores, which is known as oxygen quenching of the luminescence: higher concentrations of oxygen result in shorter lifetimes of photoluminescence, while, conversely, lower oxygen concentrations result in longer photoluminescence lifetimes. According to Henry's law, the concentration of oxygen in the paint layer is proportional to the partial pressure of oxygen of the gas above the surface. When the oxygen concentration in the gas is constant (as in ambient air), gas pressure can be related to the luminescent lifetime of the PSP luminophore [3].

2.1. Pressure Calculation

In the two-gate method, two luminescent images (I_{Gate1} and I_{Gate2}) are acquired at different times of the luminescence decay. Theoretically, using the two-gated lifetime method, the image ratio R_{on} taken at wind-on conditions can be directly related to surface pressure P as described in Equation (1).

$$R_{on} = \frac{I_{Gate1}}{I_{Gate2}} = \sum_{i=0}^{n} a_i P_{on}^i \quad (1)$$

where a_i are Stern–Volmer calibration coefficients and I_{Gate1} and I_{Gate2} represent the luminescence image intensity at each gate. The two-gate lifetime system requires a pulsed excitation light source and a fast-shutter camera. The advantage of the lifetime-based method is that the relationship between the ratio of the two gated images and pressure is, in principle, independent of the excitation light intensity distribution. However, in many cases of practical application, spatial gate ratio variations have been observed that are not related to varying surface pressures [8,17]. This can occur especially on large, coated surfaces where gate ratios show differences of several percent on different areas of the surface, but where pressures are the same. This spatial gate ratio (lifetime) variation causes critical errors in the final pressure calculation. Therefore, one single gate ratio obtained under wind-off condition R_{off} is also required to correct for these non-uniform, non-flow-related lifetime patterns. This calculation, a so-called "ratio-of-ratios" [9], can be described as in Equation (2).

$$\frac{R_{on}}{R_{off}} = \sum_{i=0}^{n} b_i \cdot \left(\frac{P_{on}}{P_{off}}\right)^i \quad (2)$$

where b_i are calibration coefficients. By using the ratio-of-ratios calculation, the effect of local lifetime variations can either be corrected for or at least their effect reduced.

2.2. Paint Composition

A typical composition of PSP which is used at DLR is described here: Platinum meso-tetra (pentafluorophenyl)porphine (PtTFPP) [23] is used as the sensor dye. This dye has a lifetime decay time of 50 μs (1/e time decay) under vacuum conditions and also a high luminescent intensity. The same polymer, poly(4-tert-butyl styrene), is used in both the active layer and the base coat to prevent any unfavorable interactions between the layers. In the base coat, Boron Nitride (BN) is mixed in as white particles to further enhance the PSP signal [24]. The mean diameter of these particles is less than 0.5 μm. The thickness of each layer is approximately 10 μm, and the surface roughness of the active layer coating is less than 1 μm in Ra. A commercially available white screen layer, based on epoxy resin, was sprayed directly onto the wind tunnel model with a thickness of about 60 μm in order to generate a homogeneous background for the succeeding layers. Figure 1 shows the different coating layers applied to the wind tunnel model surface and their corresponding paint-layer thicknesses. It should be noted at this point that the use of a separate white screen layer can be dispensed with if the wind tunnel model to be examined has a very uniform surface finish. However, if the wind tunnel model has either filled screw holes or consists of different model materials, the use of a white screen layer has been found to be highly beneficial.

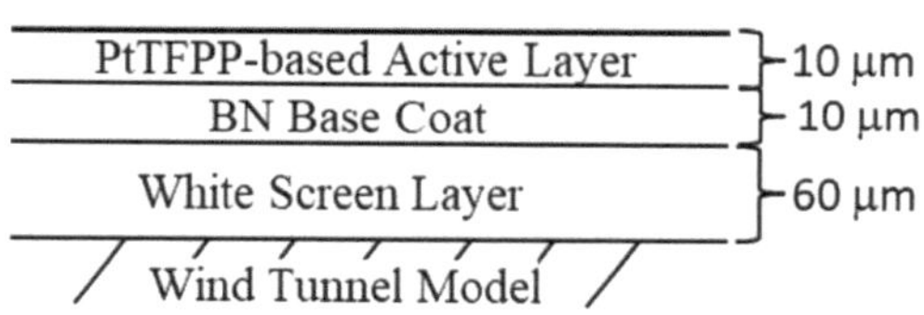

Figure 1. PSP coating layers.

2.3. Hardware

As PSP excitation light sources, UV LEDs (IL-106, HARDsoft) with a 385ET75 nm optical band-pass filter were used. The emission peak of each UV-LED system was at 390 nm. The LEDs were operated in pulsed mode, where the rise and fall times of the pulse excitation were less than 1 μs. The PSP luminescence signal was acquired using a 14-bit CCD camera (pco.2000, PCO, Kelheim, Germany), which was equipped with a 650ET100 nm optical band-pass filter. The camera was operated in on-chip accumulation mode (which is also referred to as modulation mode in the literature) to obtain an image with a high signal-to-noise ratio (SNR). In this mode, the camera shutter is opened many times (~1000) with a constant time delay relative to each LED light pulse so that PSP luminescence images are accumulated on the CCD chip via multiple exposures. During pitch–pause data acquisition, the camera was operated using one analog-to-digital converter (ADC), whereas for the pitch–traverse data acquisition, the readout speed of the camera was enhanced by using two ADCs. The camera readout speed for both data acquisition modes was set to 10 Mpx/s. For these camera settings, the expected 14-bit dynamic range of the camera could be fully used, and a camera frame rate of 3 Hz could be achieved. A faster readout of 40 Mpx/s is available for this camera but the dynamic range of the camera is then limited to 10,000 counts, see Appendix A. The gate settings for Gate1 and Gate2 of the camera are shown in Figure 2a. LED and camera timing settings employed in this test are summarized as follows: LED pulse length: 12 μs; LED pulse frequency: 5 kHz; Camera shutter width: 30 μs; Camera Gate1 start: −28.5 μs; Camera Gate2 start: +1.5 μs; $t = 0$ is defined by the end of the excitation (LED off). It is noted that the numbers of LED pulses and camera accumulations are adjusted for each test condition to obtain almost the same PSP signal level in both gated images. In Figure 2a, the luminescence lifetime excitation and decay curves for different pressures at $T = 30\ ^\circ C$ are shown, taken with the settings described above. The obtained signal intensity of Gate1 is almost two times that of Gate2 (at $P = 60$ kPa). Therefore, the number of LED pulses in Gate1 has been set to about half that in Gate2 in order to obtain about the same PSP signals in Gate1 and Gate2 images.

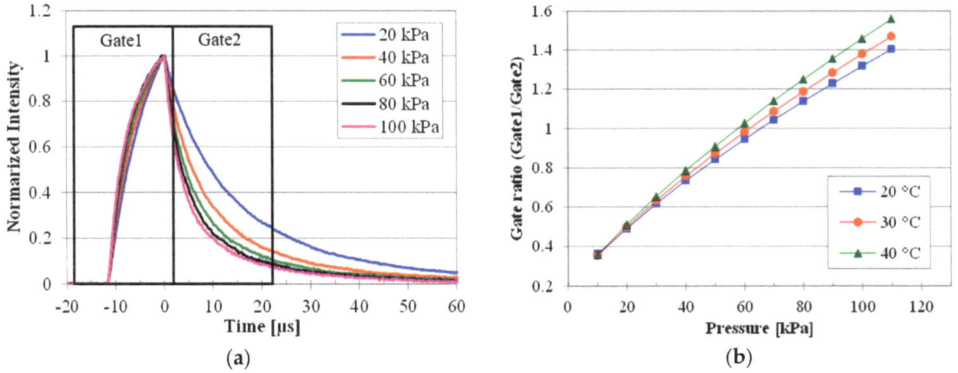

Figure 2. (**a**) Luminescence lifetime decay curves for different pressure levels at $T = 30\ ^\circ C$ and camera gate settings. (**b**) Pressure sensitivity curves for different temperatures of the lifetime-based method.

3. Application to Wind Tunnel Testing

3.1. Experimental Setup

Evaluation tests of the lifetime-based PSP were conducted in the perforated wall test section of the Transonic Wind Tunnel in Göttingen (TWG); it has a cross-section of 1 m × 1 m, and the corresponding operating range of the wind tunnel with this test section provides Mach numbers in the range M = 0.3 to 1.2 and total pressures from $P = 30$ kPa to 150 kPa [25].

Optical access is possible through glass windows from the top and from each side of the test section. In order to have sufficient excitation light available to acquire images with good SNR, three UV-LED systems were used as excitation light sources: one was located on the top wall, and two others on the starboard side of the wind tunnel model (looking in the flow direction). In Figure 3, the geometry of the model (a) and a photo of the wind tunnel model inside the test section (b) are shown. The locations of the camera and the three LEDs behind the windows of the wind tunnel wall are also shown. The photo (b) was taken with the LEDs switched on so that the model area illuminated by the LEDs can be recognized as the magenta-colored (the "color" of the luminescence) area on the wind tunnel model. The model forebody area was not illuminated with the LEDs installed here; see Section 4 later.

For image acquisition, the pco.2000 camera was installed on the top wall of the wind tunnel test section. The lens with a focal length of 20 mm allowed capturing the wing part of the model with a spatial resolution of approx. 3 pixels/mm.

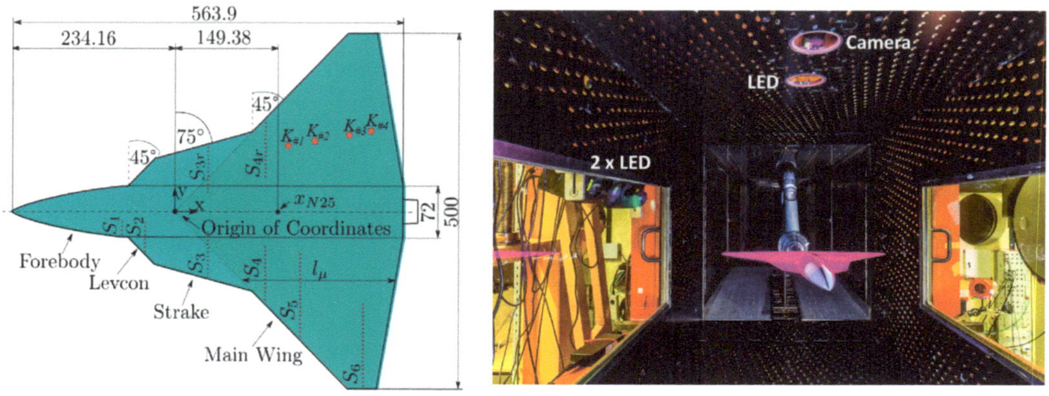

(**a**) (**b**)

Figure 3. (**a**) Geometry of DLR-F22 model [4]. (**b**) Photo of the installed DLR-F22 model in the perforated wall test section of TWG. The area illuminated by the LEDs on the wind tunnel model is shown in magenta color.

The model is a generic triple-delta wing with a flat-plate wing and a sharp leading edge (DLR-F22) with four Kulite® sensors $K_{\#1}$, $K_{\#2}$, $K_{\#3}$, and $K_{\#4}$ (Kulite, Leonia, NJ, USA) for unsteady pressure measurements, as shown in Figure 3a on the suction-side model surface. The configuration tested here is named DLR-internally as "LS1", see Figure 3a, and was designed and manufactured with a 45° "levcon" (leading edge vortex controller), a 75° strake, a 45° main wing, and a forebody. The DLR-F22 model is described in greater detail elsewhere [4]. The test conditions are summarized in Table 1 for both data acquisition cases. For the pitch–traverse case, the model was moved with a constant pitch speed of 0.1°/s.

Table 1. Test conditions investigated in the measurement campaign.

Mach Number M	Angles-of-Attack for $\beta = 0°$ α [°]	Angles-of-Yaw for $\alpha = 20°$ β [°]
0.5, 0.85, 0.95, 1.1	15 → 25 (pitch traverse)	−6 → +6 (pitch traverse)
0.5, 0.85, 0.95, 1.1	16, 20, 24 (pitch pause)	−5, 0, 5 (pitch pause)

The suction side of the model was coated with PSP using a spray gun for the different layers, as described in Section 2.2. The coating was performed after the DLR-F22 model had been fully mounted onto the wind tunnel support system and after all model-related instrumentation—temperature sensors and pressure taps—had been prepared for operation. For this purpose, a portable painting booth (commercially purchasable under the name "paintTrotter", LAGOS, Bergondo, Le Coruña, Spain) was installed; this allows safe handling of the PSP coating materials and guarantees a clean environment for the coating procedure. With this approach, the PSP coating is the final step of model preparation and thus reduces the risk of damage to the PSP coating during handling and model installation. A photo of the wind tunnel model inside the painting booth during the coating process is shown in Figure 4.

Figure 4. Model coating inside the portable painting booth.

3.2. Data Acquisition and Automation

The internal pressure in TWG can be varied so that before starting wind tunnel operation, one so-called wind-off (reference) data point without flow could be acquired to allow for the correction of inhomogeneous lifetime distributions, as has been mentioned before. To be able to make allowance for temperature differences between wind-off and wind-on conditions, one wind-off data point was also acquired after each wind tunnel run. The wind-off pressure was set near the static pressure under wind-on conditions [26].

3.2.1. Pitch–Pause Mode

Each series of data points consists of a set of different angles-of-attack or yaw angles at a constant Mach number.

The acquisition of one data point follows the procedure that was presented for the first time in [17]; a sequence of PSP image acquisition is shown in Figure 5a. The first two images are acquired with LEDs switched off to obtain dark images, followed by five times alternating acquisition of the Gate1 and Gate2 images. The alternating gate acquisition allows for taking Gate1 data before and after Gate2 data and thus compensates for possible changes related to temperature in time. Acquired images are transferred to a computer while the image acquisition sequence is in progress. At the end of the sequence, one additional dark image is acquired. In this test, five image sets of Gate1 and Gate2 images were taken and the obtained images for each gate were then averaged to obtain the final pressure image. The duration for the acquisition of one (Gate) image is approximately one second, resulting in a total measurement time for one test point of 13 seconds. Figure 5b displays an illustration of a pitch–pause run with three data points at α = 16, 20, 24°. The image acquisition sequence is carried out automatically using a measurement program developed at DLR; it is described in more detail in [26]. All measured data (PSP images, wind tunnel data, pressure tap data, etc.) are stored in a common folder during the data acquisition. After completion of each data acquisition sequence, they are then transferred to corresponding folders automatically for final storage.

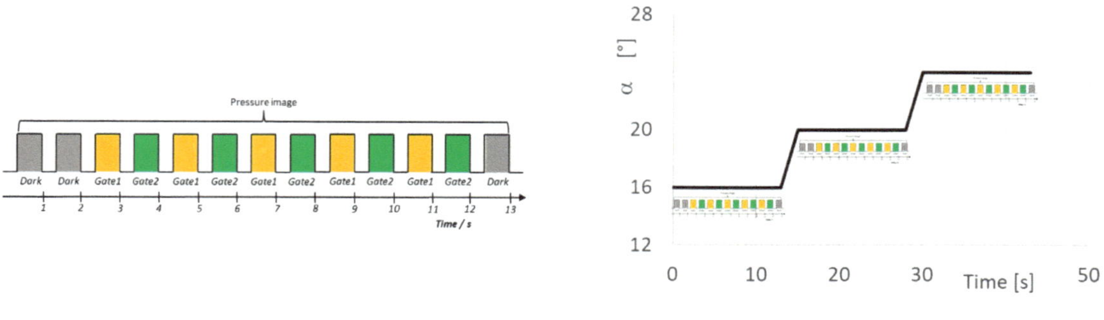

(**a**) (**b**)

Figure 5. PSP image acquisition sequences in pitch–pause mode. (**a**) At one test point (e.g., at one angle-of-attack α), it takes 13 s to acquire the one pressure image. (**b**) Example of a pitch–pause run at three different α = 16, 20, 24°.

3.2.2. Pitch–Traverse Mode

The test procedure for the pitch–traverse case (continuously moving model) can be described as follows: once the Mach number in the wind tunnel has been established and the starting position of the wind tunnel model for the pitch–traverse sweep has been reached, the acquisition procedure for the PSP images is started. Initially, dark images (with LEDs switched off) are acquired for Gate1 and Gate2 while the wind tunnel model is still kept in its start position, i.e., without starting the pitch–traverse sweep. Then, the pitch–traverse sweep for the wind tunnel model is started, see Figure 6a, and is synchronized in time with the start of alternating Gate1 and Gate2 image acquisition with a camera frequency of 3 Hz, resulting in a pressure measurement frequency (I_{Gate1}/I_{Gate2} ratios) for this lifetime-based image acquisition of 1.5 Hz, as shown in Figure 6b. Taking the pitch–traverse speed of 0.1°/s into account, one can obtain lifetime-based pressure images with an angular resolution better than 0.1°. For the given (see Table 1) angle-of-attack sweeps of α = 15→25°, approximately 300 (gated) images corresponding to 150 pressure images were acquired. Since the camera timing signal was also recorded by the wind tunnel control system while the model moved through its pitch–traverse sweep, the exact wind tunnel model position for each of the PSP pressure images could be determined.

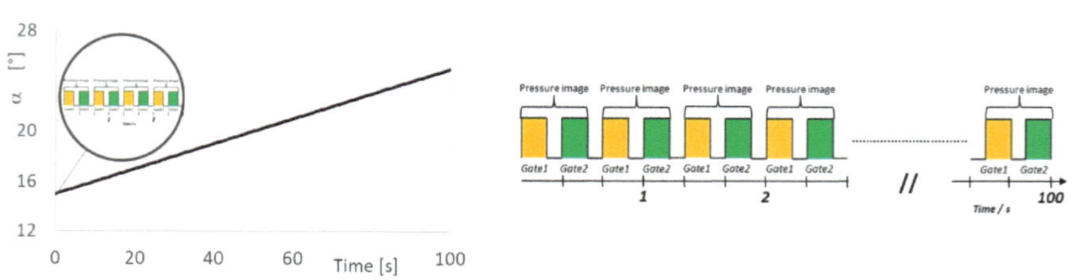

(**a**) (**b**)

Figure 6. (**a**) Example of a pitch–traverse sweep from α = 15°→25°, the total time for pitch–traverse was 100 s. Zoom shows the image acquisition in the first two seconds. (**b**) Image acquisition sequence during a pitch–traverse sweep.

3.2.3. Data Processing

Image processing starts with dark image subtraction. A gate ratio image was then calculated as ratios R of I_{Gate1} to I_{Gate2} ($R = I_{Gate1}/I_{Gate2}$). To cancel out the inhomogeneous lifetime distribution mentioned before, a ratio-of-ratios image was calculated as the ratio of R_{on} to R_{off}. Image alignment was applied between R_{on} and R_{off} to correct for model movement between wind-on and wind-off conditions. The processed PSP image was mapped onto a three-dimensional model grid by using marker positions in the images and mapping them to the coordinates of the model grid. The mapping process takes account of 11 independent parameters: translation (3), rotation (3), perspective (3), scaling (1) and lens correction (1). The value for every grid point is calculated by linearly interpolating adjacent pixels. Pressure values were calculated from the ratio images by using data from a priori calibration, which had been obtained in a special calibration chamber, where temperature and pressure can be set individually. Details on the setup and properties of the chamber can be found in [27]. A small PSP calibration coupon (an aluminum plate) with dimensions of 30 mm by 30 mm is coated at the same time as the model to ensure that the PSP coating has the same paint composition and thickness. For temperatures ranging from 20 °C to 40 °C and pressures ranging from 10 kPa to 110 kPa, the luminescent intensity is captured with a CCD camera, and the signal is averaged over an area of 40 by 40 pixels (about 10 mm × 10 mm). CCD camera, gate settings, excitation wavelength, and optical filters are the same as in the wind tunnel setup. The gate ratio values for the given pressure and temperature range are presented in Figure 2b. For the given gate settings, the PSP coating used here has a pressure sensitivity of $S_P = 0.77\%/\text{kPa}$ and a temperature sensitivity $S_T = 0.37$ at $P = 70$ kPa. A flow chart of the PSP data processing is shown in Figure 7.

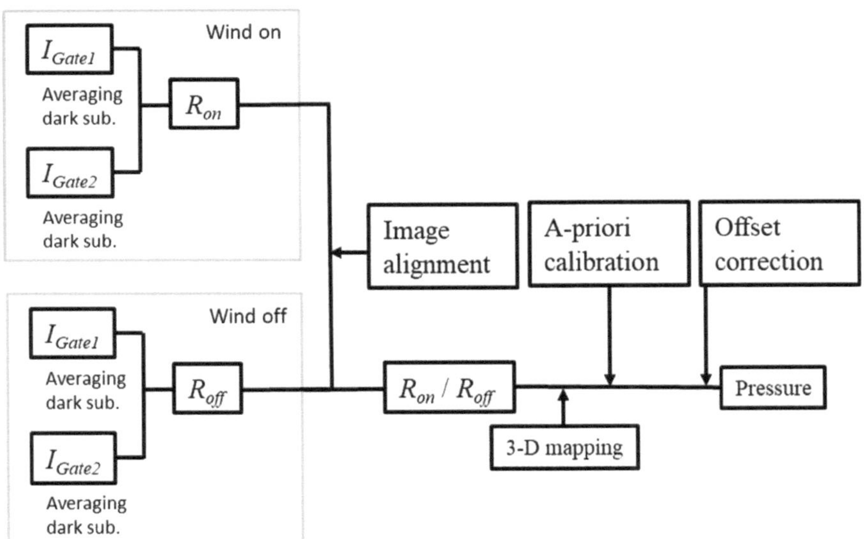

Figure 7. Flow chart of the PSP data processing.

Finally, PSP data and corresponding pressure tap and Kulite® data were compared and an "offset" correction was determined and applied to the PSP data to correct for mean differences between the two data sets. This difference is caused mainly by a temperature change between wind-on and wind-off conditions [3]. All PSP data processing was performed by the DLR in-house developed software ToPas version 6.1.0 [28].

4. Results from the Wind Tunnel Experiment

In the following images, the flow is from left to right. The same color map is used for all presented pressure distribution (pressure coefficient C_p) figures in this chapter. In all the presented images, the blue areas indicate low pressure, whereas the red areas indicate a higher pressure. Areas presented in black color could either not be acquired with the installed optical setup from the camera (forebody) or were not sufficiently excited by UV light from the installed LED systems (on the port side) to obtain usable luminescence images. The colormap applied to show the PSP results are also shown in each Figure.

By means of PSP, the shape and position of leading-edge vortices can be measured easily since the induced suction of vortical flow imprints characteristic pressure signatures (footprints) onto the wind tunnel models' surface. Some accompanying phenomena, e.g., vortex breakdown, secondary vortex structures, and vortex–shock interaction, can also be detected when PSP is applied. Thus, PSP provides valuable insights into the aforementioned phenomena and their evolution throughout a large angle-of-attack and Mach number range. The general flow structure that develops at the DLR-F22 model remains qualitatively similar for a large range of angles-of-attack and Mach numbers. For one case, M = 0.85, $\alpha = 16°$, and $\beta = 0°$, the pressure distribution measured by means of PSP is shown in Figure 8, and the herewith measured footprints of the flow field will be discussed. The flow structure is dominated by three primary vortices that originate along the leading edges of the forebody, the strake, and the main wing. These vortices are named as inboard vortex, mid-board vortex, and outboard vortex, respectively. The positions and spatial locations of the aforementioned vortices are shown in Figure 8 as a black dotted line (inboard vortex), a black dashed line (mid-board vortex), and a black long-dashed line (outboard vortex). Additionally, shocks from above the strake and main wing can clearly be recognized from the PSP image. A first shock is visible behind the kink from the levcon to the strake, and the second shock is located further downstream on the main wing. Their spatial locations are highlighted as white dash–dot lines in Figure 8. Interaction of shocks with the vortices causes shock-induced vortex breakdown that can also be easily detected from the PSP image. More details regarding the basic flow structure of the DLR-F22 model can be found in [4].

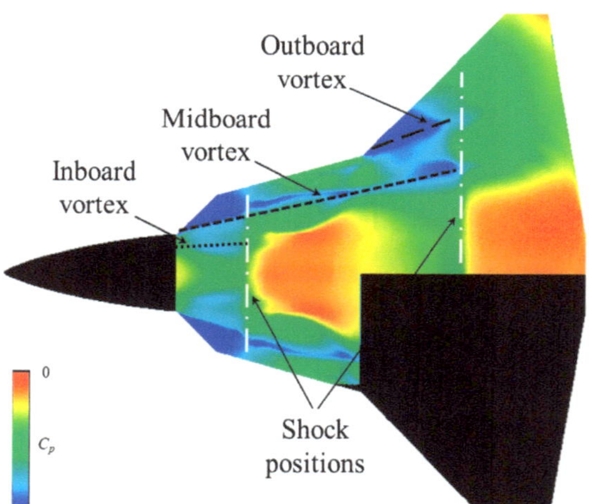

Figure 8. Pressure distribution measured by means of lifetime PSP in pitch–pause mode on the DLR-F22 model, M = 0.85, $\alpha = 16°$, $\beta = 0°$ with vortex (black) and shock (white) positions indicated by dashed/dotted lines.

There are already copious pressure data available from wind tunnel experiments using the DLR-F22 model; here, continuous lifetime PSP data acquisition by means of the on-chip accumulation technique during the pitch–traverse sweep has been carried out for the first time. As already mentioned, 150 pressure images were measured in 100 s during a pitch–traverse sweep from $\alpha = 15 \rightarrow 25°$; however, only a small selection of these 150 PSP result images will be shown and discussed here. The PSP results using the pitch–traverse mode are compared here to those that were measured in pitch–pause mode. For an image-based comparison of the measured flow topologies, PSP result images were selected for Mach numbers M = 0.85, 0.95, and 1.1 with an angle-of-attack of α = 16° and a yaw angle of β = 0°.

Figure 9 shows PSP result images for M = 0.85, with (a) showing the PSP result image using the pitch–pause mode (same as in Figure 8) and (b) the PSP result image obtained during the pitch–traverse mode. The three primary vortices (in-, mid-, and outboard vortex) are clearly visible in (a) and (b), and the shock positions can also be clearly seen in both images. In all the following discussed PSP result images, the in-, mid-, and outboard vortices, as well as the shock positions, are shown using the same assignment of black or white lines, as first introduced in Figure 8. For a better comparison of the flow topology in the PSP result images, the black and white lines for the PSP result image obtained in pitch–pause mode are copied into the PSP result image for pitch–traverse mode. This makes it possible to see immediately whether the vortex axes and shock positions have been measured at identical positions. Looking at the result images in Figure 9a,b, it is noticeable that the vortex axes of the three primary vortices and the upstream shock in the area of the levcon correspond very well locally. However, the downstream shock position in Figure 9b has slightly been shifted upstream.

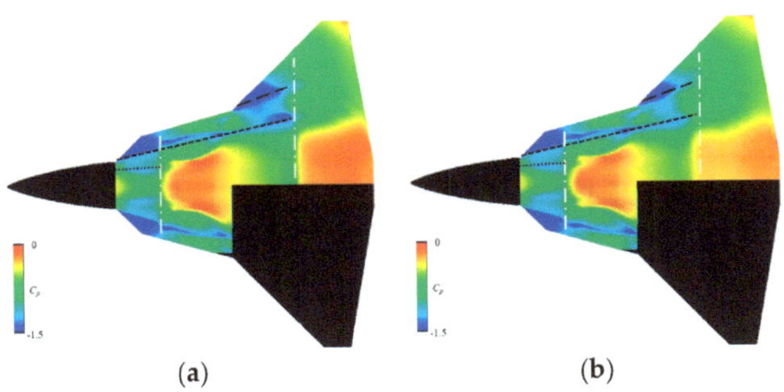

Figure 9. PSP result images obtained in pitch–pause (**a**) or pitch–traverse mode (**b**) for M = 0.85, α = 16°, β = 0°; black and white lines showing vortex and shock locations in (**a**) have been transposed directly into (**b**) to facilitate comparison.

For a better comparison of the results shown in Figure 9a,b and to determine differences arising from the two different data acquisition modes, Figure 10a shows the C_p-difference image ($\Delta C_p = C_{p,\, pitch\text{–}pause} - C_{p,\, pitch\text{–}traverse}$) obtained by subtraction of the two images, and also includes the black and white lines for orientation, as explained before. As is to be expected for a vortex- and vortex-shock-dominated unsteady flow field, pressure differences can be seen in this difference image. Maximum differences of $\Delta C_p = \pm 0.2$ can be seen in the area of the shocks above the main wing and in the area of the (outboard) vortex-shock interaction. However, outside this area on the strake, levcon, and the main wing, the essential physical flow characteristics, their local position, and also the measured pressure distributions match very well. The resulting image shown in Figure 9a has been averaged from 5 individual image pairs ($R_{on} = I_{Gate1}/I_{Gate2}$); see Section 3.2.1. Averaging of 5 image pairs allows the calculation of the root mean square (RMS) value. Here, the RMS value is calculated for those images with the highest pressure sensitivity, which are the Gate2 images, according to Equation (3).

$$\mathrm{RMS}_{Gate2} = \sqrt{\frac{\sum_{i=1}^{5} \left(I_{Gate2} - < I_{Gate2} >\right)^2}{< I_{Gate2} > \cdot 4}} \quad (3)$$

where $<I_{Gate2}>$ is the mean value of the five individual I_{Gate2} images.

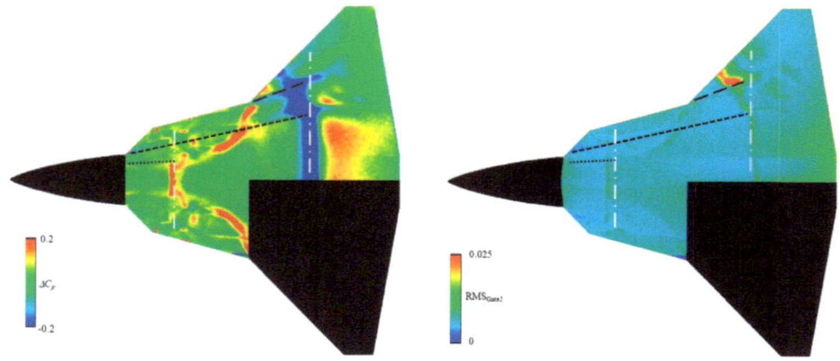

(**a**) (**b**)

Figure 10. C_p-difference image (**a**) and RMS_{Gate2}-result image (**b**) for M = 0.85, α = 16°, β = 0° (see text).

These RMS values are the largest in the region of transient or unsteady flow during the measurement acquisition time of 10 s. Figure 10b shows the RMS_{Gate2} result of the PSP result image shown in Figure 9a. The largest RMS_{Gate2} values have been measured for the outboard vortex in the area of vortex breakdown (vortex–shock interaction). Large fluctuations can also be seen in the area of the shock positions over the main wing and the levcon. The largest differences in the C_p difference image (Figure 10a) thus correlate very well with the areas of large RMS_{Gate2} values in Figure 10b.

Now, the obtained PSP results for the higher transonic Mach number of M = 0.95 will be compared. First of all, Figure 11 shows the individual PSP results for the different data acquisition modes: (a) shows the PSP result image for the pitch–pause mode, and (b) shows the PSP result image for the pitch–traverse mode. In both PSP result images, the mid- and outboard vortex structures are clearly visible. In contrast, the inboard vortex is only very weakly visible in both cases. The shock positions and the associated shock-vortex interactions, which lead to vortex breakdown, are also clearly visible in both result images. Compared to the results shown for M = 0.85 in Figure 9, it is noticeable for the M = 0.95 results that the measured suction peaks of the in-, mid-, and outboard vortices are weaker (lighter blue color in the result images). For this higher Mach number M = 0.95, the C_p-difference image has also been calculated, and this result is shown in Figure 11c. For this M = 0.95, the differences in results between both data acquisition modes are less pronounced than for M = 0.85. There are, again, local differences in the vicinity of the shock positions and the vortex breakdowns, but the spatial extent of these differences is significantly smaller for M = 0.95 compared to M = 0.85. There are also some differences very close to the trailing edge of the model, but these can be explained by impurities (oil) on the lower of the model.

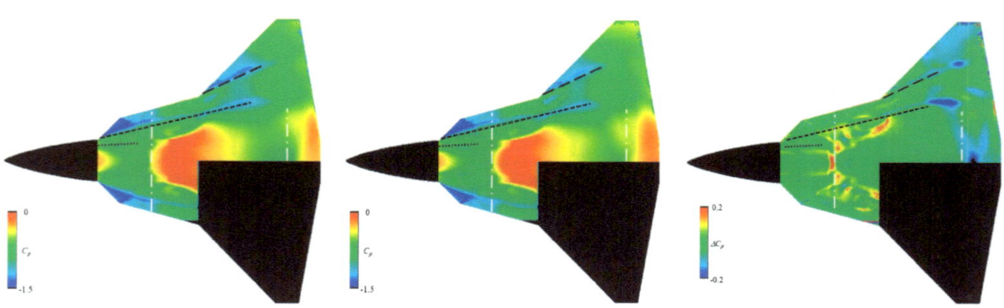

(**a**) (**b**) (**c**)

Figure 11. PSP result images obtained in pitch–pause (**a**) or pitch–traverse mode (**b**). (**c**) C_p-difference image for M = 0.95, α = 16°, β = 0°.

Finally, in Figure 12, also the individual PSP results for the supersonic Mach number M = 1.1 with α = 16°, β = 0°are presented for the two different data-acquisition modes: (a) shows the PSP result image for the pitch–pause mode and (b) shows that for the pitch–traverse case. The pressure signature of mid- and outboard vortices is clearly visible in both presented PSP results, whereas an inboard vortex cannot be seen (at least not with the chosen colormap) in the PSP results. Pressure signatures of mid- and outboard vortices are visible up to the trailing edge of the main wing. Compared to the previously discussed PSP results for the transonic Mach number, it is obvious that for the supersonic Mach number, the vortex breakdown does not occur above the main wing, and furthermore, only one shock position has been detected. Compared to the results shown for M = 0.85 and 0.95 in Figures 9 and 11, it is noticeable for the supersonic M = 1.1 PSP results that the measured suction peaks of the mid and outboard vortices are much weaker. For the highest tested Mach number M = 1.1, the C_p-difference image has also been calculated, and this ΔC_p result is shown in Figure 12c. Overall, there is very good agreement between the results shown for both data acquisition modes (a) and (b). Only in the area of the shock are there again significant pressure differences, but these are again locally very limited and restricted to the area of unsteady flow phenomena.

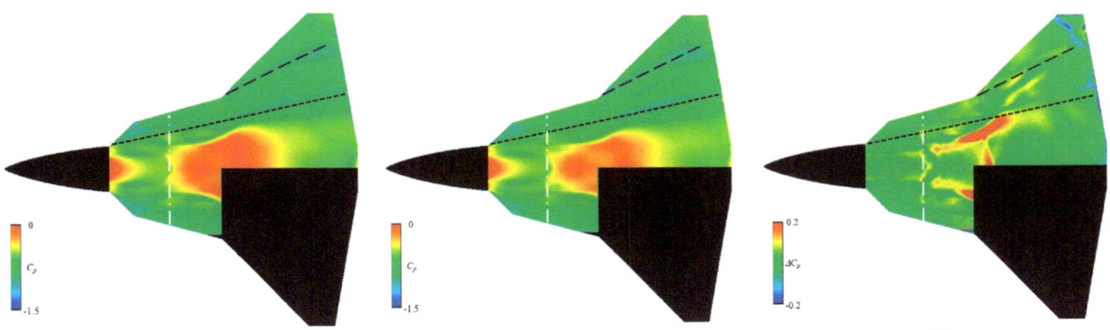

(**a**) (**b**) (**c**)

Figure 12. PSP result images obtained in pitch–pause (**a**) or pitch–traverse mode (**b**) for M = 1.1, α = 16°, β = 0°. (**c**) C_p-difference image—see text.

Error Analysis

The wind tunnel model has been equipped with conventional pressure taps for steady pressure readings and Kulite® sensors for unsteady pressure measurements, see Figure 3a. For the measurements presented here, only the Kulite® sensors for unsteady pressure measurements have been recorded during the continuous sweep of the wind tunnel model using the Dewetron® system of the wind tunnel data-acquisition system. Data acquisition using the conventional pressure taps did not take place in the measurement campaign presented here. However, since the PSP data acquisition system for the conventional pitch–pause method is identical to that presented in [17] and [26], a standard error in ΔP of 600 Pa is also to be expected here for the trans- and supersonic results presented in Figures 8, 9a, 11a and 12a.

To specify the uncertainty of the PSP results during the continuous movement of the wind tunnel model (pitch–traverse mode), the pressure distribution measured by means of PSP is considered in the local area around all the Kulite® sensors. This is possible because all PSP results are mapped onto the surface grid of the wind tunnel model, and therefore, the exact local area around the Kulite® sensors can be considered.

By way of example, error analysis has been carried out and is shown here for the case M = 0.85; see Figure 13a: for all 150 data points acquired with PSP ($15 \leq \alpha \leq 25°$, $\beta = 0°$), the PSP pressure values (in kPa) are shown on the abscissa and the corresponding Kulite® results (also in kPa) on the ordinate. The various angles of attack can be identified with the help of the symbol color map in the figure, where low to high α is coded as ranging from blue- to red-colored symbols, respectively. Fortunately, the relationship is linear over the entire measured angle-of-attack range, indicating that there is no particularly problematic angle-of-attack range that needs to be considered separately in the error analysis. The error can, therefore, be specified covering the entire data series of measurements. For this purpose, the pressure difference (PSP-to-Kulite®) has been plotted in Figure 13b as a function of the angle of attack. One can see some outliers in the range $16° \leq \alpha \leq 17°$, where larger errors of up to ±4 kPa can be seen. This may be related to the fact that the mid-board vortex, which for this angle-of-attack range propagates approximately in the direction of the Kulite® sensor locations (see Figure 9b), produces locally strong transient/unsteady pressure differences in the area of these Kulite® sensors. Statistically, this leads to a measurement error of ΔP = 790 Pa. It must be emphasized here that the PSP results can be compared with only the four available Kulite® sensors so that the statistical sample size is not particularly large. Interestingly, the result of the error analysis obtained for the pitch–traverse mode is of the same size as for the pitch–pause mode, so with some confidence, the value given here can be considered plausible.

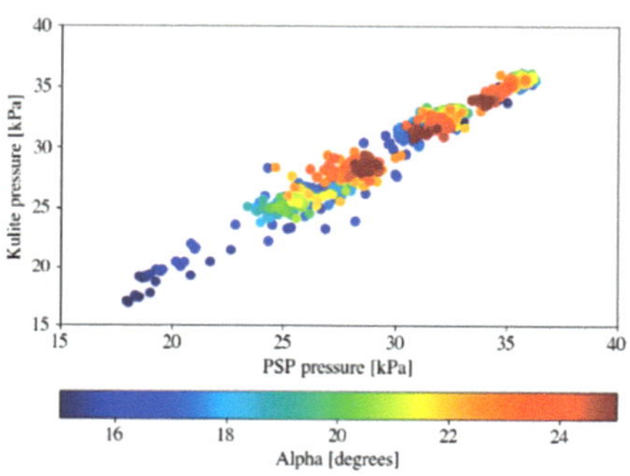

(a)

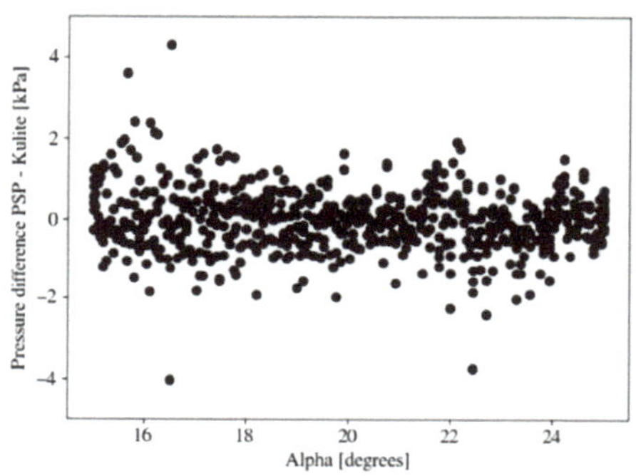

(b)

Figure 13. Error analysis for all 150 data points measured with PSP for M = 0.85 in pitch–traverse mode. (**a**) Comparison of PSP pressure values and the corresponding Kulite® results, (**b**) pressure difference (PSP-to-Kulite®).

5. Conclusions and Outlook

Lifetime-based pressure measurements by means of PSP using on-chip accumulation on the DLR-F22 triple-delta wing model have been performed in the transonic wind tunnel TWG for trans- and supersonic Mach numbers. PSP images were acquired when the wind tunnel model was moved in pitch–pause mode (standard approach) and, alternatively and for the first time, in pitch–traverse mode. The data acquisition strategies for both modes have been presented and the applied data reduction was shown. The pressure distributions obtained by PSP for the different data-acquisition approaches were shown for trans- and supersonic Mach number cases and some selected angles-of-attack and yaw angles. In general, the results for both data-acquisition modes agree very well and primary vortices and shock positions could be easily detected. When comparing the corresponding data points for the two different data acquisition modes individually, one can see small differences, for example, in locations and extent of vortex breakdown, shock position, or shock–vortex interaction. Since these measurements were performed on a model configuration at high angles-of-attack and at trans- and supersonic flow conditions, noticeable unsteady flow is to be expected and was, in fact, indeed observed in the pitch–pause pressure data, which allows RMS analysis of the pressure and thus provided an insight into the unsteadiness of the flow. By application of the PSP lifetime technique in combination with the pitch–traverse sweep of the wind tunnel model, the productivity of the data acquisition (in terms of measurement per unit time) could be increased to ≥1.5 Hz compared to <0.1 Hz for the standard pitch–pause mode, viz. an increase by a factor of 15. The accuracy of ΔP = 790 Pa achieved for PSP data acquisition in pitch–traverse mode is similar to that of pitch–pause mode (ΔP = 600 Pa). In summary, the lifetime-based PSP data acquisition technique using on-chip accumulation was applied successfully in a wind tunnel test whilst the model was moving in pitch–traverse mode during the measurement.

For future PSP tests where many model configurations need to be investigated or when a wind tunnel model is to be investigated for the first time in a wind tunnel, the application of the lifetime-based PSP measurement technique presented here in combination with continuous model movement (pitch–traverse sweep) can be recommended to enable more efficient use of tunnel runtime to be made.

Author Contributions: Conceptualization, C.K. and D.Y.; methodology, D.Y.; software, D.Y. and U.H.; validation, D.Y., U.H. and C.K.; formal analysis, C.K.; investigation, D.Y.; resources, C.K.; data curation, C.K.; writing—original draft preparation, C.K.; writing—review and editing, D.Y. and U.H.; visualization, D.Y.; supervision, C.K.; project administration, C.K.; funding acquisition, C.K. All authors have read and agreed to the published version of the manuscript.

Funding: The experiments were conducted within the framework of the DLR projects HIGHFLY and DIABOLO, which provided financial and organizational support for this research.

Institutional Review Board Statement: Not applicable.

Informed Consent Statement: Not applicable.

Data Availability Statement: All data that support the findings of this study are included within this article.

Acknowledgments: The authors would like to thank Carsten Fuchs and Tobias Kleindienst (DLR) for their support during hardware installation; Janos Agocs (DLR) for taking photos for the measurement documentation; Martin Rein, Michael Werner, and Andreas Schütte (DLR) for their support in using the DLR-F22 model and to help further improve the DLR PSP measurement technique, and also for helpful discussions and advice regarding the DLR-F22 model. Additionally, we would like to thank the TWG wind tunnel crew for their support during the measurements. We dedicate this article to our previous colleague from ETW, Martin C.N. Wright, who passed away in 2023 and who was very committed to the industrial use of PSP under cryogenic conditions in ETW. Martin would certainly have been very happy about the method for continuous measurement using lifetime PSP described in this article and would have encouraged us to possibly apply it in the ETW as well.

Conflicts of Interest: The authors declare no conflicts of interest.

Appendix A

For the sake of completeness, we would like to add a result that was not discussed in the previous results section but was also measured according to conditions given in Table 1.

PSP results for M = 0.5 were measured successfully in the pitch–pause mode only. PSP measurements during the pitch–traverse mode were also performed for M = 0.5, but the readout speed for the camera was set to 40 Mpx/s for these series and image saturation was reached at 10,000 counts already. Therefore, the expected 14-bit dynamic range of the camera could not be used and the PSP images are unusable. The results presented in Section 4 were acquired with a camera readout speed of 10 Mpx/s for which the expected 14-bit dynamic range could be fully used.

Nevertheless, the measured PSP results in pitch–pause mode can be shown here; see Figure A1. It has to be noted that the colormap for the M = 0.5 case is different than that for the other Mach numbers.

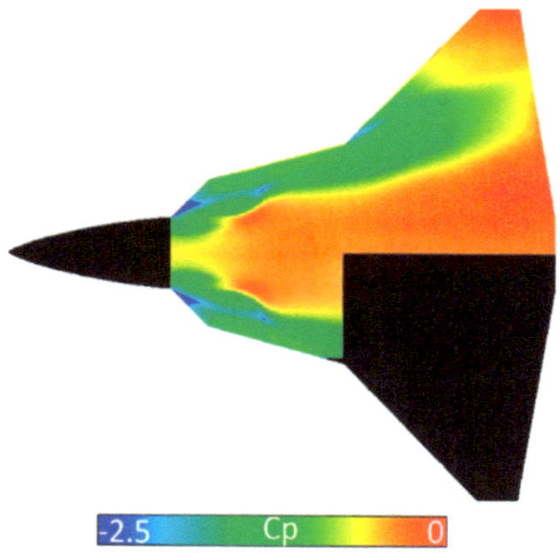

Figure A1. Pressure distribution for M = 0.5, $\alpha = 16^{\circ}$, $\beta = 0^{\circ}$ measured in pitch–pause mode.

References

1. Engler, R.H.; Hartmann, K.; Schultze, B. Aerodynamic assessment of an optical pressure measurement system (OPMS) by comparison with conventional pressure measurements in a high speed wind tunnel. In Proceedings of the ICIASF'91 Record, 14th International Congress on Instrumentation in Aerospace Simulation Facilities, Institute of Electrical and Electronics Engineers, Rockville, MD, USA, 27–31 October 1991; pp. 17–24. [CrossRef]
2. Bell, J.H.; Schairer, E.T.; Hand, L.A.; Mehta, R.D. Surface pressure measurements using luminescent coatings. *Annu. Rev. Fluid Mech.* **2001**, *33*, 155–206. [CrossRef]
3. Liu, T.; Sullivan, J.P.; Asia, K.; Klein, C.; Egami, Y. *Pressure and Temperature Sensitive Paints*, 2nd ed.; Springer: Cham, Switzerland, 2021; ISBN 978-3-030-68055-8. [CrossRef]
4. Werner, M.; Rein, M.; Richter, K.; Weiss, S. Experimental and Numerical Analysis of the Aerodynamics and Vortex Interactions on Multi-Swept Delta Wings. *CEAS Aeronaut. J.* **2023**, *14*, 927–938. [CrossRef]
5. Holmes, J.W. Analysis of radiometric, lifetime and fluorescent lifetime imaging for pressure sensitive paint. *Aeronaut. J.* **1998**, *102*, 189–194. [CrossRef]
6. Juliano, T.J.; Kumar, P.; Peng, D.; Gregory, J.W.; Crafton, J.; Fonov, S. Single-shot, lifetime-based pressure-sensitive paint for rotating blades. *Meas. Sci. Technol.* **2011**, *22*, 085403. [CrossRef]
7. Holst, G.; Kohls, O.; Klimant, I.; König, B.; Kühl, M.; Richter, T. A modular luminescence lifetime imaging system for mapping oxygen distribution in biological samples. *Sens. Actuators B Chem.* **1998**, *51*, 163–170. [CrossRef]
8. Ruyten, W.; Sellers, M.E.; Baker, W.M. Spatially Nonuniform Self-Quenching of the Pressure-Sensitive Paint PtTFPP/FIB. In Proceedings of the 47th AIAA Aerospace Sciences Meeting Including the New Horizons Forum and Aerospace Exposition, Orlando, FL, USA, 5–8 January 2009. [CrossRef]
9. Bell, J.H.; Sellers, M.E. Pressure Sensitive Paint Measurements on a Space Shuttle Model with the Luminescent Lifetime Technique. In Proceedings of the 21st International Congress on Instrumentation in Aerospace Simulation Facilities, Institute of Electrical and Electronics Engineers, Sendai, Japan, 29 August–1 September 2005; pp. 195–206. [CrossRef]
10. Ruyten, W.; Sellers, M.E. Improved Data Processing for Pressure-Sensitive Paint Measurements in an Industrial Facility. In Proceedings of the 44th AIAA Aerospace Sciences Meeting and Exhibit, Reno, NV, USA, 9–12 January 2006. [CrossRef]
11. Mitsuo, K.; Asai, K.; Takahashi, A.; Mizushima, H. Advanced Lifetime PSP Imaging System for Simultaneous Pressure and Temperature Measurements. In Proceedings of the 24th AIAA Aerodynamic Measurement Technology and Ground Testing Conference, Portland, Oregon, USA, 28 June–1 July 2004. [CrossRef]
12. Watkins, A.N.; Jordan, J.D.; Leighty, B.D. Development of Next Generation Lifetime PSP Imaging System. In Proceedings of the 20th International Congress on Instrumentation in Aerospace Simulation Facilities, Göttingen, Germany, 25–29 August 2003; pp. 372–382. [CrossRef]
13. Sugioka, Y.; Sato, H.; Nakakita, K.; Nakajima, T.; Nonomura, T.; Asia, K. In-flight visualization of shock wave on a jet aircraft wing using lifetime-based Pressure-Sensitive Paint Technique. In Proceedings of the AIAA Scitech Forum, San Diego, CA, USA, 7–11 January 2019. [CrossRef]
14. Yorita, D.; Weiss, A.; Geisler, R.; Henne, U.; Klein, C. Comparison of LED and LASER based Lifetime Pressure-Sensitive Paint Measurement Techniques. In Proceedings of the 2018 AIAA Aerospace Sciences Meeting, Kissimmee, FL, USA, 8–12 January 2018. [CrossRef]
15. Weiss, A.; Geisler, R.; Schwermer, T.; Yorita, D.; Henne, U.; Klein, C.; Raffel, M. Single-shot pressure-sensitive paint lifetime measurements on fast rotating blades using an optimized double-shutter technique. *Exp. Fluids* **2017**, *58*, 120. [CrossRef]

16. Yorita, D.; Klein, C.; Henne, U.; Ondrus, V.; Beifuß, U.; Bruse, M. Improvement of PtTFPP-based PSP for Lifetime-based Measurement and Application to Transonic Wind Tunnel Test. In Proceedings of the 16th International Symposium on Flow Visualization—ISFV16, Okinawa, Japan, 24–28 June 2014.
17. Yorita, D.; Henne, U.; Klein, C. Improvement of Lifetime-based PSP Technique for Industrial Wind Tunnel Tests. In Proceedings of the 55th AIAA Aerospace Sciences Meeting, Grapevine, TX, USA, 9–13 January 2017. [CrossRef]
18. Roberts, D. Progress at ARA's transonic wind tunnel in application of PSP for industrial testing. In Proceedings of the PSP-Workshop, AIAA SCITECH Forum, Orlando, FL, USA, 8–12 January 2024.
19. Mébarki, Y.; Benmeddour, A. Pressure-Sensitive Paint Measurements on a Moving Store in the NRC 1.5 m Blowdown Wind Tunnel. In Proceedings of the 32nd AIAA Aerodynamic Measurement Technology and Ground Testing Conference, Washington, DC, USA, 13–17 June 2016. [CrossRef]
20. Bruse, M.; (DNW—The German Dutch Wind-Tunnels, Göttingen, Germany). Personal communication, 2021.
21. Schulz, M.; (ETW—European Transonic Windtunnel, Cologne, Germany). Personal communication, 2023.
22. Ueda, Y. Notice of Discontinuation of Sony Semiconductor Corporation's Kagoshima Wafer Production Line (CCD 200 mm Line). Available online: www.electronicspecifier.com/news/analysis/ccd-200mm-wafer-line-to-be-discontinued (accessed on 6 June 2024).
23. Puklin, E.; Carlson, B.; Gouin, S.; Costin, C.; Green, E.; Ponomarev, S.; Tanji, H.; Goutermann, M. Ideality of Pressure-Sensitive Paint. I. Platinum tetra(pentafluorophenyl)porphine in fluoreacrylic polymer. *J. Appl. Polym. Sci.* **2000**, *77*, 2795–2804. [CrossRef]
24. Egami, Y.; Fujii, K.; Takagi, T.; Matsuda, Y.; Yamaguchi, H.; Niimi, T. Reduction of Temperature Effects in Pressure Sensitive Paint Measurements. *AIAA J.* **2013**, *51*, 1779–1783. [CrossRef]
25. The Transonic Wind Tunnel in Göttingen (TWG). Available online: www.dnw.aero/wind-tunnels/twg/ (accessed on 7 May 2024).
26. Henne, U.; Yorita, D.; Klein, C. Application of Lifetime-Based Pressure-Sensitive Paint for Transonic Tests on a Generic Delta Wing Planform. In *New Results in Numerical and Experimental Fluid Mechanics XII*; Dillmann, A., Ed.; Springer: Cham, Switzerland, 2020; Volume 142, pp. 287–296. [CrossRef]
27. Egami, Y.; Fey, U.; Klein, C.; Quest, J.; Ondrus, V.; Beifuss, U. Development of new two-component temperature-sensitive paint (TSP) for cryogenic testing. *Meas. Sci. Technol.* **2012**, *23*, 115301. [CrossRef]
28. Klein, C.; Engler, R.H.; Henne, U.; Sachs, W.E. Application of pressure-sensitive paint for determination of the pressure field and calculation of the forces and moments of models in a wind tunnel. *Exp. Fluids* **2005**, *39*, 475–483. [CrossRef]

photonics

Article

Improving the Two-Color Temperature Sensing Using Machine Learning Approach: $GdVO_4:Sm^{3+}$ Prepared by Solution Combustion Synthesis (SCS)

Jovana Z. Jelic [1], Aleksa Dencevski [1], Mihailo D. Rabasovic [1], Janez Krizan [2], Svetlana Savic-Sevic [1], Marko G. Nikolic [1], Myriam H. Aguirre [3,4,5], Dragutin Sevic [1,*] and Maja S. Rabasovic [1,*]

1 Institute of Physics, University of Belgrade, Pregrevica 118, 11080 Belgrade, Serbia; jelic@ipb.ac.rs (J.Z.J.); dencevski@ipb.ac.rs (A.D.); rabasovic@ipb.ac.rs (M.D.R.); savic@ipb.ac.rs (S.S.-S.); nikolic@ipb.ac.rs (M.G.N.)
2 AMI, d.o.o, 2250 Ptuj, Slovenia; janez.krizan@gmail.com
3 Department of Condensed Matter Physics, University of Zaragoza, E-50009 Zaragoza, Spain; maguirre@unizar.es
4 Instituto de Nanociencia y Materiales de Aragón, University of Zaragoza-CSIC, E-50018 Zaragoza, Spain
5 Laboratorio de Microscopías Avanzadas, Univervisity of Zaragoza, E-50018 Zaragoza, Spain
* Correspondence: sevic@ipb.ac.rs (D.S.); majap@ipb.ac.rs (M.S.R.)

Abstract: The gadolinium vanadate doped with samarium ($GdVO_4:Sm^{3+}$) nanopowder was prepared by the solution combustion synthesis (SCS) method. After synthesis, in order to achieve full crystallinity, the material was annealed in air atmosphere at 900 °C. Phase identification in the post-annealed powder samples was performed by X-ray diffraction, and morphology was investigated by high-resolution scanning electron microscope (SEM) and transmission electron microscope (TEM). Photoluminescence characterization of emission spectrum and time resolved analysis was performed using tunable laser optical parametric oscillator excitation and streak camera. In addition to samarium emission bands, a weak broad luminescence emission band of host $VO_4{}^{3-}$ was also observed by the detection system. In our earlier work, we analyzed the possibility of using the host luminescence for two-color temperature sensing, improving the method by introducing the temporal dependence in line intensity ratio measurements. Here, we showed that further improvements are possible by using the machine learning approach. To facilitate the initial data assessment, we incorporated Principal Component Analysis (PCA), t-Distributed Stochastic Neighbor Embedding (t-SNE) and Uniform Manifold Approximation and Projection (UMAP) clustering of GdVO4:Sm^{3+} spectra at various temperatures. Good predictions of temperature were obtained using deep neural networks. Performance of the deep learning network was enhanced by data augmentation technique.

Keywords: samarium-doped gadolinium vanadate nanopowder; structural and luminescent properties; lifetime; phosphor thermometry; machine learning

Citation: Jelic, J.Z.; Dencevski, A.; Rabasovic, M.D.; Krizan, J.; Savic-Sevic, S.; Nikolic, M.G.; Aguirre, M.H.; Sevic, D.; Rabasovic, M.S. Improving the Two-Color Temperature Sensing Using Machine Learning Approach: $GdVO_4:Sm^{3+}$ Prepared by Solution Combustion Synthesis (SCS). *Photonics* **2024**, *11*, 642. https://doi.org/10.3390/photonics11070642

Received: 6 June 2024
Revised: 2 July 2024
Accepted: 3 July 2024
Published: 6 July 2024

1. Introduction

In this paper, we present the results of experimental investigation of Sm^{3+}-doped $GdVO_4$ nanopowders. Nanopowder $GdVO_4$:Sm is an efficient orange-reddish light-emitting material [1]; light emission occurs due to a strong absorption of ultraviolet light by $GdVO_4$ and efficient energy transfer from vanadate groups ($VO_4{}^{3-}$) to dopants (Sm^{3-}). It is a good candidate for phosphor thermometry [1,2] Here, it is prepared by the solution combustion synthesis (SCS) method [1]. Simplicity and low cost are the main characteristics of this process. Phase identification in the post-annealed powder samples is performed by X-ray diffraction, and morphology is investigated by high-resolution scanning electron microscope (SEM) and by transmission electron microscopy (TEM). The main aim of this study is time-resolved analysis of luminescence properties of $GdVO_4:Sm^{3+}$ nanopowders. The possibility for $GdVO_4:Sm^{3+}$ usage in phosphor thermometry was analyzed in ref. [2], where temperature determination of sensing calibration curves was based on luminescence intensity ratios of samarium lines. In ref. [1], we used the intensity ratio of the host luminescence emission and samarium line to obtain temperature-sensing calibration curves, as proposed in refs. [3–6]. Calculation of temperature by intensity ratio of two luminescence peaks of different colors is called two-color thermometry. In ref. [1], the method was improved by introducing temporal dependence in the ratio measurements, as proposed in ref. [7]. Here, we propose a different approach, based on machine learning (ML). ML techniques are more and more used in the analysis of luminescence, near-infrared, and other spectral data [8–18]. Our findings demonstrate that instead of the conventional approach of identifying spectral peaks and calculating intensity ratios, it is feasible to train computer software to recognize time-resolved spectra associated with different temperatures of the thermophosphor. Traditionally, luminescence thermometric methods rely on analyzing a single spectral characteristic parameter, selected based on the properties of the thermophosphor and the researcher's intuition or experience. This approach leads to a partial utilization of spectral data, limiting measurement accuracy and thermal resolution. In contrast, the authors of [12] trained a deep learning artificial neural network to extract multiple temperature-dependent features for temperature estimation, fully utilizing temperature-dependent spectral data. However, the spectral data in ref. [12] did not include temporal information. In our research, we take a step further by incorporating time-resolved spectra into machine learning (ML)-based temperature estimation using a deep learning artificial neural network. This approach allows us incorporation of the thermophosphor's lifetime dependence into our analysis. In ref. [18], our measurements were aimed to be used by ML. We performed 50 measurements for each temperature of the sample. Upon inspecting the measured spectra in ref. [18], we realized that they have very small intensity differences, as expected. The main differences between the spectra acquired at the same temperature were contents of random noise introduced by the photon-counting operation of streak camera. On the other hand, in this study, we analyze measurements which were not obtained specifically for ML; in other words, there was not a large enough training set of training data for ML. To overcome this difficulty. we used the data augmenting techniques; see [19] and references therein.

2. Materials and Methods

$GdVO_4$ nanopowder doped with samarium ions was prepared by the solution combustion method, as described in ref. [1]. Stoichiometric quantities of starting chemicals $Sm(NO_3)_3$, $Gd(NO_3)_3{\cdot}6H_2O$, and NH_4VO_3 with the purity of 99.99% were chosen to obtain the Sm^{3+} concentration in $GdVO_4$ of 1 at.% ($Gd_{1\text{-}0.01}Sm_{0.01}VO_4$). All the chemicals and ammonium nitrate NH_4NO_3 were purchased from ABCR, and urea, $(NH_2)_2CO$, from Sigma-Aldrich Merck KGaA, Darmstadt, Germany. The dry mixture of 0.134 g (0.4 mmol) of $Sm(NO_3)_3$, 18.05 g (40 mmol) of $Gd(NO_3)_3{\cdot}6H_2O$, and 4.676 g (40 mmol) of NH_4VO_3 was combined with the mixture of 4.8 g (60 mmol) of ammonium nitrate and 3.003 g (50 mmol) of urea, which were used as organic fuels. The prepared starting reagents were combusted with aflame burner at approximately 500 °C, yielding a voluminous foamy powder in an intensive exothermic reaction. After the solution combustion synthesis, the nanopowder was annealed for 2 h, in air atmosphere, at 900 °C. Annealing has an effect on increasing the grain size of the nanopowders and it is widely used to achieve higher emission intensity.

The structure of the nanopowder was verified by X-ray diffraction analysis using a Diffractometer system: EMPYREAN, PANalytical. X-ray diffraction analysis (XRD) was performed using a X-ray Diffractometer PANalytical Empyrean, Malvern Panalytical, Malvern, United Kingdom with monochromatic CuK α radiation (γ = 0.15405980) at 45 kV and 40 mA. The diffraction data for the GdVO4-Sm nanopowder were collected over a 2θ range from 5 to 110° at a scan rate of 0.4° per minute.

The morphology of nanopowders and the size of crystallites were determined by a high-resolution scanning electron microscope (SEM) equipped with a high-brightness Schottky Field Emission gun FEGSEM, TESCAN Brno—Kohoutovice Czech Republic.

The morphology of GdVO4-Sm nanopowder was also evaluated by transmission electron microscopy (TEM), FEI Tecnai F30 Hillsboro, Oregon, United States. Photoluminescence (PL) studies reported in this work were performed using the Optical Parametric Oscillator Vibrant OPO Carlsbad, California USA, as described in ref. [1]. The output of the OPO can be continuously tuned over a spectral range from 320 nm to 475 nm. Time-resolved streak images of the emission spectrum excited by the OPO system were collected using a spectrograph (SpectraPro 2300i), Teledyne Princeton Instruments, New Jersey USA and recorded with a Hamamatsu streak camera (model C4334), Hamamatsu City, Japan. All streak camera operations were controlled by HPD-TA 8.3.0 (High-Performance Digital Temporal Analyzer) software. For measurements presented here, we used a homemade temperature control system, similar to the one in ref. [20].

To perform the machine learning (ML) analysis of the data, we utilized the Solo + Mia software package (Version 9.1, Eigenvector Research Inc, Manson, Washington USA). Solo provides a user-friendly environment, making it usable to individuals without programming expertise.

The measured spectra were initially analyzed using Principal Component Analysis (PCA) and Uniform Manifold Approximation and Projection clustering (UMAP). PCA was originally introduced in ref. [21], and for further details, we refer the reader to [22] and the referenced sources therein. Efficient implementation of t-SNE is proposed in ref. [23].

UMAP, a novel clustering technique suitable for visualizing extensive datasets, was recently proposed in ref. [24]. Ultimately, after conducting preliminary visualization experiments, we employed a deep learning artificial neural network to estimate the temperature of the heated samples.

3. Results

3.1. TEM, XRD, and SEM Study

The morphology of $GdVO_4$:Sm nanopowder was evaluated by transmission electron microscopy (TEM). TEM analysis shows the presence of slightly elongated particles formed by much smaller monocrystals as presented in Figure 1a. High-resolution TEM (HRTEM) shown in Figure 1b exposes their good crystallinity, clear lattice fringes, and the interplanar spacing of 4.77 Å, which might be associated with the (101) plane of the tetragonal $GdVO_4$ phase. Selected area electron diffraction (SAED) and fast Fourier transform (FFT) analyses revealed d values of 4.77 Å, 3.61 Å, 2.69 Å, 2.55 Å, and 1.86 Å, which match well with the (101), (200), (112), (220), and (312) crystal planes of the tetragonal $GdVO_4$ phase (JCPDS 00-017-0260). We confirmed the coexistence of a cubic Gd_2O_3 phase, specifically the (222) and (400) planes, with the corresponding d values of 3.11 Å and 2.70 Å (JCPDS 01-073-6280). In addition, the monophase composition of particles was identified by the presence of the above-mentioned crystal planes with d values corresponding well to those obtained through the X-ray diffraction (XRD) analysis of the sample presented in Figure 2.

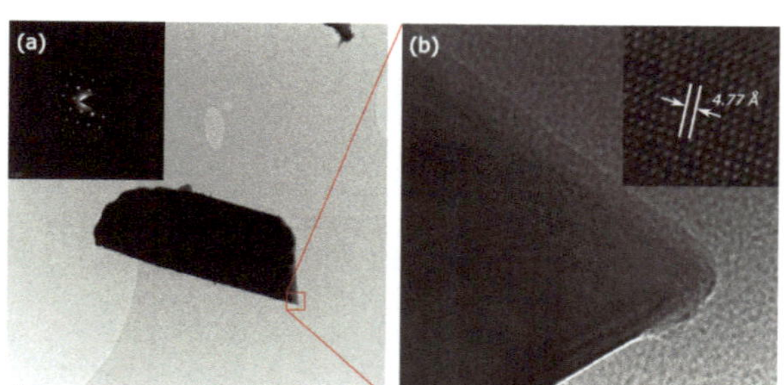

Figure 1. (**a**) TEM image of $GdVO_4$-Sm nanoparticles with the inset showing the SAED pattern. (**b**) HRTEM image of a section of the nanoparticle highlighted in the red square in (**a**). The inset displays an interplanar distance of 4.77 Å, likely corresponding to the (101) lattice plane of tetragonal $GdVO_4$.

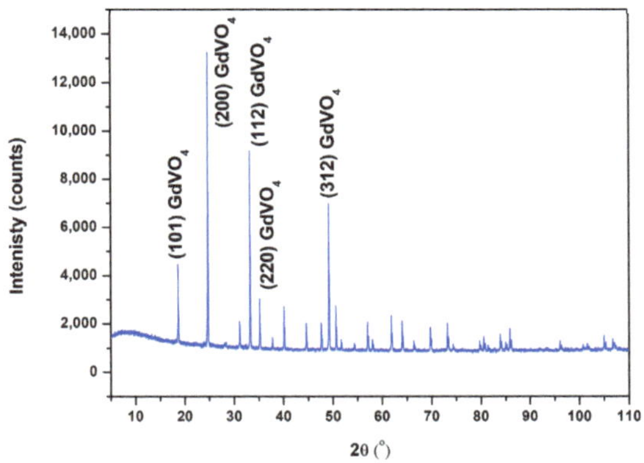

Figure 2. The XRD pattern of the $GdVO_4$-Sm nanopowder with respective Miller indices.

The particle size and morphology of the $GdVO_4$:Sm nanopowders were characterized by SEM (Figure 3). Some particles were agglomerated as clusters; however, individual spherical shaped particles are also visible in Figure 3. The estimated average particle size was about 50 nm. Looking at the SEM image (Figure 3), it is likely that the sizes of individual particles of nanopowders were about from 30 nm and up to 100 nm.

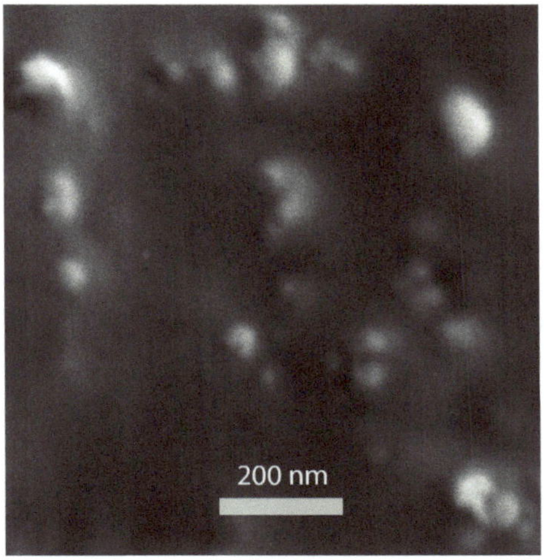

Figure 3. SEM image of $GdVO_4$:Sm nanopowder.

3.2. Photoluminescence and Lifetime Analysis

The streak image of the time-resolved photoluminescence spectrum of $GdVO_4$:Sm^{3+} using the 330 nm excitation is presented in Figure 4. The horizontal scale of the streak image corresponds to wavelength and the vertical scale shows the development of spectra in time. Images are usually presented in pseudocolor, where different colors mean different optical intensities. Using camera software, it was determined that the estimated lifetime of the most prominent samarium optical emission from the $^4G_{5/2}$ level is 0.78 ms at room temperature [1].

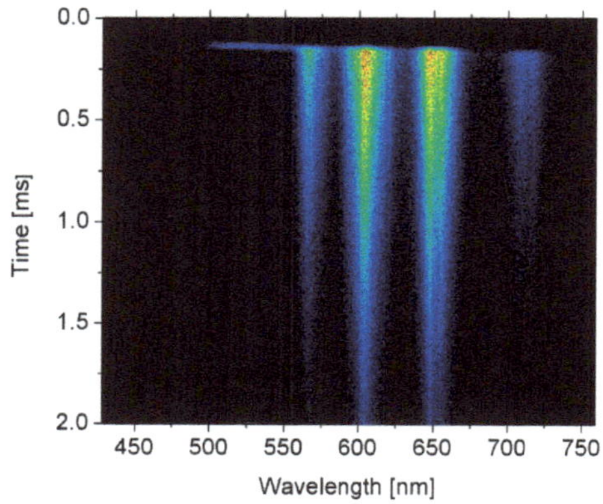

Figure 4. Streak image of the photoluminescence spectrum of $GdVO_4:Sm^{3+}$ nanophosphor (OPO excitation at 330 nm) at room temperature.

3.3. Photoluminescence and Lifetime Analysis Temperature Dependence of Photoluminescence

The new concept based on using the host luminescence for the luminescence intensity ratio method was introduced; for more details, see [4] and references therein. In our previous study [1], the method was improved by introducing temporal dependence in the intensity ratio measurements, as proposed in ref. [7]. Namely, it was possible to increase the sensitivity of the curve of intensity ratio between the host and samarium luminescence if the appropriately selected part of temporal evolution is used in calculation. This was our early attempt to improve the temperature estimation by introducing the temporal characteristics of luminescence emission, recorded by the streak camera. In ref. [25], we combined several methods for temperature measurements to obtain the best results.

The luminescence spectra of $GdVO_4$:Sm nanopowder were measured at various temperatures using OPO excitation at 330 nm and streak camera. Figure 5 shows the luminescence emission of $GdVO_4$:Sm nanopowder at two temperatures.

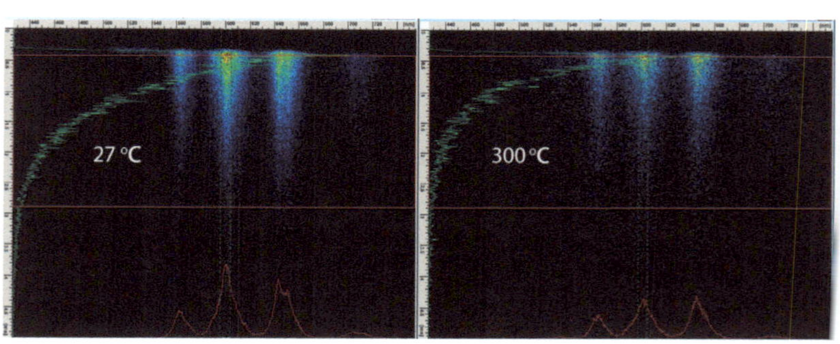

Figure 5. Streak images of the photoluminescence spectrum of $GdVO_4:Sm^{3+}$ nanophosphor at two temperatures.

Figure 6, which provides analysis of temperature dependence at eight different temperatures, is used in other analyses in this manuscript. Figure 6a shows changes in intensity ratios of luminescence peaks. Figure 6b shows and summarizes changes in luminescence lifetimes.

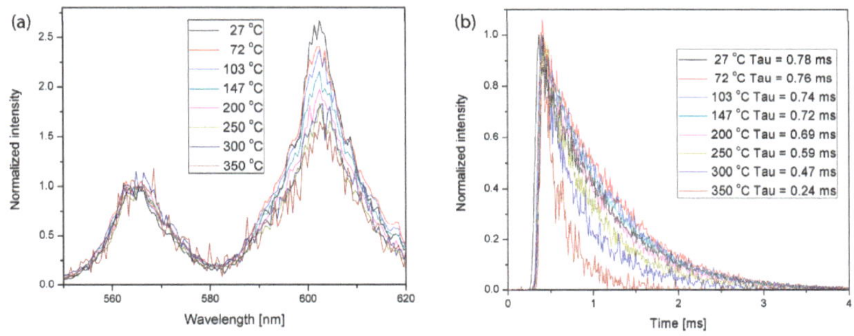

Figure 6. Temperature dependence of intensity ratios of luminescence peaks (**a**). Temperature dependence of luminescence lifetimes (**b**).

It can be seen in Figures 5 and 6 that intensity ratios between different peaks vary with changing the temperature of the sample. Moreover, lifetimes of luminescence emission decreases with the increase in temperature. So, it is intuitively clear that taking into account the intensity ratio and the lifetime of the luminescence emission provides better temperature estimation. The analysis will be further improved if the intensities of all prominent peaks are included into temperature estimation. The natural answer to the question of how to achieve all this is to use machine learning methods applied to streak images.

In this study, the spectra were measured at 640 wavelength points. The time axis of a streak image has 480 points. To alleviate the computational burden of machine learning computations, we selectively chose specific portions of the streak image that contained the majority of the spectral and time-resolved information. After some experimenting, for the ML analysis, we focused on 90 spectral points corresponding to bands between 558 nm and 568 nm, 592 nm and 612 nm, and 641 nm and 656 nm. The selected time frame consisted of 120 points. A total of 80 points were from a streak image with the time range of 100 μs, equivalent to approximately $80 \times 100/480 = 16.66$ μs, starting from 25 μs. Another 40 points dwere from a streak image (of the same sample at the same temperature) with the time range of 5 ms, equivalent to approximately $40 \times 5000/480 = 416.66$ μs, starting from 826 μs. In this way, there was sufficient information to highlight the intensity ratios and temporal decrease in luminescence. We see the omitted points as a computational burden, because in our earlier attempts, with larger ROIs, there were no increases in performance of the deep leaning network, just longer computational time.

To understand the following improvement of neural network performance, let us point out that the rows of the streak image correspond to the spectra of the image at a certain time, defined by the columns of the streak image. So, further improvement in performance was achieved when we replaced five rows of selected ROI with one row. In other words, each pixel of the new row was calculated as a mean value of five points of the same wavelength successive in time. Thus, we obtained an increased performance of the neural network (in a sense of root standard errors) because of the mean filtering of columns; moreover, we obtained calculations faster because of the reduced numbers of rows.

As a result, each sample used for ML analysis comprised $120/5 \times 90 = 24 \times 90 = 2160$ data points. To generate the input vector of 2160 data points, the pixel rows of each of the selected spectral bands in the image were packed sequentially in time row by row. In SOLO software, we used also two preprocessing techniques on neural network input: data, normalization, and data smoothing.

In order to construct the temperature calibration curve presented in ref. [1], we conducted measurements on samples at regular intervals of about 50 degrees Celsius, ranging from 27 °C to 350 °C. Measurements for temperature estimation were performed at two time ranges, 100 μs and 5 ms. However, the number of measurements was obviously insufficient for ML training. To obtain the larger training data set, we decided to use image data augmenting techniques. A nice survey on image data augmentation for deep learning is presented in ref. [19].

As a clue to augmentation of our measured spectra, we used the experience from our earlier publication where measurements were carried out with the intention to be used for ML analysis, taking 50 streak images for each temperature. The visualization of these training data in ref. [18] by PCA and UMAP showed that the measured spectra at each temperature had similar characteristics. By looking at the spectral profiles, it could be seen that they were almost the same, differing only by random noise.

Here, we use the same spectra, measured in ref. [1]. In order to obtain augmented training set for ML analysis, we used the simple technique based on the properties of the acquired streak images. Keeping in mind that the luminescence lifetime of this material is about 0.78 ms, it is easy to see that translating the region of interest of a streak image by several rows results in almost negligible intensity change and a clear difference in random noise added by the photon counting mode. After selecting the region of interest for analysis, the shifting of the vertical offset of ROI was defined by a random number generator, serving the purpose of augmenting the training data set. The procedure was coded in the C language. We first used the training set consisting of 10 samples for each temperature. The training data for each temperature were added sequentially using batch processing to obtain a training data file readable by SOLO software.

To check the usefulness of constructed training set and the actual diversity of data, we used several visualization techniques.

Principal Component Analysis (PCA) score plots serve as a valuable tool for initial data assessment and verification. In Figure 7, the scores on the first three principal components of $GdVO_4$:Sm^{3+} spectra at different temperatures are displayed. It is noticeable that spectra obtained at similar temperatures tend to cluster together, although not perfectly. Consequently, to facilitate the initial data assessment, we incorporated t-Distributed Stochastic Neighbor Embedding (t-SNE) and Uniform Manifold Approximation and Projection (UMAP) clustering of $GdVO_4$:Sm^{3+} spectra at various temperatures.

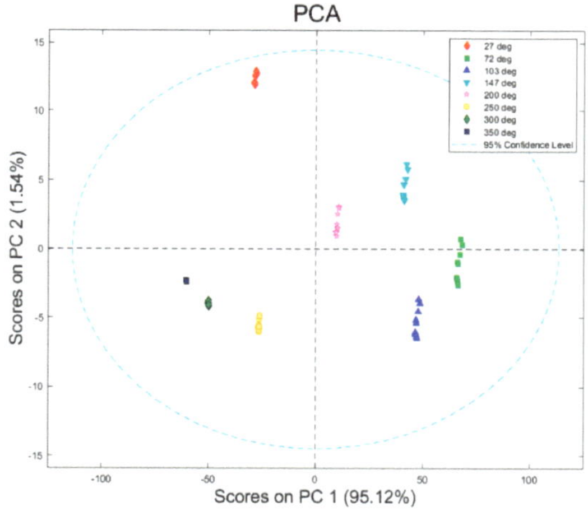

Figure 7. Scores on first two principal components of $GdVO_4$:Sm^{3+} spectra at different temperatures.

The t-SNE and UMAP clustering results of $GdVO_4$:Sm^{3+} spectra at various temperatures are depicted in Figures 8 and 9. It is obvious that the good groupings of classes are distinguishable.

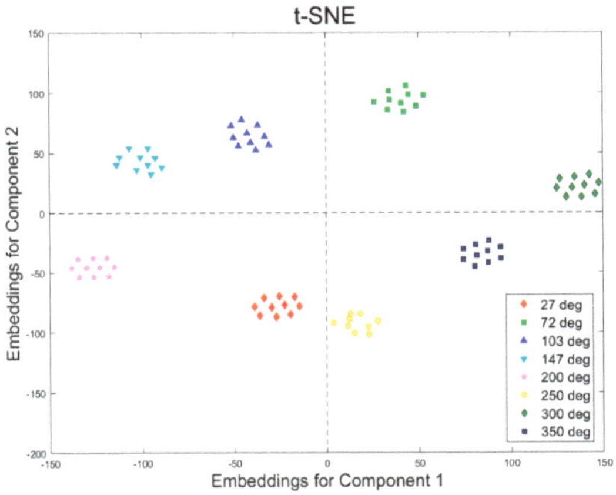

Figure 8. t-SNE component clustering of $GdVO_4:Sm^{3+}$ spectra at different temperatures.

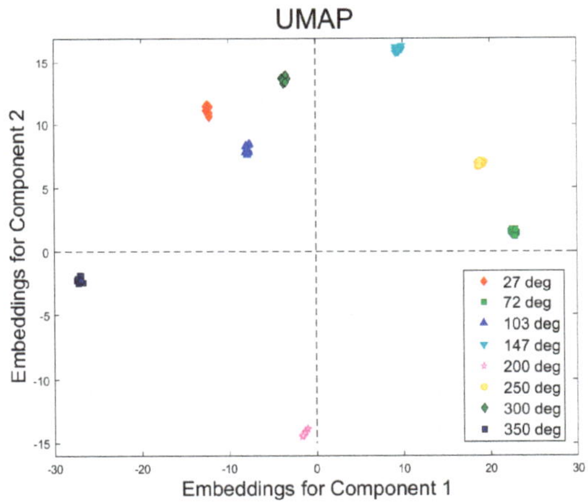

Figure 9. Uniform Manifold Approximation and Projection (UMAP) clustering of $GdVO_4:Sm^{3+}$ spectra at different temperatures.

The deep learning artificial neural network (ANNDL) implementation in SOLO can utilize either the SciKit or TensorFlow Python packages. In our case, we chose the TensorFlow Python package, which utilizes GPU acceleration for highly efficient computations.

The primary concept behind temperature estimation of thermophosphors using ANNDL involves training the network with sample spectra that have corresponding temperature measurements. During the training phase, the network iteratively minimizes errors between the calculated and predicted temperatures. Since the region of interest in the streak image of $GdVO_4:Sm^{3+}$ consists of 2160 points, the input layer of the deep learning network has 2160 nodes. Considering that this network aims to approximate two types of exponential functions—one for intensity ratios and another for luminescence lifetime—we initially considered using two hidden layers. The initial default size of a hidden layer in SOLO software is 100. However, we experimented with different numbers and sizes of hidden layers. Based on our experience, we decided that 100 nodes do not provide optimal predictions, in a sense of root mean standard errors. We gradually increased the sizes of hidden layers until the increase in neural network performance became saturated. Ultimately, we determined that a neural network with two hidden layers, each containing 540 nodes, was the optimal solution. The optimal number of epochs was 600, the batch size was 50. The output layer consisted of a single node representing the predicted temperature.

A schematic diagram illustrating the deep learning network used in this study is shown in Figure 10.

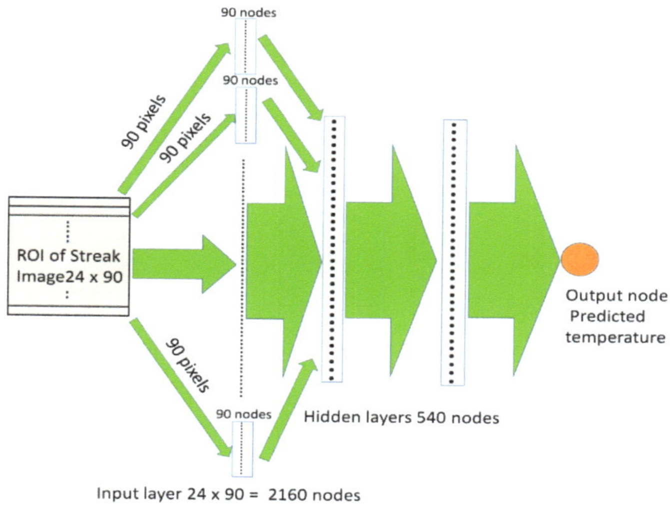

Figure 10. A schematic diagram illustrating the deep learning network used in this study. It is not possible to draw all arrows, and the shown arrows are merely symbolic. The pixel rows of the region of interest in the image are fed into the input layer as shown in the diagram.

After conducting several experiments, we determined that the *Adam* optimizer outperformed other optimizers, and that the *Relu* activation function is optimal for our case. *Adam* optimizer utilizes a default learning rate of 0.001, but it can adaptively adjust the learning rate based on the characteristics of the data. For cross-validation, we employed the Venetian blind method with a data split of 10, meaning that in each sub-validation experiment, 90% of the data was used for training and 10% for validation.

In other words, this 10% of data was used as test data, not seen by the computer in sub-validation experiment where 90% of data was used for training. To be precise, such sub-validation experiments were repeated 10 times, making test data unnecessary. Moreover, it should be noted that validation computational time is much longer than the initial construction of the neural network.

Following further experimentation, we selected a batch size of 50. After some trial and error, we decided to set the number of epochs to 600. Before processing the data, we used normalization procedure 1-Norm (area = 1) on the dataset. Then, the data were smoothed using the Savitzky–Golay filter. The width of the filter was five, with a polynomial order of six and weighted tails. These preprocessing options are built in the SOLO software.

Figure 11 shows the results of deep learning predictions of the measured temperatures when the neural network is presented with part of the spectra not seen by the computer in that validation cycle. Notably, the errors for the samples within the training set are relatively small. Validation errors are comparably larger. In comparison to the results presented in ref. [18], we observed larger maximum errors.

Based on the idea that a larger training set reduces problems in neural networks, we augmented the training set to 50 samples. Figure 12 shows the results of deep learning predictions of measured temperatures when the augmented training set was used. Moreover, optimizing the structure of the neural network resulted in decreasing of the optimal number of epochs from 600 to 100, and the batch size decreased from 50 to 12. Errors were comparable to the results presented in ref. [18].

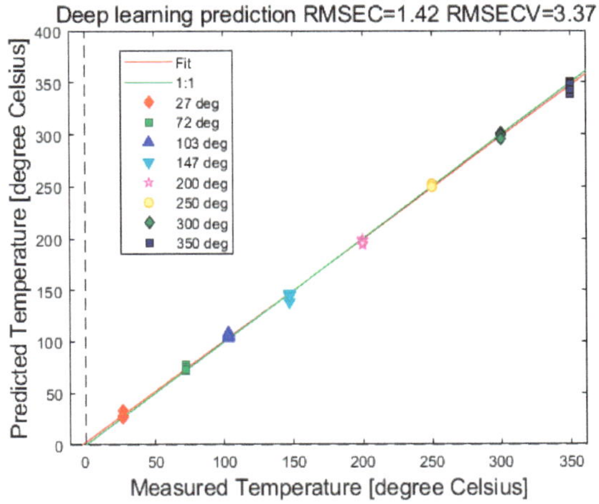

Figure 11. Plot of predicted temperatures using the training set of 10 samples for each temperature. RMSEC refers to the root mean standard error of calibration; RMSECV refers to the root mean standard error of cross-validation.

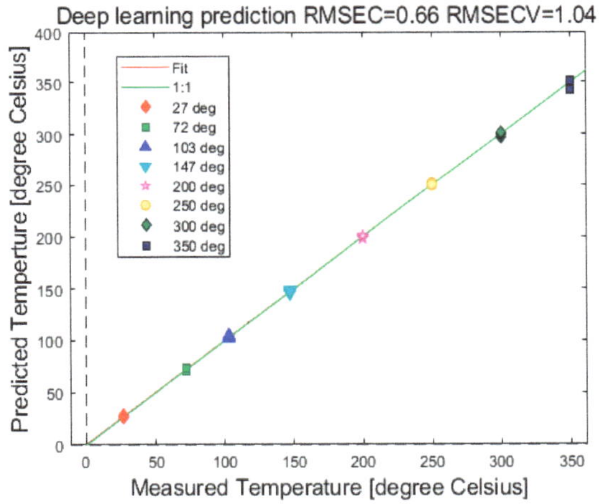

Figure 12. Plot of predicted temperatures using the training set of 50 samples for each temperature. RMSEC refers to the root mean standard error of calibration; RMSECV refers to the root mean standard error of cross-validation.

4. Discussion

In this study, the data not originally intended for ML were visualized using PCA, t-SNE, and UMAP, showing the good grouping of the data corresponding to the same temperature. Thus, in this paper, visualization techniques proved the possibility of using the original experimental data after appropriate augmentation.

In [18], for each temperature, a set of 50 spectra was measured, resulting in a total of 650 training spectra (13 temperatures multiplied by 50 spectra per temperature). Here, we analyzed the spectra corresponding to eight temperatures, and the training set initially comprised 80 (8 multiplied by 10) spectra. Using visualization techniques, we proved that the technique used for data augmentation was appropriate. However, subsequently using ANNDL to obtain temperature predictions with similar values of RMS as in ref. [18], we had to expand the training set up to 50 for each temperature. After that, the structure of ANNDL and computational times were similar to those in ref. [18]. It should be pointed out that structure (number of hidden layers and number of their nodes), number of epochs, and batch size decreased when ANNDL was trained with the training set of 50 for each sample.

In the previous section, we presented the best results achieved through a trial-and-error process. Now, we validate this educated guess using the learning curves of the neural network. The learning curves illustrating the training progress of the ANNDL model are shown in Figure 13. The curves were generated by repeatedly running the deep neural network model with an increasing number of epochs. Based on the shapes of these curves, we decided to set the number of epochs for future experiments with these data to 100 as a precaution.

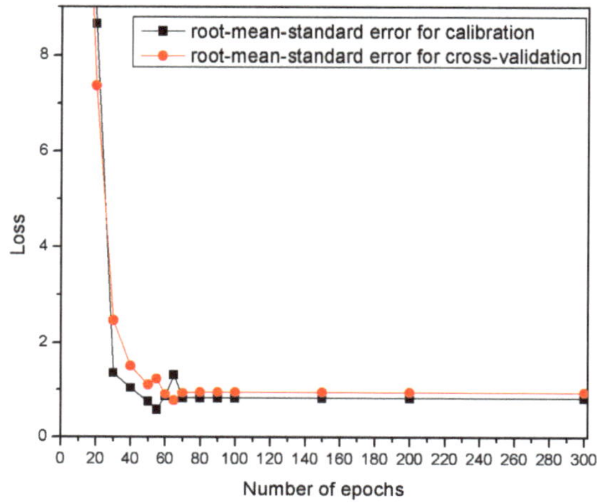

Figure 13. The learning curves depicting the training progress of the ANNDL model.

When comparing our results to those reported in ref. [26], our method yields slightly superior prediction errors. Notably, our deep neural network model achieves optimal performance after approximately 100 epochs, whereas the study in ref. [26] requires training the network for 6000 epochs.

In reference [13], the authors explore the use of linear transformations (Principal Component Analysis) and non-linear transformations (t-Distributed Stochastic Neighbor Embedding) on thermophosphor calibration datasets. Their findings show a clear advantage compared to our results when using PCA or UMAP. However, this was exactly why we chose to employ a deep learning neural network in our approach.

Compared to the results presented in ref. [12], we observed slightly larger maximum errors. We anticipated that incorporating the time development of luminescence spectra into our prediction model would enhance performance relative to the network described in ref. [12]. However, it is important to note that our temperature controller lacks manufacturer-specified resolution and accuracy. We estimate that its sporadic errors can reach up to two degrees Celsius, regardless of the temperature value. As a result, the relative error can be significant at lower temperatures. In contrast, the temperature controller used in ref. [12] has a specified temperature resolution of 0.01 degrees Celsius and a stability of ±0.05 degrees Celsius. Our main objective was to validate the concept of using the time evolution of luminescence spectra for remote temperature estimation, and the results presented here support this concept.

5. Conclusions

After phase identification and morphology of post-annealed powder samples $GdVO_4:Sm^{3+}$ using X-ray diffraction, the time-resolved analysis of this nanoposphor luminescence was conducted. The estimated lifetime of the most prominent samarium optical emission from the $^4G_{5/2}$ level is 0.726 ms at room temperature. We employed machine learning techniques to analyze the optical spectra of $GdVO_4$:Sm^{3+} thermophosphor in order to improve remote temperature measurements. Our approach deviates from the conventional method of identifying spectral peaks and calculating their intensity ratio. Instead, we trained a computer model to recognize the time-resolved spectra associated with different temperatures of the thermophosphors. This allowed us incorporation of not only the intensity ratios of all peaks but also the luminescence lifetime as an additional parameter. We showed that, for the analyzed $GdVO_4$:Sm^{3+} material, the temperature sensing is useful up to 350 °C. We used the streak camera to prove the concept. The real application of this method will be based on using the gated CCD cameras and appropriate bandwidth filters for selecting the emission region of interest. To improve the performance of the deep learning network, we used the augmenting of the training data set. In summary, our analyses proved that it is possible to use a deep learning neural network for improved temperature sensing with Sm^{3+}-doped $GdVO_4$ nanopowder.

Author Contributions: Conceptualization, D.S., M.S.R. and M.G.N.; software, D.S.; formal analysis, A.D. and J.Z.J.; investigation, D.S., M.S.R. and M.D.R.; resources, J.K., M.H.A. and S.S.-S.; writing—original draft preparation, D.S. and M.S.R.; writing—review and editing, D.S., M.S.R. and M.D.R. All authors have read and agreed to the published version of the manuscript.

Funding: This work was financially supported by funding provided by the Institute of Physics Belgrade, through the grant by the Ministry of Science, Technological Development and Innovations of the Republic of Serbia and by European Commission through Marie Skłodowska-Curie Actions H2020 RISE with ULTIMATE-I (Grant No. 101007825). Authors would like to acknowledge the access of equipment of "Servicio General de Apoyo a la Investigación (SAI), Univesidad de Zaragoza".

Data Availability Statement: The original contributions presented in the study are included in the article, further inquiries can be directed to the corresponding author/s.

Conflicts of Interest: The authors declare no conflicts of interest.

References

1. Rabasovic, M.S.; Krizan, J.; Savic-Sevic, S.; Mitric, M.; Rabasovic, M.D.; Marinkovic, B.P.; Sevic, D. Orange-Reddish Light Emitting Phosphor $GdVO_4$:Sm^{3+} Prepared by Solution Combustion Synthesis. *J. Spectrosc.* **2018**, *2018*, 1–8. [CrossRef]
2. Nikolić, M.G.; Jovanović, D.J.; Đorđević, V.; Antić, Ž.; Krsmanović, R.M.; Dramićanin, M.D. Thermographic properties of Sm^{3+} -doped $GdVO_4$ phosphor. *Phys. Scr.* **2012**, *T149*, 014063. [CrossRef]
3. Nikolić, M.G.; Antić, Ž.; Ćulubrk, S.; Nedeljković, J.M.; Dramićanin, M.D. Temperature sensing with Eu^{3+} doped TiO_2 nanoparticles. *Sens. Actuators B Chem.* **2014**, *201*, 46–50. [CrossRef]
4. Dramićanin, M.D.; Antić, Ž.; Ćulubrk, S.; Ahrenkiel, S.P.; Nedeljković, J.M. Self-referenced luminescence thermometry with Sm3+ doped TiO_2 nanoparticles. *Nanotechnology* **2014**, *25*, 485501. [CrossRef] [PubMed]
5. Lojpur, V.; Nikolić, M.G.; Jovanović, D.; Medić, M.; Antić, Ž.; Dramićanin, M.D. Luminescence thermometry with Zn_2SiO_4:Mn^{2+} powder. *Appl. Phys. Lett.* **2013**, *103*, 141912. [CrossRef]
6. Lojpur, V.; Ćulubrk, S.; Dramićanin, M.D. Ratiometric luminescence thermometry with different combinations of emissions from Eu^{3+} doped $Gd_2Ti_2O_7$ nanoparticles. *J. Lumin.* **2016**, *169*, 534–538. [CrossRef]
7. Aldén, M.; Omrane, A.; Richter, M.; Särner, G. Thermographic phosphors for thermometry: A survey of combustion applications. *Prog. Energy Combust. Sci.* **2011**, *37*, 422–461. [CrossRef]
8. Chatzidakis, M.; Botton, G.A. Towards calibration-invariant spectroscopy using deep learning. *Sci. Rep.* **2019**, *9*, 2126. [CrossRef]
9. Dramićanin, T.; Zeković, I.; Periša, J.; Dramićanin, M.D. The Parallel Factor Analysis of Beer Fluorescence. *J. Fluoresc.* **2019**, *29*, 1103–1111. [CrossRef]
10. Ranamukhaarachchi, S.A.; Peiris, R.H.; Moresoli, C. Fluorescence spectroscopy and principal component analysis of soy protein hydrolysate fractions and the potential to assess their antioxidant capacity characteristics. *Food Chem.* **2017**, *217*, 469–475. [CrossRef]
11. Weng, S.; Yuan, H.; Zhang, X.; Li, P.; Zheng, L.; Zhao, J.; Huang, L. Deep learning networks for the recognition and quantitation of surface-enhanced Raman spectroscopy. *Anal.* **2020**, *145*, 4827–4835. [CrossRef]
12. Cui, J.; Xu, W.; Yao, M.; Zheng, L.; Hu, C.; Zhang, Z.; Sun, Z. Convolutional neural networks open up horizons for luminescence thermometry. *J. Lumin.* **2023**, *256*, 119637. [CrossRef]
13. Ximendes, E.; Marin, R.; Carlos, L.D.; Jaque, D. Less is more: Dimensionality reduction as a general strategy for more precise luminescence thermometry. *Light Sci. Appl.* **2022**, *11*, 237. [CrossRef]

14. Barbon, S.; da Costa Barbon, A.P.A.; Mantovani, R.G.; Barbin, D.F. Machine Learning Applied to Near-Infrared Spectra for Chicken Meat Classification. *J. Spectrosc.* **2018**, *2018*, 1–12. [CrossRef]
15. Geronimo, B.C.; Mastelini, S.M.; Carvalho, R.H.; Barbon Júnior, S.; Barbin, D.F.; Shimokomaki, M.; Ida, E.I. Computer vision system and near-infrared spectroscopy for identification and classification of chicken with wooden breast, and physicochemical and technological characterization. *Infrared Phys. Technol.* **2019**, *96*, 303–310. [CrossRef]
16. Junior, S.B.; Mastelini, S.M.; Barbon, A.P.A.; Barbin, D.F.; Calvini, R.; Lopes, J.F.; Ulrici, A. Multi-target prediction of wheat flour quality parameters with near infrared spectroscopy. *Inf. Process. Agric.* **2020**, *7*, 342–354. [CrossRef]
17. Nolasco Perez, I.M.; Badaró, A.T.; Barbon, S.; Barbon, A.P.A.; Pollonio, M.A.R.; Barbin, D.F. Classification of Chicken Parts Using a Portable Near-Infrared (NIR) Spectrophotometer and Machine Learning. *Appl. Spectrosc.* **2018**, *72*, 1774–1780. [CrossRef] [PubMed]
18. Rabasovic, M.S.; Savic-Sevic, S.; Križan, J.; Matovic, B.; Nikolic, M.; Sevic, D. Time resolved study of temperature sensing using Gd $_2$ O $_3$:Er,Yb: Deep learning approach. *Phys. Scr.* **2023**, *98*, 116003. [CrossRef]
19. Shorten, C.; Khoshgoftaar, T.M. A survey on Image Data Augmentation for Deep Learning. *J. Big Data* **2019**, *6*, 60. [CrossRef]
20. Rabasović, M.D.; Murić, B.D.; Čelebonović, V.; Mitrić, M.; Jelenković, B.M.; Nikolić, M.G. Luminescence thermometry via the two-dopant intensity ratio of Y_2O_3:Er^{3+}, Eu^{3+}. *J. Phys. D Appl. Phys.* **2016**, *49*, 485104. [CrossRef]
21. Hotelling, H. Analysis of a complex of statistical variables into principal components. *J. Educ. Psychol.* **1933**, *24*, 417–441. [CrossRef]
22. Wise, B.M.; Gallagher, N.B.; Bro, R.; Shaver, J.M.; Windig, W.; Koch, R.S. *Chemometrics Tutorial for PLS Toolbox and Solo*; Eigenvector Research, Inc.: Wenatchee, WA, USA, 2006; ISBN 0-9761184-1-6.
23. Van der Maaten, L.; Hinton, G. Visualizing data using t-SNE. *J. Mach. Learn. Res.* **2008**, *9*, 2579.
24. McInnes, L.; Healy, J.; Saul, N.; Großberger, L. UMAP: Uniform Manifold Approximation and Projection. *J. Open Source Softw.* **2018**, *3*, 861. [CrossRef]
25. Sevic, D.; Rabasovic, M.S.; Krizan, J.; Savic-Sevic, S.; Nikolic, M.G.; Marinkovic, B.P.; Rabasovic, M.D. YVO_4:Eu^{3+} nanopowders: Multi-mode temperature sensing technique. *J. Phys. D Appl. Phys.* **2020**, *53*, 015106. [CrossRef]
26. Lewis, C.; Erikson, J.W.; Sanchez, D.A.; McClure, C.E.; Nordin, G.P.; Munro, T.R.; Colton, J.S. Use of Machine Learning with Temporal Photoluminescence Signals from CdTe Quantum Dots for Temperature Measurement in Microfluidic Devices. *ACS Appl. Nano Mater.* **2020**, *3*, 4045–4053. [CrossRef]

photonics

MDPI

Article

Smartphone-Readable Optical-Fiber Quasi-Distributed Phosphorescent Temperature Sensor

Tinko Eftimov [1,2,*], Irena Kostova [3], Samia Fouzar [1,4], Daniel Brabant [1], Kristian Nikolov [2] and Veselin Vladev [2,5]

1 Photonics Research Center, Université du Québec en Outaouais, Rue 101 St-Jean Bosco, Gatineau, QC J8X 3G5, Canada; sfouzar@cdta.dz (S.F.); dbrabant426@gmail.com (D.B.)
2 Central Laboratory of Applied Physics, Bulgarian Academy of Sciences, 61 Sanct Peterburg Blvd., 4000 Plovdiv, Bulgaria; kris7ian.nikolov@gmail.com (K.N.); v.p.vladev@abv.bg (V.V.)
3 Faculty of Chemistry, Plovdiv University, 4000 Plovdiv, Bulgaria; irena_k87@abv.bg
4 Centre de Développement des Technologies Avancées CDTA, Baba Hassen, Algiers 16303, Algeria
5 Department of Mathematics, Physics and Information Technologies, Faculty of Economics, University of Food Technologies, 26 Maritsa Blvd., 4002 Plovdiv, Bulgaria
* Correspondence: tinko.eftimov@uqo.ca

Abstract: In this paper we present the principle of operation, fabrication and performance of a phosphorescent optical-fiber quasi-distributed sensor with contactless smartphone interrogation. An array of short strong corrugated long-period gratings (C-LPG) is used as a platform to spatially locate and to excite the phosphors whose time responses are temperature-dependent. The C-LPG array was fabricated using a pulsed CO_2 laser. The quasi-distributed sensing array is excited by a UV LED and the normalized differential rise/decay time response measured by a smartphone is used as a measure of the temperature. The sensing spots have a volume smaller than 0.5 μL, can be separated by several millimeters to several meters and the interrogation can be simultaneous or in a sequence. The response and the sensitivity to temperature have been measured. The sensing array has been shown to measure abrupt and gradual temperature changes in space as well as time-dependent processes in the 0 °C to 100 °C range and with a measurement time of 1 s.

Keywords: optical fibers; quasi-distributed sensor; phosphorescence; corrugated long-period gratings; sensing arrays; smartphone interrogation; UV LED excitation

Citation: Eftimov, T.; Kostova, I.; Fouzar, S.; Brabant, D.; Nikolov, K.; Vladev, V. Smartphone-Readable Optical-Fiber Quasi-Distributed Phosphorescent Temperature Sensor. *Photonics* **2024**, *11*, 694. https://doi.org/10.3390/photonics11080694

Received: 30 June 2024
Revised: 21 July 2024
Accepted: 24 July 2024
Published: 25 July 2024

1. Introduction

The development and implementation of optical fiber-based luminescent temperature sensors has been developing for over two decades [1–3]. The use of optical fibers for the excitation of the sensing material and to lead-out the luminescence to a photo detector or spectrometer has definite advantages, such as the excitation of small-sized samples, construction of point dip sensors and isolation from electromagnetic interference. Both phosphorescent and fluorescent materials [4,5] have been proposed and tested as temperature-dependent luminescent probes. While luminescence caused by down conversion is typically used, materials with up-conversion excitation have also been proposed [6]. Both powders and polymers have been proposed and studied [7,8]. The characteristics of the luminescence that is used to retrieve the temperature can include the intensity [8], ratio of different peaks of the luminescent spectrum [5,9] and the decay time [10]. We have recently proposed [11,12] a fiber optic sensor based on the difference in the rise and decay time responses, and approach which increases the sensitivity and allows normalized signals to be used as a measure of temperature, which makes the sensor immune to power fluctuations of the source. Europium (Eu) and dysprosium (Dy)-doped strontium aluminates have been used as sensing materials since their rise and decay time responses have been studied in detail [13], especially the dependence on the excitation wavelength and on temperature [14,15]. In the above enumerated temperature sensors, either a spectrometer is used to measure the spectral changes or a photodetector to measure the total intensity changes. This means only one sensor at a time can be interrogated. In a number of cases, however, more than one sensor has to be interrogated, as is the case when we have to measure the spatial distribution of temperature and, consequently, thermal gradients and thermal flows. In such cases, the use of a smartphone as an interrogation unit that can detect both spectrally and time-dependent changes of luminescent materials is more appropriate [16]. As the smartphone can take 30 frames per second (fps), i.e., a measurement every 33.3 ms, it is perfectly suited to measure the rise and decay time responses of phosphorescent materials and perform time-gated measurements [17]. In case we need to excite more than one phosphorescent sensor for temperature measurements, the phosphor cannot be placed on the tip of a single fiber. Rather it should be deposited along the fiber which is not made to radiate light but to guide it. To solve the problem, corrugated fiber gratings can be used not as receivers of luminescence coming from outside the fiber [18] but as side-emitting excitation devices instead. Modification of the fiber surface and butt end using CO_2 pulsed lasers, including the fabrication of corrugated gratings is well known and widely used [18–20] because of its simplicity and repeatability.

In this paper we propose a quasi-distributed optical-fiber phosphorescent temperature sensor, which is an array of corrugated long-period gratings (C-LPG) that side-radiate a portion of the UV light propagating along the fiber to excite phosphorescent sites deposited upon the corrugated sections. The rise and decay time responses of the phosphorescent light of the whole spatial array is simultaneously detected by a smartphone and the temperature of each sensing point is measured individually, thus revealing the spatial distribution of temperature and its evolution in time.

Section 2 is devoted to the principle of operation of the sensor array with more details on the rise/decay responses of the phosphor, the smartphone as an interrogation instrument and the fabrication of the CLPG array.

Section 3 presents the experimental set up and the results obtained on the characterization of the individual CLPGs and the performance of the sensing array in detecting abrupt and gradual spatial as well as time-dependent temperature changes.

Section 4 presents some comments and discussion based on the results obtained.

2. Principle of Operation

2.1. Temperature-Sensitive Phosphorescent Materials

2.1.1. Strontium Aluminates

There is a large variety [2–7] of temperature-dependent phosphors that have been proposed, tested and used for the development of temperature sensors. In this paper we prefer to use Eu^{2+} and Dy^{2+}-doped strontium aluminate powder ($SrAl_2O_4$:$Eu_{0.1}$,$Dy_{0.2}$) synthesized by the combustion and solid-state reaction method [13]. Such powders have been extensively studied and their characteristics well-documented in our previous publications [12,14]. More particularly, we make use of sample #A from these previous studies because it has been found to be appropriate for temperature measurements in the 0 °C to 100 °C range [11,12,15]. Two properties of these materials are of interest: the temperature dependence of the luminescent intensity and the temperature dependence of their rise and decay time responses.

2.1.2. Temperature-Dependent Intensity Change

It is well known [4] that as temperature T increases, electrons release energy in non-radiative transitions rather than radiative recombinations thus reducing the intensity of luminescence which becomes temperature-dependent $I(T)$. The normalized intensity

$\xi(T)$ over the whole spectrum is related to the absolute temperature T through the Arrhenius equation which, in its normalized form, reads as follows [15]:

$$\xi(T) = \frac{I(T)}{I(0)} = \frac{1}{1 + A\exp\left(-\frac{\Delta E}{kT}\right)} \quad A = \frac{\Gamma_0}{\Gamma_v} \tag{1}$$

where $I(0)$ is the intensity at $T = 0$, A is the ratio of the attempt rate Γ_0 and the radiative decay rate Γ_v of the 5d state [4] and $k = 1.38 \times 10^{-23}$ J/K is the Boltzmann constant, while ΔE is the activation energy of the transition corresponding to the band gap between the lowest energy of the Eu^{2+} 4f65d1 excited level and the bottom of the conduction band. The normalized intensity $\xi(T)$ decreases slowly over a wide range of temperatures as shown elsewhere [4,15]. From the temperature dependence of the intensity $I(T)$, the activation energy and the coefficient A for the sample used in our experiments were found as in [15]: $\Delta E \approx -0.203$ eV and $A \approx 395$.

2.1.3. Rise and Decay Time Responses

The rise and decay time responses as well as the maximum intensity of the phosphorescence of Sr aluminates have been extensively studied and have been found to depend on the ambient temperature [13–15]. We consider here a phosphorescent sample excited by periodic rectangular pulses of a duty cycle

$$\eta = \frac{\tau_p}{T_p} \quad 0 \le \eta \le 1 \tag{2}$$

where T_p is the period and τ_p the duration of the ON (rise) response $R(t)$, while T_p-τ_p is that of the OFF (decay) response $D(t)$. For intensities $R_D = D_D$ measured in the dark we use the following normalized quantities [11–15]:

$$u_{ON}(t) = \frac{R(t) - R_D}{R_\infty - R_D} = 1 - A_0 t^{-\alpha_0} = 1 - \left(\frac{t}{\tau_0}\right)^{-\alpha_0}, \ \tau_0 = A_0^{\frac{1}{\alpha_0}} \text{ forthedecayresponse} \tag{3}$$

$$u_{OFF}(t) = \frac{D(t) - D_D}{D_0 - D_D} = At^{-\alpha} = \left(\frac{t}{\tau}\right)^{-\alpha}, \ \tau = A^{-1/\alpha} \text{ forthedecayresponse} \tag{4}$$

In (3) and (4), R_∞ is the total optical intensity for a sufficiently long illumination, D_0 is the total optical intensity of the whole spectrum just before switching off the excitation, R_D is the intensity measured in the dark, $R(t)$ and $D(t)$ are the emission intensities of the light emitted by the material after the excitation is switched ON (rise) and OFF (decay), t is the time in seconds, the amplitudes A, A_0 and the powers α, α_0 are constants, while τ and τ_0 are time scaling parameters. These power law approximations have been found to hold after some initial instant t_0 at which the ON response reaches the level a_0 while the OFF response reaches b_0. As shown in [14,15], the quantities b_0 and $1 - a_0$ are a measure of the relative share of phosphorescence in luminescence.

The physical meaning of these power law parameters has been shown [15,16] to be the following. For the decay (OFF) response (Figure 1b), A stands for the level from which luminescence decays. A larger $A' > A$ stands for a stronger luminescence which shifts the response to a higher level (the upper A', α dotted line in Figure 1b). The power α stands for the rate of decay. A larger $\alpha' > \alpha$ means a faster decay rate as illustrated by the A', α' dashed line in Figure 1b. For the rise (ON) response (Figure 1a), the physical meaning is reversed. A higher $A'_0 > A_0$ stands for a lower level from which luminescence starts increasing as shown by the dashed line in Figure 1a and is related to a larger share of phosphorescence while a lower level of A_0 means a higher level of fluorescence. When a time-dependent measurement of the phosphorescence response is taken the intensity until the first measurement at $t_0 = \Delta t$ is missed, the exact time response is unknown and is noted by the red lines in Figure 1a,b. By that moment, the fluorescence process is over and what is measured afterwards is the slower phosphorescence afterglow [2–4]. Therefore, the level denoted by the solid circles from which the measurable response starts represents approximately the relative share of phosphorescence.

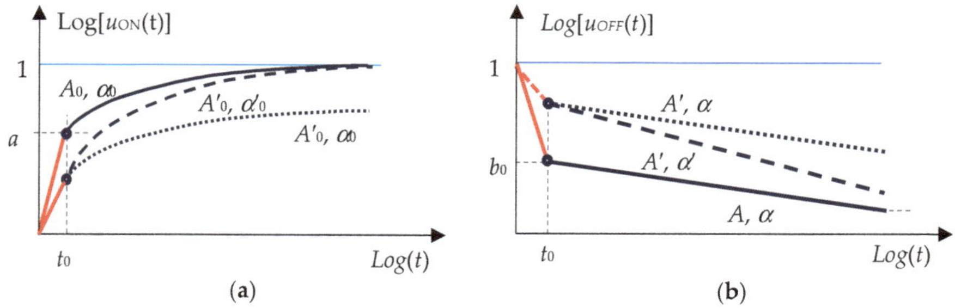

Figure 1. Double log plot of normalized time responses: (**a**) rise (ON) time response; (**b**) decay (OFF) time response for different pairs of A, a and A_0, a_0.

All of the four power law parameters from (2) have been shown [14,15] to be dependent on the temperature T, the excitation duration τ_p and period T_p and hence on the duty cycle η so we can write the following:

$$A_0 = A_0(T,\eta),\ \alpha_0 = \alpha_0(T,\eta),\ A = A(T,\eta),\ \alpha = \alpha(T,\eta) \tag{5}$$

As has been shown [15], the rise of temperature increases the relative share of phosphorescence and the rate of decay and the power parameters from (3) and (4) have been observed to increase with temperature until 60° and then slightly decrease. Figure 2 illustrates how the ON and OFF responses change as temperature increases form a lower T_0 to a higher T. As seen, the increase in temperature causes a drop of the a_0 level for the ON phase and a rise in the b_0 level for the OFF phase.

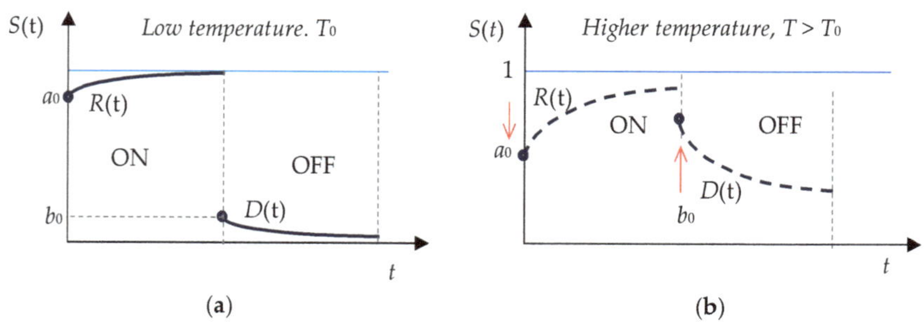

Figure 2. Double log plot of normalized time responses for temperature changes: (**a**) rise (ON) time response; (**b**) decay (OFF) time response for different pairs of A, a and A_0, a_0.

Based on the above observation [15] as a measure of the temperature, we make use of the normalized differential signal defined as follows [11]:

$$\mathrm{N} = \frac{R_i - D_i}{R_i + D_i}100\% \tag{6}$$

where R_i and D_i are the rise and decay intensities at the ith instant t_i of the measurement after the start of, correspondingly, the rise and the decay time response [11]. This quantity can be negative depending on the phosphors used. To reduce fluctuations, we can calculate the ratio not at certain moments but over certain time intervals between instant t_k and t_m, in which case the quantity becomes the following:

$$\mathrm{N} = \frac{\sum_{k}^{m} R_i - \sum_{k}^{m} D_i}{\sum_{k}^{m} R_i + \sum_{k}^{m} D_i} 100\% \tag{7}$$

As this quantity decreases with temperature, we shall make use of the quantity

$$\mathrm{N}_D = (1 - N)100\% = \left(1 - \frac{R_i - D_i}{R_i + D_i}\right)100\% = \frac{2D_i}{R_i + D_i}100\% \tag{8}$$

which is in fact the decay normalized to ON + OFF intensity and which changes in the same direction as the temperature does and is immune to power fluctuations of the source.

2.2. Smartphone-Based Interrogation of Space-Multiplexed Temperature Sensors

The detection of the temperature-dependent time responses can be performed using either a spectrometer or a simple photodetector unit. The former can take a measurement about each 200 ms and measures the whole spectrum, while the latter in less than a millisecond but measures the integral intensity. Whichever method is used, only one sensor can be measured at a time.

2.2.1. The Smartphone as a Detection Unit

Should we need to take simultaneous measurements of several phosphorescence-based sensors in order to measure spatial temperature distributions, thermal gradients and associated thermal flows, we are in need of another interrogation approach. The natural solution is the use of the smartphone (or webcam) camera which offers the following advantages:

- It is sensitive in the 400 nm to 700 nm visible range where most of the phosphors emit;
- It can measure time-dependent light emissions at rates of 30 fps (33 ms) or 60 fps (16.7 ms) which allow typical phosphorescent time responses to be tracked;
- As the camera is based on 2D CCD arrays it can simultaneously measure a considerable number of independent spatially distributed luminescent point sensors;
- Touchless remote interrogation of spatially distributed sensors;
- Data processing capabilities;
- Connectivity to wireless sensing networks using Wi-Fi and Bluetooth technology.

We, therefore, explore the possibility of developing a spatially multiplexed quasi-distributed phosphorescent temperature sensor with smartphone interrogation.

2.2.2. Possible Interrogation Schemes

Figure 3 shows the two basic schemes for the excitation of a multitude of spatially distributed temperature-sensitive phosphorescent dots that can be interrogated simultaneously by a single smartphone camera if they are within its field of view.

In the first scheme (Figure 3a), the sensing dots are illuminated from the side by a modulated UV LED whose emission wavelength is well outside the spectral range of the smartphone camera. The advantage of this approach is the simplicity of the construction, the disadvantage being the uneven illumination and need to place the sensing dots on a non-fluorescent surface.

In the second, the sensing phosphorescent dots are deposited upon the surface of an optical fiber with deliberately introduced deformations that cause parts of the guided excitation light to leave the fiber fore and excite the phosphors. The advantages of this approach are the controlled and stable excitation conditions, the freedom to attach the fibers to any type of surface and the possibility to integrate the excitation bus structure into a smart material. The disadvantage is the need to fabricate side-emitting deformations along the fiber and handling of the fiber. As the fiber can be up to tens of meters long, the fiber can be attached to large-scale structures whose surface temperatures can be monitored from a distance by a mobile phone.

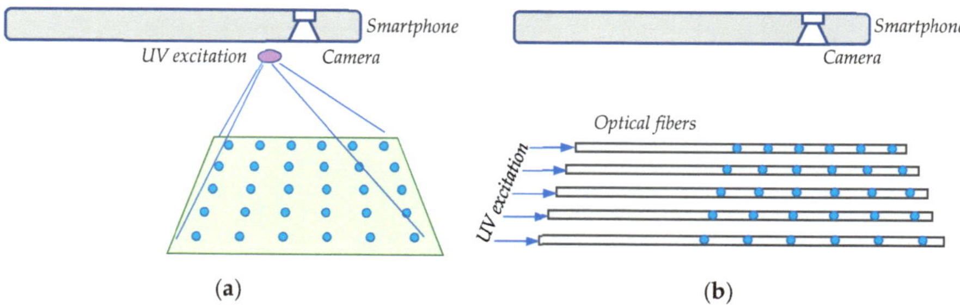

Figure 3. Two basic schemes for simultaneous interrogation of spatially distributed phosphorescent dots: (**a**) with side illumination; (**b**) illumination using array of excitation fibers.

2.3. Optical-Fiber Excitation Platform

As discussed in the introduction, a natural choice for an excitation and receiving medium is optical fiber. There are different approaches to produce a side-emitting fiber section, one of them being a taper usually produced with the help of a fusion splicer, and another being a partial chemical etching of the outer diameter. The third approach is to use a pulsed CO_2 laser to produce predesigned and controllable corrugated fiber sections with desirable dimensions and depth of dips upon the outer fiber surface. The periods are as those of long-period gratings but since the fiber is heavily multimode, the specific spectral properties of LPGs in single-mode fibers are not observed. What is of interest is the side radiation, which in traditional LPGs is not of interest. We shall refer here to these undulated fiber structures as corrugated LPGs (C-LPG). The corrugated fiber grating was shown [7] to be appropriate to accept fluorescence from the surrounding medium, which implies it can be used for the reverse function—exciting luminescence in the area around the fiber.

2.3.1. CO_2 Laser-Written Corrugated Long-Period Gratings (C-LPG)

To fabricate C-LPGs the radiation of a pulsed CO_2 laser, emitting at 10.6 µm or 9.6 µm, is focused upon the fiber surface and the absorbed radiation melts the glass. Depending on the laser power and exposure duration, a valley is formed when the melted glass solidifies. Thus, corrugated gratings of N undulations of the fiber profile can be fabricated. The corrugated section of a fiber of outer diameter D is characterized by its period Λ, the undulation width w, its thickness of the waist d and inclination angle ψ as defined in Figure 4. The width w, the depth $\Delta = D - d$ and the angle ψ of the undulations are inter-related and depend on the laser power and number of laser scans needed to achieve the desired depth. The period Λ and the number of undulations m as well as the laser power and the laser beam scanning speed are preset.

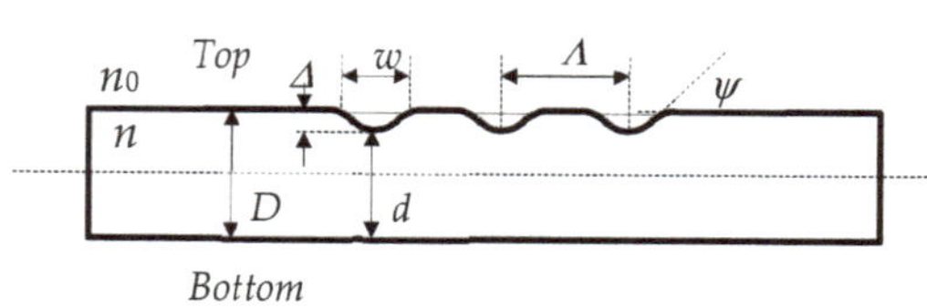

Figure 4. Schematic representation of a corrugated optical fiber.

2.3.2. Radiation Diagram of a Corrugated LPG

A simple explanation of the radiation diagram is presented in Figure 5. For the sake of simplicity, a valley of the undulated profile is represented in a triangular shape whose inclination with respect to the fiber outer surface is α. Taking into account Snell's law, we find that the rays in the fiber core propagate with respect to the axes through a maximum angle α such that

$$\sin\alpha = \frac{NA}{n} \tag{9}$$

NA being the numerical aperture and n the core refractive index. Thus for a fiber of $NA = 0.22$ and $n \approx 1.46$, $\alpha < 10°$. In the case of $\psi < 45°$, the trajectory of a ray in the fiber which is almost parallel to the fiber axis is shown in Figure 5a, from which it is evident that part of the optical power will be reflected upward from the top side while the other part will be refracted and will be irradiated downward from the bottom side.

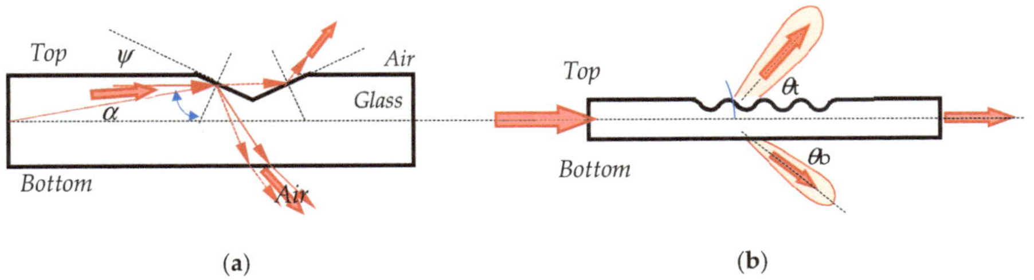

Figure 5. Radiation pattern of the corrugated fiber structure in case the modulation angle $\psi < 45°$: (**a**) ray tracing; (**b**) resultant radiation diagram.

The resultant radiation diagram is illustrated in Figure 5b. Because the corrugated surface is on the top of the fiber the radiation pattern will be in the vertical plane as shown in the figure.

3. Experiments and Results

3.1. Inscription, Characterization and Functionalization of the Corrugated LPGs

The fabrication procedure of the sensing phosphorescent fiber optic arrays is as follows:

(i) Inscription of the array of CLPGs;
(ii) Measurement of the basic parameters of the corrugated structures;
(iii) Deposition of the phosphorescent powder around the C-LPG.

We now proceed with the description of these steps.

3.1.1. Experimental Setup for C-LPG Inscription

The experimental setup to fabricate the corrugated fiber sections is shown in Figure 6a,b below. The corrugated long-period gratings were fabricated using a SYNRAD-pulsed CO_2 laser (Synrad Inc., Mukilteo, WA, USA) emitting at $\lambda = 10.6$ µm, f = 20 kHz, and at 25% of maximum power; the focal length of the scanning head was 10 cm, and the spot size was about 100 µm. Arrays of varying numbers (5 to 10) of C-LPGs spaced at several millimeters (4 mm to 10 mm) were fabricated in a 105/125 µm multimode optical fiber. The fiber was stripped, wiped clean and installed under a 3 g tension. The writing process includes consecutive scans of the focused laser beam. During each scan all of the C-LPGs of the whole array are inscribed in a row. Each subsequent scan increases the depth Δ, the length l and the angle ψ of the dip which causes the power irradiated from the fiber to increase as well. As the number of scans per C-LPG of the array can be programmed, fine tuning of the parameters (Δ, l and ψ) can be realized.

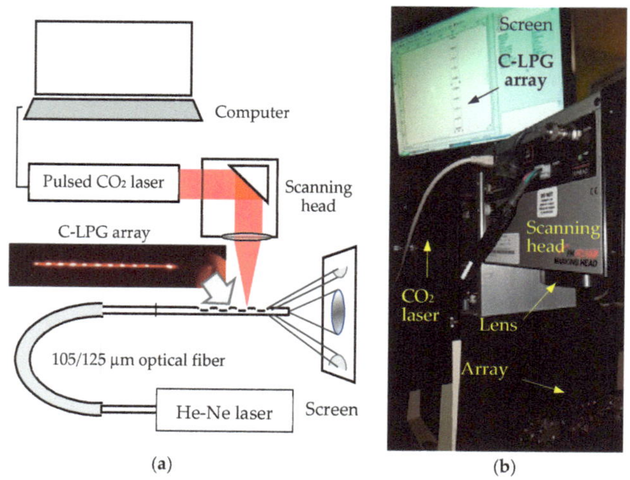

Figure 6. Experimental set up. (**a**) The inset is a photo of the N = 10 corrugated gratings spaced at 4 mm and each containing 4 dips of Λ = 250 μm. (**b**) Photo of the experimental arrangement.

3.1.2. Characteristics of the Fabricated Corrugated Structures

The basic characteristics of interest are the dimensions of the C-LPG (dip depth Δ, length l and angle ψ) as well as the top and bottom radiation angles θ_t and θ_b. In Figure 7 we show examples of two C-LPGs of different strengths as a whole and of the individual dips, photos of their resultant radiation patterns and of the functionalized structures. We clearly see that a weaker C-LPG is characterized by lower values of Δ, l, ψ and top/bottom radiation angles θ_t and θ_b. To fix the phosphorescent power to the grating surface, a micro-drop of instant glue was deposited upon the corrugated surface and then minute quantities of the phosphorescent powder were added.

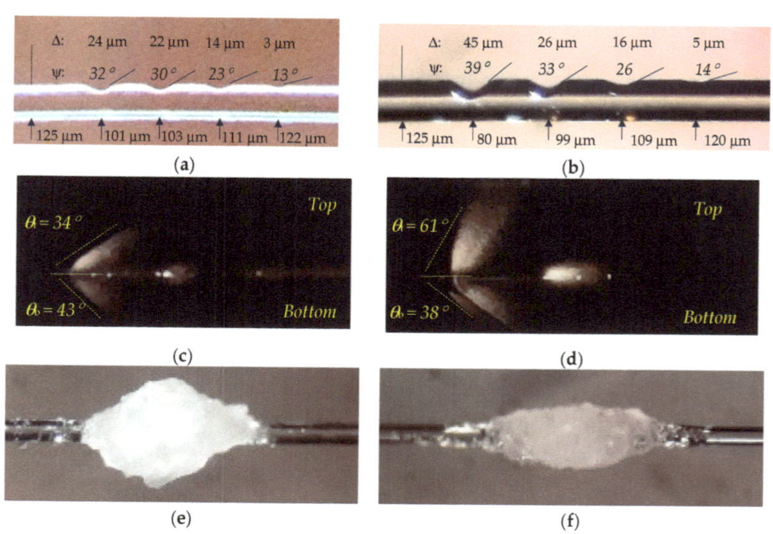

Figure 7. Photos of two single C-LPGs each containing 4 dips of Λ = 250 μm of different depths with dimensions indicated: (**a**) a weaker C-LPG-#1; (**b**) a stronger C-LPG-#2, (**c**) radiation pattern with top and bottom radiation angles for the weaker C-LPG-#1; (**d**) same as (**c**) for the stronger C-LPG-#2; (**e**) C-LPG-#1 functionalized with Sr aluminate powder; (**f**) same as (**e**) but for the C-LPG-#2.

We next fabricated an array of 10 C-LPGs each of 4 dips with Λ = 250 µm and separated by 6 mm from one another. The photos of sections of the array and the side-radiation patterns are presented in Figure 8.

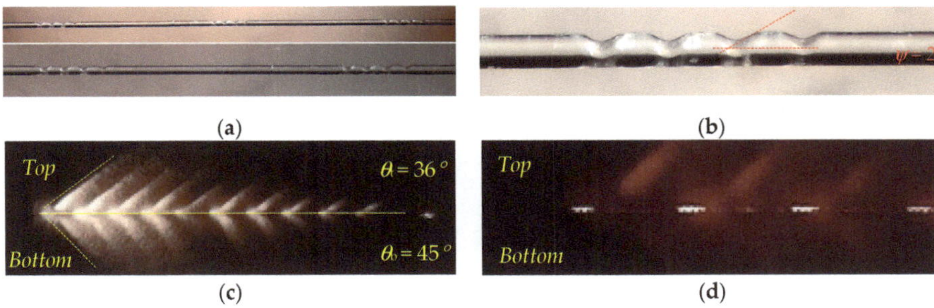

Figure 8. Photos of arrays of C-LPGs each containing 4 dips of Λ = 250 µm of the same depths: (**a**) three C-LPGs in a row and two magnified; (**b**) one C-LPG of the array magnified with angle ψ indicated; (**c**) radiation pattern of the array of 10 C-LPGs with angles θ_t and θ_b indicated; (**d**) same as (**c**) magnified with only four C-LPGs visible.

As is seen, the side-irradiated power decreases with each subsequent C-LPG. If we assume that the transmission of each C-LPG is T_0, then after the k-th grating, the resultant transmission will be

$$T = T_0^k \tag{10}$$

i.e., the optical power at the fiber output will decrease exponentially, which also means that the side-irradiated power will also decrease exponentially, which is visible in Figure 8c. There are three ways to equalize the side-irradiated power:

- To individually fine tune the dip depth and hence the radiation pattern and irradiated power;
- To excite the fiber from both ends;
- To deposit a reflector at the fiber output tip.

From the data in Figure 8a,b, we can find the experimental dependence of the dip angle ψ on the dip depth Δ, which is shown in Figure 9a. As should be expected, the dip angle increases with the depth but tends to saturate above 40 deg. On the other hand, the comparison with the side-radiation patterns shows that the radiation angled θ_t and θ_b increases with the dip angle. A more detailed study of the properties of C-LPGs in different types of fibers will be presented elsewhere.

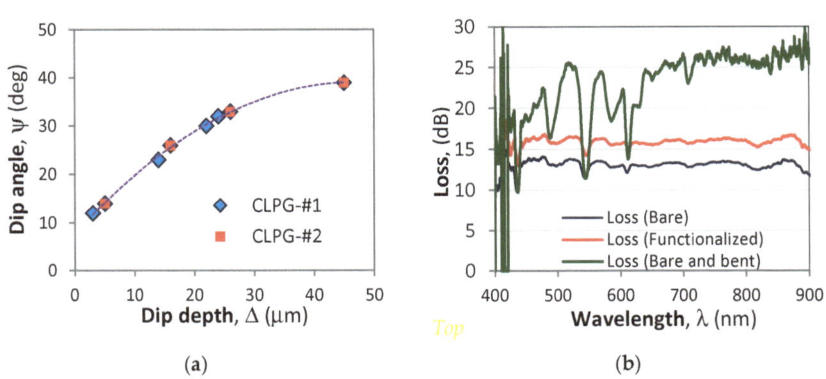

Figure 9. Characteristics of the sensing array: (**a**) dependence of the dip angle ψ on the depth Δ; (**b**) spectral losses of a bare, a functionalized and a bent bare C-LPG sensing array.

Since the dips of the C-LPG introduce losses as demonstrated above, we present in Figure 9b their spectral dependences before and after functionalization with the phosphor. As is seen, the averaged losses of the bare C-LPG are around 13 dB and increase by about 3 dB after functionalization. The reason for this rise is the increased refractive index around the corrugated surface which causes more light to leave the fiber and is then additionally scattered. In both the bare and functionalized straight C-LPG arrays the loss ripples are of the same depth and same wavelengths. i.e., they do not depend on surrounding refractive index changes as in LPGs in single-mode fibers. When bent, the depths of the loss ripples increase to more than 10 dB, but do not shift spectrally. This indicates that the losses are from side radiation and not from wavelength-dependent intermodal coupling as in LPGs.

3.2. Performance of the Quasi-Distributed Sensor

3.2.1. Measurement Set-Up

In Figure 10 we present the setup for the contactless measurement of temperature by means of a mobile phone. It consists of two thermoelectric heaters/coolers TEC_1 and TEC_2 separated by a distance Δx that can vary their temperatures T_1 and T_2 from 5 °C to 75 °C and which is controlled by two thermistors. A thin (0.5 mm) metal plate was placed to thermally connect both TECs.

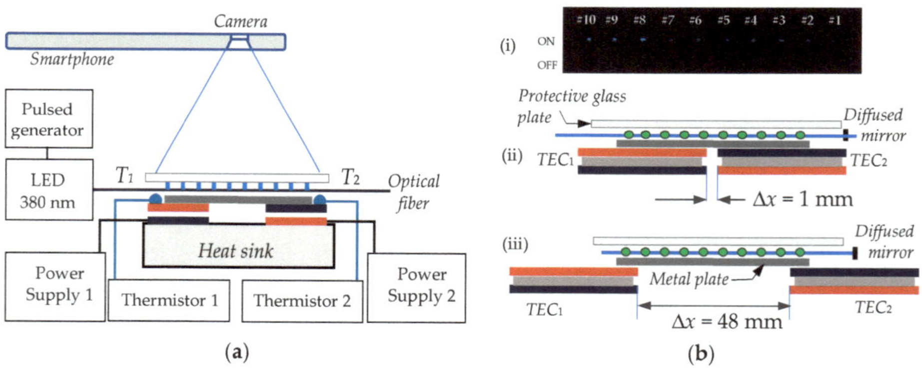

Figure 10. Smartphone interrogation of the sensor array: (**a**) experimental setup; (**b**) the TEC and array arrangement: (**i**) photo of the ON/OFF luminescent array; (**ii**) arrangement for an abrupt temperature change (Δx = 1 mm); (**iii**) arrangement for a smooth temperature change Δx = 48 mm).

The array containing 10 sensing points was placed upon the metal plate. Since the volume of each sensor was less than 0.5 mm^3, its temperature was strongly influenced by air currents. Therefore, a protective transparent glass plate was used to cover the sensing array. Since the optical power in the fiber decreases along the array (see Figure 8c), a diffused reflector created with 90% barium sulfate ($BaSO_4$) and polyvinyl alcohol (PVA) was deposited at the fiber's distal end which helped equalize the intensity of the luminescent sensors.

Three experiments were carried out. In the first, the array was placed entirely above one of the TECs and the responses of the ten sensors were taken at different temperatures in order to plot the $N(T)$ dependence and the sensitivity $S_T = \Delta N/\Delta T$.

In the second, the TECs were placed closed to each other with Δx = 1 mm. Sensing points #5 and #6 were placed exactly on either side of the separation (Figure 10b(i)).

In the third, the TECs were separated by Δx = 48 mm and sensors #1 and #10 were over the corresponding TECs (Figure 10b(ii)).

The UV LED (380 nm) was pulsed at a 1 Hz repetition rate (T = 1 s), a duty cycle of η = 50% and the mobile phone (Samsung A51) was placed at H = 15 cm above the sensing array. The frame rate of the smartphone was $F = 30$ fps, which translates into a sampling duration of $\Delta t = 33.3$ ms.

3.2.2. Results

Four consecutive experiments were performed to characterize the sensor array.

In the first experiment, the temperatures of the TEC were varied from 5 °C to 75 °C, during which a video clip was taken for at least ten ON/OFF cycles. The ON/OFF responses presenting the rise and decay curves averaged over 10 cycles for the 10th sensor is presented in Figure 11a. In this plot we clearly see the drop of phosphorescence with temperature as described by (1) and the slowing of the rise and decay responses in agreement with the comments in Section 2.1.3 and previous studies [11,15]. The red lines indicate the direction of change of the ON/OFF responses with temperature. Figure 11b presents the plots of the normalized decay N_T as a function of the temperature for the 10th sensor as calculated from the plots in Figure 11a for three different summations in (7): $i = 1$ to 5, $i = 1$ to 8 and $i = 1$ to 15. The latter means integration (summation) over the whole ON and the whole OFF durations. We see that the temperature dependences $N_D(T)$ exhibit a change in slope and hence of the sensitivities for $T > 45$ °C. The calculated sensitivities in %/°C are listed to the right of the legend in Figure 11b.

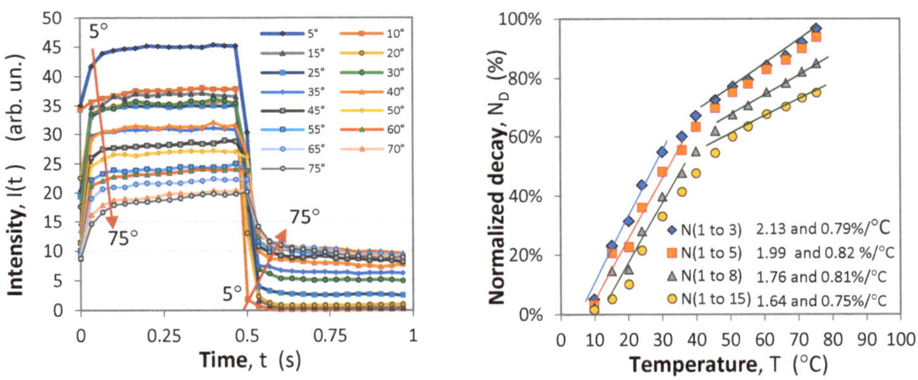

(a) (b)

Figure 11. Rise/Decay (ON/OFF) responses for temperatures from 5 °C to 75 °C: (**a**) ON/OFF time responses over a period $T = 1$ s and a duty factor $\eta = 50\%$; (**b**) temperature dependence of the normalized decay $N_T(T)$.

Comparing the sensitivities we see that they drop by a factor of two above 45 °C and if the normalized difference is calculated over the initial moments of the ON/OFF responses the sensitivity is higher. This observation implies that the frequency of the LED modulation can be doubled and the period reduced to 0.5 s.

In the second experiment, several combinations of temperature pairs (T_1, T_2) of the two TECs were applied, as shown on each inset in Figure 12a, which is a photo of the array during ON and OFF durations. We calculate the normalized difference N of each sensor ($k = 1$ to 10) along the array and plot the dependence of N on the sensor number (the individual positions are shown in Figure 10b(i)) and the results are presented in Figure 12b. The plots show that the sensor array reproduces the abrupt change in temperature and the sensors #5 and #6 reflect the transition from the hot to the cold section.

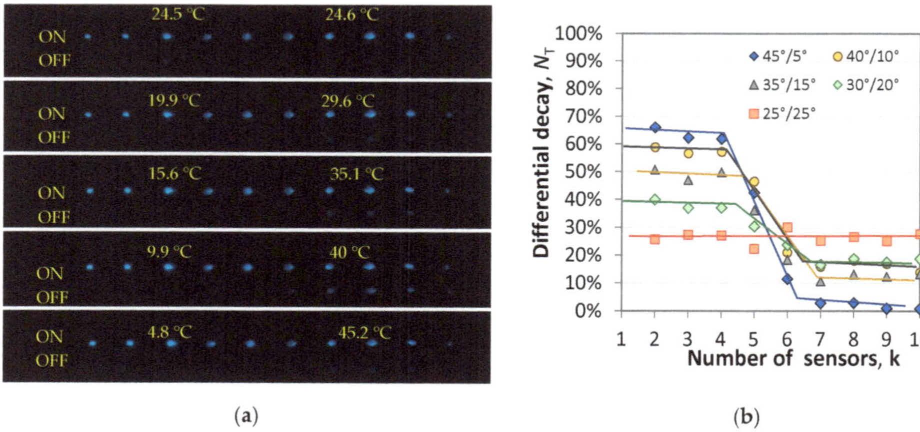

Figure 12. Rise/decay (ON/OFF) responses for different pairs of temperatures at the two TECs separated by 1 mm: (**a**) photos of the sensor array during the ON and the OFF; (**b**) normalized differential signal $N(T)$ for each sensor for an abrupt spatial change in the temperature.

In the third experiment, temperature pairs (T_1, T_2) were set with the TECs which were separated to a distance Δx = 48 mm (Figure 10b(iii)) so that sensor #1 was at the edge of the hot TEC while #10 was at the edge of the cold TEC. The responses were analogous to those from Figure 12b, but the transition was gradual in reproducing the temperature gradient along the metal plate connecting the TECs.

Figure 13 summarizes the typical response. Figure 13a compares the ON/OFF responses of sensors #10 (the coldest at 5 °C), #5 (in the middle at 40 °C) and #2 (the hottest at 75 °C), while Figure 13b presents the response of the whole array, i.e., the spatial distribution of the temperature along the metal plate. The dependence is almost linear as should be expected along a homogeneous thermally conductive medium. Figure 13c,d present the $1 - u_{ON}(t) = A_0 t^{-\alpha 0}$ and $u_{OFF}(t) = At^{-\alpha}$ expressions from (3) and (4) for sensor #1 at 75 °C and sensor #10 at the other end of the array at 5 °C. The power law fit yield A_0 = 0.332, α_0 = 0.16 for the ON and A = 0.2166, α = 0.199 for the OFF responses at 75 °C (hot end) versus A_0 = 0.1044, α_0 = 0.062 for the ON and A = 0.0188, α = 0.077 for the OFF responses at 5 °C (cold end) is in line with the reasoning in Section 2.1.3. Clearly, at high temperatures A, α, A_0 and α_0 are higher than at low temperatures, implying that at higher temperatures the relative share of phosphorescence in the overall luminescence process increases and the decay is faster as shown in [15].

In the fourth experiment, the time dependent response of the array was taken after the heaters/coolers were switched off at the end of the third experiment and the array returned to equilibrium. Figure 14a displays the evolution of the pulses of the sensor at the hot end detected by the mobile phone, and Figure 14b displays the normalized decay calculated for three different summations in (7): i = 1 to 5, i = 1 to 8 and i = 1 to 15. As is well seen, when summation is over a larger number of measurement points (i = 1 to 15) the ON and OFF phases fluctuations in the response from Figure 14b are weaker compared to the case of a smaller number of summation points (i = 1 to 5).

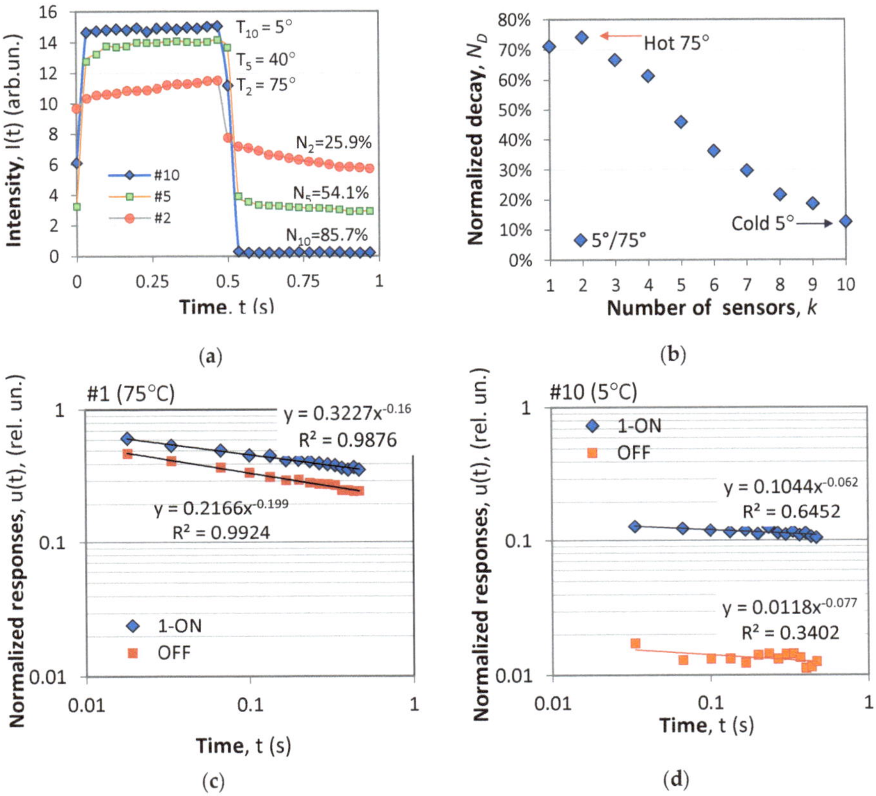

Figure 13. Rise/decay (ON/OFF) responses at the extremities and the middle of the sensing array with 5 °C and 75 °C at the TECs: (**a**) ON/OFF time responses of the 2nd, 5th and 10th sensor with the corresponding temperatures and normalized differential signals; (**b**) normalized differential signal $N(T)$ for each sensor which corresponds to the temperature gradient along the array; (**c**) $1 - u_{ON}(t)$ and $u_{OFF}(t)$ of sensor #10 at 75 °C; (**d**) same as (**c**) for sensor #1 at 5 °C.

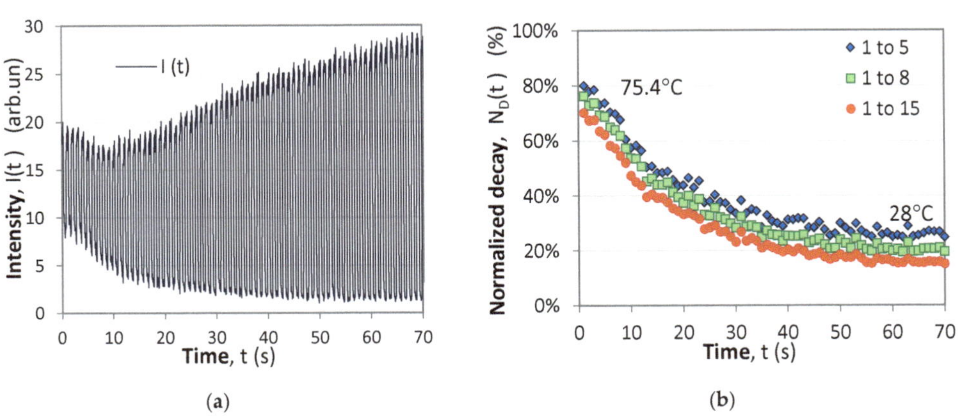

Figure 14. Temperature relaxation process of the sensor at the hot end: (**a**) time response of the pulsed modulation during the process of reaching a thermal equilibrium; (**b**) normalized decay signals for three cases of summation.

4. Discussion

The presented quasi-distributed sensor principle of operation, construction and results require some comments and discussion.

In the first place, this is a further development of the optical-fiber (OF) temperature single-sensor arrangements presented elsewhere [12] and the application of mobile phones for the contactless interrogation of phosphorescent-based sensors.

Compared to previous phosphor sensors, the present structure is characterized by much smaller volumes of the used phosphor because of the efficient C-LPG-based excitation and a better time resolution because of the use of smartphones. The volume of the sensing region is an order of magnitude smaller than the one from [11] and can be further reduced to less than 0.1 µL with an improved phosphor deposition technology and more efficient luminescence. The time response resolution, which varies between 1 s for single pulse interrogation and 10 s for averaged measurements, can easily be reduced by a factor of two with a 30 fps frame rate and to about 0.2 s for a 60 fps frame rate, which would be an order of magnitude better compared to the sensor using spectrometer detection [11,16].

Second, the proposed scheme essentially is a spatially multiplexed sensor array and can be used because of the efficient side-excitation by means of the CLPG's smaller diameter fibers. This generates two consequences. As the mass of the sensor decreases, the fiber endures a smaller bending diameter, which simplifies the handling and optical wiring of the fiber sensor. Next, since the fiber diameter is smaller, a bundle of up to 20 fibers can be powered with a single LED and hence a larger number of sensing arrays can be combined into a 2D sensing array. Such arrays can be simultaneously interrogated if the 2D sensing array is within the field of view of the smartphone camera. With an appropriate software and smartphone application, up to 20 × 10 individual points can be interrogated which implies the reconstruction of a temperature planar distribution. And hence the evaluation of heat flows and transfers. As the fiber platform can be up to tens of meters long, the individual sensors can be permanently attached to a control temperature at critical points over a fairly large area. In this case, the contactless interrogation is not simultaneous.

Third, the use of a smartphone offers the capability of contactless interrogation and the transfer of data to a larger Wi-Fi-compatible sensing network within the internet of things technology.

Fourth, the sensitivity of the sensor expressed in %/C depends on the following:

(i) The phosphorescent material (also affects the range) as shown in [11,15];
(ii) The duration over which rise and decay summation (integration) is carried out;
(iii) The duty factor η.
(iv) Fifth, the response time of the sensor essentially is affected by the following:
(v) The rise/decay times of the phosphors; so to reduce it to the order of seconds, if needed, phosphors with similar decay times must be used;
(vi) The modulation frequency (limited by the rise/decay times of the responses), whose increase will reduce the response time;
(vii) The number of cycles over which the signal is averaged to reduce fluctuations, at the expense of response time;
(viii) The frame rate of the smartphone whose increase (to 60 fps, for example) will reduce the response time.

Sixth, the temperature range of operation can be varied with a proper choice of phosphor. Phosphors having a short rise/decay time response at room temperature will have a longer response at higher temperatures and those with a long response at room temperature will have a shorter one at low temperatures. Although the principle of operation used for this sensor is the normalized differential rise/decay time response based on Equations (1) and (2), to increase the range at the expense of sensitivity and response time, only the intensity changes can be monitored. In this case the duty cycle should be closer to unity and the modulation frequency lower than 1 Hz to saturate the emission during the ON phase and monitor how the saturation level decreases with temperature following (1).

Compared to the variety of other types of temperature sensors [21–23], the quasi-distributed sensor demonstrated in this present paper has several basic features which define its specific niche:

(i) It offers contactless simultaneous interrogation using smartphones or web cameras linked via Wi-Fi and Bluetooth to the internet as a part of a broader sensing network. None of the reported temperature sensors has this feature.

(ii) It makes use of standard components of the lowest cost such as fibers and phosphorescent materials. It can also use polymer fiber. Interferometric sensors [21,22], be they Mach-Zender (MZI), Michelson (MI) or Fabry Perror (FP), require single-mode or specialty photonic crystal (PCF), holey (HOF) or microstructured (MOF) optical fibers.

(iii) A single low-cost LED can excite and a single camera can interrogate several sensing arrays. Similar is the case of a quasi-distributed FBG sensing network using wavelength multiplexing whose temperature range is up to 100 °C. In contrast, all of the non-phosphorescent types of sensors are based on considerably more costly sources and detection equipment. The fluorescence-based sensors [21] require pulsed lasers and fast electronics or spectrometers to measure the intensity ratio of different fluorescence. Practically all of the interferometric and fiber Bragg grating (FBG) sensors function in the infrared communication band (1260 nm to 1630 nm) and employ orders of magnitude more costly sources and interrogation equipment and thus make practical sense mostly for multiplexed sensing networks.

(iv) It is easy to fabricate by a high-productivity pulsed CO_2 laser technique. In contrast, FBGs are made using an order of magnitude more costly excimer or femtosecond lasers [23]. The variety of interferometric sensors need more sophisticated splicers and for the micro-cavity-based MZI and FP interferometric sensors—a femtosecond laser [23].

(v) The excitation fiber allows wiring of the sensing array along arbitrary paths to measure spatial distribution of temperature and thermal flows. This feature is similar to distributed temperature sensors (Raman, Brillouin [23] or Rayleigh type [22]) and quasi-distributed FBG sensors, which however need a special attachment in order to eliminate strain unless they measure both strain and temperature.

(vi) Depending on the phosphor used, temperatures as high as several hundred degrees can be reached, which is what most of the silica-based fiber sensors offer. The sensor proposed here, however, offers no possibility for high temperature (above several hundred degrees) measurement.

(vii) It can also be made from a polymer fiber [21], in which case the fabrication of the dips along the fiber would be even less costly. However, losses in the UV will be considerably higher so shorter fiber spans should be used.

Future research will be performed with other phosphors, types of fibers (such as 100 μm and 200 μm quartz polymer fibers) and smartphones having a 60 fps frame rate as well as the simultaneous interrogation of a larger number of arrays.

5. Conclusions

The results obtained allow the formulation of the following conclusions:

C1. An array of short and strong C-LPGs fabricated in a standard 105/125 μm multimode fiber has been proposed, developed and demonstrated as an excitation platform for a spatially quasi-distributed phosphorescent temperature sensor with contactless smartphone interrogation.

C2. The performance of the sensing array comprising 10 sensing points has been experimentally studied and has been demonstrated to correctly reproduce abrupt and gradual spatial changes in temperature caused by temperature gradients within the 5 °C to 75 °C range.

C3. Time dependent evolution of temperature with a 1 s measurement duration has been experimentally demonstrated for different integration ranges.

Author Contributions: Conceptualization, T.E.; methodology, T.E.; software, D.B. and K.N.; validation, S.F.; formal analysis, T.E.; investigation, S.F. and T.E.; resources, I.K.; data curation, K.N. and V.V.; writing—original draft preparation, T.E. and I.K.; writing—review and editing, T.E. and I.K.; visualization, T.E.; supervision, I.K.; project administration, T.E. All authors have read and agreed to the published version of the manuscript.

Funding: This research received no external funding.

Institutional Review Board Statement: Not applicable.

Informed Consent Statement: Not applicable.

Data Availability Statement: Data can be provided upon request.

Conflicts of Interest: The authors declare no conflicts of interest.

References

1. Aizawa, H.; Toba, E.; Katsumata, T.; Takahashi, J.; Matsunaga, K.; Komuro, S.; Morikawa, T. Long duration phosphor for a fiber-optic thermometer. In Proceedings of the 41st SICE Annual Conference, Osaka, Japan, 5–7 August 2002; pp. 2972–2975.
2. McSherry, M.; Fitzpatrick, C.; Lewis, E. Review of luminescent based fibre optic temperature sensors. *Sens. Rev.* **2005**, *25*, 56–62. [CrossRef]
3. Bao, Y.; Huang, Y.; Hoehler, M.; Chen, G. Review of fiber optic sensors for structural fire engineering. *Sensors* **2019**, *19*, 877. [CrossRef] [PubMed]
4. Wang, L.; Shang, Z.; Shi, M.; Cao, P.; Yang, B.; Zou, J. Preparing and testing the reliability of long afterglow $SrAl_2O_4:Eu^{2+}$, Dy^{3+} phosphor flexible films for temperature sensing. *RSC Adv.* **2020**, *10*, 11418. [CrossRef]
5. Li, J.; Chen, H.; Wang, X.; Li, P.; Lv, M.; Hao, H.; Meng, L. An optical fiber temperature sensor based on fluorescence intensity ratio in Er^{3+}/Yb^{3+} co-doped Gd_2O_3 phosphors. *Ceram. Int.* **2023**, *49*, 28913–28919. [CrossRef]
6. Zhao, Y.; Wang, X.; Zhang, Y.; Li, Y.; Yao, X. Optical temperature sensing of up-conversion luminescent materials: Fundamentals and progress. *J. Alloys Compd.* **2020**, *817*, 152691. [CrossRef]
7. Carrillo-Betancourt, R.A.; López-Camero, A.D.; Hernández-Cordero, J. Luminescent Polymer Composites for Optical Fiber Sensors. *Polymers* **2023**, *15*, 505. [CrossRef] [PubMed]
8. Miluski, P.; Dorosz, D.; Mojda, J.; Kochanowicz, M.; Dorosz, J. Luminescent Polymer Optical Fibre Sensor for Temperature Measurement. *Acta Phys. Pol. A* **2015**, *127*, 730–733. [CrossRef]
9. Li, H.; Zhang, Y.; Shao, L.; Htwe, Z.; Yuan, P. Luminescence probe for temperature sensor based on fluorescence intensity ratio. *Opt. Mater. Express* **2017**, *7*, 1077. [CrossRef]
10. Harun, S.W.; Yasin, M.; Hamzah, A.; Arof, H.; Ahmad, H. Temperature sensor based on lifetime measurement of Erbium fluorescence. *J. Optoelectr. Adv. Mater.* **2011**, *13*, 69–71.
11. Fouzar, S.; Eftimov, T.; Kostova, I.; Dimitrova, T.L.; Benmounah, A.; Lakhssassi, A. A Simple Fiber Optic Temperature Sensor for Fire Detection In Hazardous Environment Based On Differential Time Rise/Decay Phosphorescence Response. *IEEE Trans. Instr. Meas.* **2022**, *71*, 9508908. [CrossRef]
12. Fouzar, S.; Kostova, I.; Eftimov, T.A.; Nikolov, K. Phosphorescent Fibre Optic Tip Temperature Sensor. In Proceedings of the 4th International Conference on Phosphor Thermometry, Becici, Montenegro, 26–30 August 2024. *Accepted*.
13. Eftimov, T.; Kostova, I.; Arapova, A.; Patronov, G. Rise and decay time responses of Sr aluminate phosphorescent materials. *J. Lumin.* **2021**, *235*, 117985. [CrossRef]
14. Fouzar, S.; Kostova, I.; Eftimov, T.; Benmounah, A.; Ouchabane, M.; Lakhssassi, A. Excitation and spectral dependence of the rise and decay time responses of Eu^{2+} and Dy^{3+} doped strontium aluminates. *J. Mater. Sci. Mater. Electron.* **2022**, *33*, 20962–20980. [CrossRef]
15. Fouzar, S.; Eftimov, T.; Kostova, I.; Benmounah, A.; Lakhssassi, A. Effects of temperature on the time responses of strontium aluminates. *Opt. Mater.* **2021**, *122*, 111619. [CrossRef]
16. Eftimov, T.A.; Kostova, I.; Peltekov, A.; Hristova, K.; Brabant, D.; Fouzar, S.; Nikolov, K. A Study of Strontium Aluminates for All Optical Contactless Sensing Applications Using Smartphone Interrogation. *Opt. Laser Technol.* **2024**, *176*, 110965. [CrossRef]
17. Fratto, B.E.; Culver, E.L.; Davis, G.; Deans, R.; Goods, J.B.; Hwang, S.; Keller, N.K.; Lawrence, J.A.; Petty, A.R.; Swager, T.M.; et al. Leveraging a smartphone to perform time-gated luminescence measurements. *PLoS ONE* **2023**, *18*, e0293740. [CrossRef] [PubMed]
18. Chiniforooshan, Y.; Ma, J.; Bock, W.J.; Hao, W.; Wang, Z.Y. Corrugated Fiber Grating for Detection of Lead Ions in Water. *J. Light. Technol.* **2015**, *33*, 2549–2553. [CrossRef]
19. Irawan, R.; Chuan, T.S.; Meng, T.C.; Ming, T.K. Rapid Constructions of Microstructures for Optical Fiber Sensors Using a Commercial CO_2 Laser System. *Open Biomed. Eng. J.* **2008**, *2*, 28–35. [CrossRef] [PubMed]
20. Tan, C.; Zhao, L.; Chen, M.; Cheng, J.; Yin, Z.; Liu, Q.; Yang, H.; Liao, W. Physical mechanism of pulsed laser interaction with fused silica optics during CO_2 laser mitigation process. *J. Non-Cryst. Solids* **2021**, *558*, 120662. [CrossRef]
21. Szczerska, M. Temperature Sensors Based on Polymer Fiber Optic Interferometer. *Chemosensors* **2022**, *10*, 228. [CrossRef]

22. Yang, T.; Wang, X. Decoupling and Simultaneous Measurement of Nonuniform Strain and Temperature Using a Single Distributed Optical Fiber Ring. *Exp. Mech.* **2022**, *62*, 1531–1552. [CrossRef]
23. Ma, S.; Xu, Y.; Pang, Y.; Zhao, X.; Li, Y.; Qin, Z.; Liu, Z.; Lu, P.; Bao, X. Optical Fiber Sensors for High-Temperature Monitoring: A Review. *Sensors* **2022**, *22*, 5722. [CrossRef] [PubMed]

photonics

MDPI

Article

Human Vital Signs Signal Monitoring and Repairment with an Optical Fiber Sensor Based on Deep Learning

Haochun Gao [1], Qing Wang [1,*], Jing Zhou [1] and Changyuan Yu [1,2,*]

1 Department of Electrical and Electronic Engineering, The Hong Kong Polytechnic University, Hong Kong 999077, China
2 Shenzhen Research Institute, The Hong Kong Polytechnic University, Shenzhen 518057, China
* Correspondence: qing-wq.wang@connect.polyu.hk (Q.W.); changyuan.yu@polyu.edu.hk (C.Y.)

Abstract: Optical fiber sensors have been widely applied for their advantages such as small size, lightweight, and strong electronic interference robustness. Compared with current electronic sensors, optical fiber sensors perform better in measuring parameters in harsh environments, which makes them suitable for more and more applications, such as target tracing and detection and monitoring of health signs in medical services. However, due to fiber optic sensor failure, improper transmission and storage, or other reasons, missing data occur from time to time. Therefore, effective missing value processing methods are desirable as they can be used to facilitate data processing or analysis. In the present study, gated recurrent unit (GRU) interpolation is performed by using the generative adversarial network (GAN) model to process the irregular delay relationship between the data before and after the collection of incomplete vital signs data. Furthermore, a data interpolation model based on VS-E2E-GAN is proposed to reconstruct vital signs signals. The ROC curve (AUC), metrics including mean squared error (MSE), and accuracy (ACC) of experiments reach 0.901, 0.777, and 0.908, respectively, which indicates that the proposed VS-E2E-GAN model performs well in terms of vital signs data imputation and repairment, has strong robustness when compared with other works, and has potential clinical application in health monitoring, smart home, and so on.

Keywords: vital signs monitoring; optical fiber sensor; missing data; repairment; generative adversarial network; VS-E2E-GAN

Citation: Gao, H.; Wang, Q.; Zhou, J.; Yu, C. Human Vital Signs Signal Monitoring and Repairment with an Optical Fiber Sensor Based on Deep Learning. *Photonics* **2024**, *11*, 707. https://doi.org/10.3390/photonics11080707

Received: 20 June 2024
Revised: 26 July 2024
Accepted: 27 July 2024
Published: 29 July 2024

1. Introduction

Recently, the optical fiber communication industry has experienced rapid development. An optical fiber communication system is characterized by a larger channel bandwidth and more reliable information service [1–5]. At the same time, there are also various applications for optical fiber sensors, especially medical monitoring [6–9]. Currently, the incidence and mortality of various high-risk diseases show an increasing trend. With the progress in modern medicine and the shift of priority from treatment to prevention, daily assessment and detection of human health have become requisite for preventing and controlling the development of high-risk diseases [10]. For example, the vital signs of patients are an important piece of data needed for the treatment and care of patients, especially when the continuous monitoring of heart rate is required. Various vital signs, such as respiration, body temperature, blood sugar, lipids heartbeat, pulse, and blood pressure (BP), are the main health indicators of human physiological activities. Therefore, in addition to assisting in the evaluation of daily health status for individuals, monitoring of vital signs can also support the early of certain diseases like cardiovascular diseases [10,11]. Among them, heartbeat and respiratory signals are the most common ones, which are closely related to health status [12–15]. However, the method commonly used to monitor vital signs, namely, five-lead detection, requires direct contact with the patient's skin to collect vital signs data. In different situations, this measurement causes inconvenience, and it performs poorly in efficiency and accuracy, thus causing disruption to timely and effective treatment. This is a problem that can be effectively solved by using optical fiber sensors [15,16].

However, it is a common problem to encounter in practice that data missing and data anomalies result from the use of optical fiber sensors to collect vital signs signals, which is due to terminal data collectors, network communication failures, or other human factors, such as subject body flipping [17–19]. There are different methods available to perform data repair but they vary in the extent to which accuracy is improved. With regard to dealing with the outliers and missing and error values in the existing literature, there are three main methods used. The first one is the direct deletion method, which often ignores the important information in the data, and a higher missing rate influences the subsequent data analysis more significantly. The other one is the simple imputation method, which usually involves mean imputation, median imputation, and other methods of mathematical imputation, focusing only on the relationship between values. The last one includes various imputation methods based on machine learning, such as the interpolation based on the interpolation method and the *k*-nearest neighbors (KNN) based on matrix factorization. However, these methods focus narrowly on the statistical characteristics of the data itself, thus ignoring the time-series signal characteristics. Furthermore, the imputation accuracy is poor [20–24].

In an era of artificial intelligence, the technology of deep learning is being applied widely to achieve data processing and pattern recognition. Generative adversarial network (GAN) is a generative learning algorithm that has emerged in recent years [25–27]. Due to the advantage of GAN in model-free reconstruction to perform data generation tasks, it has been applied to deal with time series data generation. By following the adversarial training process, the quality of samples generated via GAN is significantly improved compared with traditional methods [28–32]. Moreover, the framework of GAN shows flexibility. The generator or discriminator can be any differentiable function, which is conducive to the further application of GAN. At present, recurrent neural networks (RNNs) are being used widely for sequence learning, with excellent achievement in digital image signal processing, video signal processing, and natural language processing (NLP) [33–39]. Some representative modules, such as gated recurrent unit (GRU) and long short-term memory (LSTM), are applicable to alleviate the problem of gradient disappearance caused by introducing gating units and applying information-locking strategies. Alternatively, extra modules can be used to solve problems in LSTM [40,41].

In the present study, to improve the completeness of the acquired heart rate (HR) and respiration rate (RR) data and the effectiveness of medical monitoring, a vital sign data imputation and end-to-end GAN-based data imputation model (VS-E2E-GAN) is proposed, which relies on the new gated recurrent unit interpolation (GRUI) module to accurately fill the missing values (missing values and outliers to be filled) in vital signs data, so as to improve the integrity and quality of data, thus compensating for the missing and incomplete signals collected by optical fiber sensors. The experimental results obtained on the real dataset demonstrate the effectiveness of the missing value imputation model of vital signs data as proposed in this paper. The main contributions of this study are summarized as follows:

(1) An optical fiber sensor based on a fiber interferometer is proposed to monitor the vital signs effectively.
(2) To deal with the missing values of vital signs data, a novel deep learning model (VS-E2E-GAN), which is based on de-noising autoencoder and GAN, is proposed to extract the distribution features of the vital signs data obtained from the optical fiber sensor.
(3) Multiple experiments are conducted by using three common evaluation metrics to verify how well the model performs, with the experimental results obtained to confirm its better imputation performance.
(4) In combination with the VS-E2E-GAN model, the optical fiber sensor is more effective in non-intrusive physiological monitoring under clinical settings.

The content of this paper is structured as follows. The related works are introduced in Section 2. In Section 3, a discussion is conducted about some methods and the proposed model structure. Section 4 elaborates on different experiments. Furthermore, we also compared the performances between the proposed model and other models. The experimental results are analyzed and researched. In Section 5, the conclusion is drawn and the outlook is outlined.

2. Related Works

In this section, a brief introduction is made to some related works, covering the optical fiber sensor in measurement and time series data processing with some deep learning techniques.

2.1. Optical Fiber Sensing and Measurement

According to the specific application requirements, optical fiber sensors can be categorized into many types, including distributed sensors, bending vector sensors, liquid level sensors, temperature sensors, pressure sensors, and so on. Based on the capability for continuous monitoring over increasing sensing distances, they are further classified as quasi-distributed (QD) optical fiber sensors and point-distributed optical fiber sensors [41]. Additionally, they can also be distinguished as interference and non-interference types based on their susceptibility to interference. There are various advantages to optical fiber sensors due to the unique properties of optical fibers compared to traditional metal wires. These include small size, eco-friendly, safety, lightweight, and reliability; resistance to electromagnetic interference and corrosion; wide applicability across different objects with minimal impact; among others. As a result of these benefits, optical fiber sensors find applications in many fields, such as industry, agriculture, aerospace, petroleum exploration, aviation, medical and health, national defense, and security [42].

Simultaneously, it demonstrates substantial potential in augmenting capabilities for micro-manufacturing, thereby finding wide-ranging applications in biomedical services and the monitoring of vital signs, particularly in the development of minimally invasive surgical instruments. Takeuchi et al. conducted a significant study in 2007 exploring the application of optical fiber sensors in pharyngeal manometry [43]. The study employed a pharynx pressure sensor based on Fabry–Perot interferometry technology (FOP-MIV, FISO technology), which exhibited a strong correlation with conventional catheter-based reference sensors (p37-4109c05, zinatics). The sensor has a compact 2.08 mm diameter conduit, measuring pressures from −30 kPa to 30 kPa at 250 Hz sampling frequency. Furthermore, its resistance to electromagnetic interference enhances its suitability for operation in environments with significant electromagnetic challenges. For instance, microbend optical fiber sensors can effectively operate in the presence of conditions such as nuclear magnetic resonance, enabling the measurement of various vital signs including HR, blood pressure (BP), and respiratory rate (RR), among others.

2.2. Time Series Data Processing

As a group of continuous series data with time tags, time series data usually reflect the laws of development and characteristics of change to things over time. The typical forms of time series data include commodity transaction prices, the temperature in a specific region at a certain time, electrocardiogram (ECG) trends, house price trends, and the solution concentration data in the process of chemical reaction. Time series prediction aims to identify the pattern of changes in the time series data by analyzing the distribution characteristics of historical time series data for quantitative prediction of the possible future values of related variables in the time series.

At the same time, time series data loss is a common problem. In different contexts of data collection, there are different causes for data loss, such as interference with data collection equipment, the population lost in social surveys like population censuses, and the limitations of existing technology like deoxyribonucleic acid (DNA) sequencing. With respect to incomplete data analysis, the missing value imputation method can be used to find a reasonable substitute value for each missing value by exploring the distribution law of the remaining data, so as to obtain a dataset with the same size and dimension as the original dataset. This method has attracted the attention of many researchers as it can not only maintain the size of the original dataset but also make reasonable inferences about the missing values. In the course of scientific research or various industrial applications such as machine learning and data mining, data preprocessing consumes more than 60% of the time and energy for researchers. Therefore, incomplete data processing is the key work to perform. For the analysis and prediction of time series, it is necessary to effectively deal with the missing values of time series data. In general, data loss mechanisms are divided into three categories, and different methods should be adopted for different loss mechanisms. These three mechanisms are detailed as follows [44]:

(1) Missing completely at random (MCAR), which means the data may be lost in some dimensions or completely. These deletions are completely random and irrelevant to any other external factors.
(2) Missing at random (MAR), which means deletion of data is related to known variables only and irrelevant to any unobservable variables.
(3) Missing not at random (MNAR), which means the missing value of the data is related to both observable and unobservable variables if the missing data are neither MCAR nor MAR.

The incompleteness of time series tends to cause deviation in the modeling of time series and even renders the results of subsequent time series data analysis incorrect. Therefore, the primary emphasis in time series research revolves around effectively addressing missing values within the data. Approaches to handling missing data are categorized into two main methods based on their data processing techniques: direct deletion and imputation. Imputation methods, in turn, branch into statistical imputation, machine learning-based imputation, and deep neural network-based imputation techniques.

2.3. Generative Adversarial Network (GAN)

The Generative Adversarial Network (GAN), introduced by Ian et al. in 2014 [45], between the generator and the discriminator is used to capture the distribution characteristics of the original dataset. In recent years, GAN has achieved outstanding performance in the field of computer vision and natural language processing, becoming a cornerstone in generating sequential data. For instance, Mogren et al. employed a C-RNN-GAN model to synthesize sequential melody data, utilizing an enhanced recurrent neural network within the model to accurately replicate musical styles [46]. Similarly, Yoon et al. explored GANs for generating time series data in 2019, focusing primarily on modeling historical trends rather than predicting future data trajectories [47].

The idea of creating a confrontation network is inspired by the "two-person zero-sum game" theory. In a space where the sum of interests is fixed, two people fight for their own benefits. The increase in benefits received by one party must be accompanied by the decrease in benefits received by the other party. The two parties constantly strive for their own benefits through competition. The whole game process can be described as that with one side's strategy determined, the other side will make the decision that is most favorable to its own side under the current conditions [48]. If both sides of the game choose to be optimal when the other side's strategy is determined, this group of strategic choices will reach Nash equilibrium. The generator (G) and discriminator (D) in the generated confrontation network represent the two sides of the game. Fighting for their own interests through the game, they learn and progress together in the game process [49].

In practice, the mutual objective of the generator and discriminator is quantified by the probability value outputted by the discriminator, which signifies the likelihood of the discriminator classifying the current input as real data. In the network training process, the generator will be interfered by the false but new samples similar to the real sample data, which will be interfered with by the discriminator by constantly generating. When these false samples are taken as the input, the discriminator should protect its own interests by outputting a probability value that is as low as possible. In the process of confrontation training between the two, the generator constantly improves its ability to generate false samples, and the discriminator constantly improves its ability to identify them. Finally, network training is completed when the "Nash equilibrium" similar to that in game theory is reached. There are no restrictions on the selection of the internal network structure of the generator and discriminator. A fully connected network (FCN) and cyclic neural network can all be treated as a generator and a discriminator, respectively.

3. Materials and Methods

This section proposes an optical fiber sensor and vital sign interpolation model. Specifically, the fiber interferometer-based optical sensor is developed to monitor vital signs. Furthermore, we construct a new model of deep learning (VS-E2E-GAN) in order to fill in the missing values in vital sign data.

3.1. Proposed Optical Fiber Sensor

In this study, ballistocardiogram (BCG) monitoring is performed non-invasively using a smart cushion. The monitoring system consists of a proportional–integral–derivative (PID) controller, a phase shifter, as well as a Mach–Zehnder interferometer (MZI) BCG monitor. The system for monitoring consists of a low-speed photodetector (PD), a DFB laser, a phase shifter, as well as the MZI-based BCG monitor. Comprising two 3 dB couplers that operate as optical splitters as well as couplers, correspondingly, the MZI is positioned atop a plastic substrate to enable smart cushion packing. The MZI along with the phase shifter have been parallelly fixed as well as located beyond the sensing area. The BCG signal is transformed by the PD into an electrical signal that splits between the CH1 and CH2 channels. The low-pass filter (LPF) processed data as an input for the PID controller is included in CH2, whereas CH1 represents the raw data. To keep the system running at a quadrature point, the controller adjusts for phase drift. A data acquisition (DAQ) card (National Instrument, USB6001) records the unprocessed information from CH1.

The two arms of the MZI design employed in this investigation are roughly 40 cm long and varied by 5 mm. Two bent arms that are placed together within a semicircle shape as well as are not overlapping make up the integrated cushion made of plastic substrate. When a subject sits upon the cushion, their body recoils from heartbeats, which causes a variation in phase within the interferometer. Via the degree of variability, this phase variation may be utilized to identify the BCG signal. Figure 1 shows the monitoring system that makes use of the optical fiber sensor. Equation (1) represents the system's optical fiber interferometer's mean output light intensity, where D is the mean output intensity of light, the interference fringes peak intensity is denoted by I_0, the total amount of output light routes is denoted by k, the variation in the phase of the signal generated by the sensor is represented by $\varphi(t)$, as well as the variation in phase brought on by shifts in the environment is indicated by $\psi(t)$.

$$I_k = D + I_0 cos[\varphi(t) - (k-1)(2\pi/3)] \tag{1}$$

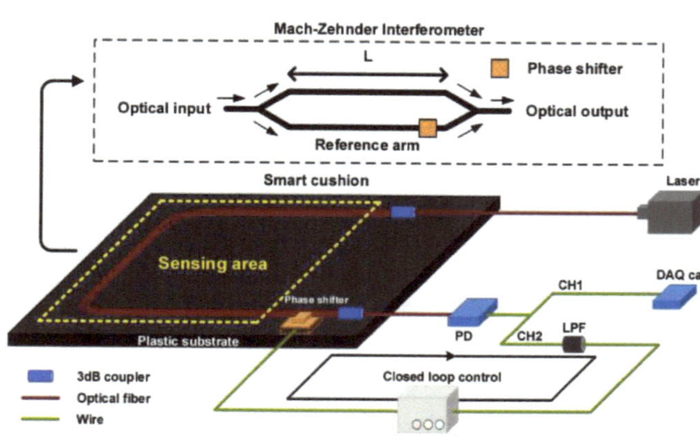

Figure 1. The proposed optical fiber sensor-based monitoring system.

3.2. Data Acquisition and Processing

As an important vital sign, RR reflects the physiological function of the lung, oxygen input, and carbon dioxide output. When abnormal RR is detected, it may be associated with many symptoms such as asthma and anemia [8]. At present, it is widely used to evaluate the physical and mental conditions of the human body, and the measurement of this parameter is the focus of vital signs detection. HR is detected by examining the characteristics of the BCG signal. The body will recoil when the heart pumps blood into the vascular system. Such pressure changes and microchanges in the human body can be detected simultaneously by using fiber-optic sensors embedded in the mattress for non-invasive detection. The original signal filtered respiration signal and filtered BCG signal as obtained by the fiber optic sensor are shown in Figure 2, in which HR and RR are shown as typical time series signals.

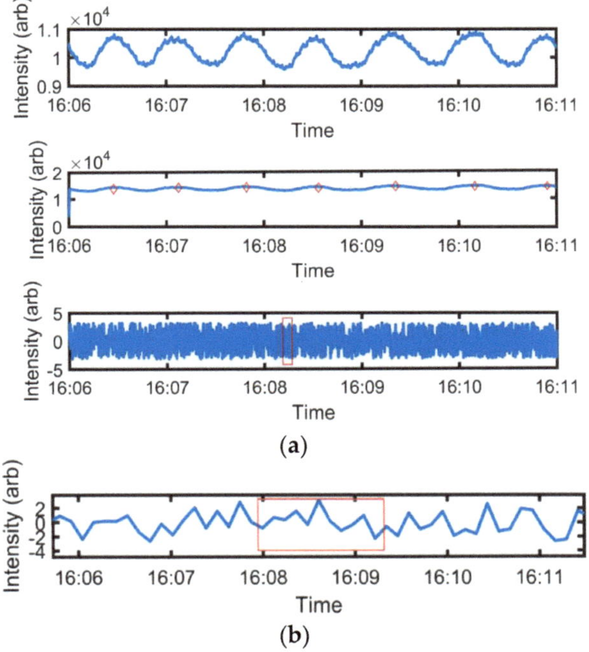

Figure 2. (**a**) The extracted heartbeat signal, respiration signal, and acquired BCG signal by the optical fiber sensor. (**b**) Detailed information of the original BCG signal framed by a red box of (**a**).

To extract HR and RR from the raw BCG signal, the RR is calculated as $(60 \times 6/12) = 30$ rpm when the raw BCG signal exhibits 6 distinct waveforms within a 12 s interval. Utilizing time-domain data and the Fast Fourier Transform (FFT) algorithm facilitates noise mitigation in the raw BCG signal through linear trend removal and filtering. Consequently, if the filtered signal exhibits 9 clear waveforms over a 6 s span, the HR can be determined as 90 bpm.

3.3. Proposed Model (VS-E2E-GAN)

In practice, it is common that sensor failure or body turnover of the subject makes the collection of vital sign signals incomplete and causes the loss of some key values. There are two main problems with the current methods of time series data missing value imputation. On the one hand, most data-imputation methods ignore the particularity of time series. On the other hand, a complete dataset is required by the time series data imputation method based on a recurrent neural network. In this paper, the VS-E2E-GAN model is proposed which can directly interpolate missing time series data. The model relies on a bidirectional gated recurrent neural network to model the time series and fully learn the implicit information contained in the remaining data. Vital signs data are usually collected continuously by sensors and show significant temporality, while the time series collected by the sensor can be lost due to the failure or interference of the sensor. Therefore, it is necessary to prevent the missing part from affecting the subsequent data analysis by processing the missing time series. Based on the observed characteristics of missing time series data, the idea of auto-encoder is introduced to propose a filled recurrent neural network model that can be applied to process time series data. However, the model requires complete time series data during training, so that it is used as a generator network to construct a GAN-based time series missing value imputation model called VS-E2E-GAN. The model starts by learning the distribution law of time series on the missing dataset. After the completion of network training, the new time series data conforming to the original data distribution are generated. Also, the network-generated time series data are used to impute missing values. The overall model structure is shown in Figure 3.

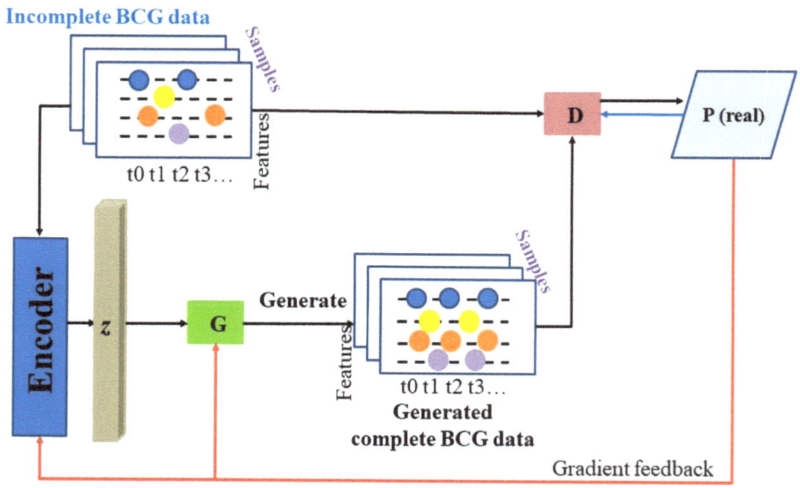

Figure 3. Architecture of the VS-E2E-GAN model.

For a given d-dimensional time series containing vacancies ($x = [x_0, x_1, \ldots, x_{n-1}]$) and corresponding time nodes ($t = [t_0, t_1, \ldots, t_{n-1}]$), the vacancies are filled in. Firstly, it is necessary to obtain the specific position of the vacancy value in the time series. To achieve this purpose, and to define the position matrix of the vacancy value (S), the 0~1 variable is used to determine whether there is any vacancy in the observation value data at the time node. For a better understanding, a three-dimensional time series with null values is assigned, with each row representing the observed value of each feature quantity at the time of 0, 1, 2, and 3, as shown in Equation (2):

$$\begin{cases} x = \begin{bmatrix} 1,5,None,8 \\ None,7,6,None \\ 4,None,None,2 \end{bmatrix} \\ S = \begin{bmatrix} 1,1,0,1 \\ 0,1,1,0 \\ 1,0,0,1 \end{bmatrix} \end{cases} \tag{2}$$

where *None* represents the data vacancy value. It can be clearly seen from above that the time delay between two adjacent effective observations changes due to the existence of *None*, which also shows the distribution characteristics between vacancy values. At the same time, different time delays have different effects on subsequent observations. If an observation is missing for some time, its effect is supposed to diminish over time. To record the time delay between two adjacent valid observations, a time delay matrix δ is introduced to describe the change in time delay. Its elements are expressed as follows:

$$\delta_{t_{i,j}} = \begin{cases} t_i - t_{i-1}, (s_{t_{i-1,j}} = 1, i > 0) \\ \delta_{t_{i-1,j}} + t_i - t_{i-1}, (s_{t_{i-1,j}} = 0, i > 0) \\ 0(i = 0) \end{cases} \tag{3}$$

The delay matrix corresponding to the aforementioned three-dimensional time series with vacancies is as follows:

$$\delta = \begin{bmatrix} 0,1,2,1 \\ 0,1,1,2 \\ 0,1,2,1 \end{bmatrix} \tag{4}$$

Since most GAN frameworks are intended mainly for image processing, they are not adaptive to time series imputation. Differently, RNN is a model applicable for processing time series, but the traditional RNN is prone to gradient explosion and gradient disappearance when a longer time series is processed. In recent years, many of its variants have emerged, such as LSTM and GRU, all of which are effective in solving this problem. To simplify the construction of GAN, it is proposed in this paper to construct the basic network structure of the generator and discriminator by adopting the GRU with higher computational efficiency and scalability. Due to the incompleteness of data and the significant variation in the time lag between two consecutive valid observations, the traditional GRU structure is not applicable to data imputation. Therefore, a GRU-based GRUI is proposed to deal with such irregular time delays more efficiently and learn implicit information from the time delays; Figure 4 shows the basic structure of GRUI.

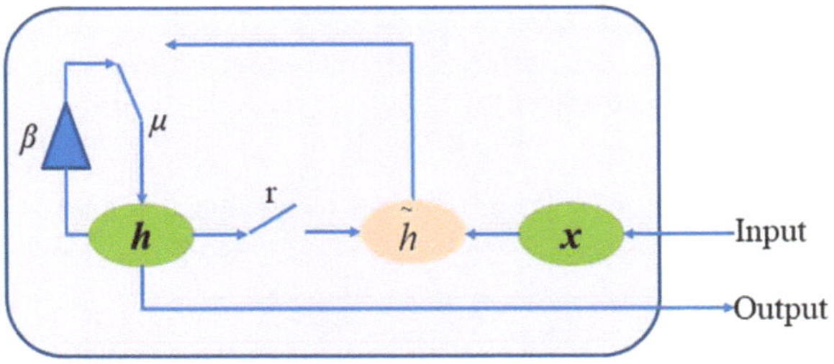

Figure 4. Basic structure of the GRUI module.

In order to reflect the impact on the observed value, the time decay vector β is introduced. The size of β is supposedly related to the delay matrix δ. If δ is large, β becomes smaller. The elements in β are expressed as follows:

$$\beta_{t_i} = \frac{1}{e^{max(0, W_\beta \delta_{t_i} + b_\beta)}} \tag{5}$$

$$\delta_{t_i} = (\delta_{t_i,j})_{1\times n} \tag{6}$$

where W_β and b_β are the parameters that need to be learned.

To ensure $\beta \in (0, 1]$, a negative exponential function is used as a constraint. Meanwhile, W_β is introduced as a full-weight matrix for the interaction of the δ matrix. After the attenuation vector is obtained, the attenuation vector β is multiplied by elements to update the hidden state of the GRU. Since data preprocessing is often conducted through the normalization method, the hidden state is updated in this study by multiplying it with the attenuation vector. Therefore, the value of the updated hidden state is not overly small. The formula used for GRUI is expressed as follows:

$$h'_{t_{i-1}} = \beta_{t_i} \odot h_{t_{i-1}} \tag{7}$$

$$\mu_{t_i} = \sigma\left(W_\mu \left[h'_{t_{i-1}}, x_{t_i}\right] + b_\mu\right) \tag{8}$$

$$r_{t_i} = \sigma\left(W_r \left[h'_{t_{i-1}}, x_{t_i}\right] + b_r\right) \tag{9}$$

$$\widetilde{h}_{t_i} = tanh(W_{\widetilde{h}} \left[r_{t_i} \odot h'_{t_{i-1}}, x_{t_i}\right] + b_{\widetilde{h}}) \tag{10}$$

$$h_{t_i} = (1 - u_{t_i}) \odot h'_{t_{i-1}} + u_{t_i} \odot \widetilde{h}_{t_i} \tag{11}$$

Since the auto-encoder is designed to reconstruct the target samples, the auto-encoder and GRUI structure in the generator are adopted in this paper. To make the auto-encoder learn more useful information, this paper adopts the idea of denoising auto-encoders, removing some original samples (destroying the original samples), and reconstructing the original samples. However, considering the small amount of data available, removing samples will further reduce the amount of data, which hinders the self-encoder from learning more information. Therefore, it is proposed in this paper to introduce the random noise η sampled from the normal distribution N (0, 0.01) into the original sample for destroying the original sample. The generator function $G(\cdot)$ is expressed as follows:

$$G(x + \eta) = x' \tag{12}$$

where x represents the real data and x' refers to the fake data.

To make the generator generate the complete time series x', which is the closest to x, the squared error loss function is incorporated into the loss function of the generator. The loss function $L_2(x)$ of the generator is expressed as follows:

$$L_2(x) = \|x \odot S - G(x + \eta) \odot S\|_2 \tag{13}$$

$$L_G = \lambda L_2(x) - D(x') \tag{14}$$

where λ represents a hyper-parameter, which is used to control the weight of the discriminator loss and variance loss, and $D(\cdot)$ refers to a function of the discriminator.

Functioning as a decoder, the network structure of the discriminator consists of a completed connected layer and a GRUI layer. The discriminator is ordered to distinguish the forged data x' from the real data x, and its output indicates the probability that the generated x' is real data. Therefore, it is necessary to find a series of parameters such that a higher probability is outputted when real data x is inputted and a lower probability is outputted when false data x' is inputted. In order to achieve this purpose, the loss function L_D of the discriminator can be defined as follows:

$$L_D = -D(x) + D(x') \tag{15}$$

Both the discriminator and the generator are trained against the min–max game according to Equation (14), which ends up leading to the equilibrium.

$$\underset{G}{min}\underset{D}{max}V(D, G) = E_{x \sim P_x(x)}(D(x)) - E_{x' \sim P_{x'}(x')}(D(x')) \tag{16}$$

where $E(\cdot)$ represents the mathematical expectation; $P_x(x)$ and $P_x{'}(x')$ denote the probability distributions of the real data x and the fake data x', respectively.

In this paper, the following methods of data quality improvement are proposed. Firstly, the -1 value is used to replace the vacancy value in the original data to obtain the marked sequence x of the vacancy value, while the corresponding position matrix S and delay matrix δ are constructed according to the distribution of the -1 value. Secondly, in the data-loading stage, the data are constructed into a three-dimensional array in various forms (batch number of samples, sequence length, sequence feature number) and inputted into the GRUI-GAN model after zero-mean normalization. Thirdly, in the generator, the sequence x and its time-delay matrix δ that are labeled and added with noise η are inputted into the cells of GRUI. After processing by the GRUI layer, a hidden state corresponding to the input sequence is generated. The state is further fed into a fully connected layer to generate a low-dimensional vector z. Then, z is sent to the next GRUI layer for processing through another fully connected layer, with all the outputs of this GRUI layer combined into a new full sequence x'. Fourthly, in the discriminator, the incomplete sequence x or the complete sequence x' and its corresponding delay matrix δ are also effectively processed by the GRUI. The last hidden state of the GRUI layer in the discriminator is fed into a fully connected layer, whose output is the probability of being discriminated as real data. Lastly, after multiple rounds of adversarial training between the discriminator and the generator, the generator can be used to get a new sequence x' that is closest to x. Then, x' is used to fill in the vacancies in x. The equation is expressed as follows:

$$x_{im} = x \odot S + (1 - S) \odot x' \tag{17}$$

where x_{im} is the x obtained after the imputation.

The imputation framework proposed in this thesis is shown in Figure 5. In order to improve the performance of the generator in fitting the original sequence, the generator is updated N times, and the discriminator is updated once in one iteration in the GAN proposed in this paper.

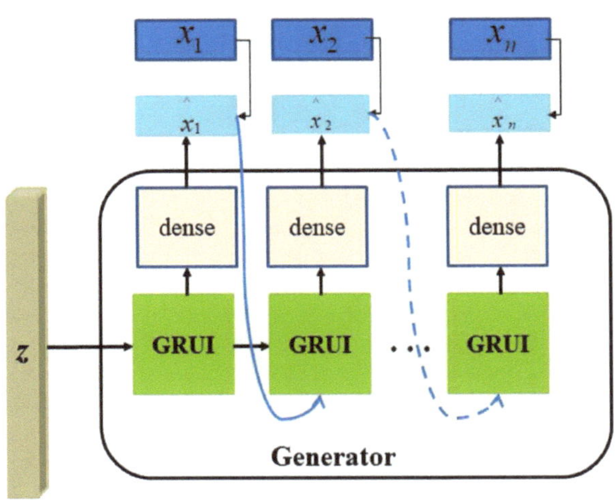

Figure 5. Basic structure of the generator.

4. Experiments and Results

This section presents a series of experiments designed to validate the VS-E2E-GAN model. It includes detailed descriptions of experimental parameters and various evaluation metrics employed. In addition, baseline methods are utilized to assess the effectiveness and performance of the proposed model.

4.1. Experiment Setup

During the experiment, the ADAM optimizer was employed with a batch size of 10 and 50 iterations. The experiments were conducted on a computational platform featuring the high-performance NVIDIA GeForce GTX1080Ti graphics card and the formidable Intel i7-8700 CPU. The model was implemented utilizing the PyTorch framework and the Python programming language.

The performance of the proposed model is assessed through the area under the ROC curve (AUC), metrics including mean squared error (MSE), and accuracy (ACC). The MSE can be defined as follows:

$$MSE = \frac{1}{n}\sum_{i=1}^{n}(\hat{y}_i - y_i)^2 \tag{13}$$

where y_i denotes the observed value and $\hat{y}_i$ signifies the predicted value.

Additionally, the proposed method is comparatively verified against several vacancy value imputation methods commonly used in traditional load forecasting (mean value imputation, piecewise linear interpolation, and KNN imputation). Given the incomplete dataset, it is impossible to directly evaluate the gap imputation accuracy of different algorithms. Also, it is inappropriate to train and test GAN models for experiments in this paper by dividing training and testing sets. Therefore, all data are trained in the proposed VS-E2E-GAN model. The training set is the entire dataset. As for the imputation performance of the VS-E2E-GAN model on the dataset after training on the entire dataset, it can be evaluated only by the subsequent prediction results. Therefore, to validate the proposed method, the experiment conducted in this paper involves the following two parts:

(1) Part of the complete data is selected from the dataset (not including missing values and not considering outliers), and the incomplete data with different proportions of random missing data for experiments are constructed to verify the imputation performance of the model trained with a complete dataset on small datasets.

(2) By using the above experimental methods, the complete incomplete dataset is obtained again, and the different complete datasets obtained by different methods are used for subsequent HR (RR) prediction. The final HR (RR) prediction results are used to indirectly verify the proposed imputation method.

4.2. Results and Analysis

The experimental results are as follows. Firstly, among all the classification results obtained by the simple binary classifier, the data filled by the proposed method lead to the best classification performance, which indicates that this method achieves the best results. Secondly, the classifier based on the simple recurrent neural network produces the best outcome of classification on the dataset filled by various methods, which shows that the recurrent neural network outperforms other networks in learning the distribution characteristics of time series data, reflecting the advantages of recurrent neural network in time series modeling. Thirdly, the GAN-based method performs better than other methods, indicating that it is achievable to produce better results by training the data filled by the GAN missing value imputation method. Furthermore, the effect of epoch and learning rate (lr) on the model performance is verified. As shown in Figure 6, the model performs best when the epoch is 30 and the lr is 0.012.

As for neural networks, the training time of neural networks is affected by the selection of software and hardware configuration, dataset size, training rounds, network layers, and other factors. For this reason, it is practically difficult to evaluate the training time. In the comparative experiment, the time efficiency of various methods was compared by fixing the hardware and software parameters, dataset size, training set, cross-validation method, test set, and other factors. The experimental results are shown in Table 1. We can see clearly from the table that the method (VS-E2E-GAN) proposed in this paper is applicable to achieve optimal imputation accuracy, because the network model is more complicated and the time cost is higher. In some tasks requiring high accuracy (such as vital signs signal imputation), the proposed method is much more effective.

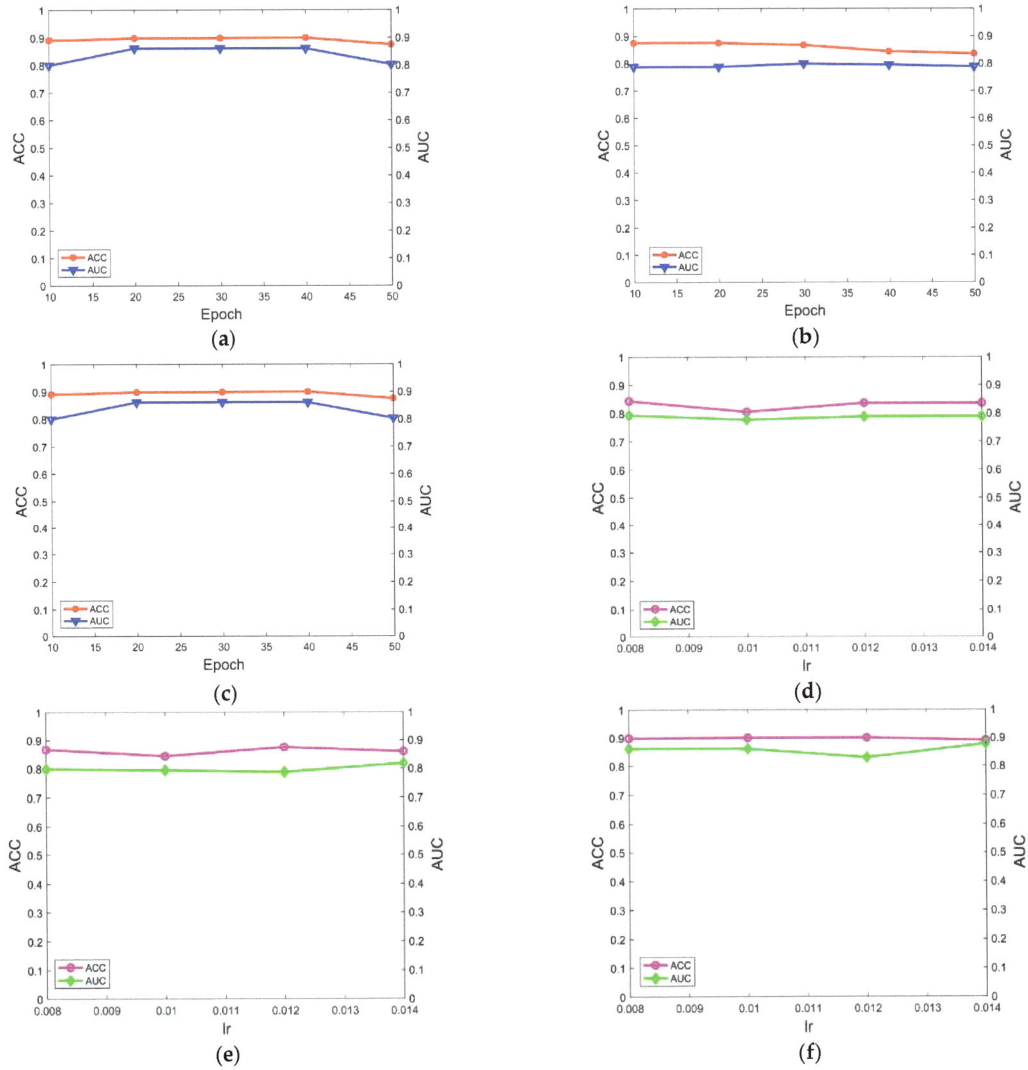

Figure 6. The model performance between different epochs and learning rates. (**a**–**c**) ACC and AUC performance with different epochs and (**d**–**f**) ACC and AUC performances with different learning rates.

Table 1. The average results of different methods.

Method	MSE	AUC	ACC	Time (min)
C-RNN-GAN [46]	0.801	0.833	0.840	17
KNN	0.835	0.799	0.832	19
TGAN [47]	0.924	0.707	0.802	18
GAN [48]	0.811	0.830	0.838	17
DU-GAN [49]	0.790	0.868	0.885	17
VS-E2E-GAN	**0.777**	**0.901**	**0.908**	**10**

5. Discussion

As information science and technology advance, computer technology has played an irreplaceable role in every aspect of life. With the popularization of global information, the Internet has given rise to a large amount of data around the world, which makes information highly valuable. However, these data suffer quality problems due to various reasons, of which the incompleteness of data is a major one. Because of a large amount of missing data, it is inevitable for incomplete data to have a negative impact on data mining or other technical applications.

There is plenty of hidden information contained in a massive amount of data, which requires the corresponding techniques for exploration, such as data mining technology. However, most data mining algorithms are incapable of processing incomplete data because the data are often incomplete in practice. Moreover, incomplete data also affect the quality of the final results. Based on optical fiber sensors, vital signs measurement is an important research topic, which is essential for the noninvasive monitoring of human health. However, in the measurement process, the measured data are often lost due to various reasons. In this paper, the imputation of incomplete data is studied for analysis, including data loss. By analyzing the particularity of time series data, a countermeasure network and a de-noising self-encoder are created. On this basis, an imputation model for the missing values of time series data (VS-E2E-GAN) is constructed to directly fill the missing dataset. We use a bidirectional gated recurrent neural network to capture the dependencies of time series data from the past to the future and from the future to the past and to fully learn the distribution patterns of time series data. Furthermore, a new generator loss function is proposed to make the model perform better in accomplishing the missing value imputation task of the time series. In case of a significant sample missing rate, the method can still get a better achievement in imputation. In addition, multiple experiments are conducted to compare them with other methods, which demonstrates that this method can achieve better results.

Vital signs signals are highly valuable for maintaining physical health. Despite some remarkable results achieved by the time series prediction model based on GAN and the time series data imputation model based on GAN as proposed in this paper, there are still some shortcomings. Therefore, improvement can be made in the future from the following perspectives

(1) The time series prediction model proposed in this paper only considers the accuracy of the design of the loss function. In the future, the loss function can be designed by combining the downstream tasks to deal with specific problems.
(2) The time series data imputation model proposed in this paper incurs substantial time costs. In the future, it is worthwhile to consider simplifying the network structure, such as model distillation.
(3) In this paper, the accuracy of the model is verified by using the real vital signs data collected by optical fiber sensors. In the future, new data can be used to validate the proposed model.

6. Conclusions

In this paper, time series data are extensively distributed in the real world, of which heart rate and respiration signals are two typical time series signals. The analysis and accurate prediction of these time series data play an important impact in production and life for people. Therefore, establishing an effective time-series signal processing model is necessary. Meanwhile, missing time series is ubiquitous, and these missing data are related to the mining of time series data. That is the reason why an urgent need to reasonably deal with the loss of time series data. In this paper, a VS-E2E-GAN-based load data imputation method is proposed, and the GRUI structure is built by introducing a time decay vector into the traditional GRU structure to understand the impact of different time intervals on the measurement results. After adversarial training, the GAN of this structure is capable to process incomplete time series data and generate higher-precision data for interpolation. In this way, the quality of data can be improved and the lack of optical fiber sensors can be solved to capture vital sign signals, which is a convenient and efficient solution to medical monitoring.

Author Contributions: Writing—original draft preparation, H.G.; writing—review and editing, Q.W.; validation, J.Z.; supervision and funding acquisition, C.Y. All authors have read and agreed to the published version of the manuscript.

Funding: The authors gratefully acknowledge the financial support of the non-wearable and noninvasive photonic sleep monitoring system-based optical fiber sensor with machine learning (HKPU 1-WZ01) and non-wearable and non-invasive photonic smart health monitoring system for atrial fibrillation diagnosis based on optical fiber sensor with machine learning (HKPU 1-CD8N).

Data Availability Statement: The original contributions presented in the study are included in the article; further inquiries can be directed to the corresponding authors.

Conflicts of Interest: The authors declare no conflicts of interest.

References

1. Zhang, J.; Jiang, W.; Zhou, J.; Zhao, X.; Huang, X.; Yu, Z.; Yi, X.; Qiu, K. An iterative BP-CNN decoder for optical fiber communication systems. *Opt. Lett.* **2023**, *48*, 2289–2292. [CrossRef] [PubMed]
2. Zhu, K.N.; Liu, S.; Wei, S.; Li, Y.; Zhao, Y.; Li, Y.; Wang, W.; Zhang, J. Physical layer secure key generation and distribution based on noise variances in optical fiber communications systems. *Opt. Laser Technol.* **2023**, *165*, 109576. [CrossRef]
3. Wang, F.; Gao, R.; Li, Z.; Liu, J.; Cui, Y.; Xu, Q.; Pan, X.; Zhu, L.; Wang, F.; Guo, D.; et al. 400 Gbit/s 4 mode transmission for IM/DD OAM mode division multiplexing optical fiber communication with a few-shot learning-based AffinityNet nonlinear equalizer. *Opt. Express.* **2023**, *31*, 22622–22634. [CrossRef] [PubMed]
4. Nagarajan, R.; Lyubomirsky, I.; Agazzi, O. Low power DSP-based transceivers for data center optical fiber communications (Invited Tutorial). *J. Light. Technol.* **2021**, *39*, 5221–5231. [CrossRef]
5. Chen, B.; Alvarado, A.; van der Heide, S.; van den Hout, M.; Hafermann, H.; Okonkwo, C. Analysis and experimental demonstration of orthant-symmetric four-dimensional 7 bit/4D-sym modulation for optical fiber communication. *J. Light. Technol.* **2021**, *39*, 2737–2753. [CrossRef]
6. Kim, C.S.; Ober, S.L.; McMurtry, M.S.; Finegan, B.A.; Inan, O.T.; Mukkamala, R.; Hahn, J.O. Ballistocardiogram: Mechanism and potential for unobtrusive cardiovascular health monitoring. *Sci. Rep.* **2016**, *6*, 31297. [CrossRef]
7. Wen, X.; Huang, Y.; Wu, X.; Zhang, B. A feasible feature extraction method for atrial fibrillation detection from BCG. *IEEE J. Biomed. Health Inform.* **2019**, *24*, 1093–1103. [CrossRef] [PubMed]
8. Wang, Q.; Zhang, Y.; Chen, G.; Chen, Z.; Hee, H.I. Assessment of heart rate and respiratory rate for perioperative infants based on ELC model. *IEEE Sens. J.* **2021**, *21*, 13685–13694. [CrossRef]
9. Chen, Z.; Hu, J.; Yu, C. Fiber senor for long-range and biomedical measurements. In Proceedings of the 2013 12th International Conference on Optical Communications and Networks (ICOCN), Chengdu, China, 26–28 July 2013.
10. Yu, C.; Xu, W.; Zhang, N.; Yu, C. Non-invasive smart health monitoring system based on optical fiber interferometers. In Proceedings of the 2017 16th International Conference on Optical Communications and Networks (ICOCN), Wuzhen, China, 7–10 August 2017.
11. Wang, Q.; Lyu, W.; Cheng, Z.; Yu, C. Noninvasive measurement of vital signs with the optical fiber sensor based on deep learning. *J. Light. Technol.* **2023**, *41*, 4452–4462. [CrossRef]
12. Chen, W.; Zhang, Y.; Yang, H.; Qiu, Y.; Li, H.; Chen, Z.; Yu, C. Non-invasive measurement of vital signs based on seven-core fiber interferometer. *IEEE Sens. J.* **2021**, *21*, 10703–10710. [CrossRef]
13. Wang, S.; Ni, X.; Li, L.; Wang, J.; Liu, Q.; Yan, Z.; Zhang, L.; Sun, Q. Noninvasive monitoring of vital signs based on highly sensitive fiber optic mattress. *IEEE Sens. J.* **2020**, *20*, 6182–6190. [CrossRef]
14. Lyu, W.; Xu, W.; Yang, F.; Chen, S.; Tan, F.; Yu, C. Non-invasive measurement for cardiac variations using a fiber optic sensor. *IEEE Photonics Technol. Lett.* **2021**, *33*, 990–993. [CrossRef]
15. Chen, S.; Tan, F.; Lyu, W.; Yu, C. Ballistocardiography monitoring system based on optical fiber interferometer aided with heartbeat segmentation algorithm. *Biomed. Opt. Express* **2020**, *11*, 5458–5469. [CrossRef] [PubMed]
16. Yang, F.; Lyu, W.; Pan, C.; Yang, S.; Tan, F.; Chen, S.; Yu, C. Contactless vital signs monitoring based on optical fiber Mach-Zehnder interferometer aided with passive homodyne demodulation methods. In Proceedings of the Asia Communications and Photonics Conference/International Conference on Information Photonics and Optical Communications 2020 (ACP/IPOC), Beijing, China, 24–27 October 2020.
17. Wang, Y.; You, M.; Zhang, Y.; Wu, S.; Zhang, Y.; Yang, H.; Zheng, T.; Chen, X.; Chen, Z.; Xie, X.; et al. Noninvasive measurement of the vital signs of cancer patients with a dual-path microbend fiber sensor. *Biomed. Opt. Express* **2022**, *13*, 982–994. [CrossRef] [PubMed]
18. Tahir, S.; Sadek, I.; Abdulrazak, B. A CNN-ELM-Based Method for Ballistocardiogram Classification in a Clinical Environment. In Proceedings of the 2021 IEEE Canadian Conference on Electrical and Computer Engineering (CCECE), Virtual Conference, 12–17 September 2021; pp. 1–6. [CrossRef]

19. Wang, Q.; Lyu, W.; Zhou, J.; Yu, C. Sleep. Condition detection and assessment with optical fiber interferometer based on machine learning. *iScience* **2023**, *26*, 107244. [CrossRef] [PubMed]
20. Zhu, C.; Chan, E.A.; Wang, Y.; Peng, W.; Guo, R.; Zhang, B.; Soci, C.; Chong, Y. Image reconstruction through a multimode fiber with a simple neural network architecture. *Sci. Rep.* **2021**, *11*, 896. [CrossRef] [PubMed]
21. Borhani, N.; Kakkava, E.; Moser, C.; Psaltis, D. Learning to see through multimode fibers. *Optica* **2018**, *5*, 960–966. [CrossRef]
22. Bahrami, M.; Forouzanfar, M. Sleep apnea detection from single-lead ECG: A comprehensive analysis of machine learning and deep learning algorithms. *IEEE Trans. Instrum. Meas.* **2022**, *71*, 4003011. [CrossRef]
23. Wang, K.; Xiao, L.; Yi, W.; Ran, S.J.; Xue, P. Experimental realization of a quantum image classifier via tensor-network-based machine learning. *Photonics Res.* **2021**, *9*, 2332–2340. [CrossRef]
24. Yan, Q.; Deng, Q.; Zhang, J.; Zhu, Y.; Yin, K.; Li, T.; Wu, D.; Jiang, T. Low-latency deep-reinforcement learning algorithm for ultrafast fiber lasers. *Photonics Res.* **2021**, *9*, 1493–1501. [CrossRef]
25. von Rueden, L.; Mayer, S.; Beckh, K.; Georgiev, B.; Giesselbach, S.; Heese, R. Informed Machine Learning—A Taxonomy and Survey of Integrating Knowledge into Learning Systems. *arXiv* **2019**, arXiv:1903.12394. [CrossRef]
26. Mohr, F.; Wever, M.; Tornede, A.; Hüllermeier, E. Predicting machine learning pipeline runtimes in the context of automated machine learning. *IEEE Trans. Pattern Anal. Mach. Intell.* **2021**, *43*, 3055–3066. [CrossRef]
27. Yang, Z.; Zhang, T.; Bozchalooi, I.S.; Darve, E. Memory-Augmented Generative Adversarial Networks for Anomaly Detection. *IEEE Trans. Neural Netw. Learn. Syst.* **2021**, *33*, 2324–2334. [CrossRef]
28. Guo, Z.; Song, J.K.; Barbastathis, G.; Glinsky, M.E.; Vaughan, C.T.; Larson, K.W.; Alpert, B.K.; Levine, Z.H. Physics-assisted generative adversarial network for X-ray tomography. *Opt. Express* **2022**, *30*, 23238–23259. [CrossRef]
29. Wang, Q.; Liu, W.; Chen, X.; Wang, X.; Chen, G.; Zhu, X. Quantification of scar collagen texture and prediction of scar development via second harmonic generation images and a generative adversarial network. *Biomed. Opt. Express* **2021**, *12*, 5305–5319. [CrossRef]
30. Hou, B.; Yan, R. Triplet-Classifier GAN for Finger-Vein Verification. *IEEE Trans. Instrum. Meas.* **2022**, *71*, 2505112. [CrossRef]
31. Wang, Q.; Liu, W.; Wang, X.; Chen, X.; Chen, G.; Wu, Q. A Spatial-Temporal Graph Model for Pronunciation Feature Prediction of Chinese Poetry. *IEEE Trans. Neural Netw. Learn. Syst.* **2022**, *34*, 10294–10308. [CrossRef]
32. An, Y.; Huang, L.; Li, J.; Leng, J.; Yang, L.; Zhou, P. Learning to decompose the modes in few-mode fibers with deep convolutional neural network. *Opt. Express* **2019**, *27*, 10127–10137. [CrossRef]
33. Rothe, S.; Zhang, Q.; Koukourakis, N.; Czarske, J. Intensity-only mode decomposition on multimode fibers using a densely connected convolutional network. *J. Light. Technol.* **2021**, *39*, 1672–1679. [CrossRef]
34. Li, H.; Chen, C.-H. Effects of Affordance State and Operation Mode on a Smart Washing Machine Touch Sensitive User Interface Design. *IEEE Sens. J.* **2021**, *21*, 21956–21967. [CrossRef]
35. Isola, P.; Zhu, J.Y.; Zhou, T.; Efros, A.A. Image-to-image translation with conditional adversarial networks. In Proceedings of the IEEE Conference on Computer Vision and Pattern Recognition, Honolulu, HI, USA, 21–26 July 2017.
36. Wang, Q.; Lyu, W.; Chen, S.; Yu, C. Non-Invasive Human Ballistocardiography Assessment Based on Deep Learning. *IEEE Sens. J.* **2023**, *23*, 13702–13710. [CrossRef]
37. Mirza, M.; Osindero, S. Conditional generative adversarial nets. *arXiv* **2014**, arXiv:1411.1784.
38. Xu, B.; Wang, N.; Chen, T.; Li, M. Empirical evaluation of rectified activations in convolutional network. *arXiv* **2015**, arXiv:1505.00853.
39. Venugopalan, J.; Chanani, N.; Maher, K.; Wang, M.D. Novel data imputation for multiple types of missing data in intensive care units. *IEEE J. Biomed. Health Inform.* **2019**, *23*, 1243–1250. [CrossRef]
40. Chen, C.; Lu, N.; Jiang, B.; Xing, Y.; Zhu, Z.H. Prediction interval estimation of aeroengine remaining useful life based on bidirectional long short-term memory network. *IEEE Trans. Instrum. Meas.* **2021**, *70*, 3527213. [CrossRef]
41. Bucaro, J.A.; Dardy, H.D.; Carome, E.F. Optical fiber acoustic sensor. *Appl. Opt.* **1977**, *16*, 1761–1762. [CrossRef]
42. Reja, M.I.; Nguyen, L.V.; Peng, L.; Ebendorff-Heidepriem, H.; Warren-Smith, S.C. Temperature-Compensated Interferometric High-Temperature Pressure Sensor Using a Pure Silica Microstructured Optical Fiber. *IEEE Trans. Instrum. Meas.* **2022**, *71*, 7002612. [CrossRef]
43. Takeuchi, S.; Tohara, H.; Kudo, H.; Otsuka, K.; Saito, H.; Uematsu, H.; Mitsubayashi, K. An optic pharyngeal manometric sensor for deglutition analysis. *Biomed. Microdevices* **2007**, *9*, 893–899. [CrossRef]
44. Miao, X.; Wu, Y.; Wang, J.; Gao, Y.; Mao, X.; Yin, J. Generative semi-supervised learning for multivariate time series imputation. *Proc. AAAI Conf. Artif. Intell.* **2021**, *35*, 8983–8991. [CrossRef]
45. Goodfellow, I.; Pouget-Abadie, J.; Mirza, M.; Xu, B.; Warde-Farley, D.; Ozair, S.; Courville, A.; Bengio, Y. Generative adversarial networks. *Commun. ACM* **2020**, *63*, 139–144. [CrossRef]
46. Mogren, O. C-RNN-GAN: Continuous recurrent neural networks with adversarial training. *arXiv* **2016**, arXiv:1611.09904.
47. Yoon, J.; Jarrett, D.; Van der Schaar, M. Time-series generative adversarial networks. *Adv. Neural Inf. Process. Syst.* **2019**, *32*, 5508–5518.

48. He, W.; He, Y.; Li, B. Generative adversarial networks with comprehensive wavelet feature for fault diagnosis of analog circuits. *IEEE Trans. Instrum. Meas.* **2020**, *69*, 6640–6650. [CrossRef]
49. Huang, Z.; Zhang, J.; Zhang, Y.; Shan, H. DU-GAN: Generative adversarial networks with dual-domain U-Net-based discriminators for low-dose CT denoising. *IEEE Trans. Instrum. Meas.* **2021**, *71*, 4500512. [CrossRef]

photonics

MDPI

Article

Highly Sensitive D-SPR Sensors with Optimized Metallic Thin Films for Bio-Analyte Detection

John Ehiabhili, Radhakrishna Prabhu * and Somasundar Kannan

School of Engineering, Robert Gordon University, Aberdeen AB10 7GJ, UK; j.ehiabhili@rgu.ac.uk (J.E.); s.kannan1@rgu.ac.uk (S.K.)
* Correspondence: r.prabhu@rgu.ac.uk; Tel.: +44-1224262252

Abstract: There is a growing need for precise and rapid detection methods in fields such as biomedical diagnostics, environmental monitoring, and chemical analysis. Surface plasmon resonance (SPR) sensors have been used for the detection and quantification of a wide range of analytes, including biomolecules, chemicals, and gases, in real-time. Despite the promising capabilities of SPR sensors, there remains a gap in creating a balance between having a large enough area to capture a significant number of analytes for detection and being small enough to ensure high sensitivity. This research aims to explore the design of a D-shaped SPR-based optical fiber sensor, focusing on the use of copper, gold, and silver thin films at optimized width and thickness of 10 μm and 45 nm, respectively to improve the sensor's performance. Employing a computational approach, this study examines the influence of the optimized width and refractive indices of metallic films on the sensor's characteristics. The 10 μm width of the metallic thin film has been found to produce an optimal balance between the sensitivity and the dynamic range of the sensor. Leveraging on the ratio of the real and imaginary parts of the dielectric constant of the thin film metal provides insight into the optical properties and sensitivity at certain wavelengths. Within an analyte refractive index range of 1.37–1.42 and a wavelength range of 650–1200 nm, results indicate that silver outperforms gold and copper at the optimized width with a wavelength sensitivity, and detection accuracy of 12,300 nmRIU^{-1}, and 3.075, respectively. By optimizing the width of the metal thin film at 10 μm, a highly sensitive D-SPR is designed, allowing for enhanced sensor detection capabilities for a wide range of bioanalytes.

Keywords: surface plasmon resonance; D-shaped sensor; metal thin film; sensitivity; optical fibre; analyte detection

Citation: Ehiabhili, J.; Prabhu, R.; Kannan, S. Highly Sensitive D-SPR Sensors with Optimized Metallic Thin Films for Bio-Analyte Detection. *Photonics* **2024**, *11*, 764. https://doi.org/10.3390/photonics11080764

Received: 29 June 2024
Revised: 10 August 2024
Accepted: 13 August 2024
Published: 15 August 2024

1. Introduction

Optical fibre sensors are a class of sensors that have attracted a lot of attention due to their versatility, compact size, and ability to operate in a wide range of environments. These sensors exploit the principle of light transmission through optical fibres, which are thin glass or plastic strands that guide light along their length [1]. Hence, they can be engineered to be sensitive to various physical, chemical, and biological parameters, making them suitable for diverse applications, including structural health monitoring, telecommunications, oil and gas, biomedicine, and environmental sensing [2]. Surface plasmon resonance (SPR) is one of the fundamental mechanisms that allows optical fibre sensors to be sensitive to different analytes. SPR is a phenomenon that occurs when light is incident at the interface between a positive and negative permittivity material, causing resonant oscillation of conduction electrons—surface plasmons [3]. SPR is a useful technique for detecting molecular interactions because of its great sensitivity to changes in the refractive index near the sensor surface, which might be influenced by the adsorption of analytes. For SPR to occur, the condition of phase matching between the incident light and the surface plasmons must be met. This excites the surface plasmons, and energy is transferred, causing a reduction in the reflected light intensity at a specific angle or wavelength, resulting in a dip in the reflected light spectrum [4]. SPR technology has emerged as an essential tool in biomedical research and diagnostics, offering a label-free, real-time, and highly sensitive method for the detection and quantification of biomolecular interactions [5]. SPR technology advancements enable multiplexed platforms for simultaneous analyte detection, accelerating drug discovery, enhancing diagnostics, and studying biomolecular mechanisms, making SPR a key tool in biomedical research and applications [6].

In SPR sensing, the selection of the sensing metal is a critical factor in the design and functionality of the sensor. The metal layer is responsible for supporting the surface plasmon waves that are sensitive to changes in the refractive index near the sensor surface, which occur due to analyte binding [7]. Plasmons are typically stimulated using a laser that matches the requirements for coupling light into the waveguide. The laser light is usually polarized to have a perpendicular component to the metal surface (p-polarized) to excite the surface plasmons. The laser light is coupled into the optical fibre using an optic coupler, which propagates through the fibre and reaches the region where the cladding has been removed and the metal film is deposited. Here, the evanescent field of the guided light interacts with the metal film, which excites surface plasmons at the metal-dielectric interface. Broadband light sources like halogen lamps, supercontinuum lasers, and tunable laser sources can be used for the stimulation of plasmons due to their wavelength variation tendency [8]. The wavelength range used in this research is 650–1200 nm; hence, the light sources stated are applicable. The power can range from a few microwatts to several milliwatts to ensure a sufficient signal-to-noise ratio while avoiding nonlinear effects. The choice of the sensing metal will significantly influence the sensor's performance characteristics, such as sensitivity, stability, and specificity [9]. The most commonly used sensing metals in SPR sensors are gold (Au) and silver (Ag), with each offering distinct advantages and limitations. Deposition techniques such as the electron beam evaporation method, pulsed laser deposition, chemical vapour deposition, and magnetron sputtering can be used to coat the metal on the optical fibre [10]. In some cases, alternative materials like graphene, aluminium (Al), or copper (Cu) might be considered for their unique plasmonic properties, though they are less common than Au and Ag. Additionally, alloying or layering multiple metals can sometimes be employed to combine the advantageous properties of different materials [11].

The incorporation of SPR with optical fibres has been the subject of extensive research due to its potential to create highly sensitive and miniaturized sensors for biomedical applications. Numerous studies have focused on developing SPR-based optical fibre sensors that exploit various fibre geometries, including U-shaped, tapered, and side-polished or D-shaped fibres, to enhance the interaction between the evanescent field and the plasmonic material [12]. The design of the D-shaped optical fibre sensor is a pivotal aspect of its functionality and performance in biomedical applications. The sensor leverages a modified optical fibre, where a portion of the cladding is removed to expose the core, creating a D-shaped profile. This design modification facilitates direct exposure of the fibre core to the external medium, enhancing the interaction between the guided light's evanescent field and the surface plasmons on the metal coating [13]. Recent advancements have demonstrated the use of D-shaped sensors in detecting changes in refractive index, temperature, and the presence of specific chemicals or biomolecules [14].

In the context of D-SPR, many studies have investigated the enhancement of sensitivity by optimizing the thickness of the metallic thin film (40–50 nm). There is no literature on the investigation of sensitivity enhancement due to the width of a metal thin film. Hence, to the best of our knowledge, this is the first study to investigate the width of a metal thin film for improving the sensitivity of a D-SPR sensor. An optimized width and thickness can maximize the interaction between the surface plasmons and the analyte, leading to higher sensitivity and better detection limits. This research work aims to design a D-shaped SPR sensor that offers high sensitivity by optimizing the width, thickness, and material of the sensing layer. The outcome of this research can lead to the development of highly sensitive, reliable, and versatile sensors suitable for a wide range of applications in healthcare, food safety, environmental monitoring, and chemical analysis. Hence, this sensor technology will provide a balance between sensitivity enhancement and detection accuracy via the optimization of the width of a D-shaped SPR sensor for optical sensing and biomedical applications.

2. Structural Design and Theoretical Simulations

For the geometrical design and numerical simulation of the performance parameters of the sensor, COMSOL Multiphysics software has been utilised. In this research work, a single-mode fibre (SMF) with a 3.1% GeO_2 silica core diameter of 9 μm and a pure silica cladding diameter of 125 μm is used. The 3D schematic diagram of the D-shaped SPR-based optical fibre sensor with coated metal thin film (copper, gold, and silver, respectively) along the axis of the sensor is shown in Figure 1a. Investigating the interaction between the evanescent waves and the external medium is essential for a D-shaped optical fibre sensor. The proposed D-shaped SPR optical fibre sensor can be fabricated through the side polishing technique or laser micro-machining to create a plane sensing area [15]. For the deposition of the metal thin films onto the plane surface of the D-shaped optical fibre, magnetron sputtering or evaporation techniques are utilized. The polishing depth between the etched cladding and the core of the optical fibre (the amount of residual cladding) can be properly controlled [16]. The metallic thin film layer, which serves as the sensing layer, is positioned at a residual cladding distance of D = 0.5 μm from the core of the fibre sensor. The width (w) and thickness (t) of the metal sensing layers were optimized at 10 μm and 45 nm, respectively. Figure 1b illustrates the 2D D-shaped SPR-based fibre design with the design parameters and thin film indicated. A D-shaped biosensor having the above structural parameters is carefully designed to support single-mode operation. Utilizing the Wave Optics Module of COMSOL Multiphysics, the finite element method (FEM) numerical technique investigates the boundary conditions, and mode analysis was used for the study of the 2D sensor design. The refractive indices of the optical fibre core and cladding, the permittivity of the metal layer, and the properties of the surrounding medium are input into the model. The model is then discretized using meshing, which divides the geometry into finite elements that can be used for numerical computation. A physics-controlled extra-fine triangular mesh was utilised to capture the details of the fibre sensor geometry and the plasmonic interactions between the metallic sensing layer and the surrounding medium. To avoid the reflection of any incident waves, regardless of their angle or frequency, an artificial absorbing layer known as a perfectly matched layer (PML) is designed to surround the simulated region. To inject an analyte into the structure, a microfluidic channel or flow cell can be integrated with the sensor. The microfluidic channel is positioned over the D-shaped region of the fibre sensor, allowing the analyte to flow directly over the metal film where the SPR occurs [17]. The analyte solution is then introduced into the channel inlet using a pump system. The analytes may be protein, glucose, heavy metals in the water sample, toxins, viruses or bacteria in the samples. However, the desired analytes in this research are viruses. To simulate the sensor's interaction with biological analytes, the refractive index at the sensor surface is varied to represent the binding of biomolecules. This allows for the simulation of changes in the SPR signal due to molecular interactions and the assessment of the sensor's sensitivity. Parameter studies are conducted to analyse the sensor's performance under various conditions. This includes sweeping through a range of wavelengths, refractive indices, and geometric parameters to explore their effects on the SPR response. The simulations help identify the optimal sensor design and operational parameters.

All fibre-based SPR sensors rely on the coupling between the core mode and the surface plasmon polariton mode under phase-matching conditions. This initiates a resonance condition, which is seen as a dip in the sensor's transmission curve. This phase-matching condition is expressed as [18]:

$$k_{SP} = \frac{\omega}{c}\left(\frac{\epsilon_m \epsilon_d}{\epsilon_m + \epsilon_d}\right)^{1/2} \tag{1}$$

where k_{SP} is the surface wave vector, ω is the angular frequency, c is the speed of light in vacuum, ϵ_m and ϵ_d are the permittivity of the metal and dielectric, respectively.

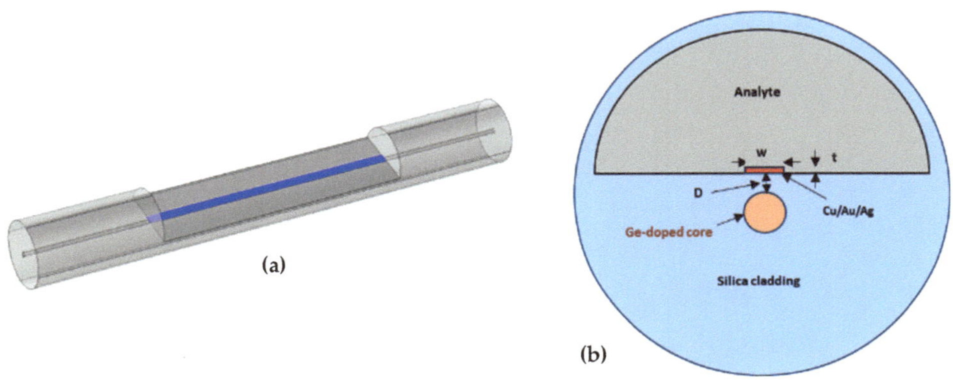

Figure 1. (**a**) 3D schematic of D-SPR sensor; (**b**) 2D sensor design showing the design parameters.

The refractive index of the silica cladding and Ge-doped core are determined using the Sellmeier equation expressed as [19]:

$$n^2(\lambda) = 1 + \frac{a_1\lambda^2}{\lambda^2 - C_1^2} + \frac{a_2\lambda^2}{\lambda^2 - C_2^2} + \frac{a_3\lambda^2}{\lambda^2 - C_3^2} \tag{2}$$

where λ is the incident light wavelength in μm; n is the wavelength-dependent refractive index of the fibre core and cladding. a_1, a_2, a_3, C_1, C_2, and C_3 are Sellmeier constants given in Table 1 [19].

Table 1. Sellmeier constants for fibre core and cladding.

Constants	a_1	a_2	a_3	C_1(μm)	C_2(μm)	C_3(μm)
Silica cladding	0.6961663	0.4079426	0.8974794	0.0684043	0.1162414	9.896161
Ge-doped core	0.7028554	0.4146307	0.8974540	0.0727723	0.1143085	9.896161

The dielectric function of a metal layer is determined using the Drude-Lorentz model and is expressed as [20]:

$$\varepsilon_m(\omega) = \varepsilon_\alpha - \frac{\omega_D^2}{\omega^2 + i\gamma\omega} + \Delta\varepsilon_p \frac{\Omega_p^2}{\Omega_p^2 - \omega^2 - i\Gamma_p\omega} \tag{3}$$

where ω_D is the plasma frequency, γ is the damping coefficient, $\Delta\varepsilon_p$ is the weighting coefficient, Ω_p is the oscillator strength, Γ_p is the spectral width, and ε_α is the interband offset for the metal [21]. The values of the Drude-Lorentz model parameters are given in Table 2 [22].

Table 2. Drude-Lorentz model parameter values.

Parameters	Ag Thin Film	Au Thin Film	Cu Thin Film
ε_α	1.55	3.5	1.25
ω_D (PHz)	2.176	2.169	1.985
γ (THz)	16.93	7.254	7.254
$\Delta\varepsilon_p$	1.0	0.35	1.39
Ω_p (PHz)	1.39	0.648	0.7133
Γ_p (PHz)	0.6529	0.1086	0.2587

The confinement loss, L_c, is calculated in dB/m using the following expression [23]:

$$L_c = 8.686 \cdot \frac{2\pi}{\lambda} \cdot I_m\left(n_{eff}\right) \tag{4}$$

where $I_m\left(n_{eff}\right)$ is the imaginary part of the effective refractive index, which is related to the attenuation or absorption of light as it propagates through the material medium. The transmission coefficient, T, can be expressed as [24]:

$$T = \exp\left(\frac{4\pi}{\lambda} \cdot \left(n_{eff}\right)_{imag} \cdot L\right) \tag{5}$$

where L is the length of the sensing region. An important parameter is the sensor's sensitivity. *Sensitivity* of a sensor is the degree of response to a change in an input signal. Sensitivity refers to the sensor's ability to detect small changes in the refractive index of the surrounding medium, which corresponds to the detection of low concentrations of analytes. Wavelength sensitivity is given as [25]:

$$S_\lambda(nm/RIU) = \frac{\Delta\lambda_{peak}}{\Delta n_a} \tag{6}$$

where $\Delta\lambda_{peak}$ is the peak shift in resonant wavelength and Δn_a is the change in the analyte's RI. Amplitude sensitivity, which refers to the change in the intensity of the transmitted light because of changes in the analyte RI in contact with the sensor, is given as [25]:

$$S_A(RIU^{-1}) = -\frac{1}{\alpha(\lambda, n_a)} \frac{\delta\alpha(\lambda, n_a)}{\delta n_a} \tag{7}$$

where $\alpha(\lambda, n_a)$ is the loss/attenuation and $\delta\alpha(\lambda, n_a)$ is the change in the loss spectrum resulting from the change in analyte RI.

Limit of Detection (LOD) is defined as the smallest quantity or concentration of an analyte that can be reliably detected by the sensor. The LOD is a critical parameter in biomedical sensing, where analytes may be present at very low concentrations.

Resolution is a measure of a sensor's ability to detect a minimum RI change in the surrounding medium. It is expressed as [26]:

$$R\ (RIU) = \frac{\Delta n_a \times \Delta\lambda_{min}}{\Delta\lambda_{peak}} \tag{8}$$

Figure of merit (FOM) and detection accuracy (*DA*) are measures of a sensor's performance obtained by combining its sensitivity, resolution, and dynamic range. It is calculated thus [26]:

$$FOM = \frac{S_\lambda}{FWHM} \tag{9}$$

$$DA = \frac{\Delta\lambda_{res}}{S_\lambda} \cdot FWHM \tag{10}$$

where *FWHM* is the full width at half-maximum of the spectra.

3. Results and Discussion

This study presents comprehensive theoretical modelling results for the optimized width and thickness of metal thin films for D-shaped SPR optical fibre sensors. The geometry of the D-shaped optical fibre sensor, including the metallic thin film width and thickness, and the length of the polished area can be optimised to maximise the overlap between the evanescent field and the plasmon field, enhancing sensitivity. Employing COMSOL Multiphysics software, the performance of sensors with a 10 μm width metal film was simulated, incorporating a 500 nm residual cladding and a thickness layer of 45 nm.

Cu, Au, and Ag were each coated on the flat surface of the D-shaped fibre sensor, and each was subject to different sensing performance measurement for analyte detection. Investigative analysis was carried out using an analyte RI range of 1.37–1.42. This falls within the range of the refractive index of most viruses and certain biomolecules. However, this research is intended for detecting the presence of viruses in biological samples.

The transmission spectrum of the phase match between the core and SPP mode of the D-shaped SPR sensor is shown in Figure 2. The grey line represents the transmission spectrum of the SPR sensor at an analyte refractive index of 1.42. The insets in Figure 2 show the electric field coupling of light between the metallic thin film and the core of the optical fibre. The electric field at the metal-dielectric interface is strengthened when energy is transferred from the core of the fibre to the interface. SPR is primarily excited by p-polarized waves because the electric field component of the p-polarized light can induce surface plasmons at the metal-dielectric interface. The x-orientation of the p-polarized wave has been considered because the electric field vector of the p-polarized light has a significant component normal to the metal surface when oriented along the x-axis. The peak is the fundamental plasmonic mode, and it occurs at the resonant wavelength due to the interaction of the evanescent field with the metallic-thin film's surface electrons. This mode is known as surface plasmon polariton (SPP) mode in the context of a planar surface or guided mode in the context of waveguides like optical fibre. This mode involves the propagation of light through the fibre core, generating an evanescent field that extends into the cladding and interacts with a metallic layer deposited on the cladding. When the conditions for SPR are met, this interaction excites surface plasmons that propagate along the metal-dielectric interface. This mode is the most sensitive to changes in the refractive index of the surrounding medium and is thus critical for sensing applications. The fundamental mode peak is typically used for sensing due to its high sensitivity.

Figure 3a shows the graph of the transmission spectra of Ag thin film for 35 nm, 40 nm, 45 nm, 50 nm, and 55 nm thickness, respectively. The dip in the transmission spectrum indicates the excitation of surface plasmons, which occurs when the momentum of the incident light matches the momentum of the surface plasmons at the metal-dielectric interface. The depth of the dip is a measure of the efficiency of this excitation process. A deeper dip typically indicates a more efficient excitation of surface plasmons, suggesting better sensitivity for sensing applications. As the thickness of the silver thin film changes, the wavelength needed to satisfy the phase matching condition also changes, leading to a shift in the SPR dip wavelength. Thus, at a metal thin film thickness of 45 nm, the coupling efficiency between the light and surface plasmons for this sensor configuration is maximized. From Figure 3b, the residual cladding distance D affects the interaction between the evanescent field of the light propagating through the fibre core and the plasmons at the metal-dielectric interface. The intensity of the evanescent field decays with distance from the fibre core. A smaller D allows for a stronger interaction between the field and the metal film, potentially leading to a more efficient excitation of surface plasmons and a deeper dip in the transmission spectrum. However, a smaller D might lead to significant absorption and damping of the plasmons due to the proximity of the metal film to the fibre core, potentially reducing the interaction length and affecting the sensor's performance.

Conversely, a larger D could result in a weaker evanescent field interaction, reducing sensitivity. In addition, the choice of D must also consider practical fabrication and stability issues to ensure a reliable sensor design. Hence, the residual cladding was optimized at 0.5 μm. Figure 4a illustrates the relationship between the wavelength at RI = 1.42 and transmission spectra in terms of the distinct width of Ag thin film. As the width increases, a remarkable dip in SPR is produced, thereby increasing the plasmonic effect. However, for a metal thin film width ≥ 10 μm, the transmission spectra are almost unchanged. The lateral extent of the evanescent field and the associated sensing capabilities are not significantly enhanced by further increases in the thin film width at these dimensions. Figure 4b shows the sensitivity of the sensor for different Ag thin film widths for a refractive index change from 1.41 to 1.42. As the width increases, the sensitivity increases. However, increasing the width of the Ag thin film beyond 10 μm results in a non-significant change in the sensitivity of the sensor. This can be attributed to light confinement within the fibre core in a single-mode regime. The interaction between the guided light and the metal thin film is maximized under this condition. Increasing the width of the thin film beyond a certain point does not significantly alter this interaction because the light remains confined within the core, and the effective area of interaction with the metal surface does not change appreciably. Hence, the width of the metal-thin film has been optimized at 10 μm.

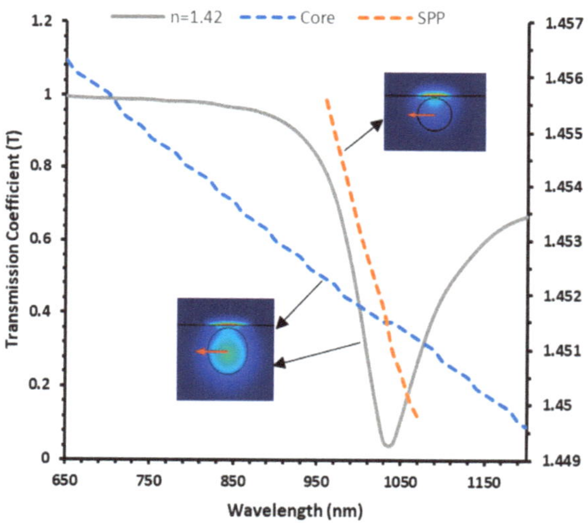

Figure 2. Phase matching between core and SPP mode of D-shaped sensor.

To understand the effect of the different metal coatings (Cu, Au, and Ag) on the D-shaped SPR sensor, the transmission spectra are plotted. Figure 5a–c shows that with increasing analyte RI (from 1.37 to 1.42) for the three-metal thin films, the SPR curve moves towards longer wavelengths, with gold showing a greater red shift compared to copper and silver. At RI of 1.42, the peak wavelengths of copper, silver, and gold were 973 nm, 977 nm, and 1035 nm, respectively. This is due to the differences in the properties of the materials. Gold has a larger free electron density and dielectric constant compared to silver and copper, leading to larger shifts in the peak wavelength [27]. Also, when the analyte refractive index increases, the resonance condition for SPR shifts because the change in refractive index alters the speed of light within the medium and, consequently, the momentum matching conditions between the incident light and the surface plasmons. This shift is due to the increased optical path length that results from a higher analyte refractive index. Figure 5d shows the nonlinear changes in the peak wavelength of the different sensors. This shows that the resonance between the core and SPP mode increases gradually with increasing analyte RI. The graph of the thin film refractive index against the wavelength using the Drude-Lorentz model is shown in Figure 6a. From the figure, silver has the lowest real part of refractive index (n) across the wavelength range. This results in a higher penetration depth of the evanescent field into the analyte region, allowing for a larger interaction volume and potentially higher sensitivity. The imaginary part of the refractive index (κ) indicates optical losses and affects the sharpness and sensitivity of the SPR signal. In the wavelength range, gold has a lower κ, indicating less absorption of light and a longer propagation length of surface plasmons. The ratio n/κ provides insight into the metal's optical properties. From Figure 6b, silver shows a high n/κ ratio, implying that it supports SPs with lower propagation losses. However, for wavelengths of about 1000 nm and above, gold has a greater n/κ ratio. This is because at these wavelengths, the κ part became significantly higher than the n part, contributing to higher optical losses.

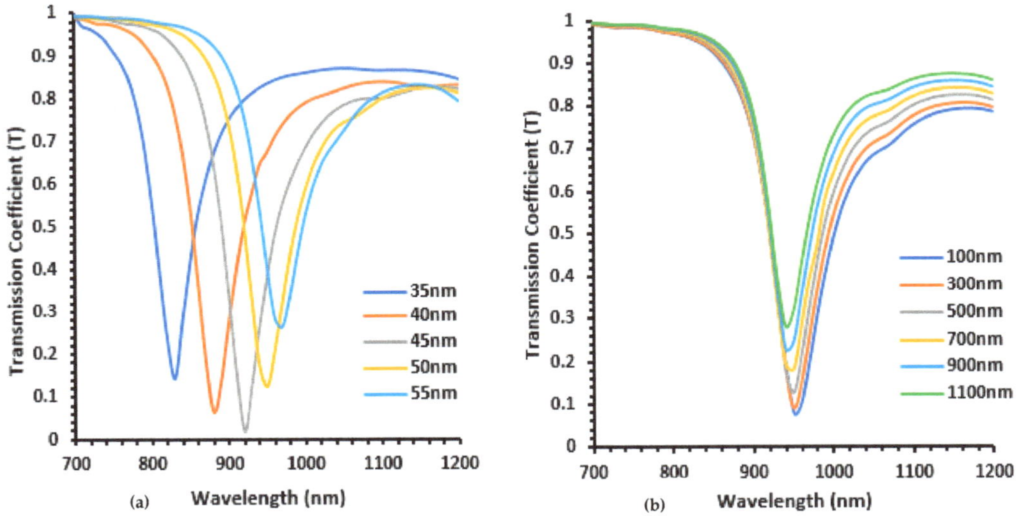

Figure 3. Transmission spectra for different (**a**) thicknesses of Ag thin film; (**b**) residual cladding thickness (D) of sensor design.

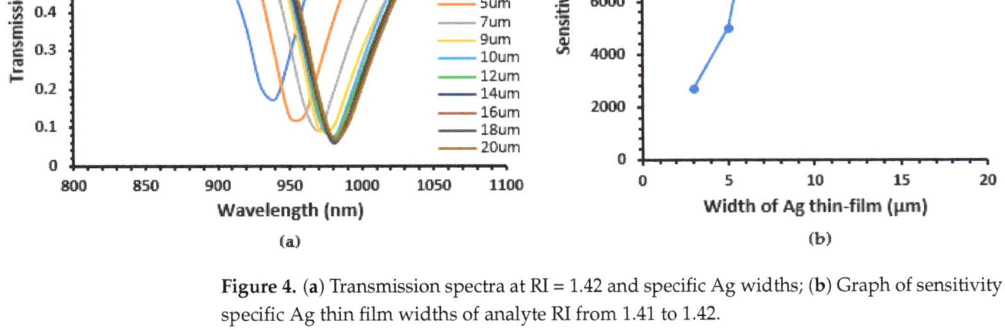

Figure 4. (**a**) Transmission spectra at RI = 1.42 and specific Ag widths; (**b**) Graph of sensitivity against specific Ag thin film widths of analyte RI from 1.41 to 1.42.

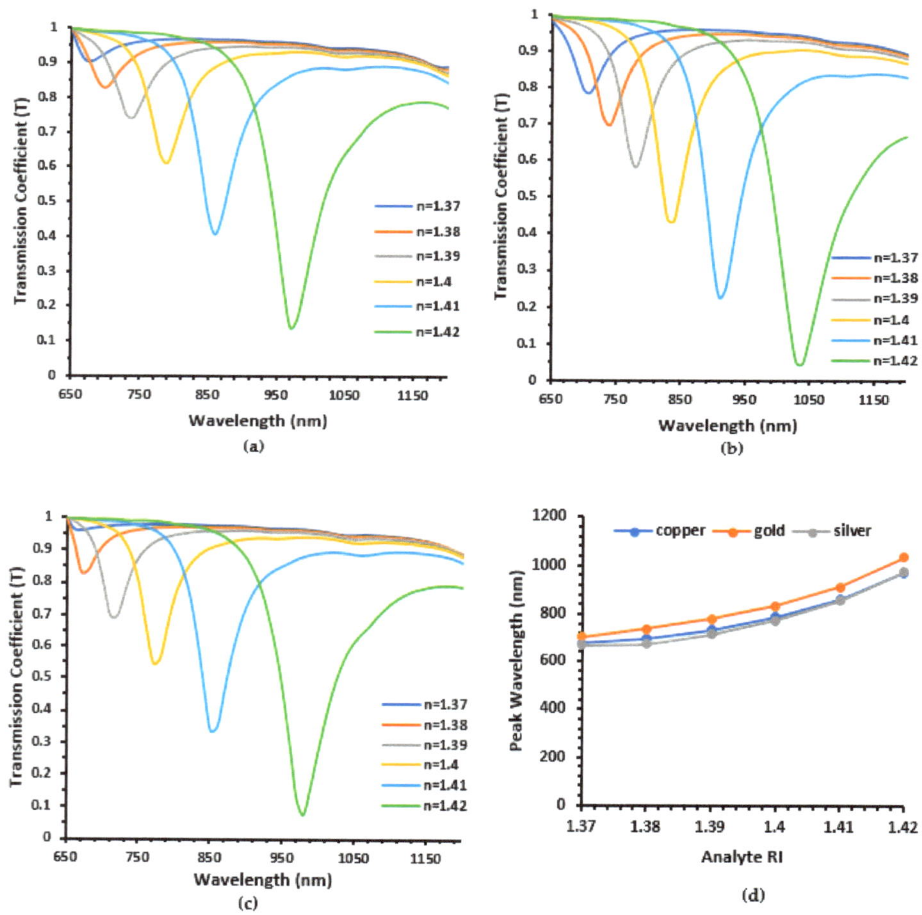

Figure 5. Transmission spectrum for (**a**) copper; (**b**) gold; (**c**) silver; (**d**) Peak wavelength as a function of analyte RI from 1.37 to 1.42.

Figure 7a–c shows the graph of the confinement loss against wavelength for copper, gold, and silver thin films, respectively. As the refractive index increases, the interaction between the guided light and the analyte changes, potentially leading to changes in the resonance conditions, deeper penetration of the evanescent field, redistribution of the optical power, and increase in absorption at the wavelength of the guiding light. All these can contribute to confinement loss increase, leading to a significant reduction in the amount of light reaching the detector, thus reducing the sensitivity. In practical situations, a high loss can degrade the SNR of the sensor design, making it more difficult to distinguish the signal from background noise. However, this loss can be minimized by developing signal processing techniques to improve the SNR. This graph, for a D-shaped optical fibre SPR-based sensor, is directly related to the amplitude sensitivity of the sensor. The height of the peak indicates the strength of the SPR interaction. For analyte RI of 1.42, silver showed the highest amplitude sensitivity of 1932.51 RIU^{-1}. For the same analyte RI, gold and copper showed amplitude sensitivities of 1454.56 RIU^{-1} and 1224.25 RIU^{-1}, respectively. The peak characteristics and the noise level in the graph are critical for assessing how effectively the sensor can detect changes in the surrounding refractive index, which is essential for applications like biosensing, chemical detection, and environmental monitoring.

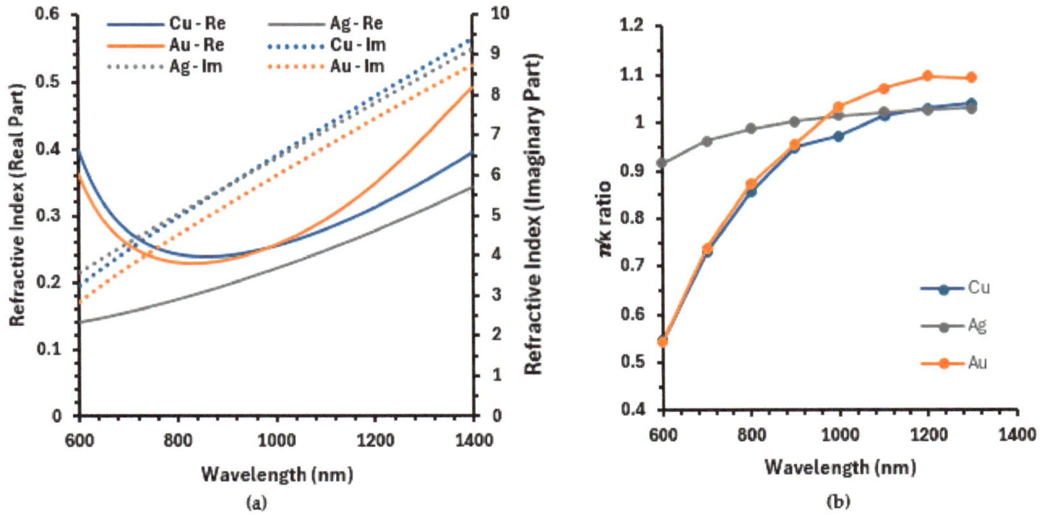

Figure 6. Graph of (**a**) real and imaginary refractive index against wavelength; (**b**) n/κ against wavelength.

Figure 8a depicts the graph of the wavelength sensitivity of the three sensor designs against the analyte refractive index. From the graph, silver has the highest wavelength sensitivity of 12,300 nmRIU^{-1} at analyte RI of 1.42, slightly better than that of gold with a wavelength sensitivity of 12,200 nmRIU^{-1}. Copper has the least wavelength sensitivity at the same RI, with 11,300 nmRIU^{-1}. This is because silver has the highest plasma frequency and the lowest optical losses in the visible and near-infrared regions among the three metals, resulting in a more pronounced and sharper SPR resonance curve, which translates to higher wavelength sensitivity. This means that small changes in the refractive index of the surrounding medium will cause a relatively large shift in the resonance wavelength, making silver an excellent choice for high-sensitivity applications. Although copper is outperformed by silver and gold thin films, it can be used in designs and specific applications where cost is a high consideration. However, silver has the tendency to oxidize and tarnish, especially when exposed to air and various chemical environments, posing a significant challenge for its practical application in sensor technologies. Applying a thin protective layer over the silver can shield it from the environment, thereby preventing or significantly slowing down the oxidation process. Materials such as graphene, silicon dioxide, or specific polymers can serve as effective protective coatings. Figure 8b shows the plot of FWHM against the analyte RI. The FWHM represents the width of the SPR resonance peak at half of its maximum intensity and is a measure of the spectral resolution and quality of the resonance. For the analyte RI considered in the design, silver showed a smaller FWHM compared to gold and copper, which indicates a sharper resonance peak. This generally corresponds to higher sensor resolution and the ability to discriminate between small changes in analyte RI. Typically, the performance of a sensor is enhanced when the sensitivity increases with a decrease in FWHM [21]. High resolution often leads to higher detection accuracy (DA), as the sensor can more precisely locate the resonance condition.

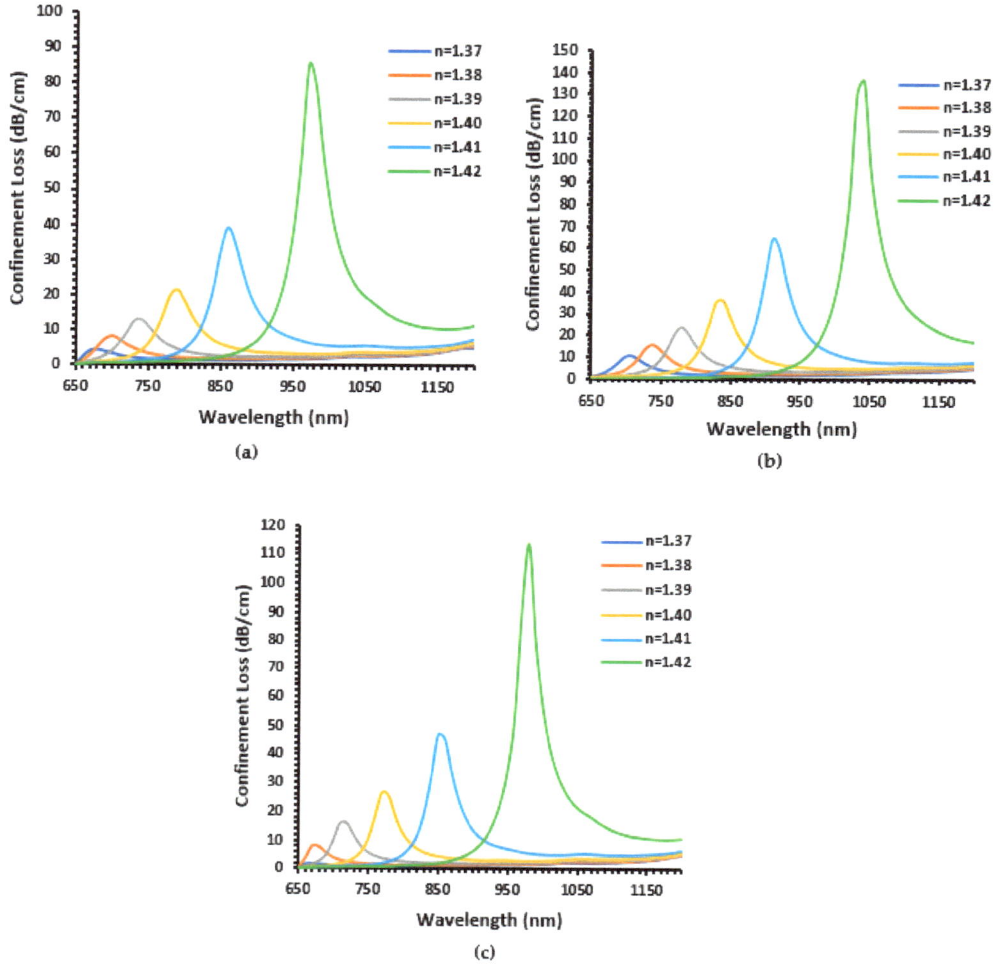

Figure 7. Confinement loss spectra for (**a**) copper; (**b**) gold; (**c**) silver thin film D-SPR sensor.

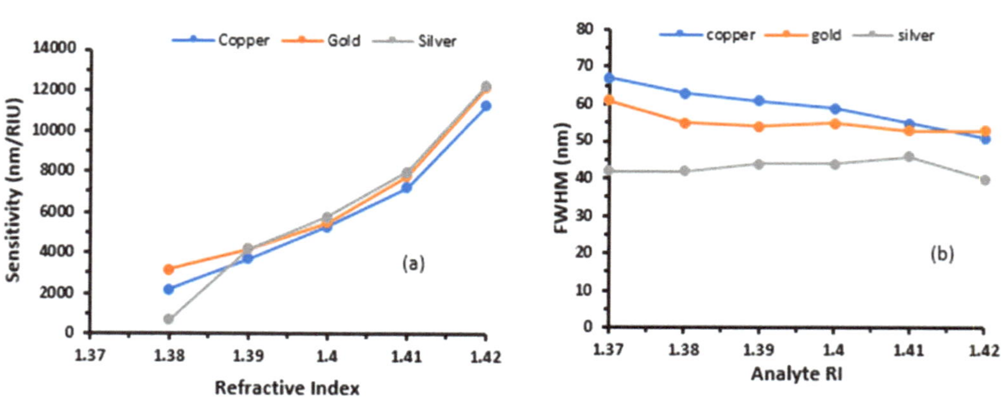

Figure 8. (**a**) Wavelength sensitivity graph; (**b**) Graph of FWHM against analyte RI.

Table 3 gives a comparison of the performance metrics for copper, gold, and silver thin film D-shaped SPR-based optical fibre sensors for analyte RI of 1.42. From the table, silver performed better than gold and copper, showing a higher wavelength and amplitude sensitivity, narrower FWHM, better FoM, resolution, and DA for a RI of 1.42.

Table 3. Comparison of Cu, Au, and Ag SPR sensor performance metrics.

	S_λ (nm/RIU)	S_A (RIU^{-1})	FWHM (nm)	FoM (RIU^{-1})	Resolution (RIU)	DA
Copper	11,300	1224.25	51	221.57	8.85×10^{-6}	2.216
Gold	12,200	1454.56	53	230.19	8.20×10^{-6}	2.302
Silver	12,300	1932.51	40	307.50	8.13×10^{-6}	3.075

The increased wavelength and amplitude sensitivity for the metal thin films as analyte RI increases is due to the increased electric field intensity at the metal-dielectric interface. To fabricate these sensors, a trade-off in wavelength sensitivity, spectral resolution, and sensor geometry must be considered for specific application requirements. A comparison of this design with existing designs from literature is presented in Table 4.

Table 4. Comparison of the performance of this D-shaped Cu, Au, and Ag optical fibre SPR sensor with other optical sensors published in literature.

Ref	Sensor Configuration	Sensing Material	Operating Wavelength (nm)	RI Range	Wavelength Sensitivity (nm/RIU)	Amplitude Sensitivity (RIU^{-1})	Resolution (RIU)
[28]	Spiral PCF-SPR	Gold	-	1.33–1.38	4600	420.4	2.17×10^{-5}
[29]	External coated PCF SPR	Gold	2550–2900	1.23–1.29	5500	333.8	7.69×10^{-6}
[30]	Hollow-core PCF	Gold	1078–2097	1.27–1.45	5653.57	590	-
[31]	Hollow-core PCF	Silver	500–800	1.33–1.37	4200	300	2.38×10^{-5}
[32]	D-shaped PCF	Gold	2900–3600	1.33–1.39	11,500	230	8.7×10^{-6}
[33]	Ag coated PCF SPR	Silver	-	1.33–1.42	11,000	1420	9.1×10^{-6}
[34]	Solid core MOF	Gold	770–850	1.36–1.39	7000	886.9	4×10^{-5}
	This work	Silver Gold Copper	650–1200	1.37–1.42	12,300 12,200 11,300	1932.51 1454.46 1224.25	8.13×10^{-6} 8.2×10^{-6} 8.85×10^{-6}

This research study considers a uniform distribution of analytes for the specified refractive index range. A non-uniform distribution of analytes means that the refractive index of analytes varies across the sensing area. This variation can affect how the evanescent field of the guided light interacts with the analytes. Since sensitivity and resonance conditions are highly dependent on the local refractive index at the sensor surface, a nonuniform distribution could lead to a broader or multi-modal response, requiring a different approach and possibly modifications in the sensor design to maintain sensitivity and accuracy.

4. Conclusions

D-shaped optical fibre SPR-based sensors have diverse advantages and are widely employed due to their high sensitivity and ease of integration with other devices. The theoretical investigation into the sensing parameters of different metallic thin films of D-shaped SPR optical fibre sensors, as presented in this study, has provided significant insights into the design and performance of SPR-based sensors. Utilising COMSOL Multiphysics for simulations, a sensor configuration employing a thin film of 45 nm thickness, with a residual cladding of 0.5 μm and an optimized sensing layer width of 10 microns has been successfully modelled. The results for the designs using three different metal thin films (Cu, Ag, and Au) showed that for an analyte RI of 1.42, silver performed best compared to gold and copper, having a wavelength sensitivity of 12,300 nmRIU^{-1}, amplitude sensitivity of 1932.51 RIU^{-1}, FWHM of 40 nm, FoM of 307.5 RIU^{-1}, resolution of 8.13×10^{-6} RIU, and DA of 3.075. This is because silver has a low real dielectric constant, which leads to a higher penetration depth of the evanescent field into the analyte region. This allows for a larger interaction volume and, thus, higher sensitivity.

The optimisation of the width and thickness of metal thin film thickness emerged as key factors in achieving a balance between sensitivity and detection accuracy. The sensor's geometry, including D-shape geometry and the design parameters, was found to significantly influence the excitation of plasmonic modes, thus affecting the sensor's sensitivity profile. The outcomes of this research suggest that these sensors could be highly effective for real-time, label-free, and selective detection of various analytes. Moreover, the insights gained from the simulations lay the groundwork for future experimental validation and the potential development of highly sensitive SPR sensor arrays for diverse applications in biochemical and environmental monitoring. The versatility and precision of the proposed sensor design hold promise for significant contributions to the realm of optical fibre sensing technology. Further research may include investigating the performance of the sensors at a higher analyte refractive index, grid structures, multilayered metal thin film, and optimising the thin film sensing lengths.

Author Contributions: Conceptualization, J.E., R.P. and S.K.; methodology, J.E. and R.P.; software, J.E., R.P. and S.K.; investigation, J.E., R.P. and S.K.; data curation, J.E. and R.P.; writing—original draft preparation, J.E.; writing—review and editing, R.P. and S.K.; visualization, J.E., R.P. and S.K.; supervision, R.P. and S.K. All authors have read and agreed to the published version of the manuscript.

Funding: This research received no funding.

Institutional Review Board Statement: Not applicable.

Informed Consent Statement: Not applicable.

Data Availability Statement: The data supporting the findings of this study are available from the corresponding author upon reasonable request.

Acknowledgments: I would like to acknowledge the support provided by the School of Engineering, Robert Gordon University in this research study.

Conflicts of Interest: The authors declare no conflicts of interest.

References

1. Correia, R.; James, S.W.; Lee, S.-W.; Morgan, S.P.; Korposh, S. Biomedical application of optical fibre sensors. *J. Opt.* **2018**, *20*, 073003. [CrossRef]
2. Johny, J.; Amos, S.; Prabhu, R. Optical fibre-based sensors for oil and gas applications. *Sensors* **2021**, *21*, 6047. [CrossRef]
3. Zeng, S.; Baillargeat, D.; Ho, H.-P.; Yong, K.-T. Nanomaterials enhanced surface plasmon resonance for biological and chemical sensing applications. *Chem. Soc. Rev.* **2014**, *43*, 3426–3452. [CrossRef]
4. Bhavsar, K.; Prabhu, R.; Pollard, P. Ultrasensitive graphene coated SPR sensor for biosensing applications. *Opt. Sens.* **2015**, *9506*, 173–178.
5. Nguyen, H.H.; Park, J.; Kang, S.; Kim, M. Surface plasmon resonance: A versatile technique for biosensor applications. *Sensors* **2015**, *15*, 10481–10510. [CrossRef] [PubMed]
6. Linman, M.J.; Abbas, A.; Cheng, Q. Interface design and multiplexed analysis with surface plasmon resonance (SPR) spectroscopy and SPR imaging. *Analyst* **2010**, *135*, 2759–2767. [CrossRef]
7. Tabasi, O.; Falamaki, C. Recent advancements in the methodologies applied for the sensitivity enhancement of surface plasmon resonance sensors. *Anal. Methods* **2018**, *10*, 3906–3925. [CrossRef]
8. Prabowo, B.A.; Purwidyantri, A.; Liu, K. Surface plasmon resonance optical sensor: A review on light source technology. *Biosensors* **2018**, *8*, 80. [CrossRef]
9. Shalabney, A.; Abdulhalim, I. Sensitivity-enhancement methods for surface plasmon sensors. *Laser Photonics Rev.* **2011**, *5*, 571–606. [CrossRef]
10. Ristau, D.; Ehlers, H. Thin film optical coatings. In *Handbook of Lasers and Optics*; Springer: Berlin/Heidelberg, Germany, 2012.
11. Hinman, S.S.; McKeating, K.S.; Cheng, Q. Surface plasmon resonance: Material and interface design for universal accessibility. *Anal. Chem.* **2018**, *90*, 19. [CrossRef]

12. Caucheteur, C.; Guo, T.; Albert, J. Review of plasmonic fiber optic biochemical sensors: Improving the limit of detection. *Anal. Bioanal. Chem.* **2015**, *407*, 3883–3897. [CrossRef] [PubMed]
13. Rifat, A.A.; Ahmed, R.; Yetisen, A.K.; Butt, H.; Sabouri, A.; Mahdiraji, G.A.; Yun, S.H.; Adikan, F.M. Photonic crystal fiber based plasmonic sensors. *Sens. Actuators B Chem.* **2017**, *243*, 311–325. [CrossRef]
14. Kadhim, R.A.; Abdul, A.K.K.; Yuan, L. Advances in surface plasmon resonance-based plastic optical fiber sensors. *IETE Technol. Rev.* **2022**, *39*, 442–459. [CrossRef]
15. Tseng, S.; Chen, C. Side-polished fibers. *Appl. Opt.* **1992**, *31*, 3438–3447. [CrossRef] [PubMed]
16. Xie, Q.; Chen, Y.; Li, X.; Yin, Z.; Wang, L.; Geng, Y.; Hong, X. Characteristics of D-shaped photonic crystal fiber surface plasmon resonance sensors with different side-polished lengths. *Appl. Opt.* **2017**, *56*, 1550–1555. [CrossRef]
17. Pathak, A.K.; Singh, V.K.; Ghosh, S.; Rahman, B.M.A. Investigation of a SPR based refractive index sensor using a single mode fiber with a large D shaped microfluidic channel. *OSA Contin.* **2019**, *2*, 3008–3018. [CrossRef]
18. Ying, Y.; Wang, J.; Hu, N.; Xu, K.; Sun, L.; Si, G. Determination of refractive index using surface plasmon resonance (SPR) and rigorous coupled wave analysis (RCWA) with a D-shaped optical fiber and a nano-gold grating. *Instrum. Sci. Technol.* **2020**, *48*, 376–385. [CrossRef]
19. Rafi, H.N.; Kaysir, M.R.; Islam, M.J. Air-hole attributed performance of photonic crystal fiber-based SPR sensors. *Sens. Bio-Sens. Res.* **2020**, *29*, 100364. [CrossRef]
20. Rioux, D.; Vallières, S.; Besner, S.; Muñoz, P.; Mazur, E.; Meunier, M. An analytic model for the dielectric function of Au, Ag, and their alloys. *Adv. Opt. Mater.* **2014**, *2*, 176–182. [CrossRef]
21. Nahid, E.; Islam, M.J.; Kaysir, M.R.; Hossain, S.S.; Rahman, M.S. Au-strip structure dependent performance investigation of D-shaped SPR sensors for biosensing applications. *Results Opt.* **2023**, *12*, 100460. [CrossRef]
22. Kravets, V.G.; Kurioz, P.Y.; Poperenko, L.V. Spectral dependence of the magnetic modulation of surface plasmon polaritons in permalloy/noble metal films. *JOSA B* **2014**, *31*, 1836–1844. [CrossRef]
23. White, T.P.; McPhedran, R.C.; de Sterke, C.M.; Botten, L.C.; Steel, M.J. Confinement losses in microstructured optical fibers. *Opt. Lett.* **2001**, *26*, 1660–1662. [CrossRef]
24. Nayak, J.K.; Jha, R. Numerical simulation on the performance analysis of a graphene-coated optical fiber plasmonic sensor at anti-crossing. *Appl. Opt.* **2017**, *56*, 3510–3517. [CrossRef] [PubMed]
25. Kadhim, R.A.; Yuan, L.; Xu, H.; Wu, J.; Wang, Z. Highly sensitive D-shaped optical fiber surface plasmon resonance refractive index sensor based on Ag-α-Fe_2O_3 grating. *IEEE Sens. J.* **2020**, *20*, 9816–9824. [CrossRef]
26. Klantsataya, E.; François, A.; Ebendorff-Heidepriem, H.; Hoffmann, P.; Monro, T.M. Surface plasmon scattering in exposed core optical fiber for enhanced resolution refractive index sensing. *Sensors* **2015**, *15*, 25090–25102. [CrossRef]
27. Indhu, A.R.; Keerthana, L.; Dharmalingam, G. Plasmonic nanotechnology for photothermal applications–an evaluation. *Beilstein J. Nanotechnol.* **2023**, *14*, 380–419. [CrossRef]
28. Rifat, A.A.; Mahdiraji, G.A.; Chow, D.M.; Shee, Y.G.; Ahmed, R.; Adikan, F.R.M. Photonic crystal fiber-based surface plasmon resonance sensor with selective analyte channels and graphene-silver deposited core. *Sensors* **2015**, *15*, 11499–11510. [CrossRef]
29. Rifat, A.A.; Hasan, M.R.; Ahmed, R.; Butt, H. Photonic crystal fiber-based plasmonic biosensor with external sensing approach. *J. Nanophotonics* **2018**, *12*, 012503. [CrossRef]
30. Liu, Y.; Li, S.; Chen, H.; Li, J.; Zhang, W.; Wang, M. Surface plasmon resonance induced high sensitivity temperature and refractive index sensor based on evanescent field enhanced photonic crystal fiber. *J. Light. Technol.* **2019**, *38*, 919–928. [CrossRef]
31. Yang, X.; Lu, Y.; Wang, M.; Yao, J. An exposed-core grapefruit fibers based surface plasmon resonance sensor. *Sensors* **2015**, *15*, 17106–17114. [CrossRef]
32. Kaur, V.; Singh, S. D-shaped photonic crystal fiber based surface plasmon resonance sensor using dual coating of metal oxide for healthcare applications. In Proceedings of the 2021 IEEE 16th Nanotechnology Materials and Devices Conference (NMDC), Vancouver, BC, Canada, 12–15 December 2021.
33. Tong, K.; Wang, F.; Wang, M.; Dang, P.; Wang, Y.; Sun, J. D-shaped photonic crystal fiber biosensor based on silver-graphene. *Optik* **2018**, *168*, 467–474. [CrossRef]
34. Popescu, V.A.; Puscas, N.N.; Perrone, G. Simulation of the sensing performance of a plasmonic biosensor based on birefringent solid-core microstructured optical fiber. *Plasmonics* **2017**, *12*, 905–911. [CrossRef]

photonics

MDPI

Article

Structural Diagnosis of Solid Rocket Motors Using Neural Networks and Embedded Optical Strain Sensors

Georgia Korompili [1], Nicholaos Cholevas [1,2], Konstantinos N. Anyfantis [2], Günter Mußbach [3] and Christos Riziotis [1,4,*]

1 Theoretical & Physical Chemistry Institute, National Hellenic Research Foundation, 11635 Athens, Greece; gkorompili@eie.gr (G.K.); cholevasnickolaos@gmail.com (N.C.)
2 School of Naval Architecture and Marine Engineering, National Technical University of Athens, 15780 Athens, Greece; kanyf@naval.ntua.gr
3 Bayern-Chemie GmbH (MBDA Germany), 84544 Aschau am Inn, Germany; guenter.mussbach@mbda-systems.de
4 Defence & Security Research Institute, University of Nicosia, CY-2417 Nicosia, Cyprus
* Correspondence: riziotis@eie.gr

Abstract: The main failures that could deteriorate the reliable operation of solid rocket motors (SRMs) and lead to catastrophic events are related to bore cracks and delamination. Current SRMs' predictive assessment and damage identification practices include time-consuming and cost-demanding destructive inspection techniques. By considering state-of-the-art optical strain sensors based on fiber Bragg gratings, a theoretical study on the use of such sensors embedded in the circumference of the composite propellant grain for damage detection is presented. Deep neural networks were considered for the accurate prediction of the presence and extent of the defects, trained using synthetic datasets derived through finite element analysis method. The evaluation of this combined approach proved highly efficient in discriminating between the healthy and the damaged condition, with an accuracy higher than 98%, and in predicting the extent of the defect with an error of 2.3 mm for the bore crack depth and 1.6° for the delamination angle (for a typical ~406 mm diameter grain) in the worst case of coexistent defects. This work suggests the basis for complete diagnosis of solid rocket motors by overcoming certain integration and performance limitations of currently employed dual bond stress and temperature sensors via the more scalable, safe, sensitive, and robust solution of fiber optic strain sensors.

Keywords: solid rocket motors; fiber Bragg gratings; optical strain sensors; finite element analysis; structural health monitoring; strain sensing; neural networks; crack; delamination

Citation: Korompili, G.; Cholevas, N.; Anyfantis, K.N.; Mußbach, G.; Riziotis, C. Structural Diagnosis of Solid Rocket Motors Using Neural Networks and Embedded Optical Strain Sensors. *Photonics* **2024**, *11*, 799. https://doi.org/10.3390/photonics11090799

Received: 25 June 2024
Revised: 23 August 2024
Accepted: 24 August 2024
Published: 27 August 2024

1. Introduction

Solid rocket motors (SRMs) stand as pivotal components across aerospace and defense applications, powering satellite launches, missile systems, and space exploration missions as solid rocket boosters (SRB). A typical schematic of an SRM structure is illustrated in Figure 1a. Despite their widespread use, SRMs face inherent vulnerabilities such as internal bore cracks and delamination, which can pose threats to their performance, reliability, and safety [1]. Timely detection of these defects is paramount to averting catastrophic failures and ensuring mission success [2].

Conventional inspection methods for SRMs typically involve visual scrutiny and labor-intensive experimental procedures on the mechanical properties of the propellant grain [3] or X-ray imaging, which can be time-consuming, expensive, and, most importantly, often destructive for the SRMs. Additionally, these testing protocols are impractical for implementation on user premises such as air bases, thus necessitating logistic efforts associated with high financial costs and equally important time delay costs. In recent years, there has been a surge in the development of structural health monitoring (SHM) systems to support condition-based maintenance (CBM) strategies for efficient inspection and monitoring of SRMs [4]. These systems particularly employ piezoelectric or dual bond stress and temperature (DBST) sensors [4–7], placed on the interface between the propellant and the casing of the SRM, as can be seen in Figure 1b, or more recently, optical fiber strain sensors [8] embedded in the grain. These sensors can detect changes in the motor's mechanical behavior, which can be indicative of the presence of internal defects [9].

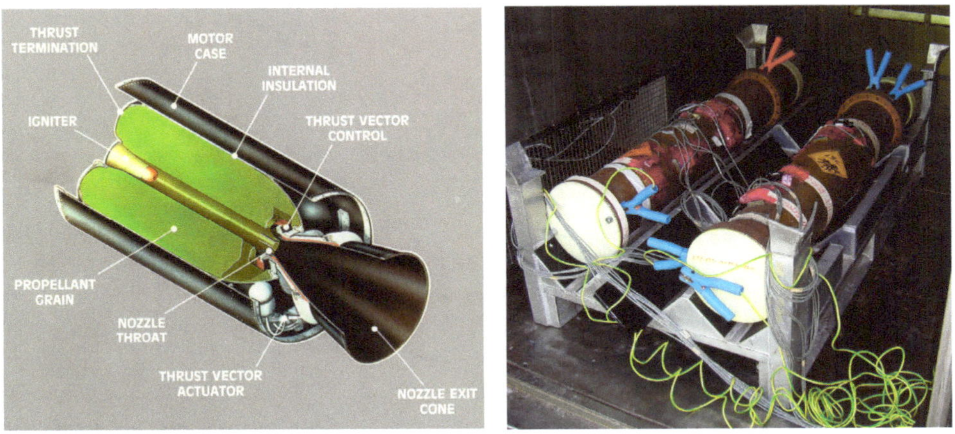

(a) (b)

Figure 1. (**a**) Schematic of a typical SRM illustrating the main components. (**b**) Figure of SRMs with embedded DBST sensors on the casing, together with the external electric wiring.

Recently, there has also been a strong interest in developing automated defect detection systems for SRMs using machine learning (ML) techniques such as neural networks (NN) [10–13] that can be trained to recognize patterns in large amounts of data. These have been successfully applied to a variety of image, signal, and feature processing applications. Particularly, concerning the SHM and diagnosing of the SRMs, there have been a few approaches evaluating the performance of the several NN-based systems in the identification of bore cracks, delamination, or both [11]. However, these approaches cover specific areas of the problem, without reporting on the potential to discriminate the healthy from the damaged state or perform complete diagnoses by predicting the type of defect and the extent of it. Moreover, the majority of the reported projects and research on the issue of SHM of SRMs refer to the use of DBST sensors [14]. To our knowledge, there is no reported work on the performance of optical fiber strain sensors [15–17], such as fiber Bragg grating (FBG)-based sensors, along with the employment of machine learning techniques for the detection of defects in SRMs. This type of optical sensor was introduced to reduce the minimum detectable defect extent to much lower values without implementing electrical wiring into the motor, as was reported in several studies [18–20]. Furthermore, optical sensors have some uniquely favorable characteristics, such as their low weight, inherent electromagnetic interference immunity, and inherent safety characteristics in explosive or flammable environments. Additionally, optical fibers can have quasi-distributed sensing capabilities by introducing multiple FBGs into a single optical fiber, while they can be interrogated remotely, from a distance through a low loss optical fiber [16,17,21] through optical transmission, meaning that the electronic-based interrogation unit and any electric power supply system can be positioned far away from the SRM. The incorporation of an optical fiber with a nominal typical diameter of 125 μm or even down to 50 μm is possible without mechanical disturbance of the propellant or degradation of the mechanical properties of the SRM system, as could be the case with DBST sensors, where invasive physical openings on the external casing surface are needed for embedding the sensors.

Another important characteristic of the fiber optic-based strain sensors is their ability to be embedded anywhere in the propellant by following proper procedures during the propellant casting process, thus providing the potential for monitoring higher deformations and higher strain changes in comparison to DBST sensors that are fixed close to the casing with limited monitored deformations.

While the capabilities of the FBG-based strain sensors appear promising in monitoring the structural integrity of the SRM [21], a diagnostic tool targeting to the detection of localized defects, such as bore cracks or delamination, based on the strain field distribution has to be re-developed to compare its performance with previously studied works that are based on stress sensors. The performance of ML, or specifically, NN-based techniques [22,23] using optical strain sensors, has been evaluated as a SHM tool in several generic diagnostic applications [23–28]. However, the field remains open to further investigation, particularly in the case of challenging SRM-related studies concerning the detection of defects or ageing and degradation of the propellant. The specific issues that need to be studied are as follows: (a) the comparison of the hoop strain field distribution, monitored using FBG-based strain sensors, with the radial stress field distribution, monitored by the currently used DBST sensors; (b) the feasible location of the recently introduced FBG-based strain sensors in the SRM and the way this location affects the performance and accuracy of detection of localized defects; and (c) the theoretically expected performance of a NN-based SRM diagnostic tool, finely tuned for the specific application and the use of optical-based strain sensors, to investigate the potential of these sensors for achieving higher accuracies in SRM health monitoring.

In this paper, we focus on the above fields of interest, and we particularly explore the feasibility of an NN-based diagnostic tool for the identification of defects in an SRM by assessing its trainability and performance. Due to the lack of experimental data, we use two-dimensional (2D) simulations of the cross section of the SRM to build a dataset of strain data in specific positions of the circular section of the motor in order to feed a two-stage deep NN-based system and to perform structural health diagnosis of the SRM. We evaluate the performance of the networks to accurately discriminate healthy from damaged states of the SRM, to identify the type of the present defect in the case of the damaged state, and to predict the spatial extent of the defects. In our analysis, we also investigate the impact of critical factors on the prediction error. These factors are defined at the simulation level and concern the angular position of the defect, the coexistence of two different defects—bore crack and delamination—and the relative angular distance between them. Our results show that deep neural networks can achieve a high accuracy in identifying both bore cracks and delamination in solid rocket motors, with an overall accuracy of over 98% on our dataset and a maximum root mean square error (RMSE) of 2.3 mm for bore cracks and 1.57°. These results were obtained using only four strain sensors in a typical SRM case with grain diameter of 406 mm. We discuss the implications of these results for the development of automated inspection systems for SRMs, and we identify several areas for future research. Overall, our work demonstrates the potential of machine learning techniques in conjunction to recently introduced optical fiber-based strain sensing capabilities in improving the safety and reliability of critical propulsion systems.

2. SRM Physical Model and Finite Element Analysis

2.1. SRM Model and Defect Definition

A typical SRM geometry is considered in this study. It mainly consists of a metallic case, an insulation layer, and the contained solid elastomeric composite material (propellant grain) itself. In general, the SRMs are structured in a cylindrical arrangement with various possible bore shapes depending on the specific SRM functional characteristics such as acceleration and thrust. Here, the simplified case of a cylindrical bore is assumed.

The entire propellant structure considered here is axisymmetric, with the bore diameter (inner diameter) being 203.2 mm and the outer diameter being 406.4 mm. The propellant material is a typical composite grain of hydroxyl-terminated polybutadiene

(HTPB)/ammonium perchlorate (HTPB/AP). The insulation is modeled as a thin layer attached to the propellant domain, with a thickness of 2.54 mm, whose material properties correspond to an ethylene propylene diene monomer (EPDM). The insulation layer is attached to a thin stainless-steel casing with a thickness of 3.125 mm. The material properties participating in the model are given in Table 1, along with the information of the inner and outer radius of the designed domains (Figure 2). They were treated as isotropic linear elastic materials. This assumption is a simplification that is not expected to affect the final results of the simulation, as the model is subjected to a stationary regime, simulating the stress and strain fields due to the transition from the initial curing and stress-free temperature, which here is set to 71 °C (or more usually in the range 50–60 °C), down to extreme conditions of −51 °C.

Table 1. Material and geometry design parameters used in the simulations.

	Material Domains		
Material Properties	**Propellant**	**Insulation**	**Casing**
Young's modulus	14.2 MPa	33.5 MPa	55.9 GPa
Coefficient of thermal expansion	9.56×10^{-5} 1/K	8.75×10^{-5} 1/K	2.16×10^{-6} 1/K
Poisson's ratio	0.499	0.499	0.3
Inner radius	101.6 mm	203.2 mm	205.74 mm
Outer radius	203.2 mm	205.74 mm	208.915 mm

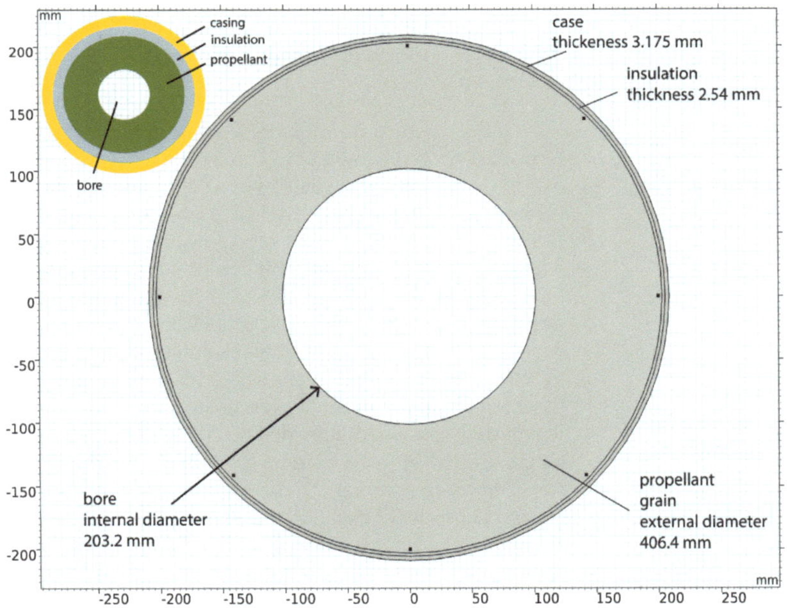

Figure 2. Plan view/cross section of the SRM for 2D analysis, together with a simplified explanatory schematic of the SRM layers (inset).

In this work, a finite element analysis (FEA), particularly the Structural Mechanics module of the COMSOL Multiphysics software (COMSOL, Inc., Burlington, MA 01803, USA), is used to study and analyze the strain distribution in the presence of thermal loads and various defects and to identify also the response of the strain sensors in the SRM under various conditions. The time and computational requirements of the study of SRM are reduced without a loss of information in a simplified two-dimensional (2D) structure of the cross-section of a center-perforated SRM, simulated using the plane strain approximation method. In Figure 2, the modeled grain structure is illustrated. For the specific case illustrated in Figure 3, we used an example of a bore crack of depth a = 22.54 mm located at 3.271° and a delamination of angle φ = 14.97° located at 117.55°. The strain field distribution in με is illustrated on the slice of the simulated cylindrical geometry of the SRM. The strain sensor locations are also depicted at 0°, 90°, 180° and 270°. The strain field distribution and the corresponding deformation is illustrated in the regions of both the bore crack and the delamination, with the deformation scale factor being set to 1.

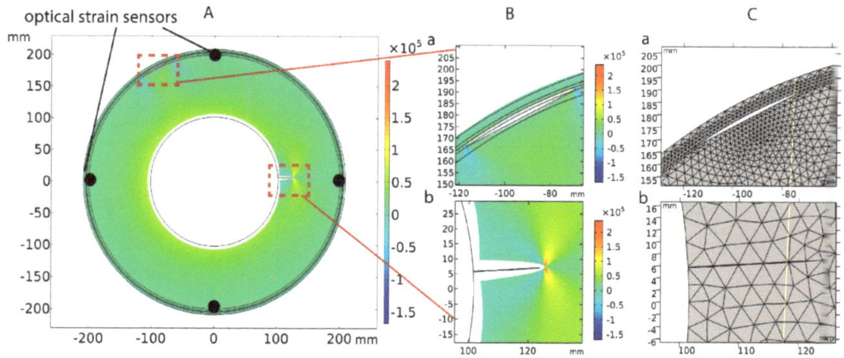

Figure 3. (**A**) Solid rocket motor geometry design, along with a bore crack and a delamination. (**B**) The strain field distribution and the simulated deformation of the grain in the region close to (**Ba**) the bore crack and (**Bb**) the delamination. (**C**) The mesh close to the delamination (**Ca**) and the bore crack (**Cb**).

In the conducted simulations, we assumed two types of possible flaws in the SRM: bore cracks and delamination. The designs of these flaws are depicted in Figure 3. In the case of the bore crack, the defect has a depth a within the range of (0.1 mm, 40 mm) with the corresponding gap width being equal to a/200. The delamination corresponds to an arc in the range φ (1°, 19°), with the gap appearing between the curved insulation boundary and a line segment between the limit points of the debonded region, which corresponds to the propellant domain boundary. Outside the debonded region, the two neighboring material domains—the insulation and the propellant—share a common interface corresponding to a curved arc with a radius equal to the outer radius of the propellant and an arc angle equal to 360° − φ. Based on the conducted calculations, the gap between the two domains after the thermal expansion is not expected to exceed 3 mm in the case of the largest debonding region angle examined. Provided that the simulations assume a homogeneously distributed temperature during the examined thermal cycle, this design can be considered accurate.

The study assumes the use of four strain sensors that are evenly distributed in the circular perimeter of the SRM, thus having an angular distance of 90° from each other and a 45° maximum angular distance from any defect in the worst-case scenario. The sensors are considered to be placed in the grain close to the interface between the propellant and the insulation, particularly with a distance of 3 mm from this interface. Such individual strain sensors are essentially multiple discrete FBGs integrated into a single optical fiber with a typical diameter of 125 μm or less. The study of the optimal number of sensors that should be placed per circular perimeter of the SRM is considered to be beyond the scope of this study; however, a short comparison with the preliminary study in the case of eight sensors and its impact on the diagnosis improvement is addressed in the Discussion section.

2.2. Hoop Strain Distribution vs. Radial Stress Distribution

Following the previously described SRM study formulation based on newly proposed strain distribution monitoring, it would be important to provide a direct comparison between the details and capabilities of strain and stress sensing, where the latter corresponds to the approach currently used in industry by DBST devices. It is expected that this comparison will highlight and demonstrate the differences and identify the advantages of strain monitoring employing optical strain sensors such as FBGs.

The key element of the present work is the use of optical-based strain sensors for the detection of defects in SRMs. Comparable studies in literature targeting SRM diagnosis and defect prediction focus on the use of stress sensors, particularly DBSTs. The transition to the recently introduced FBG-based strain sensors emerges with open questions regarding: (a) the differences between the strain field distribution and the stress field variations when a defect is present in the SRM, and (b) the consequent capabilities of the strain sensors to accurately detect the presence of a defect. Particularly, when a defect occurs in the SRM, it typically induces localized stress concentrations in the vicinity of the defect. These stress concentrations can lead to variations in the radial stress field, which DBST sensors are primarily designed to detect. However, since stress concentrations tend to dissipate more rapidly in the radial direction compared to the circumferential direction, the changes in the radial stress field may be more localized around the defect site. On the other hand, an equivalent defect can also cause significant changes in the hoop strain field along the circumference of the motor.

Hoop strain, which FBGs can measure, provides information about the circumferential deformation of the grain. The occurring defects can introduce disruptions in the structural integrity of both the casing and the grain, leading to alterations in the hoop strain distribution over a broader area compared to the localized changes in the radial stress field. However, the practical issues related to the sensor placements—specifically, the DBST sensors are placed in the insulation–casing interface, while FBGs should be placed within the propellant material—reduce the above-described effect, since the propellant is typically composed of a more homogeneous material compared to the casing material. Therefore, its structural integrity is less affected by localized defects. Figure 4 illustrates the radial stress and hoop strain field distribution in the typical radial position of the corresponding sensors. Particularly, Figure 4a depicts the residual strain in the grain 3 mm away from the insulation layer, and Figure 4b presents the residual stress in the middle of the insulation layer for three typical defects in the bore region (bore cracks) at depths of 5 mm, 10 mm, and 20 mm, respectively. In Figure 4c, we present the strain and stress field variation when studied at the same radial distance of the SRM, particularly in the middle of the insulation layer.

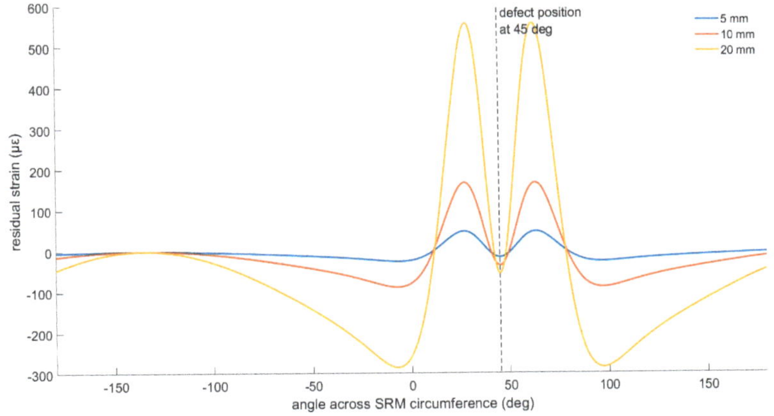

(a)

Figure 4. *Cont.*

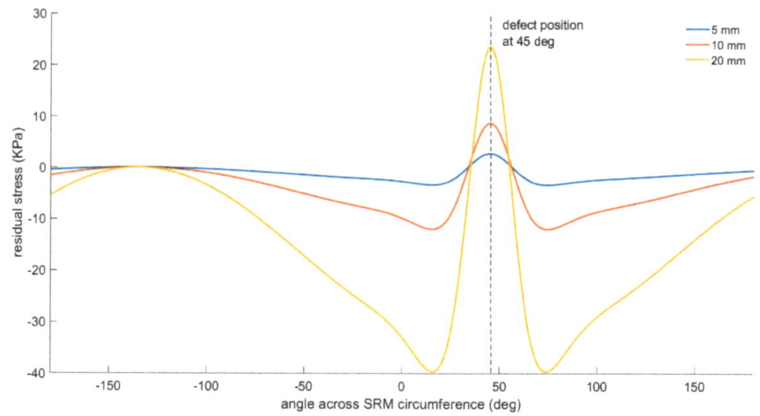

(**b**)

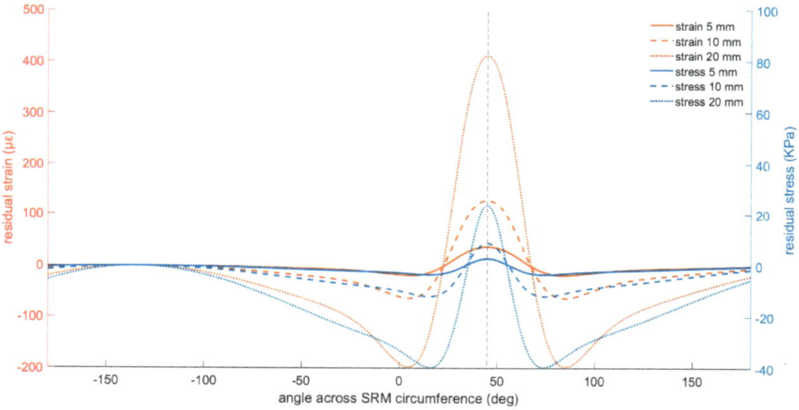

(**c**)

Figure 4. (**a**) Residual hoop strain field distribution across the SRM circumference, extracted from a simulation of a circle inside the grain, with a distance of 3 mm from the propellant–insulation circular interface for 3 typical bore cracks with depths of 5 mm, 10 mm, and 20 mm located at 45°. (**b**) Residual radial stress field distribution across the SRM circumference, extracted from a simulation of a circle located in the middle of the insulation layer for the same bore cracks. (**c**) Comparison of the strain and stress field variations produced by the same defects (bore cracks of 5 mm, 10 mm, and 20 mm located at 45°) measured at the same radial distance in the middle of the insulation layer.

Based on the study of the above graphs (Figure 4), we can conclude that: (a) a crack of double depth (10 mm compared to 5 mm) provokes a ~3.5 times higher maximum value in both the strain and the stress fields, irrespective of the location of the maxima; (b) the same happens for a bore crack depth that is four times deeper than the initial depth (20 mm compared to 5 mm crack), which provokes an ~11.5 times higher maximum value in both the stress and strain fields; (c) the variations in the hoop strain field are extended over a larger area compared to the variations in the stress field for the same crack, and they are likely to be detected even from sensors located even 90° away from the crack. This observation is further enhanced by the comparison of the absolute residual strain and stress values with the corresponding sensor resolutions, being ~4με for the FBGs and 10 KPa for the DBSTs. Particularly, the maximum change in the residual stress due to the presence of a bore crack at a 20 mm depth is 4 times the corresponding DBSTs stress resolution, while the maximum change in the residual strain exceeds 130 times the corresponding FBG strain resolution. Similar results are extracted based on the study shown in Figure 4c, where it can be seen that strain field variations are expected to be larger compared to stress field variations when measured using sensors located in the middle of the insulation layer. The strain fields tend to show more widespread variations due to the overall deformation of the material around the crack, while the stress field variations are highly localized around the crack tips. Therefore, strain measurements provide a broader indication of the defect's presence and impact.

Similarly, a delamination present in the interface between the propellant and the insulation layer produces a stress and strain field distribution shown in the following graphs based on simulation results (Figure 5). Particularly, in Figure 5a, we present the residual stress in the middle of the insulation layer, while in Figure 5b, we present the residual strain in the grain, 3 mm away from the propellant–insulation interface, provided the presence of 2°, 4°, and 8° delamination. The study of the provided data shows that the stress field variation due to the presence of a delamination of 4° is detectable by the stress sensors placed in the middle of the insulation layer only 10° away from the position of the defect (delamination position at 45°, sensor position for accurate detection at 35°). The corresponding strain field distribution presents a detectable change more than 37° away from the location of the defect (delamination position at 45°, sensor position for accurate detection at 7.5°). It is also proven that a delamination of 8° produces a maximum change in the strain field approximately 8500 times the expected strain resolution of the optical fiber strain sensors (expected to be at least 4 με), while the same delamination produces a maximum stress field change that is only110 times higher than the stress resolution of DBSTs (expected to be 10 kPa).

Based on the above observations, it was necessary to study the capabilities of SRM diagnosis and defect detection using FBG-based strain sensors, aiming at determining: (a) the minimum defect that can be detected, (b) the accuracy of defect size prediction, and (c) the corresponding accuracy of diagnosis when two types of defects are present (bore cracks and delamination). The following sections focus on these emerging issues, which comprise the main contribution of this work.

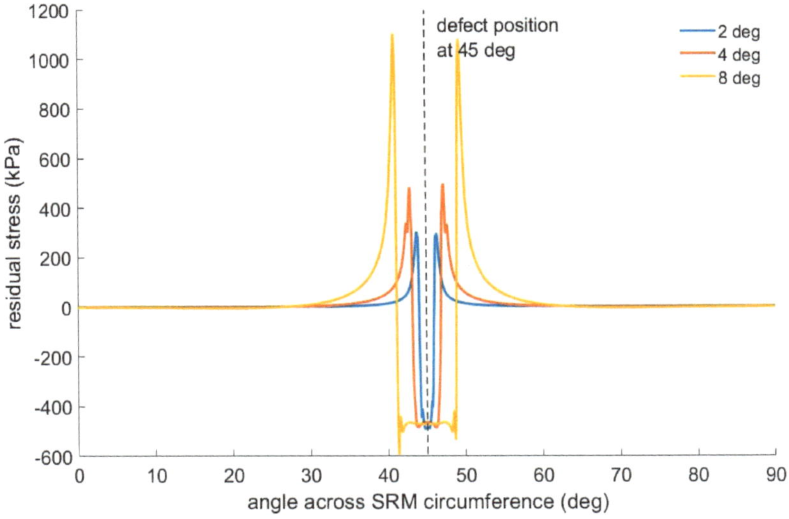

(a)

Figure 5. *Cont.*

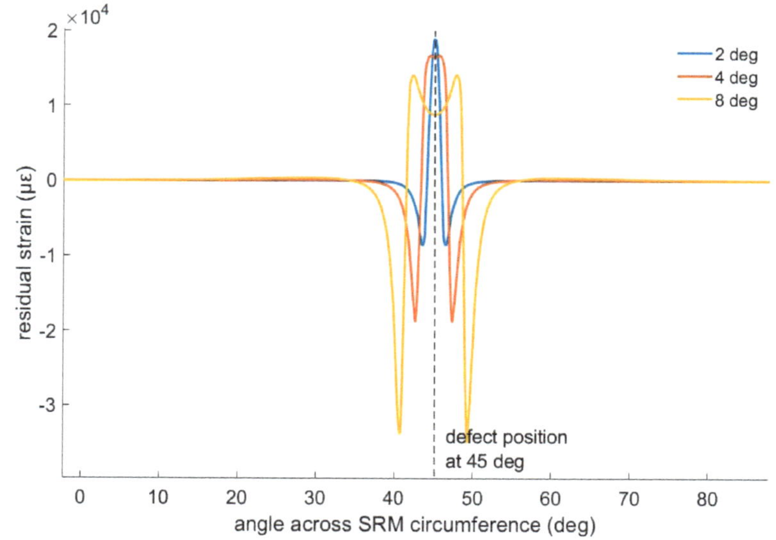

(**b**)

Figure 5. (**a**) Stress field distribution produced by 3 delamination cases of 2°, 4°, and 8° in the circumference of the SRM in the middle of the insulation layer, where DBST sensors are designed to be placed. (**b**) Corresponding strain field distribution produced by the same delamination cases in the grain, 3 mm away from the insulation–propellant interface, which is the assumed location of the optical-based strain sensors.

2.3. Healthy State and Damaged State of SRM

Discrimination between the healthy and the damaged state (Level I SHM) of the SRM is a typical problem that may be addressed using classification neural networks. The threshold values of all participating parameters, related to the extent and location of defects, correspond to the discrimination between the healthy state and the damaged state of the SRM. These are selected according to the minimum detectable flaws based on the employed optical strain sensors. Thus, all cases with a bore crack depth larger than the corresponding threshold of the minimum detectable bore crack depth belong to the damaged state, while the cases with smaller crack depths belong to the healthy state. The same applies for the delamination angle. For the determination of the minimum detectable defect extent, it is assumed in a conservative scenario that the resolution of these sensors is 4 με (which is achievable using current FBG technology), and the worst-case scenario for the position of the defect with respect to the sensing element is set to its maximum, equal to 45°, provided the use of four sensors that are evenly distributed on the SRM perimeter.

In order to define the minimum detectable bore crack, a simulation with variable bore crack depth a in the range of [0.1 mm, 15 mm] with a step of 0.1 mm for the range of [0.1 mm, 3 mm] and a step of 1 mm for the range of (3 mm, 15 mm] was performed for a bore crack located at 45°. The detection of the bore crack is assured when the absolute difference between the strain value of the sensor closest to the flaw and the most distant sensor is higher than 4 με, provided that the location of the flaw at 45°, the nearest sensor is at 0° (or equally 90°), and the most distant sensor is located at 180° (or equally 270°). Figure 6 illustrates this strain difference, where each plot is presented with reference to the corresponding strain value of the sensor located at 180°. Particularly, Figure 6 illustrates the strain field across the circular path with a radius equal to the radius of the strain sensor locations, 3 mm away from the interface between the insulation and the propellant inside the grain, which is the circular path of interest in this study. Based on the zoomed-in region illustrated in Figure 6Aa, it is depicted that a minimum depth of 2.2 mm can be detected using a sensor located at an angular distance of 45°. The same process was repeated for the delamination case, with the parameter of the delamination angle varying in the range of [1°, 10°] with a step of 1°. The minimum detectable delamination extends in a region with an angular size equal to 7°. This process is based on a rather pragmatic consideration of the healthy state depending on the accuracy of the employed sensing technology.

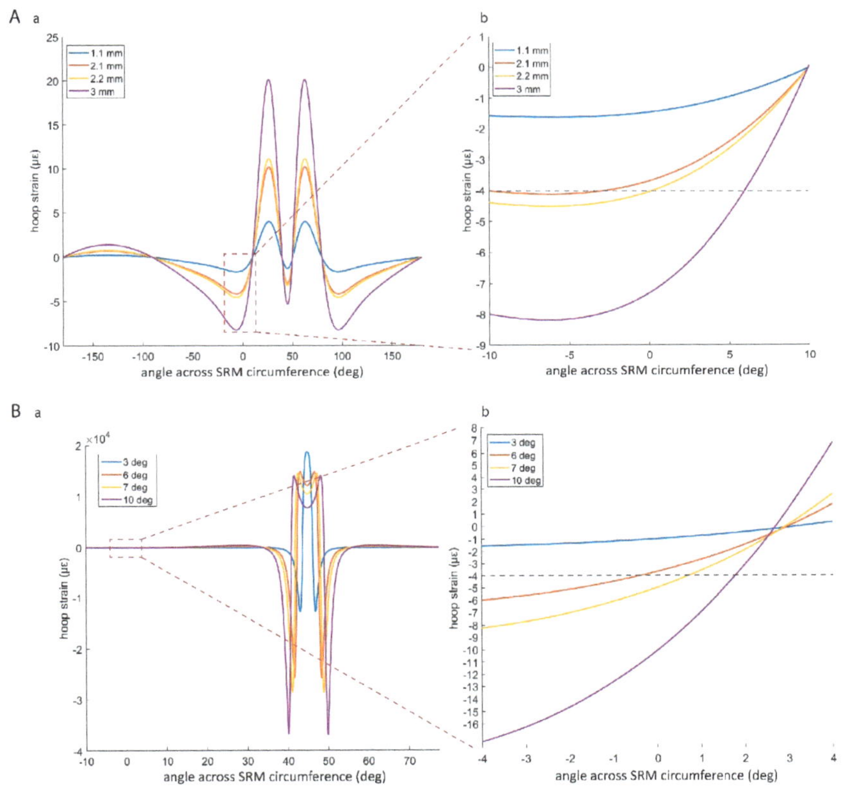

Figure 6. Strain field across the circular path of interest for different bore crack depths (**Aa**) and delamination (**Ba**) located at 45° (worst-case scenario). Zoomed regions for each of the above cases, for the bore crack (**Ab**), and for the delamination (**Bb**).

2.4. Strain Data Generation Using FEA Models

To produce an adequately large dataset for the analysis and NN training, we divided the simulations into three (3) separate cases. The first case corresponds to the presence of a bore crack only; the second case corresponds to the presence of a delamination only; and the third case corresponds to the presence of both. The presence of more than two flaws of the same or different type is considered rare and is excluded from the produced dataset. For each of the above cases, a subset of 1000 data points was produced for: (a) the "healthy condition", which corresponds to defect extent values below the threshold of the minimum detectable flaw, and (b) the "damaged condition", corresponding to defect extent values above the threshold of the minimum detectable flaw. Thus, a total of 6000 cases of different bore crack depths and delamination angles were examined. The bore crack depth and delamination values are randomly selected within the appropriate range each time.

For the above-described simulations, the position of the defects is always selected using a set of random numbers that are uniformly distributed to the appropriate range. In the case of a single bore crack, the angular position is restricted for the simulations in the range of [0°, 90°]. Using the extracted values of the strain field in the exact point location of the four sensors, and based on the circular symmetry of the SRM geometry, we extended the dataset in the remaining three quarters of the cylinder. This was performed via mutual interchange of the location of the sensors, as shown in Figure 7. Particularly, provided a bore crack at a depth located in the first quarter of the cylinder, we can extract the strain values for a bore crack of an equal depth a located in the second quarter of the cylinder (angular position range: [90°, 180°]) by mutually interchanging the location of sensors 1 and 3. The corresponding position of the bore crack is given by the appropriate shift, [180°−θ], where θ is the angular position of the initial bore crack in the range of [0°, 90°]. Respectively, by mutually interchanging the location of strain sensors 2 and 4, we can result with a bore crack of the same depth located at the fourth quarter of the cylinder; i.e., in the range of [270°, 360°]. Finally, if we interchange the sensors in locations 2 and 3 while at the same time interchanging the location of sensors 1 and 4, we extract the sensors values as if the bore crack was located in the third quarter of the cylinder; i.e., in the range of [180°, 270°). Provided that the simulations are performed for 2000 cases of bore cracks located only in the first quarter, we followed this simple strategy to multiply the dataset by a factor of four while at the same time covering the entire cylinder. The same strategy is followed for the case of delamination. In the case of coexistent bore cracks and delamination, in the simulated cases, the angular position of the bore crack is restricted in the range of [0°, 90°], corresponding to the first quarter of the cylinder, while the relative angle between the two defects is a number randomly selected in the range of [0°, 360°). In this way, it is assured that the relative angular distance between the defects can take any value in the entire cylinder. This strategy results in multiplication of the data by a factor of 4. Therefore, the final dataset used for training and validating the developed neural network-based diagnostic system consists of 24,000 cases. Moreover, this approach was proven to be equivalent to the raw data production by FEA; i.e., the process was verified by comparing the results with the strain values in the particular positions of the strain sensors when the defect is initially located in the sifted angular position. The error is negligible, lower than 0.1 με, and it is attributed to the selected mesh size of the FEA. Thus, it could be further reduced through the application of a denser mesh.

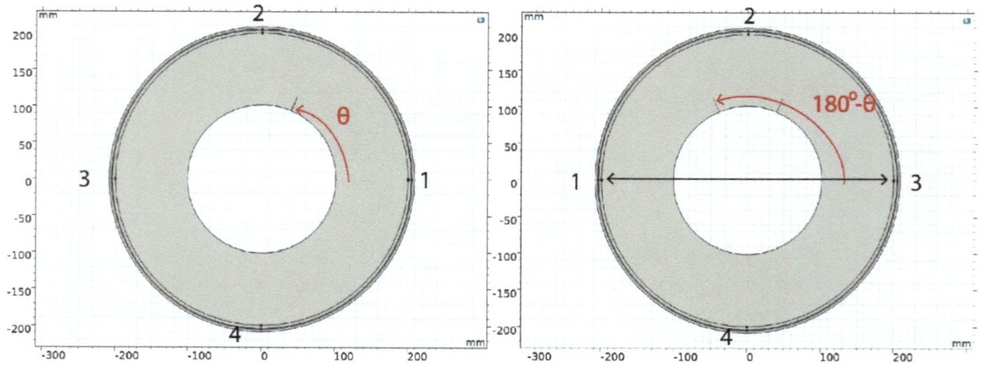

Figure 7. Description of the followed strategy of mutual interchange of sensor locations for dataset multiplication and coverage of all possible angular positions of the bore crack or delamination. In the case illustrated here, a bore crack depth initially located at an angle of θ in the range of [0°, 90°] is transferred in the second quarter of the cylinder by mutually interchanging the location of sensors 1 and 3.

3. Defect Diagnosis Using Machine Learning Techniques

3.1. Deep Neural Networks

Deep neural networks are a subcategory of machine learning techniques that can be employed to develop system diagnosis and structural health monitoring tools, provided the use of sensors readings. Deep fully connected neural networks are neural networks with multiple hidden layers between the input nodes and the output layer, which consist of fully connected (FC) layers. In FC layers, each neuron applies a linear transformation to the input vector through a weight matrix and a bias matrix. Thus, an FC layer consists of the weights and biases, along with the neurons, and it is used to connect the neurons between two different layers. In the FC layers, all the input nodes are connected to all the output nodes, meaning that a change in one input node affects all output nodes. In the following section, we present the two parts of the developed diagnostic tool, both consisting deep FC networks with the appropriate selection of the corresponding weights and bias initialization, suitable for the purposes of classification (first step of health state extraction), and regression (second step of defect extent prediction).

3.2. Classification of Health State Classes Using Deep Neural Networks

The first step of the diagnosis comprises the classification of the health state of the SRM (level I SHM) in one of the following classes: (a) healthy state in which the present flaw, either a bore crack or a delamination, is below the threshold limit, (b) damaged case with the presence of a single bore crack, (c) damaged case with the presence of a single delamination, and (d) damaged case with the presence of both a bore crack and a delamination. Figure 8 provides a high-level flowchart to illustrate the process of health state identification and defect detection. The neural network architecture is illustrated in Figure 9, while Table 2 provides a detailed description of all the layers participating in the network.

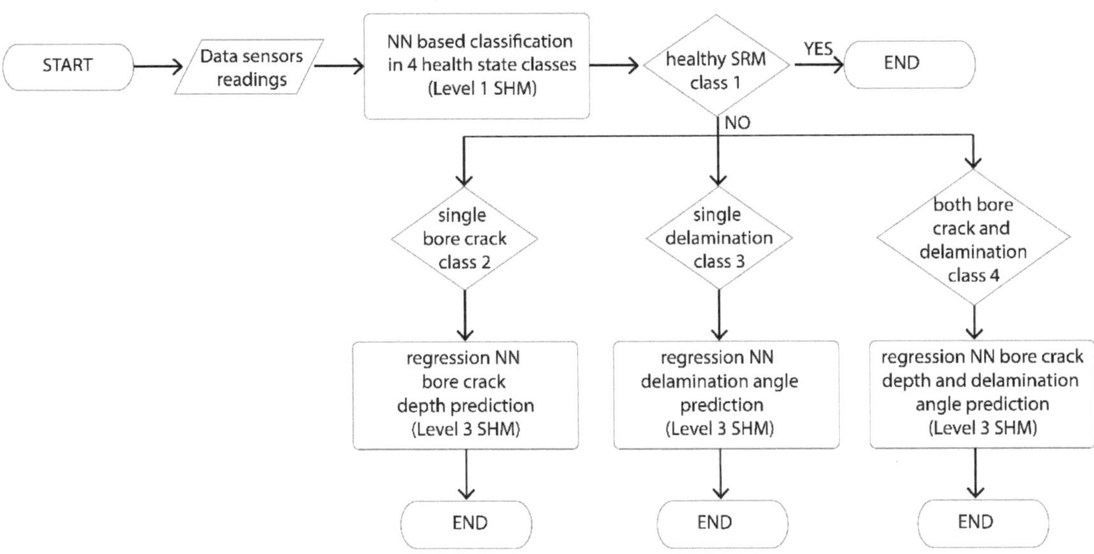

Figure 8. Flow chart depicting the 2 steps of SHM diagnosis: step 1 (Level 1 SHM), providing the classification in the four (4) classes of health state, and step 2 (Level 3 SHM), providing the quantification of defects in each of the 3 damaged classes.

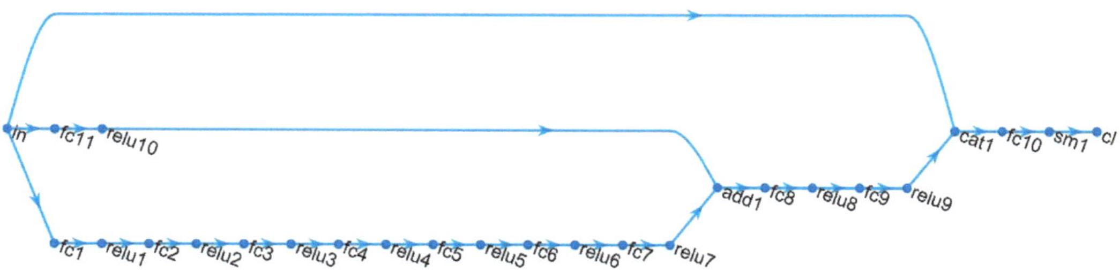

Figure 9. Architecture of the developed classification layer.

Table 2. Classification network analysis parameters.

	Name	Type	Activations	Learnables	Total Learnables
1	In	Feature input	4	-	0
2	fc1	Fully connected layer	1024	Weights 1024 × 4 Bias 1024 × 1	5120
3	relu1	ReLU	1024	-	0
4	fc2	Fully connected layer	1024	Weights 1024 × 1024 Bias 1024 × 1	1,049,600
5	relu2	ReLU	1024	-	0
6	fc3	Fully connected layer	1024	Weights 1024 × 1024 Bias 1024 × 1	1,049,600
7	relu3	ReLU	1024	-	0
8	fc4	Fully connected layer	512	Weights 512 × 1024 Bias 512 × 1	5,248,000
9	relu4	ReLU	512	-	0
10	fc5	Fully connected layer	256	Weights 256 × 512 Bias 256 × 1	131,328
11	relu5	ReLU	256	-	0
12	fc6	Fully connected layer	128	Weights 128 × 256 Bias 128 × 1	32,896
13	relu6	ReLU	128	-	0
14	fc7	Fully Connected Layer	64	Weights 64 × 128 Bias 64 × 1	8256
15	relu7	ReLU	64	-	0
16	fc11	Fully connected layer	64	Weights 64 × 4 Bias 64 × 1	320
17	relu10	ReLU	64	-	0
18	add1	Addition layer	64	-	0
19	fc8	Fully connected layer	32	Weights 32 × 64 Bias 32 × 1	2080
20	relu8	ReLU	32	-	0
21	fc9	Fully connected layer	16	Weights 16 × 32 Bias 16 × 1	528
22	relu9	ReLU	16	-	0
23	cat1	Concatenation layer	20	-	0
24	fc10	Fully connected layer	4	Weights 4 × 20 Bias 4 × 1	84
25	sm1	Softmax	4	-	0
26	cl	Classification layer	-	-	0

The selected architecture of the deep neural network contains 11 fully connected (FC) layers—denoted with the prefix 'fc'—in which all input nodes of the layer are connected to all output nodes of this specific layer. Among the 11 FC (fc) layers, 10 layers are used in a sequential structure, followed by a rectified linear unit (ReLU) layer denoted with the prefix 'ReLU'. The selected weights initializer is set to 'orthogonal', while the bias initializer is set to 'narrow–normal'. The prefixes 'add' and 'cat' are used to denote addition and concatenation layers, which are both used to connect the main branch of the neural network with the two additional branches. All the developed NNs and corresponding data handling are performed in a MATLAB environment.

The parameters significantly affecting the training process are: (a) the selected optimizer algorithm, set to ADAM (adaptive movement estimation method), which is an extended version of stochastic gradient descent (SGD) that is particularly suitable for classification purposes, (b) the batch size, which is set to 2048, resulting in 10 iterations per epoch, provided the use of 90% of the total amount of available data for the training set and 10% of the remainder for the validation set. The training is performed using a 'piecewise' learning rate schedule, which means that the learning rate starting from an initial value of 0.001 drops with a factor of 1‰ every epoch during the entire training procedure. The maximum number of epochs is set to 10,000, with a validation patience of 500 epochs to avoid overfitting. Thus, the training is stopped after 5310 epochs, reaching a validation accuracy of 98.38%. The required training time does not exceed 41 min when running on a single Intel® Core i7-10750H CPU (16 Gb RAM) on a 64-bit operating system. The plots describing the accuracy increase and loss reduction during the entire training process are given in Figure 10, showing an adequate convergence after approximately 5000 iterations.

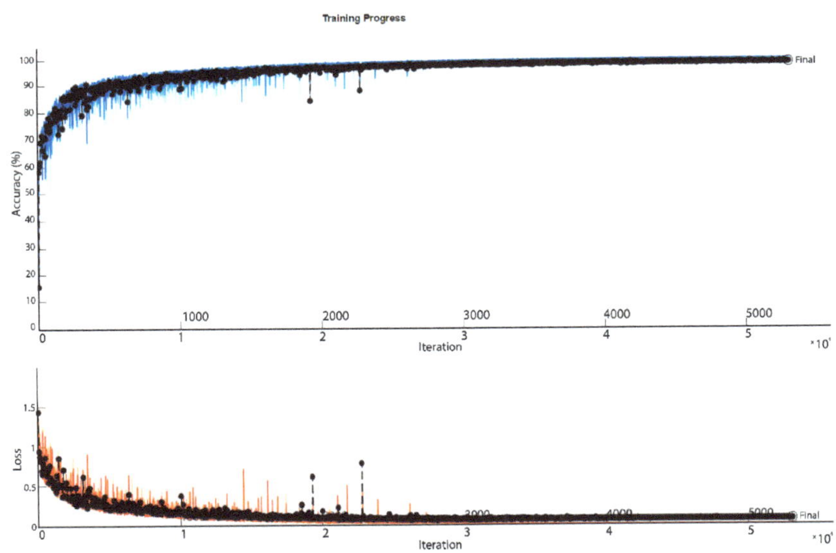

Figure 10. Accuracy and loss during the entire training process for the classifier corresponding to the prediction of the health state of the SRM. Blue and red lines indicate the training accuracy and loss respectively in a batch level, while black dots indicate the corresponding values at the end of each iteration.

The validation accuracy is presented in Table 3 per class, along with the corresponding confusion matrix (Figure 11). In the confusion matrix, the numbers in each box represent the examined cases, with the percentage over the total sum of all examined cases given below the absolute number.

The last column of the confusion matrix gives the recall, defined as the ability of a model to find all the relevant cases within a data set, mathematically given by the following formula:

$$recall = \frac{True\ positive}{True\ positive + False\ negative}, \quad (1)$$

The last line of the confusion matrix gives the precision of the model per class, defined as the ability of the classification model to identify only the relevant cases. It is described mathematically by the formula:

$$Precission = \frac{True\ positive}{True\ positive + False\ positive}, \tag{2}$$

It is proven that the false positive and false negative results are limited, with higher values given between the 'bore crack and delamination' class and the 'delamination' class. Twelve cases of 'bore crack and delamination' were falsely classified as 'delamination'. Further analysis of these false results proved that these false classification cases mainly corresponded to rather small bore cracks (bore crack depths below 12 mm and ~75% of them below 5 mm) combined with a delamination.

The accuracy of the model is mathematically calculated as follows and it is proven to be remarkably high, reaching 98.38%:

$$accuracy = \frac{True\ positive + True\ negative}{True\ positive + True\ negative + False\ positive + False\ negative}, \tag{3}$$

We additionally provide the $f1$ score per class, which is a metric of how strong a classifier is. It is given as a harmonic mean between the two other provided metrics, precision and recall, per class:

$$f1\ score = \frac{2 \times Precision \times Recall}{Precision + Recall}, \tag{4}$$

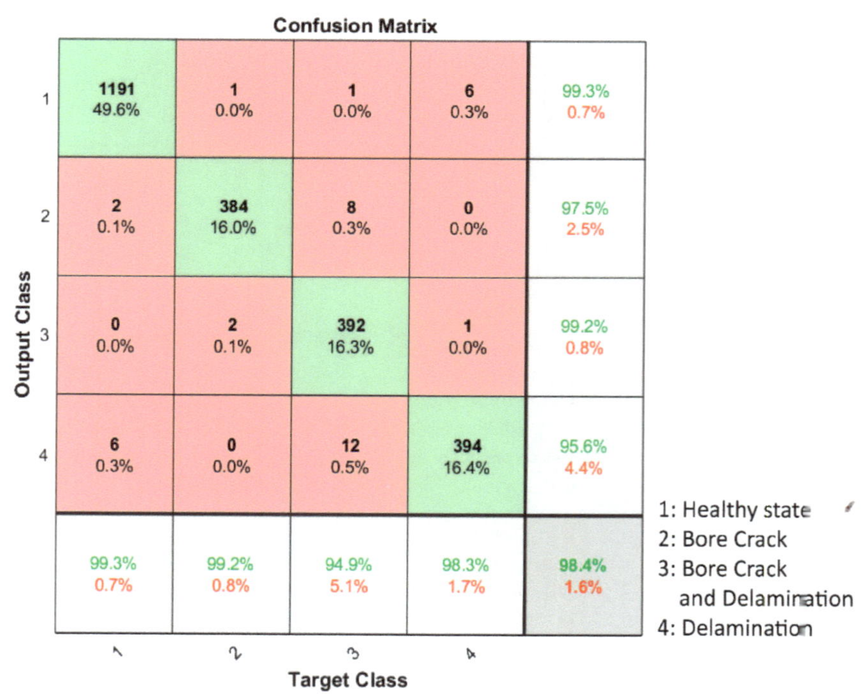

Figure 11. Confusion matrix based on the dataset used to validate the developed deep neural network. True indications appear in green color and false indications appear in red color. The same color rule applies in the boxes.

Table 3. Validation accuracy factors per health state class provided by the developed deep neural network.

	Healthy State	Bore Crack	Bore Crack and Delamination	Delamination
Accuracy			0.9838	
Precision	0.9933	0.9922	0.9492	0.9825
Recall	0.9933	0.9746	0.9924	0.9563
$f1$ score	0.9933	0.9834	0.9903	0.9692

3.3. Defect Extent Prediction Using Regression Deep Networks

The second step of the SRM diagnosis comprises a regression network built using the same architecture for each one of the above three classes of damaged states of the SRM and trained individually for each case to accurately predict the extent of the defect. The structure of the neural network follows the basic principles of the deep networks' architecture developed for classification in the previous step. The network contains a main branch with sequential use of FC layers and ReLU layers. These are combined with the raw strain sensors' data and the output of a single FC layer, along with a ReLU activation layer, using a concatenation layer and an addition layer, respectively. The analytical description of the network architecture is depicted in Figure 12, with the participating parameters being given in Table 4.

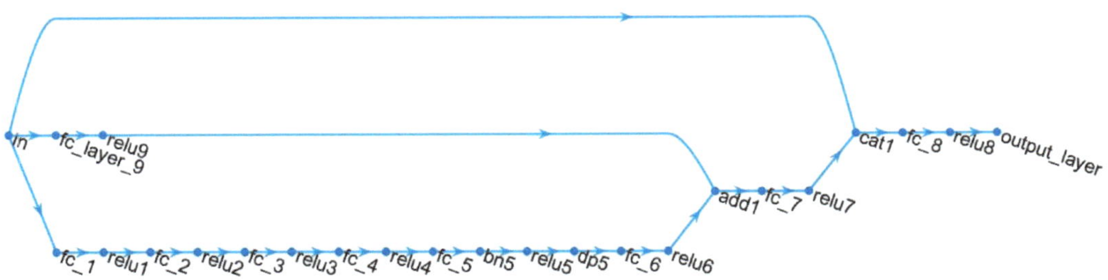

Figure 12. The architecture of the regression deep neural network developed for the prediction of the present defect in each of the damaged health state cases/classes.

In the case of the regression deep neural network, the weights initializer is set to 'He' for all FC layers. The He initialization is an initialization method used in NNs which takes into account the non-linearity of activation functions [29]. Additional batch normalization and dropout layers (20%) are used once in the entire neural network to reduce overfitting. The initial learning rate is set to 0.0005, again following a piecewise learning rate schedule with a 1‰ drop factor in every epoch. The batch size is set to a much lower value than the one in the classification due to the remarkably smaller dataset used to feed the neural network this time. The dataset for the regression network is 1/6 of the entire developed dataset, corresponding to one of the 3 cases/classes of the damaged state of the SRM. Thus, the batch size is set to 64, and the resulting number of iterations per epoch is 56. A validation patience strategy is again followed to avoid overfitting, with the corresponding parameter set to 100 epochs. Finally, the optimizer algorithm selected for the regression network is the root mean square propagation (RMSprop), which best fits regression networks, rather than the ADAM optimizer selected for the classification deep neural network. The developed network was trained individually for each case of the damaged state of the SRM; thus, all parameters related to the training process were analyzed separately for each case in the following analysis.

Table 4. Regression network analysis parameters.

	Name	Type	Activations	Learnables	Total Learnables
1	in	Feature input	4	-	0
2	fc1	Fully connected layer	1024	Weights 1024 × 4 Bias 1024 × 1	5120
3	relu1	ReLU	1024	-	0
4	fc2	Fully connected layer	1024	Weights 1024 × 1024 Bias 1024 × 1	1,049,600
5	relu2	ReLU	1024	-	0
6	fc3	Fully connected layer	1024	Weights 1024 × 1024 Bias 1024 × 1	1,049,600
7	relu3	ReLU	1024	-	0
8	fc4	Fully connected layer	512	Weights 512 × 1024 Bias 512 × 1	5,248,000
9	relu4	ReLU	512	-	0
10	fc5	Fully connected layer	256	Weights 256 × 512 Bias 256 × 1	131,328
11	bn	Batch normalization layer	256	Offset 256 × 1 Scale 256 × 1	512
12	relu5	ReLU	256	-	0
13	dp5	Dropout	256	-	0
14	fc6	Fully connected layer	32	Weights 32 × 256 Bias 32 × 1	8224
15	relu6	ReLU	32	-	0
16	fc9	Fully connected layer	4	Weights 32 × 4 Bias 32 × 1	160
17	relu9	ReLU	32	-	0
18	add1	Addition layer	32	-	0
19	fc7	Fully connected layer	16	Weights 16 × 32 Bias 16 × 1	528
20	relu7	ReLU	16	-	0
21	cat1	Concatenation layer	20	-	0
22	fc8	Fully connected layer	1 or 2	Weights 1 or 2 × 20 Bias 1 or 2 × 1	42
23	relu8	ReLU	1 or 2	-	0
24	output	Regression layer	-	-	0

3.3.1. Bore Crack Prediction

For the case of a single bore crack, the regression deep neural network is trained using 90% of the damaged cases containing a bore crack in the range of (2.2 mm, 40 mm]. The validation root mean square error (RMSE) is 0.2320 mm. A total of 428 epochs are needed to accomplish the entire training process, which corresponds to a time duration of 13 min and 50 s. In Figure 13a, the predicted values of the bore crack depth are given with respect to the true values. The fitted curve is $y = p_1 \cdot x + p_2$, with the coefficients being $p_1 = 0.9996$ (95% confidence bounds: (0.9976, 1.002)) and $p_2 = 0.07051$ (95% confidence bounds: (0.02222, 0.1188)). For the representation of the provided results, we opted for the additional use of Bland–Altman plots, giving the distribution of the error of prediction in the entire range of the targeted defect extent values to extract any dependency of the accuracy of the developed regression model on the bore crack depth. This type of plot demonstrates the error of prediction, defined as the difference $a_{true} - a_{predicted}$ over the mean of the true and predicted values of the bore crack depth, $\left(a_{true} + a_{predicted}\right)/2$. The corresponding Bland–Altman plot given in Figure 13b proves a negligible correlation between the prediction error and the depth of the bore crack. Finally, we present the distribution of the RMSE over the angular position of the bore cracks by dividing the entire angular position range (360°) into 20 bins.

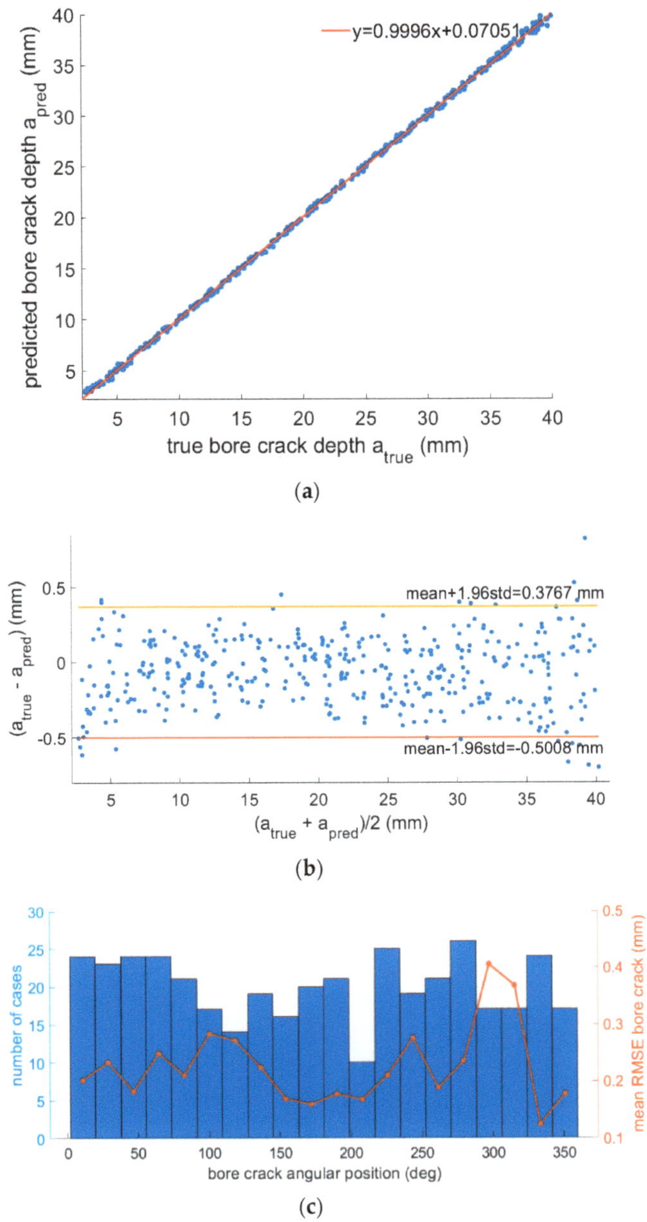

Figure 13. Validation results of the regression deep neural network trained with damaged health state cases containing a single bore crack with a depth in the range of (2.2 mm, 40 mm]. (**a**) Plot of the predicted bore crack depth with respect to the real values of the bore crack depth along with the fitting curve. (**b**) Bland–Altman plot of the same validation data. (**c**) Plot of the RMSE (right axis) with respect to the angular position of the bore crack, separated into 20 bins.

The corresponding plot given in Figure 13c shows: (a) that the number of randomly selected validation cases is uniformly distributed in the entire circular SRM cross section, given by the number of cases per bin (left axis), and (b) that the distance from the sensors does not seem to affect the prediction accuracy in the selected range of bore crack depths corresponding to the damaged health state of the SRM.

3.3.2. Delamination Prediction

In the class of a damaged health state with a single delamination, the network is trained again using the same percentage of the available data (90%), which corresponds to 3600 cases. The remaining 400 cases are used for the validation of the network. The required time for training, provided that the use of validation patience is equal to 100 epochs, does not exceed 15 min, and the provided validation RMSE is 1.2897°. In Figure 14a, the predicted values of the delamination extent are given with respect to their true values in the validation dataset. The curve fitting the combination of the predicted and true data for the angle of delamination is $y = p_1 \cdot x + p_2$, with the coefficients being $p_1 = 0.9011$ (95% confidence bounds: (0.8696, 0.9325)) and $p_2 = 1.267$ (95% confidence bounds: (0.7968, 1.737)). The Bland–Altman plot corresponding to this case and provided in Figure 14b also proves a negligible dependence of the prediction error on the angular size of the debonded region for the specified range of (7°, 19°]. iIn other words, the developed model can predict a small or a large delaminated region with approximately the same accuracy. Finally, the plot in Figure 14c gives the distribution of the RMSE over the randomly selected angular position of the delamination for all cases participating in the validation dataset, as well as the distribution of the number of examined cases over the entire SRM circular cross section. The prediction error does not seem to be strongly affected by the position of the delamination with respect to the sensors' positions, with the RMSE being held below 2° for most cases, apart from the bin (126°,144°], in which the mean RMSE reaches the highest value of approximately 4°. The plot also proves that the validation dataset comprises cases that are evenly distributed in the SRM circle and that there is no correlation between the number of cases per bin (left axis of the plot) and the mean RMSE of each bin (right axis of the plot).

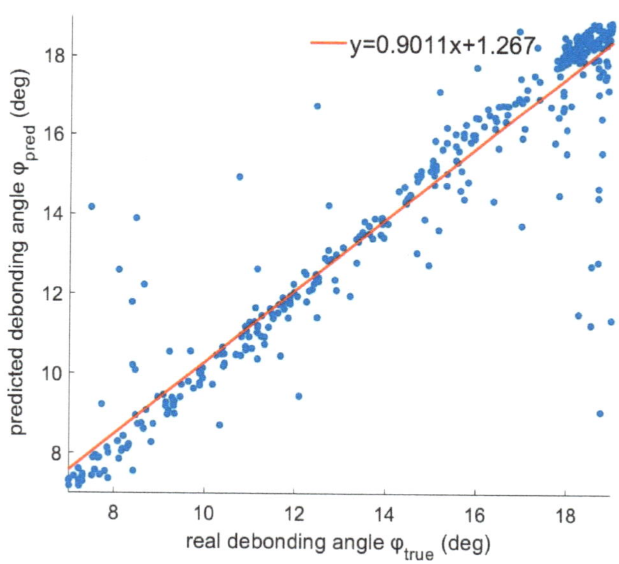

(a)

Figure 14. *Cont.*

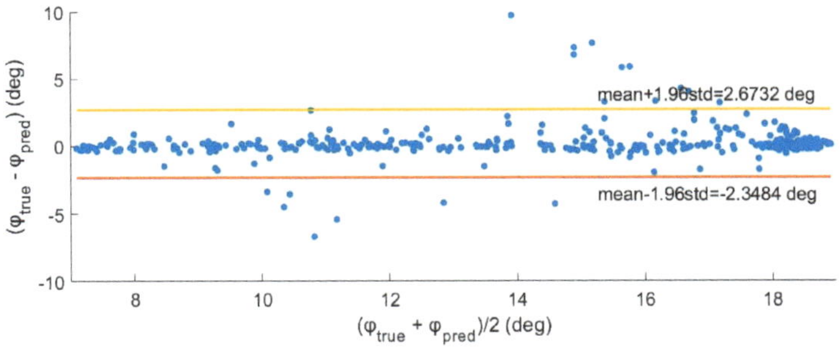

(**b**)

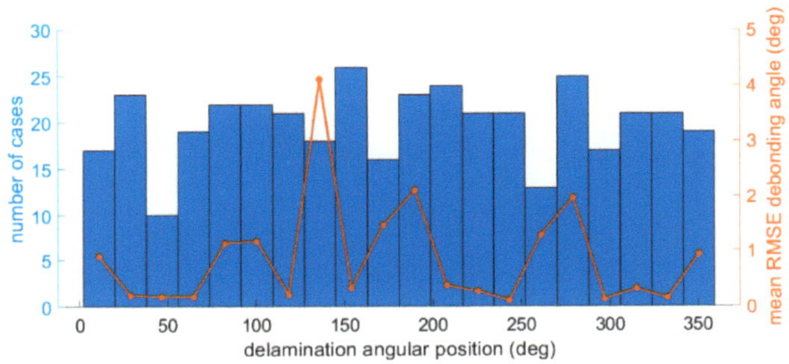

(**c**)

Figure 14. Validation results of the regression deep neural network trained using damaged health state cases containing a single delamination with an angular size in the range of (7°, 19°]. (**a**) Plot of the predicted delamination angle with respect to the real values of the delamination angle along with the fitting curve. (**b**) Bland–Altman plot of the same validation data. (**c**) Plot of the RMSE (right axis) with respect to the angular position of the delamination, separated into 20 bins.

3.3.3. Combined Cracks and Delamination Detection

The class of the damaged state of the SRM that corresponds to the coexistence of a bore crack and a delamination is the most demanding, as two different values of the extent of the defect are to be predicted. The training process stops after 1000 epochs, with the required time reaching approximately 30 min. The provided RMSE for the case of the bore crack is 2.3 mm, and for the case of the delamination, it is 1.5678°. In Figures 15a and 16a, we present the predicted values along with the true values of the bore crack depth and the delamination angle, respectively. The curve fitting the data in the predicted vs. true values plot for the bore crack is $y = p_1 \cdot x + p_2$, with the coefficients being $p_1 = 0.9573$ (95% confidence bounds: (0.9372, 0.9773)) and $p_2 = 0.0009858$ (95% confidence bounds: (0.000507, 0.001465)). The corresponding fitting curve process for the case of the delamination results in the curve $y = p_1 \cdot x + p_2$, with the coefficients being $p_1 = 0.8028$ (95% confidence bounds: (0.7647, 0.8409)) and $p_2 = 2.612$ (95% confidence bounds: (2.103, 3.122)). We also represent the corresponding Bland–Altman plots, revealing any dependency of the prediction accuracy on the extent of each defect in Figure 15 plot (b) and Figure 16 plot (b). Both Bland–Altman plots prove the negligible dependence of the prediction error on the size/extent of the defect. In this health state, we opted for presenting the variation in the RMSE with respect to the relative angle between the defects by dividing the entire range of the relative angle (180°) into 20 equally wide bins. The corresponding plots are given in Figures 15c and 16c for the case of the bore crack and the delamination, respectively. It is proven that there is no correlation between the prediction error of the bore crack depth and the relative angle between the defects, with the mean RMSE per bin reaching a maximum value of approx. 3.5 mm. In the case of the delamination, the mean RMSE per bin is held below

3°; however, it reaches its maximum value for the most distant defects, corresponding to the bin (171°,180°]; i.e., the relative angle between them is approximately equal to 180°. Additionally, we observed an increased distribution of the number of cases participating in the validation dataset that belong to the bins with narrow angular distances between the two defects. This is expected to be the case for the training dataset, and it is attributed to the fact that the angular position of both defects is uniformly distributed in the entire circle.

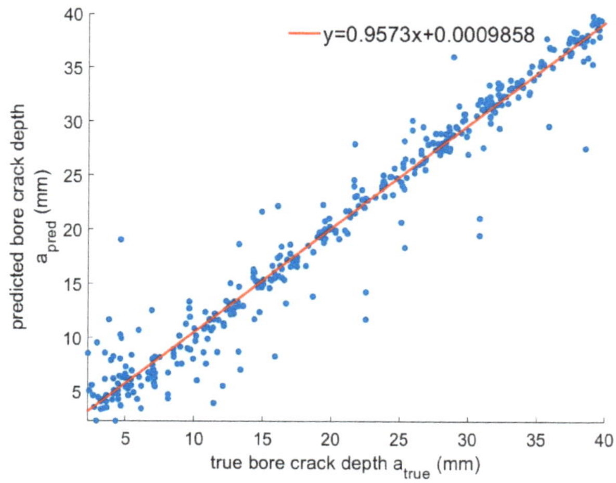

(a)

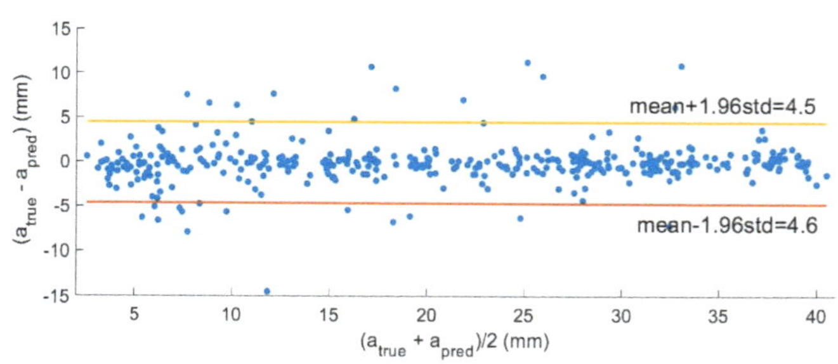

(b)

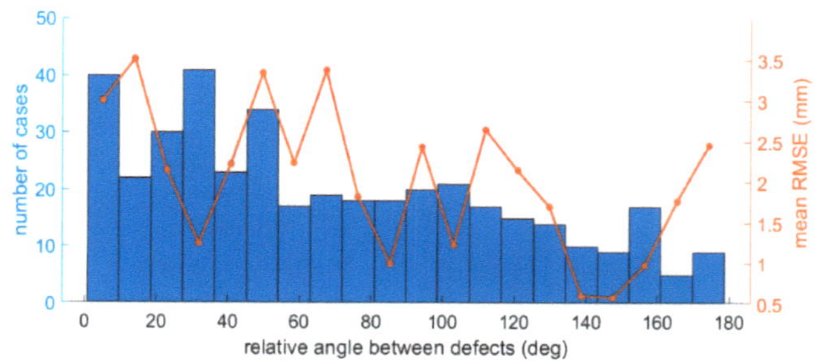

(c)

Figure 15. Validation results of the regression deep NN trained with damaged state cases containing both a bore crack and a delamination, particularly related to the prediction of the bore crack defect. (**a**) The predicted bore crack depth with respect to its real value, along with the fitting curve. (**b**) Bland–Altman plot of the same validation data; (**c**) Plot of the RMSE (right axis) and the number of cases (left axis) with respect to the relative angle between the coexisting defects, separated into 20 bins.

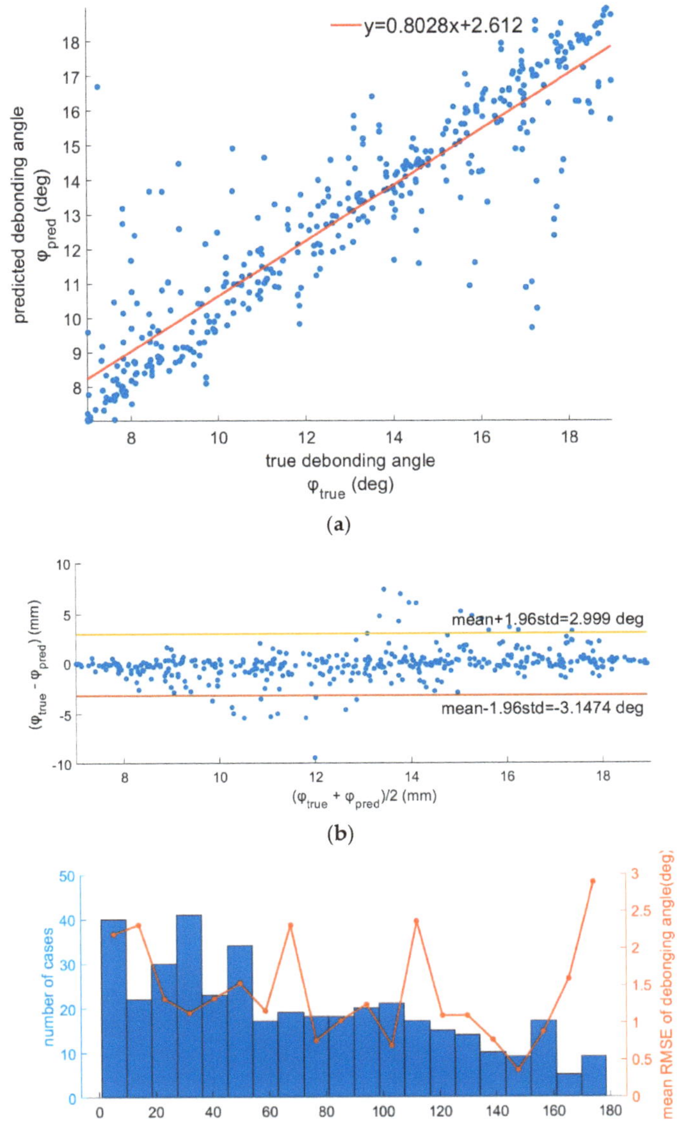

Figure 16. Validation results of the regression deep NN trained with damaged state cases containing both a bore crack and a delamination, particularly related to the prediction of the delamination. (**a**) The predicted delamination angle with respect to its real values, along with the fitting curve. (**b**) Bland–Altman plot of the same validation data. (**c**) Plot of the RMSE (right axis) and the number of cases (left axis) with respect to the relative angle between the coexisting defects, separated into 20 bins.

4. Discussion and Future Perspectives

The presented results provided a systematic study on the investigation of the combined performance of both the health state classifications, particularly in predicting the types of defects that are present in the SRM and the defect extent prediction through the corresponding regression neural network. Previous research studies have reported on some of related issues [11], but none have opted for the combination of these steps, which can lead to a complete diagnosis of the SRM health state. Particularly, Liu et al. reported on the defect extent prediction error only for the case of a coexistent bore crack and delamination [11] using an architecture of deep convolutional neural networks (CNNs) similar to the one reported here. This architecture was able to predict the delamination angle and bore crack depth in the case of coexistent defects and provided the use of DBST sensors. In this work, we introduced a NN with fully connected layers instead of a CNN; the single use of four sensors readings instead of a temperature-dependent data series as the input given to the network; and the incorporation of cases of single defect prediction along with the case of combined cracks and delamination for accurate SRM diagnosis across a wider range of defects. Particularly, in the case of coexistent bore cracks and delamination, our findings are in full compliance with previous reported works [11], additionally providing a reduced prediction error as well as a broader range of defect extent detection.

The broadening of the defect extent range, with the inclusion of smaller defects, is attributed to the employment of optical fiber strain sensors that, due to their achievable resolution, are capable of detecting smaller defects compared to well-established DBST deformation sensors. Although very limited previous work is available on optical fiber integration in composite propellants, in principle, optical fibers provide the flexibility to be incorporated anywhere in the propellant at various geometries during the propellant casting process by providing suitable bonding and glueing process with specific primers [8]. However, further investigation is required depending on the fiber material (silica, polymers) and coating. In contrast, DBST sensors can be only embedded in SRMs through the metal casing in order to reach the upper layer of insulation. Therefore, optical fiber sensors are exposed to regions of larger deformations and higher variations of the strain field, which consequently leads to the detection of smaller defects. This is proven by the comparison of the calculated threshold between the healthy and the damaged state for both the bore crack case and the delamination case, which are both much smaller than previously targeted defects [11]. We opted for assuming the use of optical strain sensors profiting from their capacity to detect much smaller defects—both bore cracks and delamination—and the performance of the developed network is also evaluated in these cases. Furthermore, optical strain sensors based on Bragg gratings on optical fibers is a dynamic research area that provides continuously enhanced sensors in terms of functionality and resolution. FBGs can be concatenated in reasonably high numbers, enabling a multipoint quasi-distributed sensor operation via monitoring without deteriorating the mechanical properties of SRMs. This is accomplished through the seamless incorporation process of FBGs within the core of a fiber with a typical diameter of 125 μm or less. Furthermore, new advances have enabled the inscription of FBGs in polymer optical fibers of a higher elasticity than typical silica fibers, thus providing higher strain range capabilities [30,31]. Based on the presented results, we demonstrated the potential of neural networks combined with optical strain sensors to perform complete SRM health state diagnoses with a high accuracy and to identify the extent of the present defect.

Our approach is limited by the use of a dataset generated through simulations of the SRM behavior in specific thermal cycles due to a lack of real data recordings. There are several assumptions related to the simulated models that are made in favor of simplification and required time compensation. These assumptions impose the main restrictions of the applications of our work. Particularly, in our finite element models (FEMs), we assume homogeneous thermal change of the entire SRM body, and all materials are treated as linear elastic. We also excluded the temperature dependence of several material parameters such as the Young's modulus to further simplify the simulation process and generate a large amount of data within a restricted time duration and a reduced computational capacity. We did not include any investigation of the material degradation that is frequent in real practice, reflecting material property changes that could result in different strain field distributions. All of above aspects are considered beyond the purposes of this work. They are treated separately in other previously published works by taking into account various material uncertainties [32], remaining open to further investigation in future works.

The present study is based on a stationary regime approach, considering only the thermal loading effect where the SRM model is subjected. Here, the transition from the initial curing and stress-free temperature of 71 °C, down to extreme conditions of −51 °C is considered. However, in realistic dynamic conditions, in addition to various environmental factors, vibrations could significantly affect the signal quality obtained by FBGs by introducing additional noise due to captured vibrations, mostly during the SRMs' transportation or pre-operation stages. Although the detailed discussion of possible vibration issues is out of the scope of the present paper, efficient solutions for vibration noise elimination are easily identifiable. One solution could be the partial cancelation of noise due to vibrations by deploying an individual FBG sensor decoupled from the propellant material itself and the resulting stress and strain effects in order to monitor only the mechanical vibrations on the SRM casing. Feeding such individual vibration information into the machine learning (ML) algorithm could significantly improve the performance of such ML-based structural diagnosis in dynamic realistic environments with the presence of vibrations.

Additional factors that need to be further investigated concern the number of sensors per circumference that achieve optimum compensation between the cost and the accuracy of the developed diagnostic system. The neural networks developed for the presented approach were also evaluated in a preliminary study with the use of eight (8) sensors, giving no noticeable increase in the provided classification accuracy or further reduction in the RMS error in the case of the regression networks. Moreover, the use of eight sensors was related to the occurrence of overfitting earlier in the training process, resulting in a final classification accuracy of 97.17% and an RMSE of 4.1 mm in bore crack depth prediction and 2.9552° for the delamination angle in the most demanding case of coexisting defects. Thus, additional techniques for overfitting reduction, such as batch normalization and dropout [33], regularization [34], and likely the reduction of the batch size, should be employed to further investigate the ways to reach higher accuracies via the increase in available strain sensors. However, it is important to note that the existence of four sensors in the circumference of the SRM possibly imposes a reasonable limit, given that in practical real monitoring applications, the entire SRM length should be covered, thus leading to issues related to the cost of the sensors and the associated interrogation complexity with the increasing number of sensing points. Thus, most studies follow the same regime concerning the number of sensors [11,18]. However, the cost and interrogation complexity of using arrays of FBG sensors compared to multiple DBST sensors would be drastically lower. Optimal numbers of sensors in a 3D structure of the SRM as well as the optimal sensors placement issue [35] that inevitably emerges are subjects to being separately investigated in future studies.

The present study is a theoretical investigation on the applicability of FBGs in SRM structural diagnosis, and a detailed study on operational scenarios as the diagnostic approaches operated in the frame of condition-based maintenance (CBM) strategies was not included. These studies are still in an initial exploratory stage in the SRM industry. However, the monitoring of propellants in SRMs needs to take place during the propellants' curing process at the manufacturing, storage, and transportation stages, as well as possibly during SRM loading before ignition and operation. During the curing process and storage, the environmental conditions change relatively slowly or remain mostly constant, respectively. Therefore, depending on the monitoring requirements, a sample rate for data acquisition could be one sample per hour, up to one sample per minute at most. During transportation and operation, the situation is significantly different for distinct application areas such as SRMs integrated in guided missiles for tactical applications or SRMs as boosters (SRBs) mainly for space exploration applications. SRM-based missiles can be loaded on aircrafts, thus leading to extreme temperature variations larger than 100 °C and also extreme accelerations >8 g during flights. For the monitoring of the impact of such fast-occurring environmental changes, a sample rate of >1 sample per second or even in the KHz range would be required. Additionally, continuous monitoring of SRBs in space applications could be required in order to confirm the structural integrity of propellants even just before the ignition stage. Widely commercially available FBG interrogators can provide sampling ratios of up to tens of KHz, which is perfectly adequate for the aforementioned applications.

The developed regression deep neural network was additionally evaluated for predicting the location of the defects present in each case. In this case, the position error was calculated, taking into account the circular geometry of the SRM cross section given by the following formula:

$$position\ error = \sqrt{\frac{\Sigma\left(360^{\circ} mod\left(Y_{validation} - Y_{prediciton}\right)\right)^2}{N}}, \quad (5)$$

where $Y_{validation}$ represents the true values of the target angular position, $Y_{prediction}$ represents the predicted values of this variable, and N is the total number of cases participating in the validation dataset. The provided RMSE was acceptable in the case of a single bore crack (RMSE 7.48°). It was severely increased in the case of a single delamination (RMSE 36.99°), while the network failed to accurately predict the position of the defects when more than one defect was present in the SRM, as the obtained RMSE was larger than 50° for both bore cracks and delamination in case of coexisting defects.

5. Conclusions

This work presented for the first time the theoretical design of a deep learning-based diagnosis system for SRMs by employing a scenario using fiber Bragg gratings-based strain sensors. The demonstrated performance exceeded the current capabilities of systems using traditional and widely employed DBST. More specifically, the use of efficient neural networks was introduced with fully connected layers, assuming a rather moderate strain resolution performance of 4 με, which is achievable using FBG-based strain sensors. This approach proves promising in providing accurate diagnosis of the SRM for the detection of defects and the extent of bore cracks and/or delamination by demonstrating an accuracy of more than 98% and predicting the extent of the defect with an error of 2.3 mm for the bore crack depth and 1.6° for the delamination angle in the worst case of coexistent defects. The higher prediction accuracy compared to previously published results, together with the unique characteristics of the optical fiber-based sensors, suggests a viable direction for a robust and reliable physical implementation for SHM systems in SRMs via the use of FBGs.

Focusing on the use of optical FBG strain sensors as a new approach as strain sensors for SRM monitoring, the following advantages can be identified and summarized:

1. The hoop strain monitoring approach is proven to be more efficient compared to radial stress monitoring, as the change range ratio to the FBG strain resolution for a given defect is at least one order of magnitude larger than the corresponding ratio in current DBST solutions.
2. Cracks or delamination defects induce a much wider spatial effect on hoop strain compared to the corresponding effect on axial/radial stress, thus easing their detection by strain point sensors compared to DBST/stress point sensors.
3. Optical fibers provide the flexibility to be integrated in various geometries and at optimal positions in the propellant grain even close to the bore, in contrast to DBSTs, which can be placed only through the metal casing and close to the insulation area, thus providing better sensitivity.
4. Optical FBG strain sensors can be seamlessly integrated in SRMs, not affecting their handling, transportation, storage, or operation, in contrast to DBSTs, where such external sensors are placed in the metal casing with external electric wiring, making monitoring difficult during the SRM life cycle.

5. FBG strain sensors are passive and can provide continuous monitoring of SRMs due to their seamless integration capability inside the SRM body, in contrast to DBSTs, which are placed outside the metal casing, limiting their monitoring to only controllable laboratory conditions.
6. As SRMs can experience drastic environmental changes during transportation, storage, or operation, in terms of acceleration, temperature, and humidity, their structural integrity needs to be continuously monitored. However, an event of a sudden defect under extreme conditions can be securely logged and registered as a permanent Bragg Grating wavelength shift even if the continuous real-time monitoring was offline. This is in contrast to DBSTs, where electric power is required to interrogate the SRM and register a measurable defect.
7. FBGs and optical solutions provide unique characteristics such as intrinsically electric safety and immunity to electromagnetic interference, which are crucial in a highly explosive propellant material system.
8. The optical fiber monitoring approach provides the capability of easy interconnection to the FBG interrogation unit using a single optical fiber output, also providing the capability of remote monitoring through a fiber optic channel or network.

Overall, it is anticipated that fiber optic strain sensors will become the preferred solution to SRM continuous monitoring, thus providing real-time data for feeding machine learning-based diagnostic approaches operating in the framework of modern condition-based maintenance (CBM) strategies.

Author Contributions: Conceptualization, C.R.; methodology, G.K. and C.R.; validation, G.K., C.R. and K.N.A.; formal analysis, G.K. and C.R.; investigation, C.R., G.K., K.N.A., N.C. and G.M.; resources, C.R. and K.N.A.; writing—original draft preparation, G.K.; writing—review and editing, G.K., C.R., K.N.A. and G.M.; visualization, G.K.; supervision, C.R.; project administration, C.R.; funding acquisition, C.R. All authors have read and agreed to the published version of the manuscript.

Funding: This research was funded by Bayern Chemie GmbH Germany in the framework of the collaborative Project RocketSens (Contract No: 2018-0720-1). Partial support was also received from the Hellenic Foundation for Research and Innovation (H.F.R.I.) under the "First Call for H.F.R.I. Research Projects to support Faculty members and Researchers and the procurement of high-cost research equipment grant" (Project Number: HFRI-FM17-640, InPhoQuC).

Institutional Review Board Statement: Not applicable.

Informed Consent Statement: Not applicable.

Data Availability Statement: The original contributions presented in this study are included in the article. Further inquiries can be directed to the corresponding author.

Acknowledgments: The authors are grateful to Ludwig Eineder of Bayern-Chemie GmbH (MBDA Germany) for his useful discussions and for providing technical insights regarding strain and stress sensing in SRMs.

Conflicts of Interest: Author Günter Mußbach was employed by the company Bayern-Chemie GmbH (MBDA Germany). The remaining authors declare that the research was conducted in the absence of any commercial or financial relationships that could be construed as a potential conflict of interest.

References

1. Dhital, D.; Lee, J.R.; Farrar, C.; Mascarenas, D. A Review of Flaws And Damage in Space Launch Vehicles: Motors and Engines. *J. Intell. Mater. Syst. Struct.* **2014**, *25*, 524–540. [CrossRef]
2. McDonald, A.J. Solid Rocket Motor Failure. *Encycl. Aerosp. Eng.* **2010**, *2*, 1123–1129.
3. Tussiwand, G.S.; Saouma, V.E.; Terzenbach, R.; de Luca, L.T. Fracture Mechanics of Composite Solid Rocket Propellant Grains: Material testing. *J. Propuls. Power* **2009**, *25*, 60–73. [CrossRef]
4. Tussiwand, G.S.; Eineder, L.; Mussbach, G.; Bohn, M.A. Non-Destructive Ageing State Determination of Solid Rocket Motors Charges. In Proceedings of the 8th European Workshop on Structural Health Monitoring EWSHM, Bilbao, Spain, 5–8 July 2016.

5. Sankar, R.A.; Anees, P.; Kumar, A.; Sriram, B.T. Structural Health Monitoring of Solid Propellant Using Embedded PVDF Sensor. In Proceedings of the International Conference on Power, Institute of Electrical and Electronics Engineers, New York, NY, USA, 18–20 January 2018. [CrossRef]
6. Lopatin, C.M. Method for Measuring the Health of Solid Rocket Propellant Using an Embedded Sensor. U.S. Patent US7652488B1, 26 January 2010.
7. Le, A.Q.; Sun, L.Z.; Miller, T.C. Detectability of Delaminations in Solid Rocket Motors with Embedded Stress Sensors. *J. Propuls. Power* **2013**, *29*, 299–304. [CrossRef]
8. Riziotis, C.; Eineder, L.; Bancallari, L.; Tussiwand, G. Fiber Optic Architectures for Strain Monitoring of Solid Rocket Motors' Propellant. *Sens. Lett.* **2013**, *11*, 1403–1407. [CrossRef]
9. Mußbach, G.; Tussiwand, G.; Buswell, J. Assessing the Ageing-State of Solid Propellant Grains in Case-Bonded Rocket Motors by Measuring Bondline Stresses. In Proceedings of the Energetic Materials. Synthesis, Characterization, Processing. Fraunhofer-Institut für Chemische Technologie (43rd International Annual Conference), Karlsruhe, Germany, 26–29 June 2012.
10. Xiao, H.-Y.; Gao, F. Fault Diagnosis of Solid Rocket Motor Based on Fuzzy Neural Network. In Proceedings of the 2012 International Conference on Electronics, Communications and Control, Yichang, China, 16–18 October 2012.
11. Liu, D.; Sun, L.; Miller, T.C. Defect Diagnosis in Solid Rocket Motors Using Sensors and Deep Learning Networks. *AIAA J.* **2021**, *59*, 276–281. [CrossRef]
12. Hoffmann, L.F.S.; Bizarria, F.C.P.; Bizarria, J.W.P. Detection of Liner Surface Defects in Solid Rocket Motors Using Multilayer Perceptron Neural Networks. *Polym. Test.* **2020**, *88*, 106559. [CrossRef]
13. Dai, J.; Li, T.; Xuan, Z.; Feng, Z. Automated Defect Analysis System for Industrial Computerized Tomography Images of Solid Rocket Motor Grains Based on YOLO-V4 Model. *Electronics* **2022**, *11*, 3215. [CrossRef]
14. Anhduong, A.; Le, Q.; Sun, L.Z.; Miller, T.C. Health Monitoring and Diagnosis of Solid Rocket Motors with Bore Cracks. *J. Aerosp. Eng.* **2016**, *29*, 04015058.
15. Ramakrishnan, M.; Rajan, G.; Semenova, Y.; Farrell, G. Overview of Fiber Optic Sensor Technologies for Strain/Temperature Sensing Applications in Composite Materials. *Sensors* **2016**, *16*, 99. [CrossRef]
16. Konstantaki, M.; Violakis, G.; Pappas, G.; Geernaert, T.; Korakas, N.; Tiriakidis, N.; Tiriakidi, T.; Tiriakidis, K.; Thienpont, H.; Berghmans, F.; et al. Monitoring of Torque Induced Strain in Composite Shafts with Embedded and Surface-Mounted Optical Fiber Bragg Gratings. *Sensors* **2021**, *21*, 2403. [CrossRef] [PubMed]
17. Filograno, M.L.; Piniotis, G.; Gikas, V.; Papavasileiou, V.; Gantes, C.; Kandyla, M.; Riziotis, C. Comparative Assessment and Experimental Validation of a Prototype Phase-Optical Time-Domain Reflectometer for Distributed Structural Health Monitoring. *J. Sens.* **2022**, *2022*, 6856784. [CrossRef]
18. Zhang, L.; Chang, X.-L.; Zhang, Y.-H.; Chen, X.-D.; Zhang, Q. On-Line Monitoring of SRM Grain Structure Based on Embedded FBG Sensor. In Proceedings of the 7th International Conference on Power Science and Engineering, EDP Sciences, Vienna, Austria, 19–21 November 2019. [CrossRef]
19. Lopatin, C.; Grinstein, D. Active Sensing for Monitoring the Properties of Solid Rocket Motor Propellant Grains. *Propellants Explos. Pyrotech.* **2015**, *40*, 295–302. [CrossRef]
20. Zhang, L.; Chang, X.; Zhang, Y.; Liu, W. Response of FBG Sensors Embedded in SRM Interface of Combustor when Subjected to Tri-Axial Normal Loadings. *IEICE Electron. Express* **2017**, *14*, 20170657.
21. Korompili, G.; Mußbach, G.; Riziotis, C. Structural Health Monitoring of Solid Rocket Motors: From Destructive Testing to Perspectives of Photonic-Based Sensing. *Instruments* **2024**, *8*, 16. [CrossRef]
22. Riziotis, C.; Vasilakos, A. Computational Intelligence in Photonics Technology and Optical Networks: A Survey and Future Perspectives. *Inf. Sci.* **2007**, *177*, 5292–5315. [CrossRef]
23. Farrar, C.; Worden, K. *Structural Health Monitoring: A Machine Learning Perspective*; John Wiley & Sons, Ltd.: Hoboken, NJ, USA, 2013.
24. Mieloszyk, M.; Krawczuk, M.; Skarbek, L.; Ostachowicz, W. An Application of Neural Network for Structural Health Monitoring of an Adaptive Wing with an Array of FBG Sensors. In *Journal of Physics: Conference Series, 9th International Conference on Damage Assessment of Structures (DAMAS 2011), St Anne's College, University of Oxford, 11–13 July 2011*; IOP Publishing: Bristol, UK. [CrossRef]
25. Kahandawa, G.C.; Epaarachchi, J.; Wang, H.; Canning, J.; Lau, K. Extraction and Processing of Real Time Strain of Embedded FBG Sensors Using a Fixed Filter FBG Circuit and an Artificial Neural Network. *Measurement* **2013**, *46*, 4045–4051. [CrossRef]
26. Sundaram, R.; Kamath, G.M.; Gupta, N.; Rao, M.S. Structural Health Monitoring of co-Cured Composite Structures Using FBG Sensors. In Proceedings of the Smart Structures and Materials 2005: SPIE Smart Structures and Materials + Nondestructive Evaluation and Health Monitoring, San Diego, CA, USA, 17 May 2005. [CrossRef]
27. Wang, D.; Zhang, W. Damage Detection Combining Principal Component Analysis and Deep Convolutional Neural Network with Dynamic Response from FBG Arrays. *J. Civ. Struct. Health Monit.* **2023**, *13*, 101–115. [CrossRef]
28. He, Y.W.M.; Ramakrishnan, K.R.; Zhang, Z. A Comparison of Machine Learning Algorithms for Assessment of Delamination in Fiber-Reinforced Polymer Composite Beams. *Struct. Health Monit.* **2021**, *20*, 1997–2012. [CrossRef]
29. Kaiming, H.; Xiangyu, Z.; Shaoqing, R.; Jian, S. Delving Deep into Rectifiers: Surpassing Human-Level Performance on ImageNet Classification. *arXiv* **2015**, arXiv:1502.01852v1. [CrossRef]

30. Lacraz, A.; Polis, M.; Theodosiou, A.; Koutsides, C.; Kalli, K. Femtosecond Laser Inscribed Bragg Gratings in Low Loss CYTOP Polymer Optical Fiber. *IEEE Photonics Technol. Lett.* **2015**, *27*, 693–696. [CrossRef]
31. Leal-Junior, A.G.; Theodosiou, A.; Díaz, C.R.; Marques, C.; Pontes, M.J.; Kalli, K.; Frizera, A. Simultaneous Measurement of Axial Strain, Bending and Torsion With a Single Fiber Bragg Grating in CYTOP Fiber. *J. Light. Technol.* **2019**, *37*, 971–980. [CrossRef]
32. Cholevas, N.; Anyfantis, K.N.; Mußbach, G.; Korompili, G.; Riziotis, C. Crack Identification in Solid Rocket Motors Through the Neyman–Pearson Detection Theory. *AIAA J.* **2023**, *61*, 2241–2254. [CrossRef]
33. Srivastava, N.; Hinton, G.; Krizhevsky, A.; Sutskever, I.; Salakhutdinov, R. Dropout: A Simple Way to Prevent Neural Networks from Overfitting. *J. Mach. Learn. Res.* **2014**, *15*, 1929–1958.
34. Nusrat, I.; Jang, S.-B. A Comparison of Regularization Techniques in Deep Neural Networks. *Symmetry* **2018**, *10*, 648. [CrossRef]
35. Liangou, T.; Katsoudas, A.; Silionis, N.; Anyfantis, K. Statistical Pattern Recognition for Optimal Sensor Placement in Damage Detection Applications. In *European Workshop on Structural Health Monitoring*; Rizzo, P., Milazzo, A., Eds.; EWSHM; Springer International Publishing: Cham, Switzerland, 2022.

photonics

MDPI

Article

Comprehensive Analysis of Optical Resonances and Sensing Performance in Metasurfaces of Silicon Nanogap Unit

Masanobu Iwanaga

Research Center for Electronic and Optical Materials, National Institute for Materials Science (NIMS), 1-1 Namiki, Tsukuba 305-0044, Japan; iwanaga.masanobu@nims.go.jp

Abstract: Metasurfaces composed of silicon nanogap units have a variety of optical resonances, including bound states in the continuum (BIC). We show comprehensive numerical results on metasurfaces of Si-nanogap units, analyze the optical resonances, and clarify optically prominent resonances as well as symmetry-forbidding resonances that are the BIC, based on the numerical analyses of optical spectra and resonant electromagnetic field distributions. Introducing asymmetry in the unit cell, the BIC become optically allowed, being identified as magnetic dipole, electric quadrupole, and magnetic quadrupole resonances. Moreover, the optical resonances are examined in terms of refractive index sensing performance. A pair of the resonances associated with electric field localization at the nanogap was found to be sensitive to the refractive index in contact with the metasurfaces. Consequently, the gap mode resonances are shown to be suitable for a wide range of refractive index sensing over 1.0–2.0.

Keywords: all-dielectric metasurface; silicon nanogap; BIC; optical sensing; refractive index

Citation: Iwanaga, M. Comprehensive Analysis of Optical Resonances and Sensing Performance in Metasurfaces of Silicon Nanogap Unit. *Photonics* **2024**, *11*, 1053. https://doi.org/10.3390/photonics11111053

Received: 14 October 2024
Revised: 4 November 2024
Accepted: 8 November 2024
Published: 10 November 2024

1. Introduction

Metasurfaces have opened the way for diverse optical properties and applications. In addition to light-wave manipulations employing the complex subwavelength structures [1–7], various types of resonances have been exploited to realize prominent effects that are helpful for practical applications such as metalenses [8–12], color splitting to pixel [13–15], liquid-crystal-based active responses [16], and biosensing based on resonance shift [17–19] and on fluorescence detection [20–23].

Among many effects in all-dielectric metasurfaces, one of the popular resonances is quasi-bound states in the continuum (qBIC). Structural asymmetry allows us to access the resonances of qBIC that are forbidden in symmetric structures. Although this concept was introduced in an electronic system almost a hundred years ago [24], a similar effect to the electronic system was shown in a nanophotonic system, which was a photonic crystal, about ten years ago [25], which stimulated nanophotonics studies on the qBIC.

As applications of the qBIC in nanophotonics, refractive index sensing [26–28] and nonlinear effects [29–31] were explored to date. On the physical limit of refractive index sensing using periodic structures, a formula for the resolution in the units of nm/refractive index unit (RIU) was recently shown [32] such that

$$\frac{\Delta\lambda}{\Delta n} = P \tag{1}$$

where λ, n, and P denote wavelength of light, refractive index of surrounding medium, and periodic length, respectively. Equation (1) indicates that the upper limit of refractive index sensing is determined by the periodic length P.

Figure 1 illustrates structural features of metasurfaces addressed in this study. In Figure 1a,b, Si-nanogap units are symmetric and asymmetric in the unit cell on the xy plane, respectively. Keeping gap width g constant, we vary y-width, y_1 and y_2, while keeping the sum of $y_1 + y_2 + g$ constant. In Figure 1b, a condition of $y_1 > y_2$ results in $-y$-direction offset of the center line in the nanogap (dotted line). We define the absolute value of the offset as α, which is an asymmetric parameter, and quantitatively characterize the asymmetry in the unit cell, and can write it down such that

$$\alpha = |y_1 - y_2|/2. \quad (2)$$

Introducing the asymmetry along only the y axis, the definition in Equation (2) is simple and explicitly expressed. Also, experimental tests in the future will be easy to conduct. Note that the asymmetric parameters for qBIC are usually normalized using a reasonable reference length to make the parameters dimensionless; for example, the offset can be normalized by the length of unit cell; however, to make the physical meaning of α clear, we simply define the α in Equation (2). Thus, the present α means the offset of the gap center from the center of unit cell; that is, $\alpha = 1$ means that the offset is 1 nm.

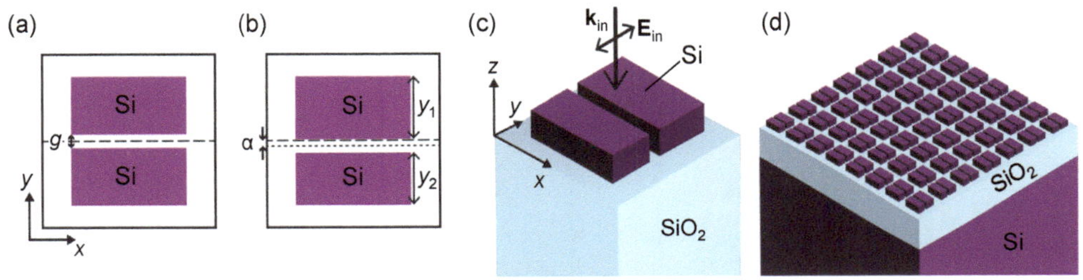

Figure 1. Schematics of metasurface structure and optical configuration in this study. (**a**) xy-section view of unit cell of metasurface of Si-nanogap unit (purple). The gap g is indicated with a both-end arrow. The center line along the x axis is drawn with a broken line, which goes across the central point of the unit cell. (**b**) Asymmetry introduced in the unit cell is indicated by a quantity α, which is offset from the center line (broken line) to the middle line (dotted line) in the gap. (**c**) Three-dimensional illustration of unit cell of metasurface of an asymmetric Si-nanogap pair. Optical configuration is shown together. The unit cells in (**a**–**c**) are set to have dimensions of 600×600 nm^2 in the xy plane. Length of the Si nanoblock along the x axis is 400 nm and the sum of $y_1 + y_2 + g$ is 400 nm. Height of the Si nanoblock is set to 200 nm. (**d**) Three-dimensional illustration of a metasurface of periodic array of asymmetric Si-nanogap units, which are assumed to be infinitely periodic in the computations.

In Figure 1c, a 3D illustration of the unit cell of the metasurface (Figure 1b) is shown together with spatial xyz coordinates and an optical configuration. Wavevector $\mathbf{k}_{\text{in}}$ denotes the wavevector of the incident plane wave, and vector $\mathbf{E}_{\text{in}}$ represents electric field (or polarization) of the incidence. In this study, we mainly study normal incidence because the asymmetry enables us to access qBIC and allows us to explore various optical resonances. The metasurface is schematically illustrated in Figure 1d, which is assumed to be infinitely periodic in the computations. In practice, the metasurface is designed to be fabricated through nanolithography on a Si-on-insulator (SOI) wafer.

2. Materials and Methods

The metasurfaces in this study are assumed to comprise dielectrics of Si and SiO_2. SOI substrates are assumed to be used as base materials to fabricate the Si metasurfaces. In particular, the thickness of the top SOI layer, the middle buried oxide layer that is SiO_2, and the base Si layer are set to 200 nm, 1000 nm, and 725 μm, respectively, in accordance with SOI substrates available in reality. The permittivity of crystalline Si was taken from the literature [33] and that of SiO_2 was set to a representative value of 2.1316 in the photon-energy range of present interest. Using these permittivities, realistic simulations were conducted.

Numerical implementation was conducted based on a method combining rigorously coupled-wave analysis (RCWA) [34], which solves Fourier-transformed Maxwell equations for infinitely periodic objects in the frequency domain, with the scattering-matrix (S-matrix) algorithm [35]. The S-matrix algorithm makes numerical calculations stable even for stacked layer structures. The combined numerical method is a frequency-domain method and suitable to precisely compute optical spectra such as reflectance and transmittance spectra of periodic structures; indeed, the method has been applied to reproduce experimental optical spectra in various periodically structured materials such as metamaterials and metasurfaces [36–41]. The code executing the RCWA and S-matrix algorithm run on supercomputers in a multiparallel implementation (MPI) manner. Optical spectra were computed in the MPI manner and therefore the net runtime was almost equal to total runtime/(number of MPI), being substantially reduced. Also, the combined method enables us to output electromagnetic (EM)-field distributions, which is helpful to understand features of optical resonances in the metasurfaces. The RCWA is mathematically exact to the Maxwell equation under infinite expansion of the Fourier series. In reality, the RCWA provides approximate solutions by truncating the Fourier series at a finite order. We set the truncation order at ± 15 and more in this study and did not see changes in the optical spectra. Note that, since we considered 2D periodic systems, the total truncation order was 31×31 and more.

3. Results

3.1. Asymmetric Metasurfaces and qBIC Modes

Figure 2a shows a series of reflectance spectra of metasurfaces of Si-nanogap units, illustrated Figure 1b–d. The reflectance is normalized and takes values in a range from 0 to 1. The reflectance spectra are displayed with offset; accordingly, the zero lines for the offset spectra are drawn with thin gray lines.

The unit cell is set to be 600×600 nm^2 in the xy plane, and the width of Si nanoblocks along the x axis is set to 400 nm. The height of Si nanoblocks along the z axis is 200 nm. The gap g between the Si nanoblocks is set to 50 nm, and the sum of $y_1 + y_2$ is set to 350 nm; consequently, the unit of Si nanoblocks has a dimension of 400×400 nm^2 in the xy plane, being set at the center of the unit cell.

Introducing the asymmetry represented by α, we can allow for the qBIC modes to appear in optical spectra, such as reflectance spectra in Figure 2a. The qBIC modes are indicated with MD, MQ, and EQ that denote magnetic dipole, magnetic quadrupole, and electric quadrupole, respectively, and are indicated with arrows. They appear for $\alpha \geq 1$ though the signals at $\alpha = 1$ are quite small. The refractive index n in the incident layer and gaps between the Si nanoblocks is set to 1.0. The Si-nanogap unit cell is anisotropic for x and y axes; accordingly, the reflectance spectra are also anisotropic. The x- and y-polarized reflectance spectra are shown from bottom to top with offset, using black-to-blue and red-to-brown colors, respectively. The parameter α varies from 0 to 25. The series of reflectance spectra exhibit definite anisotropy for the incident polarizations and various resonant responses in the shapes of the spectra. In the photon energy range over 1.41 eV, diffraction-associated resonant modes appear. We here focus on the resonances below 1.4 eV, which are diffraction-free and belong to a subwavelength range at the normal incidence.

We briefly mention a general aspect in optical systems (e.g., Figure 1c) that contain periodic structures and plain waves that are incoming and outgoing. Even when the periodic systems have inversion symmetry in the xy plane, oblique incidence to the xy plane yields asymmetry in the optical responses; consequently, qBIC modes can be observed under oblique incidence. In this study, we mainly study the normal incidence and make the plain waves symmetric in the xy plane; therefore, at the normal incidence, the qBIC modes are observed only for the metasurface of asymmetric units (i.e., $\alpha > 0$) that do not have inversion symmetry in the xy plane. Note that, under the oblique incidence, the optical systems are asymmetric for the xy plane, and that the origin of asymmetry is generally indistinguishable between the structural units and optical configurations.

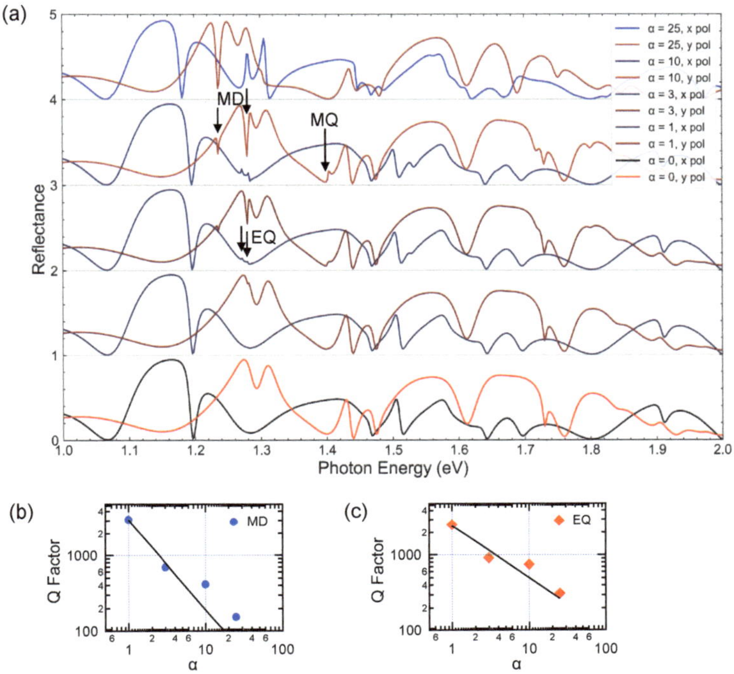

Figure 2. Optical resonances emerged in the Si-nanogap unit metasurfaces. (**a**) Computed reflectance spectra at two incident polarizations, $\mathbf{E}_{\text{in}} \parallel x$ and $\mathbf{E}_{\text{in}} \parallel y$ for various asymmetric parameters α, defined in Figure 1, and a fixed gap g of 50 nm. The reflectance spectra are shown with offset, changing color from black to blue for $\mathbf{E}_{\text{in}} \parallel x$ and from red to brown for $\mathbf{E}_{\text{in}} \parallel y$, in accordance with $\alpha = 0$–25, respectively. EQ, MD, and MQ denote electric quadrupole, magnetic dipole, and magnetic quadrupole, respectively, indicated by arrows. (**b**,**c**) Asymmetric parameter α and estimated quality (Q) factors of the MD (closed blue circles) and EQ modes (red diamonds), respectively. The Q factor is defined as Equation (3). Black lines denote fitted lines using a power function (see the text).

The reflectance spectra in Figure 2a have several peaks with broad bands at both x and y polarizations. The peaks coexist in the photon-energy range with interference coming from the multilayer structure of the SOI substrate; the interference is easy to be verified at a low energy range between 1.1 and 1.2 eV at x and y polarizations, respectively. These coexisting properties make the spectra complicated.

In addition, the spectra in Figure 2a include qBIC resonances with narrow line widths; the EQ modes are located at 1.2717 and 1.2809 eV for x polarization, the MD modes at 1.236 and 1.280 eV for y polarization, and the MQ modes at 1.403 eV for y polarization, as indicated by arrows. For smaller α, the resonant signals become smaller and finally disappear at $\alpha = 0$ for both polarizations, suggesting that these signals originate from qBIC. These resonances are quantified using quality (Q) factor, defined such that

$$Q = \frac{\omega_0}{\Delta\omega} \tag{3}$$

where ω_0 and $\Delta\omega$ denote resonant frequency and full width at the half maximum (FWHM) of the resonance in the frequency domain, respectively. In Figure 2b,c, the Q factors of MD (blue closed circle) and EQ modes (red diamond) are plotted for the parameter α, respectively. Although two MD and EQ lines appear, we chose the lower-energy resonance. The reason why the two MD modes appear is described with examining resonant EM-field distributions below. As the parameter α becomes smaller, the Q factors increase and exceed

1000 at $\alpha = 1$. The Q factor was fitted using a power function, $Q = A\alpha^n + B$, where A and B are constants, being found that $n = -1.22$ and -0.70 for the MD and EQ modes, respectively. The fitted power functions are shown with black lines. If radiative loss is simply assumed to evaluate the Q factors, a relation of $Q \propto \alpha^{-2}$ was suggested [42]. However, nonradiative losses are able to occur, due to light absorption by the constituent materials and surface roughness that could give rise to in nanofabrication processes, and substantially reduce the large Q factors over 1000 [43]. In the present numerical evaluations, the absorption loss is explicitly incorporated based on the permittivity of crystalline Si in the literature [33]. Furthermore, the high-Q modes are not isolated from the broad-band interference mode in the present metasurfaces; therefore, the EM-field confinement is considered to be reduced, though it is difficult to quantitatively evaluate the net amount of dissipation stemming from the interference mode.

Representative resonant EM-field distributions of qBIC are shown in Figure 3, where the asymmetric parameter α is chosen to be 10. EQ-mode EM-field distributions are shown in Figure 3a–c, where $|\mathbf{E}|$, $\mathrm{Re}(E_z)$, and $|\mathbf{H}|$ components are displayed, respectively. The boundaries of Si nanoblocks are shown with dotted white lines in Figure 3c. From the EM fields, several features are observed, as follows:

1. EM-field intensity is significantly enhanced on the resonance. For the incident intensities of $|\mathbf{E}_{\mathrm{in}}|^2 = 1$ and $|\mathbf{H}_{\mathrm{in}}|^2 = 1$, the maxima of $|\mathbf{E}|^2$ and $|\mathbf{H}|^2$ reach 420 and 341 in Figure 3a,c, respectively. Note that the nanogap does not contribute to the large EM-field enhancement in this mode.
2. The field pattern of $\mathrm{Re}(E_z)$ in Figure 3b indicates that the mode has quadratic oscillation on and around the Si-nanogap unit. Therefore, this mode is attributed to an EQ mode.
3. Anisotropy is introduced in the unit structure along the y axis and results in the anisotropy of the EM-field distributions, as observed in Figure 3a–c.
4. The resonant EM fields predominantly consist of near-field distributions in and around the Si-nanogap unit. As a result, the imaginary parts of EM fields contribute to the enhanced intensities. This is verified from the fact that the maximum of $|\mathrm{Re}(E_z)|$ in Figure 3b is approximately two-fold smaller than that of $|\mathbf{E}|$ in Figure 3a.

For y polarization, the qBIC modes are magnetic modes. Magnetic dipole resonance is visualized in Figure 3d–f, where $|\mathbf{E}|$, $|\mathbf{H}|$, and a snapshot of $\mathrm{Re}(H_z)$ are shown, respectively. Due to the nanogap between the two Si nanoblocks, strong localization of electric fields in the nanogap emerges, as shown in Figure 3d; accordingly, the magnetic fields are weaker at the center of the unit cell, as seen in Figure 3e; however, they exhibit a modified magnetic-dipole field distribution that is mainly localized in the Si domain, as shown in Figure 3f. Thus, the EM-field distributions retain a feature of magnetic-dipole resonance.

We here remark multipole approximations for the qBIC modes, which were reported frequently. First of all, it is to be noted that there is no analytical solution for the present metasurfaces that have complicated unit structures, in comparison with ideal spheres. Therefore, any trial to conduct multipole expansion for the metasurfaces will result in numerically approximated analyses. The issue is that it is difficult to evaluate the precision of the approximation in a quantitative manner. As shown in Figure 3d–i, the MD modes are significantly modified due to the nanogap and another mode of the gap mode coexists.As a result, if we should conduct the multipole expansion, the approximated multipolar values will be difficult to justify. Second, we already conducted and reported a multipole analysis for the Si-nanogap unit using a finite-element method [28]. A similar result is expected and is not repeated here. Instead, the information on the EM-field distributions is shown in Figure 3.

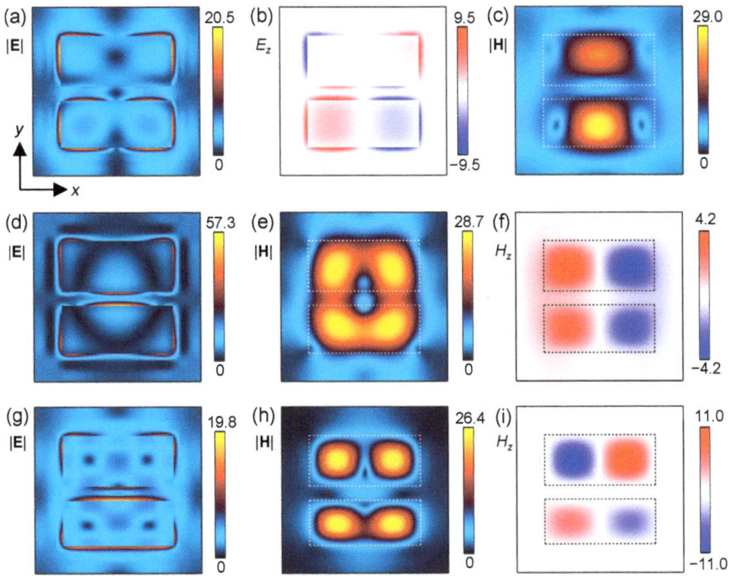

Figure 3. Resonant electromagnetic (EM)-field distributions on the qBIC resonances. (**a**) xy-section view of $|\mathbf{E}|$ distribution at 1.272 eV. The xy axes are shown together and the xy sections were taken at the half height of the Si-nanogap pair. These settings are in common with the other panels in this figure. (**b**) Snapshot of $\mathrm{Re}(E_z)$ component, corresponding to (**a**), presents a signature of electric quadrupole (EQ) mode. (**c**) $|\mathbf{H}|$ distribution, corresponding to (**a**). The incident polarization was set to be $\mathbf{E}_{\mathrm{in}} \parallel x$ in (**a–c**). (**d–f**) $|\mathbf{E}|$, $|\mathbf{H}|$, and $\mathrm{Re}(H_z)$ distributions at 1.236 eV, respectively. The magnetic-field distribution shows a signature of magnetic dipole (MD) mode. (**g–i**) $|\mathbf{E}|$, $|\mathbf{H}|$, and $\mathrm{Re}(H_z)$ distributions at 1.402 eV, respectively. The magnetic-field distribution shows a signature of magnetic quadrupole (MQ) mode. The incident polarization was set to be $\mathbf{E}_{\mathrm{in}} \parallel y$ in (**d–i**). The color bars indicate values of resonantly enhanced EM fields, when the absolute values of incident fields were set to unity, that is, $|\mathbf{E}_{\mathrm{in}}| = 1$ and $|\mathbf{H}_{\mathrm{in}}| = 1$.

In Figure 3d, the electric field is strongly localized in the nanogap and takes the maximum intensity at the center of x positions. In other words, the electric field is a single node in the nanogap along the x axis. We note that another MD mode in Figure 2a is located at a higher photon energy than this MD mode and has two maxima on the electric-field intensity in the nanogap; that is, it has two nodes in the nanogap. Thus, the two MD modes correspond to the first and second electric-field nodes in the nanogap. We mention that the third node is not induced because the length of the Si nanoblocks along the x axis is not enough. A similar type of resonances appears later (Section 3.2).

Magnetic quadrupole resonance is visualized in Figure 3g–i that shows $|\mathbf{E}|$, $|\mathbf{H}|$, and a snapshot of $\mathrm{Re}(H_z)$, respectively. The electric and magnetic field distributions show four minimum and maximum points in the Si domain, respectively, which indicates quadrupole. This feature is confirmed from the $\mathrm{Re}(H_z)$ distribution in Figure 3i.

Summing up Figure 3 briefly, the qBIC modes are found to be EQ, MD, and MQ modes. They have resonantly enhanced EM fields, that reach hundred-to-thousand-fold intensity, compared to that of incidence.

3.2. Resonant Shift for Refractive Index

Two series of numerically calculated reflectance spectra of metasurfaces of asymmetric Si-nanogap units are shown in Figure 4a, where the asymmetric parameter α was set to 10 in the unit cell. We note that the spectra for $n = 1.0$ are identical to those of $\alpha = 10$ in Figure 2a and that $y_1 = 185$ nm, $y_2 = 165$ nm, and $g = 50$ nm.

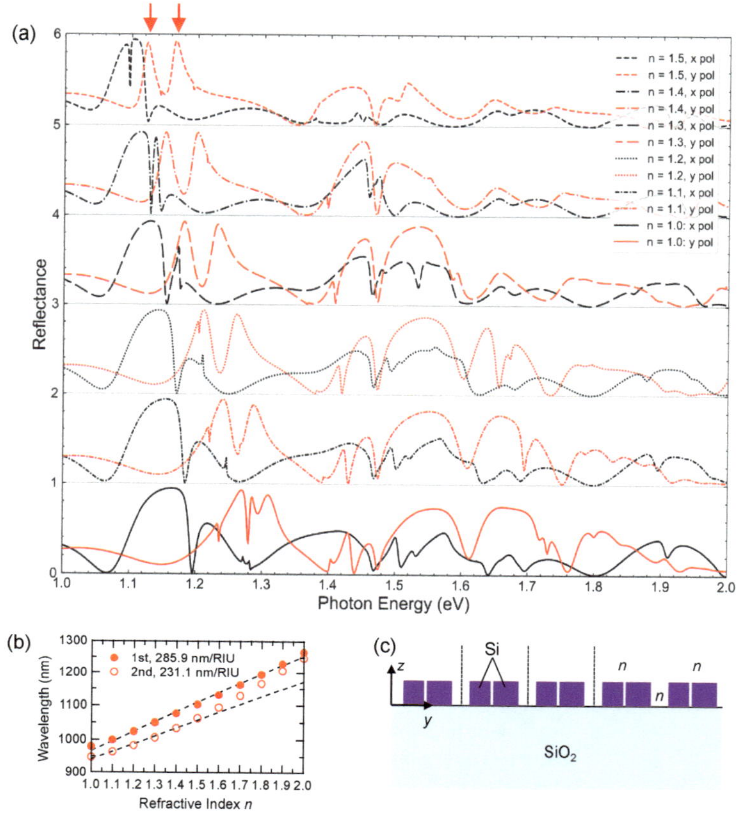

Figure 4. (**a**) Reflectance spectra dependent on refractive index n in the medium contacting the metasurface. Black and red curves show x- and y-polarized spectra, respectively, which are displayed with offset. Red arrows indicate a pair of resonances responsive to the index n over a wide range of $n = 1.0$–2.0. (**b**) Resonance shift of the pair of resonances. The first and second ones are shown with closed and open red circles, respectively. The slopes, fitted using a linear function and shown with dashed black lines, approximating performance as optical sensors in the units of nm/RIU. See more details in the text. (**c**) Schematic of a yz-section-view metasurface and medium of refractive index n. Dotted lines section the unit cell along the y axis.

Refractometric responses of the two resonances for $n = 1.5$, indicated by red arrows in Figure 4a, are plotted for the refractive index n over a range from 1.0 to 2.0 in Figure 4b, where the lower and higher energy peaks are shown with closed and open red circles, respectively. The wide range of the refractive index enables us to detect diverse materials from gas to liquid. Note that the vertical axis in Figure 4b is in the units of nm, which is inversely proportional to eV. The peaks were fitted using a linear function; dashed black lines represent the fitted line, reproducing the peak profiles over $n = 1.0$–1.5 in a good approximation; the first and second peaks plotted for the wavelength show refractometric responses of 285.9 and 231.1 nm/RIU, respectively. One of the advantages in the range of $n = 1.0$–1.5 is the approximately linear responsivity to the wide range.

Generally, refractometric responses can be fitted using quadratic functions [32] because dispersion relations between wavenumber and photon energy are mostly quadratic in 2D periodic systems. Indeed, the two peak profiles in Figure 4b can be traced for the whole range of $n = 1.0$–2.0 using quadratic functions.Thus, quantitative tracing of the two peaks is possible. Other various nanostructures including metasurfaces have been proposed as refractive-index sensors [26–28,32,43]; however, almost all of them do not guarantee such a wide-range responsivity and do have a narrow range of target, and are limited to a narrower range of targets than the present metasurface.

The medium of refractive index n is assumed to exist in the incident layer and gaps between the Si nanoblocks, as drawn in Figure 4c. Practically, liquid or gas media are assumed to be the target of refractive-index sensing. In addition, large molecules such as biomolecules can be detected as resonant shift, similarly to the conventional surface plasmon resonance method [44].

Figure 5 shows EM-field distributions at the two reflectance peaks at y polarization, indicated by the red arrows in Figure 4a. The peaks correspond to resonances responsive to the surrounding refractive index n. In Figure 5, the index n was set to 1.5.

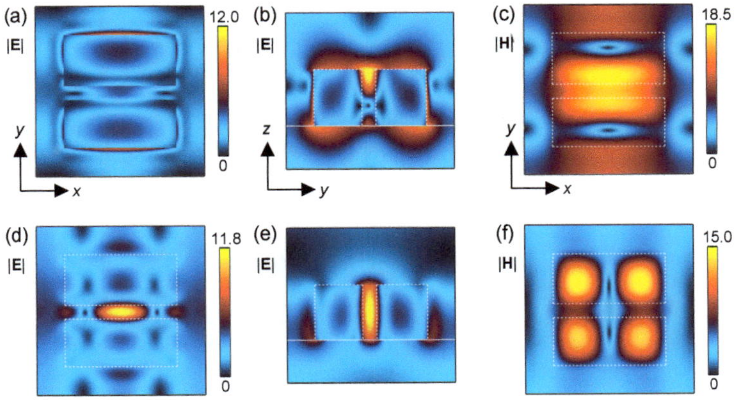

Figure 5. Resonant EM-field distributions on the two resonances responsive to surrounding refractive index, which is set to $n = 1.5$. (**a**,**b**) xy- and yz-section views of $|\mathbf{E}|$ distributions, respectively. The yz section cuts across the center of Si nanoblocks in the x direction. (**c**) $|\mathbf{H}|$ distribution in xy-section view. The photon energy is 1.124 eV in (**a**–**c**). (**d**–**f**) $|\mathbf{E}|$ and $|\mathbf{H}|$ distributions at 1.167 eV, displayed in a similar manner to (**a**–**c**). The incident polarization was set to $\mathbf{E}_{\text{in}} \parallel y$ in common. These xy sections are taken at the half-height of the Si-nanogap pair. The yz sections are through the center of the unit cell.

The first resonance at 1.124 eV is visualized in Figure 5a–c. The $|\mathbf{E}|$ component is presented in xy- and yz-section views, respectively. The color bar indicates the absolute values in common. A feature observed in the yz-section view is that the enhanced electric fields localize at the upper side of the nanogap. Since the xy section was taken at the half height of the Si nanoblocks, the enhanced electric fields are not seen; in contrast, the magnetic fields are enhanced on the xy section.

The second resonance at 1.167 eV is visualized in Figure 5d–f. The way of presentation is similar to that in Figure 5a–c. Obviously, the electric fields are enhanced at the nanogap. The magnetic fields are mostly localized in the Si nanoblocks.

From the enhanced electric-field distributions, the refractometric response in Figure 4a is most likely to originate from the locally enhanced electric fields at the nanogap in Figure 5, which are considered to enhance responses susceptible to the refractive index. Indeed, as shown in Figure A1, in a metasurface of symmetric Si nanoblock unit without any nanogap, resonances corresponding to the two resonances in the asymmetric metasurface (Figure 4) are not observed, and the refractometric response becomes small and is not found for a wide range of $n = 1.0$–1.5.

3.3. Nanogap Modes with Hugely Enhanced Electric Fields

Reflectance spectra of an asymmetric metasurface with $\alpha = 5$ and $g = 10$ nm are shown in Figure 6a. The sum of $y_1 + y_2 + g$ is set to 400 nm. The unit cell is the same dimension as that for Figure 2. The setting of incident wave is in common with that for Figure 3.

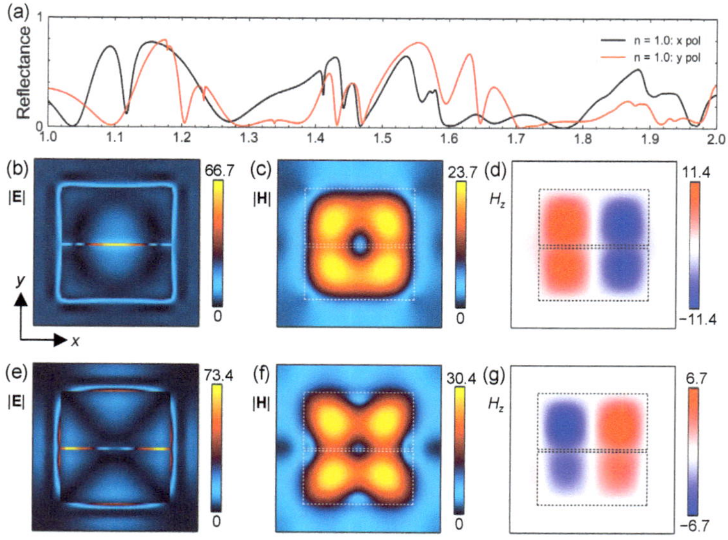

Figure 6. Optical resonances in an asymmetric metasurface of narrow gap $g = 10$ nm. (**a**) Reflectance spectra at x and y polarizations for refractive index $n = 1.0$, shown with black and red curves, respectively. (**b–d**) xy-section views of $|\mathbf{E}|$, $|\mathbf{H}|$, and snapshot of $\mathrm{Re}(H_z)$ at 1.178 eV, respectively. (**e–g**) xy-section views of $|\mathbf{E}|$, $|\mathbf{H}|$, and snapshot of $\mathrm{Re}(H_z)$ at 1.233 eV, respectively. The incident polarization was set to $\mathbf{E}_{\text{in}} \parallel y$ in common. The xy sections were taken at the half-height of the Si-nanogap pair.

The reflectance spectrum in Figure 6a presents definite anisotropy for the incident x and y polarizations, shown with black and red curves, respectively. High-Q modes with narrow line widths appear at 1.178 and 1.233 eV at the y polarization, suggesting that they are qBIC modes. We examine the features in terms of the EM-field distributions in the following.

In Figure 6b–d, the resonant EM fields, $|\mathbf{E}|$, $|\mathbf{H}|$, and $\mathrm{Re}(H_z)$, are visualized, respectively, at the condition of the left arrow in Figure 4a, that is, 1.178 eV and y polarization. The EM fields are shown in a xy-section view of the unit cell; the section was taken at the half height of the Si-nanogap unit. The refractive index in the medium contacting the metasurface is set to 1.5. Evidently, strong localization of the electric field takes place at the Si nanogap, reaching the maximum of $|\mathbf{E}| = 66.7$ (or intensity $|\mathbf{E}|^2 = 4448.9$). In particular, the electric field localizes around the center of the nanogap. The magnetic field mainly distributes in the Si nanoblocks (Figure 6b) and the major component is found to be H_z (Figure 6d). The magnetic-field distribution indicates that this resonance is basically an MD mode. Introducing the nanogap, the electric field is able to localize inside the nanogap.

In Figure 6e–g, the resonant EM fields correspond to the condition of the right arrow in Figure 4a, that is, 1.233 eV and y polarization. The most prominent feature is seen at the electric-field distribution that has two maxima points regarding $|\mathbf{E}|$ in the nanogap. The maxima mean that the intensity $|\mathbf{E}|^2$ exceeds 5387, which is a huge enhancement, compared to the incidence of $|\mathbf{E}_{\text{in}}|^2 = 1$. The magnetic field is qualitatively similar to that in Figure 6b,c. Thus, the two resonances in Figure 6 are basically MD modes associated with the first and second localized electric-field localizations in the nanogap.

This asymmetric metasurface of 10 nm gap can be compared to symmetric metasurfaces of Si-nanogap unit (Figure A2), which has a 10 nm gap. In the symmetric case, there is no qBIC mode, in contrast to the asymmetric case in Figure 6. Further details are described in Appendix A.

Regarding refractometric responses, the interaction volume with the surrounding medium becomes small in this asymmetric metasurface. Consequently, the responsiveness is similar to that in Figure A2, being quantitatively smaller than that in Figure 4a.

4. Discussion

Except for the qBIC modes, the metasurfaces have prominent optical resonances in the reflectance spectra. When the parameter $\alpha = 10$, it was found that the two gap modes emerge and are susceptible linearly to a wide range of refractive index of $n = 1.0$–1.5 (Figure 4a) and quadratically to a further wide range of $n = 1.0$–2.0 (Figure 4b); in the linear range, the responses were evaluated to be 286 and 231 nm/RIU. The two peaks of reflectance enable us to clearly identify the refractometric responses in the multi-peak reflectance spectra. Such a wide-range response has hardly been ensured for other metasurfaces, which are often limited to sensing for narrow refractive-index ranges [17,26–28,45–49]. Table 1 selectively lists the reported quantities based on Si-based nanostructures. Other platforms such as plasmonic structures were already addressed in other references [32,46–50].

Table 1. Comparison of Si metasurfaces and lattice in terms of refractive-index sensing. MSF, Exp, and R denotes metasurface, experiment, and reflectance, respectively. C band means a telecommunication band of 1530–1580 nm. Detected n ranges without any indication resulted from simulations.

Structure	Feature	Detected n	Response	Reference
Si MSF	2D array of 50 nm height pellets. Shift at 840–850 nm. 227 nm/RIU	1.33–1.35, Exp	Linear	[17]
Si MSF	qBIC at C band. Asymmetric Si bars. 440 nm/RIU	1.0–1.1	Linear	[26]
Si MSF	2D array of 450 nm height pairs. Shift at 1330–1600 nm. 612 nm/RIU	1.3–1.7	Linear	[27]
Si MSF	qBIC MD mode at C band. 30 nm nanogap. 258 nm/RIU	1.0–1.4	Linear	[28]
Si MSF	qBIC MD mode at C band. 30 nm nanogap. 317 nm/RIU	1.33–1.36, Exp	Linear	[28]
Si lattice	1D line and space. Large R change at 900–1000 nm	1.0–1.5 (wide)	Quadratic	[32]
Si lattice	2D square nanoblocks. Large R change at 745–790 nm	1.0–1.5 (wide)	Quadratic	[32]
Si MSF	qBIC MD mode at C band. Long-short bar pair. 231 nm/RIU	1.33–1.36, Exp	Linear	[43]
Si MSF	2D ring and bar unit. Shift at 1340–1360 nm. 289 nm/RIU	1.40–1.44, Exp	Linear	[45]
Si MSF	2D cylinders with thin spokes. Shift at 1381.99–1382.82 nm. 166 nm/RIU	1.000–1.005	Linear	[46]
Si lattice	1D grating of pedestal unit. Shift at 1480–1555 nm. 536 nm/RIU	1.33–1.47	Linear	[47]
Si MSF	qBIC at C band. 2D bar-pair unit. Shift at 1340–1360 nm. 501 nm/RIU	1.33–1.48	Linear	[48]
Si MSF	qBIC. Asymmetric Dimer. Shift at 710–760 nm. 140 nm/RIU	1.4–1.6, Exp	Linear	[49]

Table 1. *Cont.*

Structure	Feature	Detected n	Response	Reference
Si MSF	Asymmetric Si-nanogap unit. Shift at 950–1120 nm. 286, 231 nm/RIU	1.0–1.5 (wide)	Linear	This study
Si MSF	Asymmetric Si-nanogap unit. Shift at 950–1270 nm. Traceable double R peaks	1.0–2.0 (wide)	Quadratic	This study

As listed in Table 1, refractive index sensing was frequently studied for narrow ranges of the refractive index. In recent papers [26–28,43,48,49], qBIC modes were explored for refractive index sensing. Although the motives were to find highly sensitive refractive index sensors, the responses in the units of nm/RIU are competing with other resonant nanostructures. Indeed, the physical limit of refractive index sensors comprising periodic structures is given by Equation (1), indicating that the limit is proportional to the periodic length and that larger periodic lengths tend to exhibit larger sensitivity in nm/RIU. The performance of the qBIC-based sensors [26–28,43,48,49] is 20–70%, compared to the physical limit. Thus, there is a room to seek for better performance. Indeed, a 1D grating structure of a large periodic length of 820 nm showed an improved value of the sensitivity [47], compared with the qBIC-based sensors. The present metasurface sensors have a feature that they are responsive to a wide range of refractive indices, showing a practical potential. Other refractive index sensors using plasmonic lattices, optical fibers, and THz absorbers were discussed in a previous publication [32], where 569.1 nm/RIU was shown using a 1D plasmonic lattice and means 94.9% realization of the physical limit.

Regarding qBIC resonances, it is an experimental issue to obtain high-Q qBIC modes because nanofabricated metasurfaces have inevitable scattering loss that originates from roughness at the outermost surface of the nanostructures, even when Si is etched through a BOSCH process that allows for high-contrast, deep reactive ion etching of Si [43]; in fact, we refer to the fact that the experimental limitation for high-Q modes exists at $Q \approx 3000$ in the Si-nanogap metasurfaces. In the structures described in this study, Q factors over 1000 will be a practical goal when experimentally pursuing the high-Q qBIC modes in the present Si-nanogap metasurfaces.

5. Conclusions

Metasurfaces of Si-nanogap units were numerically examined in this study. Introducing asymmetry in the unit cell, qBIC modes became detectable. As the asymmetric parameter α became larger, the qBIC signals in the reflectance spectra became evident. This is a feature of qBIC. Examining the resonant EM-field distributions, the EQ mode was observed at x polarization, and the MD and MQ modes were observed at y polarizations (Figure 2a). The MD modes were modified into two modes in accordance with the first and second electric-field nodes in the nanogap (Figure 6). Adjusting the gap g and the α, a huge electric-field-intensity enhancement exceeding 5000 was found. This is an outstanding feature in this all-dielectric metasurface. Considering application for refractive index sensing, the present metasurfaces are responsive to a very wide range of refractive indices from 1.0 to 2.0 (Figure 4). The linear approximation for sensitivity holds for $n = 1.0$–1.5, whereas the whole range of $n = 1.0$–2.0 is reproduced using quadratic functions. The changes in reflectance spectra are traceable and calibrated by measuring the pair of two reflectance peaks. Such a feature responding to the wide n range, to the best of our knowledge, has not been shown in other systems so far.

Funding: This study was partially supported by JSPS KAKENHI Grant Number JP24K01389.

Institutional Review Board Statement: Not applicable.

Informed Consent Statement: Not applicable.

Data Availability Statement: Data in this study are accessible upon reasonable request to the author.

Acknowledgments: The author thanks Keisuke Watanabe for discussion on Si-nanogap structures. Numerical implementations were conducted on supercomputers SX-AOBA in Cyberscience Center, Tohoku University, Japan.

Conflicts of Interest: The author declares no conflicts of interest.

Abbreviations

The following abbreviations are used in this manuscript:

qBIC	quasi-bound state in the continuum
SOI	Si-on-insulator
RCWA	rigrously coupled-wave analysis
S matrix	scattering matrix
MPI	multiparallel implementation
EM	electromagnetic
EQ	electric quadrupole
MD	magnetic dipole
MQ	magnetic quadrupole
FWHM	full-wave half maximum
Q factor	quality factor

Appendix A. Reflectance Spectra of Symmetric Metasurfaces

Here, we show two illustrative results using metasurfaces of symmetric units. In Figure A1, the unit has a single square Si nanoblock. In Figure A2, the unit has a symmetric Si pair. The refractometric responses are shown for comparison with Figure 4a.

Figure A1a shows the unit cell in a xy-section view, which is set to a dimension of 600×600 nm^2. The square Si nanoblock is $400 \times 400 \times 200$ nm^3 in the unit cell located on the SiO_2 and Si base substrate, similarly to the unit cell in Figure 1.

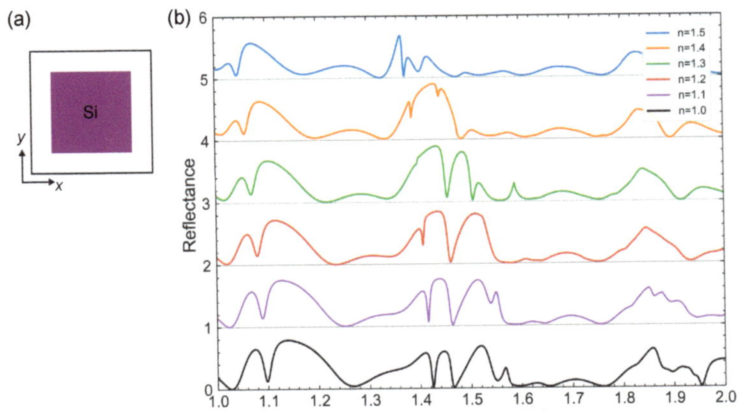

Figure A1. Reflectance spectra of metasurface of a single symmetric Si-nanoblock unit. The spectra depend on refractive index n. (**a**) Unit cell in xy-section view. (**b**) Reflectance spectra dependent on refractive index n in the contacting medium. They are displayed with offset.

Reflectance spectra for different refractive index n in the medium contacting the metasurface are shown in Figure A1b. The spectra exhibit small responsiveness to the refractive index, compared with the asymmetric metasurface in Figure 4a. This result indicates that the unit of a simple Si nanoblock is unsuitable for refractive index sensing.

Metasurface of symmetric Si-nanogap unit is tested in a configuration of refractive-index sensing. The unit structure has two Si nanoblocks of $400 \times 200 \times 200$ nm^3, and the gap is set to 10 nm, as illustrated in Figure A2a. The Si-nanoblock pair is placed at the center of unit cell of 600×610 nm^2 in the xy plane and is assumed to be on the SiO_2 layer, similarly to Figure 1. The refractive index n changes from 1.0 to 1.5 in the medium in contact with the metasurface in Figure A2b.

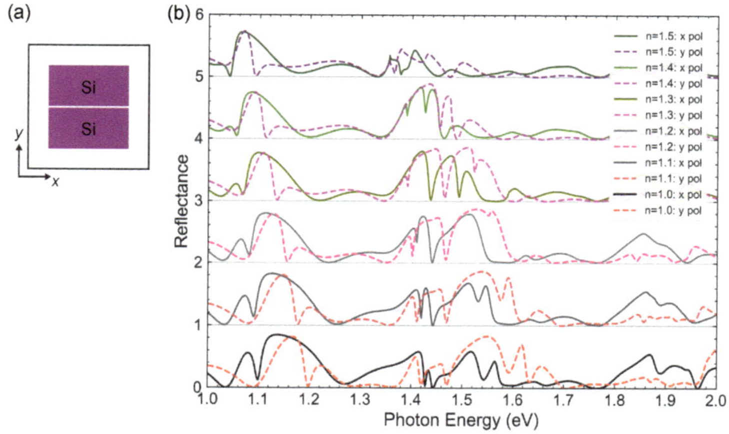

Figure A2. Reflectance spectra of metasurface of symmetric Si-nanogap unit. The spectra depend on refractive index n. (**a**) Unit cell in xy-section view. (**b**) Reflectance spectra dependent on refractive index n in the contacting medium. Solid and dashed curves indicate x- and y-polarized spectra, respectively. They are displayed with offset.

Response to refractive index n is evaluated for the y-polarization resonance that appears at 1.170 eV for $n = 1.0$, as shown with the red dashed curve in Figure A2. It turned out that the sensitivity is 193.6 nm/RIU, which is approximately 10% smaller than that of the asymmetric metasurface in Figure 4. This reduction probably comes from less enhanced EM fields in this symmetric Si-pair metasurface than that in the asymmetric metasurface (Figure 6); in particular, the maximum of $|\mathbf{E}|$ component in the nanogap of this symmetric metasurface is 10.9, which is approximately 16.3% of the maximum value in Figure 6a.

References

1. Yu, N.; Genevet, P.; Kats, M.A.; Aieta, F.; Tetienne, J.P.; Capasso, F.; Gaburro, Z. Light Propagation with Phase Discontinuities: Generalized Laws of Reflection and Refraction. *Science* **2011**, *334*, 333–337. [CrossRef]
2. Ni, X.; Kildishev, A.V.; Shalaev, V.M. Metasurface holograms for visible light. *Nat. Commun.* **2013**, *4*, 2807. [CrossRef]
3. Zheng, G.; Mühlenbernd, H.; Kenney, M.; Li, G.; Zentgraf, T.; Zhang, S. Metasurface holograms reaching 80% efficiency. *Nat. Nanotechnol.* **2015**, *10*, 308–312. [CrossRef]
4. Arbabi, A.; Horie, Y.; Bagheri, M.; Faraon, A. Dielectric metasurfaces for complete control of phase and polarization with subwavelength spatial resolution and high transmission. *Nat. Nanotechnol.* **2015**, *10*, 937–943. [CrossRef]
5. Sell, D.; Yang, J.; Doshay, S.; Yang, R.; Fan, J.A. Large-Angle, Multifunctional Metagratings Based on Freeform Multimode Geometries. *Nano Lett.* **2017**, *17*, 3752–3757. [CrossRef] [PubMed]
6. Fan, Z.; Shcherbakov, M.R.; Allen, M.; Allen, J.; Wenner, B.; Shvets, G. Perfect Diffraction with Multiresonant Bianisotropic Metagratings. *ACS Photonics* **2018**, *5*, 4303–4311. [CrossRef]
7. Chen, W.T.; Zhu, A.Y.; Sisler, J.; Huang, Y.W.; Yousef, K.M.A.; Lee, E.; Qiu, C.W.; Capasso, F. Broadband Achromatic Metasurface-Refractive Optics. *Nano Lett.* **2018**, *18*, 7801–7808. [CrossRef]
8. Khorasaninejad, M.; Chen, W.T.; Devlin, R.C.; Oh, J.; Zhu, A.Y.; Capasso, F. Metalenses at visible wavelengths: Diffraction-limited focusing and subwavelength resolution imaging. *Science* **2016**, *352*, 1190–1194. [CrossRef]
9. She, A.; Zhang, S.; Shian, S.; Clarke, D.R.; Capasso, F. Large area metalenses: Design, characterization, and mass manufacturing. *Opt. Express* **2018**, *26*, 1573–1585. [CrossRef] [PubMed]
10. Callewaert, F.; Velev, V.; Jiang, S.; Sahakian, A.V.; Kumar, P.; Aydin, K. Inverse-designed stretchable metalens with tunable focal distance. *Appl. Phys. Lett.* **2018**, *112*, 091102. [CrossRef]
11. Aiello, M.D.; Backer, A.S.; Sapon, A.J.; Smits, J.; Perreault, J.D.; Llull, P.; Acosta, V.M. Achromatic Varifocal Metalens for the Visible Spectrum. *ACS Photonics* **2019**, *6*, 2432–2440. [CrossRef]

12. Meem, M.; Banerji, S.; Pies, C.; Oberbiermann, T.; Majumder, A.; Sensale-Rodriguez, B.; Menon, R. Large-area, high-numerical-aperture multi-level diffractive lens via inverse design. *Optica* **2020**, *7*, 252–253. [CrossRef]
13. Horie, Y.; Han, S.; Lee, J.Y.; Kim, J.; Kim, Y.; Arbabi, A.; Shin, C.; Shi, L.; Arbabi, E.; Kamali, S.M.; et al. Visible Wavelength Color Filters Using Dielectric Subwavelength Gratings for Backside-Illuminated CMOS Image Sensor Technologies. *Nano Lett.* **2017**, *17*, 3159–3164. [CrossRef]
14. Chen, B.H.; Wu, P.C.; Su, V.C.; Lai, Y.C.; Chu, C.H.; Lee, I.C.; Chen, J.W.; Chen, Y.H.; Lan, Y.C.; Kuan, C.H.; et al. GaN Metalens for Pixel-Level Full-Color Routing at Visible Light. *Nano Lett.* **2017**, *17*, 6345–6352. [CrossRef]
15. Miyata, M.; Nakajima, M.; Hashimoto, T. High-Sensitivity Color Imaging Using Pixel-Scale Color Splitters Based on Dielectric Metasurfaces. *ACS Photonics* **2019**, *6*, 1442–1450. [CrossRef]
16. Zheng, R.; Wei, Y.; Zhang, Z.C.; Wang, Z.Y.; Ma, L.L.; Wang, Y.; Huang, L.; Lu, Y.Q. Stimuli-responsive active materials for dynamic control of light field. *Responsive Mater.* **2023**, *1*, e20230017. [CrossRef]
17. Yavas, O.; Svedendahl, M.; Dobosz, P.; Sanz, V.; Quidant, R. On-a-chip Biosensing Based on All-Dielectric Nanoresonators. *Nano Lett.* **2017**, *17*, 4421–4426. [CrossRef]
18. Tittl, A.; Leitis, A.; Liu, M.; Yesilkoy, F.; Choi, D.Y.; Neshev, D.N.; Kivshar, Y.S.; Altug, H. Imaging-based molecular barcoding with pixelated dielectric metasurfaces. *Science* **2018**, *360*, 1105–1109. [CrossRef]
19. Zhou, J.; Tao, F.; Zhu, J.; Lin, S.; Wang, Z.; Wang, X.; Ou, J.Y.; Li, Y.; Liu, Q.H. Portable tumor biosensing of serum by plasmonic biochips in combination with nanoimprint and microfluid. *Nanophotonics* **2019**, *8*, 307–316. [CrossRef]
20. Iwanaga, M. All-Dielectric Metasurface Fluorescence Biosensors for High-Sensitivity Antibody/Antigen Detection. *ACS Nano* **2020**, *14*, 17458–17467. [CrossRef] [PubMed]
21. Iwanaga, M. Robust Detection of Cancer Markers in Human Serums Using All-Dielectric Metasurface Biosensors. *Biosensors* **2023**, *13*, 377. [CrossRef] [PubMed]
22. Iwanaga, M.; Hironaka, T.; Ikeda, N.; Sugasawa, T.; Takekoshi, K. Metasurface Biosensors Enabling Single-Molecule Sensing of Cell-Free DNA. *Nano Lett.* **2023**, *23*, 5755–5761. [CrossRef] [PubMed]
23. Fu, Q.; Iwanaga, M.; Tang, Y. Metasurface Platform Incorporating Aggregation Induced Emission Based Biosensor for Enhanced Human Serum Albumin Detection. *Adv. Opt. Mater.* **2024**, *12*, 2400868. [CrossRef]
24. Von Neumann, J.; Wigner, E. Über merkwürdige diskrete Eigenwerte. *Phys. Z.* **1929**, *30*, 465–467.
25. Hsu, C.W.; Zhen, B.; Lee, J.; Chua, S.L.; Johnson, S.G.; Joannopoulos, J.D.; Soljačić, M. Observation of trapped light within the radiation continuum. *Nature* **2013**, *499*, 188–191. [CrossRef]
26. Ndao, A.; Hsu, L.; Cai, W.; Ha, J.; Park, J.; Contractor, R.; Lo, Y.; Kanté, B. Differentiating and quantifying exosome secretion from a single cell using quasi-bound states in the continuum. *Nanophotonics* **2020**, *9*, 1081–1086. [CrossRef]
27. Hsiao, H.H.; Hsu, Y.C.; Liu, A.Y.; Hsieh, J.C.; Lin, Y.H. Ultrasensitive Refractive Index Sensing Based on the Quasi-Bound States in the Continuum of All-Dielectric Metasurfaces. *Adv. Opt. Mater.* **2022**, *10*, 2200812. [CrossRef]
28. Watanabe, K.; Iwanaga, M. Nanogap enhancement of the refractometric sensitivity at quasi-bound states in the continuum in all-dielectric metasurfaces. *Nanophotonics* **2023**, *12*, 99–109. [CrossRef]
29. Kodigala, A.; Lepetit, T.; Gu, Q.; Bahari, B.; Fainman, Y.; Kanté, B. Lasing action from photonic bound states in continuum. *Nature* **2017**, *541*, 196–199. [CrossRef]
30. Ha, S.T.; Fu, Y.H.; Emani, N.K.; Pan, Z.; Bakker, R.M.; Paniagua-Domínguez, R.; Kuznetsov, A.I. Directional lasing in resonant semiconductor nanoantenna arrays. *Nat. Nanotechnol.* **2018**, *13*, 1042–1047. [CrossRef]
31. Liu, Z.; Xu, Y.; Lin, Y.; Xiang, J.; Feng, T.; Cao, Q.; Li, J.; Lan, S.; Liu, J. High-*Q* Quasibound States in the Continuum for Nonlinear Metasurfaces. *Phys. Rev. Lett.* **2019**, *123*, 253901. [CrossRef] [PubMed]
32. Iwanaga, M. A Design Strategy for Surface Nanostructures to Realize Sensitive Refractive-Index Optical Sensors. *Nanomaterials* **2023**, *13*, 3081. [CrossRef] [PubMed]
33. Palik, E.D. *Handbook of Optical Constants of Solids II*; Academic: San Diego, CA, USA, 1991.
34. Li, L. New formulation of the Fourier modal method for crossed surface-relief gratings. *J. Opt. Soc. Am. A* **1997**, *14*, 2758–2767. [CrossRef]
35. Li, L. Formulation and comparison of two recursive matrix algorithm for modeling layered diffraction gratings. *J. Opt. Soc. Am. A* **1996**, *13*, 1024–1035. [CrossRef]
36. Christ, A.; Tikhodeev, S.G.; Gippius, N.A.; Kuhl, J.; Giessen, H. Waveguide-Plasmon Polaritons: Strong Coupling of Photonic and Electronic Resonances in a Metallic Photonic Crystal Slab. *Phys. Rev. Lett.* **2003**, *91*, 183901. [CrossRef]
37. Christ, A.; Zentgraf, T.; Tikhodeev, S.G.; Gippius, N.A.; Kuhl, J.; Giessen, H. Controlling the interaction between localized and delocalized surface plasmon modes: Experiment and numerical calculations. *Phys. Rev. B* **2006**, *74*, 155435. [CrossRef]
38. Christ, A.; Ekinci, Y.; Solak, H.H.; Gippius, N.A.; Tikhodeev, S.G.; Martin, O.J.F. Controlling the Fano interference in a plasmonic lattice. *Phys. Rev. B* **2007**, *76*, 201405. [CrossRef]
39. Iwanaga, M. Polarization-selective transmission in stacked two-dimensional complementary plasmonic crystal slabs. *Appl. Phys. Lett.* **2010**, *96*, 083106. [CrossRef]
40. Iwanaga, M.; Choi, B. Heteroplasmon Hybridization in Stacked Complementary Plasmo–Photonic Crystals. *Nano Lett.* **2015**, *15*, 1904–1910. [CrossRef]
41. Iwanaga, M. All-Dielectric Metasurfaces with High-Fluorescence-Enhancing Capability. *Appl. Sci.* **2018**, *8*, 1328. [CrossRef]

42. Koshelev, K.; Lepeshov, S.; Liu, M.; Bogdanov, A.; Kivshar, Y. Asymmetric Metasurfaces with High-Q Resonances Governed by Bound States in the Continuum. *Phys. Rev. Lett.* **2018**, *121*, 193903. [CrossRef] [PubMed]
43. Watanabe, K.; Iwanaga, M. Optimum asymmetry for nanofabricated refractometric sensors at quasi-bound states in the continuum. *Appl. Phys. Lett.* **2024**, *124*, 111705. [CrossRef]
44. Homola, J.; Koudela, I.; Yee, S.S. Surface plasmon resonance sensors based on diffraction gratings and prism couplers: Sensitivity comparison. *Sens. Actuator B-Chem.* **1999**, *54*, 16–24. [CrossRef]
45. Yang, Y.; Kravchenko, I.I.; Briggs, D.P.; Valentine, J. All-dielectric metasurface analogue of electromagnetically induced transparency. *Nat. Commun.* **2017**, *5*, 5753. [CrossRef]
46. Tognazzi, A.; Rocco, D.; Gandolfi, M.; Locatelli, A.; Carletti, L.; De Angelis, C. High Quality Factor Silicon Membrane Metasurface for Intensity-Based Refractive Index Sensing. *Optics* **2021**, *2*, 193–199. [CrossRef]
47. Beliaev, L.Y.; Stounbjerg, P.G.; Finco, G.; Bunea, A.I.; Malureanu, R.; Lindvold, L.R.; Takayama, O.; Andersen, P.E.; Lavrinenko, A.V. Pedestal High-Contrast Gratings for Biosensing. *Nanomaterials* **2022**, *12*, 1748. [CrossRef]
48. Liu, Z.; Guo, T.; Tan, Q.; Hu, Z.; Sun, Y.; Fan, H.; Zhang, Z.; Jin, Y.; He, S. Phase Interrogation Sensor Based on All-Dielectric BIC Metasurface. *Nano Lett.* **2023**, *23*, 10441–10448. [CrossRef]
49. Van Loon, T.; Liang, M.; Delplace, T.; Maes, B.; Murai, S.; Zijlstra, P.; Rivas, J.G. Refractive index sensing using quasi-bound states in the continuum in silicon metasurfaces. *Opt. Express* **2024**, *32*, 14289–14299. [CrossRef]
50. Qin, J.; Jiang, S.; Wang, Z.; Cheng, X.; Li, B.; Shi, Y.; Tsai, D.P.; Liu, A.Q.; Huang, W.; Zhu, W. Metasurface Micro/Nano-Optical Sensors: Principles and Applications. *ACS Nano* **2022**, *16*, 11598–11618. [CrossRef]

hv *photonics*

MDPI

Article

A Novel Demodulation Algorithm for Micro-Displacement Measurement Based on FMCW Sinusoidal Modulation

Zhen Xu [1,2], Yongjie Wang [1,2,*], Zhenqiang Li [1,2], Gaochao Li [1,2], Ke Li [1,2], Hongtao Zhang [1,2] and Fang Li [1,2,*]

1 Laboratory of Optoelectronic Systems, Institute of Semiconductors, Chinese Academy of Sciences, Beijing 100083, China; xuzhen22@semi.ac.cn (Z.X.); lizhenqiang@semi.ac.cn (Z.L.); ligaochao96@semi.ac.cn (G.L.); like@semi.ac.cn (K.L.); zhanghongtao@semi.ac.cn (H.Z.)

2 College of Materials Science and Opto-Electronic Technology, University of Chinese Academy of Sciences, Beijing 100049, China

* Correspondence: wyj@semi.ac.cn (Y.W.); lifang@semi.ac.cn (F.L.)

Abstract: Frequency-modulated continuous wave (FMCW) interferometry, an emerging laser interferometry technology, can be applied in the field of fibre-optic sensing to achieve high-precision micro-displacement measurements. To address nonlinearity issues in laser frequency modulation and localisation deviations of feature points in traditional algorithms, this paper proposes a demodulation algorithm suitable for sinusoidal frequency modulation schemes, incorporating the principle of orthogonal phase-locked amplification. The algorithm includes signal preprocessing, phase-locked amplification, error correction, and phase calculation. Experimental results show that the system achieves a measurement error standard deviation of 3.23 nanometres for static targets. The displacement measurement error at 100 μm is 0.057% F.S., and the linearity between the measured values and the actual displacement values is 0.99997. Compared with conventional methods, the approach introduced in this paper eliminates the need for separate nonlinear corrections of the current-to-optical frequency relationship and minimises the issue of feature point localization deviations, showing significant potential for practical applications.

Keywords: fibre-optic sensors; interferometric sensors; micro-displacement; frequency-modulated continuous wave; sinusoidal wave frequency modulation; phase demodulation

Citation: Xu, Z.; Wang, Y.; Li, Z.; Li, G.; Li, K.; Zhang, H.; Li, F. A Novel Demodulation Algorithm for Micro-Displacement Measurement Based on FMCW Sinusoidal Modulation. *Photonics* **2024**, *11*, 1196. https://doi.org/10.3390/photonics11121196

Received: 16 November 2024
Revised: 9 December 2024
Accepted: 16 December 2024
Published: 20 December 2024

1. Introduction

The rapid advancement of modern technology and industry has led to an increasing demand for micro-displacement measurement across various fields, including aerospace, precision engineering, and marine fibre-optic sensing [1–3]. Laser displacement measurement technology offers several advantages, such as non-contact operation, ease of use, high measurement accuracy, and strong anti-interference capability [4]. Frequency-modulated continuous wave (FMCW) laser interferometry is a non-contact laser displacement measurement method known for its high precision, lightweight design, extensive dynamic measurement range, and high sensitivity [5–8]. Consequently, it has been widely applied in the manufacturing and research of various large-scale precision machining equipment [9–11].

An FMCW laser interferometry system modulates a distributed feedback (DFB) laser to emit light waves with time-varying optical frequencies. These light waves pass through a Fabry–Pérot (F–P) interferometer, generating a beat signal that contains relative displacement information. By demodulating the phase variation in this beat signal, displacement information can be obtained [12–14].

In existing measurement schemes, optical frequency modulation is typically performed using sawtooth or triangular waveforms. Due to the nonlinear relationship between the laser's current and optical frequency tuning, significant deviations occur between the actual modulation curve and the ideal curve, a primary source of error. Therefore, in practical measurements, auxiliary equipment is often required for nonlinear correction.

For example, in 2009, Peter A. Roos et al. [15] achieved the linearisation of ultra-wideband laser frequencies using fibre-optic self-heterodyne techniques, reducing the frequency error to 170 kHz at a modulation bandwidth of 5 THz. In 2011, Barber et al. [16] used optical frequency combs to perform frequency modulation nonlinearity correction on tuneable lasers, reducing the frequency error to 60 kHz within the same bandwidth. In 2022, Lu Cheng et al. [17] demonstrated a distance extraction method based on frequency modulation nonlinear kernel functions, achieving a ranging accuracy of 1.9 μm for targets at a distance of 5 m.

In terms of signal processing, existing FMCW interferometric phase demodulation algorithms rely on feature points as initial phase characteristic values for phase demodulation, such as maxima, zero-crossings, and centroid points. During dynamic displacement measurements, these methods often suffer from inaccurate point localisation, leading to errors. Although increasing the hardware sampling frequency can mitigate this error, this typically slows down measurement speeds [18]. For example, in 2020, Bin Sun et al. [19] used the average zero-crossing position as the initial phase characteristic value for phase demodulation; their contact displacement sensor achieved a standard deviation of static error of 8.9 nm at a displacement of 12 mm. In 2023, Mengchao Yan et al. [20] used the average position of centroid points as the initial phase characteristic value, thereby reducing random errors in displacement measurement.

Compared with linear frequency modulation schemes, sinusoidal frequency modulation signals are easier to generate [21] and the interference signals they produce do not have dead zones. However, the beat signal produced by sinusoidal modulation has a non-constant frequency, making it challenging to capture the initial phase variation, and there is currently a lack of effective demodulation algorithms.

In this paper, by incorporating the principle of orthogonal phase-locked amplification, we propose, for the first time, a new displacement demodulation method under sinusoidal frequency modulation. This method involves the shaping and filtering of the beat interference signal, segmenting it according to the modulation period, passing it through a phase-locked amplifier, performing elliptical fitting and correction, and finally calculating the phase to demodulate the target displacement.

2. Principle of Sinusoidal FMCW Interference for Displacement Measurement

FMCW refers to the continuous modulation of light waves by frequency or angular frequency [22]. A sine wave is used to modulate a coherent light source, emitting two modulated light waves that propagate along separate paths and eventually meet at a point in space, forming interference.

Figure 1 shows the frequency relationship among the reference light, signal light, and beat signal. The solid line represents the frequency of the reference light; the dashed line represents the frequency of the signal light; and the dotted line represents the frequency of the interference beat signal. Based on the waveform in Figure 1, the three curves can be expressed as follows [21]:

$$\begin{cases} \omega_1(t) = \frac{\Delta\omega}{2}\sin(\omega_m t) + \omega_0 \\ \omega_2(t) = \frac{\Delta\omega}{2}\sin[\omega_m(t-\tau)] + \omega_0 \\ \omega_b(t) = \frac{\Delta\omega\omega_m\tau}{2}\cos(\omega_m t) \end{cases} \quad (1)$$

In this equation, $\Delta\omega$ represents the peak-to-peak value of the modulated angular frequency, ω_m is the modulation angular frequency, ω_0 is the centre angular frequency, and τ represents the delay time between the reference and signal light. The beat signal generated by the interference of the signal and reference light can be expressed as follows:

$$I(\tau,t) = I_0\left\{1 + V cos\left[\frac{\Delta\omega\tau}{2} sin\ (\omega_m t) + \omega_0\tau\right]\right\} \quad (2)$$

In this equation, I_0 represents the average intensity of the beat signal and V represents the contrast of the beat signal. The waveform of the beat signal is shown in Figure 2.

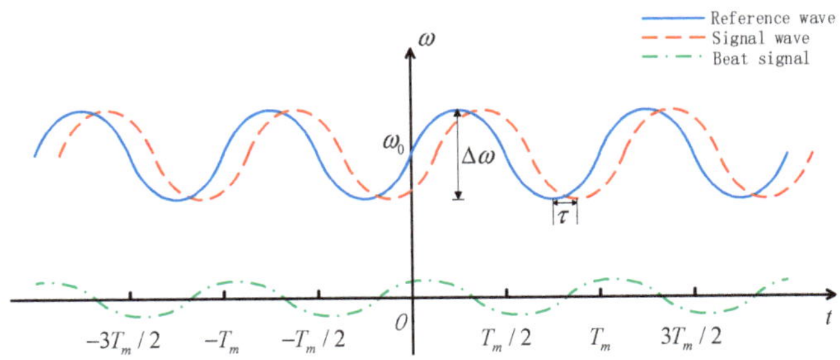

Figure 1. Frequency relationship between the reference wave, signal wave, and beat signal during sinusoidal wave modulation.

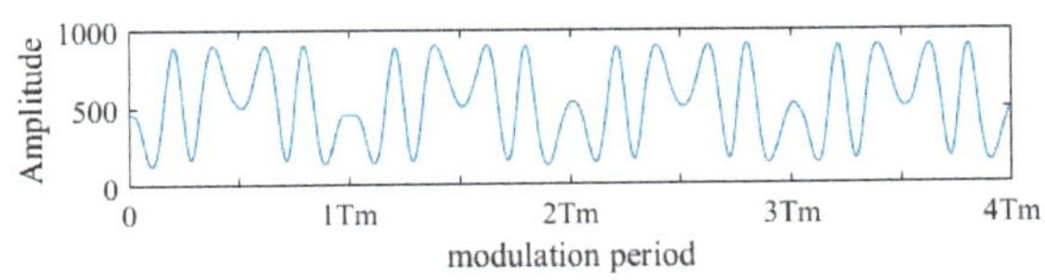

Figure 2. Waveforms of beat signals from a real optical sinusoidalwave FMCW interferometer.

By substituting the optical path difference formula represented by $OPD = c\tau$ into Equation (2), we can obtain an expression that relates the optical path difference to frequency, phase, and other parameters as follows:

$$I(OPD,t) = I_0\left\{1 + V\cos\left[\frac{\Delta\omega \cdot OPD}{2c}\sin\left(\omega_m t\right) + \frac{2\pi}{\lambda_0}OPD\right]\right\} \quad (3)$$

where c represents the speed of light propagating in a vacuum. Since the optical path difference in the F–P interferometer is represented by $OPD = 2nd$, the initial phase ϕ_{b0} of the beat signal can be expressed as follows:

$$\phi_{b0} = \omega_0\tau = \frac{2\pi}{\lambda_0}OPD = \frac{4\pi nd}{\lambda_0} \quad (4)$$

where n is the refractive index of air and d is the distance between the partial reflector and the collimator. Therefore, the relative displacement Δd of the target can be expressed as follows:

$$\Delta d = \frac{\lambda_0}{4\pi n}\Delta\phi_{b0} \quad (5)$$

where $\Delta\phi_{b0}$ represents the change in initial phase between adjacent modulation cycles. The total displacement of the target can be determined by summing Δd over the entire displacement process.

3. Principle of the Demodulation Algorithm

3.1. Signal Processing Flow

Based on the principle of orthogonal lock-in amplification, this paper proposes a phase demodulation method based on sinusoidal-wave optical frequency-modulated continuous-wave interference. The algorithm's principle is shown in Figure 3.

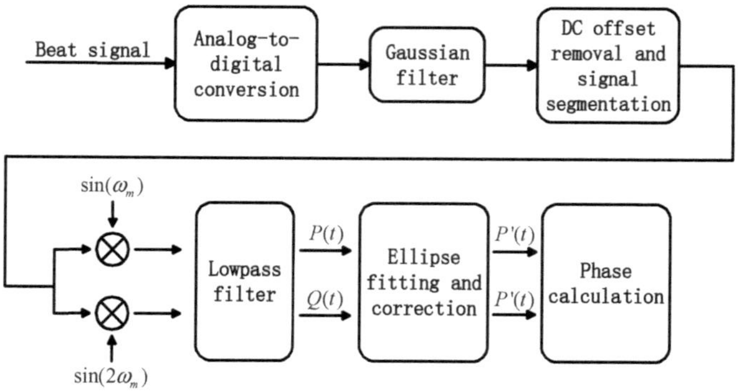

Figure 3. Signal processing flow, consisting of four parts: signal preprocessing, lock-in amplification, ellipse fitting and error correction, and phase calculation.

During signal preprocessing, the beat frequency signal is converted into a digital signal, and a Gaussian filter is applied to reduce signal noise. The signal is then segmented according to the frequency modulation cycle. The beat frequency signal is rewritten as follows based on Equation (2):

$$I(t) = A + Bcos[Csin\,(\omega_m t + \Delta\theta) + \phi_{b0}(t)] \tag{6}$$

where A represents the DC bias of the beat frequency signal, B is the AC amplitude, C is the phase modulation depth, $\Delta\theta$ is the carrier phase delay caused by factors like photoelectric signal transmission and digital-to-analogue conversion, and $\phi_{b0}(t)$ represents the function of ϕ_{b0} changing with time t during the displacement process. Expanding Equation (7) using the Bessel function can be expressed as follows:

$$\begin{aligned} I(t) = A \quad &+ B\left\{ J_0(C) + 2\sum_{n=1}^{\infty} J_{2n}(C)\cos[2n(\omega_m t + \Delta\theta)] \right\} cos\phi_{b0}(t) \\ &- B\left\{ 2\sum_{n=1}^{\infty} J_{2n-1}(C)\cos[(2n-1)(\omega_m t + \Delta\theta)] \right\} sin\phi_{b0}(t) \end{aligned} \tag{7}$$

After expanding the beat signal $I(t)$ using Bessel functions, multiple harmonics with frequency intervals of ω_m appear. The amplitudes of each harmonic component are given by the values of first-kind Bessel functions of different orders. By multiplying the preprocessed beat signal $I(t)$ with the carrier signal $\sin(\omega_m t)$ and its second harmonic $\sin(2\omega_m t)$ and applying a low-pass digital filter to remove high-frequency components, two orthogonal signals $P(t)$ and $Q(t)$ are obtained as follows:

$$\begin{cases} P(t) = LPF[I(t) \times \sin(\omega_m t)] = -BJ_1(C)cos\Delta\theta \cdot sin\phi_{b0}(t) \\ Q(t) = LPF[I(t) \times \sin(2\omega_m t)] = -BJ_2(C)cos2\Delta\theta \cdot cos\phi_{b0}(t) \end{cases} \tag{8}$$

In this context, LPF denotes the low-pass filtering operation. Under ideal conditions, $C = 2.63$, $\Delta\theta = 0$, at which point,

$$J_1(C) = J_2(C),\ \phi_{b0}(t) = \arctan\left[\frac{P(t)}{Q(t)}\right] \tag{9}$$

Next, based on Equation (5), the relative displacement for a single measurement cycle can be calculated. By accumulating the relative displacements of each measurement cycle, the total displacement of the target can be obtained.

3.2. Error Correction Algorithm

The primary sources of error in the measurement system include (1) phase modulation depth $C \neq 2.63$, (2) carrier phase delay $\Delta\theta \neq 0$, (3) a lack of quadrature, and (4) deviations in the actual optical frequency modulation curve from an ideal sine wave. From Equation (8), it is observed that errors (1) and (2) cause a shift in the amplitude ratio of the two orthogonal signals, while error (3) results in the phase difference between the two orthogonal signals deviating from 90°. Error (4) introduces multiple harmonics into the modulation curve of the actual optical frequency, as shown in Equation (11).

$$\omega_1(t) \approx \omega_0 + \frac{\Delta\omega}{2}\sin(\omega_m t) + c_1\sin(2\omega_m t + \varphi_1) + c_2\sin(3\omega_m t + \varphi_2) + \ldots \quad (10)$$

This results in a tilted ellipse in the Lissajous figure, which can be compensated by elliptic correction. If the nonlinearities are temporally stable, they contribute no additional measurement uncertainty [23].

In summary, the actual orthogonal signals are now defined as follows:

$$\begin{cases} P(t) = Rsin\phi_{b0}(t) + p \\ Q(t) = \frac{R}{r}cos[\phi_{b0}(t) + \alpha] + q \end{cases} \quad (11)$$

where $\frac{1}{r}$ is the channel gain ratio, p and q are the DC offsets, and α is the orthogonal phase offset. Under ideal conditions, $\frac{1}{r} = 1$ and $\alpha, p, q = 0$, at which point $(P(t), Q(t))$ always lies on a circle of radius R. However, due to the aforementioned errors, the endpoint of $(P(t), Q(t))$ will be distributed on a distorted ellipse, with the equation given by

$$AP^2(t) + BQ^2(t) + CP(t)Q(t) + DP(t) + EQ(t) = R^2 \quad (12)$$

Next, the coefficients A, B, C, D, E are obtained through ellipse fitting, followed by inversely calculating the error terms [24–26]:

$$\begin{cases} \alpha = arcsinC(4AB)^{-\frac{1}{2}} \\ r = \left(\frac{B}{A}\right)^{\frac{1}{2}} \\ p = \frac{2BD-EC}{C^2-4AB} \\ q = \frac{2AE-DC}{C^2-4AB} \end{cases} \quad (13)$$

By inverting Equation (11), the following corrected orthogonal signals $P'(t)$ and $Q'(t)$ are obtained:

$$\begin{cases} P'(t) = P(t) - p \\ Q'(t) = \frac{1}{cos\alpha}[(P(t) - p)sin\alpha + r(Q(t) - q)] \end{cases} \quad (14)$$

4. Experimental Setup

To verify the displacement demodulation capabilities of the proposed method in practical engineering, an FMCW laser interferometry system was built for experimental testing. The system primarily consists of a DFB laser, fibre circulator, fibre collimator, total reflection mirror, and photodetector. The system setup is shown in Figure 4. Initially, the system uses an arbitrary signal generator to output a sinusoidal signal to modulate the DFB laser. The laser emits light waves with a time-varying frequency, which are transmitted along the fibre through Port 1 of the circulator to Port 2, reaching the collimator coated with a 30% reflective film. The interference beat signal occurs within the interferometer, which is composed of a partially reflective film and a total reflection mirror. This signal is received by the photodetector via Port 3 of the circulator and then processed and demodulated by the host computer.

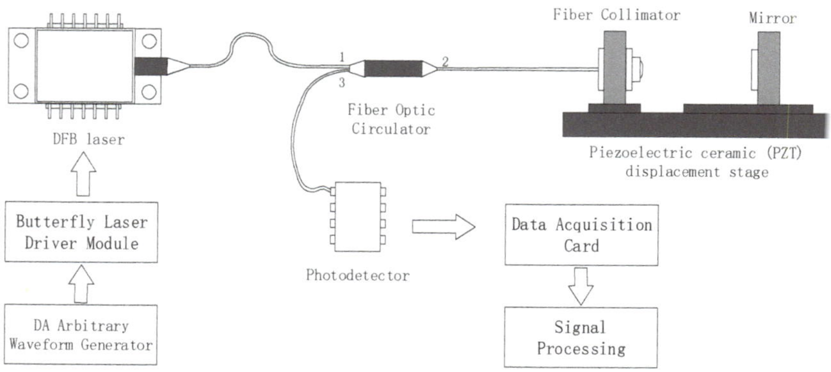

Figure 4. Diagram of the displacement measurement system structure.

The system uses a high-performance DFB laser with a central wavelength of 1550 nm, a threshold current of 9.5 mA, and a maximum operating current of 300 mA. Since different currents have varying effects on laser wavelength modulation, it is essential to select a current range with good optical frequency linearity and stable wavelength output for modulation. Based on test results for the DFB laser, a high-performance signal generator was used to output a sine wave signal within a current range of 205–215 mA to modulate the DFB laser at a frequency of 10 kHz. A TEC temperature control module maintains the temperature of the DFB laser, reducing the optical frequency drift caused by temperature changes. The measurement optical path consists of a collimator coated with a partially reflective film and a total reflection mirror mounted on a high-precision displacement stage. The optimal working range of the collimator is 46–50 mm. The high-precision translation stage used in this experiment has a range of 50 mm, a repeatable displacement accuracy of 2 nm, and a displacement speed of 0.1–3 mm/s, thus meeting the experimental requirements. The DFB laser, photodetector, fibre circulator, and translation stage are fixed along the optical path on an air-floating platform to minimise the influence of environmental factors such as vibrations. The physical diagram of the system is shown in Figure 5.

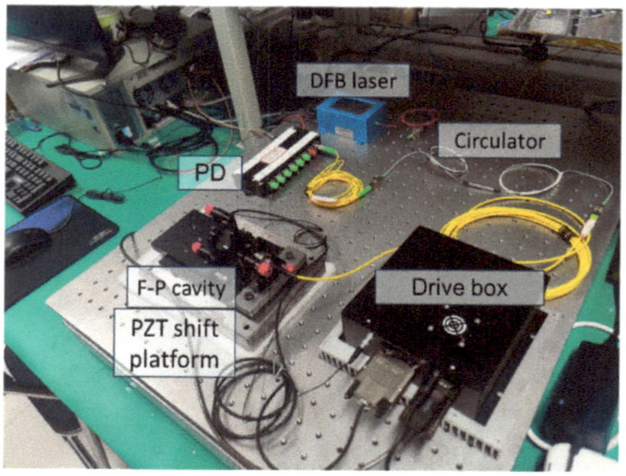

Figure 5. Physical diagram of the displacement measurement system structure.

5. Results

This section may be divided into subheadings. It provides a concise and precise description of the experimental results, their interpretation, and conclusions.

5.1. Static Stability Test

To test the static stability of the system, displacement measurements were performed on a stationary target fixed to an optical platform. To minimise the influence of environmental factors on the measurement results, an air-floating optical platform was used to isolate the measurement optical path from external vibrations in a quiet environment. During the measurement, the system recorded displacement every 0.2 s, continuously recording for 800 s. A scatter plot and distribution histogram of the static error are shown in Figure 6. The experimental data show that the standard deviation of the measurement error for the static target is 3.23 nm, with a normal distribution of random error, demonstrating the system's good stability.

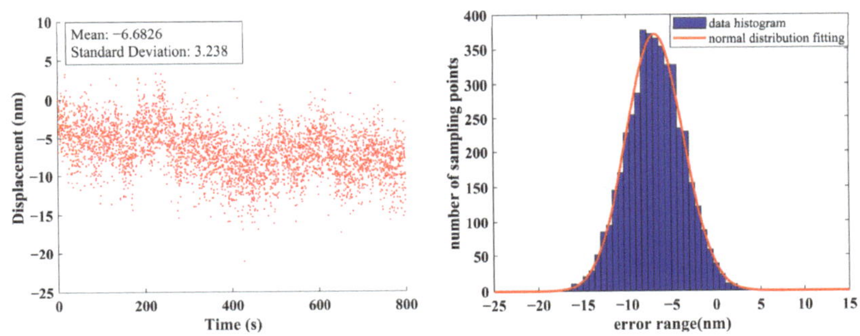

(a) (b)

Figure 6. Static displacement error test results: (**a**) scatter plot of error variation over time; (**b**) error distribution histogram.

5.2. Repeatability Displacement Test

To test the accuracy of the designed displacement measurement system, a micron-level displacement repeatability experiment was conducted. The experimental procedure was as follows: (1) Before measurement, the translation stage was driven to move linearly by 100 μm, and data from two orthogonal signals were collected. Initial correction parameters for the orthogonal components were obtained through ellipse fitting, with continuous updates in subsequent experiments. (2) The translation stage was moved to the zero point of the guide rail, and the system began measurement. The target underwent a uniform displacement of 100 μm at a speed of 0.5 mm/s. After completing the displacement, experimental data were recorded at a sampling rate of 20 kHz. This process was repeated 40 times, and the displacement variation curve is shown in Figure 7a. The bottom right corner of the image illustrates that the piezoelectric ceramic translation stage generates an "oscillating peak" at the moment each movement stops. To represent a single measurement result, the 100th sample point after the peak was taken as the measurement value once the system had stabilised. (3) The displacement error was calculated by subtracting the actual displacement value from the measured value for each of the 40 displacements, as shown in Figure 7b. The results indicate that the average measurement error is 8.6 nm, with a standard deviation of 24.36 nm. The maximum displacement error among the 40 measurements was 57 nm, which corresponds to a full-scale error of 0.057% F.S. at 100 μm.

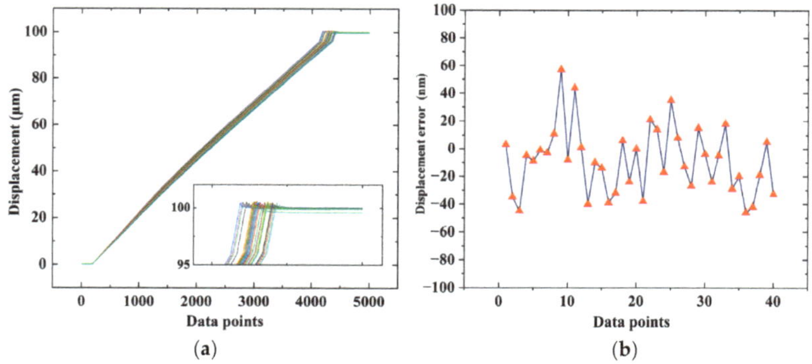

Figure 7. Displacement demodulation results for the displacement stage when the preset value is 100 μm on the basis of 40 measurements: (**a**) linear displacement; (**b**) error distribution.

5.3. Linearity Test

To determine the linearity between the system's measurement values and the actual displacement values, multiple stepping experiments were conducted on the measurement system. The experimental steps are as follows: (1) The translation stage was moved to the zero point of the guide rail, and the system began measurement. (2) The step size of the translation stage was set to 5 μm, which was used as the actual displacement value. During the test, the target was allowed to rest for one second after each 5-micrometre step, with a total of 20 steps taken, and experimental data were recorded. The relationship between the measurement values and time is shown in Figure 8a. (3) The data from Figure 8a were then sampled at equal intervals to extract 20 data points as the measurement values after each step. The relationship between the measurement values and the number of steps is shown in Figure 8b. After calculation, the linear regression R^2 value between the measurement values and the actual displacement values was 0.99997.

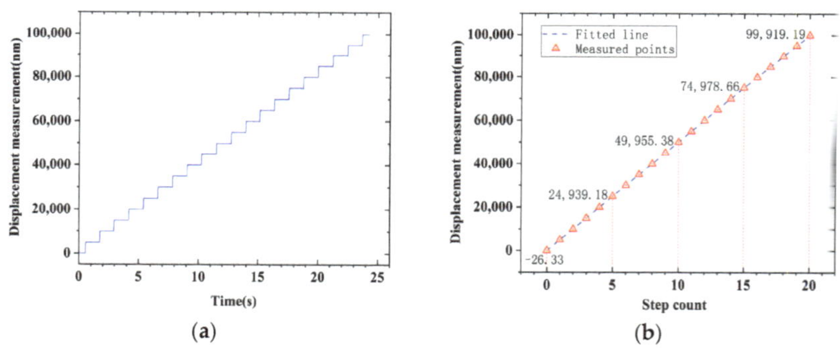

Figure 8. Results of 20 steps of 5 μm for the target: (**a**) relationship between measurement values and time during the stepping process; (**b**) linear fit between the measurement values of 20 steps and the actual displacement.

5.4. Data Correction

The test target is controlled by a displacement stage to move at a constant speed of 0.5 mm/s over a distance of 100 μm. The data from the two orthogonal signals generated during the displacement process are saved. Ellipse fitting is performed on both the original dataset and calibrated dataset, with the results being shown in Figure 9.

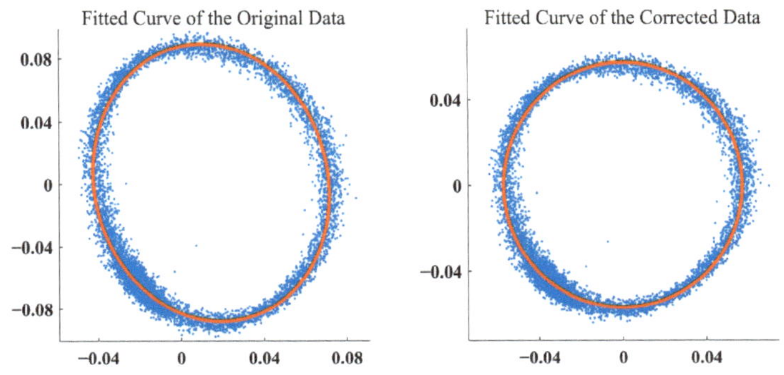

(a) (b)

Figure 9. Plots of uncorrected (**a**) and corrected (**b**) experimental datasets.

To represent the degree of closeness between the overall distribution of the dataset and the ideal circle, the goodness of fit is defined as

$$R^2 = 1 - \frac{\sum_i^N \left[\sqrt{(x_i - u)^2 + (y_i - v)^2} - \hat{r}\right]^2}{\sum_i^N \left[\sqrt{(x_i - \overline{x})^2 + (y_i - \overline{y})^2} - \hat{r}\right]^2} \quad (15)$$

where N is the number of samples, (x_i, y_i) are the sample points, $(\overline{x}, \overline{y})$ is the sample mean, and (u, v) and $\hat{r}$ are the centre and radius of the ideal circle, respectively.

The results show that after algorithmic calibration, $R^2 = 0.9974$.

6. Discussion

Most existing FMCW displacement measurement systems use a linear frequency modulation scheme and rely heavily on phase information from specific feature points in the demodulation algorithm, such as peak positioning methods. However, the beat signal has a very small slope near the peak, making these systems prone to positioning errors during actual measurements, which affects the results. The sinusoidal frequency modulation scheme used in this study eliminates the need for a separate nonlinear correction while also avoiding the problem of extracting information from only a few feature points in the algorithm. The experimental results show that the standard deviation of the static error in the current measurement system reaches 3.23 nm, the measurement error at 100 μm is 0.057% F.S., and the linearity between the measured values and the actual displacement values is 0.99997. Additionally, we speculate that since the phase detection accuracy of this method does not rely on high sampling rates, it is possible to reduce the number of sample points per cycle, allocating more hardware resources to increase the modulation frequency, thereby improving the maximum measurement speed of the system.

As measurement results may be influenced by hardware structure, environmental conditions, and human factors, the displacement measurement system constructed in this study can still be further optimised. Next, I will analyse the static drift issue of this measurement system.

In the static sampling process shown in Figure 6a, the interferometer's zero point exhibits a certain degree of negative drift, which causes the statistical distribution of the data to be directly affected by the sampling duration. Extending the sampling time may increase the static standard deviation. To investigate the relationship between the sampling duration and static standard deviation, I extended the static measurement time to one hour and calculated the static standard deviation for different sampling durations, as shown in

Figure 10. The results indicate a certain linear relationship between the sampling duration and the magnitude of the static error, though this relationship weakens over time.

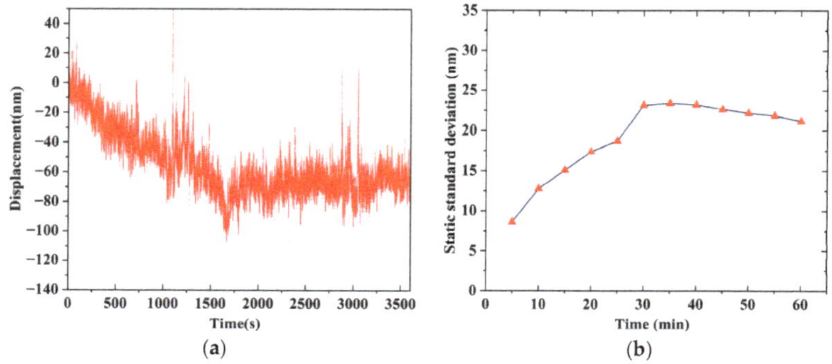

Figure 10. (**a**) Distribution of static displacement errors within one hour; (**b**) relationship between static standard deviation and time.

It is hypothesised that the static measurement error is influenced not only by the negative drift but also by environmental disturbances, such as micro-vibrations of the table, changes in the refractive index of the medium, and other factors. Therefore, the relationship between the sampling duration and static standard deviation is not entirely linear. Further research is needed in the future to conduct more specific theoretical analysis and validation.

Additionally, system optimisation includes the selection of DFB lasers and collimators, vibration isolation measures, and enhanced error compensation. In the future, we plan to implement a more precise micro-displacement calibration scheme to provide a solid experimental foundation while continuing to optimise the system's hardware structure and error compensation algorithm to achieve higher performance metrics.

7. Conclusions

This paper proposes a micro-displacement measurement method based on FMCW laser interferometry, as suitable for sinusoidal frequency modulation schemes. This method eliminates the need for the separate nonlinear correction of the current–optical frequency relationship and increases the utilisation of signal sampling points. In this study, we conducted a theoretical analysis, built a measurement system, and performed an experimental validation.

The results indicate that the standard deviation of the static error in the current measurement system reaches 3.23 nm, exhibits a measurement error of 0.057% F.S. at 100 µm, and the linearity between the measured values and the actual displacement values is 0.99997. This method meets the high-precision dynamic measurement requirements in fields such as ultra-precision machining and sensor design. Additionally, it features a high sweep rate and efficient sampling point utilisation, presenting promising potential for high-speed displacement demodulation applications.

Author Contributions: Methodology, Z.X.; writing, Z.X. and G.L.; supervision, F.L.; software, Z.X. and Z.L.; project administration, Y.W.; conceptualization and validation, H.Z. and K.L.; funding acquisition, Y.W. All authors have read and agreed to the published version of the manuscript.

Funding: This research was funded by the National Key R&D Program of China, grant number 2022YFB3205300, and the National Natural Science Foundation of China, grant number 42275194.

Institutional Review Board Statement: Not applicable.

Informed Consent Statement: Not applicable.

Data Availability Statement: Data are contained within the article.

Conflicts of Interest: The authors declare no conflicts of interest.

References

1. Fernández, R.; Durana, G.; Aldabaldetreku, G.; Przysowa, R.; García, I. Analysis of the shaft behaviour in aircraft engines using tip clearance data and custom designed laser sensors. *Proc. SPIE* **2021**, *11772*, 1177224. [CrossRef]
2. Lin, Z.; Ji, L.; Wang, W. Precision machining of single crystal diamond cutting tool via picosecond laser irradiation. *Int. J. Refract. Met. Hard Mater.* **2023**, *114*, 106226. [CrossRef]
3. Li, G.; Wang, Y.; Shi, A. Review of seawater fiber optic salinity sensors based on the refractive index detection principle. *Sensors* **2023**, *23*, 2187. [CrossRef] [PubMed]
4. Zhang, Y.; Zhang, T.; Liu, X. Research progress and application of air refractive index measurement technology based on laser interference. *Proc. SPIE* **2021**, *11915*, 1191502. [CrossRef]
5. Xu, Z.; Sun, X.; Yu, F.; Chen, K.; Pan, S. Frequency-modulated continuous-wave laser ranging using low-duty-cycle signals for the applications of real-time super-resolved ranging. *Opt. Lett.* **2021**, *46*, 258–261. [CrossRef] [PubMed]
6. Hariyama, T.; Sandborn, P.A.M.; Watanabe, M.; Wu, M.C. High-accuracy range-sensing system based on FMCW using low-cost VCSEL. *Opt. Express* **2018**, *26*, 9285–9297. [CrossRef] [PubMed]
7. Shi, G.; Wang, W.; Zhang, F. Precision improvement of frequency-modulated continuous-wave laser ranging system with two auxiliary interferometers. *Opt. Commun.* **2018**, *411*, 152–157. [CrossRef]
8. Fan, B.; Xu, L. Research on the frequency modulation continuous wave ranging method based on spectral feature matching. *Proc. SPIE* **2020**, *11427*, 1142719. [CrossRef]
9. Wu, J.; Zhao, S.; Zhang, J. A continuous phase-frequency modulation technique and adaptive pump noise cancellation method of continuous-wave signal over mud telemetry channel. *Digit. Signal Process* **2021**, *117*, 103147. [CrossRef]
10. Shi, G.; Zhang, F.; Qu, X. High-resolution frequency-modulated continuous-wave laser ranging for precision distance metrology applications. *Opt. Eng.* **2014**, *53*, 122402. [CrossRef]
11. Blanche, J.; Lewis, H.; Couples, G.D. Dynamic fluid ingress detection in geomaterials using K-Band frequency modulated continuous wave radar. *IEEE Access* **2020**, *8*, 111027–111041. [CrossRef]
12. Zheng, J. Reflectometric fiber optic frequency-modulated continuous-wave interferometric displacement sensor. *Opt. Eng.* **2005**, *44*, 124404. [CrossRef]
13. Zheng, J. Optical frequency-modulated continuous-wave interferometers. *Appl. Opt.* **2006**, *45*, 2723–2730. [CrossRef]
14. Zheng, J. Single-mode birefringent fiber frequency-modulated continuous- wave interferometric strain sensor. *IEEE Sens. J.* **2010**, *10*, 281–285. [CrossRef]
15. Roos, P.A.; Reibel, R.R.; Berg, T.; Kaylor, B.; Barber, Z.W.; Babbitt, W.R. Ultrabroadband optical chirp linearization for precision metrology applications. *Opt. Lett.* **2009**, *34*, 3692–3694. [CrossRef]
16. Barber, Z.W.; Giorgetta, F.R.; Roos, P.A.; Coddington, L.; Dahl, J.R.; Reibel, R.R.; Greenfield, N.; Newbury, N.R. Characterization of an actively linearized ultrabroadband chirped laser with a fiber-laser optical frequency comb. *Opt. Lett.* **2011**, *36*, 1152–1154. [CrossRef] [PubMed]
17. Lu, C.; Yu, Z.; Liu, G. A high-precision range extraction method using an FM nonlinear kernel function for DFB-array-based FMCW lidar. *Opt. Commun.* **2022**, *504*, 127469. [CrossRef]
18. Sheng, Q. Research on the Displacement Measurement System of Frequency-Modulated Continuous Wave Laser Interferometry. Master's Thesis, Xi'an Technological University, Xi'an, China, 2022. [CrossRef]
19. Sun, B.; Zheng, G.; Zhang, X. Application of contact laser interferometry in precise displacement measurement. *Measurement* **2021**, *174*, 108959. [CrossRef]
20. Zhang, D.; Yan, M.; Shi, A. Frequency-modulated continuous-wave laser interferometry displacement sensor based on centroid peak-finding. *Opto-Electron. Eng.* **2023**, *50*, 220315. [CrossRef]
21. Zheng, J. Analysis of optical frequency-modulated continuous-wave interference. *Appl. Opt.* **2004**, *43*, 4189–4198. [CrossRef]
22. Zhuang, J.; Li, X.; Li, S. Frequency-modulated microwave generation with feedback stabilization using an optically injected semiconductor laser. *Opt. Lett.* **2016**, *41*, 5764–5767. [CrossRef] [PubMed]
23. Meiners-Hagen, K.; Burgarth, V.; Abou-Zeid, A. Profilometry with a multi-wavelength diode laser interferometer. *Meas. Sci. Technol.* **2004**, *15*, 741. [CrossRef]
24. Heydemann, P.L.M. Determination and correction of quadrature fringe measurement errors in interferometers. *Appl. Opt.* **1981**, *20*, 3382–3384. [CrossRef] [PubMed]
25. Fu, H.; Wang, Y.; Hu, P.; Tan, J.; Fan, Z. Nonlinear Errors Resulting from Ghost Reflection and Its Coupling with Optical Mixing in Heterodyne Laser Interferometers. *Sensors* **2018**, *18*, 758. [CrossRef]
26. Ferrero, B.; Bellon, L. Harmonic Calibration of Quadrature Phase Interferometry. *EPL* **2022**, *139*, 55002. [CrossRef]

photonics

Article

Study of Point Scanning Detection Mechanisms for Vibration Signals with Wavefront Sensors

Quan Luo [1,†], Hongsheng Luo [1,†], Yiyou Fan [1], Guihan Wu [2], Haoming Chen [1], Yun Pan [1], Wei Jiang [2] and Jinshan Su [1,*]

1 Key Laboratory of Vibration Signal Capture and Intelligent Processing, School of Electronic Engineering, Yili Normal University, 448 Jiefang Road, Yining 835000, China
2 College of Engineering & Applied Sciences, Key Laboratory of Intelligent Optical Sensing and Manipulation (Nanjing University), Ministry of Education, Nanjing University, 163 Xianlin Ave, Nanjing 210023, China
* Correspondence: 14222431@ylnu.edu.cn
† These authors contributed equally to this work.

Abstract: Seismic wave laser remote sensing is extensively employed in seismic monitoring and resource exploitation. This work establishes a vibration signal point scanning detection system utilizing wavefront sensors, leveraging their high resolution, array detection capabilities, and the independent detection of each microlens based on research into seismic wave laser remote sensing detection. The experiments validate that each microlens of the wavefront sensor possesses autonomous detecting capabilities, enabling the sensor to scan and identify points of vibrational signals. This work also significantly improved the scanning efficiency by increasing the diameter of the scanning spot.

Keywords: vibration signal; wavefront sensor; point scan

1. Introduction

Seismic wave exploration is the fundamental method of geophysical investigation. Analyzing seismic wave propagation properties in subterranean media elucidates details of the geological structure [1,2]. In recent years, seismic wave laser remote sensing detection technology has advanced swiftly, employing laser-based non-contact monitoring of ground vibration waveforms characterized by high resolution and sensitivity [3]. Due to variations in wave velocity when seismic waves traverse various geological layers, optical remote sensing technology may more precisely analyze seismic waveforms and reveal subsurface structures [4,5]. Analyzing the seismic wave signals in a region can identify its geological features, including rock types, faults, and folds, and assess the distribution and reserves of minerals, oil, and natural gas, providing a foundation for energy development [6,7]. Advancements in high-resolution optical remote sensing technologies have enabled scholars to detect seismic waves and capture minor vibration signals remotely [8,9]. This technology surpasses the conventional seismic research techniques, which is particularly advantageous for oil and gas development. The traditional exploration methods exhibit difficulties in deep and diverse terrain due to the rising requirements for high resolutions, rapid sampling, and weak signal identification. Optical remote sensing, in conjunction with remote sensing [10], GIS [11], and satellite data [12], has markedly enhanced the efficiency and precision of resource exploitation, emerging as a novel option for geophysical research. Compared to the conventional approaches, laser remote sensing technology offers reduced costs and enhanced efficiency in detecting areas with mineral potential, facilitating advancements in deep geological structure study and resource development.

Received: 28 November 2024
Revised: 10 January 2025
Accepted: 15 January 2025
Published: 16 January 2025

Citation: Luo, Q.; Luo, H.; Fan, Y.; Wu, G.; Chen, H.; Pan, Y.; Jiang, W.; Su, J. Study of Point Scanning Detection Mechanisms for Vibration Signals with Wavefront Sensors. *Photonics* **2025**, *12*, 78. https://doi.org/10.3390/photonics12010078

Laser remote sensing detection technology offers advantages such as non-contact measurement, minimal constraints, enhanced efficiency, and long-range detection, making it extensively utilized to detect object vibrations. To address the issues of remote laser echo signal attenuation and vibration waveform analysis, researchers have investigated the application of He–Ne laser interferometers [13] and Michelson interferometers [14] for measuring vibration signals. Nonetheless, these technologies depend on intricate optical systems and the configuration of reference and measuring arms, which possess certain limits. Silvio Bianchi [15] suggested a novel approach for sensing surface vibration signals without employing an interferometer, utilizing speckle patterns generated by laser irradiation on rough surfaces. This approach enables the efficient detection of vibration signals across several hundred meters. Numerous techniques exist for detecting ground vibration signals based on phase, including wavefront sensors [16–18] and shear interferometry [19,20]. Jorge Ares [21] et al. employed a Shack–Hartmann wavefront sensor as a positional sensing apparatus to precisely ascertain the location of point or substantially extended planar objects.

Compared to conventional seismic wave laser remote sensing technology, employing wavefront sensors for detecting ground vibration signals offers substantial advantages. The wavefront sensor eliminates the need for an additional reference and measuring arm, resulting in a more compact and easily deployable device. Secondly, mechanical vibration information can be precisely acquired by detecting the phase shift of the wavefront. Moreover, the wavefront sensor possesses array detection capabilities, augmenting its detection precision and scope of application. In 2023, several researchers employed wavefront sensors to identify ground vibration signals, developed a seismic wave laser remote sensing detection system utilizing wavefront sensors [22], and successfully acquired ground vibration information [23]. The existing technology primarily gathers vibration information from a single site; it cannot analyze regional vibration signals and suffers from low detection efficiency. This work proposes a method for laser remote sensing detection of regional vibration signals, utilizing a microlens array's non-interference properties and array detection capabilities in a wavefront sensor, as indicated by the research in Ref. [22]. Laser point scanning detection is employed to achieve a measurement method characterized by high precision and high resolution for regional vibration signals, thereby improving detection efficiency. The first section introduces the working principle of the point scanning detection system and the changes in wavefront during vibration. Section 2 introduces wavefront reconstruction using the zonal method, the calculation of wavefront phase, and wavefront crosstalk. Section 3 introduces single-point vibration detection, which determines the relationship between amplitude and phase and then performs point scanning on an area to obtain the vibration information of that area. In order to quickly detect the vibration information in an area, the area of the detection spot is expanded for scanning.

2. Vibration Signal Laser Point Scanning Surface Detecting System and Operational Mechanism

During an earthquake, a longitudinal wave is produced, causing the wavefront of the incident light to be altered by the vertical ground vibrations, thereby transmitting pertinent characteristic information of the seismic wave through the backscattered light of the laser. A laser beam irradiates vertically or obliquely in the target area of the item being measured within a wavefront measurement system. The object's surface irregularities will cause light to scatter and reflect. The laser's reflected or scattered light is concentrated and imaged by a microlens from a different angle. The vibration-related information is acquired by reconstructing the wavefront via imaging the microlens.

The work presents the design of a remote sensing detecting system for scanning vibration signal points. The system primarily consists of a transmitter and a receiver, as demonstrated in

Figure 1. The transmitter comprises a laser, a fiber optic collimator, and a telescope. The laser produces a continuous wave at a wavelength of 635 nm with constant power. The laser output from a fiber passes through fiber-optic collimator 1, then telescope 2 to ensure consistent angle and power irradiation to the target region 3. The receiver consists of telescope 5, filter 4, and wavefront sensor 6. The telescope captures the reflected and dispersed light from target area 3 and eliminates extraneous environmental light using filter 4 to enhance the contrast of the reflected light. This is a bandpass filter that enables light with a wavelength range from 600 nm to 700 nm to pass through, with a center wavelength of 650 nm. The light subsequently strikes the wavefront sensor 6. The wavefront sensor is an instrument capable of accurately detecting wavefront distortion in optical systems. The assembly consists of a microlens array, with each microlens possessing independent detection capabilities. The lens size of the wavefront sensor is 146 microns, the pitch is 150 microns, the focal length is 4 millimeters, the camera is CMOS type, and the software is wavefront sensor software (version number:18183-D03). The microlens array segments the sub-aperture of the incident light spot and concentrates the image in each detection window 8 within the array.

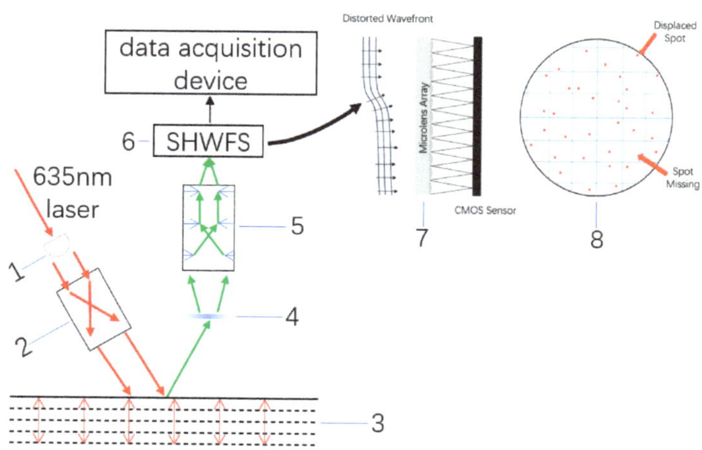

Figure 1. Vibration signal point scanning detection system. 1 Fiber optic collimator; 2 telescope; 3 ground target; 4 filter; 5 telescope system; 6 Shack–Hartmann wavefront sensor; 7 microlens array; 8 detection window.

This system detects wavefront distortion caused by vibration, which causes the centroid of the light spot to deviate from the reference centroid coordinates and cause phase changes. By analyzing the phase changes, we obtain vibration information. Based on this principle, we use this system to detect micro-vibrations in the target area. Understanding the correlation between the phase information of the target vibration and the displacement of the receiving spot in the CMOS sensor pixel is essential. When object 3 vibrates, the centroid offset ΔS of the wavefront sensor transmits the vibration amplitude ΔZ. The literature [22] states that $\Delta Z = \mu \Delta S$, where μ is a proportional coefficient. Light propagation in the medium between the transmitting and receiving optics is described by the well-known Fourier optics principles [24]. Note that, for this initial study with a short distance (about 10 m) to the target, the non-ideal effect due to atmosphere is neglected. When not vibrating, the wavefront is a plane; when vibrating, the wavefront changes from a plane to a concave surface, as shown in Figure 2. In Figure 2, we simplify the analysis by assuming the ground is flat when not vibrating. This assumption is made to illustrate the basic relationship between the transmitted and reflected wavefronts. In reality, uneven ground may combine with other non-ideal factors to influence the wavefront's shape.

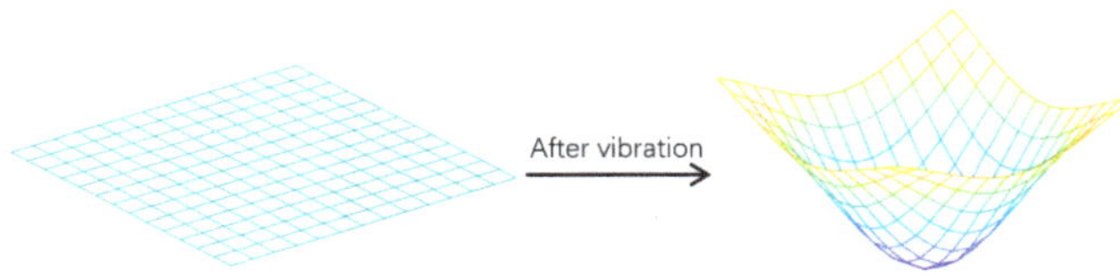

Figure 2. Vibration wavefront.

Shack–Hartmann wavefront sensor (SHWFS) disaggregates the incident wavefront into numerous minuscule spots via a microlens array, with each spot's centroid position indicating the local wavefront's tilt angle. Ideally, the microlens produces a perfect spot on the focal plane. However, due to wavefront distortion, the centroid of the spot will wander from its optimal position, as illustrated in Figure 3. The wavefront distortion can be deduced by examining the sub-aperture spot imaging of the microlens array. When an ideal plane wave is incident, the spot distribution remains uniform; nevertheless, wavefront distortion will result in a displacement of the spot. The wavefront slopes S_x and S_y can be determined by the centroid offsets Δx and Δy ratio to the focal length f of the microlens, facilitating further analysis of the wavefront's shape and distortion. Figure 4 shows the wavefront sensor used for research.

The amplitude of wavefront distortion in the z direction can be acquired using the wavefront sensor software, as shown in Figure 5a. Simultaneously, the wavefront sensor software may visually monitor the wavefront slope of the complete microlens array in high-speed sampling mode, as illustrated in Figure 5b. When the object under measurement vibrates, the position of the spot and the angle of reflection produced by laser irradiation on the object will vary, resulting in wavefront distortion. The variations in wavefront distortion amplitude and slope precisely correlate with the phase information of object vibrations, hence facilitating precise detection and assessment of vibrations.

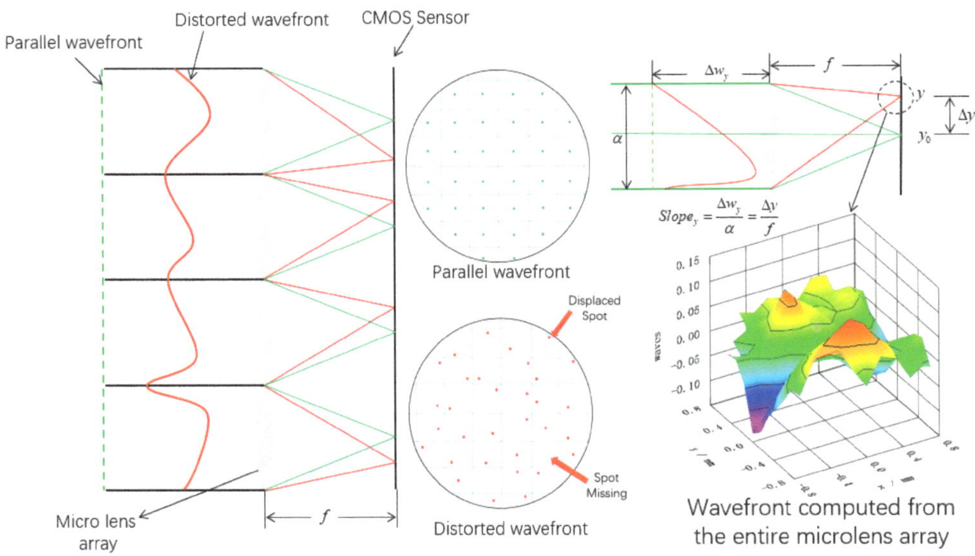

Figure 3. Distortion of the wavefront in the microlens array of the wavefront sensor. Upon entry into the parallel wavefront, all centroid points of the spots are located at the center of the microlens array. Upon entering the deformed wavefront, the spot centroid may shift from its center or vanish.

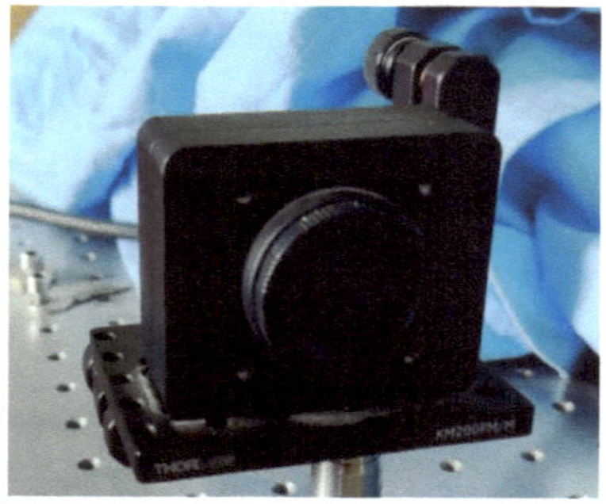

Figure 4. Shack–Hartmann wavefront sensor.

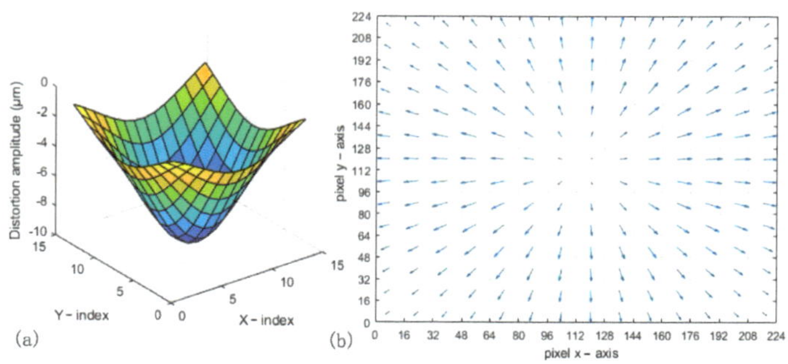

Figure 5. Example wavefront distortion. (**a**) Amplitude of wavefront distortion in the z direction; (**b**) slope of the wavefront of the microlens array.

3. Theoretical Analysis of Vibration Signal Laser Array Remote Sensing Detection System

3.1. Zonal Wavefront Reconstruction

The zonal method is a frequently employed waveform reconstruction algorithm in SHWFS [25–27]. The approach partitions the entire wavefront into multiple small sections, with the wavefront information of each region being measured by individual microlens from the microlens array. Initially, the wavefront slopes of each minor region are computed, followed by the execution of a double integral of these slopes across two spatial dimensions, subsequently reconstructing the wavefront height of each minor region. The reconstructed wavefronts of each region are combined to derive the three-dimensional shape of the complete wavefront. The zonal method offers considerable benefits. This method successfully decreases computational complexity and markedly enhances reconstruction speed by individually reconstructing the wavefront of each microlens region. Secondly, the area approach may effectively address wavefronts with significant local distortion, particularly in uneven wavefront distortion. This approach is extensively employed in practical applications for rapid and efficient wavefront reconstruction [28].

In the vibration signal point scanning detection system, Southwell's reconstruction method is employed to reconstruct the wavefront using the zonal method, as illustrated in Figure 6. The wavefront reconstruction result W' is derived from the findings in the literature [29]:

$$W' = VD^{+}U^{T}CS \tag{1}$$

Here, W' denotes the wavefront data after reconstruction; S signifies the wavefront slope data acquired from the wavefront sensor; $VD^{+}U^{T}$ indicates the inversion of the echo derived from the slope data using the generalized inverse; C serves to transform the slope data into wavefront height. Normalization is employed to adjust to the actual grid distance ds:

$$W = reshape\left(W', n, n\right)/ds \tag{2}$$

In the above formula, W' is the result of wavefront reconstruction, which is stored in the form of a one-dimensional vector. Reshape (W',n,n) rearranges this one-dimensional vector into an $n \times n$ two-dimensional matrix, representing the height data of the wavefront. $/ds$ normalizes each element of the wavefront to adapt to the actual grid distance. The formula above enables the derivation of the wavefront's three-dimensional shape.

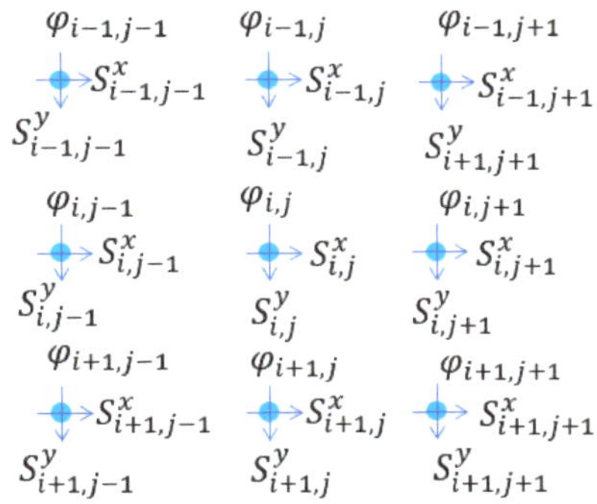

Figure 6. Southwell model (circle represents reconstruction point; arrow represents measurement point).

3.2. Wavefront Phase

The slope data from SHWFS can be utilized to rebuild the wavefront phase [30–41]. Assuming that the wavefront slope matrices in the x and y directions have been obtained from the Shack–Hartmann sensor, the wavefront slopes in the x and y directions are defined as

$$S_x(x,y) = \frac{\partial W(x,y)}{\partial x}, S_y(x,y) = \frac{\partial W(x,y)}{\partial y} \tag{3}$$

Among them, the wavefront slope is S_x (x,y) in the x-direction, and $S_y(x,y)$ is the wavefront slope in the y-direction; W(x,y) is a two-dimensional function representing the wavefront phase, which is inversely obtained from the known slopes S_x and S_y. The wavefront phase $W(x,y)$ can be determined by resolving these equations. From these two partial derivative equations, we can obtain the Poisson equation that the wavefront phase $W(x,y)$ must satisfy:

$$\nabla^2 W(x,y) = \frac{\partial S_x(x,y)}{\partial x} + \frac{\partial S_y(x,y)}{\partial y} \tag{4}$$

The issue of resolving the phase is converted into the Poisson equation. The first derivatives ($\frac{\partial S_x}{\partial x}$ and $\frac{\partial S_y}{\partial y}$) are second-order derivatives used to convert slope data into phase. $\frac{\partial S_x}{\partial x}$ and $\frac{\partial S_y}{\partial y}$ reflect the change in wavefront curvature by differentiating the slope data, thus being associated with the Laplacian operator of wavefront phase. In the Poisson equation for wavefront phase reconstruction, the right-hand side of the equation is not equal to zero. In the general Poisson equation, there is a source term $f(x,y)$ on the right, which can be expressed as the derivative of the wavefront slope. In the wavefront reconstruction problem, this source term is $\frac{\partial S_x}{\partial x} + \frac{\partial S_y}{\partial y}$. If the right side is equal to zero, it means that the wavefront is a smooth and undisturbed wavefront, without any phase change. This obviously does not apply to reconstructing the actual wavefront phase with disturbances, so the right side is not equal to zero.

Fourier transform in two-dimensional space can quickly solve Laplace equation and efficiently and accurately reconstruct wavefront phase. To resolve this Poisson equation, we can employ Fourier transform, which converts the differential operation into an algebraic one. In the frequency domain, the Poisson equation can be expressed as

$$F\left(\nabla^2 W(x,y)\right) = k_x^2 \cdot F(W) + k_y^2 \cdot F(W) \tag{5}$$

Here, F represents the two-dimensional Fourier transform, while k_x and k_y denote the spatial frequencies in the frequency domain, respectively. Consequently, the Fourier transform $W_{ft}(k_x, k_y)$ of the wavefront phase in the frequency domain can be articulated as

$$W_{ft}(k_x, k_y) = \frac{-i\left[k_x \cdot F(S_x) + k_y \cdot F(S_y)\right]}{k_x^2 + k_y^2 + \varepsilon} \tag{6}$$

In the frequency domain, the value at zero frequency will render the denominator zero. To prevent the inaccuracy of zero removal, a minimum value ε is incorporated into the denominator to enhance calculation stability. Upon acquiring $W_{ft}(k_x, k_y)$, the wavefront phase expression $W(x,y)$ in the spatial domain is derived through the inverse Fourier transform as follows:

$$W(x,y) = F^{-1}\left[W_{ft}(k_x, k_y)\right] \tag{7}$$

The slope of the vibration wavefront in Figure 2 is shown in Figure 7. Consequently, the phase alteration of the wavefront is derived utilizing the vibration wavefront slope S depicted in Figure 2, as illustrated in Figure 8.

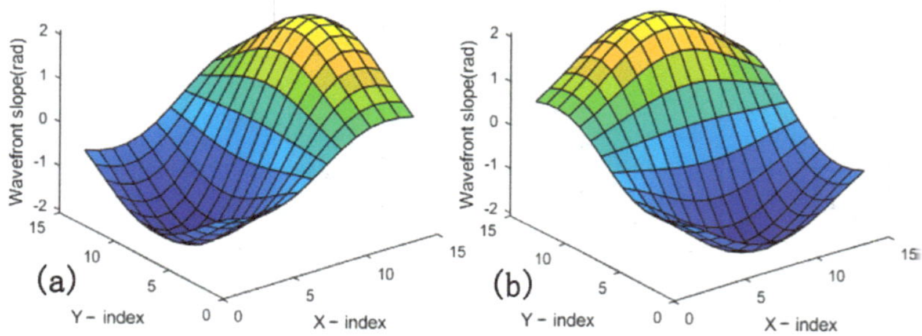

Figure 7. The wavefront slope in Figure 2. (**a**) The wavefront slope in the x-direction; (**b**) the wavefront slope in the y-direction.

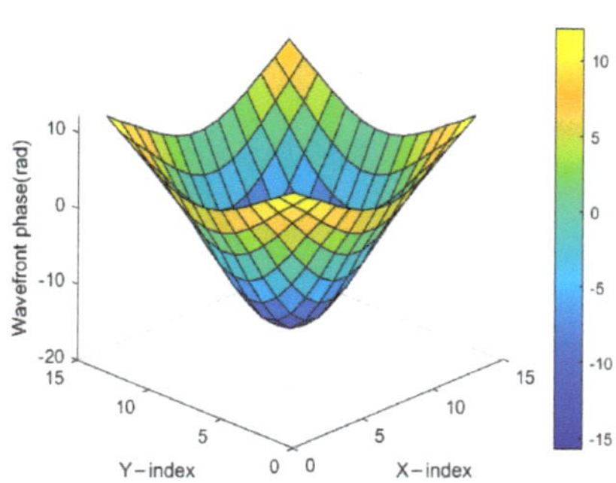

Figure 8. Wavefront phase for an illustrative example.

The phase transition of SHWFS during vibration enables the establishment of the correlation between amplitude and phase, as demonstrated in the subsequent formula:

$$\Delta\varphi = k\Delta d + c \tag{8}$$

where $\Delta\varphi$ denotes the phase alteration and Δd signifies the amplitude variation. In the absence of vibration, the laser echo detected by the wavefront sensor would induce a minor displacement of the spot centroid on the microlens array, resulting in a slight alteration of the wavefront slope and the emergence of a constant term in the equation.

3.3. *Wavefront Crosstalk*

In the SHWFS employing the zonal method for wavefront reconstruction, the problem of wavefront crosstalk frequently results in measurement inaccuracies. The zonal method segments the detection area into multiple small pieces, each concentrated by an individual microlens, and enables the measurement of the local wavefront slope. When the spot's area is beyond the optimal range, it encroaches upon the neighboring microlens region, resulting in the erroneous reception of target area information. This phenomenon causes crosstalk in the wavefront information, hence compromising measurement precision. This issue typically arises from high irradiation power, resulting in an overload of the spot power received by the sensor, hence compromising the accuracy of wavefront reconstruction. The spot's dimensions, the microlens spacing, and the distribution of spot energy are critical elements influencing the precision of wavefront reconstruction. A considerable spot or high energy concentration will amplify measurement error and induce crosstalk between neighboring regions when employing the zonal method, thereby diminishing reconstruction accuracy.

3.3.1. Spot Size

The focal length of each microlens is denoted as f, the diameter as d, and the detector is positioned on the focal plane of the microlens, meaning that the distance from the detector to the microlens array is f. The local tilt angles of the wavefront on the microlens plane are denoted as θ_x and θ_y, and the spot offsets can be articulated as

$$\Delta x = f \cdot \theta_x, \Delta y = f \cdot \theta_y \tag{9}$$

It deviates from its center when the spot is sufficiently small and concentrated within a single microlens. Ideally, this offset can be accurately measured. Nonetheless, the detector's restricted resolution and the spot's defocusing effect may contribute to actual measurement inaccuracies, impacting wavefront reconstruction accuracy.

Consider a scenario where a laser-emitted spot, with a radius r_1, is formed at a certain distance d from the telescope. As the light passes through the telescope, which acts as a converging lens, the spot is focused to form a smaller spot with a radius r_2, which can be changed to a different value as needed, as shown in Figure 9.

$$\frac{1}{F'} = \frac{1}{d} + \frac{1}{d'} \tag{10}$$

where F' is the focal length of the telescope, d is the object distance, and d' is the image distance. As the spot image moves through the focal plane, its size varies depending on whether it is positioned inside or outside the focal length, thus leading to a different spot size at the focal distance. The magnification M is the ratio of the image size to the object size:

$$M = \frac{r_2}{r_1} = \frac{d'}{d} \tag{11}$$

By integrating Formulas (10) and (11) above, one can derive the correlation between the spot radius and the receiving distance as follows:

$$r_2 = r_1 \cdot \frac{F'}{d} \tag{12}$$

Among them, r_1 denotes the radius of the circular spot captured by the telescope objective, r_2 represents the radius of the spot once the telescope is focused, d indicates the receiving distance of the telescope (the distance from the spot to the objective), and F' signifies the focal length of the telescope objective.

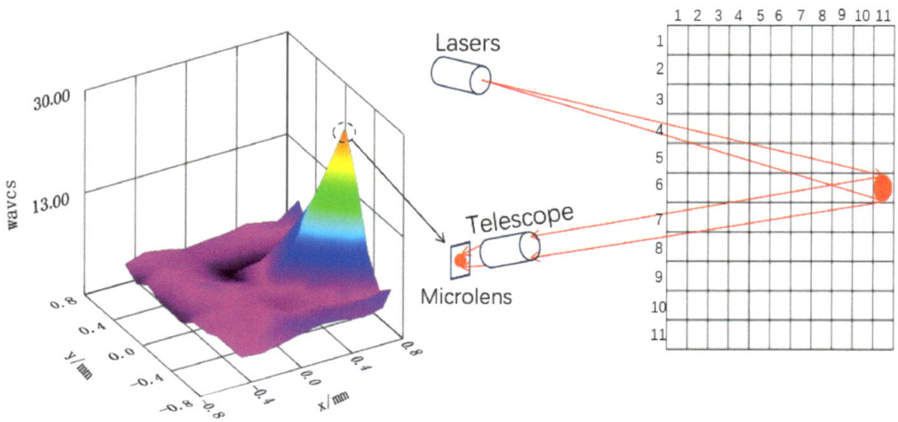

Figure 9. The target area is divided into 11 × 11 sub-regions, and the laser is focused on one of the sub-regions. The spot signal of the sub-region is received by the telescope and focused on the corresponding microlens in the microlens array.

3.3.2. Microlens Spacing

Assuming that the spacing of the microlens array is p when the diameter of the spot exceeds the spacing p, the spot may be simultaneously captured by many nearby microlenses. Designate the spot radius as r and the spacing as p. When the spot obscures the neighboring microlens, the position of the spot recorded by the adjacent microlens will coincide, leading to an inaccuracy in the computation of the spot centroid, which subsequently induces crosstalk. The local tilt angle θ of the wavefront will influence the measurement accuracy as the spot traverses many microlenses. Crosstalk shift can be represented by the following formula:

$$D_{cs} = r - p \cdot \sin\theta \tag{13}$$

When the spot is huge and encompasses several microlenses, this phenomenon will considerably impact the precision of wavefront reconstruction.

3.3.3. Spot Energy Distribution

A Gaussian function can define the energy distribution of the spot. The intensity distribution of the spot on the microlens plane is presumed to be

$$I(x,y) = I_0 \left[-\frac{(x-x_0)^2 + (y-y_0)^2}{2\sigma^2} \right] \tag{14}$$

Among these parameters, I_0 denotes the most incredible intensity at the spot center, (x_0, y_0) indicates the coordinates of the spot center, and σ signifies the standard deviation of the spot, representing its size. A significant local alteration in the wavefront may cause the intensity of the spot on certain microlenses to be insufficient or undetected, leading to heightened local reconstruction error and crosstalk.

4. The Experiment and Results of Analysis of Vibration Signal Laser Point Scanning Detection System

We experimented with a 70 mW laser at a distance of 10 m. The automatic shutter management of the wavefront sensor camera enables it to process optical input power across a broad dynamic range, with sensitivity significantly influenced by wavelength. The Thorlabs WFS-20-5C Shack–Hartmann wavefront sensor features a high sampling rate of 800 frames per second and a wavefront accuracy of $\lambda/30$ rms at 633 nm. The divergence angle of the laser is 0.25 degrees. This experiment employs a Shack–Hartmann wavefront sensor to detect ground vibration signals. In the single-point detection experiment of vibration signals, a controlled shaking table is employed as a vibration source to replicate the features of longitudinal waves in nature by varying the frequencies and amplitudes. Simultaneously, in the vibration signal point scanning experiment, we employ the vibration motor as the vibration source for testing. The vibration of the tent can usually be described by the displacement function $\Delta z(x, y; t)$. To isolate the reciprocal effect of the vibration source, we affixed the random amplitude vibration motor to the soft white fabric, as indicated in Figure 10. In the experiment, we employ a laser to illuminate the vibration source and capture the laser echo signal reflected from it using a wavefront sensor.

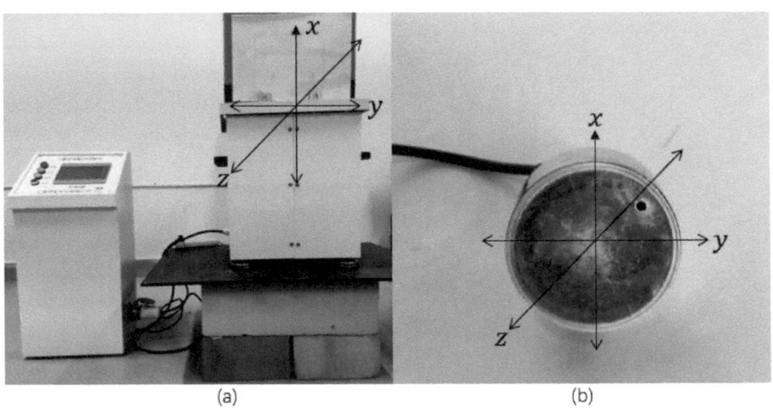

Figure 10. (**a**) Controllable vibration table; (**b**) vibration motor. The black arrow indicates the direction of vibration.

We employ the wavefront sensor software to document the wavefront variations and monitor the real-time alterations in the incident wavefront. Simultaneously, we compute and exhibit the microlens alterations via the LabView software platform, facilitating additional analysis of the wavefront sensor's variation upon excitation of the vibration source. The program's flow is illustrated in Appendix A. Utilizing the architecture and recording approaches of the system above, we can precisely monitor and analyze the small alterations in the wavefront sensor induced by the vibration source's oscillations. The wavefront slope of the wavefront sensor deviates during the activation of the vibration source, with the deviation occurring in both horizontal and vertical orientations. Detecting longitudinal seismic waves results in a very significant offset in the y-direction and a comparatively minor offset in the x-direction [23]. The vibrometer concurrently detects the vibration signal.

During the test of the reception point of an individual microlens, the laser echo signal is concentrated on a microlens inside the microlens array via a telescope. The high-speed sampling mode (800 frames per second) in wavefront sensor software is a feature that enables real-time acquisition and analysis of wavefront data. This mode captures wavefront data at a high sampling rate, enabling the system to monitor rapidly changing wavefronts in real time, especially suitable for dynamic scenes or fast-moving objects. The Beam View mode in wavefront sensors is a display mode used to observe and analyze the distribution of light beams on the sensor. In this mode, the output signal of the wavefront sensor will be displayed in the form of a beam, usually as an image or a visualized light spot. This mode enables intuitive observation of the characteristics of the beam, such as the shape, size, and position of the light spot. In Beam View mode, the various features of the beam can determine the alignment of the optical system, especially whether there are issues such as focal shift, beam expansion, and skewness. We focused the laser echo signal onto the 6 × 6 microlens in the wavefront sensor microlens array using a telescope and enabled the high-speed sampling mode of the wavefront sensor. In the 6 × 6 green square (corresponding to a 6 × 6 microlens), a spot centroid and its location will be displayed, as shown in Figure 11a. Due to the fact that the laser echo signal is only focused on the 6 × 6 microlens, although there are weak signals around the microlenses, they are not displayed. In the Beam View mode of the wavefront sensor, it is observed that the target microlens exhibits a spot signal. At the same time, numerous neighboring microlenses also display comparatively weaker spot signals, as illustrated in Figure 11b. This indicates that the relationship between the laser echo signal and the microlens array is not a perfect 'one-to-one' correspondence but rather a 'one-to-many' complex relationship.

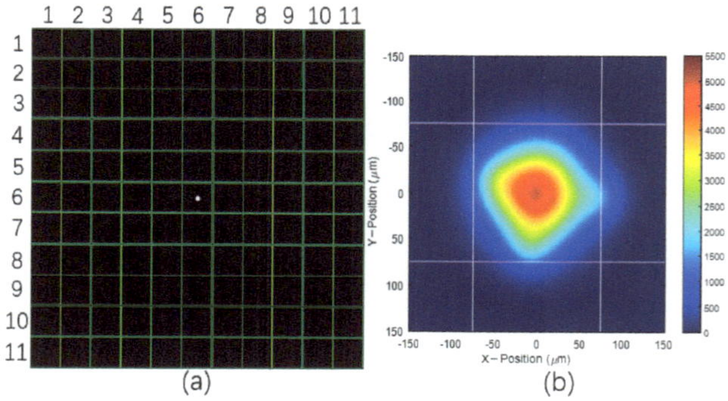

Figure 11. Single microlens receiving spot in microlens array. (**a**) The spot signal of 6 × 6 microlens in high-speed sampling mode; (**b**) the spot signal of 6 × 6 microlens in Beam View mode.

The laser creates a spot on the vibration table, and the echo signal is directed to the 6 × 6 microlens of the wavefront sensor via the telescope. The wavefront slopes (S_x and S_y) of the centroid of the 6 × 6 microlens array are acquired using LabView software, and the related wavefront picture is rebuilt utilizing the zonal method. Figure 12a–d illustrate that, as the vibration amplitude escalates, the extent of wavefront distortion correspondingly intensifies, resulting in mild disturbances to the microlenses around the 6 × 6 microlens array. In the presence of vibration signals of varying amplitudes inside a particular area, the laser echo signal will transmit the vibrational information of that area. The wavefront sensor acquires and evaluates the associated vibration information by capturing these laser echo signals. Analyzing the wavefront phase alterations at varying amplitudes elucidates the correlation between amplitude and phase, as illustrated in Figure 12e. The wavefront's phase is altered with changes in amplitude, enabling the derivation of the amplitude variation law using phase change measurement, as indicated in the subsequent formula:

$$\Delta\varphi = k\Delta d + c, k = 0.0143(rad/mm), c = 0.0354 \quad (15)$$

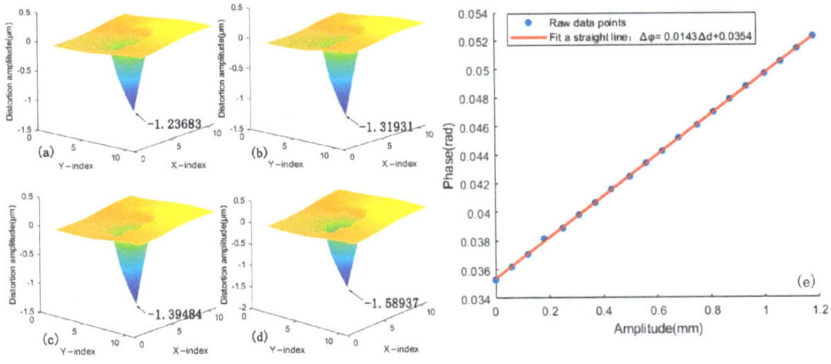

Figure 12. The distortion amplitude is indicated for various vibration amplitudes of the 6 × 6 microlens wavefront. (**a**) Amplitude: 0.06 mm, phase: 0.036195 rad; (**b**) amplitude: 0.31 mm, phase: 0.039825 rad; (**c**) amplitude: 0.75 mm, phase: 0.04607 rad; (**d**) amplitude: 1.18 mm, phase: 0.052289 rad; (**e**) vibration amplitude and phase change.

According to the tests above, we partition a 1.2 m × 1.2 m target into 11 × 11 sub-regions, commencing from the upper left corner, designating the initial area as 1 × 1 coordinates, and sequentially constructing the array. A vibration source is randomly positioned within 11 × 11 sub-regions behind the target. The laser sequentially irradiates each sub-region, employing two pan-tilt mechanisms to regulate the laser head and telescope for point scanning throughout each sub-region, as illustrated in Figure 13. Point scanning is conducted for each sub-region to acquire the wavefront slope variations regarding the respective microlenses. During the detection process, we first use pan-tilt 1 to project the collimated laser onto the target area. Then, we use a clamp to fix telescope 1 and manually adjust its objective lens and eyepiece to adjust the spot area on the target area. Then, we use pan-tilt 2 to adjust the receiving angle of telescope 2 and further adjust the objective lens and eyepiece of telescope 2 to ensure the best receiving effect. Subsequently, through the high-speed sampling interface of the wavefront sensor, we manually adjust the position of the wavefront sensor so that the laser echo signal can be accurately focused on the corresponding lens in the wavefront sensor microlens array, thereby achieving vibration signal point scanning detection. Figure 14 shows the physical system. The field of view of the telescope is $16.4° - 4.1°$, and the distance between the telescope and the spot is 10 m. Therefore, the 10 cm spot size is fully within the field of view of the telescope.

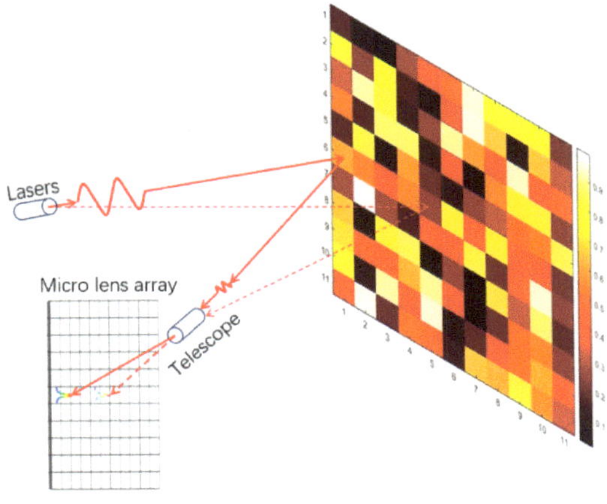

Figure 13. Diagram illustrating the scanning of regional vibration signal sites.

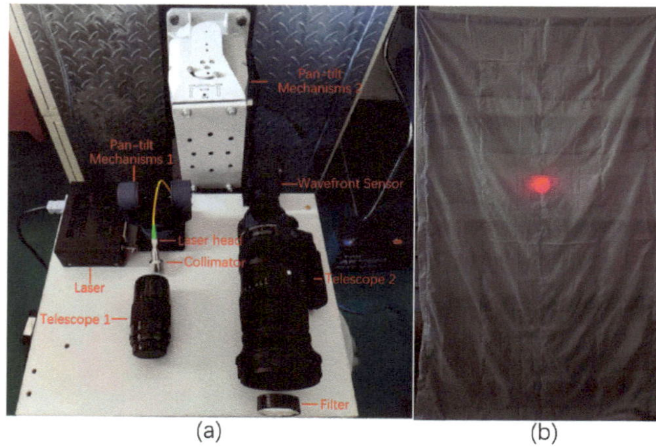

Figure 14. (**a**) Physical image of point scanning detection system; (**b**) a light spot with a diameter of 10 cm formed in the target area.

The generated spot is directed onto the wavefront sensor using a telescope, with each spot corresponding individually to the microlens array in the sensor. Figure 15a–h illustrate the outcomes of point scanning over all sub-locations, demonstrating that the microlens array in various regions captures the spot, further substantiating the one-to-one correlation between the microlens array and the target area. During single-point detection across several regions, the application of excitation vibrations by the vibration source to the sub-regions of the target object results in synchronous distortion of the laser echo wavefront. The variation in the wavefront gradient calculates the phase alteration during vibration, and the associated vibration amplitude is computed. The anticipated amplitude of Figure 15b is 1.08 mm, that of Figure 15d is 0.86 mm, the amplitude of Figure 15f is 0.88 mm, and the amplitude of Figure 15h is 1.3 mm. The minor root mean square error is 0.00045 for the amplitude measured by the vibrometer.

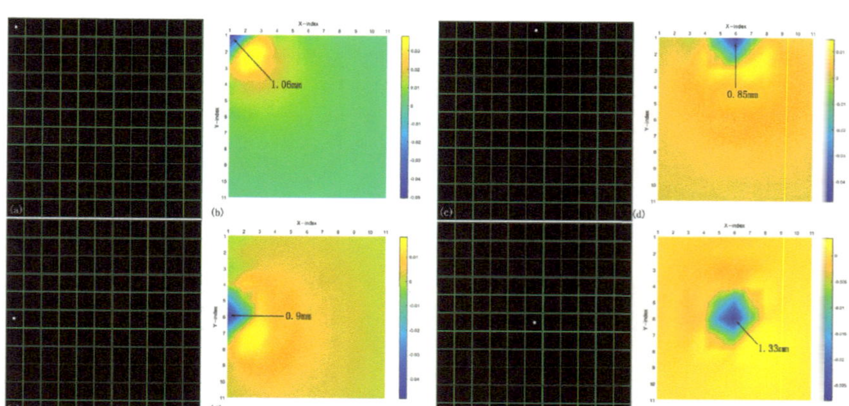

Figure 15. Relationship between microlens and wavefront when laser irradiates various locations. The measurement values of the vibrometer are marked in millimeters in the figure. (**a**) Focus the laser echo signal onto the 1th × 1th microlens; (**b**) The wavefront phase change of the 1th × 1th microlens; (**c**) Focus the laser echo signal onto the 6th × 1th microlens; (**d**) The wavefront phase change of the 6th × 1th microlens; (**e**) Focus the laser echo signal onto the 1th × 6th microlens; (**f**) The wavefront phase change of the 1th × 6th microlens; (**g**) Focus the laser echo signal onto the 6th × 6th microlens; (**h**) The wavefront phase change of the 6th × 6th microlenses.

In order to achieve faster detection of vibration information in an area, the designated area of 1.2 m × 1.2 m is segmented into a 2 × 2 grid, and a laser beam is directed to four sections to create a spot with a diameter of 60 cm. The telescope is utilized to position the laser echo signal onto the microlens array; in the high-speed sampling mode of the wavefront sensor, the microlens array is viewed to capture the spot centroid, as illustrated in Figure 16.

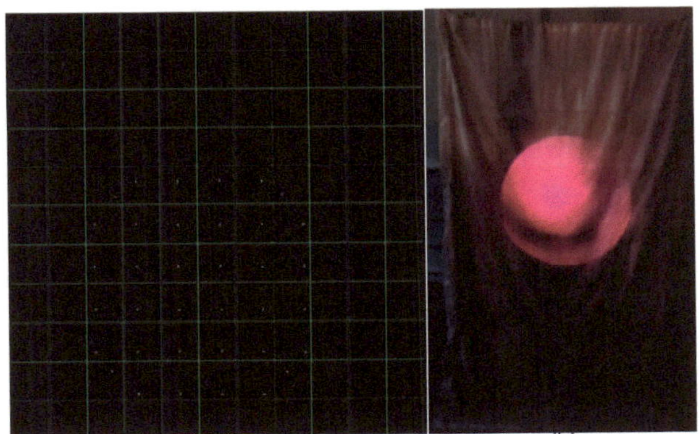

Figure 16. (**a**) Multiple microlenses in high-speed sampling mode for wavefront sensors to receive spot signals; (**b**) a light spot with a diameter of 60 cm formed in the target area.

Based on the detection in the area of Figure 16, we randomly arranged multiple vibration sources in the target area of 1.2 m × 1.2 m. The coordinates of the vibration source are marked in Figure 17a,c,e,g. The laser beam shines a 60 cm diameter spot through a telescope onto multiple sub-regions of the target area. In the high-speed sampling mode of the wavefront sensor software, we observed that 36 microlenses received spot signals, with two detection windows showing significant wavefront distortion, as shown in Figure 17. Through LabView software (version number: LabView2020.0 32-bit), we can accurately locate the microlens detection window with significant wavefront distortion, thereby obtaining the corresponding changes in the wavefront slope of the microlens and calculating the wavefront phase. Figure 17b shows the information of the two vibrations detected in the first sub-region. Subsequently, we scanned the next sub-region and successfully obtained the vibration information of that region, as shown in Figure 17d,f,h. For completeness, we include a case where the two sources are not parallel to the horizontal or vertical axis. The wavefront slope of Figure 17 is shown in Appendix B. This indicates that increasing the spot area of the target region can effectively detect the vibration information of the sub-region, thereby improving the detection efficiency. The minimum root mean square error between the vibration amplitude calculated by phase change and the amplitude recorded by the vibrometer is 0.00049112. Figure 18 shows the amplitude and phase changes in surface scanning. The amplitude generated by the vibration sources in Figure 18 is random, so the abscissa of the eight vibration sources does not increase proportionally. The insufficient accuracy and sensitivity of the vibration meter, as well as the failure to firmly adhere to the vibration source, may lead to inaccurate measurements and cause significant errors.

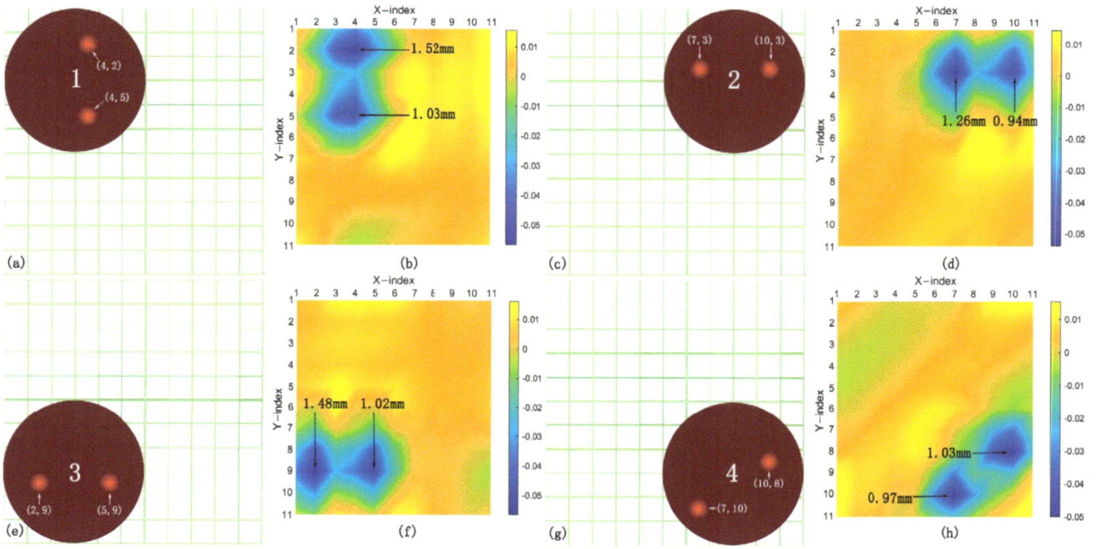

Figure 17. Detection outcomes of extensive spots in the target area. The amplitude value, measured in millimeters, is illustrated in the image. (**a**) Project laser to the region 1; (**b**) The wavefront phase change corresponding to the region 1; (**c**) Project the laser at the region 2; (**d**) The wavefront phase change corresponding to the region 2; (**e**) Project the laser at the region 3; (**f**) The wavefront phase change corresponding to the region 3; (**g**) Project the laser at the region 4; (**h**) The wavefront phase change corresponding to the region 4.

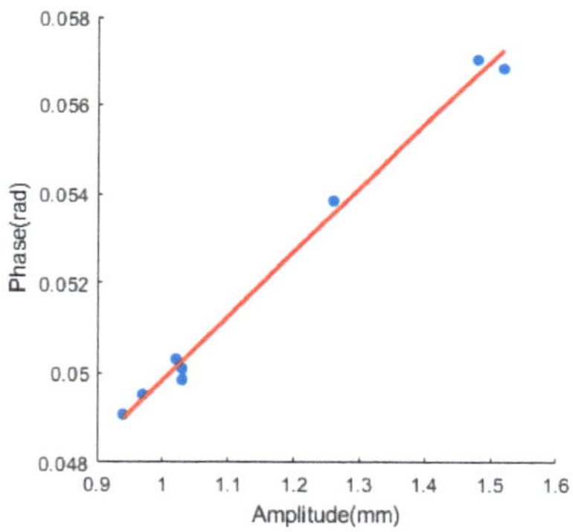

Figure 18. The amplitude and phase changes of surface scanning.

Figure 17 illustrates the variations in the vibration amplitude of the source correspondingly alter the wavefront phase. This indicates that the vibration signal laser point scanning detection system can efficiently identify vibrations across many target regions and can simultaneously detect the vibration information of multiple sub-areas. By examining the wavefront distortion across varying vibration amplitudes, one may deduce the vibration properties of each sub-region and their impact on the overall wavefront. This method confirms that the laser point scanning detection system possesses excellent sensitivity and resolution in intricate situations, enabling precise capture of local vibration information, hence offering robust support for vibration signal identification. Note that focal plane wavefront sensing is used here as an initial approach, although pupil plane wavefront sensing may also be explored for advanced functionality in the future [42].

5. Conclusions

A vibration signal point scanning detection system utilizing Shack–Hartmann wavefront sensor methods for seismic wave laser remote sensing detection has been established. The system's transmitter leverages the benefits of short laser wavelengths, elevated detection sensitivity, and high measurement resolution for information collecting. The technique of point scanning detection for extensive ground vibration signals is investigated by utilizing the small size, high precision, high resolution, and high sensitivity of the Shack–Hartmann wavefront sensor. This research provides a vibration signal point scanning detection method that enables remote, non-contact, and effective data collection, addressing the limitations of the current laser remote sensing seismic wave detection technology. This method diverges from conventional single-point spot detection by including the whole surface of the spot. Based on the fundamental principle of the wavefront sensor, we analyze and verify the variations in the vibration of the spot across different locations. Upon excitation of the vibration source, the microlens array within the wavefront sensor captures the ground vibration signal. The telescope system focuses the laser echo signal in the target area onto the wavefront sensor, with the microlens array's sub-aperture corresponding to the received light spot. The experimental results indicate that, at a fixed vibration frequency, each microlens' wavefront slope within the array varies with different amplitudes. Moreover, by increasing the diameter of the scanning spot and detecting multiple sub-regions, the scanning efficiency has been significantly improved. The point scanning method concurrently detects the vibrations of various target sub-regions, successfully capturing the vibration wavefront. This outcome confirms the validity of our suggested vibration signal point scanning system and demonstrates its capability to detect vibration information across various regions within an extensive range.

In contrast to the conventional single-point spot detection method, the system has markedly enhanced the detection range and comprehensively analyzed the response of each sub-aperture within the microlens array. This paper presents experiments on low-frequency and small-amplitude vibrations. The system's features are analyzed, and vibration signal point scanning detection and data analysis are performed. With promising seismic wave detection and exploration applications, it can monitor and record minute ground vibrations in real time. Nonetheless, the vibration signal point scanning detection system established in this study faces the issue of laser power attenuation during the detection of seismic waves. To attain more excellent distance detection in the future, we must prioritize the utilization of high-power lasers. Moreover, noise interference constitutes a significant concern. Future studies may explore the application of neural networks to eliminate noise caused by environmental interference.

Author Contributions: Conceptualization, Q.L. and H.L.; methodology, Q.L. and J.S.; software, H.L.; validation, Q.L., G.W. and Y.P.; formal analysis, Q.L., G.W., Y.F. and J.S.; investigation, J.S.; resources, W.J.; data curation, Q.L., G.W. and J.S.; writing—original draft preparation, Q.L. and H.L.; writing—review and editing, J.S. and W.J.; visualization, Q.L., G.W. and H.C.; supervision, J.S. and W.J.; project administration, J.S.; funding acquisition, J.S. All authors have read and agreed to the published version of the manuscript.

Funding: The National Natural Science Foundation of China (Grant No. 62165015).

Institutional Review Board Statement: Not applicable.

Informed Consent Statement: Not applicable.

Data Availability Statement: The data that support the findings of this study are available from the corresponding author upon reasonable request.

Conflicts of Interest: The authors declare no conflicts of interest.

Appendix A

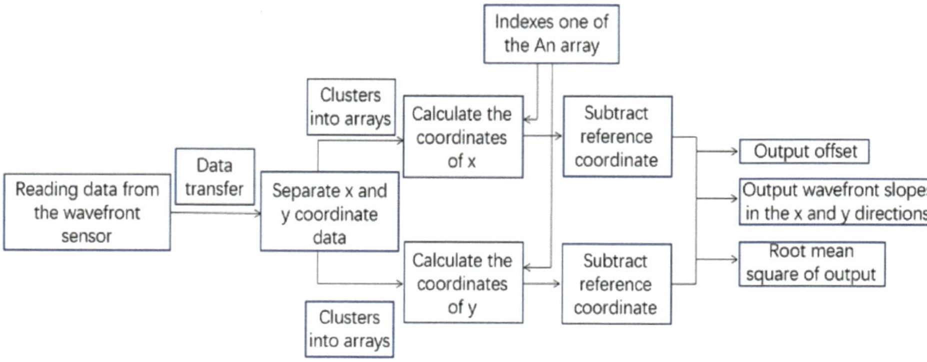

Figure A1. Calculation process.

Appendix B

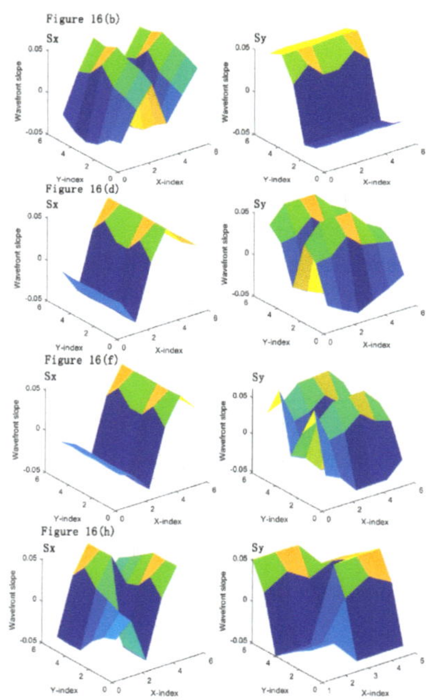

Figure A2. The wavefront slope in Figure 16.

References

1. Scott, W.; Martin, J.; Larison, G. Experimental model for a seismic landmine detection system. *IEEE Trans. Geosci. Remote Sens.* **2001**, *39*, 1155–1164. [CrossRef]
2. Polychronopoulou, K.; Lois, A.; Draganov, D. Body-wave passive seismic interferometry revisited: mining exploration using the body waves of local microearthquakes. *Geophys. Prospect.* **2020**, *68*, 232–253. [CrossRef]
3. Berni, A.J. Remote sensing of seismic vibrations by laser Doppler interferometry. *Geophysics* **1994**, *59*, 1856–1867. [CrossRef]
4. Yu, J.; Xu, P.; Yu, Z.; Wen, K.; Yang, J.; Wang, Y.; Qin, Y. Principles and Applications of Seismic Monitoring Based on Submarine Optical Cable. *Sensors* **2023**, *23*, 5600. [CrossRef]
5. Fernández-Ruiz, M.R.; Soto, M.A.; Williams, E.F.; Martin-Lopez, S.; Zhan, Z.; Gonzalez-Herraez, M.; Martins, H.F. Distributed acoustic sensing for seismic activity monitoring. *APL Photonics* **2020**, *5*, 030901. [CrossRef]

6. Mondol, N.H. Seismic Exploration. In *Petroleum Geoscience*; Springer: Berlin/Heidelberg, Germany, 2015; pp. 427–454. [CrossRef]
7. Yin, X.; Zhen, D.; Zhao, S.; Zhang, T.; Jia, M.; Iqbal, I.; Zhang, L. A seismic acquisition technique for the exploration of unconventional gas: An example from southeastern Shanxi Province, China. *Pet. Sci. Technol.* **2024**, *42*, 768–784. [CrossRef]
8. Wu, C.; Li, X.; Chen, W.; Li, X. A Review of Geological Applications of High-Spatial-Resolution Remote Sensing Data. *J. Circuits, Syst. Comput.* **2020**, *29*, 2030006. [CrossRef]
9. Shroll, R.M.; St. Peter, B.; Richtsmeier, S.; Tannian, B.; Jensen, E.; Kielkopf, J.; Pereira, W.E. Remote optical detection of ground vibrations. In Proceedings of the Infrared Remote Sensing and Instrumentation XXIII, San Diego, CA, USA, 9–13 August 2015; p. 96080K. [CrossRef]
10. Ozcan Tatar, C.; Cabuk, S.N.; Ozturk, Y.; Senyel Kurkcuoglu, M.A.; Ozenen-Kavlak, M.; Bilge Ozturk, G.; Dabanli, A.; Kucukpehlivan, T.; Cabuk, A. Impacts of earthquake damage on commercial life: A RS and GIS based case study of Kahramanmaraş earthquakes. *Int. J. Disaster Risk Reduct.* **2024**, *107*, 104464. [CrossRef]
11. Li, Q.; Zhang, J.; Jiang, H. Incorporating multi-source remote sensing in the detection of earthquake-damaged buildings based on logistic regression modelling. *Heliyon* **2024**, *10*, e32851. [CrossRef]
12. Fuchs, M.J.; Rexer, M.; Schaider, F. Detection and analysis of seismic induced GNSS station motion in a North American network following the 2017 Chiapas earthquake. *J. Geodyn.* **2022**, *149*, 101881. [CrossRef]
13. Bruce, R.A.; Fitzpatrick, G.L. Remote vibration measurement of rough surfaces by laser interferometry. *Appl. Opt.* **1975**, *14*, 1621. [CrossRef] [PubMed]
14. Yan, B.; Shi, H.; Wang, P.; Tong, Q.; Su, J. Optoelectric characteristics of laser remote sensing for measurement of ground vibration. *J. Appl. Remote. Sens.* **2021**, *15*, 014505. [CrossRef]
15. Bianchi, S. Vibration detection by observation of speckle patterns. *Appl. Opt.* **2014**, *53*, 931. [CrossRef] [PubMed]
16. Li, X.; Zhao, L.P.; Fang, Z.P.; Asundi, A.; Yin, X.M. Surface measurement with Shack-Hartmann wavefront sensing technology. In Proceedings of the Ninth International Symposium on Laser Metrology, Singapore, 30 June–2 July 2008. [CrossRef]
17. Yi, S.; Xiang, J.; Zhou, M.; Wu, Z.; Yang, L.; Yu, Z. Angle-based wavefront sensing enabled by the near fields of flat optics. *Nat. Commun.* **2021**, *12*, 6002. [CrossRef]
18. Riley, Z.B.; Hagen, B.; Ehrhardt, D.A. Aero-Optical Measurements of the Response of a Thin Panel at Mach 6. *AIAA J.* **2022**, *60*, 1522–1536. [CrossRef]
19. Zhu, H.; Li, G.; Wang, J. Flow-induced vibration of a circular cylinder with splitter plates placed upstream and downstream individually and simultaneously. *Appl. Ocean. Res.* **2020**, *97*, 102084. [CrossRef]
20. Chen, Z.S.; Wang, S.; Jiang, X. Effect of wake interference on vibration response of dual tandem flexible pipe. *Ocean. Eng.* **2023**, *269*, 113497. [CrossRef]
21. Ares, J.; Mancebo, T.; Bará, S. Position and displacement sensing with Shack–Hartmann wave-front sensors. *Appl. Opt.* **2000**, *39*, 1511. [CrossRef]
22. Shi, H.; Wang, P.; Su, J.. A remote sensing study of Wavefront-Sensor-Based Measurements of Seismic Wave. *Appl. Laser* **2023**, *43*, 111–117. [CrossRef]
23. Luo, Q.; Luo, H.; Wu, G.; Ji, X.; Su, J.; Jiang, W. Laser Remote Sensing of Seismic Wave with Sub-Millimeter Scale Amplitude Based on Doppler Characteristics Extracted from Wavefront Sensor. *Photonics* **2024**, *11*, 204. [CrossRef]
24. Goodman, J.W. *Introduction to Fourier Optics*; Roberts and Company Publishers: Greenwood Village, CO, USA, 2005.
25. Panagopoulou, S.I.; Neal, D.P. Zonal Matrix Iterative Method for Wavefront Reconstruction From Gradient Measurements. *J. Refract. Surg.* **2005**, *21*, S563–S569. [CrossRef] [PubMed]
26. Liu, S.; Zhong, H.; Li, Y.; Liu, K. Fast and Highly Accurate Zonal Wavefront Reconstruction from Multi-Directional Slope and Curvature Information Using Subregion Cancelation. *Appl. Sci.* **2024**, *14*, 3476. [CrossRef]
27. Huang, L.; Xue, J.; Gao, B.; Zuo, C.; Idir, M. Zonal wavefront reconstruction in quadrilateral geometry for phase measuring deflectometry. *Appl. Opt.* **2017**, *56*, 5139. [CrossRef]
28. Wang, K.; Xu, K. A Review on Wavefront Reconstruction Methods. In Proceedings of the 2021 4th International Conference on Information Systems and Computer Aided Education, New York, NY, USA, 24–26 September 2021; pp. 1528–1531. [CrossRef]
29. Southwell, W. Wave-front estimation from wave-front slope measurements. *J. Opt. Soc. Am.* **1980**, *70*, 998. [CrossRef]
30. Freischlad, K.; Koliopoulos, C.L. Wavefront Reconstruction from Noisy Slope or Difference Data Using the Discrete Fourier Transform. In Proceedings of the 1985 Technical Symposium East, Arlington, TX, USA, 10–12 April 1986; p. 74. [CrossRef]
31. Poyneer, L.A. Advanced techniques for Fourier transform wavefront reconstruction. In Proceedings of the Astronomical Telescopes and Instrumentation, Waikoloa, HI, USA, 22–28 August 2002; p. 1023. [CrossRef]
32. Li, T.; Huang, L.; Gong, M. Wavefront sensing for a nonuniform intensity laser beam by Shack–Hartmann sensor with modified Fourier domain centroiding. *Opt. Eng.* **2014**, *53*, 044101. [CrossRef]
33. Shevkunov, I.A.; Balbekin, N.S.; Petrov, N.V. Comparison of digital holography and iterative phase retrieval methods for wavefront reconstruction. In Proceedings of the Holography, Diffractive Optics, and Applications VI, Beijing, China, 9–11 October 2014; p. 927128. [CrossRef]

34. Bond, C.Z.; Correia, C.M.; Sauvage, J.F.; Neichel, B.; Fusco, T. Iterative wave-front reconstruction in the Fourier domain. *Opt. Express* **2017**, *25*, 11452. [CrossRef]
35. Ribak, E.N.; Carmon, Y.; Talmi, A.; Glazer, O.; Srour, O.; Zon, N. Full wave front reconstruction in the Fourier domain. In Proceedings of the Advances in Adaptive Optics II, Orlando, FL, USA, 24–31 May 2006; p. 627254. [CrossRef]
36. Poyneer, L.A.; Gavel, D.T.; Brase, J.M. Fast wave-front reconstruction in large adaptive optics systems with use of the Fourier transform. *J. Opt. Soc. Am. A* **2002**, *19*, 2100. [CrossRef]
37. Goldberg, K.A.; Bokor, J. Fourier-transform method of phase-shift determination. *Appl. Opt.* **2001**, *40*, 2886. [CrossRef]
38. Meng, X.F.; Peng, X.; Cai, L.Z.; Li, A.M.; Guo, J.P.; Wang, Y.R. Wavefront reconstruction and three-dimensional shape measurement by two-step dc-term-suppressed phase-shifted intensities. *Opt. Lett.* **2009**, *34*, 1210. [CrossRef]
39. Katkovnik, V.; Bioucas-Dias, J. Wavefront reconstruction in phase-shifting interferometry via sparse coding of amplitude and absolute phase. *J. Opt. Soc. Am. A* **2014**, *31*, 1801. [CrossRef]
40. Poyneer, L.A.; Véran, J.P. Optimal modal Fourier-transform wavefront control. *J. Opt. Soc. Am. A* **2005**, *22*, 1515. [CrossRef] [PubMed]
41. Fauvarque, O.; Janin-Potiron, P.; Correia, C.; Brûlé, Y.; Neichel, B.; Chambouleyron, V.; Sauvage, J.F.; Fusco, T. Kernel formalism applied to Fourier-based wave-front sensing in presence of residual phases. *J. Opt. Soc. Am. A* **2019**, *36*, 1241. [CrossRef] [PubMed]
42. Ma, X.; Wang, J. The research of wavefront sensor based on focal plane and pupil plane. *Optik* **2016**, *127*, 2688–2693. [CrossRef]

photonics

Article

An FBG-Based Hard Landing Monitoring System: Assessment for Drops on Different Soils

Angela Brindisi [1,*], Cristian Vendittozzi [2], Lidia Travascio [3], Marika Belardo [4], Michele Ignarra [5], Vincenzo Fiorillo [5] and Antonio Concilio [1]

1 Department of Adaptive Structures, CIRA (The Italian Aerospace Research Centre), 81043 Capua, Italy; a.concilio@cira.it
2 Department of Chemical Engineering Materials Environment, Sapienza University of Rome, 00184 Rome, Italy; cristian.vendittozzi@uniroma1.it
3 Department of Reliability and Safety of Critical Infrastructure, CIRA (The Italian Aerospace Research Centre), 81043 Capua, Italy; l.travascio@cira.it
4 Department of Airframe Design and Dynamics, CIRA (The Italian Aerospace Research Centre), 81043 Capua, Italy; m.belardo@cira.it
5 Aerospace Structure Crash Facility, CIRA (The Italian Aerospace Research Centre), 81043 Capua, Italy; m.ignarra@cira.it (M.I.); v.fiorillo@cira.it (V.F.)
* Correspondence: a.brindisi@cira.it

Abstract: This study aims to develop an integrated monitoring system using a fiber Bragg grating sensor network to record the structural response of a landing gear system under operational loads to detect hard landing conditions on soils with different absorbing characteristics and to differentiate between soil types during landings. This paper refers to drop tests carried out at a drop tower of the test article, an integrated leaf spring landing gear with fiber Bragg grating sensors, measuring strain to evaluate landings from different heights on different soil types: hard soil, sand, and gravel. Cross-correlation and fast Fourier transform analyses can help to assess the repeatability of the impact tests, to assess the developed system as very reliable in detecting landing conditions and ensure very low error in the accuracy of the sensor placement, or to assess whether different impacts under different conditions produce consistent responses.

Keywords: landing gear; drop test; hard impact; integrated sensor network; strain detection; fiber Bragg grating (FBG); fiber optics; aircraft

Received: 10 January 2025
Revised: 18 February 2025
Accepted: 20 February 2025
Published: 26 February 2025

Citation: Brindisi, A.; Vendittozzi, C.; Travascio, L.; Belardo, M.; Ignarra, M.; Fiorillo, V.; Concilio, A. An FBG-Based Hard Landing Monitoring System: Assessment for Drops on Different Soils. *Photonics* **2025**, *12*, 197. https://doi.org/10.3390/photonics12030197

1. Introduction

Landing gear, a vital component in aircraft and helicopters, traditionally employs safe-life design methods [1]. This approach ensures structural integrity throughout the component's expected lifespan but limits the potential for advanced health monitoring techniques [2]. Unlike fail-safe systems, where continuous monitoring is crucial, landing gear monitoring primarily focuses on verifying actual loads against oversized design loads to assess the remaining lifespan [3]. This approach offers several advantages [4]; in fact, by accurately determining actual loads, operators can increase the operational lifespan of landing gear components. Improved monitoring can identify potential issues early, reducing the risk of unexpected failures, and by extending service life and reducing maintenance costs, operators can achieve substantial economic benefits [5–7].

To achieve these goals, various monitoring techniques have been explored, such as sensor integration into the structure. Incorporating sensors or sensor networks into landing gear systems can provide valuable data for tasks, such as hard landing (HL)

detection, flight management, weight distribution, and vibration analysis. Researchers have developed specialized methods to accurately measure loads experienced during landing gear operation. Overcoming technical challenges, engineers have created practical and reliable hardware solutions for implementing monitoring systems in real-world aircraft. By combining these approaches, the aviation industry aims to optimize landing gear performance, improve safety, and reduce operational costs.

This work is part of the "More Electric Aircraft—Smart Landing Gear (MELA-SLG)" project, which aims to develop a Smart Landing Gear (SLG) system. This system would integrate a sensory network of fiber Bragg grating (FBG) optical sensors to monitor the structural response of the "Intelligent Gear" under operational loads, with the ultimate aim of assisting the pilot during flight operations. The SLG system is designed to provide information on weight on wheel, hard lading detection, and weight and balance of the aircraft.

A useful application of the SLG is the weight on wheel (WoW). Historically, WoW systems have primarily relied on the shock absorbers' stroke movement to determine whether an aircraft's wheels are loaded. Inductive sensors, capable of detecting displacement, are commonly used in these systems.

In [8], a novel WoW system architecture based on strain gauges was introduced. This approach offers several advantages over traditional methods. Strain gauges can detect smaller changes in stress, potentially providing a more accurate and timely indication of wheel loading. Moreover, this system demonstrated the ability to detect wheel contact at a lower threshold compared to traditional systems. By measuring shear stresses, strain gauges can potentially provide valuable insights into the distribution of loads on the landing gear. Another useful application is the weight and balance (W&B) system that can estimate the total weight and the position of the center of gravity of an aircraft after the loading phase, including passengers, baggage, and so on, and before taking off. A solution was proposed in [9].

A wheel dynamometer is a specialized device used to measure the forces acting on an aircraft's wheels during ground testing. This information is essential for understanding the performance and safety of the landing gear system [10].

Modern wheel dynamometers often incorporate strain gauges, which can accurately measure the forces exerted on the wheel. Wireless communication technologies enable these systems to transmit data to both on-board and ground-based systems [11].

Wheel dynamometers have various applications in aircraft testing, including HL detection, operational loads monitoring (OLM), and vibration analysis. Identifying excessive forces during landing can help assess the structural integrity of the aircraft.

OLM systems use data from wheel dynamometers to evaluate the remaining useful life of critical components [12]. Analysing vibrations during taxiing can provide insights into the dynamic behaviour of the landing gear.

Wheel dynamometers, equipped with advanced technologies like strain gauges and wireless communication, play a crucial role in ensuring the safety and performance of aircraft. By accurately measuring wheel loads, these devices provide valuable data for assessing structural integrity, optimizing maintenance schedules, and improving overall aircraft safety.

An HL occurs when an aircraft makes abnormal contact with the runway. This can be caused by various factors, including adverse weather conditions, mechanical failures, excessive weight, pilot errors, and other unforeseen circumstances.

Currently, there is no definitive, in-flight method for directly determining whether a landing gear has been overloaded during landing or ground maneuvers. Airlines and aircraft manufacturers rely on a combination of pilot assessment and recorded data to evaluate potential HL events [13].

To address this challenge, researchers and engineers have explored various approaches for direct load measurement in landing gear structures. Battery-powered strain gauge data logger: This system integrates strain gauges into the landing gear structure to collect data without requiring external power sources [14].

Laser deflection measurement: This method uses laser-based sensors to measure the deflection of the landing gear structure, providing insights into the loads experienced [15].

Instrumented load pin: Incorporating load pins equipped with sensors allows for direct measurement of the forces acting on the landing gear [5,14].

A recent study proposed a capacitive transducer-based HL recorder for helicopters. This device can be used to monitor the condition of the helicopter's landing gear and detect potential damage or overloading [16]. While current methods for HL detection have limitations, emerging technologies offer promising solutions. By implementing direct load measurement systems, the aviation industry can improve safety, reduce maintenance costs, and enhance our understanding of landing gear behavior during critical events.

Several early attempts at HL detection focused on monitoring the shock absorber's behavior. Stroke displacement: Measuring the displacement of the shock absorber proved to be unreliable, as it was not directly correlated with the severity of an HL [17]. Fluid pressure: Monitoring the fluid pressure within the shock absorber provided a more promising approach. However, this method requires significant modifications to the shock absorber's architecture, making it expensive and sometimes inaccessible. Piezoelectric accelerometer: While sensitive to accelerations, this approach could lead to false alarms, particularly for fighters that experience high vertical accelerations during flight. Pilot's physiological detection: Relying on the pilot's subjective perception of accelerations was deemed unreliable due to its subjective nature and limited time frame. More recent approaches have explored the use of heuristic algorithms that analyze flight parameters and sensor data to estimate impact loads. These algorithms offer a more data-driven and objective approach to HL detection [18].

The development of reliable hard-landing detection systems has been challenging due to the complexity of the phenomenon and the limitations of existing technologies. While various methods have been proposed, none have proven to be entirely satisfactory. Future research may focus on combining multiple approaches and leveraging advanced data analysis techniques to improve the accuracy and reliability of HL detection systems.

Traditional load cells have faced limitations in size, weight, and calibration issues when used for monitoring landing gear loads. Additionally, transmitting large volumes of data from the aircraft to the ground for analysis presents significant challenges. Integrating a sensor network into a landing gear requires careful consideration of several factors [19,20]. These are the following: Limited space: the available space within a landing gear is constrained, necessitating compact and lightweight sensor designs; harsh operating conditions: landing gears are subjected to severe conditions, including shock loads, vibrations, temperature extremes, and environmental factors; maintainability and reliability: the sensor network should not compromise the landing gear's overall maintainability and reliability. Fiber Bragg grating sensors have emerged as a promising solution for structural health monitoring (SHM) applications [21,22]. Their small size, resistance to harsh environments, and ability to measure strain and deformation make them well-suited for landing gear monitoring [23,24] by overcoming the limitations of traditional methods. The proposed approach demonstrates the potential of FBG technology in this critical application area.

FBG sensors have been successfully applied in a variety of engineering fields, including railway infrastructure monitoring, where they are used to measure mechanical stress in overhead contact wires [25], and in automotive modal testing, where they provide insights into strain distribution and vibrational analysis of composite materials [26]. These applications demonstrate the versatility of FBG sensors for real-time strain monitoring in dynamic environments.

A key challenge in impact strain measurement is the ability to capture high-frequency dynamic events. Hard landings induce rapid strain variations, requiring a high-speed interrogation system capable of resolving short-duration impact loads. Studies on ultrasonically embedded FBG sensors have demonstrated that FBGs can measure strain at frequencies up to 10 kHz, making them well-suited for impact and vibration analysis in aerospace applications [27]. Furthermore, modal analysis in automotive structures has validated the feasibility of FBG-based strain sensors for high-impact mechanical loads, providing a strong foundation for their use in landing gear SHM [26].

FBG-based monitoring systems have also been implemented in railway applications, where sensors were embedded into overhead contact wire clamps to track strain variations due to mechanical loading and environmental factors [25]. Additionally, impact testing using FBG sensors in composite and metallic structures has demonstrated their effectiveness in detecting structural responses to mechanical stress [28]. Numerical models analyzing strain values in composite materials with embedded FBGs further validate the use of FBG sensors in complex structural applications, particularly in cases where uniaxial stress assumptions affect sensor accuracy [29]. These applications highlight the potential of FBG sensors in assessing transient impact loads and structural integrity in high-risk environments.

Accurate data processing techniques are also essential for FBG-based monitoring systems. The interrogation of FBG signals can be affected by signal noise and spectral distortions, reducing measurement accuracy. To address this, advanced signal processing methods, such as center-of-mass algorithms, have been developed to improve wavelength resolution and strain sensitivity [28]. Additionally, finite element analysis (FEA) has been widely used to validate FBG sensor measurements under simulated loading conditions, enhancing confidence in their real-world application [27].

This paper presents drop tests conducted on a test article (TA), an integrated leaf spring landing gear with strain sensors, using a drop tower. The experiments validate a methodology for measuring the dynamic strain field upon ground impact using FBG sensors embedded along an optical fiber. This study aims to detect HL conditions, defined by the exceedance of specific vertical velocities.

Preliminary results from previous experiments on the same TA and drop tower have been presented by the authors in prior research [30].

The objective of the previous campaign was to verify the potential of a landing gear monitoring system based on FBG technology to distinguish an HL or to operate as an active weight-on-wheel system [31]. The present study further investigates the capability of the FBG array installed on the TA to detect HL conditions on soils with different absorbing characteristics and differentiate between soil types during landing.

This objective is pursued through the results of a test campaign in which different types of soil were arranged at the base of the drop tower. The effects of soil type on the deformation of the TA were then compared. To evaluate the feasibility of detecting HL events on different terrains, the system underwent a series of drop tests on surfaces of varying consistencies, including hard (steel plate), intermediate (gravel), and soft (sand).

The present study provides additional empirical validation of the proposed methodology; in this paper, cross-correlation analysis can help quantify the repeatability of the impact test or assess whether different impacts under different conditions produce consistent responses and fast Fourier analysis confirms the repeatability of the experiments.

The remainder of this paper is organized as follows: Section 2 describes the methodology and experimental setup, including sensor configuration and data acquisition procedures. Section 3 presents the results of impact strain measurements, followed by a discussion in Section 4 on the implications of numerical validation techniques for improving FBG sensor performance. Finally, Section 5 concludes the study with recommendations for future applications of SHM in aviation.

2. Materials and Methods

2.1. The FBG Sensor Network

FBGs are a well-established technology for measuring structural deformations [14,21] and offer several advantages for integrating into landing gear structures such as high sensitivity, electromagnetic immunity, minimal harness, and minimum weight. In fact, they can accurately measure small strains and deformations, are not affected by electromagnetic interference, have a small diameter, reducing the need for bulky cabling, and are lightweight, making them suitable for aircraft applications. Moreover, FBGs can be embedded into structures with minimal impact on their integrity, and multiple FBG sensors can be connected to a single interrogation unit, reducing the number of required cables. They operate as optical filters that reflect light at a specific wavelength, known as the Bragg wavelength. This wavelength is determined by the grating period and the effective refractive index of the fiber core.

The Bragg condition describes the relationship between the Bragg wavelength (λ_B), the grating period Λ, and the effective refractive index n_{eff} [21], $\lambda_B = 2n_{eff}\Lambda$. The strain dependence of an FBG can be calculated by differentiating the Bragg condition:

$$\frac{\Delta\lambda_B}{\lambda_B} = \left(1 + \frac{1}{n_{eff}}\frac{\partial n_{eff}}{\partial \varepsilon}\right)\Delta\varepsilon = (1 + p_e)\Delta\varepsilon = \beta_\varepsilon \Delta\varepsilon \quad (1)$$

This equation shows that the change in Bragg wavelength is directly related to the strain (ε) applied to the FBG. By monitoring the shift in Bragg wavelength, it is possible to accurately measure the strain variation ($\Delta\varepsilon$) experienced by the landing gear structure.

In this equation, β_ε represents the strain sensitivity of the Bragg grating at a given temperature, and p_e is the photo-elastic constant (variation of the core refraction index vs. the axial stress). For most silica optical fibers, the effect of core doping (typically, germanium) can be neglected. Theoretical mechanical strain sensitivity at a given temperature may be expressed as

$$S_{theor} = \frac{\Delta\lambda_B}{\Delta\varepsilon} = \beta_\varepsilon \lambda_B \quad (2)$$

2.2. The Test Article: Leaf Spring Landing Gear

The TA is a 1:3 replica of a Piper PA-18 Super Cub leaf-spring landing gear (Figure 1), a single-engine high-wing tourism light aircraft. The TA's max dimensions are 770 (width) by 325 mm (height), and the weight is 2.97 kg. The elastic architecture is particularly prone to absorbing dynamic loads at the impact.

On each leg of the TA, an FBG strain sensor is integrated as shown in Figure 2. Additionally, in Figure 2 are shown two conventional strain gauges (designated as SG3 and SG4, with the numerical value referring to the reference FBG), which were employed to substantiate the data collected by the FBGs, as detailed in Section 3.2.

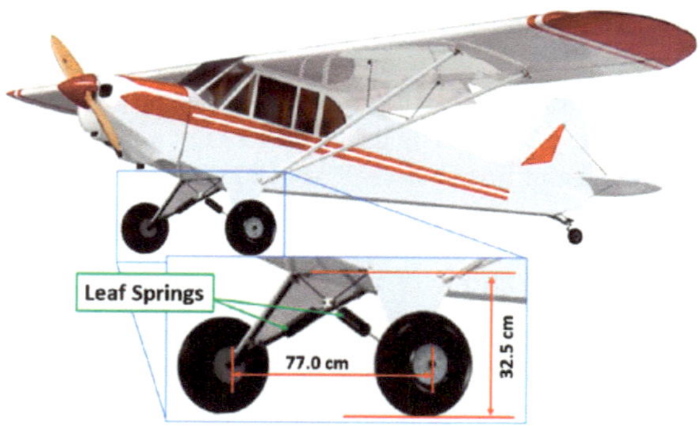

Figure 1. The Test Article.

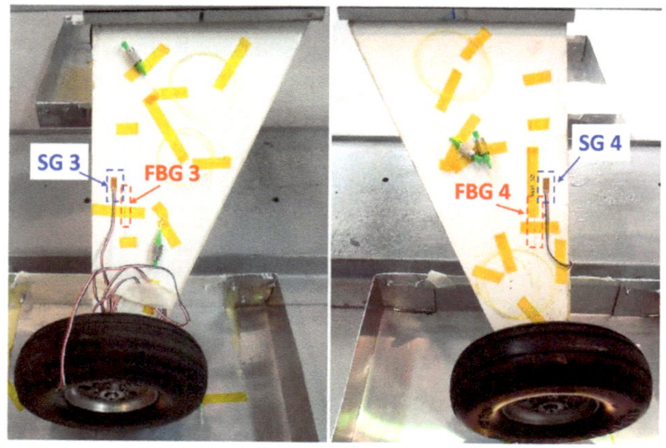

Figure 2. The test article: the two legs of the LG are integrated with strain sensors.

2.3. The Test Set-Up (The Tower and the Soils)

Impact tests were executed utilizing a small drop tower (Figure 3a). Two metal containers were then deployed at ground level in the region where the wheels were expected to make contact with the ground (Figure 3b). The dimensions of the containers were 380 by 400 by 70 mm. These containers were used to lay out the three different impact grounds: soft with sand, intermediate with gravel, and hard, i.e., the free surface of the bottom of the containers. Within each container, an FBG was installed to detect the impact time (Figure 3c). It is imperative to note that both FBGs were placed at a safe distance from the impact area (Figure 3c), as their sole function is to detect impacts. The objective is not to quantify the level of strain recorded by the sensors on the plates during impacts; rather, it is to ascertain the moment at which the TA wheels make contact with the ground (i.e., the bottom of the container) or even the instant of the detachment of the wheels during the lift phase. This is done to facilitate a comparison with the recorded signals.

Two measurement chains, each consisting of two gratings, were prepared and integrated into the TA and the test rig. Two of the gratings (FBG1 and FBG2, Figure 3c) were attached to the containers deployed at the base of the drop tower to detect the instant of impact, while the other two (FBG3 and FBG4, Figure 2) were attached to the legs of the TA to monitor the structure's deformation. The sensors were bonded to the outer surfaces of the steel containers and the legs of the TA using HBM Z70 cyanoacrylate adhesive. The two chains were connected to two separate channels of the Micron Optics' SM130 Acquisition

System, with an acquisition rate of 1000 Hz for the entire test campaign. Table 1 summarises the wavelengths and positions of each FBG.

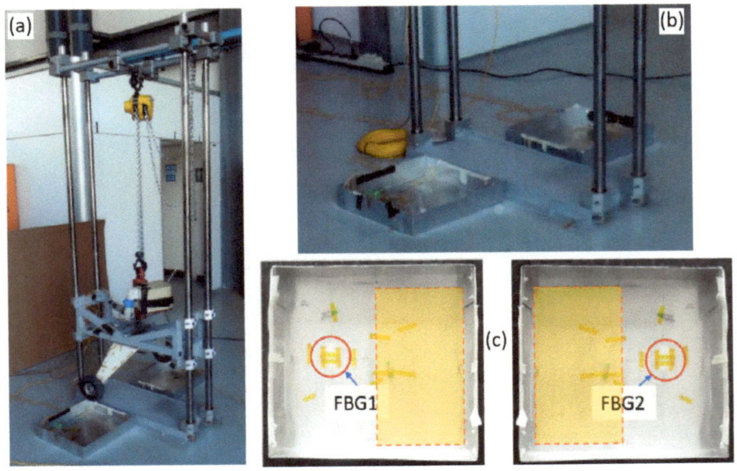

Figure 3. The test article. Installation details: (**a**) the test rig and the TA installed on the slit at a suitable drop height; (**b**) the drop tower with two containers deployed at its base; (**c**) detail over the two containers deployed at the drop tower base, with an FBG sensor installed in each of them. In the two containers, the impact areas of the wheels of the TA are highlighted in yellow.

Table 1. Measurement chain configuration.

Sensor Reference Number	WL (nm)	Position	Chain/Channel
FBG 1	1,530,254	Left Container	Ch1
FBG 2	1,540,333	Right Container	Ch1
FBG 3	1,549,921	Left Leg	Ch2
FBG 4	1,570,187	Right Leg	Ch2

To facilitate a more effective comparison of the data, the recorded wavelength shifts were converted to microstrain ($\mu\varepsilon$, dimensionless) using a set of equations (based on Equations (1) and (2)) embedded in the ENLIGHT measurement analysis software included with the Micron Optics sensing instruments.

3. Experimental Tests

3.1. Test Description

The tests performed consisted of a series of drops of the TA from different heights at the drop tower on different types of soils.

Three different impact heights (0.10, 0.15, and 0.20 m) (Figure 4) were considered, corresponding to three different impact velocities, as reported in Table 2.

Table 2. Drop height (h) and related impact velocity (v).

Drop Height, h (m)	Impact Velocity, v (m/s)
0.10	1.40
0.15	1.71
0.20	1.98

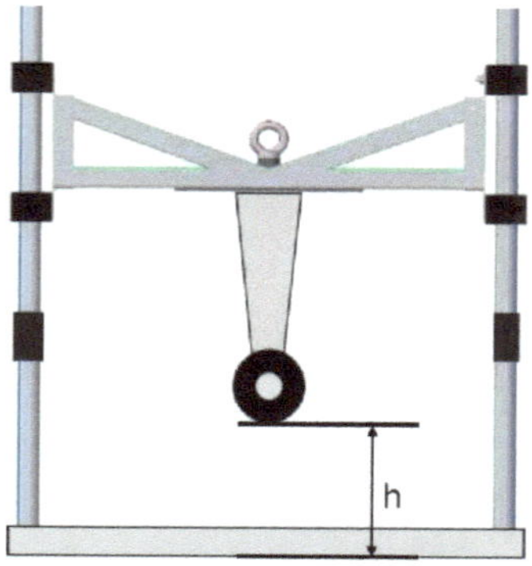

Figure 4. Free-fall height, h.

Impacts occurred on three soil types: rigid soil (a steel plate fixed on a concrete floor), sand, and gravel, as shown in Figure 5.

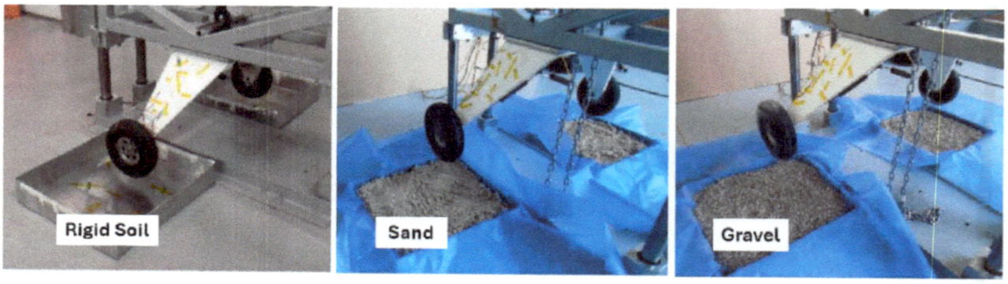

Figure 5. The three different impact soils: rigid (**left**), sand (**center**), and gravel (**right**).

The tests are performed by adopting the following procedure: For each soil, drop tests are carried out at three different heights. The first set of drops is carried out on the rigid soil, followed by the drop tests on the sand and gravel. For each set of tests, drops are performed from increasing heights, starting from 0.1 m. At each drop, the landing gear is placed at a height *h* from the soil, as shown in Figure 4. The height is measured to the lowest point of the TA, i.e., the bottom point of the wheel. The single experiment is then conducted by letting the landing gear drop freely, after having activated the release system; the mechanical guidance system allows the TA to follow a pure vertical descent, up to hitting the ground. When the landing gear is placed at a certain height, it is suspended. FBG sensors on the landing gear legs and the ground containers are zeroed. In other words, the reference is set as the rest condition. Indeed, weight force and thermal effects could lead the strain sensors to some deviations.

3.2. Support Data

During each test, two accelerations were measured on the drop tower slide and two strains on the test article by classical instrumentation (accelerometers and strain gauges, respectively), for a total of two acquisition channels. The acquisition frequency band was set at 1650 Hz, while the sampling frequency was set at 10 kHz, according to the applicable SAE standards. Such data were essential to give a prompt validation of the measures retrieved by the fiber optic system. In Figures 6–8, as an example case, the time history of the signals representative of each cited sensor is reported, regarding the test featuring the test article dropped by 0.20 m on the three available surfaces (rigid, sand, and gravel).

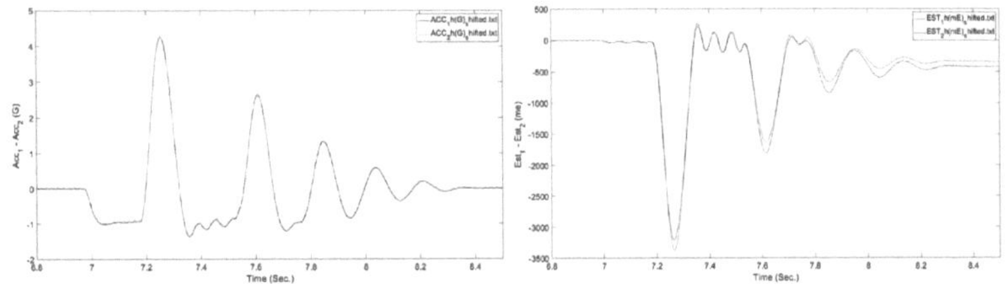

Figure 6. Time histories of the reference sensors during a 0.20 m drop test over a rigid surface: accelerations (**left**) and deformations (**right**).

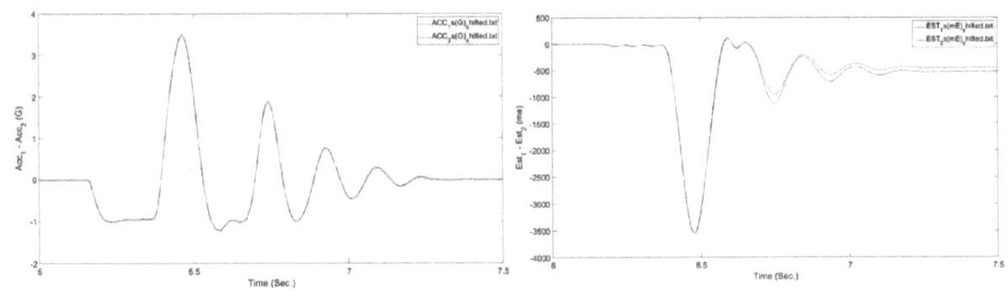

Figure 7. Time histories of the reference sensors during a 0.20 m drop test over sand: accelerations (**left**) and deformations (**right**).

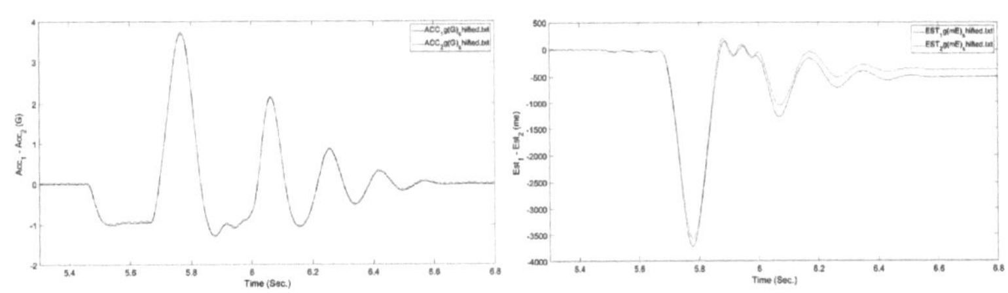

Figure 8. Time histories of the reference sensors during a 0.20 m drop test over gravel: accelerations (**left**) and deformations (**right**).

These simple diagrams evidence some peculiar properties. First of all, it can be seen that the two pieces of information are strictly correlated to each other. Impact times, max excursion, rebound and free air movement, and so on are distinguishable among the couples of signals. Accelerations are instead characterized by two more features at the very beginning of the test. After the first rest segment (acceleration, a = 0), it gradually moves to the value of 1 g free-fall. The regular growth is related to the characteristic nature of the release system, which is not capable of releasing the test article immediately, but after a few seconds. Then, the free-fall follows, up to the touchdown, where a sharp variation of the acceleration does occur. The rest of the tracks are very similar; among the three drop tests, it is also easy to distinguish among the different impact surfaces, providing more (sand, gravel in sequence) or less damping (rigid plate), intended as energy dissipation capability.

4. Results

4.1. Test Results: Strain Measurements

The strain measurements retrieved by the FBG sensors placed on the soil and the falling TA show a behavior that is repeated for different drops from different heights and on different soils. The image in Figure 9 shows the strain measurements retrieved by the FBG sensors placed on the containers (indicated as 1 and 2) and on the legs of the TA (indicated as 3 and 4) vs. time. The signals are taken during the drop and the impact of the TA from a 0.10 m height on rigid soil.

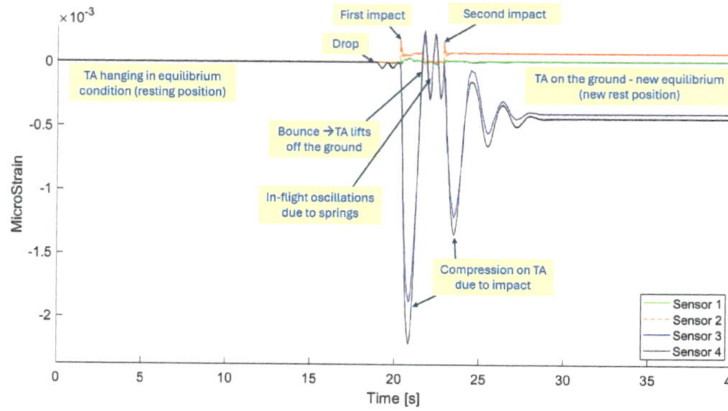

Figure 9. Deformation measurements of the FBG sensors for a drop test from 0.10 m on a rigid impact surface: in the first phase, the test article is suspended; then, there is the free-fall frame, up to the impact; a rebound then occurs, followed by an ascent and descent phase, up to the second impact, and finally a smooth oscillation until rest.

As the landing gear is released, a shortfall occurs, followed by the impact to the ground, a first rebound leading to a short ascent, and so on, until the relaxation of the whole structural system. In the drop phase, the two FBGs placed on the containers do not record any signal, while the other two deployed on the landing gear start feeling the inertial force and, above all, the vibration field associated with the free fall and the sliding movement over the guides. At the impact, the sensors on the containers are dynamically excited by the impulse caused by the landing gear hitting the base, while the sensors on the landing gear retrieve the dynamic deformations generated by the same impact forces on its legs. As the landing gear rebounds and detaches from the soil, the sensors continue to record a dynamic motion, associated with the landing gear vibrations. It is relevant to say that both these oscillations refer mainly to the first eigenmode of the landing gear structure featuring two different constraints: the first, at the impact, referring to a 2-point simple supported structure, and the second, to a free structure. At the second impact, the signal resembles the first path again. The sequence is finally interrupted as the landing gear does not rebound anymore and slowly loses its energy until rest. Such a sequence is illustrated in the diagram reported in Figure 9.

In the first phase, the TA is hung on the test rig at rest, and the 4 FBG sensors record no signal coherently to the zeroing set at the start. At the impact with the ground, the sensors on the containers record the event with a sharp peak deformation, while the sensors on the trolley record the compression progress, occurring on the two legs of the landing gear with negative strain, following the way they were deployed. Since the containers are much more rigid than the TA, the strains therein measured are significantly lower. In this frame, the time history of the deformation of the containers is significantly regular, since no other relevant phenomenon occurs. Instead, the strain recorded over the landing gear shows another history: first, there is a regular compression, describing the typical movement of something elastic touching the ground and continuing its movement downwards, until the elastic forces exhaust the inertial ones. Then, the elastic forces recall the mass upwards, causing its detachment from the ground. This process occurs at a frequency about the first resonance of the object, simply supported on its wheels. As the TA loses contact with the soil, its free movement in the air starts, and it is possible to detect an insurgent vibration at a higher frequency than before. This is very close to the first frequency of the free object (i.e., without constraints to the ground). This phase is visible in the diagrams and terminates with its new contact with the ground (complete rebound). At that point, the cycle is repeated, with the difference that in this latter case, the structure does not leave the ground anymore but continues its oscillations until rest. Such oscillations occur at the same frequency as before, featuring again the landing gear simply supported on its wheels. It can be remarked that the resting phase concludes with a perceptible displacement from the zero line for all the FBG strain sensors, since the weight affects both the TA, imposing its deformation, and the containers, where the landing gear weight is transferred. A discernible discrepancy in the strain information between the two legs and the two containers is evident, attributable to variations in position for the left and right leg of the TA, as well as an imperfectly even distribution of the masses. The container exhibited discrepancies in its position, the distribution of the applied load (as the TA did not strike the ground at the same relative location as the two containers), minor variations in its structural design, and its asymmetrical deployment on the ground. Nevertheless, the retrieved data are sufficient for the aims of this investigation. The images in Figure 10 show the results of the deformation measurements for the case of drops on rigid soil from the impact heights of 0.15 and 0.20 m, respectively. The result of the drop from 0.10 m on the same ground has already been shown in Figure 9. The trend documented by the same sensor in the case of drops from three distinct heights exhibits a high degree of similarity. As might be expected, the magnitude of the peaks increases in proportion to the height of the drops. In instances of impacts at 0.15 and 0.20 m, as well as at 0.10 m, two complete rebounds are observed. That is to say, following the initial drop, the system rebounds to the ground. Before reaching a state of complete rest, during which dynamic energy is fully dissipated, the system undergoes a subsequent, shorter detachment from the ground. This phenomenon can be discerned by the presence of the two densest peaks between the largest and widest peaks, which are indicative of on-ground oscillations.

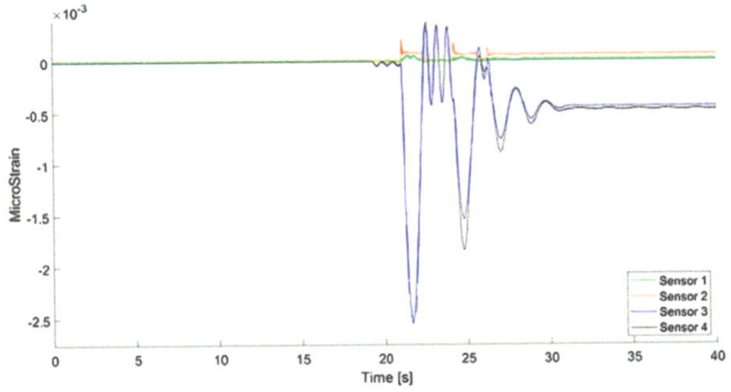

(a)

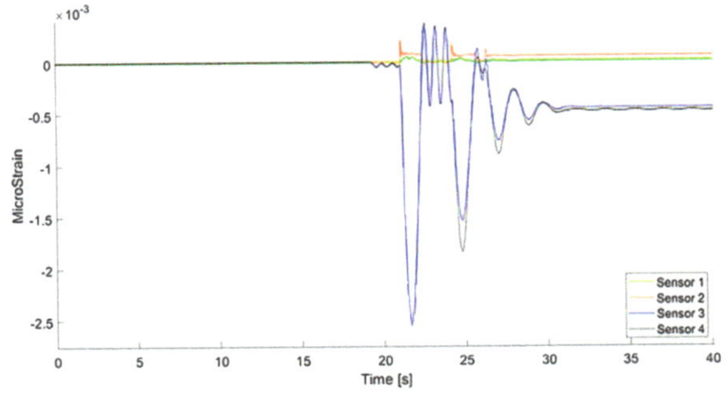

(b)

Figure 10. Outcome of drop tests from (**a**) 0. 15 and (**b**) 0.20 m, featuring an impact on a rigid surface.

In the images reported in Figure 11, the deformation measures by the FBG sensors for drops from 0.10, 0.15, and 0.20 m on sand (right) and gravel (left) are shown, respectively. The behaviors are similar to the previous ones on rigid soil but the signals result damped significantly. For instance, it retrieved just one bounce in all the cases, but also one (0.10 m, sand) when the TA never leaves the ground after the impact. It should be observed that such an energy dissipation process mainly affects the signal coming from the sensors deployed on the containers and not the landing gear response. Sand and gravel are distributed over the surface of the containers to directly influence their vibration, but they have a minimal role in the TA dynamics. The ground materials have a poor influence on the nominal constraints of the structure under investigation. This is trivial in the case of the free movement in the air, where there is not even a minimal contact between the two items (sand or gravel vs. the landing gear legs). Instead, when the TA touches the ground, the only interface between the two elements is at the level of the wheels. Therefore, the only way for the materials deployed in the containers to contribute to the damping is to intervene in the wheels' movement during the oscillation; this is a very small, almost negligible, contribution. While the strains measured at rest result in the same magnitude, some difference in the absolute values may be perceived. This means that the landing gear does not assume the same configuration after the different drops; this is very important as the system is intended to detect the HL condition, i.e., quantify the overall energy entering the TA at the touchdown. It follows that a suitable dense sensor network shall be sized to ensure coherent information for those aims. The design of such a map is one of the arguments that shall be faced in the next stages of the activities.

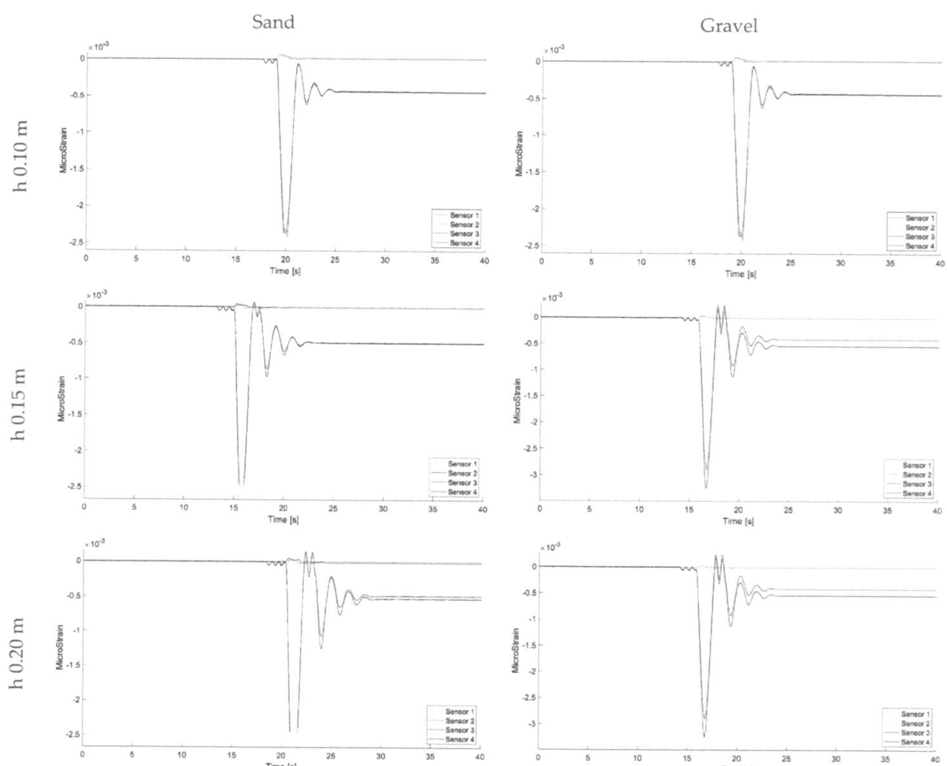

Figure 11. On the **left** column, from **top** to **bottom**, are the 0.10, 0.15, and 0.20 drops on sand; on the **right** column are the 0.10, 0.15, and 0.20 drops on gravel.

4.2. Test Results: Signal Correlation

It may be of certain interest to compare the different responses of the proposed architecture, the sensor network integrated into the TA, for a series of drops, to verify the repeatability of the tests, in case of drops with the same characteristics, or the coherence of the whole system in case of drops under different conditions, such as different heights or different soils. In this paper, cross-correlation (CC) analysis can help quantify the repeatability of the impact test or assess whether different impacts under different conditions produce consistent responses.

A high value of the CC coefficient indicates that the signals are similar. Figure 12 shows the plots of the signals from the two FBGs, sensor S3 on the left leg and sensor S4 on the right leg of the TA, recorded during three consecutive drops from 0.10 m on hard soil. The figures show the signals of the first drop (S3.1) with the blue dashed curve, of the second drop (S3.2) with the red continuous curve, and of the third drop (S3.3) with the black dashed curve. Tables 3 and 4 show the correlation coefficient matrices of the signals for these drops. A strong similarity between repeated drops of the same height appears as a high CC coefficient. The maximum value is 1, which means that the two signals are identical, as is the case on the matrix diagonal where the signals of each sensor are correlated with themselves. Variability between impacts with the same height and on the same soil can indicate possible internal or external noise, such as changes in the structure during the impacts or environmental factors. Moreover, the sensors integrated into the TA may have errors in the positions or orientations; hence, variations in the strain wave propagation paths and arrival times at each sensor could result in low correlation coefficient values.

Table 3. Matrix of correlation coefficients for three consecutive drops from h 0.10 m on rigid soil, related to sensor 3.

	S3.1	**S3.2**	**S3.3**
S3.1	1.000	0.995	0.986
S3.2	0.995	1.000	0.996
S3.3	0.986	0.996	1.000

Table 4. Matrix of correlation coefficients for three consecutive drops from h 0.10 m on rigid soil, related to sensor 4.

	S4.1	**S4.2**	**S4.3**
S4.1	1.000	0.995	0.989
S4.2	0.995	1.000	0.997
S4.3	0.989	0.997	1.000

The reported CC of the signals related to different drops from the same height and on the same soil is generally very high, indicating a high correlation between repeated events in the same conditions, hence assessing the developed system as very reliable in detecting landings and assuring very low error in the accuracy of the sensor placement. Nonetheless, the signal of the sensor S3, particularly in the third drop, shows a trend slightly different from the two previous drops (see Figure 12a). This deviation occurs mostly in the latter part of the test, after 7 s, and is most evident when the TA is stationary on the ground, after 11 s. The third drop was expected to stabilize around the same equilibrium value, but instead, it is slightly more compressed. When the TA was checked after the last drop, it was observed that the left leg was slightly deformed at the constraint between the TA and the wheel, resulting in a difference in height of approximately 5 mm between the right and left legs and consequently a different pressure on the two wheels. The TA load weighed more on the left leg. A slight deformation of the structure resulted in a slight lowering of the CC value.

A second CC analysis was executed by comparing the drop signals of a sensor for any given height on three distinct surfaces. The ensuing results are displayed in Figure 13. Figure 13a illustrates the signals of sensor S3 from the given height of 0.10 m on the three different soils; Figure 13b is from 0.15 m, and Figure 13c is from 0.20 m. The corresponding CC matrices are provided in Tables 5–7.

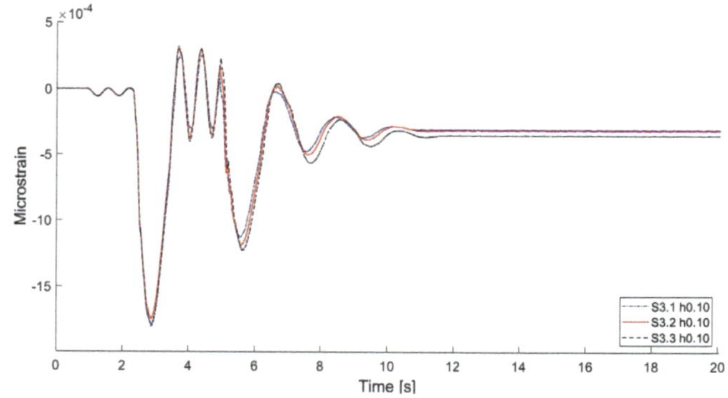

(**a**)

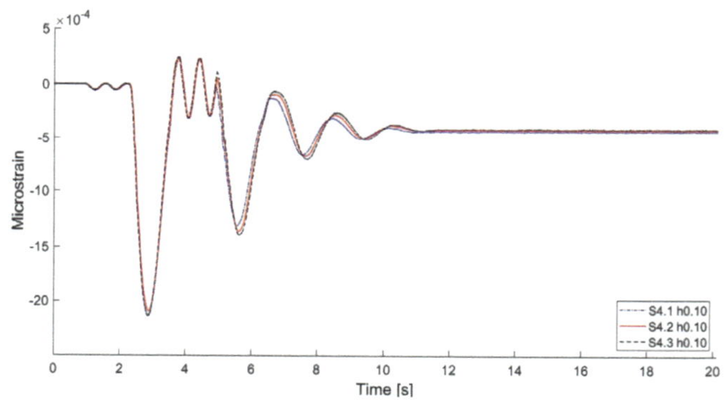

(**b**)

Figure 12. Correlation graphs between three consecutive drops from 0.10 m on rigid soil related to (**a**) sensor S3 and (**b**) sensor S4. The first drop is indicated with the blue dashed curve, the second drop with the red continuous curve, and the third drop with the black dashed curve.

Table 5. Matrix of correlation coefficient signals of sensor S3 for drops from the same height of 0.10 m for three different soils: rigid soil, sand, and gravel.

	Rigid Soil	**Sand**	**Gravel**
Rigid Soil	1.000	0.437	0.477
Sand	0.437	1.000	0.989
Gravel	0.477	0.989	1.000

Table 6. Matrix of correlation coefficient signals of sensor S3 for drops from the same height of 0.15 m for three different soils: rigid soil, sand, and gravel.

	Rigid Soil	**Sand**	**Gravel**
Rigid Soil	1.000	0.695	0.721
Sand	0.695	1.000	0.987
Gravel	0.721	0.987	1.000

Table 7. Matrix of correlation coefficient signals of sensor S3 for drops from the same height of 0.20 m for three different soils: rigid soil, sand, and gravel.

	Rigid Soil	Sand	Gravel
Rigid Soil	1.000	0.696	0.743
Sand	0.696	1.000	0.977
Gravel	0.743	0.977	1.000

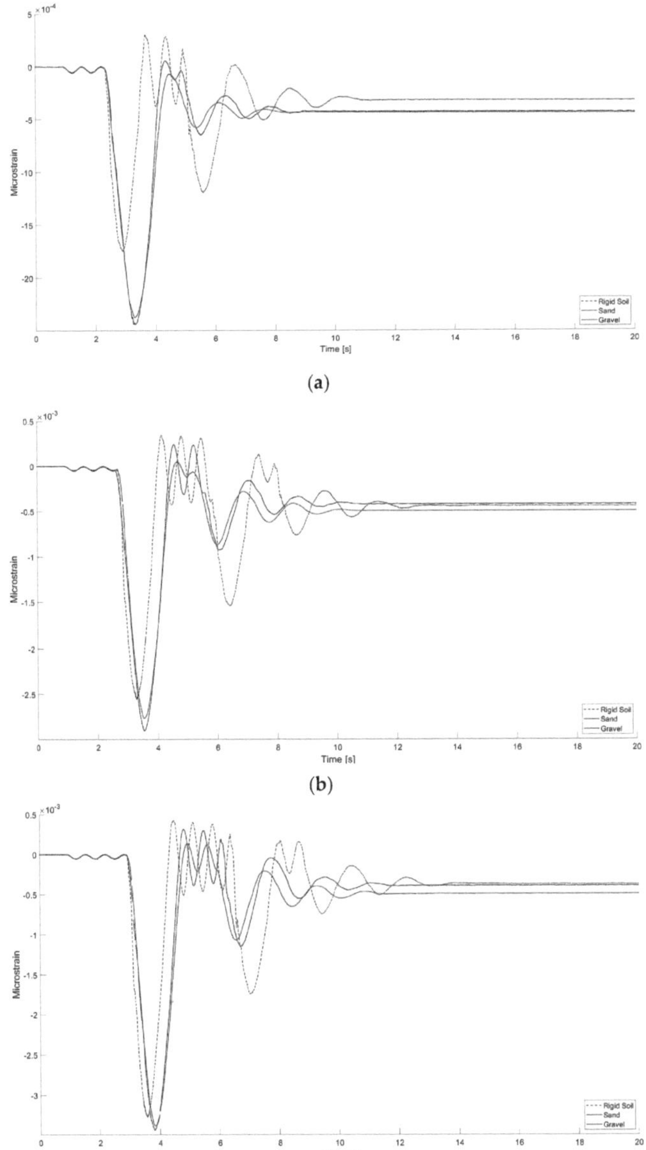

Figure 13. Correlation graphs between signals of sensor S3 for drops from a given height on the three different soils. From top to bottom: (**a**) drops from 0.10 m, (**b**) drops from 0.15 m, and (**c**) drops from 0.20 m.

The CC matrices in Tables 5–7 show a low correlation between signals related to different soils, suggesting that the strain waveforms captured by the FBGs, although similar in general appearance, are different. Impacts on different soils, which absorb the impact energy differently, generate different strain magnitudes, frequencies, and temporal patterns. The material and surface properties of the ground significantly influence the distribution and propagation of the impact forces. Some soils may exhibit nonlinear behavior, such as soft soils like sand that compress more on higher impacts, or uneven soils like gravel that compress unevenly, displace on impact, or resist more, to the same or greater extent than rigid soils. This can result in distinct deformation patterns that are less correlated between different elevations or soils. In addition, soil materials with different damping properties may dissipate energy at different rates, resulting in different characteristics of signal variation.

A third CC analysis compared the signals of a sensor for impacts on any given type of soil from three different heights.

The results are shown in Figure 14a for the drop on hard ground, Figure 14b for the drop on sand, and finally Figure 14c for the drop on gravel. The corresponding CC matrices are given in Tables 8–10. The CC matrices in Tables 8–10 show a low correlation between signals related to different heights of the drops. Impacts from different heights produce mainly different strain magnitudes and temporal patterns. This could be due to different energy levels associated with each height.

Table 8. Matrix of correlation coefficients for three drops on rigid soil from three different heights: h 0.10 m, 0.15, and 0.20.

	h0.10	h0.15	h0.20
h0.10	1.000	0.253	−0.033
h0.15	0.253	1.000	0.527
h0.20	−0.033	0.527	1.000

Table 9. Matrix of correlation coefficients for three drops on gravel from three different heights: h 0.10 m, 0.15, and 0.20.

	h0.10	h0.15	h0.20
h0.10	1.000	0.839	0.502
h0.15	0.839	1.000	0.765
h0.20	0.502	0.765	1.000

Table 10. Matrix of correlation coefficients for three drops on the same soil (sand) from three different heights (h 0.10 m, 0.15, and 0.20).

	h0.10	h0.15	h0.20
h0.10	1.000	0.853	0.568
h0.15	0.853	1.000	0.811
h0.20	0.568	0.811	1.000

The findings from the two most recent correlation analyses demonstrate that the developed system can discern impacts from varying heights, resulting in deformations of differing magnitude, frequency, and time course, particularly on different terrains. Additionally, they underscore the significant influence that the characteristics of diverse impact surfaces exert on the distribution and propagation of impact forces.

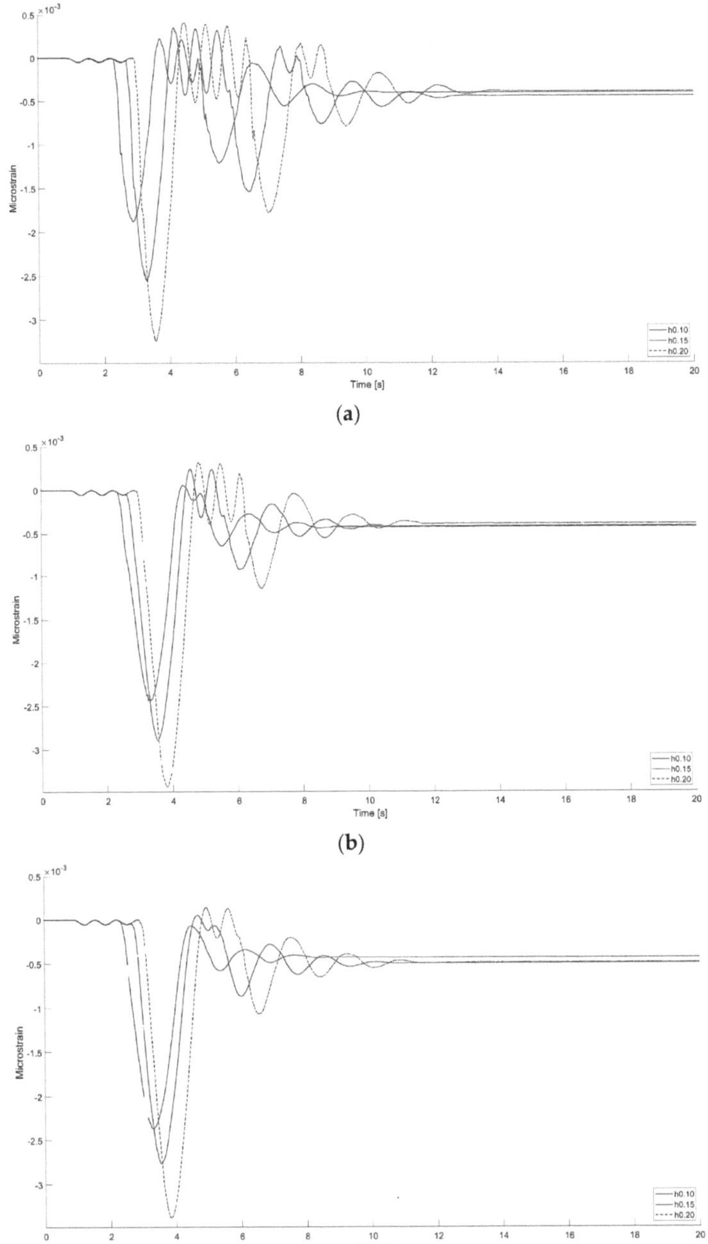

Figure 14. Correlation graphs between signals of sensor S3 for drops on the same soil from three heights. From top to bottom: (**a**) drops on rigid soil, (**b**) drops on sand, and (**c**) drops on gravel.

4.3. Test Results: Fast Fourier Transform Analysis

Without loss of generality, a study was carried out on three specific signals to extrapolate what kind of detailed information can be extracted from a fast Fourier transform (FFT) analysis and to what extent. The three signals refer to three consecutive drops, recorded by sensor S3 from a height of 0.10 m on hard ground. The three signals clearly show a drop at 0.9 s, an impact at 2.3 s, a rebound at 2.8 s, and free flight until 5.0 s. During this time, the signal displays three large amplitude oscillations. There is a second impact at 5.1 s, followed by a series of rebounds until the dynamics come to a complete stop at 11 s, as shown in Figure 15. As in the previous cases, the sampling rate is 1.0 kHz. An obvious consideration is that the timing of the phenomenon is very narrow. Therefore, an acquisition aimed at one of the different phases would lead to a very coarse discretization in the frequency domain, regardless of the time resolution. The elementary relations may be recalled:

$$\Delta f = 1/T \quad (3)$$

where Δf is the frequency resolution, and T is the observation time, while

$$F = 1/\Delta t \quad (4)$$

where Δt is the temporal resolution, and F is the observable frequency range. A series of graphically synthesized analyses was then carried out in the upper figure. Details are given in Table 11.

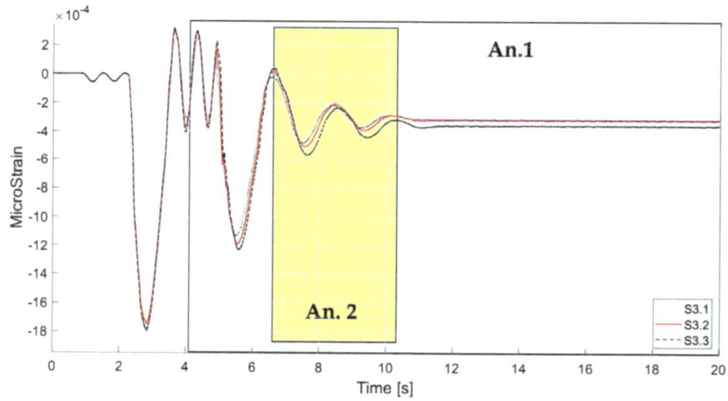

Figure 15. Detailed analysis schematic and an indication of the two different analysis windows, whose description is reported in Table 11. An1 (0–16,384 ms) and An2 (2500–6596 ms).

Table 11. Description of the FFT analysis carried out on the signal reported in Figure 15.

ID	t Start (s)	t End (s)	Samples	Δf (Hz)	n → (2^n)	Description
1.1	0.000	16.384	16384	0.061	14	The analysis is carried out from the first acquisition (t = 0), aiming at covering the whole phenomenon and extending the observation time at the furthest n exponent (2n)
1.2	2.500	6.596	4096	0.24	12	The focus of the analysis is on the three in-flight oscillations between the first and second impacts, extending the observation period at the furthest n exponent (2n)

Figures 16 and 17 show the FFT results in the two examined windows. Four distinct peaks are confirmed in the various representations at the frequencies reported in Table 12, which also shows the range of variation in each analysis.

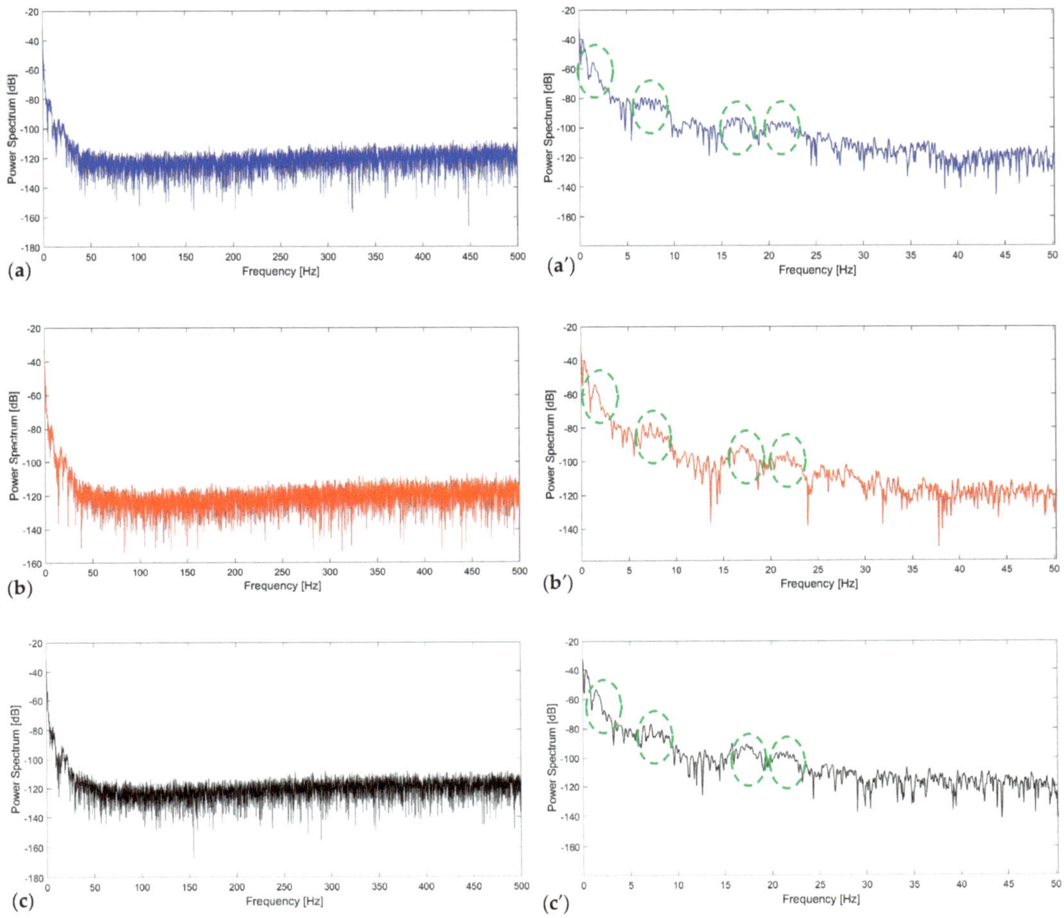

Figure 16. Analysis no 1. FFT analyses of the signal frames are defined in Table 11. The repeated peaks are evident, even though it is not possible to distinguish among the system responses in the supported (grounded) and floating (free) conditions. The FFT analyses are related to three drops from a height of 0.10 m on a rigid soil recorded by sensor no. 3. On the left, (**a**–**c**) are the complete frequency range from 0 to 500 Hz; on the right, (**a′**–**c′**) are zoomed in from 0 to 50 Hz.

Table 12. Main frequencies (eigenfrequencies): the outcome of the FFT of the signal reported in Figure 15.

Freq ID	Freq (Hz)
F1	0.2–0.9
F2	6.5–7.5
F3	16.0–17.5
F4	21.5–22.5

In this phase of the study, the FFT is not meant to verify the modal analysis characteristics; such an investigation deserves its own devoted publication. In this case, a further complication is associated with the co-existence of two different constraints in the observation period (namely, free–free (during floating) and simply supported (during the landing moments)). Instead, the FFT was intended as a check to verify the repeatability of the experiments. To avoid any influence on the elaboration process, it was chosen to identify two different time windows, where the contribution of the different boundary conditions could have influenced the transformed time signals in the frequency domain differently. Unfortunately, as already cited in the text, it was not possible to reduce the time window too much, otherwise the frequency resolution would have been very poor (several Hz), attaining the same magnitude of the characteristic structural system eigenvalues. Though these precautions were taken, the high decay of the signal made the peaks little discernible from the background noise, making the analysis difficult but not impossible to perform. A look at Figures 16 and 17 permits easy highlighting of four different peaks, repeated all along the different analyses at the same levels. As can be observed, the background noise (the line outside the marked circles) appears much more irregular in itself and somehow different in the various FFTs.

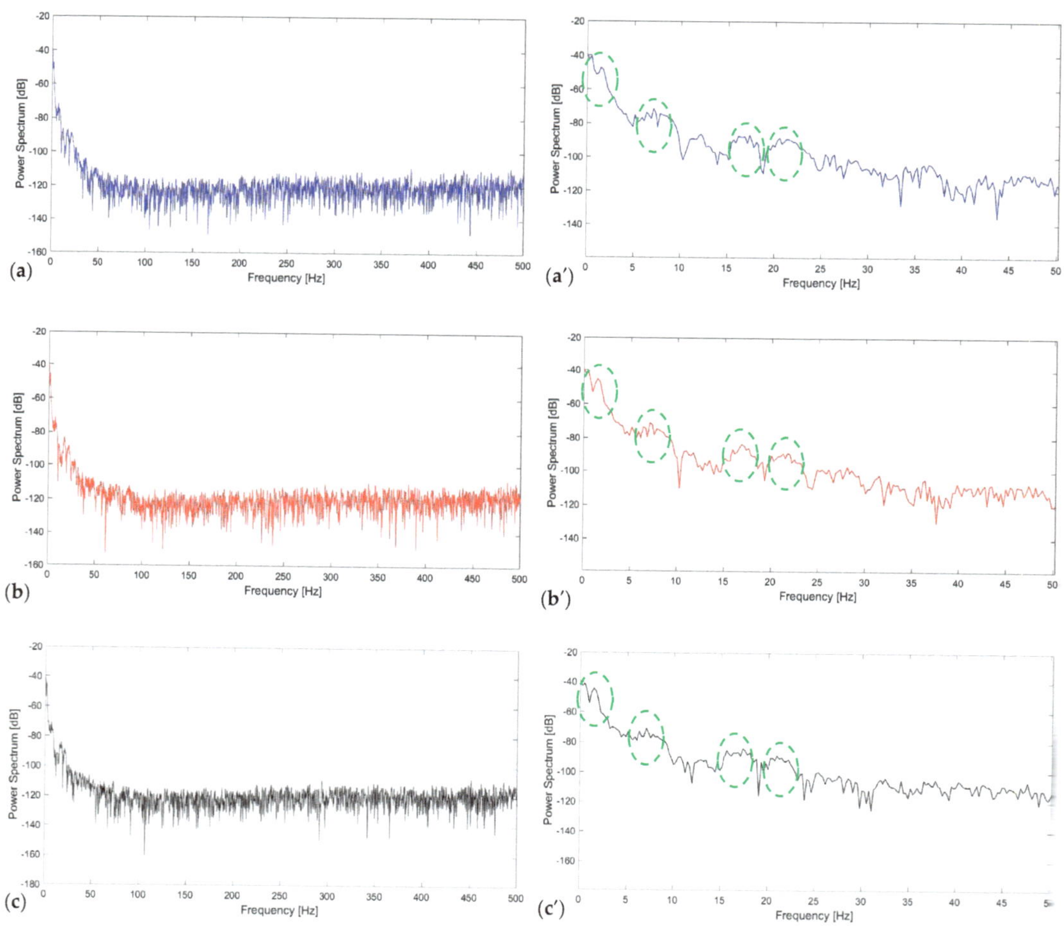

Figure 17. Analysis no 2. The FFT analyses are related to three drops from a height of 0.10 m on a rigid soil recorded by sensor no. 3. On the left, (**a**–**c**) are the complete frequency range from 0 to 500 Hz; on the right, (**a′**–**c′**) are zoomed in from 0 to 50 Hz.

The results are consistent: the identified eigenmodes are repeated in the three drops from the same height on the same soil, characterizing the TA both in the ground contact and in the rebound flight phase.

5. Conclusions

The objective of the activity presented here is to develop a Smart Landing Gear system capable of recording the structural response when subjected to operational loads during landings and to detect hard landing conditions on various soils with different absorbing characteristics. The SLG is an integrated system with a sensory network based on the use of fiber Bragg grating optical sensors. The focus of the experimental activities described here concerned the landings from different impact heights on different types of soils: hard soil, sand, and gravel.

This paper refers to drop tests carried out at a drop tower with the test article, an integrated leaf spring landing gear with strain sensors. The experiments aimed to validate a methodology to measure the dynamic strain field during the impact on the ground and to assess the repeatability and consistency of the acquired signals of the SLG when undergoing landings on different soils.

The strain measurements retrieved by the FBG sensors placed on the soil and the falling TA show a behavior that is repeated for different drops from different heights and on different soils.

Cross-correlation analysis is used to assess the correlation between signals for a series of the same drops, all with the same characteristics, to verify the repeatability of the tests or assess whether different impacts under different conditions produce consistent responses and to test the coherence of the whole system.

The reported CC of the signals related to different drops from the same height and on the same soil is generally very high, indicating a high correlation between repeated events in the same conditions, hence assessing the developed system as very reliable in detecting landings and assuring very low error in the accuracy of the sensor placement. Impacts from different heights produce mainly different strain magnitudes and temporal patterns. This could be due to different energy levels associated with each height. Impacts on different soils, which absorb the impact energy differently, generate different strain magnitudes, frequencies, and temporal patterns. Correlation analyses demonstrate that impacts from varying heights result in deformations of differing magnitude, frequency, and time course, particularly on different terrains. Additionally, they underscore the significant influence that the characteristics of diverse impact surfaces exert on the distribution and propagation of impact forces.

Finally, a study was carried out on three specific signals, a series of the same drops with the same characteristics, to extrapolate what kind of detailed information can be extracted from a fast Fourier transform analysis and to what extent. The identified eigenmodes from the FFT analysis are repeated in the signals related to the three drops from the same height on the same soil, showing the consistency of the results.

Author Contributions: Conceptualization, A.B.; methodology, C.V. and A.B.; software, C.V.; validation, A.B. and C.V.; formal analysis, A.C. and C.V.; investigation, M.I. and V.F.; resources, L.T. and M.I.; data curation, C.V., A.B. and V.F.; writing—original draft preparation, A.B. and C.V.; writing—review and editing, A.C.; visualization, A.B. and C.V.; supervision, L.T.; project administration, M.B.; funding acquisition, A.C. All authors have read and agreed to the published version of the manuscript.

Funding: PRORA, the (Italian) National Program for Aerospace Research; MELA-SLG (More Electric Aircraft—Smart Landing Gear).

Institutional Review Board Statement: Not applicable.

Informed Consent Statement: Not applicable.

Data Availability Statement: The data are not publicly available due to confidentiality reasons.

Conflicts of Interest: The authors declare no conflicts of interest.

References

1. Howe, D. *Aircraft Loading and Structural Layout*; Professional Engineering Publishing Limited: London, UK; Bury Edmunds, UK, 2004; ISSN 1743-386X.
2. *SAE Aerospace Information Report*; Landing Gear Structural Health Monitoring; AIR6168; SAE International: Warrendale, PA, USA, 2012.
3. Phillips, P.; Diston, D.; Starr, A. Perspectives on the commercial development of landing gear health monitoring systems. *Transp. Res. Part C* **2011**, *19*, 1339–1352. [CrossRef]
4. McGehee, J.R.; Carden, H.D. *Analytical Investigation of Landing Dynamics of a Large Airplane with a Load-Control System in Main Landing Gear*; NASA Technical Paper TP-1555; NASA Langley Research Center: Hampton, VA, USA, 1979.
5. Forrest, C.; Wiser, D. Landing gear structural health monitoring (SHM). *Procedia Struct. Integr.* **2017**, *5*, 1153–1159. [CrossRef]
6. Cowan, S.J.; Cox, R.L.; Slusher, H.W.; Jinadsa, S. Airplane HL Indication System. U.S. Patent No 6,767,075, 27 July 2004.
7. Pritchard, J. Overview of Landing Gear Dynamics. *J. Aircr.* **2001**, *38*, 1. [CrossRef]
8. Tripero, A.; Arevalo, J. Weight on wheel system based on strain gauges. In Proceedings of the ICAS 2012 28th International Congress of the Aeronautical Sciences, Brisbane, Australia, 28 September 2012.
9. Brindisi, A.; Ameduri, S.; Concilio, A.; Ciminello, M.; Iele, M.L.A.; Consales, M.; Cusano, A. A multi-scaled demonstrator for aircraft weight and balance measurements based on FBG sensors: Design rationale and experimental characterization. *Measurement* **2019**, *141*, 113–123. [CrossRef]
10. Pytka, J.; Jozwik, J.; Lyszczyk, T. Embedded Wheel Force Sensor For Aircraft Landing Gear Testing. In Proceedings of the 5th IEEE International Workshop on Metrology for Aerospace, Rome, Italy, 20–22 June 2018.
11. Jebacek, I.; Horak, M. Measuring of a Nose Landing Gear Load during Take-Off and Landing. In Proceedings of the 21st International Conference Engineering Mechanics 2015, Svratka, Czech Republic, 11–14 May 2015; pp. 120–121.
12. Gorinevsky, D.; Gordon, G.A.; Beard, S.; Kumar, A.; Chang, F.-K. Design of Integrated SHM System for Commercial Aircraft Applications. In Proceedings of the 5th International Workshop on Structural Health Monitoring, Stanford, CA, USA, 12–14 September 2005.
13. Baird, B.W. Overload Detection/Health Monitoring Landing Gear Sensor System Proposal. Master's Thesis, Ryerson University, Toronto, ON, Canada, 2008; p. 1157.
14. Schmidt, R.K. Monitoring of aircraft landing gear structure. *J. Aircr.* **2008**, *112*, 275–278. [CrossRef]
15. Schmidt, R.K.; El-Samid, N.A. Structural Deflection and Load Measuring Device. Patent WO 2006/024146 A1, 9 March 2006.
16. Sagi, O.T.; Maynard, D.; Enikov, E. Capacitive Transducer for Condition Based Maintenance After Harsh Landing Events. In Proceedings of the AUTOTESTCON, Baltimore, MD, USA, 12–15 September 2011; pp. 286–291.
17. Robert, J. Finance, Device for Detecting Hard Landings. U.S. Patent US4278219A, 14 July 1981.
18. Davis, C.L.; Hagelin, J.S.; LePere, J.R.; Earle, D.P.; Reuter, R.J.; Akdeniz, A.; Haugse, E.D.; Anderson, D.M. Hard Landing Detection. U.S. Patent US 2008/011450.6 A1, 15 May 2008.
19. Anì, T.; Ting, Z.Z.; Tao, C.J. A technique of landing gear loads calibration with strain gauges. In Proceedings of the 27th International Congress of Aeronautical Sciences, Nice, France, 19–24 September 2010.
20. Schmidt, K.R. System and Method for Determining Aircraft Hard Landing Events from Inertial and Aircraft Reference Frame Data. U.S. Patent US20070008187A1, 15 September 2009.
21. Majumder, M.; Gangopadhyay, T.K.; Chakraborty, A.K.; Dasgupta, K.; Bhattacharya, D.K. Fibre Bragg gratings in structural health monitoring—Present Status and Applications. *Sens. Actuators A Phys.* **2008**, *47*, 150–164. [CrossRef]
22. Lopez-Higuera, L.M.; Cobo, L.R.; Incera, A.Q.; Cobo, A. Fiber Optic sensors in structural health monitoring. *J. Light Technol.* **2011**, *29*, 587–708. [CrossRef]
23. Briancourt, S.; Seymour, T. Aircraft Landing Gear Load Sensor. U.S. Patent US 8,235,326 B2, 7 August 2012.
24. Rodriguez, G.; Sandberg, R.L.; Lalone, B.M.; Marshall, B.R.; Grover, M.; Stevens, G.; Udd, E. High pressure sensing and dynamics using high speed fiber Bragg grating interrogation systems. In *Fiber Optic Sensors and Applications XI*; SPIE: Bellingham, WA, USA, 2014. [CrossRef]
25. Ciro Zuleta, E.; Lupi, C.; Felli, F.; Paris, C.; Vendittozzi, C. Railway overhead contact wire monitoring system by means of FBG sensors. *Fract. Struct. Integr.* **2021**, *15*, 246–258. [CrossRef]
26. Falcetelli, F.; Martini, A.; Di Sante, R.; Troncossi, M. Strain Modal Testing with Fiber Bragg Gratings for Automotive Applications. *Sensors* **2022**, *22*, 946. [CrossRef]
27. Zhao, J.; Dong, W.; Hinds, T.; Li, Y.; Splain, Z.; Zhong, S.; Wang, Q.; Bajaj, N.; To, A.; Ahmed, M.; et al. Embedded Fiber Bragg Grating (FBG) Sensors Fabricated by Ultrasonic Additive Manufacturing for High-Frequency Dynamic Strain Measurements. *IEEE Sens. J.* **2024**, *24*, 2853–2862. [CrossRef]
28. Iodice, M.; Striano, V.; Cappuccino, G.; Palumbo, A.; Cocorullo, G. Fiber Bragg grating sensors-based system for strain measurements. In Proceedings of the 2005 IEEE/LEOS Workshop on Fibres and Optical Passive Components, Palermo, Italy, 22–24 June 2005; pp. 307–312. [CrossRef]

29. Kosheleva, N.A.; Matveenko, V.; Serovaev, G.; Fedorov, A. Numerical analysis of the strain values obtained by FBG embedded in a composite material using assumptions about uniaxial stress state of the optical fiber and capillary on the Bragg grating. *Fract. Struct. Integr.* **2019**, *13*, 177–189. [CrossRef]
30. Brindisi, A.; Vendittozzi, C.; Travascio, L.; Di Palma, L.; Ignarra, M.; Fiorillo, V.; Concilio, A. A Preliminary Assessment of an FBG-Based Hard Landing Monitoring System. *Photonics* **2021**, *8*, 450. [CrossRef]
31. Brindisi, A.; Vendittozzi, C.; Travascio, L.; Di Palma, L.; Belardo, M.; Ignarra, M.; Fiorillo, V.; Concilio, A. Preliminary Assessment of an FBG-Based Landing Gear Weight on Wheel System. *Actuators* **2022**, *11*, 191. [CrossRef]

photonics

Article

Detection of Vibration Signal by Plane Scanning Utilizing Wavefront Sensors

Yun Pan †, Quan Luo †, Yiyou Fan, Haoming Chen, Hongsheng Luo and Jinshan Su *

Key Laboratory of Vibration Signal Capture and Intelligent Processing, School of Electronic Engineering, Yili Normal University, 448 Jiefang Road, Yining 835000, China
* Correspondence: 14222431@ylnu.edu.cn
† These authors contributed equally to this work.

Abstract: Laser remote sensing of seismic waves is extensively utilized in earthquake monitoring and resource exploitation. This article leverages wavefront sensors' high resolution and array detection capabilities to effectively conduct planar scanning of target areas in seismic wave laser remote sensing research, thereby properly acquiring regional vibration data. The error margin is between 1% and 2% relative to the source information.

Keywords: vibration signal; wavefront sensor; plane scanning

1. Introduction

As the core means of resource exploration and environmental protection [1,2], seismic wave exploration technology can accurately identify key geological elements such as lithologic interface, fault structure, and fold shape by analyzing the coupling response characteristics of elastic wave field and geological medium and provide quantitative geophysical parameters such as wave impedance inversion and seismic data quantitative interpretation technology for mineral resource assessment [3–6]. Through travel time tomography and a full waveform inversion algorithm [7], this technology can build a three-dimensional velocity structure model and realize the three-dimensional characterization of oil and gas storage spatial distribution and physical parameters [8]. However, the traditional seismic exploration has significant exploration blind areas in high altitude no man's land and complex structural zones. At present, it is urgent to develop the aerial seismic wave laser remote sensing system to achieve large-scale, high-resolution seismic field aerial monitoring.

In recent years, seismic wave laser remote sensing technology has rapidly developed, using high-resolution [9], non-contact laser measurements to accurately analyze seismic waveforms, identify geological structures [10,11], and evaluate the distribution of minerals, oil, and natural gas [12,13]. Compared to traditional methods, this technology can detect weak vibrations from a distance [14,15], breaking through the limitations of complex terrain and deep exploration. Combining remote sensing [16], GIS [17], and satellite data [18], laser remote sensing has improved exploration efficiency and accuracy, identified resource potential areas at lower costs, and promoted deep geological research and energy development.

Laser remote sensing detection technology has the advantages of non-contact measurement, fewer restrictions, higher efficiency and long-distance detection, and is widely used in object vibration detection. In order to meet the challenges of long-range laser echo signal attenuation and vibration waveform analysis, researchers have explored the use of the He-Ne laser interferometer [19] and Michelson interferometer [20] to measure vibration

Received: 13 February 2025
Revised: 10 April 2025
Accepted: 12 April 2025
Published: 14 April 2025

Citation: Pan, Y.; Luo, Q.; Fan, Y.; Chen, H.; Luo, H.; Su, J. Detection of Vibration Signal by Plane Scanning Utilizing Wavefront Sensors. *Photonics* **2025**, *12*, 377. https://doi.org/10.3390/photonics12040377

signals. However, these technologies depend on the complex optical system and the setting of the reference arm and the measurement arm, which has certain limitations. Therefore, Silvio Bianchi [21] proposed a new method without using an interferometer to sense the surface vibration signal by using the speckle pattern formed after laser irradiating the rough surface. This method can effectively detect the vibration signal within a distance of several hundred meters. There are many methods to detect ground vibration signal based on phase, such as wavefront sensor [22–24], shear interference [25,26], and so on. Jorge ares et al. [27] used the Shack–Hartmann wavefront sensor as a position sensing device to accurately determine the position of a point or moderately extended plane object.

Compared to traditional seismic wave laser remote sensing technology, wavefront sensors have significant advantages in detecting ground vibrations. Firstly, it eliminates the need for reference arms and measurement arms, making the device more compact and easy to deploy. Secondly, by measuring the phase change of the wavefront, precise mechanical vibration information can be obtained. In addition, wavefront sensors have array detection capabilities, which improve detection accuracy and application range. In 2023, researchers constructed a seismic wave laser remote sensing system based on wavefront sensors [28] and successfully detected ground vibration signals [29]. However, existing methods mainly collect single point vibration information, which makes it difficult to process regional vibration signals and has limited detection efficiency. Therefore, based on the research in reference [28], this article proposes a region vibration detection method based on laser plane scanning, utilizing the array detection characteristics of wavefront sensors, to achieve high-precision and high-resolution vibration measurement.

2. Vibration Signal Laser Plane Scanning Detection System and Its Operational Mechanism

When an earthquake occurs, longitudinal waves cause vertical ground vibrations, which, in turn, change the wavefront of the incident light, causing the backscattered light of the laser to carry the characteristic information of seismic waves. In wavefront measurement systems, laser irradiates the target area in a vertical or oblique manner, and the small fluctuations on the measured surface cause light scattering and reflection. Subsequently, the micro lens array converges and images these reflected or scattered light, and extracts vibration information through wavefront reconstruction to achieve accurate measurement of seismic wave characteristics.

This article proposes a vibration signal remote sensing detection system based on planar scanning that consists of a transmitting end and a receiving end, as shown in Figure 1. The transmitting end includes a laser, collimator 1, and telescope 2. The system uses a gradient refractive index fiber collimator to emit a laser beam with a divergence angle of 0.25°, and its oxide glass material collimating lens achieves collimated output with a beam quality factor $M^2 < 1.1$ through a 0.15 numerical aperture. The laser emits a continuous laser with a wavelength of 635 nm and is collimated by a tail fiber type fiber collimator 1 (equipped with a gradient refractive index GRIN lens) to ensure stable illumination of the target area 3. The receiving end consists of telescope 5, filter 4, and wavefront sensor 6. This is a bandpass filter that can filter out 97% of stray light within the wavelength range of 624–750 nm. The telescope is responsible for collecting reflected and scattered light from target area 3, while filter 4 is used to remove stray light from the environment to improve the utilization of echo signals. Subsequently, the light passing through the filter is incident on the wavefront sensor 6, which is composed of a micro lens array 7 and can accurately detect wavefront distortion in the optical system. The micro lens array divides the incident light spot into multiple sub apertures and focuses the imaging onto the light spot detection window 8 in the array. The system extracts vibration information by measuring the offset of the light spot and uses laser remote sensing technology to detect micro vibrations in the target area. We study the relationship between the phase information of the target vibration and the offset of the received light spot on the CMOS sensor. When the target object 3 vibrates, the displacement ΔS of the center of mass of the light spot recorded by the wavefront sensor carries vibration information ΔZ. According to reference [28], the relationship between the two satisfies the equation $\Delta Z = \mu\Delta S$, where μ is the proportionality coefficient. Figure 2 shows the changes in wavefront during the vibration process. In Figure 2, we simplify the analysis by assuming the ground is flat when not vibrating. This assumption is made to illustrate the basic relationship between the transmitted and reflected wavefronts. In reality, uneven ground may combine with other non-ideal factors to influence the wavefront's shape.

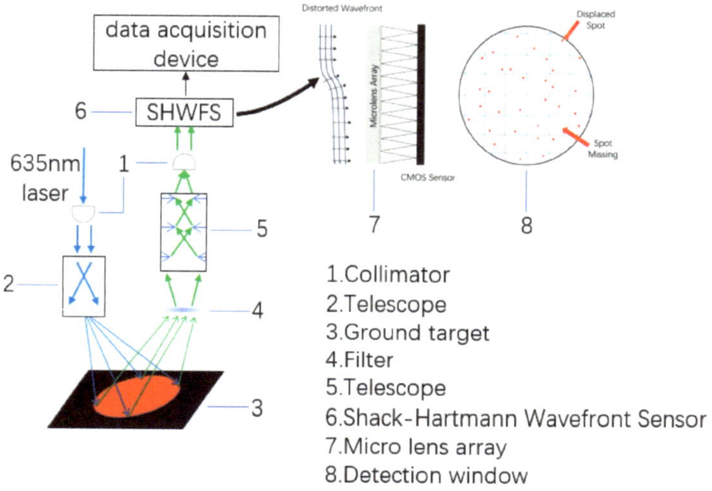

Figure 1. Vibration Signal Plane Scanning Detection System.

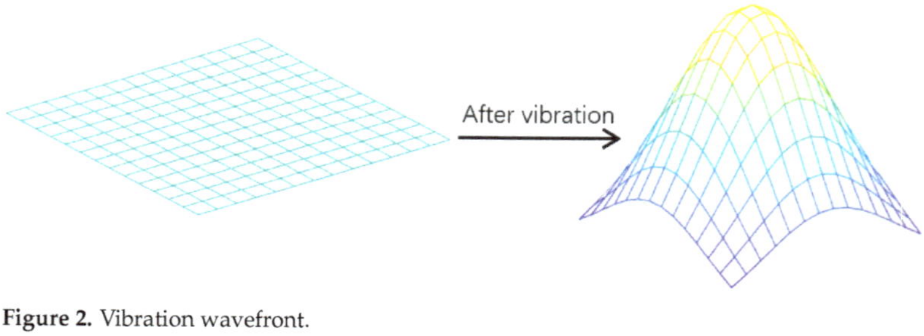

Figure 2. Vibration wavefront.

In an ideal situation, when a parallel wavefront is irradiated onto a micro lens array, each micro lens will focus the incident light, forming a uniformly distributed light spot, and the centroid position of these light spots should be located at the predetermined center position. Therefore, the distribution of light spots formed by the micro lens array is regular and symmetrical, with each light spot representing a local area of the wavefront, reflecting the degree of wavefront tilt in that area. However, when the wavefront undergoes distortion, the incident light no longer maintains a parallel state, and local areas of the wavefront become tilted or deformed. At this point, the micro lens array will convert these wavefront changes into corresponding spot offsets. The difference between the centroid shift of the light spot caused by distorted wavefront and the ideal position can be used to quantify the degree of wavefront distortion,as shown in Figure 3.

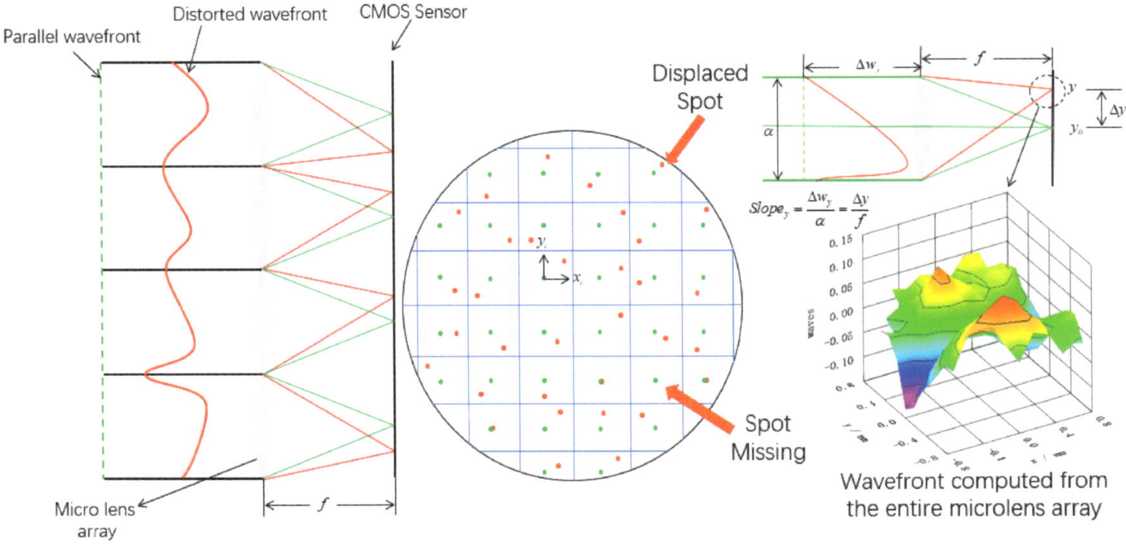

Figure 3. Wavefront aberration of the microlens array in the wavefront sensor. The green dot represents the center of mass at the origin when a parallel wavefront enters the wavefront sensor. The red dot represents a shift in the center of mass when a distorted wavefront is detected. In the case of a plane wavefront, all spot centroids align at the center of the microlens array. However, when a warped wavefront enters the sensor, the centroids of the spots may shift away from the center or even disappear entirely.

3. Theoretical Analysis of Vibration Signal Laser Plane Scanning Detection System

3.1. Wavefront Reconstruction Using Zonal Methods

This article adopts the region method proposed in references [30–34] for wavefront reconstruction. We used the wavefront slope S corresponding to the vibration wavefront shown in Figure 2 for reconstruction and successfully obtained the corresponding wavefront distribution, as shown in Figure 4.

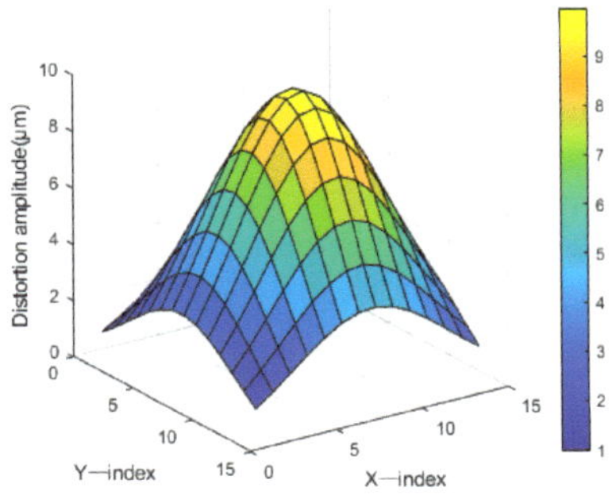

Figure 4. Wavefront rebuilt using the zonal approach.

3.2. Wavefront Phase

To calculate the phase from the reconstructed wavefront, the following formula can be used [35]:

$$\Delta\varphi = \frac{2\pi}{\lambda}|\Delta W| \tag{1}$$

Here, $\varphi(x, y)$ is the phase, $W(x, y)$; and λ is the wavelength. Therefore, the phase change of the wavefront was obtained using the wavefront in Figure 4, as shown in Figure 5. Wavefront sensors can capture wavefront distortion caused by vibration and derive amplitude changes by measuring changes in wavefront phase. When using Shack–Hartmann sensors, the micro lens array records the slope changes in various directions and then calculates the phase distortion of the wavefront. By analyzing the phase change of the wavefront sensor during vibration, the relationship between amplitude and phase can be established, and the formula is as follows:

$$\Delta\varphi = k\Delta d + c \tag{2}$$

Here, $\Delta\varphi$ represents phase change, and Δd is the amplitude variation. In the absence of vibration, the laser echo received by the wavefront sensor will cause a slight shift in the centroid of the light spot on the micro lens array, resulting in a slight change in the wavefront slope. The constant term c in the formula reflects this shift.

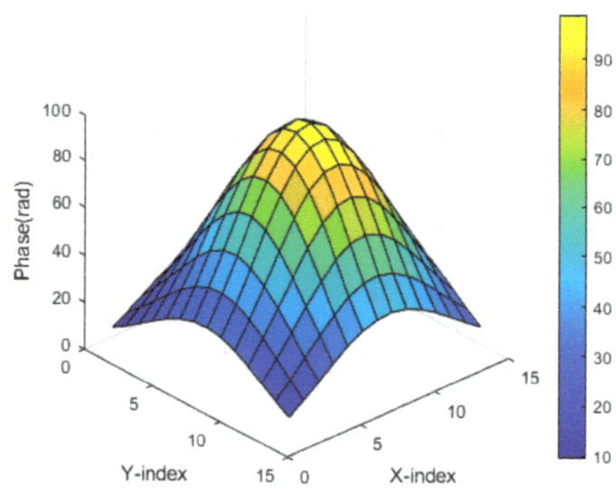

Figure 5. Wavefront phase.

This article uses the region method to reconstruct the wavefront of a certain region, obtaining a global wavefront. However, it is impossible to obtain each microlens' wavefront contribution directly. To solve this problem, we separate the overall wavefront and remove the PSF connections between each microlens, thereby obtaining the wavefront phase changes of each microlens during vibration. Through this method, the vibration situation of each subregion can be visually observed, as shown in Figure 6. This method effectively separates and analyzes the vibration characteristics of each subregion, revealing the spatial distribution and variation patterns of vibration.

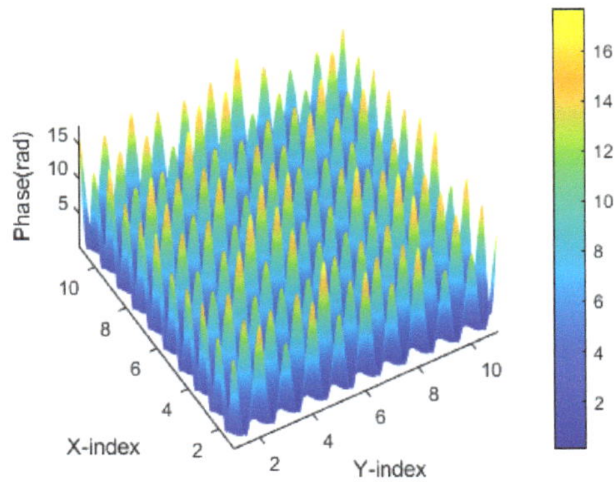

Figure 6. Wavefront rebuilt using the finite element approach.

3.3. Laser Attenuation

A laser has geometric attenuation and atmospheric attenuation in atmospheric transmission [36,37]. As the propagation distance of the laser beam increases, its light intensity gradually decreases due to the expansion of the beam. Geometric attenuation describes the decrease in laser intensity during propagation due to the expansion of the beam. At the propagation distance z, the beam's diameter increases with distance, resulting in a decrease in light intensity per unit area. Assuming the divergence angle of the beam is θ, the diameter $d(z)$ and area $A(z)$ of the light spot at the propagation distance z is

$$w(\theta) = ztan(\theta) \tag{3}$$

$$d(z) = 2ztan(\theta) \tag{4}$$

$$A(z) = \pi[\frac{d(z)}{2}]^2 = \pi z^2 tan^2(\theta) \tag{5}$$

As the propagation distance of the laser beam increases, its light intensity gradually decreases due to the expansion of the beam. Assuming the initial light intensity of the laser, since the light intensity is inversely proportional to the area, the geometric attenuation can be expressed as

$$I_{geo}(z) = k\frac{I_0}{w^2(z)}exp\left(-\frac{2r^2}{w^2(z)}\right) \tag{6}$$

Among them, k is a constant, and $w(z)$ is the beam's radius. When a laser passes through the atmosphere, it is also subject to scattering and absorption by the atmosphere, resulting in a decrease in light intensity. Assuming the absorption and scattering coefficients of the atmosphere are α, the effect of atmospheric attenuation can be described by the following formula:

$$I_{atm}(z) = I_0 e^{-\alpha z} \tag{7}$$

When there are both geometric and atmospheric attenuation, light intensity attenuation is the combined effect of both. Assuming the initial light intensity, we can combine the attenuation models of both to obtain the overall attenuation formula:

$$I(z) = I_{geo}(z) * I_{atm}(z) \tag{8}$$

Substitute the above two formulas into

$$I(z) = \left[I_0 e^{-\alpha z}\right] * \left[k\frac{I_0}{w^2(z)}exp\left(-\frac{2r^2}{w^2(z)}\right)\right] \tag{9}$$

Therefore, the overall light intensity attenuation model is

$$I(z) = k\frac{I_0^2 e^{-\alpha z}}{w^2(z)} exp\left(-\frac{2r^2}{w^2(z)}\right) \quad (10)$$

Here, I_0 is the initial light intensity, z is the propagation distance, θ is the divergence angle of the laser, and α is the atmospheric attenuation coefficient.

4. Experimental and Result Analysis of the Vibration Signal Laser Plane Scanning Detection System

We use a laser with a wavelength of 635 nm and an output power of 120 mW at a distance of 5 m. In order to reduce the interference of ambient light and increase the reception effect of the laser echo signal of the wavefront sensor, we choose to carry out the experiment at night. The temperature of the laboratory is 14 °C, the humidity is 23%, and the change in the surrounding environment is very small. Figure 7 shows the sensor used for research, and Table 1 shows the detailed parameters of this model of sensor. The sensor has high sampling rate and automatic shutter control function, which can adapt to different optical power inputs, and its sensitivity is significantly affected by wavelength, making it suitable for detecting ground vibration signals. In the single point detection experiment, we used a controllable vibration table to simulate longitudinal wave characteristics by adjusting frequency and amplitude, Figure 8a shows a diagram of a controllable vibration table; In the surface scanning experiment, a vibration motor is used as the vibration source, and the interference of the vibration source is isolated by a soft white cloth. Figure 8b shows a diagram of vibration motor. The laser irradiates the surface of the vibration source, and the wavefront sensor receives the reflected echo light and extracts vibration information. To monitor the wavefront changes of the wavefront sensor in real time, we use Wavefront Sensor software to record the wavefront dynamics and use LabView software to calculate and display the changes in individual microlenses in order to deeply analyze the response of the wavefront when excited by the vibration source. The specific program flow is detailed in Appendix A. The system configuration and data recording method ensure accurate monitoring and analysis of small wavefront changes caused by vibration sources. Figure 9a shows the deviation in wavefront slope in the horizontal and vertical directions when the vibration source vibrates. When longitudinal waves are detected, the deviation in the y direction is larger, while the deviation in the x direction is smaller. Meanwhile, Figure 9b shows the vibration signal recorded using a vibrometer.

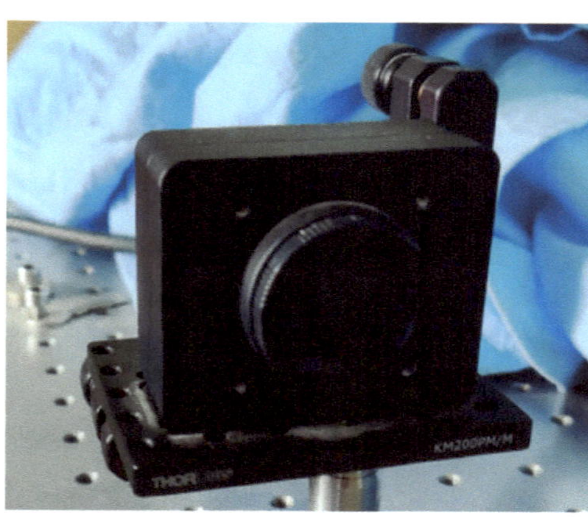

Figure 7. Shack–Hartmann wavefront sensor.

Table 1. Detailed parameters of the Shack–Hartmann wavefront sensor.

Parameter Name	Parameter Data
Wavelength Range/nm	300–1100
Lenslet Pitch/μm	150
Lens Diameter/μm	146
Wavefront Accuracy	$\lambda/30$ rms @ 633 nm
Wavefront Sensitivity	$\lambda/100$ rms @ 633 nm
Wavefront Dynamic Range	>100λ@ 633 nm
Pixel size/μm	5 × 5
Framerate/(f·s^{-1})	23–880

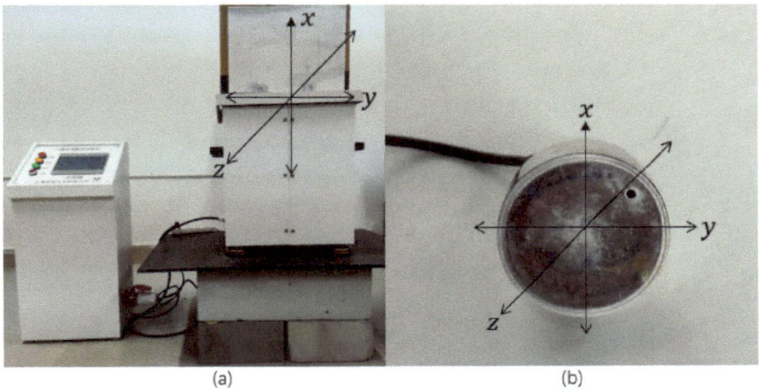

Figure 8. (**a**) Controllable vibration table; (**b**) Vibration motor.

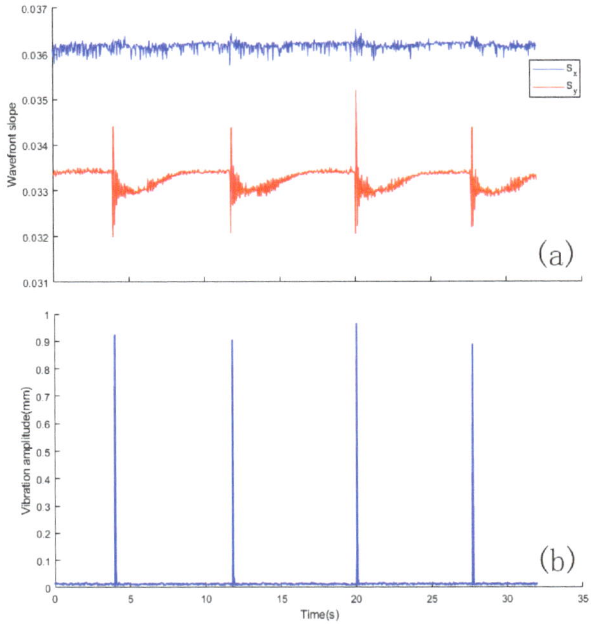

Figure 9. When the vibration source is excited, the wavefront slope changes. (**a**) When vibrating, the wavefront slope of the microlens is in the horizontal direction, denoted as S_x, and the wavefront slope in the vertical direction is denoted as S_y; (**b**) The vibration signal is detected by the vibration meter.

When testing the light spot received by a single microlens, the laser echo signal is focused onto a certain microlens in the microlens array through a telescope. In the beam view mode of the wavefront sensor, in addition to the target microlens, several adjacent microlenses around it will also display weak spot signals, as illustrated in Figure 10. This indicates that the relationship between laser echo signals and microlens arrays is not simply "one-to-one" but presents more complex "one to many" characteristics.

We tested the attenuation of laser power, and this effect usually shows a certain pattern. We used a 635 nm continuous laser to irradiate the target area separately and measured the laser power using an optical power meter, as shown in Table 2. The test results indicate that the laser power shows an exponential decay trend with increasing irradiation distance. Table 2 is only applicable for small-scale laboratory testing. When the receiving end is a wavefront sensor, the attenuation of laser power will significantly affect the accuracy of surface scanning detection, which is reflected in the reduction in signal amplitude, which leads to the reduction in the contrast and signal-to-noise ratio of the wavefront speckle pattern received by the detector, resulting in the inaccurate estimation of speckle location error and wavefront gradient and further causing the accumulation of wavefront reconstruction error; especially when the sub aperture signal is too weak or missing, local wavefront information will be missing or distorted. In addition, the nonuniformity of energy attenuation in spatial distribution will also destroy the spatial response consistency of the system, which will reduce the geometric reconstruction accuracy of the whole surface scanning, especially in the long distance, edge area or weak reflection surface.

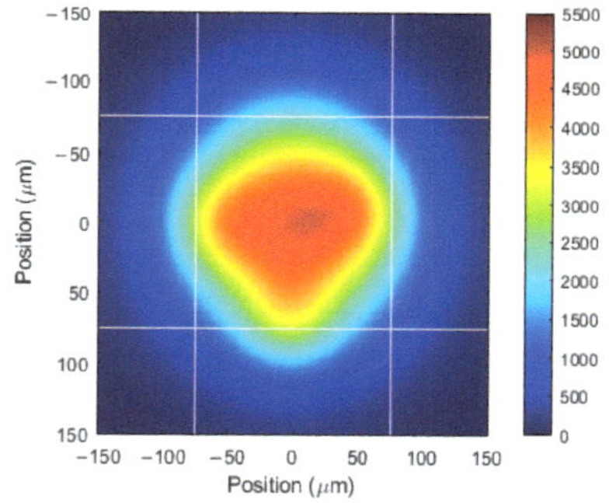

Figure 10. A single microlens in the microlens array receives the light spot, and the microlens with coordinates of 6 × 6 and adjacent microlenses also receive the light spot signal.

Table 2. Laser reflection power of different diameter light spots.

Distance Between Laser and Target (cm)	Laser Output Power (μw)	Spot Diameter (cm)	Spot Center Optical Power (nw)	Laser Power at the Edge of the Spot (nw)	Receiving End Power (nw)
183	464	45	752.8	102.8	129.92
	326	62	280.53	38.7	91.28
	10.5	114	2.7	0.4	0.194
314	1200	45	662	334	336.01
	446	62	128	63	64.88
	274	114	23	11	11.72

We mark the 11 × 11 microlenses of the wavefront sensor as 1 × 1 coordinates from the upper left corner, as shown in Figure 11. The laser forms a spot surface on the shaking table, and the echo signal is focused on the microlens with the coordinates of 6 × 6 of the wavefront sensor through the telescope. The wavefront slope and wavefront images of the microlens with coordinates of 6 × 6 and adjacent microlenses were obtained by using LabVIEW software, and the corresponding wavefront images were reconstructed by using the region method. As shown in Figure 12, with the increase in vibration amplitude, the degree of wavefront distortion increases accordingly, and the microlens around the microlens with coordinates of 6 × 6 is also slightly disturbed. Changes in amplitude and wavefront phase are shown in Appendix B. When there are vibration signals with different amplitudes in a certain area, the laser echo signal will carry the vibration information of the area. By receiving these laser echo signals, the wavefront sensor obtains and analyzes the corresponding vibration information. By analyzing the wavefront phase changes under different amplitudes, the relationship between amplitude and phase can be revealed, as shown in Figure 13. The wavefront phase shifts with the change in amplitude, which makes it possible to deduce the variation law of amplitude by measuring the phase change, as shown in the following formula:

$$\Delta\varphi = k\Delta d + c, k = 4.8485(\text{mm/rad}), c = 11.9836 \quad (11)$$

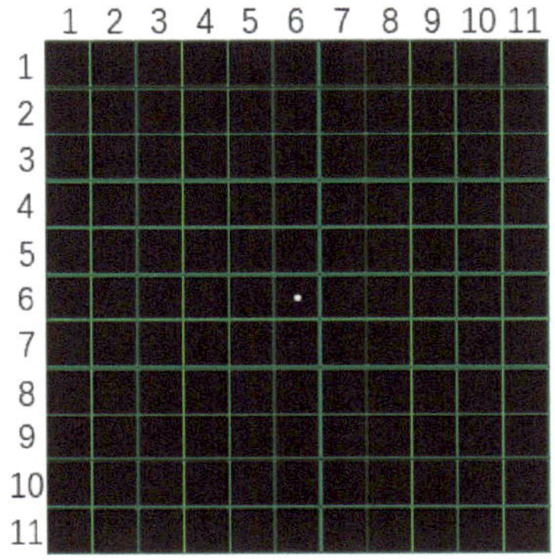

Figure 11. Vibration amplitude and phase variation.

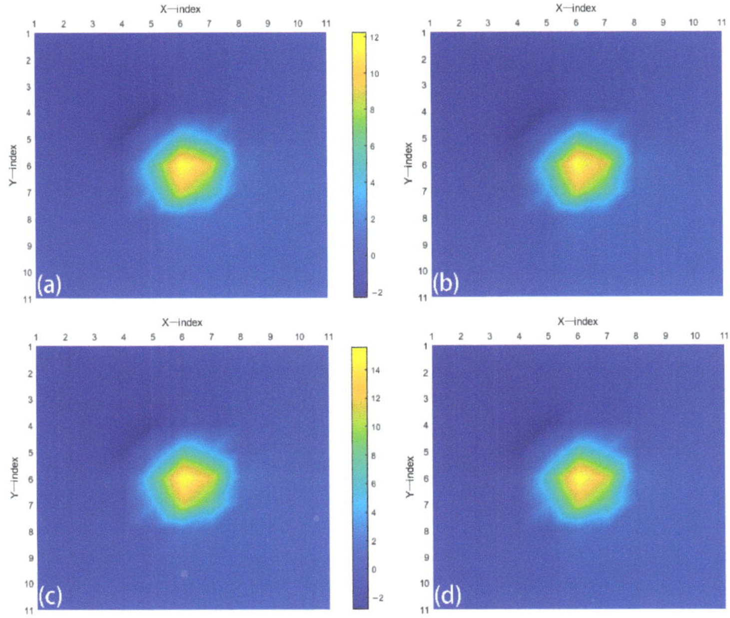

Figure 12. Wavefront of microlens with coordinates of 6 × 6 at different amplitudes. (**a**) Amplitude: 0.06 mm, Phase: 12.26 rad; (**b**) Amplitude: 0.31 mm, Phase: 13.49 rad; (**c**) Amplitude: 0.75 m, 15.60 rad; (**d**) Amplitude: 1.18 mm, phase: 17.70 rad. The error is ±0.05 rad.

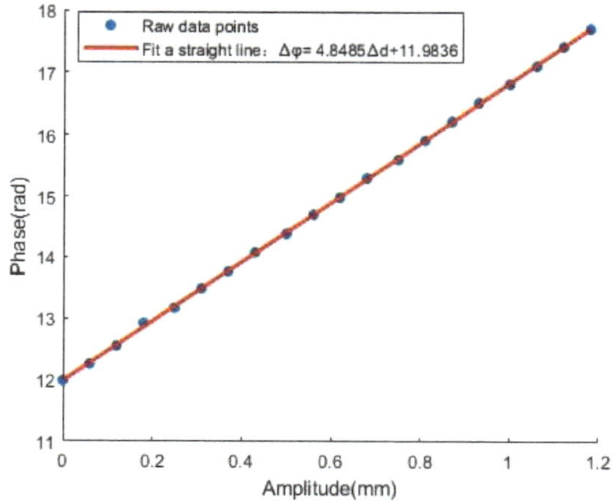

Figure 13. Vibration amplitude and phase variation.

Based on the above experiment, we divided a 1.2 m × 1.2 m target object into 11 × 11 sub regions, starting from the upper left corner, with the first region marked as 1 × 1 coordinate, and established an array in order. We randomly arranged multiple vibration sources within the 11 × 11 sub regions of the target object. We perform planar scanning on each subregion to obtain the corresponding changes in the wavefront slope of the microlens, as seen in Figure 14. A one meter diameter light spot is formed on the target area, and the laser echo signal is focused onto a microlens array through a telescope. Thirty-two microlenses in the microlens array receive the signal. In the target area, there are 4 sub areas with very small spot signals that cannot cover most of their area. Therefore, there are 4 microlenses in the microlens array that have no signal. We extract the changes in wavefront slope during the vibration of 32 microlenses and reconstruct the corresponding wavefront phase changes for each subregion. We calculate the corresponding vibration amplitude based on the wavefront phase changes in each subregion.Therefore, only one system is needed to obtain vibration information of multiple sub regions within the target area. Figure 15 shows the physical device of the object and the 1 m light spot formed in the target area. The field of view of the telescope is 4.1°–39.6°, and the distance between the telescope and the spot is 5 m. Therefore, the 100 cm spot size is fully within the field of view of the telescope. During the detection process, we first use pan-tilt 1 to project the collimated laser onto the target area. Then, we use a clamp to fix telescope 1 and manually adjust its objective lens and eyepiece to adjust the spot area on the target area. Then, we use pan-tilt 2 to adjust the receiving angle of telescope 2 and further adjust the objective lens and eyepiece of telescope 2 to ensure the best receiving effect. Subsequently, through the high-speed sampling interface of the wavefront sensor, we manually adjust the position of the wavefront sensor so that the laser echo signal can be accurately focused on the corresponding lens in the wavefront sensor microlens array, thereby achieving vibration signal point scanning detection. The spatial resolution of the system in Figure 15a is 146 μm.

We divide the 1.2 m × 1.2 m target area into 11 × 11 sub areas; divide the target area into four areas, each including 36 sub areas of the target area; and place vibration sources within the 36 sub areas. Using a telescope, a laser beam is directed onto the target object to form a spot with a diameter of 60 cm, while detecting 36 sub regions within the target area. However, due to the shape of the spot, only 32 subregions can actually be detected. The laser echo signal is focused by a telescope and forms a spot surface on the microlens array of the wavefront sensor. In the high-speed sampling mode of the wavefront sensor, the 32 microlenses of the microlens array received laser echo signals within the target area, as shown in Figure 16. Subsequently, the LabView software was used to accurately locate the microlenses in the array that correspond to the target subregion, thereby obtaining the wavefront slope changes in the corresponding microlenses and calculating the corresponding wavefront phase. As shown in Figure 16, due to the inconsistent vibration amplitudes in the 32 sub regions of the target area, the wavefront slope of each sub region in the 32 sub regions is different.

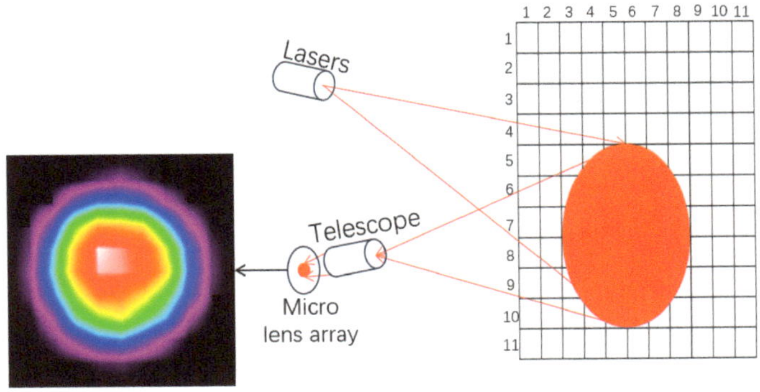

Figure 14. Illuminate the designated area with a laser to create a light spot surface over several sub-regions, subsequently capturing the light spot surface signal via a telescope and concentrating it onto a microlens array.

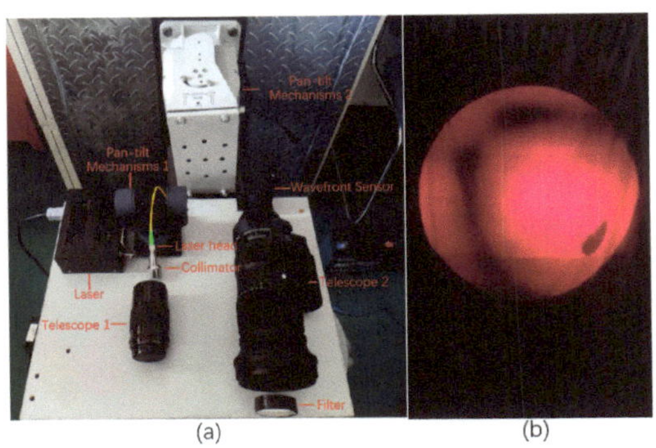

Figure 15. (**a**) Physical image of plane scanning detection system; (**b**) A light spot with a diameter of 1 m formed in the target area.

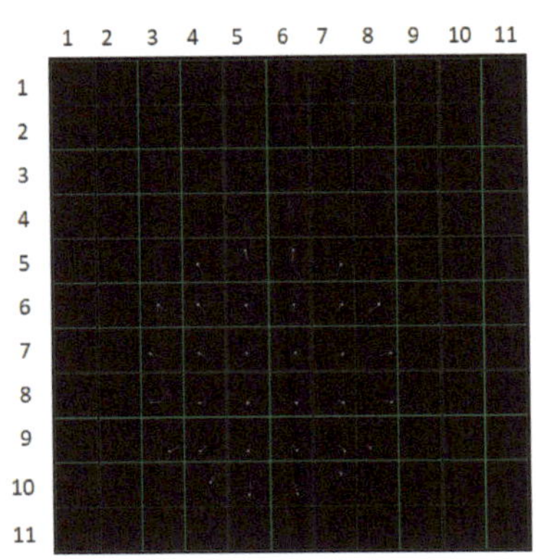

Figure 16. Multiple microlenses receiving spot signals.

We sample a region multiple times. We selected 8 microlens spot centroid changes from Figure 16, as shown in Figure 17. Figure 17 shows the centroid displacement of the light spot in eight sub regions of the target area. The vibration amplitudes of different subregions in the target area are inconsistent, and the vibration amplitudes displayed on the microlens array of the wavefront sensor are also inconsistent. Due to the use of a vibration motor, the orientation of the echo laser received by the wavefront sensor will undergo a slight change, and the center of mass of the illuminated spot will continue to move. As shown in Table 3, the offset of the center of mass of each microlens in the microlens array is different due to the inconsistent amplitude of the target subregion.

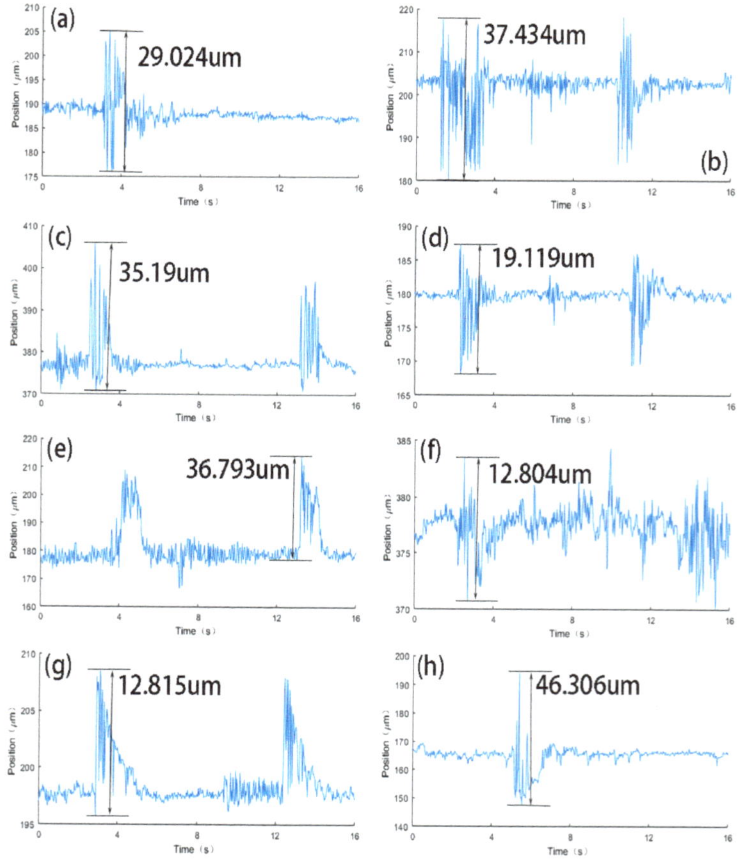

Figure 17. The spot centroid offset of eight sub regions in the target region. (**a**) The focal spot centroid offset of the microlens with coordinates of (6,5) is 29.024 μm; (**b**) The focal spot centroid offset of the microlens with coordinates of (6,8) is 37.434 μm; (**c**) The focal spot centroid offset of the microlens with coordinates of (7,4) is 35.19 μm; (**d**) The focal spot centroid offset of the microlens with coordinates of (7,7) is 19.119 μm; (**e**) The focal spot centroid offset of the microlens with coordinates of (8,3) is 36.793 μm; (**f**) The focal spot centroid offset of the microlens with coordinates of (8,6) is 12.804 μm; (**g**) The focal spot centroid offset of the microlens with coordinates of (9,5) is 12.815 μm; (**h**) The focal spot centroid offset of the microlens with coordinates of (10,7) is 46.306 μm.

Table 3. Spot center-of-mass offsets for eight subregions.

Serial Number	Coordinate	Center of Mass Offset (μm)
1	(6,5)	29.024
2	(6,8)	37.434
3	(7,4)	35.19
4	(7,7)	19.119
5	(8,3)	36.793
6	(8,6)	12.804
7	(9,5)	12.815
8	(10,7)	46.306

According to the detection in the vicinity of Figure 16, the target area of 1.2 m × 1.2 m is partitioned into 11 × 11 sub-regions, placing all sub areas in the target area with vibration sources. The whole target area is divided into a 4 × 4 area, and each area is detected by plane scanning. Utilizing a telescope, the laser beam creates a spot with a diameter of 60 cm on the target area, concurrently identifying the 32 sub-regions within that area. The telescope concentrates the laser echo signal, creating a light spot on the microlens array of the wavefront sensor. Thereafter, the microlens associated with the designated sub-region in the microlens array is accurately identified using LabView software, facilitating the acquisition of the wavefront slope variation of the respective microlens and the computation of the related wavefront phase, as illustrated in Figure 18. Figure 18 illustrates that variations in the magnitude of the vibration source correspondingly alter the wavefront phase. This signifies that the vibration signal laser planar scanning detection system can efficiently identify the vibrational states of several small locations inside the designated zone. The calculated vibration amplitude via phase changes exhibits a relative inaccuracy of 1% to 2% compared to the amplitude recorded by the vibration meter, as illustrated in Figure 19.

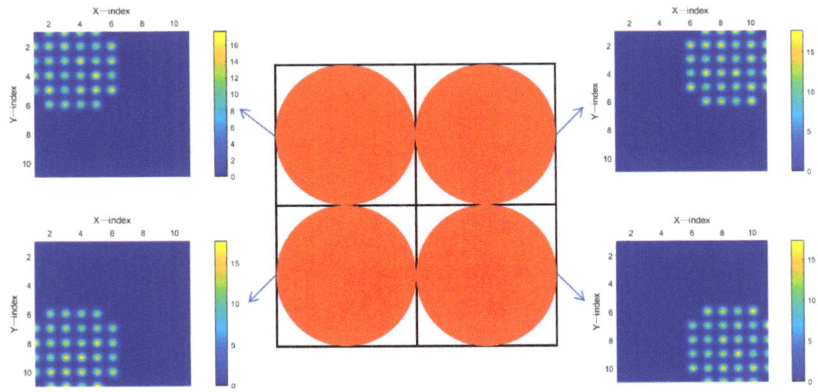

Figure 18. Divide the target area into a 4 × 4 region and perform planar scanning detection.

In order to improve scanning efficiency, we further expanded the spot area. Similarly, a circular spot with a diameter of 1 m can only cover 69 sub areas of the target area. The 1.2 m × 1.2 m target area is divided into 11 × 11 sub areas, and 69 sub areas are selected to place vibration sources. The laser beam uses a telescope to form a spot with a diameter of 1 m on the target area and covers 69 sub areas in the target area. The laser echo signal is focused on the microlens array through the telescope, and 69 microlenses on the microlens array receive the signal, as shown in Figure 20. We use LabVIEW software to obtain the wavefront slope of 69 microlenses and calculate the wavefront phase change of each microlens so as to obtain the amplitude of the corresponding sub region in the target region. At present, the detection of a 1.2 m × 1.2 m target area is only achieved in a relatively stable laboratory environment. Due to the lack of high-power lasers and telescopes with large viewing angles, it is temporarily impossible to achieve large-scale scanning.

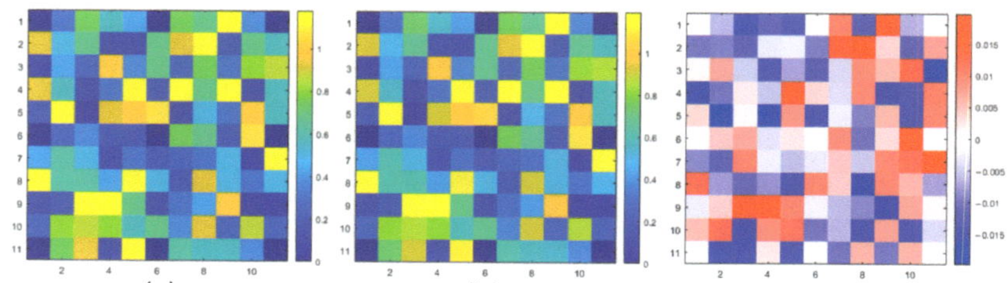

Figure 19. Error in one area. (**a**) Theoretical value; (**b**) Measured value; (**c**) Error.

The telescope concentrates the laser echo signal, creating a light spot on the microlens array of the wavefront sensor. LabView software is used to precisely identify the microlens associated with the designated sub-area and acquire the phase variation of the vibrating wavefront through the alterations in wavefront slope induced by vibration. This enables the prediction of amplitude values based on the phase change. Figure 21 illustrates the outcomes of identifying a region with three vibration sources utilizing a 69 microlens, exhibiting a relative inaccuracy of 1% to 2%.

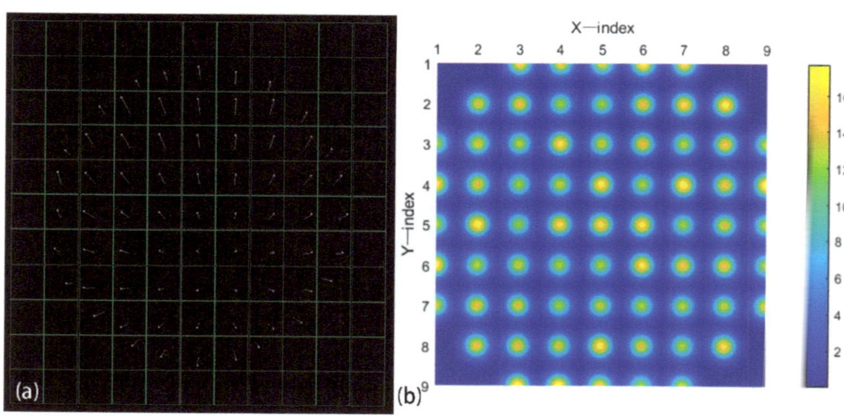

Figure 20. (**a**) Multiple microlenses receiving spot signals; (**b**) The wavefront phase variation of 81 microlenses.

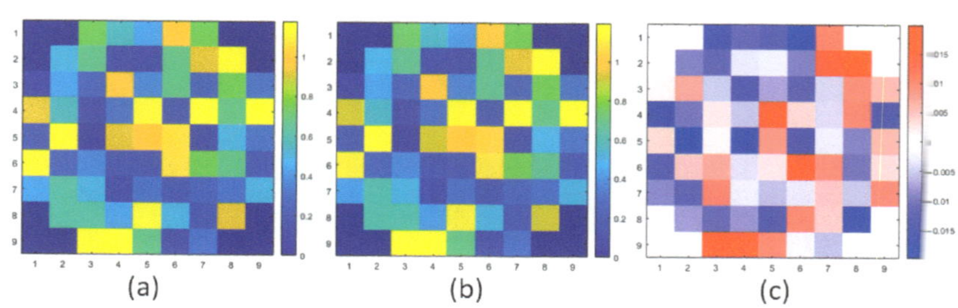

Figure 21. Error of scanning target area. (**a**) Theoretical value; (**b**) Measured value; (**c**) Error.

This research employs the zonal approach for wavefront reconstruction of a specified region, deriving the wavefront distortion associated with the microlenses of each sub-region and facilitating an intuitive analysis of the vibrational conditions of each sub-region, as illustrated in Figure 21. This method enables the effective separation and analysis of the vibration characteristics of each sub-region, elucidating the spatial distribution and variation patterns in the vibrations. In the experiment, we initially process the wavefront data obtained from the wavefront sensor, employing the zonal approach to divide them into several sub-regions. By reconstructing the wavefront of each sub-region, we can achieve higher-resolution vibration images, which are essential for recognizing local vibration sources and comprehending vibration transmission mechanisms.

This article is based on the characteristics of high resolution and sensitivity of wavefront sensors to detect low-frequency vibration signals, as shown in Figure 22. As shown in Figure 22, as the frequency of the controllable vibration table gradually increases, the frequency of the detected signal also increases.

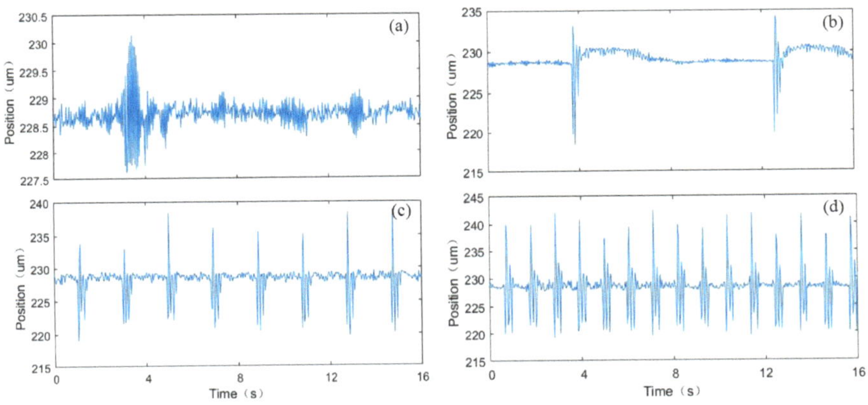

Figure 22. Low frequency vibration signal. (**a**) 0.01 Hz; (**b**) 0.1 Hz; (**c**) 0.5 Hz; (**d**) 1 Hz.

We conducted tests on the maximum and minimum vibration amplitudes using a controllable vibration table. The vibration frequency of the vibration table was set to 0.1 Hz, and the wavefront sensor was set to high-speed sampling mode (sampling rate: 700–880 fps). Therefore, this paper achieved the detection of large amplitude vibration with a vibration amplitude of 5.94 mm and small amplitude vibration with a vibration amplitude of 0.06 mm, as shown in Figure 23. However, due to the insufficient sampling rate of the wavefront sensor, it is unable to detect high-frequency vibration signals.

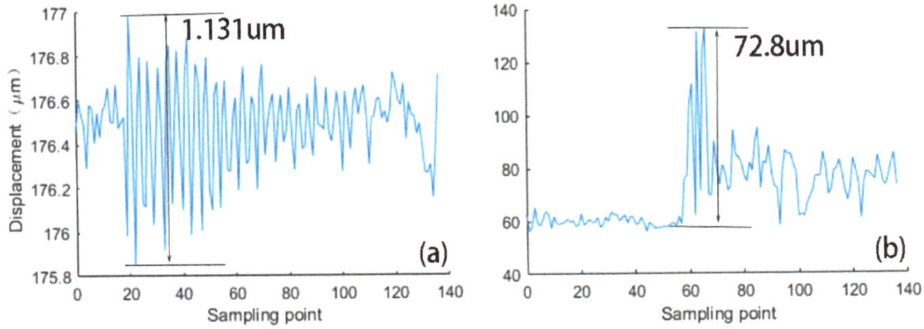

Figure 23. Maximum and minimum amplitudes and corresponding center-of-mass offsets of the wave front. (**a**) Minimum amplitude 0.06 mm; (**b**) maximum amplitude 5.94 mm.

The detection accuracy has been markedly enhanced relative to the conventional laser triangulation approach by utilizing minor wavefront alterations to identify a specific area's vibrational signals. The error in detection by wavefront distortion ranges from 1.06% to 2%, but the error in detection utilizing the laser triangulation method spans from 1.54% to 3.5%, as illustrated in Table 4. Wavefront sensors can accurately detect phase alterations in light waves, enabling more sensitive assessments of minute vibrations. In contrast to the laser triangulation method, which depends on variations in the light spot's position, wavefront sensors can efficiently mitigate environmental noise and external disturbances by examining the comprehensive phase information of the light wave, thereby yielding more stable and dependable vibration monitoring outcomes. Moreover, wavefront sensors possess benefits in spatial resolution, facilitating high-density sampling across extensive areas, hence, rendering vibration analysis more thorough and intricate.

Although the seismic wave laser remote sensing surface scanning method based on the wavefront sensor shows good application prospects in laboratory conditions, it still faces many challenges in the actual environment. Atmospheric turbulence will cause the phase distortion of the laser wavefront, interfere with the measurement accuracy, and its influence fluctuates significantly with the change in altitude, time, and meteorological conditions. In addition, aerosols such as dust and haze will aggravate the scattering and attenuation of laser signals. Uneven distribution of precipitation and humidity will also lead to signal absorption, defocusing or deviation, and even complete signal masking in severe cases, thus affecting the stability and reliability of measurement results. In terms of environmental stability, field deployment faces multiple challenges. Extensive temperature changes lead to laser wavelength drift, optical system thermal expansion mismatch, and detector performance fluctuations. Especially in the presence of a significant thermal gradient, the spatial refractive index change will further distort the wavefront shape. In addition, if the common environmental vibration on site is not effectively isolated, it is easy to introduce artifacts, affecting the measurement accuracy and consistency. Surface complexity is also a major obstacle in practical application. Different from the idealized reflector in the laboratory, the reflectivity of natural surface materials is significantly different. The low reflectivity area easily cause weak echo signals, and the high reflectivity surface may cause detector saturation or artifacts. The surface roughness will also weaken the coherence of the reflected wavefront and reduce the reconstruction accuracy. In addition, surface dynamic changes such as vegetation shaking and water flow disturbance may introduce interference similar to the characteristics of seismic signals, which increases the difficulty of effective identification and classification.

Table 4. Detection results using wavefront and laser triangulation.

Serial Number	Measurement Value of Vibration Meter (mm)	Wavefront (mm)	Laser Triangulation (mm)
1	0.12	0.1182	0.1158
2	0.25	0.2542	0.2449
3	0.37	0.3753	0.3651
4	0.5	0.5097	0.4911
5	0.62	0.6076	0.6358
6	0.81	0.8186	0.8255
7	0.93	0.9168	0.9474
8	1	1.0158	0.9833
9	1.06	1.0737	1.0413
10	1.18	1.1659	1.1982

5. Conclusions

This paper investigates the seismic wave laser remote sensing detection technology based on the Shack–Hartmann wavefront sensor and proposes a novel method for detecting regional vibration signals through planar scanning. This method fully utilizes the short wavelength, high sensitivity, and high resolution characteristics of lasers, combined with the compactness, accuracy, and sensitivity of wavefront sensors, to explore the detection of large-scale ground vibration signals. Unlike traditional single point spot detection, this article extends the detection range and enables wide area vibration signal acquisition. By analyzing the working principle and experimental data of the wavefront sensor, the differences in the vibration of the light spot in different regions were verified, and the vibration wavefront data of multiple regions were successfully obtained. Compared with traditional methods, this system significantly expands the detection range and can monitor and record small ground vibrations in real time, demonstrating its potential for application in seismic wave detection and exploration. However, the laser power attenuation in the experiment limited the long-range detection capability, and in the future, it is necessary to consider using high-power lasers to improve the detection distance.

Author Contributions: Conceptualization, Y.P. and Q.L.; methodology, Q.L. and J.S.; software, H.L.; validation, Q.L. and Y.P.; formal analysis, Q.L., Y.F., and J.S.; investigation, J.S.; resources, J.S.; data curation, Q.L., and J.S.; writing—original draft preparation, Q.L. and H.L.; writing—review and editing, J.S.; visualization, Q.L. and H.C.; supervision, J.S.; project administration, J.S.; funding acquisition, J.S. All authors have read and agreed to the published version of the manuscript.

Funding: The national natural science foundation of China (Grant No. 62165015).

Institutional Review Board Statement: The study did not require ethical approval.

Informed Consent Statement: The study did not require ethical approval.

Data Availability Statement: The data that support the findings of this study are available from the corresponding author upon reasonable request.

Conflicts of Interest: The authors declare no conflicts of interest.

Appendix A

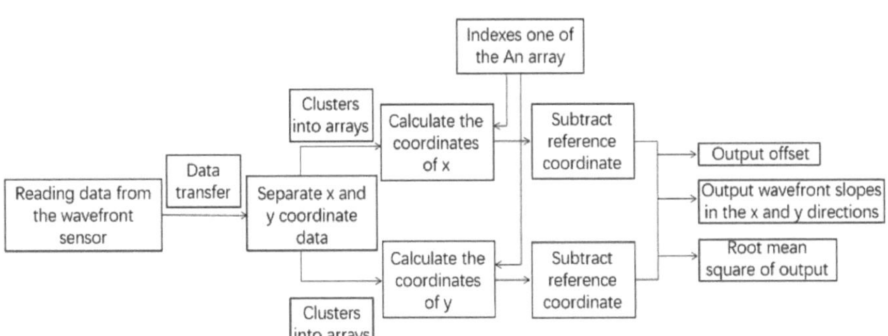

Figure A1. Calculation process.

Appendix B

According to Rodier's description, the microlens array can be regarded as a phase grating, while shwfs can be regarded as a phase grating interferometer, and its phase grating function $G(x, y)$ is a biperiodic function that represents the distribution of the entire wavefront on the microlens array, and its expression is

$$G(x, y) = \frac{1}{p^2} \sum_{n=-\infty}^{+\infty} \sum_{m=-\infty}^{+\infty} C_{n,m}(t) e^{\frac{2i\pi}{p}(nx+my)} \quad \text{(A1)}$$

Including:

$$C_{n,m} = \text{FT}\left[\prod_{p,p}(x, y) e^{\frac{i\pi\left(x^2+y^2\right)}{\lambda f_{\text{ML}}}}\right]_{\frac{n}{p}, \frac{m}{p}} \quad \text{(A2)}$$

where t is time; m, n is the position coordinate of the spot measured by the wavefront sensor; u, v is the displacement variation of the measured spot on a single microlens; P is the distance between microlenses; f_{ML} is the focal length of the microlens; λ is the laser wavelength; (x, y) is the coordinate of any point; FT means Fourier transform of the value at the point $\left(\frac{n}{p}, \frac{m}{p}\right)$; $C_{n,m}$ is the imaging result of a single microlens measured on the wavefront sensor. $C_{n,m}$ can write

$$C_{n,m} = \psi_{\text{ML}}\left(\frac{n}{p}, \frac{m}{p}\right) = \left[\frac{\sin(n\pi)}{n\pi} \frac{\sin(m\pi)}{m\pi}\right] \times e^{i\pi\lambda f_{\text{ML}}\left(u^2+v^2\right)} \quad \text{(A3)}$$

According to the imaging diagram of the distorted laser wavefront with object vibration information on a single microlens described in Figure 3, taking the imaging principle of the distorted wavefront in the y direction as an example, the object vibration amplitude is ΔZ_y, similarly, the object vibration amplitude in the x direction is ΔZ_x, from which we can see

$$\tan\alpha = \frac{\Delta Z_y}{\Delta y} = \frac{\delta y}{f_{\text{ML}}} \quad \text{(A4)}$$

$$\tan\beta = \frac{\Delta Z_x}{\Delta x} = \frac{\delta x}{f_{\text{ML}}} \quad \text{(A5)}$$

This shows that the slope of the laser wavefront can be approximately obtained by calculating the displacement of the measured spot, and the measured spot can reflect the vibration information of the object. The imaging formula of the distorted wavefront can be obtained by substituting formula (A4) and (A5) into formula (A3):

$$C_{n,m} = \left[\frac{\sin\left(\pi n + \frac{x'}{tg\alpha}\right)}{n\pi} \frac{\sin\left(\pi m + \frac{y'}{tg\beta}\right)}{m\pi}\right] \times e^{i\pi\lambda f_{\text{ML}}\left[\left(\pi n+\frac{x'}{tg\alpha}\right)^2+\left(\pi m+\frac{y'}{tg\beta}\right)^2\right]} \quad \text{(A6)}$$

where (x', y') is the coordinate of any point on the object surface. Equation (A6) shows that the offset of the light spot contains the distortion information of the beam wavefront, that is, the phase information, and the distortion of the laser wavefront is caused by the vibration of the object. By analyzing and measuring the offset distance of the light spot, we can analyze the change in the wavefront phase and the ground vibration information that causes this change.

References

1. Scott, W.; Martin, J.; Larison, G. Experimental model for a seismic landmine detection system. *IEEE Trans. Geosci. Remote Sens.* **2001**, *39*, 1155–1164. [CrossRef]
2. Polychronopoulou, K.; Lois, A.; Draganov, D. Body-wave passive seismic interferometry revisited: Mining exploration using the body waves of local microearthquakes. *Geophys. Prospect.* **2020**, *68*, 232–253. [CrossRef]
3. Komatitsch, D.; Ritsema, J.; Tromp, J. The Spectral-Element Method, Beowulf Computing, and Global Seismology. *Science* **2002**, *298*, 1737–1742. [CrossRef] [PubMed]

4. Bruno, P.P.G. Seismic Exploration Methods for Structural Studies and for Active Fault Characterization: A Review. *Appl. Sci.* **2023**, *13*, 9473. [CrossRef]
5. Bishop, J.W.; Fee, D.; Modrak, R.; Tape, C.; Kim, K. Spectral Element Modeling of Acoustic to Seismic Coupling Over Topography. *J. Geophys. Res. Solid Earth* **2022**, *127*. [CrossRef]
6. Purnomo, E.W.; Abdul Latiff, A.H.; Elsaadany, M.M.A.A. Predicting Reservoir Petrophysical Geobodies from Seismic Data Using Enhanced Extended Elastic Impedance Inversion. *Appl. Sci.* **2023**, *13*, 4755. [CrossRef]
7. Dutta, G.; Saragoussi, E.; Cheng, X.; Zhai, Y.; Parekh, C.; Vigh, D. Time-lapse velocity changes revealed with joint 4D full waveform inversion. In Proceedings of the 84th EAGE Annual Conference & Exhibition. European Association of Geoscientists & Engineers, Vienna, Austria, 5–8 June 2023, pp. 1–5. [CrossRef]
8. Mardan, A.; Giroux, B.; Fabien-Ouellet, G.; Saberi, M.R. Monitoring fluid saturation in reservoirs using time-lapse full-waveform inversion. *Geophys. Prospect.* **2023**, *71*, 1012–1029. [CrossRef]
9. Berni, A.J. Remote sensing of seismic vibrations by laser Doppler interferometry. *Geophysics* **1994**, *59*, 1856–1867. [CrossRef]
10. Yu, J.; Xu, P.; Yu, Z.; Wen, K.; Yang, J.; Wang, Y.; Qin, Y. Principles and Applications of Seismic Monitoring Based on Submarine Optical Cable. *Sensors* **2023**, *23*, 5600. [CrossRef]
11. Fernández-Ruiz, M.R.; Soto, M.A.; Williams, E.F.; Martin-Lopez, S.; Zhan, Z.; Gonzalez-Herraez, M.; Martins, H.F. Distributed acoustic sensing for seismic activity monitoring. *APL Photonics* **2020**, *5*, 030901. [CrossRef]
12. Nanda, N.C. *Analysing Seismic Attributes*; Springer: Cham, Switzerland, 2021; pp. 205–222. [CrossRef]
13. Nanda, N.C. *Seismic Wave and Rock-Fluid Properties*; Springer: Cham, Switzerland, 2021; pp. 3–23. [CrossRef]
14. Wu, C.; Li, X.; Chen, W.; Li, X. A Review of Geological Applications of High-Spatial-Resolution Remote Sensing Data. *J. Circuits Syst. Comput.* **2020**, *29*, 2030006. [CrossRef]
15. Shroll, R.M.; St. Peter, B.; Richtsmeier, S.; Tannian, B.; Jensen, E.; Kielkopf, J.; Pereira, W.E. Remote optical detection of ground vibrations. In *Infrared Remote Sensing and Instrumentation XXIII*; SPIE: Bellingham, WA, USA, 2015; p. 96080K. [CrossRef]
16. Ozcan Tatar, C.; Cabuk, S.N.; Ozturk, Y.; Senyel Kurkcuoglu, M.A.; Ozenen-Kavlak, M.; Bilge Ozturk, G.; Dabanli, A.; Kucukpehlivan, T.; Cabuk, A. Impacts of earthquake damage on commercial life: A RS and GIS based case study of Kahramanmaraş earthquakes. *Int. J. Disaster Risk Reduct.* **2024**, *107*, 104464. [CrossRef]
17. Li, Q.; Zhang, J.; Jiang, H. Incorporating multi-source remote sensing in the detection of earthquake-damaged buildings based on logistic regression modelling. *Heliyon* **2024**, *10*, e32851. [CrossRef] [PubMed]
18. Fuchs, M.J.; Rexer, M.; Schaider, F. Detection and analysis of seismic induced GNSS station motion in a North American network following the 2017 Chiapas earthquake. *J. Geodyn.* **2022**, *149*, 101881. [CrossRef]
19. Bruce, R.A.; Fitzpatrick, G.L. Remote vibration measurement of rough surfaces by laser interferometry. *Appl. Opt.* **1975**, *14*, 1621. [CrossRef]
20. Yan, B.; Shi, H.; Wang, P.; Tong, Q.; Su, J. Optoelectric characteristics of laser remote sensing for measurement of ground vibration. *J. Appl. Remote Sens.* **2021**, *15*, 014505. [CrossRef]
21. Bianchi, S. Vibration detection by observation of speckle patterns. *Appl. Opt.* **2014**, *53*, 931. [CrossRef]
22. Li, X.; Zhao, L.P.; Fang, Z.P.; Asundi, A.; Yin, X.M. Surface measurement with Shack-Hartmann wavefront sensing technology. In *Ninth International Symposium on Laser Metrology*; SPIE: Bellingham, WA, USA, 2008; p. 715515. [CrossRef]
23. Yi, S.; Xiang, J.; Zhou, M.; Wu, Z.; Yang, L.; Yu, Z. Angle-based wavefront sensing enabled by the near fields of flat optics. *Nat. Commun.* **2021**, *12*, 6002. [CrossRef]
24. Riley, Z.B.; Hagen, B.; Ehrhardt, D.A. Aero-Optical Measurements of the Response of a Thin Panel at Mach 6. *AIAA J.* **2022**, *60*, 1522–1536. [CrossRef]
25. Zhu, H.; Li, G.; Wang, J. Flow-induced vibration of a circular cylinder with splitter plates placed upstream and downstream individually and simultaneously. *Appl. Ocean. Res.* **2020**, *97*, 102084. [CrossRef]
26. Chen, Z.S.; Wang, S.; Jiang, X. Effect of wake interference on vibration response of dual tandem flexible pipe. *Ocean Eng.* **2023**, *269*, 113497. [CrossRef]
27. Ares, J.; Mancebo, T.; Bará, S. Position and displacement sensing with Shack–Hartmann wave-front sensors. *Appl. Opt.* **2000**, *39*, 1511. [CrossRef] [PubMed]
28. Su, H.S.W. A Remote Sensing Study of Wavefront-Sensor-Based Measurements of Seismic Wave. *Appl. Laser* **2023**, *43*, 111–117. [CrossRef]
29. Luo, Q.; Luo, H.; Wu, G.; Ji, X.; Su, J.; Jiang, W. Laser Remote Sensing of Seismic Wave with Sub-Millimeter Scale Amplitude Based on Doppler Characteristics Extracted from Wavefront Sensor. *Photonics* **2024**, *11*, 204. [CrossRef]
30. Panagopoulou, S.I.; Neal, D.P. Zonal Matrix Iterative Method for Wavefront Reconstruction From Gradient Measurements. *J. Refract. Surg.* **2005**, *21*, S563–S569. [CrossRef]
31. Liu, S.; Zhong, H.; Li, Y.; Liu, K. Fast and Highly Accurate Zonal Wavefront Reconstruction from Multi-Directional Slope and Curvature Information Using Subregion Cancelation. *Appl. Sci.* **2024**, *14*, 3476. [CrossRef]

32. Huang, L.; Xue, J.; Gao, B.; Zuo, C.; Idir, M. Zonal wavefront reconstruction in quadrilateral geometry for phase measuring deflectometry. *Appl. Opt.* **2017**, *56*, 5139. [CrossRef]
33. Wang, K.; Xu, K. A Review on Wavefront Reconstruction Methods. In Proceedings of the 2021 4th International Conference on Information Systems and Computer Aided Education, New York, NY, USA, 24–26 September 2021; pp. 1528–1531. [CrossRef]
34. Southwell, W. Wave-front estimation from wave-front slope measurements. *J. Opt. Soc. Am.* **1980**, *70*, 998. [CrossRef]
35. Schulze, C.; Naidoo, D.; Flamm, D.; Schmidt, O.A.; Forbes, A.; Duparré, M. Wavefront reconstruction by modal decomposition. *Opt. Express* **2012**, *20*, 19714. [CrossRef]
36. Al-Gailani, S.; Mohammad, A.; Shaddad, R. Enhancement of free space optical link in heavy rain attenuation using multiple beam concept. *Optik* **2013**, *124*, 4798–4801. [CrossRef]
37. Kim, I.I.; McArthur, B.; Korevaar, E.J. Comparison of laser beam propagation at 785 nm and 1550 nm in fog and haze for optical wireless communications. In *Optical Wireless Communications III*; SPIE: Bellingham, WA, USA, 2001; pp. 26–37. [CrossRef]

photonics

Review

Overview of Addressed Fiber Bragg Structures' Development

Timur Agliullin [1], German Il'In [2], Artem Kuznetsov [1], Rinat Misbakhov [1], Rustam Misbakhov [3], Gennady Morozov [1], Oleg Morozov [1,*], Ilnur Nureev [1] and Airat Sakhabutdinov [1]

1 Department of Radiophotonics and Microwave Technologies, Kazan National Research Technical University Named after A.N. Tupolev-KAI, 10 K. Marx St., Kazan 420111, Russia
2 Department of Electronic and Quantum Means of Information Communication, Kazan National Research Technical University Named after A.N. Tupolev-KAI, 10 K. Marx St., Kazan 420111, Russia
3 Almetyevsk Branch, Kazan National Research Technical University Named after A.N. Tupolev-KAI, 9b Stroiteli Avenue, Almetyevsk 423400, Russia
* Correspondence: ogmorozov@kai.ru

Abstract: An addressed fiber Bragg structure (AFBS) is a special type of fiber Bragg grating simultaneously performing the functions of a two-frequency radiation shaper and a sensitive element. An AFBS forms a two-frequency optical spectral response at its output, the difference frequency of which is invariant to measured physical fields and is referred to as the address frequency of the AFBS. Each of the AFBSs in the system has its own address frequency; therefore, a number of such structures can be interrogated simultaneously enabling the addressed multiplexing. In this article, we provide an overview of the theory and technology of AFBS, including the structures with three or more spectral components with various combinations of difference frequencies, both symmetrical and asymmetric. The subjects of interrogation of AFBSs, their fabrication and calibration are discussed as well. We also consider a wide range of applications in which AFBS can be used, covering such areas as oil and gas production, power engineering, transport, medicine, etc. In addition, the prospects for the further development of AFBS are proposed that mitigate the shortcomings of the current AFBSs' state of the art and open up new possibilities of their application.

Keywords: addressed fiber Bragg structure; interrogation of addressed fiber Bragg structures; fabrication of addressed fiber Bragg structures; fiber Bragg grating sensor implementation; microwave photonics

Citation: Agliullin, T.; Il'In, G.; Kuznetsov, A.; Misbakhov, R.; Misbakhov, R.; Morozov, G.; Morozov, O.; Nureev, I.; Sakhabutdinov, A. Overview of Addressed Fiber Bragg Structures' Development. *Photonics* **2023**, *10*, 175. https://doi.org/10.3390/photonics10020175

Received: 9 January 2023
Revised: 27 January 2023
Accepted: 5 February 2023
Published: 7 February 2023

1. Introduction

Fiber Bragg gratings (FBG) have acquired widespread application since their introduction in the late 1970s [1], especially as photonic sensing elements for the measurement of various physical fields. Their attractive properties, such as small footprint and low weight, immunity to electromagnetic disturbances, high sensitivity and possibility of multiplexing several FBGs into a single system, provide advantages in numerous areas, including aerospace [2], automotive [3], biomedical [4], civil engineering, oil and gas industries [5], etc.

Several techniques for FBG multiplexing and interrogation have been developed to date. The most common approaches, such as wavelength [6], frequency [7], time [8] and spatial [9] division multiplexing, are implemented using complex and costly optoelectronic devices, such as spectrum analyzers, tunable Fabry–Perot interferometers, diffraction gratings, etc. Another problem of the traditional interrogating methods is the lack of sensor addressability, which leads to interrogation errors when spectrum overlapping takes place. In order to mitigate this issue, optical spectrum-coded FBG interrogation methods were proposed [10,11], in which the sensors are interrogated in real time according to autocorrelation between the sensor spectra and its code signature, thus allowing several FBGs to be distinguished within the same spectral range.

A different approach was proposed in which FBG performs a triple function: besides sensing, it acts as a two-frequency radiation shaper as well as enables address-based multiplexing. Such FBGs are referred to as addressed fiber Bragg structures (AFBS) [12,13]. An AFBS is a type of FBG, the spectral response of which has two narrow components (notches). When an AFBS is connected to a wideband optical source, it forms an output radiation consisting of two narrowband frequencies, the difference between which is called the address frequency and belongs to the microwave range (GHz). The address frequency is invariant to the AFBS central wavelength shift when it is exposed to strain or temperature variations. Therefore, the address frequency is used as a distinguishing parameter, which makes it possible to interrogate several AFBSs even if their central wavelengths coincide.

The usage of AFBS significantly simplifies the interrogation scheme compared with the abovementioned optoelectronic methods, as it requires only a broadband light source, an optical filter with a predefined frequency response with an inclined profile and a photodetector.

The concept of AFBS was subsequently expanded to include the structures with three or more spectral components forming two or more address frequencies, which are also known as multi-addressed fiber Bragg structures (MAFBS) [14]. The increased number of address frequencies allows the enhancement of the accuracy of the central wavelength determination as well as the expansion of the sensor capacity of the system.

The current paper presents a comprehensive classification of addressed fiber Bragg structures, including both AFBSs and MAFBSs with various relative positions of the spectral address components. The theoretical and technological aspects of AFBS implementation are also discussed. An overview of a wide range of AFBS applications is given along with the directions of further AFBS development.

2. Classifications of Addressed Fiber Bragg Structures

The following classifications consider AFBSs with up to three spectral components. AFBSs with four or more address components can be classified in the same manner. Two approaches to the formation of AFBS have been proposed: the introduction of two or more phase π-shifts into the periodic structure of FBG ($N\pi$-FBG, where N is the number of phase shifts, Figure 1) [15] and the sequential recording of several ultra-narrowband FBGs with different central wavelengths ($N\lambda$-FBG, Figure 2) [16]. For the former type, the transmitted radiation is used for interrogation, while for the latter type, the reflected light is utilized. Thus, the reflecting and transmitting AFBSs constitute the first AFBS classification.

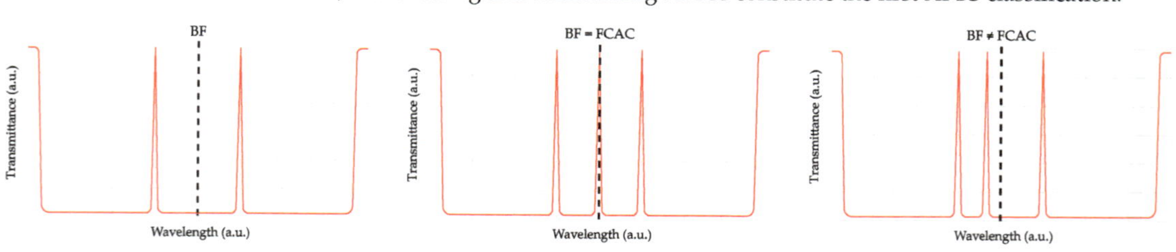

(a) (b) (c)

Figure 1. Transmitted spectrum of AFBS ($N\pi$-FBG type): (**a**) double-component AFBS (DCAFBS) with the Bragg frequency (BF) in between the frequencies of lateral address components (FLACs); (**b**) symmetric triple-component AFBS (STCAFBS) with the BF coinciding with the frequency of the central address component (FCAC); (**c**) asymmetric triple-component AFBS (ATCAFBS) with the BF not coinciding with the FCAC.

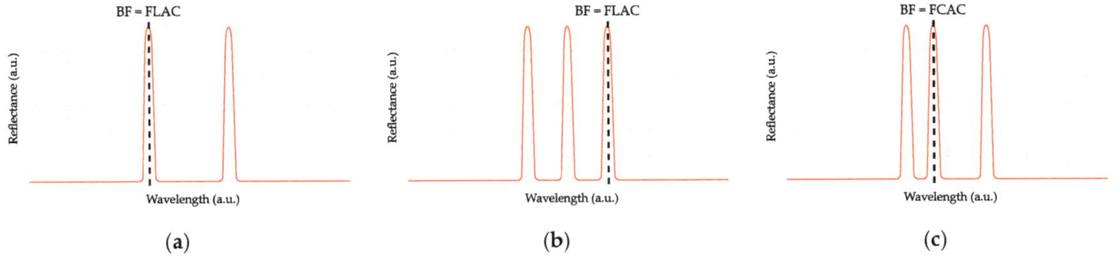

Figure 2. Reflected spectrum of AFBS ($N\lambda$-FBG type) with: (**a**) double-component AFBS (DCAFBS) with the Bragg frequency (BF) coinciding with the left frequency of lateral address component (FLAC); (**b**) symmetric triple-component AFBS (STCAFBS) with the BF coinciding with the right FLAC; (**c**) asymmetric triple-component AFBS (ATCAFBS) with the BF coinciding with the frequency of the central address component (FCAC).

The second classification of AFBS is according to the number of address frequencies: a single-addressed AFBS (Figures 1a and 2a), which is a double-component AFBS (DCAFBS), and a two-addressed AFBS (Figures 1b and 2b), which is a symmetrical triple-component AFBS (STCAFBS) with lateral address components spaced at the same address frequency from the central address component. A three-addressed AFBS (Figures 1c and 2c) is an asymmetric triple-component AFBS (ATCAFBS) with lateral address components spaced at different address frequencies from the central address component.

The third classification is the classification according to the coincidence of the frequency of the central address component and the Bragg frequency of the entire structure as a whole. The possibility of Bragg frequency definition for the entire structure as a whole follows from the invariance of the position of the address components when physical fields are applied to the structure. The Bragg frequency can be defined to be in the middle between the frequencies of lateral address components (FLACs) (Figure 1a) coinciding with the frequency of the central address component (FCAC), as in the case of STCAFBS (Figure 1b), or it cannot coincide with FCAC, as in the case of ATCAFBS (Figure 1c).

Alternatively, the Bragg wavelength can be defined to be coinciding with the FLAC of the DCAFBS (Figure 2a), or the STCAFBS (Figure 2b), or to be coinciding with the FCAC of the ATCAFBS (Figure 2c).

3. Interrogation of Addressed Fiber Bragg Structures

Typical schemes for AFBS interrogation are presented in Figure 3. The scheme for the transmitting $N\pi$-FBG AFBS is shown in Figure 3A, while Figure 3B represents the one for the reflecting $N\lambda$-FBG structures. In the schemes below, double-component AFBSs are used as an example. The schemes for the structures with three or more components are designed in the same way.

The schemes operate as follows. An optical source (1) generates a wideband optical radiation (insertion a), the bandwidth of which covers the whole range of AFBS components' wavelength shifts. The radiation passes through N addressed structures 2.1–2.N connected either in parallel using fiber-optic splitter nine and combiner ten (in the case of transmitting AFBS, Figure 3A), or sequentially (in the case of reflecting AFBS, Figure 3B). At the output of each AFBS, a radiation with two spectral components is formed, the spacing between which corresponds to the address frequency of the AFBS and is unique for each sensor in the system. The combined multi-frequency radiation from all the AFBSs (insertions b and c) is divided by means of a fiber-optic splitter six into two measuring and one reference channels. In each measuring channel, the radiation passes through an optical filter (3.1 and 3.2) with a linear inclined frequency response, which modifies the amplitudes of the frequency components of the AFBSs according to its known frequency response. The key difference between the measuring channels is that the optical filters 3.1 and 3.2 have different known temperature sensitivities of their spectral responses. The optical radiation in all the channels is received by the corresponding photodetectors 4.1, 4.2 and 7, at the output of which the electric beating signals are generated at the address frequencies of the AFBSs. The electric signals are digitized using ADCs 5.1, 5.2 and 8, and the subsequent calculations of the AFBS central wavelengths are carried out for the ratio of the power of each measuring channel to the power in the reference channel. This eliminates the influence of the optical source fluctuations on the AFBS interrogation process.

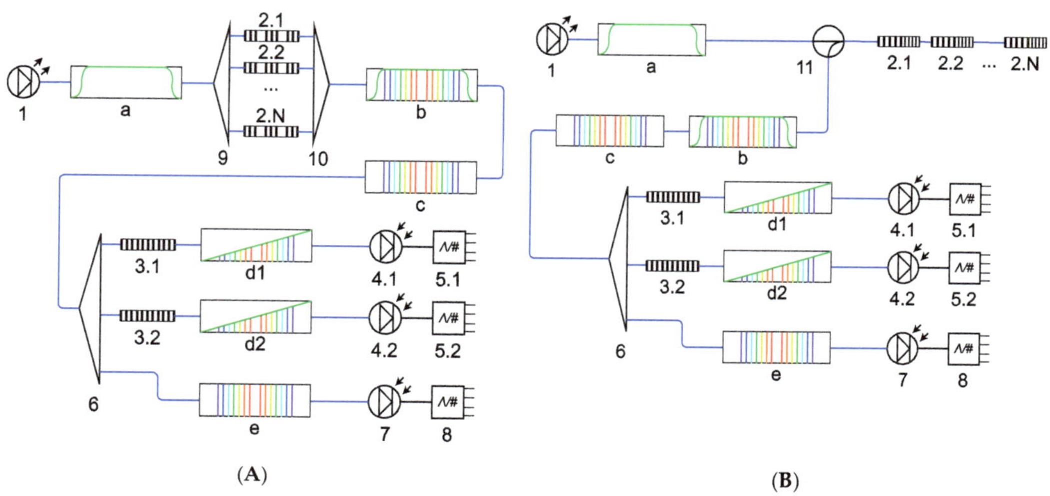

Figure 3. Interrogation schemes for AFBSs of: (**A**) transmitting type; (**B**) reflecting type; (1) wideband light source, (2.1)–(2.N) AFBS sensors, (3.1) and (3.2) optical filters with linear inclined frequency responses, (4.1) and (4.2) photodetectors of the measuring channels, (5.1) and (5.2) ADCs of measuring channels, (6) and (9) fiber-optic splitters, (7) photodetector of the reference channel, (8) ADC of the reference channel, (10) fiber-optic combiner, (11) fiber-optic circulator; inserted diagrams: (a) spectrum of the wideband light source, (b) and (c) spectra of light propagated through the AFBS sensors, (d1) and (d2) spectra of AFBS sensors at the output of the optical filter, (e) spectra of AFBSs in the reference channel; blue connection lines represent optical fibers, black connection lines represent electrical wires.

One of the key components of the AFBS interrogation scheme is the optical filter with a linear inclined frequency response. Such filters can be fabricated based on an FBG with the known spectral response having linear slopes. The deviation from the linear approximation of the optical filter frequency response is one of the main components of the measurement error, since the position of the AFBS spectrum is determined relative to it. Therefore, for the given FBG-based filter, the acceptable range of the AFBS spectral components shifts is determined, in which the deviation of the filter spectrum from the linear approximation does not exceed the desired value, as it is discussed by the authors in [17].

Figure 4 illustrates the principle of AFBS interrogation. Colored lines in Figure 4a denote the different variants of spectral positions of the two AFBSs relative to the spectral response of the optical filter, and in Figure 4b the corresponding RF spectra at the photodetector output are indicated in the same colors.

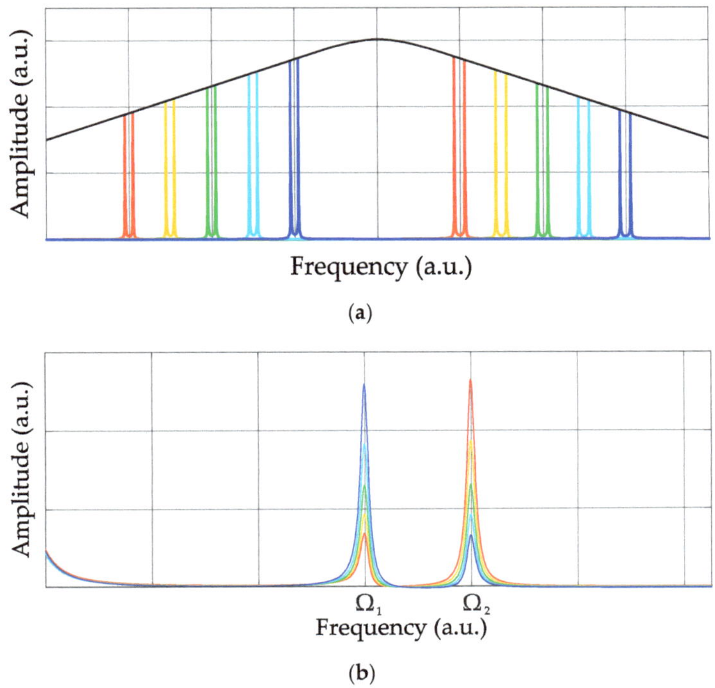

Figure 4. Interrogation of two double-component AFBSs with the address frequencies Ω_1 and Ω_2, respectively, with different variants of Bragg frequency: (**a**) spectral positions of AFBSs (colored lines) relative to the spectral response of the optical filter (black line); (**b**) corresponding spectra of the beating signal at the output of the photodetector.

As it follows from the principle of AFBS interrogation, in order to correctly determine the AFBS central wavelength, it is necessary to take into account the temperature drifts of the optical filters 3.1 and 3.2. For this reason, the filters are located close to each other in the system layout so that their temperature is assumed to be the same. Therefore, knowing the difference between the center wavelengths of the same AFBS determined using the filters, it is possible to calculate their temperature using the pre-defined temperature characteristics of the filters [17]. After that, the estimated value of temperature is used to calculate the absolute value of the center frequency of the filter, based on which the correction to the center frequency of the AFBS is determined.

Consider the output optical radiation of the i-th AFBS, which is represented as a sum of two harmonic oscillations spaced by the address frequency Ω_i:

$$E_i(t) = A_i e^{j\omega_i t + \varphi_{Ai}} + B_i e^{j(\omega_i + \Omega_i)t + \varphi_{Bi}}, \quad (1)$$

where A_i and B_i are the amplitudes of the AFBS spectral components passed through the optical filter ((3.1) or (3.2) in Figure 3); ω_i is the frequency of the "left" spectral component of the i-th AFBS; Ω_i is the address frequency; and φ_{Ai} and φ_{Bi} are the initial phases, which can be unequal, but their difference is constant over time.

The luminous power received by the photodetector from N double-component AFBSs can be expressed by multiplying the Equation (1) with its complex conjugate:

$$\begin{aligned} P(t) &= \left(\sum_{i=1}^{N} E_i(t)\right)\left(\overline{\sum_{i=1}^{N} E_i(t)}\right) = \left(\sum_{i=1}^{N}\left(A_i e^{j\omega_i t+\varphi_{Ai}} + B_i e^{j(\omega_i+\Omega_i)t+\varphi_{Bi}}\right)\right)\left(\sum_{k=1}^{N}\left(A_k e^{-(j\omega_k t+\varphi_{Ak})} + B_k e^{-(j(\omega_k+\Omega_k)t+\varphi_{Bk})}\right)\right) = \\ &= \sum_{i=1}^{N}(A_i^2+B_i^2) + 2\sum_{i=1}^{N} A_i B_i \cos(\Omega_i t+\varphi_{Ai}-\varphi_{Bi}) + 2\sum_{i=1}^{N}\sum_{k=i+1}^{N}\begin{pmatrix} A_iA_k\cos((\omega_i-\omega_k)t+\varphi_{Ai}-\varphi_{Ak})+ \\ A_iB_k\cos((\omega_i-\omega_k-\Omega_k)t+\varphi_{Ai}-\varphi_{Bk})+ \\ B_iA_k\cos((\omega_i-\omega_k+\Omega_i)t+\varphi_{Bi}-\varphi_{Ak})+ \\ B_iB_k\cos((\omega_i-\omega_k+\Omega_i-\Omega_k)t+\varphi_{Bi}-\varphi_{Bk}) \end{pmatrix}. \end{aligned} \quad (2)$$

Thus, the oscillation of the amplitude of the electrical signal of the photodetector at the address frequency of the AFBS Ω_i is proportional to the amplitudes of the AFBS optical spectral components A_i and B_i, which are defined by the parameters u (the slope) and v (the intercept) of the linear function describing the inclined frequency response of the optical filter ((3.1) or (3.2) in Figure 3):

$$A_i = L_0 \cdot (u \cdot \omega_i + v), B_i = L_0 \cdot (u \cdot (\omega_i + \Omega_i) + v), \quad (3)$$

where L_0 is the initial amplitude of the AFBS optical spectral components at the input of the filter with inclined linear frequency response. By measuring the amplitude of the photodetector output signal at the address frequency Ω_i, it is possible to define the central frequency shift (or the frequency of the left spectral component ω_i) of the AFBS relative to the inclined frequency response of the optical filter. However, due to the appearance of the additional frequency components in the last sum of Equation (2), the filtering of the electrical signal at the address frequencies is required.

By assuming that $B_i = A_i + L_0 \cdot u \cdot \Omega_i$ (which follows from (3)) and by filtering the photodetector output signal at the address frequency, the system of equations for the calculation of the AFBS spectral components' positions is obtained:

$$\sum_{i=1}^{N}\sum_{k=1}^{N}\begin{pmatrix} A_iA_k \cdot F(\Omega_j, \omega_i-\omega_k)+ \\ A_i(A_k+u_k\Omega_k)\cdot F(\Omega_j, \omega_i-\omega_k-\Omega_k)+ \\ A_k(A_i+u_i\Omega_i)\cdot F(\Omega_j, \omega_i-\omega_k+\Omega_i)+ \\ (A_i+u_i\Omega_i)(A_k+u_k\Omega_k)\cdot F(\Omega_j, \omega_i-\omega_k+\Omega_i-\Omega_k) \end{pmatrix} = D_j, \forall j = \overline{1,N}, \quad (4)$$

where the function $F(\Omega, \omega)$ describes the frequency response of the bandpass filter of the address frequency. The system of Equation (4) in the variable ω_i is solved numerically (for example, using the Levenberg–Marquardt or the Newton–Raphson algorithms), taking the previously calculated value of ω_i as the initial conditions.

In most cases, the error of the AFBS spectral position calculation does not exceed 0.1 pm even in multi-sensor systems comprising double-component structures [18]. However, in certain cases of the AFBS spectral components' relative positions, when the components of different AFBSs coincide or the third summand in (2) coincides with the address frequency of any of the structures, the error can reach 2 pm [18]. This issue can be mitigated with the usage of the addressed structures having three or more spectral components forming two or more address frequencies [14].

As is known, fiber Bragg gratings are sensitive both to strain and temperature at the same time; therefore, FBG-based sensor systems generally include at least one FBG isolated from any physical influence except for the temperature in order to perform thermal compensation of the other sensors. Thus, it is necessary to define a method for the combined calibration of strain and temperature sensors. In the work [19], the following procedure is described, considering a combination of strain and temperature sensors as an example. The value of strain ε causing the central wavelength shift $\Delta\lambda_\varepsilon$ of an AFBS can be expressed as a third power polynomial [20]:

$$\varepsilon = a_3 \cdot \Delta\lambda_\varepsilon^3 + a_2 \cdot \Delta\lambda_\varepsilon^2 + a_1 \cdot \Delta\lambda_\varepsilon + a_0, \quad (5)$$

where a_i, $i = 1 \ldots 3$ are certain coefficients. In its turn, the value of temperature T causing the wavelength shift $\Delta\lambda_T$ of the sensor is defined as a second power polynomial [20]:

$$T = c_2 \cdot \Delta\lambda_T^2 + c_1 \cdot \Delta\lambda_T + c_0, \tag{6}$$

where c_i, $i = 1, 2$ belong to the other set of coefficients. In order to take into account the thermal wavelength shift of the strain sensor, the coefficients $a_0 \ldots a_3$ in (5) can be expressed as the functions of temperature, similarly to (6):

$$\varepsilon = a_3(T) \cdot \Delta\lambda_\varepsilon^3 + a_2(T) \cdot \Delta\lambda_\varepsilon^2 + a_1(T) \cdot \Delta\lambda_\varepsilon + a_0(T), \tag{7}$$

where

$$a_n(T) = c_{2,n} \cdot \Delta\lambda_T^2 + c_{1,n} \cdot \Delta\lambda_T + c_{0,n}, n = \{0,1,2,3\}. \tag{8}$$

The resulting strain dependence on the central wavelength shifts ($\Delta\lambda_T$, $\Delta\lambda_\varepsilon$) can be expressed as follows:

$$\varepsilon = F(\Delta\lambda_T, \Delta\lambda_\varepsilon, c_{m,n}) = \sum_{m=0}^{2} \sum_{n=0}^{3} c_{m,n} \cdot \Delta\lambda_T^m \cdot \Delta\lambda_\varepsilon^n \,. \tag{9}$$

During the sensor calibration, the dataset $\{\Delta\lambda_{Ti}, \Delta\lambda_{\varepsilon i}, T_i, \varepsilon_i\}$ is obtained, in which $\Delta\lambda_{Ti}$ is the central wavelength shift of the temperature sensor, $\Delta\lambda_{\varepsilon i}$ is the central wavelength shift of the strain sensor, T_i and ε_i are the values of temperature and strain induced by the thermal chamber and the translation stage, respectively, and i is the number of test measurements. The unknown coefficients $\{c_{m,n}\}$ of the approximating surface (9) are defined from the conditions of the minimal deviation of the measured dataset $\{\Delta\lambda_{Ti}, \Delta\lambda_{\varepsilon i}, T_i, \varepsilon_i\}$ from this approximating surface. The unknown coefficients $\{c_{m,n}\}$ are calculated using the least squares method so that the surface (9) follows the strain sensor behavior as precisely as possible at various strain–temperature combinations [19], i.e., the condition is met:

$$\Phi = \sum_{i=1}^{N} (\varepsilon - \varepsilon_i)^2 = \sum_{i=1}^{N} (F(\Delta\lambda_T, \Delta\lambda_\varepsilon, c_{m,n}) - \varepsilon_i)^2 \to \min. \tag{10}$$

The minimum condition of (10) requires all partial derivatives of the function Φ with respect to all of the $\{c_{m,n}\}$ to be equal to zero:

$$\frac{\partial\Phi(c_{m,n})}{\partial c_{m,n}} = 2\sum_{i=1}^{N}\left[(F(\Delta\lambda_T, \Delta\lambda_\varepsilon, c_{m,n}) - \varepsilon_i) \cdot \frac{\partial F(\Delta\lambda_T, \Delta\lambda_\varepsilon, c_{m,n})}{\partial c_{m,n}}\right] = 0 \; \forall m \in \{2,1,0\} \cup n \in \{3,2,1,0\}. \tag{11}$$

Thus, the system of 12 equations is obtained, by solving which the 12 unknown coefficients $\{c_{m,n}\}$ are calculated.

Challenges and limitations. The main limitation to the number of AFBS sensors that can be simultaneously interrogated in the system is the maximum operating frequency of the photodetector, i.e., the maximum address frequency, at which the beating signal can be generated by the photodetector. One of the main components of the measurement error is the deviation of the optical filter frequency response from its linear approximation, since the amplitudes of the AFBS spectral components used for the calculation of the AFBS spectral position relative to the optical filter are defined by the parameters of the linear function describing the inclined frequency response of the filter. Another challenge is the necessity to ensure the uniformity of strain and temperature impact on the AFBS sensing element in order to maintain its address frequency unchanged.

The references regarding the subject of AFBS interrogation are listed in Table 1.

Table 1. Works related to the AFBS interrogation.

References	Subject
[17]	Requirements to the optical filter with linear frequency response; compensation of the thermal drift of the optical filter.
[18]	Accuracy estimation of the AFBS Bragg wavelength determination.
[14]	Interrogation of multi-addressed fiber Bragg structures.
[19,20]	Combined calibration of strain and temperature sensors.

4. Fabrication Methods of Addressed Fiber Bragg Structures

This section considers the fabrication techniques for transmitting ($N\pi$-FBG) and reflecting ($N\lambda$-FBG) AFBS separately. As an example, the fabrication of asymmetric triple-component AFBSs is discussed.

4.1. Fabrication of Transmitting AFBS

For the fabrication of the transmitting addressed structures with ultra-narrowband transparency windows in their spectra, the following techniques were chosen: the technique involving the spectral overlapping of two identical FBGs [21], the technology based on a stepped phase mask [22] and the method based on changing the geometry of an optical fiber using an electric arc of a welding machine [23]. The works [21,23] use conventional phase masks to record FBGs, while a special mask with a thickness difference of 2300 nm was utilized in [22], using which a phase shift was formed at the place of the mask thickness step.

Based on the abovementioned methods, the following technique for $N\pi$-FBG fabrication has been established [24]. In a single-mode optical fiber (for example, SMF-28), an FBG is recorded using conventional holographic recording schemes based on the Lloyd's interferometer [21]. After that, the fiber is shifted transversely to the recording beam using a high-precision translation stage, and another FBG is recorded. The displacement and the beam size are chosen in such a way that the FBGs are superimposed; thus, a phase shift is created and a 2π-FBG-type structure with length L_1 is formed. For more accurate control of the shift value, the scheme can use a Michelson interferometer based on bulk optical elements [21]. Then, in accordance with the technique presented in [23], the fiber is displaced by a distance equal to the first 2π-FBG length (L_1) and the distance between the gratings (Δl), after which the second 2π-FBG with the length $L_2 \neq L_1$ is recorded using the technology [21]. Since the recording conditions remain unchanged, the characteristics of both 2π-FBGs will be basically the same, except for a slight difference in the bandwidth of their spectra. At the next step, the fiber is placed in a splicer, and the electric arc induces a phase difference between the radiation reflected from both structures, thus creating a 3π-FBG structure, as shown in Figure 5. The bandwidth of the AFBS transparency windows created using the presented method is 30–35 MHz [23].

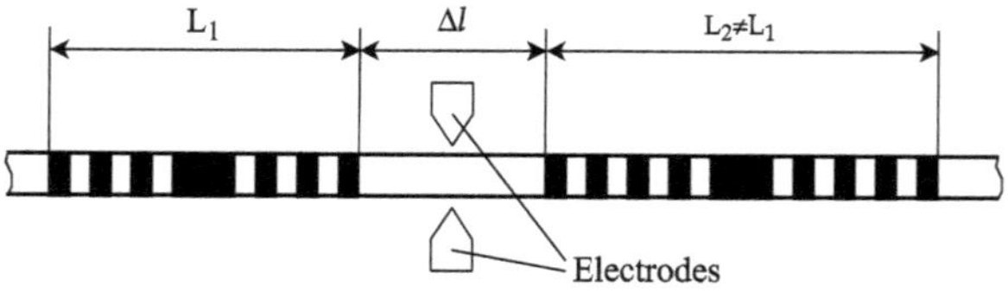

Figure 5. The structure of an asymmetric 3π-FBG fabricated using a combination of techniques in [21,23].

The work [25] studies the influence of the $N\pi$-FBG structure parameters on the spectral response of such structures. Thus, an increase in the induced refractive index and the length of the outermost uniform sections of the AFBS has the greatest effect on reducing the width of the transparency windows. Asymmetric change of address frequencies are achieved by varying the phase shifts. The findings of [25] can be used to obtain the required spectral characteristics of the $N\pi$-FBG structure.

4.2. Fabrication of Reflecting AFBS

The reflecting $N\lambda$-FBG addressed structures can be fabricated by sequential recording of N conventional FBGs with different Bragg wavelengths using a standard setup based on a 244 nm ultraviolet laser and the nanometer translation stages, such as STANDA-8MT173 [26,27].

The work [28] presents a method suitable for $N\lambda$-FBG fabrication using strain of an optical fiber and a phase mask. A translation stage is used to move the phase mask, while a spring induces tension on the optical fiber. Using the presented method, two identical FBGs were recorded with the difference in the Bragg wavelength of 1 nm and a bandwidth of 0.3 nm [28]. It is possible to obtain narrower grating bandwidths by increasing their lengths, as we demonstrate in Figure 6, where the spectral response of a 2λ-FBG with the component bandwidths of 110 pm is shown.

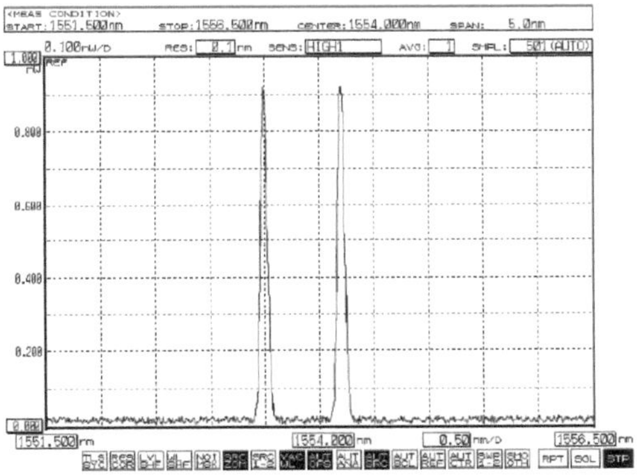

Figure 6. Reflectance spectrum of a 2λ-FBG recorded at KNRTU-KAI.

Another method for $N\lambda$-FBG fabrication uses structured grating technology by alternating sections with an induced grating and "empty" sections of the fiber [29]. During the recording, the phase mask is fixed, and the fiber is displaced with an accuracy of 10 nm. At certain fiber positions, the recording ultraviolet beam is turned off. The parameters of the resulting $N\lambda$-FBG structure are adjusted by varying the displacement step of the translation stage. In [29], a cascaded connection of structured FBGs with estimated bandwidths of 1 pm was proposed.

Thus, it is recommended to use the $N\lambda$-FBG structures recorded with the conventional phase mask method in the applications with a small number of sensors, while the technology of structured FBG recording should be used for the sensors applied in high-resolution measurement systems with a large number of sensors due to the narrower bandwidths of the spectral components.

Challenges and limitations. The lowest address frequency of an AFBS is determined mainly by the width of its spectral components; therefore, it is preferable to ensure the smallest bandwidth of such components of the fabricated AFBS. The latter is achieved by increasing the induced refractive index in the case of transmitting AFBS, and by decreasing the induced refractive index and increasing the FBG length at the same time in the case of reflecting AFBS. The former is limited by the photosensitivity of the optical fiber, while the latter is restricted by the maximum length of the sensing element suitable for the particular application.

The references regarding the subject of AFBS fabrication are listed in Table 2.

Table 2. Works related to AFBS fabrication.

References	Subject
[21]	Recording of phase-shifted FBGs by physical overlapping.
[22]	Recording of phase-shifted FBGs using stepped phase mask.
[23]	Recording of phase-shifted FBGs by changing the geometry of an optical fiber using an electric arc.
[24]	$N\pi$-FBG fabrication using a combination of FBG physical overlapping and changing the fiber geometry techniques.
[25]	Influence of the $N\pi$-FBG structure parameters on its spectral response.
[26,27]	Sequential recording of FBGs.
[28]	Reflecting AFBS recording using strain of an optical fiber and a phase mask.
[29]	Reflecting AFBS recording using structured technology by alternating sections with an induced FBG and "empty" fiber sections.

5. Application of Addressed Fiber Bragg Structures

In this section, several examples of AFBS application are discussed. Generally, the AFBSs can be used in various systems instead of conventional fiber Bragg grating sensors, providing such advantages as a simplified interrogation scheme and enhanced metrological performance, including higher measurement resolution and rate. It must be noted that the $N\lambda$-FBG structures are commonly longer than the $N\pi$-FBGs, which limits the applicability of the former in the areas where the sensor length is restricted. On the other hand, the $N\lambda$-FBGs can be connected sequentially and recorded in the same optical fiber, thereby providing more flexibility in the system layout in comparison with the $N\pi$-FBGs that require parallel connection.

5.1. Power Engineering

The features of fiber-optic sensors, namely immunity to electromagnetic interference, ability to operate in harsh environments, suitability for remote sensing, intrinsic galvanic isolation, etc., make them advantageous in various monitoring systems for power engineering [30]. An example of such application is the vibration monitoring of underground power transmission lines, which is necessary for detecting destructive vibrations and preventing the damages. The systems for simultaneous vibration and temperature measurement are of particular interest for power transmission lines monitoring. Such distributed acoustic sensor (DAS) systems can be based on a distributed array of a large number of FBGs. However, the common disadvantages of the DAS systems based on an array of ultra-weak FBGs [31,32] are the high cost of optoelectronic interrogators and the complexity of the interference device, its calibration and control. In order to eliminate these issues, the work [33] introduced the microwave-photonic DAS system based on $N\lambda$-FBG addressed structures.

Another proposed application of AFBS in power engineering is the measurement of the relative humidity of switchgear devices [34]. Relative humidity (RH) is an important parameter that is used to determine the possibility of the intensity increase in partial discharges, formation of an arc, etc. RH sensors can be developed using an FBG with hygroscopic coating instead of the standard silicone covering. The most widely used coating material for RH sensors is polyimide [35,36], which provides the sensitivity of 2–6 pm/%. Figure 7 presents an RH sensor developed in [34] consisting of three 2π-FBG AFBSs: $AFBS_1$ is a strain sensor with polyimide coating (binding areas are shown in checkered filling, and flexible zones are highlighted with oblique filling), $AFBS_2$ with etched cladding acts as a refractometer sensor, while $AFBS_3$ acts as a temperature sensor.

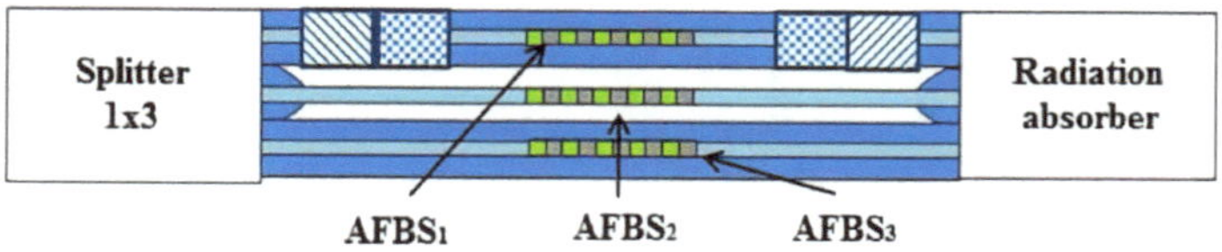

Figure 7. Scheme of an RH sensor based on three AFBSs [34].

As shown in [34], the Bragg wavelength of the $AFBS_1$ with polyimide coating linearly increases with the increase in the RH at constant temperature, due to the strain effect caused by the expansion of the polyimide when it absorbs the moisture.

In addition to RH measurement, the proposed sensor provides measurements of condensed moisture, which are also important in high-voltage power applications, by means of a refractometric sensor with an etched cladding ($AFBS_2$). The $AFBS_2$ is sensitive to the refractive index of the environment due to the etched cladding. The increase in the environmental refractive index causes the increase in the Bragg wavelength of the $AFBS_2$ [37], thereby enabling the sensor to detect the condensed moisture. The sensor's temperature measurement range was $-60 \ldots +180$ °C with an error of ± 0.1 °C, and it demonstrated the sensitivity of RH measurement of 6 pm/% in the range of 30% ... 80% [34]. By multiplexing a number of such RH sensors, it is possible to create a multi-sensor system for distributed monitoring of RH, the amount of condensed moisture, the level of partial discharges, the temperature of buses, contacts and other elements of power systems.

The references regarding the subject of AFBS application in power engineering are listed in Table 3.

Table 3. Works related to the AFBS application in power engineering.

References	Subject
[30]	Overview of FBG applications in power industry.
[31,32]	Distributed acoustic sensor system based on an array of ultra-weak FBGs.
[33]	Distributed acoustic sensor system based on AFBS.
[34]	Measurement of relative humidity of switchgear devices using AFBS.
[35,36]	Measurement of relative humidity using FBG with polyimide coating.
[37]	Detection of environmental refraction index change using FBG with etched cladding.

5.2. Oil and Gas Industry

Fiber-optic sensors are widely used in the oil and gas industries due to their inherent advantages, namely the ability to operate in harsh environments, including at temperatures exceeding 250 °C, the possibility to create systems for distributed and remote measurements over long distances, as well as the lack of electrical power supply in sensing locations [5]. Distributed fiber-optic sensor systems based on Raman and Brillouin scattering [38,39] have been used for thermal monitoring, by means of which, for example, pipeline leak detection can be performed. In addition, pipeline structures are prone to damage due to ground movement caused by earthquakes, erosion, etc., and the strain measurement technique based on Brillouin scattering has been used to monitor the movement of the pipeline [39].

The work [40] proposes a combined sensor system for simultaneous local and distributed strain and temperature measurements [41] for downhole telemetry. The scheme of the system is presented in Figure 8 [40].

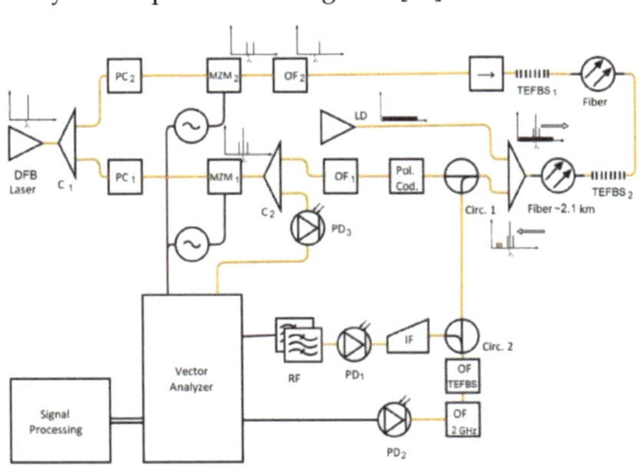

Figure 8. Scheme of combined sensor system for local and distributed downhole monitoring [40].

The system includes a DFB laser generating continuous narrowband radiation, which is divided into two branches by means of fiber-optic coupler C_1. In one of the branches, a continuous signal for Brillouin pumping is formed, while in the other branch, a continuous temperature probe is created. A wideband optical radiation from another source (LD) is directed to the addressed structures $TEFBS_1$ and $TEFBS_2$ (referred to as double-component AFBS (DCAFBS) in the current paper), which serve as local temperature sensors. The radiation from LD then enters the fiber from the side of the pump arm after the circulator Circ. 1. The polarization controller PC_1 and the Mach–Zehnder modulator MZM_1 controlled by a vector network analyzer (vector analyzer) are utilized for sinusoidal modulation of the continuous radiation intensity from the DFB laser. Noise is removed by an optical filter OF_1, and a polarization coder (pol. cod.) depolarizes the pump signal and allows the avoidance of fluctuations caused by polarization in Brillouin amplification. In the second branch, the optical radiation is modulated by another modulator MZM_2 in accordance with a microwave signal generator to create a probing signal. After that, optical filter OF_2 selects the low-frequency probe (Stokes component), removes noise as well as the secondary carrier, and also selects a high-frequency probe sideband (anti-Stokes component). The pump laser radiation passes through an optical circulator (Circ. 1) and enters a ~2.1-km standard single-mode fiber (SMF-28), which is also used to select both the probe signal and the reflected components at the pump frequency. Both radiation components are directed to the optical circulator (Circ. 2), after which the addressed structures' components are selected with the filter OF TEFBS and are transmitted through the filter with inclined frequency response (IF) and received by the photodetector PD_1, while the reflected pump light filtered by 2 GHz OF is received by PD_2. The work [40] reported that the system allows one to simultaneously and independently carry out distributed and local temperature measurements with the resolution of 0.01 °C, providing the benefits of addressed interrogation of local sensors.

The references regarding the subject of AFBS application in the oil and gas industries are listed in Table 4.

Table 4. Works related to the AFBS application in oil and gas industries.

References	Subject
[5]	Overview of applications of fiber-optic sensors in oil and gas industries.
[38,39]	Thermal and strain monitoring of pipelines using distributed fiber-optic sensors.
[40]	AFBS-based system for combined local and distributed sensing for downhole telemetry.
[41]	Combined sensor system based on Brillouin optical frequency-domain analysis and FBG for simultaneous temperature/strain measurements.

5.3. Automotive Engineering

The significantly reduced costs of AFBS interrogation systems due to their significant simplification (as it was described in Section 3), in comparison with the traditional opto-electronic FBG interrogators, open up opportunities for FBG-based sensor application in areas where it has not been economically viable, such as in automotive engineering.

One of the examples of such AFBS applications is the instrumentation of the load-sensing wheel hub bearings [13]. The load-sensing bearing is a promising type of the automotive sensing components that enhance the efficiency of various active safety systems of vehicles [42]. For instance, when the vehicle brakes or accelerates, the load sensing wheel hub bearings detect the maximum wheel load, and the wheel slip ratio is corrected in such a way as to maintain the peak value of the wheel load, therefore enhancing the effectiveness of the wheel–road interaction [43].

The load-sensing wheel hub bearing operates by measuring the strain of the bearing outer ring, which is used to estimate the load applied to the bearing. The research conducted in [13] demonstrated the feasibility of AFBS usage as sensing elements in automotive load-sensing bearings. Figure 9 shows an experimental setup for static load tests including a prototype bearing with two AFBS sensors. The first AFBS acts as a strain sensor detecting the tangential deformation of the bearing outer ring, and the second AFBS (not shown in Figure 9) is isolated from strain and serves for thermal compensation of the strain sensor.

Figure 9. Automotive load-sensing bearing with AFBS sensor during static load testing [13].

In this application, minimal sensor length is required, therefore a 2π-FBG structure was used with a typical length of 5–7 mm, since the $N\pi$-FBGs are generally shorter in comparison with $N\lambda$-FBGs. This is due to the fact that in order to ensure the narrow bandwidth of the FBG spectrum, its induced refractive index must be low, which in its turn necessitates the increase in the FBG length to keep the resulting FBG reflectance sufficiently high.

Another development direction of the automotive sensor systems are the so-called "intelligent tires", which provide real-time measurements of tire grip parameters, including tire–road friction, contact patch dimensions, loads, etc., in various driving conditions [44]. Fiber-optic sensors also possess benefits in intelligent tire instrumentation due to their small footprint, low weight, flexibility and the possibility to be embedded in the tire structure. Figure 10 shows an FBG-based strain sensor attached to the inner surface of a prototype intelligent tire [44].

Figure 10. FBG-based strain sensor attached to the inner surface of the tire [44].

The works [19,45] proposed the usage of AFBSs instead of conventional FBGs to measure the strain and temperature of intelligent tires. An additional advantage of the simplified interrogation scheme of AFBS, besides the reduced costs, is the possibility to create a compact and vibration-proof interrogator suitable for in-wheel installation. In the proposed system, a flexible ring model [46] is used to analyze the tangential deformation of the tire.

The works regarding the subject of AFBS application in automotive engineering are listed in Table 5.

Table 5. Works related to the AFBS application in automotive engineering.

Reference	Subject
[13]	Instrumentation of load-sensing wheel hub bearing using AFBS.
[42]	Application of load-sensing wheel hub bearings for vehicle dynamics control.
[44]	Instrumentation of "intelligent" tires using FBGs.
[19,45]	Instrumentation of "intelligent" tires using AFBSs.

5.4. Medicine

Fiber-optic sensors also offer benefits in a range of biomedical applications due to their small cross-section, lightness, flexibility and chemical resistance, which makes them minimally invasive and suitable for in vivo measurements (i.e., measurements directly inside a patient) [4]. Pressure measurement is an important asset in various medical procedures, including cardiovascular, urologic diagnostics and the monitoring of invasive treatments. A significant research effort is dedicated to the development of catheters for high-resolution manometry, which presupposes the spatial resolution of pressure measurement of 1–2 cm [47].

Addressed fiber Bragg structures were utilized in a catheter for esophagus sphincters monitoring proposed in [48]. The scheme of the catheter placement is shown in Figure 11.

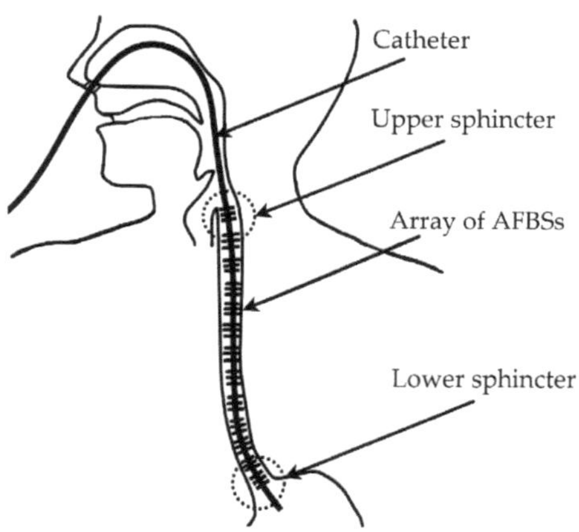

Figure 11. Scheme of catheter placement for esophagus sphincters monitoring, according to [48].

The catheter is instrumented with an array of 2π-FBG addressed structures that measure pressure caused by the sphincters of the esophagus. The interrogation of AFBSs is implemented according to the scheme described in Section 3 of the current paper, which provides benefits in terms of implementation costs in comparison with the conventional FBG sensors. The experimental investigation of the catheter was carried out taking the values of pressure measured by certified insufflator for reference. The maximum deviation of the pressure values measured with the catheter was $\pm 0.1\%$, which meets the requirements for medical applications [48]. In addition, a catheter with a larger number of AFBS sensors (up to 72) was proposed for intestinal peristalsis monitoring [48].

The references related to the subject of AFBS applications in medicine are listed in Table 6.

Table 6. Works related to the AFBS application in medicine.

Reference	Subject
[4]	Application of fiber-optic pressure sensors in medicine.
[47]	Instrumentation of high-resolution catheters using FBGs.
[48]	Instrumentation of high-resolution catheters using AFBSs.

5.5. Environmental Monitoring

A significant research effort is now dedicated to the development of various sensors for environmental monitoring, particularly for concentration measurements of greenhouse gases. An attractive solution for simultaneous gas, temperature and pressure measurement is based on the combination of FBG and a Fabry–Perot resonator (FPR). Such combined fiber-optic sensors (CFOSs) consist of an FPR formed of a thin film at the end face of the optical fiber with FBG near it [49]. The film is made of a polymer, the refractive index of which reversibly changes depending on the gas concentration. The spectral response of the FPR is sensitive to various environmental parameters, including the changes of gas concentration, temperature and pressure, while the FBG central wavelength shift depends mainly on the temperature, which enables the simultaneous measurement of temperature and gas concentration using the CFOS.

The usage of AFBS instead of conventional FBG in CFOS has the advantage of a simplified interrogation scheme as well as the possibility to include several AFBSs with identical central wavelength and addressable interrogation. The scheme of the addressed combined fiber-optic sensor (ACFOS) incorporating a 2λ-FBG structure is presented in Figure 12 [49]. The thickness of the film h and its material are selected based on the type of gas under test. For instance, PEI/PVA coating is used for CO_2 concentration measurements [50]; polyaniline/Co_3O_4–for CO [51]; $LuPc_2$–for NO_2 [52]; PDMS/PMMA–for NH_3 [53]; Cryptophane A–for CH_4 [54] and others. Humidity measurement can be performed using PVA coating [55].

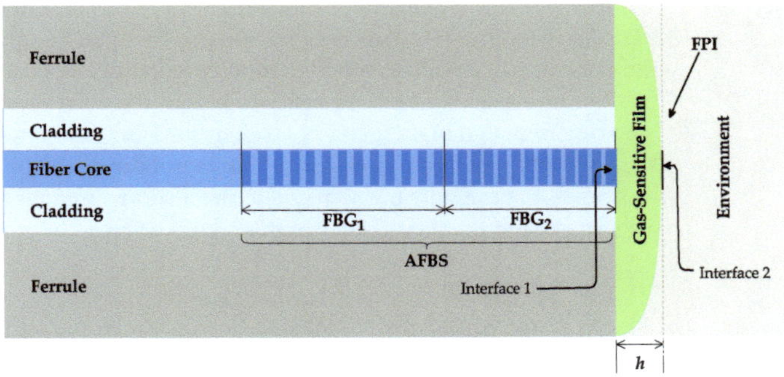

Figure 12. The structure of ACFOS [49].

The ACFOS spectrum is a superposition of the AFBS and FPR spectra, as shown in Figure 13; therefore, it is necessary to separate the FBG and the FPR spectra in order to successfully interrogate the combined sensor, which can be achieved using Karhunen–Loeve transform [56]. According to estimations, the sensor system based on ACFOS allows the measuring the gas concentrations in the range of 10−90% with an error of 0.1−0.5% [49].

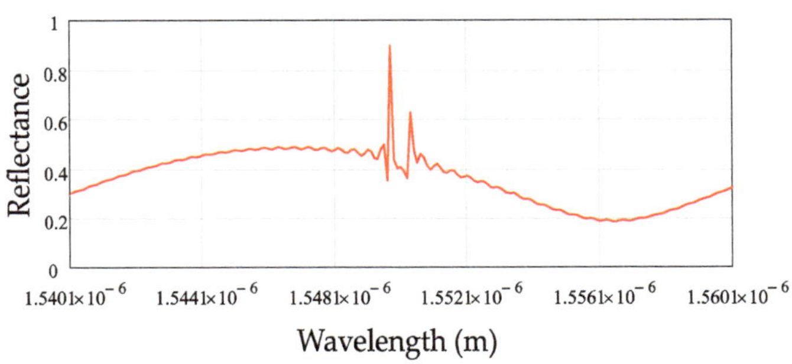

Figure 13. ACFOS reflectance spectrum [49].

The works regarding the subject of AFBS application in environmental monitoring are listed in Table 7.

Table 7. Works related to AFBS application in environmental monitoring.

References	Subject
[49]	AFBS usage in combined sensors for gas concentration measurement.
[50]	Fiber-optic Fabry–Perot interferometer based on PEI/PVA coating for CO_2 concentration measurement.
[51]	Fiber-optic Fabry–Perot interferometer based on polyaniline/Co_3O_4 coating for CO concentration measurement.
[52]	Fiber-optic Fabry–Perot interferometer based on $LuPc_2$ coating for NO_2 concentration measurement.
[53]	Fiber-optic Fabry–Perot interferometer based on PDMS/PMMA coating for NH_3 concentration measurement.
[54]	Fiber-optic Fabry–Perot interferometer based on Cryptophane A coating for CH_4 concentration measurement.
[55]	Fiber-optic Fabry–Perot interferometer for humidity measurement.
[56]	Interrogation of combined sensors using Karhunen–Loeve transform.

6. Development Prospects of Addressed Fiber Bragg Structures

Based on the theoretical and implementation background of the addressed fiber Bragg structures reviewed above, several shortcomings of the current AFBS techniques can be identified. The disadvantage of the $N\lambda$-FBG addressed structures, in general, is that they have significantly greater length in comparison with the $N\pi$-FBGs, which makes it problematic to ensure the uniform temperature and strain variations of all the FBGs constituting the $N\lambda$-FBG in order to maintain the unchanged address frequencies. In addition, some applications require the usage of AFBSs with high difference frequencies between their spectral components (up to hundreds of GHz). In order to obtain the corresponding beating frequency of the electrical signal, a high-speed photodetector is required, which significantly increases the total cost of the system. However, a number of directions of AFBS further development can be drawn that mitigate the abovementioned shortcomings of the current AFBSs' state of the art and expand the possibilities of their application.

6.1. Moiré Recording of $N\lambda$-FBG

As stated before, the $N\lambda$-FBG addressed structures are formed by sequential recording of several ultra-narrowband FBGs with different central wavelengths. In order to provide ultra-narrowband spectral response, it is necessary to increase the FBG length, and the sequential recording of a number of such FBGs results in the total length of $N\lambda$-FBG reaching centimeters or even dozens of centimeters. However, in order to maintain the unchanged address frequency, it is crucial to ensure that all the FBGs belonging to the same AFBS are subjected to the same strain or temperature impact, which is problematic in cases of extensive structure length. This issue can be mitigated by using the moiré recording technique [57,58]. During the moiré fabrication, the first FBG is written using the conventional recording method, such as the one using phase mask, then the fiber is stretched, and the second FBG is written directly over the first grating. Stretching the fiber between the recordings changes the length (and the period) of the initial grating with respect to the second. Since both FBGs are located in the same section of the optical fiber, they are always subjected to the same temperature and strain, therefore the address frequency of such 2λ-FBG remains invariant.

6.2. Transverse Load Sensing Using 2π-FBG

Fiber Bragg gratings can be used not only to measure axial strain of the optical fiber but also to measure the transverse load applied to it, which is possible due to the induced-birefringence effects of the FBGs [59]. Axial strain results in linear shift of the FBG central wavelength, while the transverse load causes an additional birefringence in the optical fiber, creating two different Bragg wavelengths corresponding to each of the polarization modes. This effect can be utilized to measure transverse load since the wavelength difference between the two Bragg wavelengths has a linear relationship with the applied load. However, due to the significantly lower sensitivity to the transverse load in comparison with the axial load sensitivity, this limits the measurement resolution of the transverse load of the conventional FBGs, which have much wider spectral bandwidth than the wavelength shift induced by the birefringence. For this reason, the usage of the FBG with phase shift provides more accurate measurements of the transverse load because of the narrowness of the transmission window of such FBG [60].

If a 2π-FBG structure is subjected to the transverse load, then the output optical radiation includes four components:

$$I(t) = I_{-1-1}\cos(\omega_{-1-1}t) + I_{-1+1}\cos(\omega_{-1+1}t) + I_{+1-1}\cos(\omega_{+1-1}t) + I_{+1+1}\cos(\omega_{+1+1}t), \quad (12)$$

where ω_{-1-1}, ω_{-1+1}, ω_{+1-1}, ω_{+1+1} and I_{-1-1}, I_{-1+1}, I_{+1-1} and I_{+1+1} are the frequencies and the amplitudes of the spectral components of the optical signal formed due to the induced-birefringence effect. The address frequency Ω is equal to the difference between the maximum beating frequency and the half of the minimal beating frequency:

$$\Omega = (\omega_{+1+1} - \omega_{-1-1}) - (\omega_{+1+1} - \omega_{+1-1})/2. \quad (13)$$

The frequency resulting from the transverse load corresponds to the minimal beating frequency. Therefore, the main requirement of the 2π-FBG for the transverse load sensing is the minimal beating frequency at the maximum transverse load being lower than half of the address frequency, i.e., $(\omega_{+1+1} - \omega_{+1-1}) < \Omega$.

6.3. Address Frequency Downshifting

In some applications, it is necessary to use the AFBS with high address frequency to ensure the sufficient difference of the spectral components' amplitudes. An example of such application is the probing of the spectral response of the Fabry–Perot interferometer (FPI) in order to define its position relative to the AFBS spectrum, which can also be used in gas concentration measurements. In Figure 14, the dashed line represents a spectral response of the Fabry–Perot interferometer, while the solid lines denote the spectral components of the AFBS. In order to provide the sufficient difference of the AFBS spectral components' amplitudes, the difference frequency Ω_3 must be rather high, reaching hundreds of GHz, due to the slow variation of the FPI amplitude over frequency. Therefore, the usage of the double-component AFBS in this application is problematic since it requires the usage of a high-frequency photodetector to generate the beating frequency of hundreds of GHz. This issue can be solved by using the AFBS with four spectral components, as shown in Figure 14. The difference frequency between the first and the second component (Ω_1), as well as the third and the fourth one (Ω_2) are in the range of several GHz, which is significantly lower than Ω_3. Using the amplitudes of the beating frequencies Ω_1 and Ω_2 and knowing the difference frequency Ω_3, it is possible to define the position of the FPI spectrum relative to the AFBS spectral response, using a comparatively low-frequency photodetector.

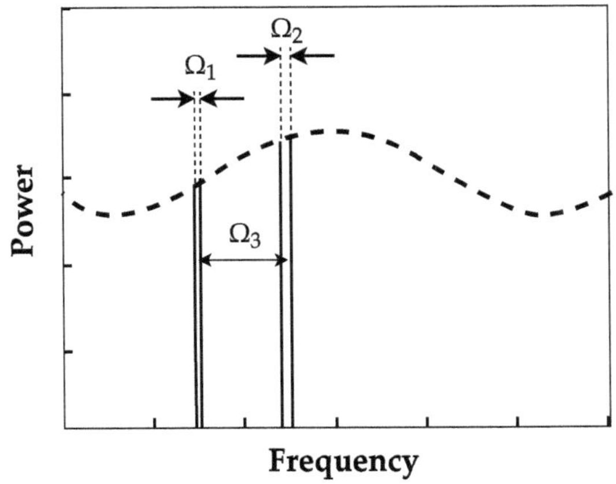

Figure 14. Probing of the Fabry–Perot interferometer using AFBS.

The proposed "downshifting" of the address frequency can also be implemented in the cases when it is required to separate the spectral components by high frequency in order to ensure that they do not coincide with the components of another AFBS.

7. Conclusions

The current paper is dedicated to comprehensive review of the works on the subject of addressed fiber Bragg structures (AFBS) covering theoretical background, interrogation principles, fabrication, calibration, as well as implementation examples. Generally, AFBS can be used in all the applications of conventional FBG-based sensors offering a simplified interrogation scheme, which can provide additional benefits in some areas, such as increased measurement rate and resolution. Based on the presented review, the directions of AFBS further development are proposed, including the moiré recording of the $N\lambda$-FBG structures and transverse load sensing using AFBSs with phase shifts.

Author Contributions: T.A. and A.S. drafted the manuscript. O.M. supervised the work. T.A., G.I., A.K., R.M. (Rinat Misbakhov), R.M. (Rustam Misbakhov), G.M., O.M., I.N. and A.S. commented, edited and reviewed the manuscript. All authors have read and agreed to the published version of the manuscript.

Funding: This work was financially supported by the Ministry of Science and Higher Education as part of the "Priority 2030" program.

Institutional Review Board Statement: Not applicable.

Informed Consent Statement: Not applicable.

Data Availability Statement: The data presented in this study are available on request from the corresponding author. The data are not publicly available due to rules of our contract conditions with our customer.

Conflicts of Interest: The authors declare no conflict of interest.

References

1. Hill, K.O.; Fujii, Y.; Johnson, D.C.; Kawasaki, B.S. Photosensitivity in Optical Fiber Waveguides: Application to Reflection Filter Fabrication. *Appl. Phys. Lett.* **1978**, *32*, 647–649. [CrossRef]
2. Hegde, G.; Asokan, S.; Hegde, G. Fiber Bragg Grating Sensors for Aerospace Applications: A Review. *ISSS J. Micro Smart Syst.* **2022**, *11*, 257–275. [CrossRef]
3. Falcetelli, F.; Martini, A.; Di Sante, R.; Troncossi, M. Strain Modal Testing with Fiber Bragg Gratings for Automotive Applications. *Sensors* **2022**, *22*, 946. [CrossRef] [PubMed]
4. Poeggel, S.; Tosi, D.; Duraibabu, D.; Leen, G.; McGrath, D.; Lewis, E. Optical Fibre Pressure Sensors in Medical Applications. *Sensors* **2015**, *15*, 17115–17148. [CrossRef]
5. Baldwin, C. 8—Fiber Optic Sensors in the Oil and Gas Industry: Current and Future Applications. In *Opto-Mechanical Fiber Optic Sensors*; Alemohammad, H., Ed.; Butterworth-Heinemann: Oxford, UK, 2018; pp. 211–236. ISBN 978-0-12-803131-5.
6. Measures, R.M.; Melle, S.; Liu, K. Wavelength Demodulated Bragg Grating Fiber Optic Sensing Systems for Addressing Smart Structure Critical Issues. *Smart Mater. Struct.* **1992**, *1*, 36–44. [CrossRef]
7. Matveenko, V.P.; Shardakov, I.N.; Voronkov, A.A.; Kosheleva, N.A.; Lobanov, D.S.; Serovaev, G.S.; Spaskova, E.M.; Shipunov, G.S. Measurement of Strains by Optical Fiber Bragg Grating Sensors Embedded into Polymer Composite Material. *Struct. Control Health Monit.* **2018**, *25*, e2118. [CrossRef]
8. Davis, M.A.; Bellemore, D.G.; Kersey, A.D. Structural Strain Mapping Using a Wavelength/Time Division Addressed Fiber Bragg Grating Array. In Proceedings of the Second European Conference on Smart Structures and Materials; McDonach, A., Gardiner, P.T., McEwen, R.S., Culshaw, B., Eds.; International Society for Optics and Photonics: Glasgow, UK, 1994; Volume 2361, pp. 342–345.
9. Rao, Y.J.; Kalli, K.; Brady, G.; Webb, D.J.; Jackson, D.A.; Zhang, L.; Bennion, I. Spatially-Multiplexed Fibre-Optic Bragg Grating Strain and Temperature Sensor System Based on Interferometric Wavelength-Shift Detection. *Electron. Lett.* **1995**, *31*, 1009–1010. [CrossRef]
10. Triana, A.; Pastor, D.; Varón, M. A Code Division Design Strategy for Multiplexing Fiber Bragg Grating Sensing Networks. *Sensors* **2017**, *17*, 2508. [CrossRef]
11. Triana, C.A.; Pastor, D.; Varón, M. Optical Code Division Multiplexing in the Design of Encoded Fiber Bragg Grating Sensors. *Opt. Pura Apl.* **2016**, *49*, 17–28. [CrossRef]
12. Morozov, O.G.; Sakhabutdinov, A.J. Addressed fiber bragg structures in quasi-distributed microwave-photonic sensor systems. *Comput. Opt.* **2019**, *43*, 535–543. [CrossRef]

13. Agliullin, T.; Gubaidullin, R.; Sakhabutdinov, A.; Morozov, O.; Kuznetsov, A.; Ivanov, V. Addressed Fiber Bragg Structures in Load-Sensing Wheel Hub Bearings. *Sensors* **2020**, *20*, 6191. [CrossRef]
14. Morozov, O.; Sakhabutdinov, A.; Anfinogentov, V.; Misbakhov, R.; Kuznetsov, A.; Agliullin, T. Multi-Addressed Fiber Bragg Structures for Microwave-Photonic Sensor Systems. *Sensors* **2020**, *20*, 2693. [CrossRef]
15. Morozov, O.G.; Sakhabutdinov, A.Z.; Nureev, I.I.; Misbakhov, R.S. Modelling and Record Technologies of Address Fiber Bragg Structures Based on Gratings with Two Symmetrical Pi-Phase Shifts. *J. Phys. Conf. Ser.* **2019**, *1368*, 022048. [CrossRef]
16. Gubaidullin, R.R.; Agliullin, T.A.; Nureev, I.I.; Sahabutdinov, A.Z.; Ivanov, V. Application of Gaussian Function for Modeling Two-Frequency Radiation from Addressed FBG. In Proceedings of the 2020 Systems of Signals Generating and Processing in the Field of on Board Communications, Moscow, Russia, 19–20 March 2020; IEEE: New York, NY, USA, 2020; pp. 1–4.
17. Sakhabutdinov, A.Z.; Morozov, O.G.; Agliullin, T.A.; Gubaidullin, R.R.; Karimov, K.G. Radiophotonic Sensor System Based on Multi-Addressed Fiber Bragg Structures. In *Proceedings of the Optical Technologies for Telecommunications 2021*; SPIE: Bellingham, WA, USA, 2022; Volume 12295, pp. 32–37.
18. Agliullin, T.A.; Gubaidullin, R.R.; Sakhabutdinov, A.Z. Multi-Sensory Strain Measurement Using Addressed Fiber Bragg Structures in Load Sensing Bearings. In Proceedings of the 2021 Systems of Signals Generating and Processing in the Field of on Board Communications, Moscow, Russia, 16–18 March 2021; IEEE: New York, NY, USA, 2021; pp. 1–5.
19. Agliullin, T.A.; Gubaidullin, R.R.; Morozov, O.G.; Sahabutdinov, A.Z.; Ivanov, V. Tire Strain Measurement System Based on Addressed FBG-Structures. In Proceedings of the 2019 Systems of Signals Generating and Processing in the Field of on Board Communications, Moscow, Russia, 20–21 March 2019; IEEE: New York, NY, USA, 2019; pp. 1–5.
20. Sahabutdinov, A.Z.; Kuznetsov, A.Z.; Nureev, I.I.; Morozov, O.G.; Faskhutdinov, L.M.; Petrov, A.V.; Kuchev, S.M. Calibration of Combined Pressure and Temperature Sensors. *Int. J. Appl. Eng. Res.* **2015**, *10*, 44948–44957.
21. Chehura, E.; James, S.W.; Tatam, R.P. A Simple Method for Fabricating Phase-Shifted Fibre Bragg Gratings with Flexible Choice of Centre Wavelength. In Proceedings of the 20th International Conference on Optical Fibre Sensors, Edinburgh, UK, 5–9 October 2009; SPIE: Bellingham, WA, USA, 2009; Volume 7503, pp. 1043–1046.
22. Martinez, C.; Ferdinand, P. Analysis of Phase-Shifted Fiber Bragg Gratings Written with Phase Plates. *Appl. Opt. AO* **1999**, *38*, 3223–3228. [CrossRef]
23. Novikova, V.; Varzhel, S.; Dmitriev, A.; Zalesskaya, Y.; Idrisov, R. Spectral Characteristics Study of Phase-Shifted Fiber Bragg Gratings under Pressure Applied Perpendicular to Fiber Axis. *J. Sci. Tech. Inf. Technol. Mech. Opt.* **2018**, *117*, 751–757. [CrossRef]
24. Morozov, O.G.; Sakhabutdinov, A.Z.; Nureev, I.I.; Kuznetsov, A.A.; Misbakhov, R.S. Holographic methods for recording of addressed fiber Bragg structures. In Proceedings of the XVII International Conference on Holography and Applied Optical Technology, Moscow, Russia, 8–9 September 2020; BMSTU: Moscow, Russia, 2020; pp. 30–39.
25. Agliullin, T.A.; Gubaidullin, R.R.; Kuznetsov, A.A.; Sakhabutdinov, A.Z.; Nureev, I.I.; Morozov, O.G.; Sharafiev, U.F. Control of the Fine Structure of Three-Component Addressed Bragg Measuring Converters Sequentially Written in the Optical Fiber of a Quasi-Distributed Microwave-Photonic Sensor System. *Vestn. Volga State Univ. Technol. Ser. Radio Eng. Infocommun. Syst.* **2020**, *4*, 72–81. [CrossRef]
26. Tremblay, G.; Gillet, J.-N.; Sheng, Y.; Bernier, M.; Paul-Hus, G. Optimizing Fiber Bragg Gratings Using a Genetic Algorithm With Fabrication-Constraint Encoding. *J. Light. Technol. JLT* **2005**, *23*, 4382. [CrossRef]
27. Kashyap, R.; Mckee, P.F.; Armes, D. UV Written Reflection Grating Structures in Photosensitive Optical Fibres Using Phase-Shifted Phase Masks. *Electron. Lett.* **1994**, *30*, 1977–1978. [CrossRef]
28. Jiang, L.; Zhang, D.; Zhang, W.; Kai, G.; Dong, X. Fabrication of Dual-Wavelength Fiber Bragg Grating with a Longitudinal Stretch. *Front. Phys. China* **2006**, *1*, 108–111. [CrossRef]
29. Pisco, M.; Iadicicco, A.; Campopiano, S.; Cutolo, A.; Cusano, A. Micro-Structured Chirped Fiber Bragg Gratings: Toward New Spatial Encoded Fiber Optic Sensors. In Proceedings of the Third European Workshop on Optical Fibre Sensors, Napoli, Italy, 4–6 July 2007; SPIE: Bellingham, WA, USA, 2007; Volume 6619, pp. 460–463.
30. Allil, R.C.S.B.; Werneck, M.M.; Ribeiro, B.A.; de Nazaré, F.V.B. *Application of Fiber Bragg Grating Sensors in Power Industry*; IntechOpen: London, UK, 2013; ISBN 978-953-51-1131-3.
31. Daley, T.M.; Freifeld, B.M.; Ajo-Franklin, J.; Dou, S.; Pevzner, R.; Shulakova, V.; Kashikar, S.; Miller, D.E.; Goetz, J.; Henninges, J.; et al. Field Testing of Fiber-Optic Distributed Acoustic Sensing (DAS) for Subsurface Seismic Monitoring. *Lead. Edge* **2013**, *32*, 699–706. [CrossRef]
32. Wang, C.; Shang, Y.; Liu, X.-H.; Wang, C.; Yu, H.-H.; Jiang, D.-S.; Peng, G.-D. Distributed OTDR-Interferometric Sensing Network with Identical Ultra-Weak Fiber Bragg Gratings. *Opt. Express OE* **2015**, *23*, 29038–29046. [CrossRef] [PubMed]
33. Misbakhov, R.S. DAS Quazi-Distributed System Combined with Temperature Measuring on Addressed Fiber Bragg Gratings. In Proceedings of the Optical Technologies for Telecommunications 2019, Kazan, Russia, 18–22 November 2019; SPIE: Bellingham, WA, USA, 2020; Volume 11516, pp. 196–203.
34. Misbakhov, R.S.; Vasev, A.N.; Sakhabutdinov, A.Z.; Nureev, I.I.; Morozov, O.G.; Lipatnikov, K.A.; Vasilets, A.A. Address Fiber Optical Sensor for Relative Humidity Measuring in a Switchgear. In Proceedings of the Optical Technologies for Telecommunications 2019, Kazan, Russia, 18–22 November 2019; SPIE: Bellingham, WA, USA, 2020; Volume 11516, p. 1151612.
35. Yeo, T.L.; Sun, T.; Grattan, K.T.V.; Parry, D.; Lade, R.; Powell, B.D. Polymer-Coated Fiber Bragg Grating for Relative Humidity Sensing. *IEEE Sens. J.* **2005**, *5*, 1082–1089. [CrossRef]

36. Huang, X.F.; Sheng, D.R.; Cen, K.F.; Zhou, H. Low-Cost Relative Humidity Sensor Based on Thermoplastic Polyimide-Coated Fiber Bragg Grating. *Sens. Actuators B Chem.* **2007**, *127*, 518–524. [CrossRef]
37. Liang, W.; Huang, Y.; Xu, Y.; Lee, R.K.; Yariv, A. Highly Sensitive Fiber Bragg Grating Refractive Index Sensors. *Appl. Phys. Lett.* **2005**, *86*, 151122. [CrossRef]
38. Hartog, A.H. Distributed Fiber-Optic Sensors. In *Optical Fiber Sensor Technology*; Grattan, K.T.V., Meggitt, B.T., Eds.; Optical and Quantum Electronics Series; Springer Netherlands: Dordrecht, The Netherlands, 1995; pp. 347–382. ISBN 978-94-011-1210-9.
39. Inaudi, D.; Glisic, B. Long-Range Pipeline Monitoring by Distributed Fiber Optic Sensing. *J. Press. Vessel Technol.* **2009**, *132*, 011701. [CrossRef]
40. Nureev, I.I.; Agliullin, T.A.; Kadushkin, V.V.; Kurbiev, I.Y.; Proskuriakov, A.D. Combined Brillouin Sensor System for Simultaneous Local and Distributed Temperature and Strain Measurements for Downhole Telemetry. *IOP Conf. Ser. Mater. Sci. Eng.* **2020**, *734*, 012135. [CrossRef]
41. Zaidi, F.; Krebber, K. Advanced Hybrid BOFDA/FBG Sensor System for Simultaneously Point-Wise and Distributed Temperature/Strain Measurements. In Proceedings of the 23rd International Conference on Optical Fibre Sensors, Santander, Spain, 2–6 June 2014; SPIE: Bellingham, WA, USA, 2014; Volume 9157, pp. 885–888.
42. Kerst, S.; Shyrokau, B.; Holweg, E. Reconstruction of Wheel Forces Using an Intelligent Bearing. *SAE Int. J. Passeng. Cars–Electron. Electr. Syst.* **2016**, *9*, 196–203. [CrossRef]
43. Dinçmen, E.; Güvenç, B.A.; Acarman, T. Extremum-Seeking Control of ABS Braking in Road Vehicles With Lateral Force Improvement. *IEEE Trans. Control Syst. Technol.* **2014**, *22*, 230–237. [CrossRef]
44. Roveri, N.; Pepe, G.; Mezzani, F.; Carcaterra, A.; Culla, A.; Milana, S. OPTYRE—Real Time Estimation of Rolling Resistance for Intelligent Tyres. *Sensors* **2019**, *19*, 5119. [CrossRef]
45. Gubaidullin, R.; Agliullin, T.; Sakhabutdinov, A.J. Fiber Optic Tire Speed Sensor. In Proceedings of the 2021 Systems of Signals Generating and Processing in the Field of on Board Communications, Moscow, Russia, 16–18 March 2021; IEEE: New York, NY, USA, 2021; p. 7.
46. Xiong, Y.; Tuononen, A. Rolling Deformation of Truck Tires: Measurement and Analysis Using a Tire Sensing Approach. *J. Terramechanics* **2015**, *61*, 33–42. [CrossRef]
47. Bueley, C.; Wild, P.M. A Flexible, Highly Sensitive Catheter for High Resolution Manometry Based on in-Fibre Bragg Gratings. *Meas. Sci. Technol.* **2013**, *24*, 094003. [CrossRef]
48. Purtov, V.; Sakhabutdinov, A.; Nureev, I.; Morozov, O.; Agliullin, A. Few and Multisensor Catheters for the High-Resolution Manometry on the Basis of Addressed Fiber Bragg Gratings. *AP* **2018**, *5*, 310–335. [CrossRef]
49. Morozov, O.; Tunakova, Y.; Hussein, S.M.R.H.; Shagidullin, A.; Agliullin, T.; Kuznetsov, A.; Valeev, B.; Lipatnikov, K.; Anfinogentov, V.; Sakhabutdinov, A. Addressed Combined Fiber-Optic Sensors as Key Element of Multisensor Greenhouse Gas Monitoring Systems. *Sensors* **2022**, *22*, 4827. [CrossRef]
50. Ma, W.; Wang, R.; Rong, Q.; Shao, Z.; Zhang, W.; Guo, T.; Wang, J.; Qiao, X. CO_2 Gas Sensing Using Optical Fiber Fabry–Perot Interferometer Based on Polyethyleneimine/Poly(Vinyl Alcohol) Coating. *IEEE Photonics J.* **2017**, *9*, 6802808. [CrossRef]
51. Peng, J.; Feng, W.; Yang, X.; Huang, G.; Liu, S. Dual Fabry–Pérot Interferometric Carbon Monoxide Sensor Based on the PANI/Co $_3$ O $_4$ Sensitive Membrane-Coated Fibre Tip. *Z. Für Nat. A* **2019**, *74*, 101–107. [CrossRef]
52. Debliquy, M.; Bueno, A.; Caucheteur, C.; Megret, P.; Olivier, M.-G.; Lahem, D.; Bouvet, M. Nitrogen Dioxide Sensor Based on Optical Fiber Coated with a Porous Silica Matrix Incorporating Lutetium Bisphthalocyanine. In Proceedings of the 2013 Seventh International Conference on Sensing Technology (ICST), Wellington, New Zealand, 3–5 December 2013; IEEE: New York, NY, USA, 2013; pp. 655–659.
53. Kanawade, R.; Kumar, A.; Pawar, D.; Late, D.; Mondal, S.; Sinha, R.K. Fiber Optic Fabry–Perot Interferometer Sensor: An Efficient and Fast Approach for Ammonia Gas Sensing. *J. Opt. Soc. Am. B JOSAB* **2019**, *36*, 684–689. [CrossRef]
54. Haitao, Y.; Xiaoyan, Z.; Hui, H.; Zhiqiang, Z.; Zhi, Z.; Huijuan, L.; Daofu, H.; Ming, W. Optical Fiber Methane Gas Sensor Based on Fabry-Perot Cavity. *CJL* **2012**, *39*, s208004. [CrossRef]
55. Su, D.; Qiao, X.; Rong, Q.; Sun, H.; Zhang, J.; Bai, Z.; Du, Y.; Feng, D.; Wang, Y.; Hu, M.; et al. A Fiber Fabry–Perot Interferometer Based on a PVA Coating for Humidity Measurement. *Opt. Commun.* **2013**, *311*, 107–110. [CrossRef]
56. Tosi, D. Advanced Interrogation of Fiber-Optic Bragg Grating and Fabry-Perot Sensors with KLT Analysis. *Sensors* **2015**, *15*, 27470–27492. [CrossRef]
57. Legoubin, S.; Fertein, E.; Douay, M.; Bernage, P.; Niay, P.; Bayon, F.; Georges, T. Formation of Moiré Grating in Core of Germanosilicate Fibre by Transverse Holographic Double Exposure Method. *Electron. Lett.* **1991**, *27*, 1945–1947. [CrossRef]
58. Gillooly, A.M.; Dobb, H.; Zhang, L.; Bennion, I. Distributed Load Sensor by Use of a Chirped Moiré Fiber Bragg Grating. *Appl. Opt. AO* **2004**, *43*, 6454–6457. [CrossRef] [PubMed]

59. Gafsi, R.; El-Sherif, M.A. Analysis of Induced-Birefringence Effects on Fiber Bragg Gratings. *Opt. Fiber Technol.* **2000**, *6*, 299–323. [CrossRef]
60. Wang, Y.; Li, N.; Huang, X.; Wang, M. Fiber Optic Transverse Load Sensor Based on Polarization Properties of π-Phase-Shifted Fiber Bragg Grating. *Opt. Commun.* **2015**, *342*, 152–156. [CrossRef]

photonics

MDPI

Review

Recent Progress in Photonic Crystal Devices and Their Applications: A Review

Rahul Kumar Gangwar [1], Akhilesh Kumar Pathak [2,*] and Santosh Kumar [3,*]

1 Department of Physics, Rajdhani College, University of Delhi, M. G. Marg, Raja Garden, New Delhi 110015, India; rahul0889@gmail.com
2 Center for Smart Structures and Materials, Department of Mechanical Engineering, Northwestern University, Evanston, IL 60208, USA
3 Shandong Key Laboratory of Optical Communication Science and Technology, School of Physics Science and Information Technology, Liaocheng University, Liaocheng 252059, China
* Correspondence: akhilesh.pathak@northwestern.edu (A.K.P.); santosh@lcu.edu.cn (S.K.)

Abstract: The research field of photonic crystals (PhCs) remains active on a global scale. PhCs, which are periodic optical nanostructures with the characteristics of excellent light field confinement and numerous varying degrees of freedom, provide a solid foundation for controlling the movement of light. Periodic variation of the index of refraction in two or three spatial dimensions with a substantial high-to-low ratio generates a number of intriguing phenomena and enables a variety of potential functionalities. Recently, intriguing devices based on PhCs, such as Y-branches, small-diameter bent waveguides, and miniature resonator cavities, have been proposed and extensively utilized. PhC waveguides are considered ideal candidates for a variety of applications, such as in power splitters, logic gates, sensing and communication fields, etc. These exceptional characteristics may facilitate the development of a dense integrated circuit. However, PhC technology is still relatively new and therefore requires additional effort to fully exploit it. This paper reviews the most popular and essential optical components based on PhCs, including power splitters, modulators, polarization maintaining devices, sensors, and lasers, to summarize the most recent developments relating this hot topic. These devices have superior performance and a smaller footprint compared to conventional photonic devices.

Keywords: photonic crystals; photonic band gap; photonic crystal waveguide; photonic crystal fiber; photonic crystal sensors; photonic crystal beam splitter; lasers

Citation: Gangwar, R.K.; Pathak, A.K.; Kumar, S. Recent Progress in Photonic Crystal Devices and Their Applications: A Review. *Photonics* **2023**, *10*, 1199. https://doi.org/10.3390/photonics10111199

Received: 5 October 2023
Revised: 22 October 2023
Accepted: 26 October 2023
Published: 27 October 2023

1. Introduction

Photonic crystals (PhCs) were discovered in 1987 and were first reported by Yablonovitch [1] and John [2]. Within PhCs, electromagnetic waves scatter. For some wavelengths, destructive interference occurs, resulting in the formation of a photonic bandgap (PBG), which is similar to the energy bandgap of electron waves in a semiconductor. It may be able to influence the propagation of light due to the likelihood of producing a PBG. PhCs are used in a number of notable applications, including self-collimation [3], negative refraction [4], optical diodes [5], and light bending [6].

One-dimensional (1D), two-dimensional (2D), and three-dimensional (3D) photonic crystals represent distinct dimensions of periodic structures designed to control the flow of light. In 1D photonic crystals, refractive index variations occur along a single direction, enabling precise control over light in that dimension. Two-dimensional photonic crystals expand this control to two dimensions, offering more complex light manipulation and enabling the creation of planar structures with applications in waveguides and microcavities. In contrast, 3D photonic crystals extend refractive index variations across all three spatial dimensions, granting comprehensive control over light propagation in volumetric structures. These dimensionality differences lead to variations in band gap properties, complexity of design and fabrication, and suitability for diverse applications, making each type of photonic crystal well-suited for specific optical and photonic device requirements. Although the development of 3D PhCs first attracted a lot of attention, planar 2D PhC devices are now favored because of their comparatively simple production procedure. Many 3D structural traits are present in 2D PhC structures. These factors have sparked an increase in research and development for sub-micrometer and micrometer-sized active and passive photonic elements like logic gates, polarizers, polarization maintaining waveguides (WGs), lasers, and solar absorbers, among others.

PhCs are a good example of a resolution that encourages great performance in sensing applications. PhCs provide strong optical confinement to a very tiny volume, enabling the identification of chemical species classified according to their nanoscale size. The use of PhC WG elements in microfluidic [7] and biochemical sensing [8] has lately come under investigation. In sensing applications, a number of photonic structure designs based on various platforms have been extensively researched and used [9–11].

A 1D PhC consists of a periodic variation of the refractive index in the direction of light propagation. However, it provides a homogeneous medium in the other two directions [12]. The refractive index varies in two directions for 2D PhCs, but not in the third direction. This can be realized just by introducing a hole in a high-refractive-index material, e.g., silicon [13]. Three-dimensional PhCs can be fabricated by varying the refractive index of the material in all three directions, such as the stack of spheres made of some dielectric material kept in air [14]. The propagation of light in a periodic structure can be studied by the periodic arrangement of atoms, such as the propagation of electrons. PhCs are frequently mentioned in theories like energy bands, the Bloch theorem, and Brillouin zones. One-dimensional PhCs, also known as multilayers, lack a complete PBG, while 3D PhC fabrication is difficult because of their narrow lattice constant. However, 2D PhCs comprise a full PBG and a simpler fabrication step compared to 3D PhCs, making them more appealing to researchers. A 2D PhC is composed of an air pore arranged in a dielectric substrate or a cylindrical rod made of dielectric material submerged in air. The PBG of a PhC can be tuned by altering the radius of the rod, the lattice constant, and the refractive index of the dielectric. The resonant cavities are considered one of the potential candidates in order to create a tunable optical filter for dense wavelength–division multiplexing (DWDM) systems [15–17]. For DWDM systems, the resonant cavities with a high quality factor (Q-factor) filter the desired wavelengths with a suitable bandwidth. The adjustable filters can also be formed by introducing composition improvements to these cavities. In 2010, Rostami et al. utilized this technique to develop a demultiplexer [18]. The advantages of PhC cavities for optical interconnects can be summed up in one number: the resonator's interference order. The free spectral range (FSR), and thus the capacitance, increase by reducing this number. The conceptual limit of this theory is a PhC resonator, which often serves as the fundamental cavity mode. Due to their wide FSR, single-mode (SM) PhC cavities with various cavity resonances can be constructed in large arrays. When the PhC interacts with the incoming light, it gets reflected from each interface. A high reflection is witnessed for the particular wavelength of light when this condition is fulfilled.

Two dimensional PhCs, which are periodic in two directions and have a photonic band gap for light propagating in the plane of periodicity, are extensively used in a diverse range of applications, such as high Q-cavity devices for efficient light confinement [19], slow light optical waveguide devices [20], and optical waveguides with sharp bands and low losses for photonic integration [21]. Photonic crystal waveguides (PCWs), a class of 2-D PhC-based devices, offer an intriguing framework for making use of slow light effects. The large group index, which is a result of in-plane back-and-forth scattering at optical frequencies at the boundaries of photonic transmission bands, is where the slow light effects in PCWs arise [22].

This review is organized in the following manner: Section 2 contains basic ideas of the PhC structure; Section 3 describes the PhC-based polarization maintaining WGs; Section 4 covers the literature showing the role of PhC in the development of solar cells; Section 5 covers the potential of PhC in sensing; Section 6 discusses the PhC-based logic gates; and Section 7 presents an overview of the PhC-based amplifiers. Section 8 discusses the future prospects and challenges related to PhC-based devices, and finally, an overall conclusion is drawn in Section 9.

2. Photonic Crystal Structures

In the past decade, a strong effort has been made to control and manipulate light propagation. The propagation of light in a periodic material has been a burning topic of research for a large span of time. John [1] and Yablonovitch [2] separately reported the concept of PhC in 1987, utilizing the concept of semiconductor crystals along with the analogy of an electronic band gap. In general, PhCs can be defined as a periodic dielectric material with periodicity in one, two, or all three orthogonal directions; therefore, they are known as 1D PhCs, 2D PhCs, and 3D PhCs. Among these three structures, a 2D PhC combines most of the interesting optical characteristics with the availability and accessibility of fabrication, and thus has wider research applications. It took a century to discover 2D photonic band gap (PBG) materials after the identification of 1D band gaps. The 2D system shows most of the significant properties of PhCs, ranging from nontrivial Brillouin zones to topological sensitivity, even with a minimal index contrast. They hold great potential for validating the most widely favored PhC devices. Understanding 2D PhCs hinges on recognizing that the fields in two dimensions can be divided into two polarizations by virtue of their symmetry: transverse electric (TE) and transverse magnetic (TM). In the TE state, the electric field resides within the plane, while the magnetic field extends perpendicularly to it. Conversely, in the TM state, the magnetic field lies within the (xy) plane, while the electric field is perpendicular (z). Within 2D PhCs, a diverse array of configurations is possible, thanks to variations in permittivity along two directions, with the medium remaining uniform in the third direction. Utilizing PhCs with 2D periodic lattices significantly broadens the scope for engineering photonic band structures. Two-dimensional PhCs offer a broad range of configurations due to the periodic variation of dielectric materials along two directions, while the third direction remains uniform. Two-dimensional PhCs can be categorized into two types: (i) the air holes in materials of a high refractive index and (ii) the rods of materials of a high refractive index. The former type of 2D PhC can be easily fabricated just by etching holes periodically in high-dielectric materials such as gallium arsenide (GaAs), silicon (Si), germanium (Ge), etc. [3]. Because of the periodic variation of the refractive index, the PhC possesses a PBG, which is defined as a range of frequencies that is prohibited from propagating inside the crystal. This particular unique feature has been exploited widely in order to form waveguides by introducing line defects in the PhC structure. The line defect in PhC introduces a guided mode in PBG, and it can therefore be utilized to guide light from one point to another. The allowed frequencies in the PBG are guided via defect and out-of-plane confinement based on the total internal reflection (TIR) phenomenon. PhC WGs fabricated by removing holes from a PhC structure are shown in Figure 1a. The dispersion relation diagram of the PhC WG (considering 400 nm diameter holes arranged periodically in a hexagonal manner with a lattice constant of 500 nm) is calculated using the plane wave expansion method and is shown in Figure 1b. The grey-shaded region in Figure 1b represents the leaky zone, while the black lines exhibit the guided modes [4].

A silicon-on-insulator (SOI) is generally preferred for fabricating PhC WGs due to its favorable optical characteristic, which appears at 1.55 μm and complementary metal–oxide semiconductor (CMOS) fabrication compatibility for cost-effective and high-volume production. An SOI comprises a thin silicon guiding layer with a refractive index (RI) of 3.45 on a thick buried oxide cladding layer with an RI of 1.45, which provides a strong confinement of the light. The high RI of the silicon guiding layer makes the SOI an excellent choice to fabricate the PhC structure with a higher PBG, which is often required in various applications. The thickness of the guiding layer plays a crucial role in determining the properties of the PhC slab. If the guiding layer is too thin, then the PhC slab will provide weak confinement, as the guided mode will be very close to the light cone. On the other hand, the thick slab generates higher-order modes, which may destroy the PBG. The optimum thickness of the PhC slab satisfying the single-mode condition can be defined as follows [4]:

$$h_{max} = \frac{\lambda_0}{2\pi} \frac{1}{\sqrt{n^2 - n_2^2}} \left[\pi + \tan^{-1} \left(\frac{\sqrt{n_2^2 - n_1^2}}{\sqrt{n^2 - n_2^2}} \right) \right] \tag{1}$$

where λ_0, n, n_1, and n_2 represent the free-space wavelength, RI of the guiding layer, and upper cladding and lower cladding, respectively. According to the above Equation (1), the maximum optimized thickness of the guiding layer for a PhC on an SOI is h_{max} = 273 nm, operating at a wavelength of 1550 nm. Therefore, the guiding layer thickness must be smaller than 273 nm to satisfy the single-mode condition, which will increase the transmission.

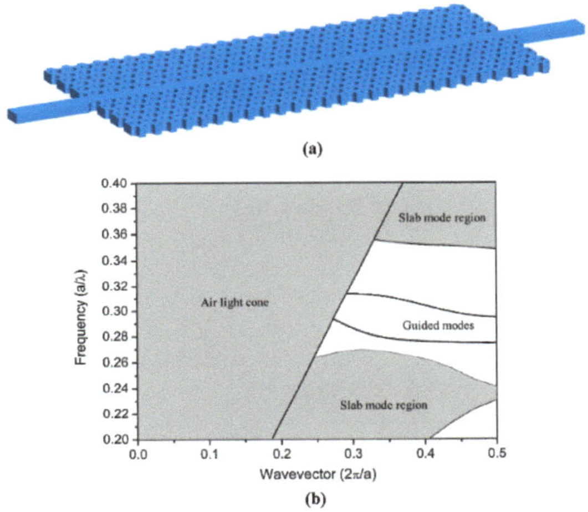

Figure 1. (**a**) Schematic of a W1 PhC WG slab. (**b**) Corresponding photonic band diagram. Reprinted/adapted with permission from [4].

The 2D photonic crystal slab plays a crucial role among label-free optical biosensing methods. The 2D photonic crystal slab biosensing holds significant promise for the miniaturization, integration, and simultaneous detection of multiple targets when integrated with microfluidic chips. Furthermore, the slow light effect inherent in 2D PhC slabs improves light–matter interactions, while ultra-high quality (Q) factor cavities, proportional to the cavity volume (V), enable the continuous reduction in detection limits through innovative design. These characteristics collectively render the 2D PhC slab a favorable option for the fabrication of optical biosensor transducers. The 2D PhC slab biosensor generally utilizes the evanescent field, which is confined within the waveguide, in order to detect the molecular interactions. This evanescent field behaves as an electromagnetic field that appears along the surface of the waveguide during total internal reflection. It has the ability to interact with nearby materials, forming an external region sensitive to changes in the refractive index. When the measurand attaches to receptors on the waveguide's surface, the accumulation of molecules with varying refractive indexes changes the external refractive index, perturbs the evanescent field, and subsequently influences the behavior of the propagated guided light within the waveguide. The sensitivity of the devices is formally quantified as the ratio of the minimum shift in the resonance wavelength to the alteration in the refractive index, i.e., $S = \Delta\lambda/\Delta n$.

Creating a 2D PhC biosensor typically involves a two-step process. First, one must devise an appropriate lattice structure to establish the frequency position of the band gap and the desired mode type. The prevailing approach for this is the plane wave expansion

(PWE) method. Second, the sensor's design needs to incorporate a practical sensing component, which often entails the introduction of one or several defect states to capture alterations in the spectrum caused by changes in the refractive index of the slab surface due to biomolecule interactions. The finite-difference time-domain (FDTD) method is the most commonly employed technique for this purpose.

3. PhC-Based Polarization Maintaining WGs

The polarization dependency of PhC structures is one of its distinguishing properties, which has been used to develop numerous polarization-maintaining devices such as polarizers. Polarizers are very helpful in optical systems because they filter out undesirable polarized light for specialized applications. The polarizer's operation is based on the polarization-dependent propagation of a PhC waveguide (WG), which differs from standard polarizers. It is founded on the specific features of two-dimensional (2D) PBG materials. Unpolarized light may be split into two components: transverse electric (TE), which have an E-field parallel to the periodic plane; and transverse magnetic (TM), which have a magnetic field parallel to the periodic plane. In 2D PBG materials, TE and TM polarized light propagations can occur independently of one another. It is important to note that the band structures and PBGs of the TE and TM polarizations are distinct. Both passive and active PBGs are possible. The Maxwell equations may be used to completely explain how passive or active PBGs interact with light. Similar to how Bragg reflection and diffraction gratings interact with and confine light, PBGs use multiple scattering and diffraction to do the same. The local EM field may be significantly impacted by PBG defects. Consequently, it has been demonstrated that these PBGs possess peculiar optical resonances, and their characteristics may be customized by picking the suitable bandgap and generating defects. The TE and TM polarizations propagate independently of one another in 2D PBG materials. In other words, the PBGs and band structures for the TE and TM polarizations are distinct. Figure 2 depicts typical band structures of a 2D PBG crystal projected in the direction of incoming light, from which some general characteristics of EM wave propagation may be suggested. The frequency of an EM wave and the band structure of the PBG crystal both affect its transmission. Both TE and TM waves can transmit in areas where the TE and TM bands overlap. Both TE and TM wave propagations are prohibited in the area where TE and TM PBGs intersect. The propagations of the TE and TM polarizations in 2D PBGs are independent of each other, or we can say that the TE and TM polarizations have their own band structures and PBGs [23].

Chun et al. proposed a PhC waveguide-based polarization-sensitive polarizer. Here, a 2D PBG crystal consisting of dielectric rods arranged in a square lattice form is designed and used to make a polarizer. A polarizer with a 40 μm long 1D PhC waveguide with a 40 dB extinction ratio has been reported by Zhao et al. [24]. Similarly, polarizers with improved extinction ratios have been designed using the 1D PhC waveguide structure [25–27]. Apart from this, a transverse magnetic (TM) polarizer based on the 1D PhC structure has been proposed with a high extinction ratio of 8.5 dB/μm by Kim et al. [28]. Recently, Pandey et al. presented a brief report on the tunable dispersion characteristics of TE and TM modes in a ternary polymer PhC composed of graphene, silicon, and polymer as nanolayers. By employing a TMM, it has been investigated that the structure shows novel ideas of tunability with variations in the incident angle and polymer thickness in which the Bloch waves are affected [29]. The performance of some photonic devices using TM-polarized light is astounding [30]. Additionally, TM-polarized light is useful for sensing applications because, in contrast to TE-polarized light, its evanescent field penetrates farther into the top and bottom cladding.

Additionally, TM-polarized light is used in polarization multiplexing devices to enable the majority of the channel capacity. As a result, it is recommended that TM-polarization pass devices are essential for effectively developing photonic-integrated circuits. In order to profit from the large index contrast of silicon-to-silicon dioxide, strong structural birefringence, considerable optical nonlinearities, and well-established CMOS manufacturing capabilities, a bulk of integrated circuits is being constructed on an SOI platform [31].

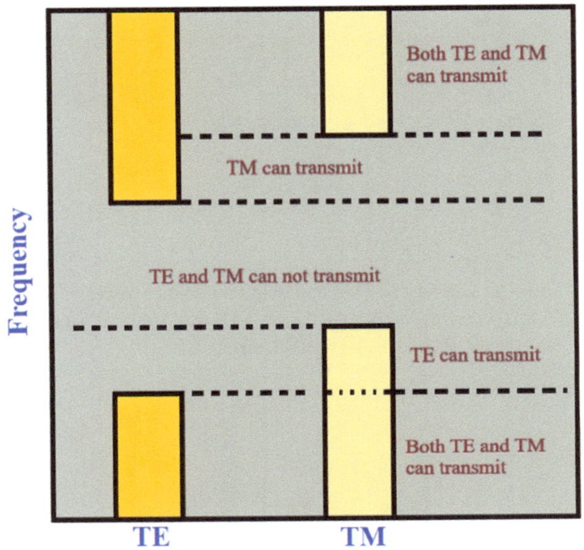

Figure 2. Photonic band structures are projected along the direction of incident light for a 2D PBG crystal.

Recently, Jerakani et al. proposed an ultra-compact polarizer by using the linear birefringence effect of the uniaxial anisotropic material. They proposed a 1D PhC structure having an anisotropic defect layer, which is used to break the degeneracy of the TE and TM modes of resonance [32]. Figure 3 shows the 1D PhC structure used for the polarizer which has alternating YVO4 and SiO_2 layers. The TE and TM band diagrams of the proposed structure are shown in Figure 4, respectively.

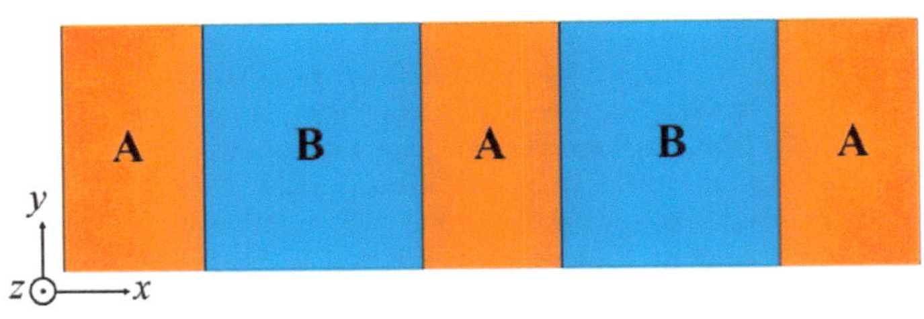

Figure 3. A 1D PhC is composed of alternating anisotropic (A) and isotropic (B) materials. The periodicity is assumed along the x-direction. Reprinted/adapted with permission from [32].

There are several polarization devices presented by the researcher that are based on photonic crystals [33–40]. In [41], Kang et al. show that by tuning the relative phase of multiple inputs for coherent control, it is possible to attain coherent complete polarization conversion. They experimentally demonstrated flexible polarization control in the terahertz region using a silicon photonic crystal slab. The schematic diagram of the photonic crystal slab is shown in Figure 5.

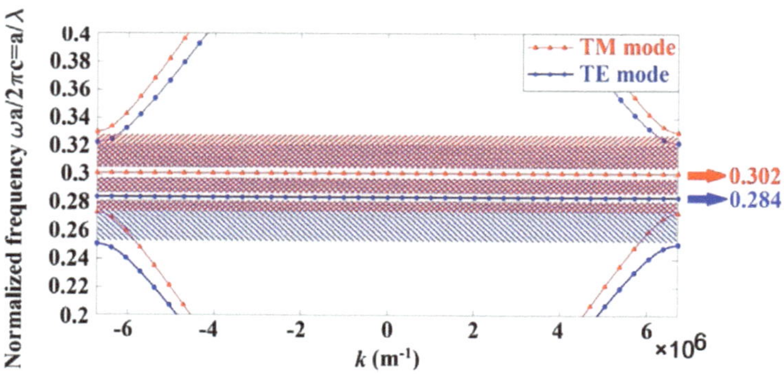

Figure 4. The TE and TM band diagrams of the structure, which is made of alternating YVO4 and SiO_2 layers as the high-index layer and low-index layer materials, and a single-defect layer of YVO4. Reprinted/adapted with permission from [32].

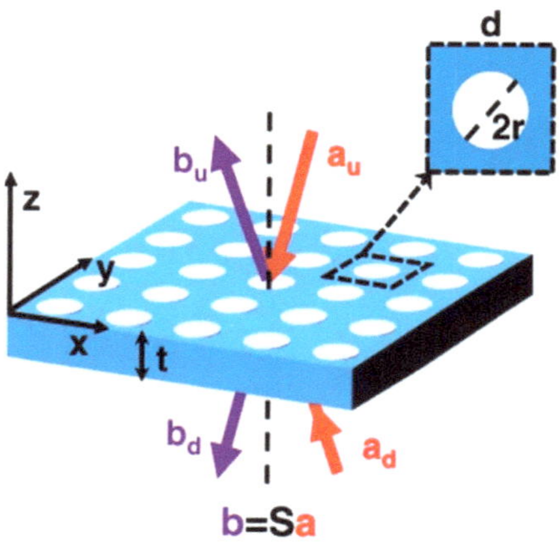

Figure 5. Schematic of the photonic crystal slab. Reprinted/adapted with permission from [41].

Also, Xintao shows 2D valley photonic crystals with dual-polarization by simultaneously opening two frequency degenerate Dirac cones. The opposite-valley Chern numbers of the two band gaps with distinct polarizations are supported by opposite-phase vortex distributions of the bulk modes and the opposing Berry curvatures. Due to this circumstance, the bulk and edge modes, which are located in opposing valleys, exhibit polarization-dependent refraction. In this instance, they also illustrated topological valley transfer that is polarization-independent [42].

The fundamental parts of integrated photonics and communication systems include polarization beam splitters (PBSs) [43,44] and optical power beam splitters (BSs) [45]. While a PBS offers a distinct routing of light in the two polarization states known as the TE and TM [46,47], a BS separates the entering light into many pieces at the output [48,49]. Two-dimensional PhCs can be used as polarizing beam splitters, another important device necessary for electromagnetic wave polarization control, because their band gap profiles are polarization-dependent. Such devices were created for a number of reasons, one of which is that, unlike traditional polarizing beam splitters, they do not separate polarizations using Brewster's angle. Since retroreflection is a possibility, the angle between the wave vectors of the transmitted and reflected beams can theoretically range from 0° to 180°. Beams produced by a high-quality polarization splitter must be almost entirely polarized. This means that the TM and TE band gaps must be separated in order for a 2D photonic crystal to act as a polarizer. In this scenario, the photonic crystal can function as a polarizer in the area of the spectrum where the waves of one polarization are totally transmitted, but the waves of the other polarization are virtually entirely reflected. Although the band gap profiles for various polarizations often do not match completely for a 2D photonic crystal due to the factors previously addressed, they may not be sufficiently separated to create highly polarized beams. Because of this, some crystals might not possess the required efficiency and/or practical bandwidth to be employed as polarizing beam splitters. Because of their diminutive size, PBSs built on PhCs are very appealing [50,51]. Light passing through a PhC could come into contact with a refractive index that is negative or positive. The three factors of refractive index contrast, angle of incidence, and slab thickness interact to produce the two refraction situations. By effectively utilizing these qualities, simple and effective optical modules to steer the light are likely to be realized. As an illustration, a passive optical device that acts as a PBS is shown, with the polarization of the TM and TE waves twisted in the opposite positive and negative directions.

In 2003, Solli et al. experimentally demonstrated PhC-based polarizers and polarizing beam splitters. They used a hexagonal-type PhC structure and obtained the transmission and reflection characteristics of the proposed beam splitter [52]. The polarization extinction ratio (PER), a crucial indicator of merit, is used to gauge PBS efficiency. In 2021, Butt et al. proposed and examined a 2D PhC heterostructure compact design for a PBS and light steering applications. Given their small size and great attractiveness, the PER that the design provides is >30 dB [53,54]. There are several unique designs of PhC-based beam splitters that have been proposed and studied extensively [55–58].

Recently, by using a 2D heterostructured PhC, a small PBS device has been quantitatively studied. The PhC formations are made using a special method that allows PhC-1 to transport light with self-collimated TE- and TM-polarization. The design of PhC-2 creates a PBG for TE-polarized light. However, an interference-free transmission of TM-polarized light is possible. When these two unique PhC formations are merged at an angle of 45°, the self-collimated TE-polarized light interferes with the PC-2 and is reflected at a steep 90° bend, while the self-collimated TM-polarized light may flow through both forms since it does not encounter any PBG on its path. Therefore, in a small space, both polarizations are separated at a 90° angle [59]. The two optimum PC forms should be merged for the polarization division-multiplexing system so that the light with the TE and TM polarizations can divide at 90 degrees. The construction of a PBS, which may be used as an on-chip PDM system, results from the unique union of PC-1 and PC-2. As seen in Figure 6, both structures are united at a 45-degree angle. The normalized electric field mapping of TE-, TM-, and TE + TM-polarized light at 1500 nm is shown in Figure 7.

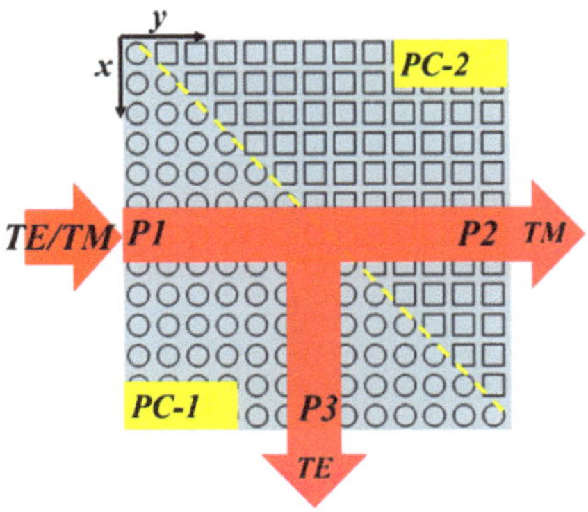

Figure 6. Graphical representation of heterostructure-PhC on-chip polarization division multiplexing. Reprinted/adapted with permission from [59].

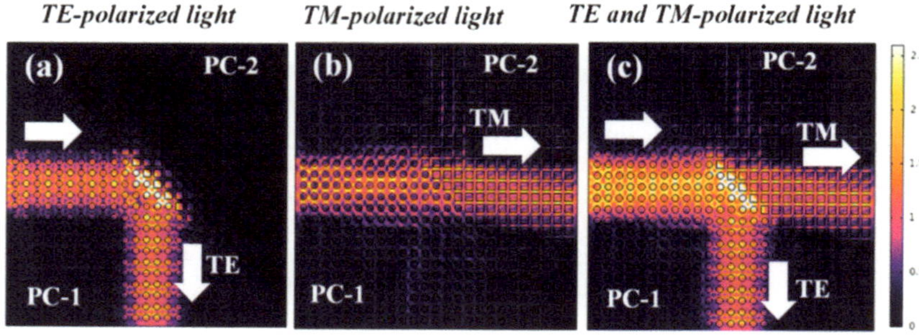

Figure 7. Norm of electric field mapping at 1550 nm for (**a**) light with TE polarization; (**b**) light with TM polarization; and (**c**) TE + TM polarization of light. Reprinted/adapted with permission from [59].

4. PhC-Based Solar Cells

The recent year witnessed a wide exploitation of PhCs in solar cells due to their unique optical properties, i.e., PBG and slow photon effect. The introduction of PhCs can successfully modulate the distribution and propagation of photons in solar cells. Therefore, the key to improving the efficiency of photoelectric conversion can only happen through a full understanding of the optical characteristics of PhCs. At present, PhC structures are being utilized in quantum-dot-sensitized solar cells (QDSSCs) and dye-sensitized solar cells (DSSCs), but comparatively few in the field of perovskite solar cells (PSCs) [60]. This section summarizes the applications of various types of PhCs in sensitized solar cells and silicon solar cells, and their possible limitations and advantages are analyzed.

One of the most promising ways to improve the performance of a thin film solar cell is to improve light trapping [61]. Light trapping in the thin film is usually improved via a surfaced metal back reflector that allows for the dispersion of light inside the absorber layer and therefore significantly increases the optical path length of the photons, which improves the light trapping inside the thin solar cells. PhCs can be integrated along with a suitable absorber, which helps in selective solar absorption. By varying the structural parameters and moving the PBG, the absorption cut-off can be tuned. Additionally, PhCs can significantly boost absorption by tuning the quality factor [62]. In 2019, Cen et al. reported a numerical analysis and exhibited a perfect absorption of monolayer graphene, showing a high quality factor and high sensitivity [63]. The author studied the single-band absorption of graphene via impedance-matching and the critical coupling effect. The analysis exhibited a high Q-factor of 664.2, achieved by varying the ratio between the structural period and the radii of the elliptic cylinder air hole. A PhC offers the benefit of incorporating diffraction, where the photon momentum (k) can be dispersed away from the specular path with $k = k_i + g$, where g is a reciprocal lattice vector and k_i is the incident wave vector.

In 2012, Mavrokefalos et al. reported on the light absorption within the crystalline silicon, which was significantly improved by inverting the nanopyramid light-trapping technique for crystalline silicon thin films, as shown in Figure 8a [64]. In this reported work, the author proposed an inverted nanopyramid light-trapping technique for crystalline silicon thin films, fabricated at a wafer scale using a cost-effective, wet, etching procedure that significantly enhances the absorption of light within the crystalline silicon thin films. Furthermore, the author also validated that the absorptivity of crystalline silicon thin films with a thicknesses less than 10 μm is significantly comparable to that of planar crystalline silicon wafers with a thicknesses greater than 300 μm, as shown in Figure 8b. From the investigation, it was observed that these structures can not only significantly minimize the processing costs, but also provide the improved efficiency required to enable viable crystalline silicon thin film solar cells in the future.

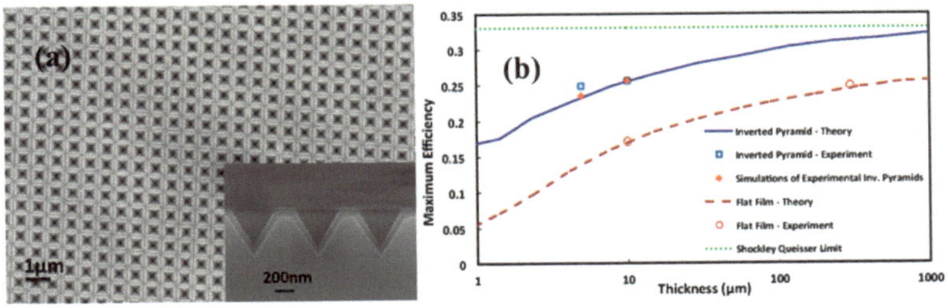

Figure 8. (**a**) SEM image of inverted pyramid solar cells. (**b**) Maximum efficiency at a normal incidence as a function of thickness for simulated inverted nanopyramids (blue solid line) and a planar film (brown dashed line), with 90 and 63 nm SiNx antireflection coatings, respectively. Reprinted/adapted with permission from [64]).

In 2014, Yu et al. reported semitransparent polymer solar cells (STPSCs) utilizing a PhC reflector to achieve 5% conversion efficiency [65]. The 1D PhC comprises multiple optimized layer WO_3/LiF pairs, as shown in Figure 9. This optimized 1D PhC reflected light back into the STPSCs because of the PBG. The STPSCs with eight pairs of 1D PhCs exhibited an attractive performance, along with a short-circuit current density (J_{sc}) and power conversion efficiency (PCE) of 9.76 mA/cm^2 and 5.16%, respectively. A maximum enhancement of 20.2% in J_{sc} was achieved, while the PCE was obtained to be increased by ~21.7%.

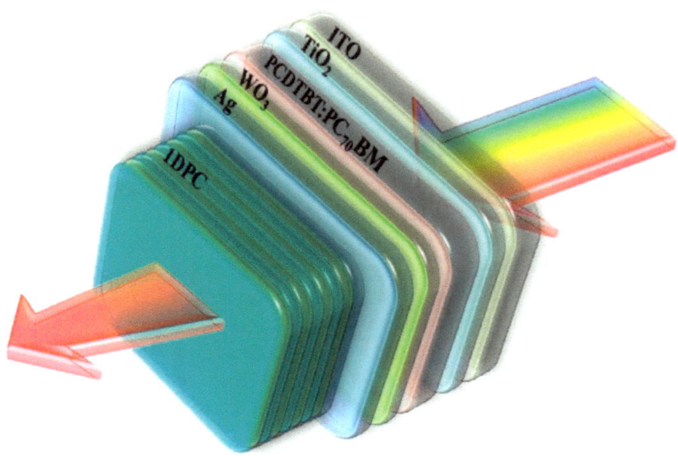

Figure 9. Device architecture of an ST−PSC with eight pairs of 1D PhCs. Reprinted/adopted with permission from [65].

In 2015, Sergey et al. reported a numerical solution of Maxwell's equations, near-perfect solar light trapping and absorption over the 300–1100 nm wavelength band in PhC architectures, amenable to fabrication using the wet-etching technique, and requiring less than 10 μm of crystalline silicon. The PhC comprises a square lattice of inverted pyramids with sides consisting of several (111) silicon facets, and the pyramid center-to-center spacing was used, varying from 1.3 to 2.5 μm. On the wet-etched slab with a height of 10 μm and a lattice constant of 2.5 μm, the author observed a maximum photo current density of 42.5 mA/cm^2, falling not far from 43.5 mA/cm^2, corresponding to 100% solar absorption. The author also investigated that these wet-etched structures required double the volume of silicon compared to the overall numerically optimum PhC structure. In the same year of 2015, Zhang et al. introduced TiO_2/SiO_2 1D PhCs as active layers into photonic solar cells [66]. To maintain a high reflectivity, the TiO_2 and SiO_2 microstructure layers were optimized in order to reduce the thickness of the photonic crystal layer. On the other hand, the author utilized the photonic crystal structure color to visualize and realize multicolor, non-transparent photonic solar cells with blue, blue–green, and green hues. At the same instance, the PhC reduces thermal gain and improves the reflectance of the incident light at a specific frequency, which is highly beneficial for the enhancement of the conversion efficiency for practical applications. In 2016, Ramos et al. utilized angular physical deposition to accurately control the growth of nanorods [67]. They reported the preparation of three types of 1D PhCs that can be used as photonic solar cell photoanodes by controlling the angle of physical deposition. With the reported design, the author achieved conversion efficiencies of 6.39%, 10.94%, and 12.03%, respectively.

In 2017, polymer solar cells were reported as an important category of the next generation of new thin-film solar cells. This technology has gained huge attention in recent years, as its energy conversion efficiency was reported to be more than 13% [68]. Following the results, Shi et al. reported a 1D PhC with various Bragg reflection wavelengths and implemented it in translucent polymer solar cells [69]. The bulk heterojunction active layer of the translucent polymer solar cell was 90 nm thick (PTB7-Th: $PC_{71}BM$). Considering the advantages of this high-performance thin film device structure, the author fabricated semitransparent photonic solar cells by introducing an ultra-thin transparent silver cathode by replacing a thick silver mirror cathode, resulting in high-performance, semitransparent, photonic solar cells with a conversion energy of over 6% and an average visible transmittance between 20% and 30%. In the same year of 2017, Gupta and Janyani improved the performance by using a 2D PhC-based thin layer heterojunction gallium arsenide (GaAs) solar cell with a periodic pattern [70]. In their reported work, the authors compared two structures: (i) having a PhC structure extended from the top transparent, conducting oxide to the p-AlGaAs window layer placed above the GaAs material; and (ii) the PhC structure etched only in a transparent conducting oxide. A comparative analysis of the structure with the planer cell and Lambertian light trapping limits was also performed, and a significant increment in the efficiency was observed for thinner active layers, showing the advantage of a wavelength scale. The 'Lambertian' term refers to a type of surface or material that exhibits diffuse reflection, meaning it scatters incident light in all directions uniformly. In the context of solar cells, 'Lambertian' surfaces are often used to enhance light trapping and absorption within the cell structure. The results exhibited that the PhC structure (etched inside p-AlGaAs)-based solar cell exceeded the efficiency of the double-antireflection coating (ARC)-based structure by more than 18% for a 50 nm thin active layer cell. Later, in 2018, Zhang et al. proposed a numerical analysis of a multi-junction solar cell structure [71]. The designed structure was a combination of grating and a 1D PhC considered to be made of crystalline silicon and silicon dioxide. It uses gratings to increase its effective optical path. The design was optimized carefully, leading to remarkable efficiency for the cells with very thin silicon active layers due to the significantly enhanced absorption. In 2019, Bhattacharya et al. achieved 29% power conversion efficiencies by utilizing thin, crystalline silicon, following the efficacy of wave-interference-based light trapping and carrier transport in a parabolic−pore PhC [72]. In the reported work, the author demonstrated the geometry of a simple PhC solar cell which enables the power conversion efficiency of a single-junction c-Si solar cell to exceed the previously reported theoretical limits.

In the most recent article published in 2022, Morsy and Saleh reported another PhC solar cell utilizing GaAs to develop nano-solar cells [73]. A nano-solar cell based on a 2D PhC ARC trapping layer and a 2D-graded-index GaAs active layer was reported and numerically investigated using the finite-difference time domain (FDTD). The proposed cell absorption and conversion efficiency were analyzed as a function of the cell layer thickness in comparison with the Lambertian absorption and cell efficiency limits. All the simulations were performed in the range of 300 to 1100 nm using the FDTD technique, where the 2D PhC structure was shown by indium tin oxide nanorods in an air background. The study exhibited that the design of a 100 μm^2 nano-solar cell achieved a 39.2% conversion efficiency. In addition to these structures, several other structures were also reported in the past decades, which are summarized in Table 1.

Table 1. A summary of the PhC nanostructures utilized in photocatalytic degradation, PEC water splitting, and solar cells to improve their solar conversion efficiencies [74].

Photonic Nanostructure	Material	Photocurrent Density	Power Conversion Efficiency	Reference
Microcavity	Fe_2O_3	0.35 at 1.23 V_{RHE}	-	[75]
	Cu_2O/ZnO		0.51	[76]
	TiO_2		-	[77]
	Polymer solar cells		4.5	[78]
	Semitransparent polymer	-	24–27%	[79]
	$WO_3/Ag/WO_3$	-	8.37%	[80]
PhC	TiO_2/ZrO_2	-	-	[81]
	$BiVO_4$	1.35 at 1.2 V vs. Ag/AgCl	-	[82]
	Mo:$BiVO_4$	2 at 1.0 V vs. Ag/AgCl	-	[83]
	Bi_2WO_6	-	-	[84]
	WO_3	2.5 at 1.0 V vs. SCE	-	[85]
	TiO_2 nanodisk	-	18.7%.	[86]
	Silicon	-	31%	[87]
	CdS/TiO_2	4.84 at 0 V vs. Ag/AgC	-	[88]
PhC + Plasmon	Au/TiO_2	-	-	[89]
	Au/TiO_2	-	-	[90]
	$Au/BiVO_4$	3.1 at 1.23 V_{RHE}	-	[91]
	TiO_2/Au	-	4.67	[92]

From the above discussion, we can clearly see improvement in the performance of various types of PhC solar cells. However, can also be observed that the technology still faces a huge research gap due to the limited exploitation of the features of PhCs. To overcome various challenges, researchers should fully understand the PhC light control performance, the selection of the appropriate PhC structure, and the illumination direction. At present, photonic crystal structures are widely used in dye-sensitive solar cells and quantum-dot-sensitive solar cells, but relatively few in the field of perovskite solar cells. It is worth noting that although PhCs have several excellent optical control properties, it is difficult to achieve the thickness requirements of a perovskite solar cell because of the limitations of the preparation technology and structure of PhCs.

5. Photonic-Crystal-Based Sensors

Because of the growing demand for sensing devices, PhCs emerge as a potential tool in healthcare [93,94], physical parameter monitoring [95,96], defense [97], food quality control [98], and aerospace [99]. The surface of PhCs is made of a periodically modulated dielectric material that can be modified in order to create a PBG, which helps prevent light from propagating at particular wavelengths [100]. Consequently, the local optical modes of a PhC can be utilized as a label-free and highly sensitive platform for monitoring chemical and physical changes in the surrounding environment. In order to evaluate the sensing performance of such devices, several parameters were investigated such as sensitivity, response time, selectivity, stability, and the limit of detection (LoD) [101].

In their initial days, PhCs were popular among gas-sensing applications, as they required light to be modulated through the waveguide and recorded on the detector. PhCs based gas sensors were considered the safest way to detect hazardous gas without compromising sensitivity and selectivity compared to electrochemical, chemiresistive, and ratiometric sensors. Gas sensors based on PhCs are reported for the mid-infrared range for various toxic gases; for example, CH_4, CO_2, and CO exhibited absorption lines in the mid-IR wavelength region. A high-precision gas index sensor based on a PhC air–slot cavity was proposed, offering a sensitivity of 510 nm/RIU [102]. In 2012, Jun reported that a surface plasmon resonance (SPR) nanocavity antenna array for gas sensing application achieved a maximum sensitivity of 3200 nm/RIU [103]. A guided-mode resonance gas sensor with a sensitivity of 748 nm/RIU was reported [104]. In addition to gas sensing applications,

PhCs also gained huge popularity in chemical- and biological-sensing applications. From a technical point of view, the optical sensors based on PhCs, including PhC optical fiber, also known as PCF, and integrated planar PhCs, are suitable for label-free detection and multiplexing. For example, in 2012, Zou et al. reported large-scale chip-integrated PhC sensor microarrays, and demonstrated their application in biosensing on an SOI-based platform [105]. These days, integrated PhC-based optical sensors are considered the most popular class of photonic sensors, which are generally utilized to monitor physical and chemical parameters. In 2013, Lu et al. also fabricated an integrated temperature sensor based on an enhanced pyroelectric PhC [106]. The sensing response of the devices was achieved with the pyroelectric effect of lithium–niobate, in which a suitable air–membrane PhC cavity has been fabricated. The wavelength position of the cavity mode is tuned at 11.5 nm for a temperature variation of only 32 °C. The reported device exhibited an average temperature sensitivity of 0.359 nm/°C for an active length of around ~5.2 µm. The main advantages of these sensing configurations are their high light confinement, excellent sensitivity, and selectivity in the sensing mechanism.

In 2014, Shafiee et al. utilized nanostructured PhCs for the label-free detection and quantification of intact viruses (HIV-1) from biologically relevant samples [107]. Figure 10 illustrates the schematic of a nanostructured PhC HIV sensor platform. In the proposed scheme, the PhCs resonantly reflect a very narrow wavelength band during illumination with a broadband light source. The surface of PhC-adsorbed biotarge induces a shift in the resonance spectral peak, which is detectable with a <10 pm wavelength resolution. The process allows for the successful detection of a small number of viruses that sparsely populate the transducer surface. In the reported work, the authors successfully detected HIV-1 in serum and phosphate-buffered saline (PBS) samples with viral loads varying from 10^4 to 10^8 copies/mL.

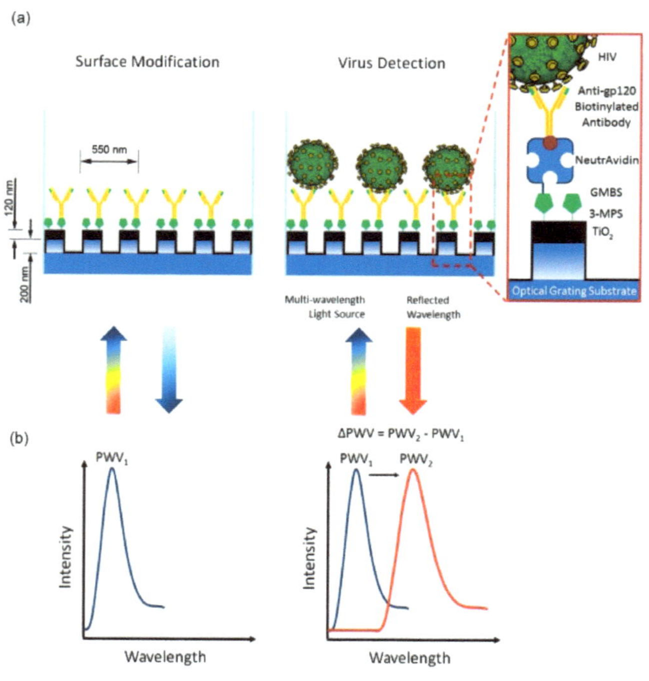

Figure 10. (**a**) The bottom surface of PhC biosensor microplate wells is composed of a nanostructured sub-wavelength grating that is coated with TiO_2. (**b**) Binding events within the close vicinity of the sensing area change the bulk index of refraction; thus, the peak wavelength value (PWV) of the reflected light is altered. Reprinted/adopted with permission from [107].

In 2015, Bouzidi et al. reported a tiny and portable gas-sensing system using 1D PhCs [108]. The sensor comprises a 1D PhC formed using alternate layers of magnesium fluoride (MgF_2) and silicon (Si), with a vacant layer in the middle. The empty layer was later filled with the polluted air, which had a refractive index closest to that of pure air, varying from 1.00 to 1.01. Transmission spectra were used here for the real-time detection of gaseous environments. The peak frequency of the transmission spectra was observed to be sensitive to the air–gas mixture with a high sensitivity of 700 nm/RIU. In 2016, Farhat et al. reported a surface-plasmon-resonance-based liquid crystal photonic crystal fiber (PCF) sensor for a temperature sensor [109]. The reported work briefly exploits the coupling characteristics between the fundamental mode inside the designed PCF infiltrated with a nematic liquid core and the plasmonic mode on the surface of a nanogold wire. The designed sensor exhibited a high sensitivity of 10 nm/°C for temperatures ranging from 30 to 50 °C. In 2017, Liu et al. reported an refractive index sensor using high quality factor bound states in free-space-coupled PhC slabs [110]. Figure 11 illustrates the schematic diagram and key parameters of the proposed design, with angle and polarization defined for the incident light beam. The reported paper demonstrates a numerical investigation over single and coupled bi-layer PhC slabs that possess a simultaneously high S and Q near the bound states in the continuum. The reported article validated the performance of the device experimentally and numerically. The numerical analysis exhibited S > 800 nm/RIU and Q > 107 for refractive index sensing in the 1400–1600 nm optical wavelength band, while experimentally, the value achieved 94 nm/RIU and 1.2×10^4 for S and Q, respectively.

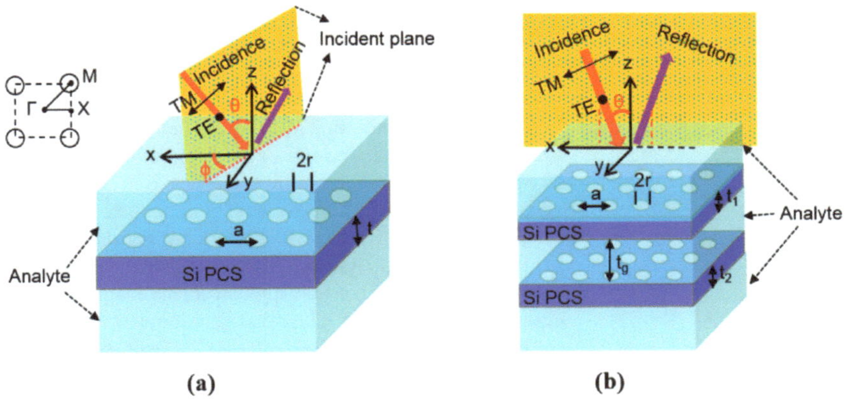

Figure 11. Schematic and key design parameters of the proposed sensor structures, with angle and polarization defined for the incident light beam: (**a**) single-layer PhC slab, and (**b**) coupled bi-layer PhC slab. Reprinted/adopted with permission from [110].

In 2018, Maurya et al. proposed a 2D layered nanomaterial (graphene, $MoSe_2$, MoS_2, WSe_2 and WS_2)-decorated 1D PhC refractive index sensor based on the Kretschmann configuration [111]. The 2D nanomaterial shows an excellent sensing response due to its superior adsorption, as well as absorption properties. The reported design comprises alternate layers of 2D nanomaterials and silicon, as shown in Figure 12. The performance of the design, including sensitivity, quality factor, resolution, and evanescent field penetration depth, was investigated and compared with a 1D PhC decorated with polymethyl methacrylate (PMMA) in place of silicon. An increased shift in resonance angle was obtained along with a high quality factor by replacing PMMA with silicon, but at the cost of poor resolution.

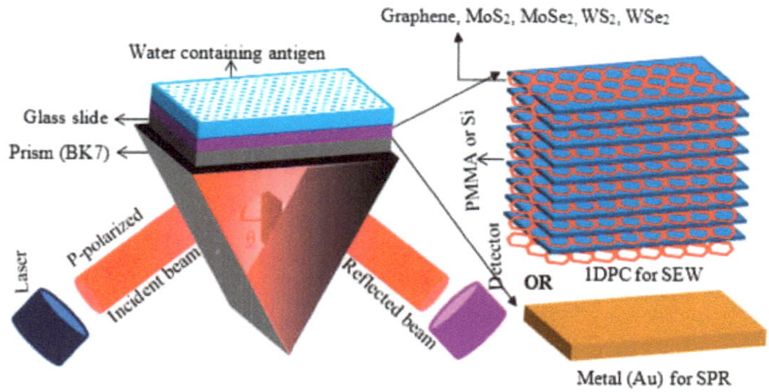

Figure 12. Schematic of 2D-material-decorated 1D PhC refractive index sensor. Reprinted/adopted with permission from [111]).

In 2019, Snapp et al. reported a colloidal PhC strain sensor integrated with a graphene phototransducer [112]. In the reported work, the author used a colloidal solution to develop a PhC which changes color in the presence of externally applied strain. This hybrid sensing system allows for a direct visual perception of strain, while strain quantification via electrical measurement of the hybrid system outperforms that of crumpled graphene strain sensors by more than 100 times. In 2020, Resende et al. developed a PhC sensor for the label-free and target-specific detection of urinary venous thromboembolism biomarkers using molecularly imprinted polymers [113]. The molecularly imprinted photonic polymer (MIPP) was achieved by manipulating an imprinted polymer on highly ordered silica nanoparticles assembled via vertical deposition. Owing to the hierarchical sensing configuration, the resulting MIPP exhibited optical properties that changed upon rebinding of the target analyte. The sensor exhibited a linear response, varying from 0.2 ng mL^{-1} to 22 ng mL^{-1}, and showed a lower limit of detection of 0.13 ng mL^{-1}. In the most recent article published in 2022, Kawasaki et al. reported the application of PhCs for monitoring SARS-CoV-2 in saliva to tackle the global pandemic [114]. The sensor comprises an imprinted photonic crystal thin film functionalized with an anti-SARS-CoV-2 spike protein antibody, enabling label-free, highly sensitive, and selective detection using a smartphone-equipped optical setup. The sensing response of the sensor was examined with artificial saliva, and a specific and quantitative detection of the spike protein was successfully obtained, with a low LOD of 429 fg/mL. Additionally, the author also provides an IoT-based, user-friendly interface to store the result data and return it immediately to the patient. The schematic of the reported sensors is shown in Figure 13.

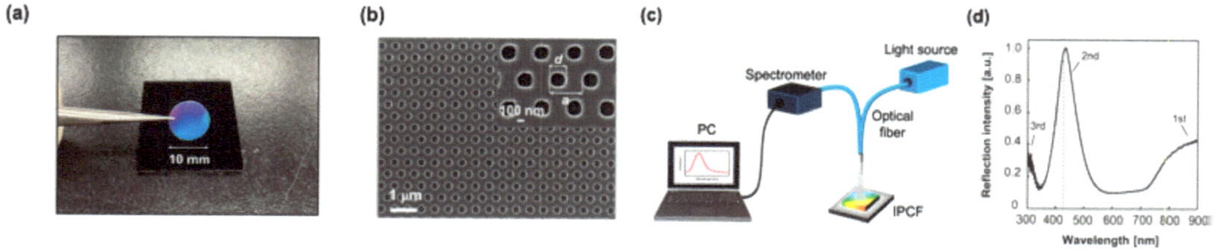

Figure 13. Schematic of the reported PhC sensor for SARS-CoV-2 showing (**a**) image of an IPCF chip, (**b**) SEM image of the fabricated PhC surface, (**c**) experimental setup used to characterize the sensor, and (**d**) reflection spectrum of the sensor. Reprinted/adapted with permission from [114]).

Sensor applications based on PhC also hold significant promise in the realm of liquid-based sensing. In 2011, Pu et al. reported such a study and explored the influence of temperature on the band gap properties of magnetic fluid-based PhCs [115]. In the reported article, a method was developed to extract temperature-dependent structural information from existing data, yielding crucial parameters for band diagram calculations. The research reveals that temperature has a limited impact on the mid-frequencies and positions of forbidden bands, but it can alter the number of forbidden bands at specific magnetic field strengths. Later, in 2016, Su et al. numerically designed a novel PhC magnetic field sensor, featuring a shoulder-coupled resonant cavity filled with magnetic fluid [116]. By tracking the changes in the resonant wavelength, the sensor achieves magnetic field sensing, with greater sensitivity and improved performance obtained through specific infiltration schemes, resulting in an optimal structure with eight air holes filled with magnetic fluid. PhCs' inherent ability to seamlessly interact with liquids opens doors to several practical applications. Such devices can be employed to monitor and analyze a broad range of materials in liquid form, from biological molecules and pathogens to environmental contaminants. This versatility positions PhC-based sensors as powerful tools for advancing fields like medical diagnostics, environmental monitoring, and food safety, where precision and sensitivity are paramount. The ease with which photonic crystals can be infiltrated with liquids underscores their potential to revolutionize liquid-based sensing, offering a path to more effective and efficient detection and analysis methods. In addition to several advantages, such as its small size, immunity to electromagnetic interference, high sensitivity, strong light confinement, etc., it also possesses a limitation in terms of its fabrication and implementation. The sensing material is disadvantaged by poor adhesion during its deposition on a PhC [97]. In some cases, the author fabricated the Panda fiber by integrating the sensing material into the silica cladding and monitoring the birefringence to detect hydrogen leakage [117]. Overall, the sensing devices based on PhC offer a huge possibility, ranging from biomolecules to gas detection, and can hence be considered a potential candidate for the next generation of sensing. In addition to the PhC sensors we reviewed above, there are some other sensors that have been reported in the past decades, which are summarized in Table 2.

Table 2. A summary of various PhC-based sensors, along with their sensing response and detection range.

Type of PhC	Sensing Material	Target	Detection Range	Sensitivity	Reference
		Gas Sensors			
D shape fiber	Ag/Porous TiO_2	Isopropanol	20–100%	1.35 nm/%	[118]
PhC long-period grating	-	Methane	0–3.5%	6.39 nm/%	[119]
Microtube PhC	-	Propene	33 vol%	8.6–10.1%	[120]
hybrid porous core PCF	-	Methane	-	21.2%	[121]
PC slab (PCS)	-	Carbon di oxide	0–2130 ppm	-	[122]
2D defect PCS	Polymer	Hexane and ethanol	0–3 ppm	123.4 nm/RIU	[123]
PhC cavity	-	Tetrahydrofuran	0–31.5 mmol.L^{-1}	0.128 nm·mmol.L^{-1}	[124]
PCF	NA	Sulfur Dioxide	1.3406–1.3388	83.64%	[125]
PCF	-	Nitrous Oxide and Benzene	1.295–1.3	73.822% 76.422%	[126]
Periodic PCS	Graphene	gas	-	-	[127]
2D PhC	-	Sulfuric Acid and Hydrogen Peroxide	0–90%	575–600 nm/RIU	[128]
		Physical sensor			
Bragg grating PhC	-	Temperature sensor	5 °C–160 °C	0.082 nm/°C	[129]
Dual-core PhC fiber	Fe_3O_4	Magnetic field sensor	89.9–271 Oe	305.8 pm/Oe	[130]
Elliptical hole PhC Fiber	Platinum	Temperature sensor	20–160 °C	77 nm/°C	[131]

Table 2. *Cont.*

Type of PhC	Sensing Material	Target	Detection Range	Sensitivity	Reference
		Gas Sensors			
MMF-PCF	-	Strain sensor	0–1600 με	~−14.89 pm/με	[132]
D shaped PCF	Au film	Magnetic field and temperature sensor	5 °C−65 °C 30 Oe~270 Oe	−1.25 nm/°C 0.21 nm/Oe	[133]
PhC	Self-adhesive hydrogel	Underwater motion detection	0–200%	≈2.09 nm/%	[134]
2D PhC	GaAs	Pressure sensor	0–5 GPa	17.00 nm/GPa	[135]
PhC micro ring resonator	-	Pressure sensor	0–0.5 GPa	25.26 nm/GPa	[136]
2D PhC	-	Pressure Sensor	0–6 GPa	3.1 nm/GPa	[137]
PCF	Silver	Temperature sensor	15–35 °C	2 nm/°C	[138]
		Refractive index sensor			
D-shaped PhC Fiber	Gold film	Refractive index	1.43–1.46	7700 nm/RIU	[139]
Dual drilled channel	Gold nanowire	Refractive index	1.31–1.40	90,500 nm/RIU	[140]
H shaped fiber	Gold nanowire	Refractive index	1.33–1.40	4000 nm/RIU	[141]
Tapered PCF	-	Bio-chemical	1.30–1.32	722.3 nm/RIU	[142]
Hollow core	Cu nanowire	Refractive index	1.33–1.38	12,400 nm/RIU	[143]
Hollow core PCF	-	Poisonous chemical	-	92.08%	[144]
2D PhC	-	Refractive index	1.33–1.481	347.99–473.38 nm/RIU	[145]
Single channel PCF	-	Petrol adulteration	1.415–1.440	20,161.2 nm/RIU	[146]
1D defect PhC	Superconducting material	Biosensing	0–50%	68.18–85.22 nm/RIU	[147]
1D PhC	-	Hemoglobin	10–150 g/L	1916.77 nm/RIU	[148]
2D PCF	Au film	Protein	0–10 μg/L	10^{-3} μg/mL	[149]

6. PhC-Based Logic Gates

All-optical systems have garnered significant interest among researchers in recent years due to their exceptional applications in fiber optic communication, optical computing, and optical signal processing [150]. For the realization of all-optical devices, PhCs are offered a new avenue [151–153]. These devices utilize the periodic arrangement of crystals with varying refractive indexes to regulate the passage of light. PhCs have great qualities such as strong light confinement, compact size and high-speed light propagation, quick response time, wide bandwidth, and low-power dissipation [150]. Light confinement is strong in silicon PhC waveguides, making them appropriate for the design of combinational and logical optical circuits. The periodic arrangement of atoms in a crystal gives rise to energy bandgaps in PhCs. Another intriguing phenomenon that can be produced with PhCs is the prevention of light transmission via optical switching techniques [154]. In this way, researchers [155] have proposed a half-subtractor using PhCs as a possible structure for usage in arithmetic logic circuits. To calculate addresses and increment or decrement operators, a half-subtractor plays a crucial role. The most fundamental combinational logic circuit utilized in digital electronics is the half-subtractor. This circuit allows for the subtraction of two binary digits and creates two output bits (difference and borrow). Similarly, Rao et al. [156] have designed the all-optical MUX using square lattice-type silicon rods and a combination of T- and Y-shaped waveguides. Here, silicon rods have been used for back reflection that also reduces the power loss. This structure consists of a silicon rod with a diameter of 0.2a, a refractive index (RI) of 3.4, and a lattice constant of 0.6 μm. Using silicon junction rods with radii of r_3 = 0.3a and r_4 = 0.25a at junctions J1 and J2, respectively, prevents undesirable back reflections at the input ports. A PC-based all-optical MUX is shown in Figure 14. The design is composed of two T-junctions and one Y-junction, together with three horizontal and two vertical waveguides. Two horizontal waveguides on the right side of the device are utilized as the reference (R) and select (S) input ports, while another horizontal waveguide on the left side serves as output port Y. A waveguide on the top is utilized for input port A, while a waveguide on the bottom is used for input port B. The suggested all-optical device is based on the phenomenon of beam interference, as shown in Figure 15. The theory of wave optics states that interference occurs within waveguides. Depending on the beginning phase and path covered by incoming light signals, a constructive or destructive interference would ensue.

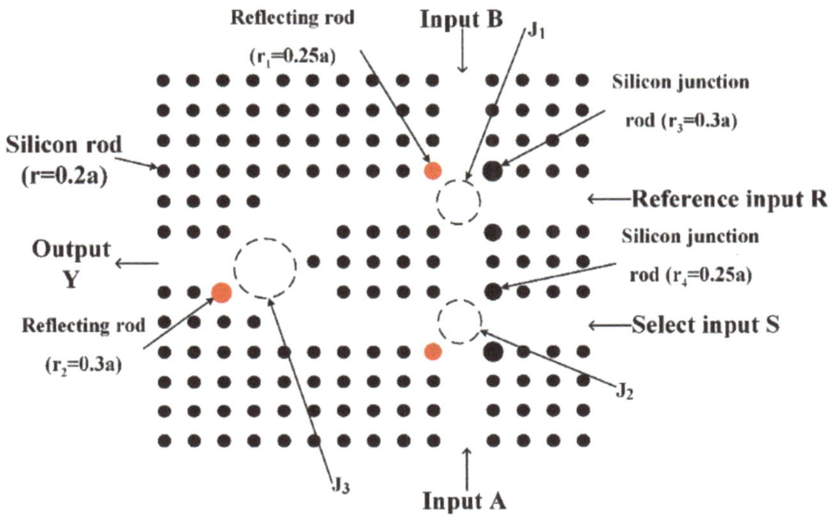

Figure 14. Diagram of an optical MUX with T- and Y-shaped PhCs [156].

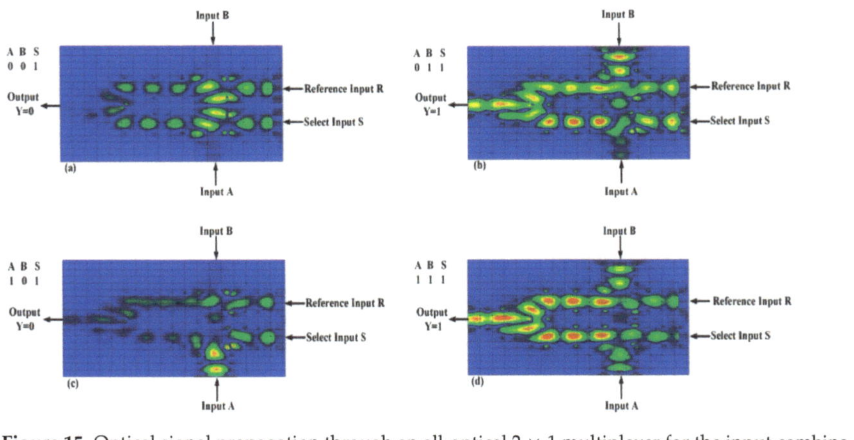

Figure 15. Optical signal propagation through an all-optical 2 × 1 multiplexer for the input combinations (**a**) 00, (**b**) 01, (**c**)10, and (**d**) 11 when S = 1. Reprinted/adapted with permission from [156].

Rao et al. [153] also proposed a reversible logic gate for a lossless digital system. It is very impressive because it lowers heat generation and power consumption in a photonic-integrated system. They have designed the Toffoli and Feynman logic gates using 2D PhCs for low-power integrated circuits. These optical reversible gates are designed without introducing any nonlinear materials, and the design is validated through FDTD-based commercial software in the wavelength range of 1550 nm. The crosstalk ratio, insertion loss, and transmittance of 32.5 dB, −0.04 dB, and 0.9 are obtained in the case of the Toffoli logic gate. Similarly, the crosstalk ratio, insertion loss, and transmittance of 12.4 dB, −0.015 dB, and 0.96 dB are found in the case of the Feynman logic gate. Thereafter, Rachana et al. [157] proposed a 3-input AND logic gate using a T-shaped, silicon-based, 2D photonic crystal.

There are various applications of this gate such as error correction, performance recognition, code conversion, arithmetic operations, and encryption/decryption. The size of the 3-input AND logic gate is 8.4 μm × 5.4 μm; that is a reasonable size. The performance analysis of this gate has been performed at 1.55 μm wavelength using the FDTD method. The best and worst CRs were found to be 24.533 dB and 8.6 dB, respectively. Also, its minimum and maximum transmission efficiency were found to be 19.6% and 142%, respectively. The reaction time of the gate is 18 fs and the bit rate is 55.5 Tbps, that provides high-speed optical signal processing. Overall, this gate provides a high CR and consumes minimal space.

Rao et al. [151] also proposed all-optical photonic devices for large-scale integration and high-speed optical computing. They have designed and analyzed the NOR, NAND, and XNOR logic gates using a single structure. This is possible with the help of optimized phase changes in the applied optical signal. The size is 7.2 μm × 5.4 μm, and the CR is 14.3 dB, 17.59 dB, and 10.52 dB of the proposed NOR, NAND, and XNOR logic gates.

The same research group [152] has also used a photonic crystal to design the ultra-compact, all-optical D flip–flop by optimizing the radius of the silicon rod and those refractive indexes. The proposed D flip–flop helps reduce the complexity of digital circuits and enhance optical computing and networking. The optical interference principle has been utilized in the T-shaped waveguide to design the D flip–flop at a wavelength of 1550 nm. This design is compact and operates at a low power level; the insertion loss is 0.87 dB, the CR is 25 dB, and the transmission ratios are more than 96%.

As we know, in order to accomplish several operations in optical communication technology, all-optical logic gates are necessary components. Thus, researchers of [158] have also presented a revolutionary design for optical NAND, NOR, and XNOR logic gates based on the optical interference phenomenon. The proposed topology for all-optical logic gates is realized with 2D photonic crystal-based, Y-shaped waveguides made of silicon rods with an air background. The proposed configuration for NAND, NOR, and XNOR logic gates occupies a smaller area of 9 μm × 9 μm and delivers a contrast ratio of 21.2 dB, 18 dB, and 17.16 dB at a wavelength of 1.53 μm, with less insertion loss of 0.011 dB and a greater-than-98% transmission ratio. Consequently, the proposed logic gates may be relevant to optical processors and photonic-integrated devices.

NNX logic gates are built utilizing square lattice rods based on PhCs in an air background. This structure consists of Y-shaped and T-shaped waveguides, with a silicon rod radius of 0.2a and a refractive index (RI) of 3.44, where a = 0.6 m. These waveguides are created by introducing structural line faults. At the junctions J_1 and J_2, reflecting rods r_1 = 0.12 μm and r_2 = 0.138 μm, and junction rods r_j = 0.18 μm are employed to prevent unwanted back reflections into the input ports. One Y-junction and one T-junction with three horizontal waveguides and one vertical waveguide comprise the design. The waveguides located on the left side of the device serve as input ports A and B, and the distance between these two input waveguides is 5a, allowing for the loss-free confinement of light within the waveguides. The vertical waveguide at the top of the structure serves as the reference input port R, while the horizontal waveguide on the right side of the structure serves as the output port Y. This design functions flawlessly as optical NAND, NOR, and XNOR logic gates across the wavelength range of 1.53–1.57 μm [158]. Similarly, as indicated in Figure 16, all-optical AND, OR, and XOR logic gates using square lattice silicon rods with an air background can be developed.

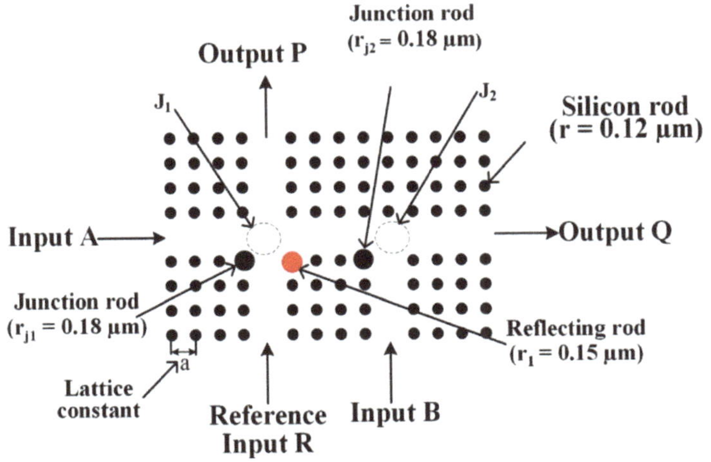

Figure 16. Schematic of AOX gates with T-shaped photonic crystals. Reprinted/adapted with permission from [158].

7. PhC-Based Amplifiers

A form of optical amplifier that uses photonic crystals to improve and regulate the amplification of light signals is known as a photonic crystal-based optical amplifier. In the fields of photonics and optical communication, these amplifiers are very important. Photonic crystals are periodic formations with alternating high- and low-refractive index areas that may control light propagation at specified wavelengths. Wavelength selectivity, enhanced light–matter interaction, efficient amplification, compact size, tunable gain, and reduced noise are some of the main benefits and features of photonic crystal-based optical amplifiers. Photonic bandgaps, which stop some light wavelengths from propagating, are built into photonic crystals. The photonic crystal's characteristics may be carefully engineered to produce amplifiers that are incredibly selective for certain wavelengths or small bands of wavelengths. Applications for photonic crystal-based optical amplifiers in optical communication include WDM systems, fiber optic network signal amplification, and the development of small and effective optical amplification components for use in integrated photonics.

Many optoelectronic devices, such as semiconductor lasers [159], optical waveguides [160], semiconductor optical amplifiers (SOAs) [161], and all-optical switches [162], have successfully benefited from the usage of PhCs in recent years. Semiconductor optical amplifiers (SOAs) include a variety of special qualities that provide difficulties for those who want to utilize them as linear amplifiers, while also creating considerable opportunity for quick, all-optical signal processing. All-optical gates can be built with semiconductor optical amplifiers as wavelength converters and OTDM demultiplexers [163].

Due to PBGs' flexibility, PCs have drawn significant interest in the realms of optical equipment, biosensing, anti-counterfeiting tools, and even quantum communications. Due to the effects of light collection and amplification, PCs in particular are frequently utilized in the present to amplify signals. They are suitable for constructing extremely compact optical devices because of their capacity. Therefore, any optical devices that are compatible with photonic-integrated circuits (PICs) may be designed on PCs [164,165]. A new device known as a photonic crystal semiconductor optical amplifier (PhC–SOA) is created by altering the semiconductor optical amplifier device utilizing a PhC structure [166,167]. This gadget makes use of SOAs, as well as the benefits of PhC architecture. The SOA length can be drastically reduced because of the improved interaction between light and matter in PC structures [168–173]. The advantages of the optical current amplifier over conventional current-sensing technologies are less electrical interference, incredibly quick reaction times, excellent computation accuracy, and reduced sensor load and length. Other than that, optical sensing and communications are the main uses for optical amplifiers [174].

In [174], with adjusting the radius of the circular photonic structure from 340 nm to 390 nm at a pressure range from 0 to 10 GPa, where the lattice space retains 800 nm, the germanium-based PC waveguide (L-shape) defect (15 × 15) of the electric distribution field is calculated. By using peak electric field modeling work, the input and output power of the electric field are carefully analyzed in relation to the incoming signal at 1.55 µm for a germanium-based photonic crystal waveguide (L-shape defect). The relationship between input power and output power demonstrates that under different pressures, the output power is amplified in relation to the radius increment. A schematic diagram of the L−defect-based PhC waveguide and distribution of the electric field are shown in Figure 17a,b, respectively.

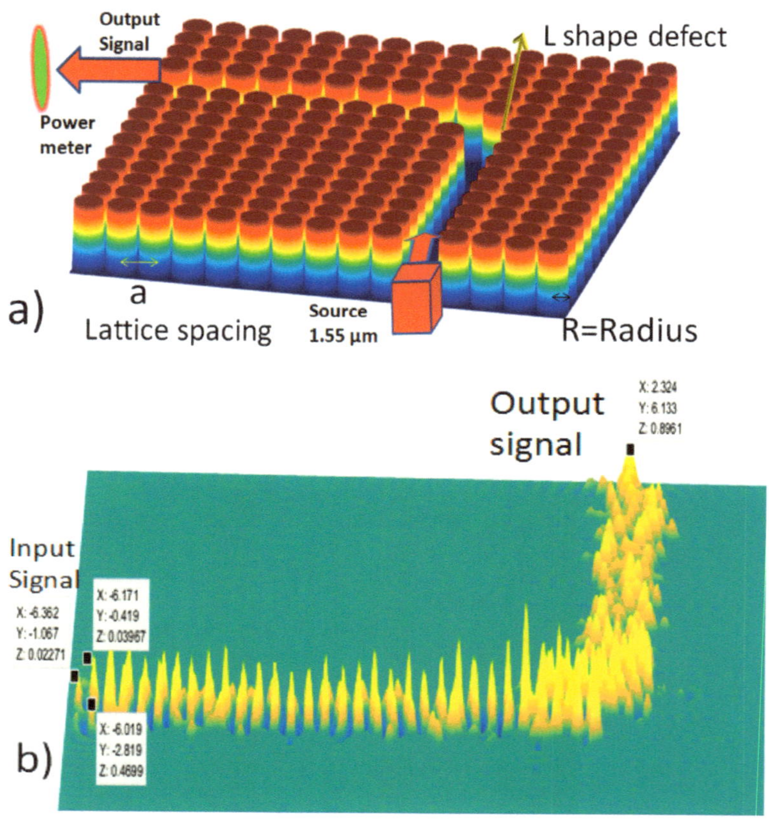

Figure 17. (**a**) L-defect-based photonic crystal waveguide at a pressure of 10 GPa. (**b**). Electric field distribution of a 0.345 µm radius at an applied pressure of 10 GPa. Reprinted/adapted with permission from [174].

Concerning PhC devices relying on miniaturized resonant cavities, random lasers (RLs) are also a very important application of PhCs [175–177]. RLs have garnered considerable interest over several decades due to their captivating features, including the presence of multiple emission spikes in the spectrum, a wide emission angle, and low spatial coherence. These RLs have been observed in a variety of gain media, encompassing disordered micro/nanostructures in semiconductors, polymers, organic materials, and liquid crystals [178–180]. In the pursuit of engineering new advanced materials, natural photonic crystals have served as a wellspring of inspiration for diverse material designs. For instance, dielectric materials emulating the photonic crystals found in butterfly wings have been harnessed for vapor sensors [181], the creation of structural colors [182], and infrared detection systems [183]. ZrO_2-inverse opal thin films, mimicking photonic crystal structures, have exhibited remarkable catalytic performance [184]. Moreover, the intricate patterns on Lepidoptera wings have captivated scientists in their quest to harness laser actions. Wang et al. accomplished random lasing by coating ZnO nanoparticles as a gain medium on the surface of butterfly wings [185], and Zhang et al. ventured into creating a liquid gain waveguide through biological scattering [186]. To explore the fascinating realm of photonics, Hakan et al. introduced a semiclassical multimode laser theory [187], while Oleg et al. contributed to this field [188]. Based on Fabry–Perot or waveguide resonance, the ridges on the wings were primarily responsible for the lasing action in these studies.

Over the next ten years, it is anticipated that technology for making 3D PhCs, which are currently thought to be more challenging to make than 2D crystals, will also advance. Three-dimensional crystals are expected to undergo entirely new levels of light control that will also enable an intricate "complete control of fields".

8. Future Prospects

Over the next 10 years, it is anticipated that nanoprocessing technology will progress significantly, and that the development of increasingly dependable and accurate devices will continue. Two-dimensional PhC slabs provide the potential for significant improvements in Si-based systems and advancements in integration with electronic circuitry. Combined optical and electrical circuits with capabilities like optical switching, tuning, and delay capability may be expected to make further advancements. It is anticipated that optical components will make up the majority of these circuits, and that optical/electronic chips will be created. Without a doubt, the size and power requirements of such devices will be tens to hundreds of times lower than they are at the moment. It is also possible to anticipate advancements in a wide range of applications, such as next-generation, miniaturized, multiple-wavelength light sources with active functionality, single-photon light sources, optical memory functionalities that demand a high Q-value, supersensitive sensors, etc. Band-edge lasers are anticipated to permit surface emission across a sizable region with a single wavelength, single light polarization, and single spot, and the beam pattern will be completely programmable by adjusting the photonic crystal structure. Band-edge lasers are anticipated to be used in a variety of disciplines in the future, including information processing, communications, and bio-related fields. It is anticipated that significant advancements will be achieved in extremely efficient emission diodes, display technologies, and the integration of PhCs into organic EL and blue LED devices.

In the realm of solar cell technology, integrating the concept of a photonic crystal with a slow light effect holds immense promise [189–192]. The integration of slow light effects in photonic crystal-based solar cells marks a groundbreaking development in renewable energy technology. Photonic crystals, with their engineered periodic structures, can manipulate the speed and propagation of light, allowing it to linger within the material. This prolonged interaction period offers several advantages, such as improved light absorption, reduced energy losses, and enhanced absorption of specific wavelengths relevant to solar energy conversion. Furthermore, the slow light effect enhances the absorption of weakly absorbed or low-energy photons, which often go untapped in conventional solar cells. This innovation not only boosts the overall efficiency of the solar cell, but also opens up opportunities for the use of thinner, more cost-effective photovoltaic materials. As we continue to harness the potential of slow light in photonic crystals, we move closer to creating solar cells that are not only more efficient, but also more sustainable and economically viable.

Over the next ten years, it is anticipated that technology for making 3D PhCs, which are currently thought to be more challenging to make than 2D crystals, will also advance. Three-dimensional crystals are expected to undergo entirely new levels of light control that will also enable intricate "complete control of fields".

9. Conclusions

In this article, we reviewed several photonic devices and their applications in beam splitters, polarizers, logic gates, solar cells, and sensors, along with their structural design and response. The small, cost-effective, and miniature device in which the propagation of electromagnetic fields can be modified makes it a highly versatile system that can be applied to a wide scale of our daily lives. Obviously, the experts and early researchers who are interested in this field can not only see the unique features and flexibilities in the structural design of such PhC-based devices, but also broaden their thoughts and understanding to further exploit more applications. The states of the art of the development of PhCs are summarized, providing a basic idea to the readers about the significance of this research topic and its history. PhCs can be combined with an appropriate absorber to aid in the selective absorption of solar energy. In recent years, several absorbers have been proposed and are being utilized to achieve high performance and durability. From the above summary, one can notice that strong optical confinement can be achieved even in a miniature system, allowing for the detection of the surrounding environment, including stress, temperature, chemical analytes, etc. Alongside that, such devices are being utilized on a large scale in biomedical fields for the selective detection of proteins and immunoassays. Several highly sensitive sensors based on various PhC configurations have been reviewed and summarized here. Owing to the growth of scientific research in the field of MIP-based sensors, we sincerely apologize to researchers for overlooking their important contributions to the field; however, we tried our best to include most of the article in the manuscript.

Author Contributions: All authors have equally contributed to the preparation of the manuscript. All authors have read and agreed to the published version of the manuscript.

Funding: This work was supported by the Double-Hundred Talent Plan of the Shandong Province, China; Liaocheng University (318052341); and Science and Technology Support Plan for Youth Innovation of Colleges and Universities of the Shandong Province of China (2022KJ107).

Acknowledgments: S. Kumar acknowledges aid from the Double-Hundred Talent Plan of the Shandong Province, China.

Conflicts of Interest: The authors declare no conflict of interest.

References

1. Yablonovitch, E. Inhibited spontaneous emission in solid-state physics and electronics. *Phys. Rev. Lett.* **1987**, *58*, 2059–2062. [CrossRef]
2. John, S. Strong localization of photons in certain disordered dielectric superlattices. *Phys. Rev. Lett.* **1987**, *58*, 2486–2489. [CrossRef]
3. Noori, M.; Soroosh, M.; Baghban, H. Self-Collimation in Photonic Crystals: Applications and Opportunities. *Ann. Der Phys.* **2018**, *530*, 1700049. [CrossRef]
4. Dutta, H.S.; Goyal, A.K.; Srivastava, V.; Pal, S. Coupling light in photonic crystal waveguides: A review. *Photonics Nanostructures-Fundam. Appl.* **2016**, *20*, 41–58. [CrossRef]
5. Liu, B.; Liu, Y.-F.; Jia, C.; He, X.-D. All-optical diode structure based on asymmetrical coupling by a micro-cavity and FP cavity at two sides of photonic crystal waveguide. *AIP Adv.* **2016**, *6*, 065316. [CrossRef]
6. Zhang, Y.; Li, B. Photonic crystal-based bending waveguides for optical interconnections. *Opt. Express* **2006**, *14*, 5723–5732. [CrossRef] [PubMed]
7. Shen, W.; Li, M.; Ye, C.; Jiang, L.; Song, Y. Direct-writing colloidal photonic crystal microfluidic chips by inkjet printing for label-free protein detection. *Lab Chip* **2012**, *12*, 3089–3095. [CrossRef]
8. Zhang, Y.; Zhao, Y.; Zhou, T.; Wu, Q. Applications and developments of on-chip biochemical sensors based on optofluidic photonic crystal cavities. *Lab Chip* **2018**, *18*, 57–74. [CrossRef]
9. Butt, M.A.; Kazanskiy, N.L.; Khonina, S.N. Label-free detection of ambient refractive index based on plasmonic Bragg gratings embedded resonator cavity sensor. *J. Mod. Opt.* **2019**, *66*, 1920–1925. [CrossRef]
10. Butt, M.A.; Khonina, S.N.; Kazanskiy, N.L. An array of nano-dots loaded MIM square ring resonator with enhanced sensitivity at NIR wavelength range. *Optik* **2020**, *202*, 163655. [CrossRef]
11. Butt, M.A.; Khonina, S.N.; Kazanskiy, N.L. Highly sensitive refractive index sensor based on hybrid plasmonic waveguide microring resonator. *Waves Random Complex Media* **2020**, *30*, 292–299. [CrossRef]
12. Shen, H.; Wang, Z.; Wu, Y.; Yang, B. One-dimensional photonic crystals: Fabrication, responsiveness and emerging applications in 3D construction. *RSC Adv.* **2016**, *6*, 4505–4520. [CrossRef]

13. Yu, S.-P.; Muniz, J.A.; Hung, C.-L.; Kimble, H.J. Two-dimensional photonic crystals for engineering atom–light interactions. *Proc. Natl. Acad. Sci. USA* **2019**, *116*, 12743–12751. [CrossRef] [PubMed]
14. Tajiri, T.; Takahashi, S.; Ota, Y.; Watanabe, K.; Iwamoto, S.; Arakawa, Y. Three-dimensional photonic crystal simultaneously integrating a nanocavity laser and waveguides. *Optica* **2019**, *6*, 296–299. [CrossRef]
15. Kumari, S.; Gupta, S. Study of Stress Effect on Optical Performance of Surface-Corrugated Hybrid Plasmonic IBG Filter. *Plasmonics* **2022**, *17*, 339–348. [CrossRef]
16. Kumari, S.; Gupta, S. Simulation Study of Stress Effect on Performance and Design Methodology of Proposed Si/SiGe Integrated Bragg Grating Filter. *IEEE Trans. Device Mater. Reliab.* **2021**, *21*, 569–578. [CrossRef]
17. Clementi, M.; Barone, A.; Fromherz, T.; Gerace, D.; Galli, M. Selective tuning of optical modes in a silicon comb-like photonic crystal cavity. *Nanophotonics* **2019**, *9*, 205–210. [CrossRef]
18. Rostami, A.; Nazari, F.; Banaei, H.A.; Bahrami, A. A novel proposal for DWDM demultiplexer design using modified-T photonic crystal structure. *Photonics Nanostructures Fundam. Appl.* **2010**, *8*, 14–22. [CrossRef]
19. Akahane, Y.; Asano, T.; Song, B.S.; Noda, S. High-Q photonic nanocavity in a two-dimensional photonic crystal. *Nature* **2003**, *425*, 944–947. [CrossRef]
20. Andreani, L.C.; Gerace, D. Light-matter interaction in photonic crystal slabs. *Phys. Status Solidi B* **2007**, *234*, 3528–3539. [CrossRef]
21. Mekis, A.; Chen, J.C.; Kurland, I.; Fan, S.; Villeneuve, P.R.; Joannopoulos, J.D. High transmission through sharp bends in photonic crystal waveguides. *Phys. Rev. Lett.* **1996**, *77*, 3787–3790. [CrossRef]
22. Krauss, T.F. Slow light in photonic crystal waveguides. *J. Phys. D Appl. Phys.* **2007**, *40*, 2666–2670. [CrossRef]
23. Chun, Z.; Feng, Q.; Jun, W.; Jian, Z. A novel polarizer made from two-dimensional photonic bandgap materials. *Chin. Phys. Lett.* **2001**, *18*, 1082–1084. [CrossRef]
24. Zhao, D.; Shi, B.; Jiang, Z.; Fan, Y.; Wang, X. Silicon-based optical waveguide polarizer using photonic band gap. *Appl. Phys. Lett.* **2002**, *81*, 409–411. [CrossRef]
25. Sinha, R.K.; Kalra, Y. Design of optical waveguide polarizer using photonic band gap. *Opt. Express* **2006**, *14*, 10790–10794. [CrossRef] [PubMed]
26. Sinha, R.K.; Kalra, Y. Design of a photonic band gap polarizer. *Opt. Eng.* **2006**, *45*, 110503. [CrossRef]
27. Cui, Y.; Wu, Q.; Schonbrun, E.; Tinker, M.; Lee, J.-B.; Park, W. Silicon-based 2-D slab photonic crystal TM polarizer at telecommunication wavelength. *IEEE Photon. Technol. Lett.* **2008**, *20*, 641–643. [CrossRef]
28. Kim, D.W.; Lee, M.H.; Kim, Y.; Kim, K.H. Ultracompact transverse magnetic mode-pass filter based on one-dimensional photonic crystals with subwavelength structures. *Opt. Express* **2016**, *24*, 21560–21565. [CrossRef] [PubMed]
29. Pandey, G.N.; Kumar, N.; Singh, P.; Thapa, K.B. Analysis of Photonic Band Structure Tunability for TE and TM Modes in a Silicon and Polymer Based Ternary Photonic Crystal for Visible Range Devices. *Silicon* **2022**, *14*, 11659–11666. [CrossRef]
30. Wang, B.; Blaize, S.; Salas-Montiel, R. Nanoscale plasmonic TM-pass polarizer integrated on silicon photonics. *Nanoscale* **2019**, *11*, 20685–20692. [CrossRef]
31. Butt, M.A.; Khonina, S.N.; Kazanskiy, N.L. A highly sensitive design of subwavelength grating double-slot waveguide microring resonator. *Laser Phys. Lett.* **2020**, *17*, 076201. [CrossRef]
32. Jervakani, A.T.; Darki, B.S. An ultracompact optical polarizer based on the one-dimensional photonic crystals containing anisotropic layers. *Opt. Commun.* **2023**, *526*, 128884. [CrossRef]
33. Alam, M.; Aitchsion, J.S.; Mojahedi, M. Compact hybrid TM-pass polarizer for silicon-on-insulator platform. *Appl. Opt.* **2011**, *50*, 2294–2298. [CrossRef] [PubMed]
34. Kurt, H.; Yilmaz, D.; Akosman, A.E.; Ozbay, E. Asymmetric light propagation in chirped photonic crystal waveguides. *Opt. Express* **2012**, *20*, 20635–20646. [CrossRef]
35. Alam, M.Z.; Aitchison, J.S.; Mojahedi, M. Compact and silicon-on-insulatorcompatible hybrid plasmonic TE-pass polarizer. *Opt. Lett.* **2012**, *37*, 55–57. [CrossRef]
36. Zhang, J.; Cassan, E.; Zhang, X. Wideband and Compact TE-pass/TM-stop polarizer based on a hybrid plasmonic bragg grating for silicon photonics. *J. Light. Technol.* **2014**, *32*, 1383–1386. [CrossRef]
37. Choi, K.-H.; Huh, J.; Cui, Y.; Trivedi, K.; Hu, W.; Ju, B.-K.; Lee, J.-B. One-step combined-nanolithography and photolithograp for a 2D photonic crystal TM polarizer. *Micromachines* **2014**, *5*, 228–238. [CrossRef]
38. Guan, X.; Chen, P.; Chen, S.; Xu, P.; Shi, Y.; Dai, D. Low-loss ultracompact transversemagnetic-pass polarizer with a silicon subwavelength grating waveguide. *Opt. Lett.* **2014**, *39*, 4514–4517. [CrossRef]
39. Prakash, C.; Sen, M.; Mondal, H.; Goswami, K. Design and optimization of a TE-pass polarization filter based on a slotted photonic crystal waveguide. *J. Opt. Soc. Am. B* **2018**, *35*, 1791–1798. [CrossRef]
40. Butt, M.A. Numerical investigation of a small footprint plasmonic Bragg grating structure with a high extinction ratio. *Photonics Lett. Pol.* **2020**, *12*, 82–84. [CrossRef]
41. Kang, M.; Zhang, Z.; Wu, T.; Zhang, X.; Xu, Q.; Krasnok, A.; Han, J. Coherent full polarization control based on bound states in the continuum. *Nat. Commun.* **2022**, *13*, 4536. [CrossRef] [PubMed]
42. He, X.-T.; Liu, J.-W.; Shi, F.-L.; Shen, K.; Chen, W.-J.; Chen, X.-D.; Dong, J.-W. Dual-polarization two-dimensional valley photonic crystals. *Sci. China Phys. Mech. Astron.* **2022**, *65*, 284212. [CrossRef]
43. Chen, N.; Zhang, X.; Lu, X.; Zhang, Z.; Mu, Z.; Chang, M. Numerical investigation of a short polarization beam splitter based on dual-core photonic crystal fiber with As_2S_3 layer. *Micromachines* **2020**, *11*, 706. [CrossRef] [PubMed]

44. Wen, K.; Wang, R.; Wang, J.; Li, J. Polarization beam splitter based on photonic crystal fibers. *Front. Optoelectron. China* **2009**, *2*, 76–80. [CrossRef]
45. Park, I.; Lee, H.-S.; Kim, H.-J.; Moon, K.-M.; Lee, S.-G.; Beom-Hoan, O.; Park, S.-G.; Lee, E.-H. Photonic crystal power-splitter based on directional coupling. *Opt. Express* **2004**, *12*, 3599–3604. [CrossRef]
46. Butt, M.; Khonina, S.; Kazanskiy, N. Compact design of a polarization beam splitter based on silicon-on-insulator platform. *Laser Phys.* **2018**, *28*, 116202. [CrossRef]
47. Butt, M.; Khonina, S.; Kazanskiy, N. Optical elements based on silicon photonics. *Comput. Opt.* **2019**, *43*, 1079–1083. [CrossRef]
48. Butt, M.A.; Kozlova, E.S. Multiport optical power splitter design based on coupled mode theory. *J. Phys. Conf. Ser.* **2019**, *1368*, 022006. [CrossRef]
49. Butt, M.A.; Khonina, S.N.; Kazanskiy, N.L. Ultra-short lossless plasmonic power splitter design based on metal-insulator-metal waveguide. *Laser Phys.* **2019**, *30*, 016201. [CrossRef]
50. Lu, Z.; Tang, Y.; Shen, Y.; Liu, X.; Zi, L. Polarization beam splitting in two dimensional photonic crystals based on negative refraction. *Phys. Lett. A* **2005**, *346*, 243–247. [CrossRef]
51. Hou, J.; Wang, L.; Yang, C.; Wang, B.; Chen, S. Compact high extinction ratio asymmetric polarization beam splitter of periodic rods waveguide. *Appl. Opt.* **2015**, *54*, 10277–10282. [CrossRef] [PubMed]
52. Solli, D.R.; McCormick, C.F.; Chiao, R.Y.; Hickmann, J.M. Photonic crystal polarizers and polarizing beam splitters. *J. Appl. Phys.* **2003**, *93*, 125. [CrossRef]
53. Butt, M.A.; Kazanskiy, N.L. Two-dimensional photonic crystal heterostructure for light steering and TM-polarization maintaining applications. *Laser Phys.* **2021**, *31*, 036201. [CrossRef]
54. Butt, M.A.; Khonina, S.N.; Kazanskiy, N.L. 2D-Photonic crystal heterostructures for the realization of compact photonic devices. *Photonics Nanostructures Fundam. Appl.* **2021**, *44*, 100903. [CrossRef]
55. Dan, L.; Citrin, D.S.; Hu, S. Compact high-performance polarization beam splitter based on a silicon photonic crystal heterojunction. *Opt. Mater.* **2020**, *109*, 110256.
56. Rafiee, E.; Emami, F.; Negahdari, R. Design of a novel nano plasmonic-dielectric photonic crystal power splitter suitable for photonic integrated circuits. *Optik* **2018**, *172*, 234–240. [CrossRef]
57. Razmi, H.; Soroosh, M.; Kavian, Y.S. A new proposal for ultra-compact polarization independent power splitter based on photonic crystal structures. *J. Opt. Commun.* **2017**, *39*, 375–379. [CrossRef]
58. Ren, C.; Wang, L.; Kang, F. Adjustable unidirectional beam splitters in two dimensional photonic crystals. *Opt. Quantum Electron.* **2020**, *52*, 475. [CrossRef]
59. Kazanskiy, N.L.; Butt, M.A.; Khonina, S.N. 2D-Heterostructure Photonic Crystal Formation for On-Chip Polarization Division Multiplexing. *Photonics* **2021**, *8*, 313. [CrossRef]
60. Liu, W.; Ma, H.; Walsh, A. Advance in photonic crystal solar cells. *Renew. Sustain. Energy Rev.* **2019**, *116*, 109436. [CrossRef]
61. Butt, M.A.; Kazansky, N.L. Narrowband perfect metasurface absorber based on impedance matching. *Photonics Lett. Pol.* **2020**, *12*, 88–90. [CrossRef]
62. Nawi, N.; Majlis, B.Y.; Mahdi, M.A.; De La Rue, R.M.; Lonĉar, M.; Md Zain, A.R. Enhancement and reproducibility of high quality factor, one-dimensional photonic crystal/photonic wire (1D PhC/PhW) microcavities. *J. Eur. Opt. Soc. Publ.* **2018**, *14*, 6. [CrossRef]
63. Cen, C.; Chen, Z.; Xu, D.; Jiang, L.; Chen, X.; Yi, Z.; Wu, P.; Li, G.; Yi, Y. High Quality Factor, High Sensitivity Metamaterial Graphene—Perfect Absorber Based on Critical Coupling Theory and Impedance Matching. *Nanomaterials* **2020**, *10*, 95. [CrossRef] [PubMed]
64. Mavrokefalos, A.; Han, S.E.; Yerci, S.; Branham, M.S.; Chen, G. Efficient Light Trapping in Inverted Nanopyramid Thin Crystalline Silicon Membranes for Solar Cell Applications. *Nano Lett.* **2012**, *12*, 2792–2796. [CrossRef] [PubMed]
65. Yu, W.; Shen, L.; Shen, P.; Long, Y.; Sun, H.; Chen, W.; Ruan, S. Semitransparent Polymer Solar Cells with 5% Power Conversion Efficiency Using Photonic Crystal Reflector. *ACS Appl. Mater. Interfaces* **2014**, *6*, 599–605. [CrossRef] [PubMed]
66. Zhang, W.; Anaya, M.; Lozano, G.; Calvo, M.E.; Johnston, M.B.; Míguez, H.; Snaith, H.J. Highly Efficient Perovskite Solar Cells with Tunable Structural Color. *Nano Lett.* **2015**, *15*, 1698–1702. [CrossRef]
67. Ramos, F.J.; Oliva-Ramírez, M.; Nazeeruddin, M.K.; Graetzel, M.; González-Elipe, A.R.; Ahmad, S. Light management: Porous 1-dimensional nanocolumnar structures as effective photonic crystals for perovskite solar cells. *J. Mater. Chem. A* **2015**, *4*, 4962–4970. [CrossRef]
68. Zhao, W.; Li, S.; Yao, H.; Zhang, S.; Zhang, Y.; Yang, B.; Hou, J. Molecular Optimization Enables over 13% Efficiency in Organic Solar Cells. *J. Am. Chem. Soc.* **2017**, *139*, 7148–7151. [CrossRef]
69. Shi, H.; Xia, R.; Sun, C.; Xiao, J.; Wu, Z.; Huang, F.; Yip, H.-L.; Cao, Y. Synergic Interface and Optical Engineering for High-Performance Semitransparent Polymer Solar Cells. *Adv. Energy Mater.* **2017**, *7*, 1701121. [CrossRef]
70. Gupta, N.D.; Janyani, V. Lambertian and photonic light trapping analysis with thickness for GaAs solar cells based on 2D periodic pattern. *IET Optoelectron.* **2017**, *11*, 217–224. [CrossRef]
71. Zheng, G.; Xu, L.; Pei, S.; Su, W. Theoretical analysis of performance improvement of thin film solar cells with two-dimensional photonic crystals and distributed Bragg reflector textured structures. *Opt. Commun.* **2013**, *309*, 318–322. [CrossRef]
72. Bhattacharya, S.; John, S. Designing High-Efficiency Thin Silicon Solar Cells Using Parabolic-Pore Photonic Crystals. *Phys. Rev. Appl.* **2018**, *9*, 044009. [CrossRef]

73. Morsy, M.A.; Saleh, K. Efficiency Enhancement of GaAs Nano Solar Cell Based on 2D Photonic Crystal Trapping Layer and 2D Index Modulation Layer. *IEEE Access* **2022**, *10*, 44147–44158. [CrossRef]
74. Zheng, X.; Zhang, L. Photonic nanostructures for solar energy conversion. *Energy Environ. Sci.* **2016**, *9*, 2511–2532. [CrossRef]
75. Dotan, H.; Kfir, O.; Sharlin, E.; Blank, O.; Gross, M.; Dumchin, I.; Ankonina, G.; Rothschild, A. Resonant light trapping in ultrathin films for water splitting. *Nat. Mater.* **2013**, *12*, 158–164. [CrossRef]
76. Cheng, K.; Li, Q.; Meng, J.; Han, X.; Wu, Y.; Wang, S.; Qian, L.; Du, Z. Interface engineering for efficient charge collection in Cu2O/ZnO heterojunction solar cells with ordered ZnO cavity-like nanopatterns. *Sol. Energy Mater. Sol. Cells* **2013**, *116*, 120–125. [CrossRef]
77. Yan, T.; Yuan, R.; Li, W.; You, J. Origin of superior photocatalytic activity in rutile TiO_2 hierarchical microspheres: The dominate role of "microcavity" structure. *Appl. Catal. A Gen.* **2014**, *478*, 204–210. [CrossRef]
78. An, C.J.; Cho, C.; Choi, J.K.; Park, J.-M.; Jin, M.L.; Lee, J.-Y.; Jung, H.-T. Highly Efficient Top-Illuminated Flexible Polymer Solar Cells with a Nanopatterned 3D Microresonant Cavity. *Small* **2014**, *10*, 1278–1283. [CrossRef]
79. Zheng, W.; Luo, X.; Zhang, Y.; Ye, C.; Qin, A.; Cao, Y.; Hou, L. Efficient Low-Cost All-Flexible Microcavity Semitransparent Polymer Solar Cells Enabled by Polymer Flexible One-Dimensional Photonic Crystals. *ACS Appl. Mater. Interfaces* **2020**, *12*, 23190–23198. [CrossRef]
80. Zhong, J.; Liang, W.; Long, Y. Simultaneously improving efficiency and transparency of semitransparent organic solar cells by constructing semitransparent microcavity. *Optik* **2018**, *171*, 706–714. [CrossRef]
81. Wang, C.; Geng, A.; Guo, Y.; Jiang, S.; Qu, X.; Li, L. A novel preparation of three-dimensionally ordered macroporous M/Ti (M = Zr or Ta) mixed oxide nanoparticles with enhanced photocatalytic activity. *J. Colloid Interface Sci.* **2006**, *301*, 236–247. [CrossRef] [PubMed]
82. Zhou, M.; Wu, H.B.; Bao, J.; Liang, L.; Lou, X.W.D.; Xie, Y. Ordered Macroporous BiVO 4 Architectures with Controllable Dual Porosity for Efficient Solar Water Splitting. *Angew. Chem. Int. Ed.* **2013**, *52*, 8579–8583. [CrossRef]
83. Zhou, M.; Bao, J.; Xu, Y.; Zhang, J.; Xie, J.; Guan, M.; Wang, C.; Wen, L.; Lei, Y.; Xie, Y. Photoelectrodes Based upon Mo:BiVO 4 Inverse Opals for Photoelectrochemical Water Splitting. *ACS Nano* **2014**, *8*, 7088–7098. [CrossRef] [PubMed]
84. Zhang, L.; Baumanis, C.; Robben, L.; Kandiel, T.; Bahnemann, D. Bi_2WO_6 Inverse Opals: Facile Fabrication and Efficient Visible-Light-Driven Photocatalytic and Photoelectrochemical Water-Splitting Activity. *Small* **2011**, *7*, 2714–2720. [CrossRef] [PubMed]
85. Chen, X.; Ye, J.; Ouyang, S.; Kako, T.; Li, Z.; Zou, Z. Enhanced Incident Photon-to-Electron Conversion Efficiency of Tungsten Trioxide Photoanodes Based on 3D-Photonic Crystal Design. *ACS Nano* **2011**, *5*, 4310–4318. [CrossRef] [PubMed]
86. Choi, D.H.; Nam, S.K.; Jung, K.; Moon, J.H. 2D photonic crystal nanodisk array as electron transport layer for highly efficient perovskite solar cells. *Nano Energy* **2019**, *56*, 365–372. [CrossRef]
87. Bhattacharya, S.; John, S. Beyond 30% Conversion Efficiency in Silicon Solar Cells: A Numerical Demonstration. *Sci. Rep.* **2019**, *9*, 12482–12515. [CrossRef]
88. Cheng, C.; Karuturi, S.K.; Liu, L.; Liu, J.; Li, H.; Su, L.T.; Tok, A.I.Y.; Fan, H.J. Quantum-Dot-Sensitized TiO_2 Inverse Opals for Photoelectrochemical Hydrogen Generation. *Small* **2012**, *8*, 37–42. [CrossRef]
89. Lu, Y.; Yu, H.; Chen, S.; Quan, X.; Zhao, H. Integrating Plasmonic Nanoparticles with TiO_2 Photonic Crystal for Enhancement of Visible-Light-Driven Photocatalysis. *Environ. Sci. Technol.* **2012**, *46*, 1724–1730. [CrossRef]
90. Dinh, C.-T.; Yen, H.; Kleitz, F.; Do, T.-O. Three-Dimensional Ordered Assembly of Thin-Shell Au/TiO_2 Hollow Nanospheres for Enhanced Visible-Light-Driven Photocatalysis. *Angew. Chem. Int. Ed.* **2014**, *53*, 6618–6623. [CrossRef]
91. Zhang, L.; Lin, C.; Valev, V.K.; Reisner, E.; Steiner, U.; Baumberg, J.J. Plasmonic Enhancement in $BiVO_4$ Photonic Crystals for Efficient Water Splitting. *Small* **2014**, *10*, 3970–3978. [CrossRef] [PubMed]
92. Li, H.; Hu, X.; Hong, W.; Cai, F.; Tang, Q.; Zhao, B.; Zhang, D.; Cheng, P. Photonic crystal coupled plasmonic nanoparticle array for resonant enhancement of light harvesting and power conversion. *Phys. Chem. Chem. Phys.* **2012**, *14*, 14334. [CrossRef] [PubMed]
93. Gangwar, R.K.; Qin, J.; Wang, X. Porous Silicon–Based Microring Resonator for Temperature and Cancer Cell Detection. *Front. Phys.* **2022**, *10*. [CrossRef]
94. Pathak, A.K.; Viphavakit, C. A Review on All-optical Fiber-based VOC sensors: Heading Towards the Development of Promising Technology. *Sens. Actuators A Phys.* **2022**, *338*, 113455. [CrossRef]
95. Kumari, S.; Gupta, S. Cladding stress induced performance variation of silicon MMI coupler. *Photonics Nanostructures Fundam. Appl.* **2019**, *33*, 55–65. [CrossRef]
96. Kumari, S.; Gupta, S. Performance Estimation of Hybrid Plasmonic Waveguide in Presence of Stress. *Plasmonics* **2021**, *16*, 359–370. [CrossRef]
97. Rahman, B.M.A.; Viphavakit, C.; Chitaree, R.; Ghosh, S.; Pathak, A.K.; Verma, S.; Sakda, N. Optical Fiber, Nanomaterial, and THz-Metasurface-Mediated Nano-Biosensors: A Review. *Biosensors* **2022**, *12*, 42. [CrossRef]
98. Malinin, A.V.; Zanishevskaja, A.A.; Tuchin, V.V.; Skibina, Y.S.; Silokhin, I.Y. *Photonic Crystal Fibers for Food Quality Analysis*; Popp, J., Drexler, W., Tuchin, V.V., Matthews, D.L., Eds.; 2012; p. 842746. Available online: https://www.spiedigitallibrary.org/conference-proceedings-of-spie/8427/842746/Photonic-crystal-fibers-for-food-quality-analysis/10.1117/12.924096.short (accessed on 4 October 2023).

99. Falconi, M.C.; Palma, G.; Starecki, F.; Nazabal, V.; Troles, J.; Adam, J.L.; Taccheo, S.; Ferrari, M.; Prudenzano, F. Novel pumping schemes of Mid-IR photonic crystal fiber lasers for aerospace applications. In Proceedings of the 2016 18th International Conference on Transparent Optical Networks (ICTON), Trento, Italy, 10–14 July 2016; pp. 1–5.
100. Mashev, L.; Popov, E. Diffraction efficiency anomalies of multicoated dielectric gratings. *Opt. Commun.* **1984**, *51*, 131–136. [CrossRef]
101. Zhou, Y.; Zhang, M.; Guo, Z.; Miao, L.; Han, S.-T.; Wang, Z.; Zhang, X.; Zhang, H.; Peng, Z. Recent advances in black phosphorus-based photonics, electronics, sensors and energy devices. *Mater. Horiz.* **2017**, *4*, 997–1019. [CrossRef]
102. Jágerská, J.; Zhang, H.; Diao, Z.; Le Thomas, N.; Houdré, R. Refractive index sensing with an air-slot photonic crystal nanocavity. *Opt. Lett.* **2010**, *35*, 2523. [CrossRef]
103. Zhao, H.-J. High sensitivity refractive index gas sensing enhanced by surface plasmon resonance with nano-cavity antenna array. *Chin. Phys. B* **2012**, *21*, 087104. [CrossRef]
104. Wang, L.; Sang, T.; Li, J.; Zhou, J.; Wang, B.; Wang, Y. High-sensitive transmission type of gas sensor based on guided-mode resonance in coupled gratings. *J. Mod. Opt.* **2018**, *65*, 1601–1608. [CrossRef]
105. Zu, P.; Chan, C.C.; Jin, Y.; Gong, T.; Zhang, Y.; Chen, L.H.; Dong, X. A Temperature-Insensitive Twist Sensor by Using Low-Birefringence Photonic-Crystal-Fiber-Based Sagnac Interferometer. *IEEE Photonics Technol. Lett.* **2011**, *23*, 920–922. [CrossRef]
106. Lu, H.; Sadani, B.; Ulliac, G.; Guyot, C.; Courjal, N.; Collet, M.; Baida, F.I.; Bernal, M.-P. Integrated temperature sensor based on an enhanced pyroelectric photonic crystal. *Opt. Express* **2013**, *21*, 16311–16318. [CrossRef] [PubMed]
107. Shafiee, H.; Lidstone, E.A.; Jahangir, M.; Inci, F.; Hanhauser, E.; Henrich, T.J.; Kuritzkes, D.R.; Cunningham, B.T.; Demirci, U. Nanostructured Optical Photonic Crystal Biosensor for HIV Viral Load Measurement. *Sci. Rep.* **2015**, *4*, 4116. [CrossRef]
108. Bouzidi, A.; Bria, D.; Akjouj, A.; Pennec, Y.; Djafari-Rouhani, B. A tiny gas-sensor system based on 1D photonic crystal. *J. Phys. D Appl. Phys.* **2015**, *48*, 495102. [CrossRef]
109. Hameed, M.F.O.; Azab, M.Y.; Heikal, A.M.; El-Hefnawy, S.M.; Obayya, S.S.A. Highly Sensitive Plasmonic Photonic Crystal Temperature Sensor Filled With Liquid Crystal. *IEEE Photonics Technol. Lett.* **2016**, *28*, 59–62. [CrossRef]
110. Liu, Y.; Zhou, W.; Sun, Y. Optical Refractive Index Sensing Based on High-Q Bound States in the Continuum in Free-Space Coupled Photonic Crystal Slabs. *Sensors* **2017**, *17*, 1861. [CrossRef]
111. Maurya, J.; François, A.; Prajapati, Y. Two-Dimensional Layered Nanomaterial-Based One-Dimensional Photonic Crystal Refractive Index Sensor. *Sensors* **2018**, *18*, 857. [CrossRef]
112. Snapp, P.; Kang, P.; Leem, J.; Nam, S. Colloidal Photonic Crystal Strain Sensor Integrated with Deformable Graphene Phototransducer. *Adv. Funct. Mater.* **2019**, *29*, 1902216. [CrossRef]
113. Resende, S.; Frasco, M.F.; Sales, M.G.F. A biomimetic photonic crystal sensor for label-free detection of urinary venous thromboembolism biomarker. *Sens. Actuators B Chem.* **2020**, *312*, 127947. [CrossRef]
114. Kawasaki, D.; Yamada, H.; Sueyoshi, K.; Hisamoto, H.; Endo, T. Imprinted Photonic Crystal-Film-Based Smartphone-Compatible Label-Free Optical Sensor for SARS-CoV-2 Testing. *Biosensors* **2022**, *12*, 200. [CrossRef] [PubMed]
115. Pu, S.; Bai, X.; Wang, L. Temperature dependence of photonic crystals based on thermoresponsive magnetic fluids. *J. Magn. Magn. Mater.* **2011**, *323*, 2866–2871. [CrossRef]
116. Su, D.; Pu, S.; Mao, L.; Wang, Z.; Qian, K. A Photonic Crystal Magnetic Field Sensor Using a Shoulder-Coupled Resonant Cavity Infiltrated with Magnetic Fluid. *Sensors* **2016**, *16*, 2157. [CrossRef]
117. Aazi, M.; Kudinova, M.; Kinet, D.; Auguste, J.-L.; Delépine-Lesoille, S.; Mégret, P.; Humbert, G. Impact of H_2 gas on disruptive birefringence optical fibers with embedded Palladium particles for developing robust sensors. *J. Phys. Photonics* **2020**, *2*, 014005. [CrossRef]
118. Pathak, A.K.; Viphavakit, C. VOC Biomarker Monitoring for Diabetes Through Exhaled Breath Using Ag/P-TiO_2 Composite Plasmonic Sensor. *IEEE Sens. J.* **2021**, *21*, 22631–22637. [CrossRef]
119. Liu, H.; Zhang, Y.; Chen, C.; Bai, B.; Shao, Q.; Wang, H.; Zhang, W.; Chen, C.; Tang, S. Transverse-Stress Compensated Methane Sensor Based on Long-Period Grating in Photonic Crystal Fiber. *IEEE Access* **2019**, *7*, 175522–175530. [CrossRef]
120. Kraeh, C.; Martinez-Hurtado, J.L.; Popescu, A.; Hedler, H.; Finley, J.J. Slow light enhanced gas sensing in photonic crystals. *Opt. Mater.* **2018**, *76*, 106–110. [CrossRef]
121. Sardar, M.R.; Faisal, M. Methane Gas Sensor Based on Microstructured Highly Sensitive Hybrid Porous Core Photonic Crystal Fiber. *J. Sens. Technol.* **2019**, *9*, 12–26. [CrossRef]
122. Chang, Y.; Hasan, D.; Dong, B.; Wei, J.; Ma, Y.; Zhou, G.; Ang, K.W.; Lee, C. Surface-Enhanced Infrared Absorption-Based CO_2 Sensor using Photonic Crystal Slab. In Proceedings of the 2019 IEEE 32nd International Conference on Micro Electro Mechanical Systems (MEMS), Seoul, Republic of Korea, 27–31 January 2019; pp. 141–144.
123. Liu, Y.; Wang, S.; Biswas, P.; Palit, P.; Zhou, W.; Sun, Y. Optofluidic vapor sensing with free-space coupled 2D photonic crystal slabs. *Sci. Rep.* **2019**, *9*, 4209. [CrossRef]
124. Wang, X.; Lv, J.; E, S.; Han, B.; Zhang, Y. Theoretical Design and Simulation Optimization of Photonic Crystal Cavity for Tetrahydrofuran Vapor Sensing. *Phys. Status Solidi* **2019**, *256*, 1900221. [CrossRef]
125. Britto, E.C.; Nizar, S.M.; Krishnan, P. A Highly Sensitive Photonic Crystal Fiber Gas Sensor for the Detection of Sulfur Dioxide. *Silicon* **2022**, *14*, 12665–12674. [CrossRef]

126. Loga, K.; Nizar, S.M.; Caroline, B.E. Design of Photonic Crystal Fiber Based Gas Sensor for Sensing Nitrous Oxide (N_2O) and Benzene (C_6H_6) at Atmospheric Level. In Proceedings of the 2022 International Conference on Smart Technologies and Systems for Next Generation Computing (ICSTSN), Villupuram, India, 25–26 March 2022; pp. 1–6.
127. Ghods, M.M.; Afsahi, M. Ultra-sensitive absorption-based gas detecting using graphene-covered periodic photonic crystal slabs resonating under critical coupling condition at mid-infrared frequencies. *IEEE Trans. Nanobioscience* **2022**, *22*, 143–148. [CrossRef] [PubMed]
128. Zegadi, R.; Zegadi, A.; Zebiri, C.; Mosbah, S.; Mekki, S.; Bouknia, M.L.; Bendjedi, H. Enhanced 2D Photonic Crystal Sensor for High Sensitivity Sulfuric Acid (H_2SO_4) and Hydrogen Peroxide (H_2O_2) Detection. *Silicon* **2022**, *14*, 11001–11006. [CrossRef]
129. Klimov, N.N.; Mittal, S.; Berger, M.; Ahmed, Z. On-chip silicon waveguide Bragg grating photonic temperature sensor. *Opt. Lett.* **2015**, *40*, 3934–3936. [CrossRef]
130. Gangwar, R.K.; Bhardwaj, V.; Singh, V.K. Magnetic field sensor based on selectively magnetic fluid infiltrated dual-core photonic crystal fiber. *Opt. Eng.* **2016**, *55*, 026111. [CrossRef]
131. Kumar Vyas, A.; Kumar Gangwar, R.; Kumar, S. Elliptical air hole PCF-based low-cost sensor for refractive index and temperature detection: Design and analysis. *Opt. Fiber Technol.* **2022**, *73*, 103060. [CrossRef]
132. Yang, F.; Wang, Z.K.; Wang, D.N. A highly sensitive optical fiber strain sensor based on cascaded multimode fiber and photonic crystal fiber. *Opt. Fiber Technol.* **2019**, *47*, 102–106. [CrossRef]
133. Ying, Y.; Hu, N.; Si, G.; Xu, K.; Liu, N.; Zhao, J.-Z. Magnetic field and temperature sensor based on D-shaped photonic crystal fiber. *Optik* **2019**, *176*, 309–314. [CrossRef]
134. Li, Q.; Liu, S.; Wang, J.; Mbola, N.M.; Meng, Z.; Wang, X.; Xue, M. A biocompatible, self-adhesive, and stretchable photonic crystal sensor for underwater motion detection. *J. Mater. Chem. C* **2022**, *10*, 9025–9034. [CrossRef]
135. Dinodiya, S.; Bhargava, A. A Comparative Analysis of Pressure Sensing Parameters for Two Dimensional Photonic Crystal Sensors Based on Si and GaAs. *Silicon* **2022**, *14*, 4611–4618. [CrossRef]
136. Biswas, U.; Rakshit, J.K.; Suthar, B.; Kumar, D.; Nayak, C. Modeling and analysis of all-optical pressure sensor using photonic crystal based micro ring resonator. *Int. J. Numer. Model. Electron. Netw. Devices Fields* **2022**, *35*, e2962. [CrossRef]
137. Gupta, P.K.; Tripathi, S.; Paltani, P.P. Optical Add/Drop Filter and Pressure Sensor Design Using Two-Dimensional Photonic Crystal. In Proceedings of the 2022 Workshop on Recent Advances in Photonics (WRAP), Mumbai, India, 4–6 March 2022; pp. 1–2.
138. Meng, X.; Zhao, Y.; Li, J.; Guo, Y.; Li, S.; Guo, H.; Du, H.; Li, Z.; Li, M.; Feng, Y.; et al. High-performance plasmonic sensor based on photonic crystal fiber for refractive index and temperature sensing. *Infrared Phys. Technol.* **2022**, *122*, 104036. [CrossRef]
139. Gangwar, R.K.; Singh, V.K. Highly Sensitive Surface Plasmon Resonance Based D-Shaped Photonic Crystal Fiber Refractive Index Sensor. *Plasmonics* **2017**, *12*, 1367–1372. [CrossRef]
140. Pathak, A.K.; Rahman, B.M.A.; Viphavakit, C. Nanowire Embedded Micro-Drilled Dual-Channel Approach to Develop Highly Sensitive Biosensor. *IEEE Photonics Technol. Lett.* **2022**, *34*, 707–710. [CrossRef]
141. Mustafa, Q.M.S.; Kadhim, R.A.; Kalankesh, H.V.; Pourziad, A. Dual-channel plasmonic refractive index sensor based on H-shaped optical fibres for multi-analyte detection. *J. Opt.* **2022**, *52*, 1171–1179. [CrossRef]
142. Minkovich, V.P.; Sotsky, A.B. Tapered photonic crystal fibers coated with ultra-thin films for highly sensitive bio-chemical sensing. *J. Eur. Opt. Soc. Publ.* **2019**, *15*, 7. [CrossRef]
143. Pathak, A.K.; Ghosh, S.; Gangwar, R.K.; Rahman, B.M.A.; Singh, V.K. Metal Nanowire Assisted Hollow Core Fiber Sensor for an Efficient Detection of Small Refractive Index Change of Measurand Liquid. *Plasmonics* **2019**, *14*, 1823–1830. [CrossRef]
144. Mahmud, S.S.; Islam, M.T.; Atiqullah, S.M. Poisonous Chemical Sensing Using Highly Sensitive Terahertz Photonic Crystal Fiber Sensor. In Proceedings of the 2019 4th International Conference on Electrical Information and Communication Technology (EICT), Khulna, Bangladesh, 20–22 December 2019; pp. 1–5.
145. Mohammed, N.A.; Hamed, M.M.; Khalaf, A.A.M.; Alsayyari, A.; El-Rabaie, S. High-sensitivity ultra-quality factor and remarkable compact blood components biomedical sensor based on nanocavity coupled photonic crystal. *Results Phys.* **2019**, *14*, 102478. [CrossRef]
146. De, M.; Pathak, A.K.; Singh, V.K. Single channel photonic crystal fiber based high sensitive petrol adulteration detection sensor. *Optik* **2019**, *183*, 539–546. [CrossRef]
147. Ramanujam, N.R.; El-Khozondar, H.J.; Dhasarathan, V.; Taya, S.A.; Aly, A.H. Design of one dimensional defect based photonic crystal by composited superconducting material for bio sensing applications. *Phys. B Condens. Matter* **2019**, *572*, 42–55. [CrossRef]
148. Panda, A.; Pukhrambam, P.D.; Dadure, P.; Pakray, P. Application of Machine Learning for Accurate Detection of Hemoglobin Concentrations Employing Defected 1D Photonic Crystal. *Silicon* **2022**, *14*, 12203–12212. [CrossRef]
149. Chen, Y.-T.; Liao, Y.-Y.; Chen, C.-C.; Hsiao, H.-H.; Huang, J.-J. Surface plasmons coupled two-dimensional photonic crystal biosensors for Epstein-Barr virus protein detection. *Sens. Actuators B Chem.* **2019**, *291*, 81–88. [CrossRef]
150. Mostafa, T.S.; Mohammed, N.A.; El-Rabaie, E.-S.M. Ultra-High bit rate all-optical AND/OR logic gates based on photonic crystal with multi-wavelength simultaneous operation. *J. Mod. Opt.* **2019**, *66*, 1005–1016. [CrossRef]
151. Rao, D.G.S.; Swarnakar, S.; Kumar, S. Performance analysis of all-optical NAND, NOR, and XNOR logic gates using photonic crystal waveguide for optical computing applications. *Opt. Eng.* **2020**, *59*, 057101. [CrossRef]
152. Rao, D.G.S.; Palacharla, V.; Swarnakar, S.; Kumar, S. Design of all-optical D flip-flop using photonic crystal waveguides for optical computing and networking. *Appl. Opt.* **2020**, *59*, 7139–7143. [CrossRef]

153. Rao, D.G.S.; Swarnakar, S.; Kumar, S. Design of all-optical reversible logic gates using photonic crystal waveguides for optical computing and photonic integrated circuits. *Appl. Opt.* **2020**, *59*, 11003–11012. [CrossRef]
154. Krauss, T.F.; Rue, R.M.D.L.; Brand, S. Two-dimensional photonic-bandgap structures operating at near-infrared wavelengths. *Nature* **1996**, *383*, 699–702. [CrossRef]
155. Swarnakar, S.; Kumar, S.; Sharma, S. Design of all-optical half-subtractor circuit device using 2-D principle of photonic crystal waveguides. *J. Opt. Commun.* **2019**, *40*, 195–203.
156. Rao, D.G.S.; Swarnakar, S.; Kumar, S. Design of photonic crystal based compact all-optical 2 × 1 multiplexer for optical processing devices. *Microelectron. J.* **2021**, *112*, 105046. [CrossRef]
157. Rachana, M.; Swarnakar, S.; Krishna, S.V.; Kumar, S. Design and analysis of an optical three-input AND gate using a photonic crystal fiber. *Appl. Opt.* **2022**, *61*, 77–83. [CrossRef] [PubMed]
158. Rao, D.G.S.; Swarnakar, S.; Palacharla, V.; Raju, K.S.R.; Kumar, S. Design of all-optical AND, OR, and XOR logic gates using photonic crystals for switching applications. *Photonic Netw. Commun.* **2021**, *41*, 109–118. [CrossRef]
159. Painter, O.; Lee, R.K.; Scherer, A.; Yariv, A.; O'Brien, J.D.; Dapkus, P.D.; Kim, I. Two-dimensional photonic bandgap defect mode laser. *Science* **1999**, *284*, 1819–1821. [CrossRef] [PubMed]
160. Adibi, A.; Lee, R.; Xu, Y.; Yariv, A.; Scherer, A. Design of photonic crystal optical waveguides with single-mode propagation in the photonic bandgap. *Electron. Lett.* **2000**, *36*, 1376–1378. [CrossRef]
161. Mizuta, E.; Watanabe, H.; Baba, T. All semiconductor low-photonic crystal waveguide for semiconductor optical amplifier. *Jpn. J. Appl. Phys.* **2006**, *45*, 6116–6120. [CrossRef]
162. Nozaki, K.; Tanabe, T.; Shinya, A.; Matsuo, S.; Sato, T.; Taniyama, H.; Notomi, M. Sub-femtojoule all-optical switching using a photoniccrystal nanocavity. *Nat. Photon.* **2010**, *4*, 477–483.
163. MÃ, J.; Berg, T.W. The Dynamics of Semiconductor Optical Amplifiers: Modeling and Applications. *Opt. Photonics News* **2003**, *14*, 42–48.
164. Krauss, T.F.; Wilson, R.; Baets, R.; Bogaerts, W.; Kristensen, M.; Borel, P.I.; Frandsen, L.H.; Thorhauge, M.; Tromborg, B. Lavrinenko, Photonic Integrated Circuits using Crystal Optics, PICCO. In Proceedings of the 11th European Conference on Integrated Optics, ECIO'03, Prague, Czech Republic, 2–4 April 2003; pp. 115–120.
165. Shih, M.-H. *Photonic Crystal Devices for Photonic Integrated Circuits;* ProQuest Dissertations Publishing: Ann Arbor, MI, USA, 2006.
166. Mandal, M.; De, P.; Lakshan, S.; Sarfaraj, M.N.; Hazra, S.; Dey, A.; Mukhopadhyay, S. A review of electro-optic, semiconductor optical amplifier and photonic crystal-based optical switches for application in quantum computing. *J. Opt.* **2023**, *52*, 603–611.
167. Cao, T.; Ho, Y.L.D.; Heard, P.J.; Barry, L.P.; Kelly, A.E.; Cryan, M.J. Fabrication and measurement of a photonic crystal waveguide integrated with a semiconductor optical amplifier. *J. Opt. Soc. Am. B* **2009**, *26*, 768–777. [CrossRef]
168. Zhang, Y.; Zheng, W.; Qi, A.; Qu, H.; Chen, L. Optical amplification based on slow light effects in the photonic crystal waveguide. *Microw. Opt. Technol. Lett.* **2011**, *53*, 2997–3001. [CrossRef]
169. Chen, Y.; Mørk, J. Simulation of nonlinear gain saturation in active photonic crystal waveguides. In Proceedings of the Photonics in Switching, PS, 2012 International Conference on IEEE, Ajaccio, France, 11–14 September 2012; pp. 1–3.
170. Kotb, A.; Zoiros, K.E. Performance analysis of all-optical XOR gate with photonic crystal semiconductor optical amplifier-assisted Mach-Zehnder interferometer at 160 Gb/s. *Opt. Commun.* **2017**, *402*, 511–517. [CrossRef]
171. Kotb, A.; Zoiros, K.E.; Guo, C. 160 Gb/s photonic crystal semiconductor optical amplifiers-based all-optical logic NAND gate. *Photon. Netw. Commun.* **2018**, *36*, 246–255. [CrossRef]
172. Kotb, A.; Zoiros, K.E.; Guo, C. Ultrafast performance of all-optical AND and OR logic operations at 160 Gb/s using using photonic crystal semiconductor optical amplifer. *Opt. Laser Technol.* **2019**, *119*, 1–10. [CrossRef]
173. Nosratpour, A.; Razaghi, M.; Darvish, G. Computational study of pulse propagation in photonic crystal semiconductor optical amplifier. *J. Nanophoton.* **2018**, *12*, 036015.
174. Mishra, C.S. Design and analysis of L shape photonic waveguide for realisation of power amplifier application. *Opt. Mater.* **2022**, *127*, 112298. [CrossRef]
175. Capocefalo, A.; Quintiero, E.; Bianco, M.; Zizzari, A.; Gentilini, S.; Conti, C.; Arima, V.; Viola, I.; Ghofraniha, N. Random Laser Spectral Fingerprinting of Lithographed Microstructures. *Adv. Mater. Technol.* **2021**, *6*, 2001037. [CrossRef]
176. Massaro, L.M.; Gentilini, S.; Portone, A.; Camposeo, A.; Pisignano, D.; Conti, C.; Ghofraniha, N. Heterogeneous Random Laser with Switching Activity Visualized by Replica Symmetry Breaking Maps. *ACS Photonics* **2021**, *8*, 376–383. [CrossRef]
177. Chen, S.-W.; Lu, J.-Y.; Hung, B.-Y.; Chiesa, M.; Tung, P.-H.; Lin, J.-H.; Yang, T.C.-K. Random lasers from photonic crystal wings of butterfly and moth for speckle-free imaging. *Opt. Express* **2021**, *29*, 2065–2076. [CrossRef]
178. Ford, A.D.; Morris, S.C.; Coles, H.J. Polymer whispering gallery mode lasers. *Opt. Mater.* **1998**, *9*, 12–17.
179. Gao, H.; Zhao, Y.G.; Ho, S.T.; Seelig, E.W.; Wang, Q.H.; Chang, R.P.H. Random Laser Action in Semiconductor Powder. *Phys. Rev. Lett.* **1999**, *82*, 2278–2281.
180. Kuehne, A.J.C.; Gather, M.C. Organic Lasers: Recent Developments on Materials, Device Geometries, and Fabrication Techniques. *Chem. Rev.* **2016**, *116*, 12823–12864. [CrossRef] [PubMed]
181. Potyrailo, R.A.; Starkey, T.A.; Vukusic, P.; Ghiradell, H.; Vasudev, M.; Bunning, T.; Naik, R.R.; Tang, Z.; Larsen, M.; Deng, T.; et al. Discovery of the surface polarity gradient on iridescent Morpho butterfly scales reveals a mechanism of their selective vapor response. *Proc. Natl. Acad. Sci. USA* **2013**, *110*, 15567–15572. [CrossRef] [PubMed]

182. Ruiz-Clavijo, A.; Tsurimaki, Y.; Caballero-Calero, O.; Ni, G.; Chen, G.; Boriskina, S.V.; Martín-González, M. Engineering a Full Gamut of Structural Colors in All-Dielectric Mesoporous Network Metamaterials. *ACS Photonics* **2018**, *5*, 2120–2128. [CrossRef]
183. Zhang, F.; Shen, Q.; Shi, X.; Li, S.; Wang, W.; Luo, Z.; He, G.; Zhang, P.; Tao, P.; Song, C.; et al. Infrared Detection Based on Localized Modification of Morpho Butterfly Wings. *Adv. Mater.* **2015**, *27*, 1077–1082. [CrossRef]
184. Waterhouse, G.I.N.; Chen, W.-T.; Chan, A.; Sun-Waterhouse, D. Achieving Color and Function with Structure: Optical and Catalytic Support Properties of ZrO_2 Inverse Opal Thin Films. *ACS Omega* **2018**, *3*, 9658–9674. [CrossRef] [PubMed]
185. Wang, C.-S.; Chang, T.-Y.; Lin, T.-Y.; Chen, Y.-F. Biologically inspired flexible quasi-single-mode random laser: An integration of Pieris canidia butterfly wing and semiconductors. *Sci. Rep.* **2015**, *4*, 6736. [CrossRef]
186. Zhang, H.; Feng, G.; Wang, S.; Yang, C.; Yin, J.; Zhou, S. Coherent random lasing from liquid waveguide gain channels with biological scatters. *Appl. Phys. Lett.* **2014**, *105*, 253702. [CrossRef]
187. Türeci, H.E.; Ge, L.; Rotter, S.; Stone, A.D. Strong Interactions in multimode random lasers. *Science* **2008**, *320*, 643–646. [CrossRef]
188. Zaitsev, O.; Deych, L. Recent developments in the theory of multimode random lasers. *J. Opt.* **2010**, *12*, 024001. [CrossRef]
189. Lei, W.; Pu, S. Tunable Slow Light with Large Bandwidth and Low-dispersion in Photonic Crystal Waveguide Infiltrated with Magnetic Fluids. *J. Magn.* **2015**, *20*, 110–113. [CrossRef]
190. Pu, S.; Dong, S.; Huang, J. Tunable slow light based on magnetic-fluid-infiltrated photonic crystal waveguides. *J. Opt.* **2014**, *16*, 045102. [CrossRef]
191. Pu, S.; Wang, H.; Wang, N.; Zeng, X. Extremely large bandwidth and ultralow-dispersion slow light in photonic crystal waveguides with magnetically controllability. *Appl. Phys. B* **2013**, *112*, 223–229. [CrossRef]
192. Pu, S.; Wang, H.; Wang, N.; Zeng, X. Tunable flat band slow light in reconfigurable photonic crystal waveguides based on magnetic fluids. *Opt. Commun.* **2013**, *311*, 16–19. [CrossRef]

photonics

MDPI

Review

Advances in Optical Fiber Speckle Sensing: A Comprehensive Review

Ivan Chapalo [1,2], Andreas Stylianou [3], Patrice Mégret [1] and Antreas Theodosiou [4,5,*]

1 Electromagnetism and Telecom Department, University of Mons, 7000 Mons, Belgium; ivan.chapalo@umons.ac.be (I.C.); patrice.megret@umons.ac.be (P.M.)
2 Foundation for Research and Technology Hellas (FORTH), Institute of Electronic Structure and Laser (IESL), 70013 Heraklion, Greece
3 School of Sciences, European University Cyprus/EUC Research Centre, 1516 Nicosia, Cyprus; an.stylianou@euc.ac.cy
4 Photonics Research Center (PRC), Lumoscribe Ltd., 8201 Paphos, Cyprus
5 Institute of Photonics and Electronics (UFE), The Czech Academy of Sciences, Chaberska 57, 8-Kobylisi, 18200 Prague, Czech Republic
* Correspondence: theodosiou.antreas@lumoscribe.com

Abstract: Optical fiber sensors have been studied, developed, and already used in the industry for more than 50 years due to their multiplexing capabilities, lightweight design, compact form factors, and electromagnetic field immunity. The scientific community continuously studies new materials, schemes, and architectures aiming to improve existing technologies. Navigating through diverse sensor technologies, including interferometry, intensity variation, nonlinear effects, and grating-based sensors, fiber specklegram sensors (FSSs) emerge as promising alternatives due to their simplicity and low cost. This review paper, emphasizing the potential of FSSs, contributes insights to the present state and future prospects for FSSs, providing a holistic view of advancements propelling FSSs to new frontiers of innovation. Subsequent sections explore recent research, technological trends, and emerging applications, contributing to a deeper understanding of the intricacies shaping the future of FFS sensor technologies.

Keywords: optical fiber sensors; optical fibers; interferometric sensors; multimode optical fibers; sensing; image processing; demodulation algorithms; future sensing solutions; fiber specklegram sensors; speckle interferometry

Citation: Chapalo, I.; Stylianou, A.; Mégret, P.; Theodosiou, A. Advances in Optical Fiber Speckle Sensing: A Comprehensive Review. *Photonics* **2024**, *11*, 299. https://doi.org/10.3390/photonics11040299

Received: 2 March 2024
Revised: 20 March 2024
Accepted: 21 March 2024
Published: 26 March 2024

1. Introduction

In the dynamic landscape of sensing technologies, the ubiquity of optical fiber sensors has propelled them to the forefront of scientific and industrial applications. Their widespread use is underpinned by distinctive advantages, such as multiplexing capabilities, lightweight construction, compact form factors, and immunity to electromagnetic fields [1–3]. Optical fiber sensors have become indispensable tools across an array of disciplines, demonstrating their efficacy in medical diagnostics [4,5], structural health monitoring [6–9], robotics [10,11], gait analysis [12], and the detection of chemical compounds [13]. The adaptability and versatility of optical fiber sensors have prompted researchers to explore various technological approaches, each contributing to the ongoing evolution of this field. A plethora of optical fiber sensor technologies have been proposed, capitalizing on distinct effects on optical fibers to achieve specific sensing outcomes. These include interferometry [14], intensity variation [15], nonlinear effects [2,16], grating-based sensors [17], and fiber specklegram sensors (FSSs) as a type of interferometric sensor [18]. The diversity within these approaches underscores the richness of possibilities in optical fiber sensing, allowing for customization to suit the unique requirements of different applications.

The dichotomy between multimode and single-mode optical fibers (SMF) adds another layer of complexity to the sensor landscape. On the one hand, multimode optical fibers

(MMFs) are accompanied by drawbacks such as modal dispersion, modal noise, and modal behavior complexity. On the other hand, MMFs offer practical advantages such as their ease of coupling. Moreover, multimode light propagation allows for increasing information capacity by applying spatial division multiplexing techniques [19] and opens the possibility for mode-related fiber sensing techniques [18]. Conversely, single-mode optical fibers, designed for the propagation of only one mode of light, provide higher fidelity in signal transmission but may be constrained in applications that demand greater modal diversity. Striking a balance between these considerations is crucial in developing optical fiber sensors tailored to specific use cases.

The inherent complexity of optical fiber sensors demands a nuanced understanding of each sensor's approach. For instance, sensors based on fiber nonlinear effects eliminate the need for structural modifications to the fiber, making them suitable for seamless integration into existing optical fiber communication lines [20]. However, the requirement for bulky and expensive equipment for interrogation poses challenges in practical implementation and may limit spatial resolution [21]. Fiber Bragg grating (FBG) sensors, a significant category within optical fiber sensors, provide solutions for quasi-distributed sensing applications but introduce complexities related to both sensor interrogation and the specialized and sometimes expensive equipment needed for grating inscription and interrogation [22–24]. Intensity-variation-based sensors offer a cost-effective alternative, analyzing power variations concerning a specific measurement [25]. Nevertheless, their low sensitivity and susceptibility to light source power variation necessitate additional compensation techniques to maintain accuracy and reliability [26].

In response to the quest for a sensor that amalgamates reliability with key parameters like compactness, lightweight design, simplicity, and cost-effectiveness, FSSs have emerged as a promising alternative in the optical fiber sensing landscape [18,27]. Operating by analyzing the specklegram created on the end facet of a multimode fiber illuminated with a coherent light source, FSSs correlate this speckle pattern with external fiber perturbations (EFPs). Notably, FSSs exhibit potentially multipoint capabilities through the use of beams with different angular separations or wavelengths [28], offering a portable and low-cost interrogation system that stands out in contrast to bulkier and more expensive alternatives.

As technology continues its relentless march forward, the intricate interplay between different types of optical fiber sensors, the challenges associated with their designs, and the continuous innovations in sensing methodologies form the crux of an exciting and dynamic field of research. This review aims to delve deeper into the promising field of FSSs, providing insights into the present state and future prospects of current technology. Through a nuanced examination of the complexities and advancements, this paper seeks to contribute to a deeper understanding of the intricacies that shape the future potential of FSSs. Further sections explore recent research findings, technological trends, and emerging applications, offering a holistic view of the advancements propelling FSSs into prospects with improved sensing capabilities.

2. Brief Historical Background and Terminology

The first research works related to fiber speckles appeared in the 1970s: in 1976, the investigations of speckle patterns at MMF outputs were presented in a number of papers [29,30]. Immediately after, the first works dedicated to acoustic sensors based on intermodal interference were published in 1977–1979 [31–33]. FSSs have been attracting noticeable attention from researchers due to their simplicity and low cost: the basic scheme required just a coherent light source, a multimode fiber, and a photodetector (PD) or CCD/camera matrix (Figure 1a). Moreover, with their interferometric nature, FSSs demonstrated high sensitivity to external fiber perturbations (EFPs); however, it was estimated as 10–1000 times less depending on the number of interfering modes when compared to other interferometric schemes such as two-arm interferometers [32,33]. Along with advantages, FSSs also exhibited several limitations, the main limitations of which were the strong nonlinearity of a transfer function and signal fading. Therefore, a significant number of research works have focused on overcoming these problems (see Section 4), while other papers have addressed FSS applications (see Section 5). It should be mentioned that in different works, FSSs—sensors based on intermodal interference and analysis of the speckle pattern—are also referred to as mode–mode interferometers [32,34], multimode interferometers [35,36], statistical-mode sensors [37], and intermodal fiber interferometers [38–40].

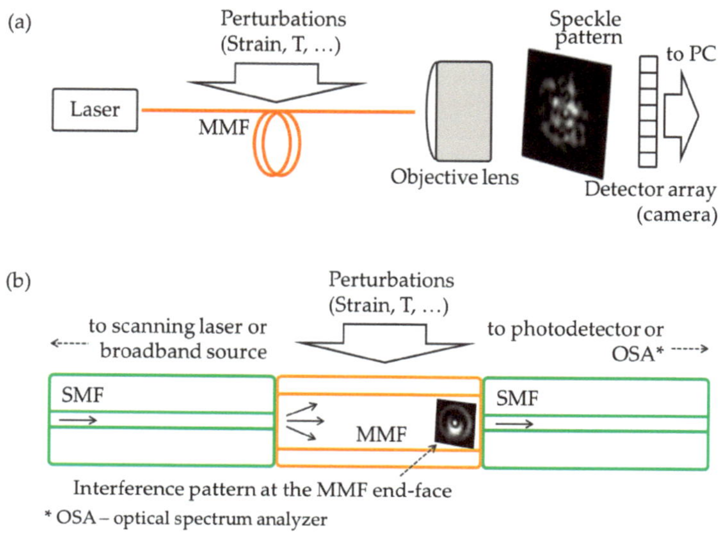

Figure 1. Basic schematic of (**a**) fiber specklegram sensor (intermodal fiber interferometer) and (**b**) interferometric SMS sensor.

In the 2000s, another direction of intermodal interferometric sensors was launched [41]. The basic scheme of a sensor consists of a short piece of MMF coaxially aligned and sandwiched between two SMFs. The input SMF excites a certain number of modes in the MMF. The interferometric pattern formed at the MMF output facet is spatially filtered by the output SMF. Thus, the output signal is proportional to the light intensity "accepted" by the output SMF according to the interference pattern formed at the MMF facet. Unlike "traditional" FSSs, which in the basic configuration utilize a coherent light source with a static wavelength, this type of sensor operates in the spectral domain: either a scanning laser at the input and a PD at the output, either a broadband source at the input and an optical spectrum analyzer (OSA) or a spectrometer at the output have been utilized in this case. Since the interferometric signal is sensitive to both EFPs and laser frequency (wavelength), the output OSA spectrum (or the PD signal formed by a wavelength scan of the scanning laser) represents an interferometric signal of a complex shape, which transforms under external perturbations of the MMF. Thus, tracking a certain interferometric maximum or minimum in a consecutively updating spectrum allows the measurement of EFPs. According to the basic structure of this type of sensor, in the literature, they are usually referred to as single-mode–multimode–single-mode (SMS) sensors (Figure 1b). It should be noted that in SMS sensor designs, the MMF section can be replaced by various specialty fibers providing unique properties and sensitivity to a desired measurand. For example, a standard MMF can serve as a temperature sensor, while a section of a coreless fiber allows for the sensing of refractive index. SMS sensors are attractive as simple and effective point sensors of various physical quantities; however, the necessity of using a scanning source or an OSA noticeably increases the price of these devices. In the literature, there is a wide variety of proposed and investigated sensor configurations, which are well summarized and categorized in several reviews [42–47].

Let us conventionally list the main differences between "classic" fiber interferometric FSSs and interferometric SMS sensors. In SMS sensors, very short lengths of the MMF (typically from centimeters to several meters) are usually used, while FSSs utilize lengths up to hundreds of meters. In a basic SMS scheme (in the ideal case), several modes with an exclusively circularly symmetric field distribution ($PL_{0\mu}$ modes) are excited due to a central launch of the MMF; therefore, an interference pattern is formed as a ring-shaped structure. By contrast, in FSSs, because of their long fiber length, even a central launch leads to the formation of a speckle pattern due to power redistribution to other modes including modes with a non-zero azimuthal number (due to mode coupling). Moreover, it is usually assumed that the power is distributed among a large number of modes, so the speckle pattern consists of many interference spots. The significant fiber length in FSSs also leads to the introduction of random phase shifts to modes' phases (see Section 3). SMS sensors operate in a spectral domain, while FSSs typically work at static wavelengths. In FSSs, the entire speckle pattern is often used for signal processing, while SMS sensors use only a part (usually central) of the interferometric pattern, "accepted" by the output SMF. Taking the above into account, the theory and sensor's simulation approaches are sometimes different or focused on different aspects. For instance, in the case of SMS sensors, researchers often involve the effect of self-focusing [48]. Another example is a calculation of launching conditions in a broad spectral range in the case of SMS sensors [49], thus assuming a change in the modal power distribution (MPD) with wavelength during sensor operation (for FSSs, it is desirable to know the MPD only at the operating wavelength). Therefore, it should be emphasized that following the title of this review, in the next section, we provide a theoretical background based on a "traditional" specklegram approach.

Finally, it should be mentioned that the term "SMS", usually associated with intermodal interferometric sensors, also simply describes a fiber structure, and therefore sensors operating on principles other than interferometry are often called SMS sensors as well. This can cause a certain confusion in terminology. For example, a microbend sensor, presented in [50], consists of an SMS structure but operates as an intensity sensor with a modulated MPD.

3. Fundamentals of Optical Fiber Speckle Sensing

Optical fibers are characterized by various parameters. Among them, the parameters responsible for modal properties are mostly fiber core diameter, relative refractive index difference of the core and cladding (refractive index contrast), wavelength, and the refractive index profile of the core (the standard ones are step-index (SI) and graded-index (GI) profiles) [51]. Unlike single-mode fibers, multimode fibers (MMFs) support the propagation of many modes. Each fiber mode is characterized by its field distribution $E(r,\phi)$ in the fiber cross-section and propagation constant β, where r and ϕ are the radius and the angle of the polar coordinates in the fiber cross-section. For standard fibers, the weakly guiding approximation is applied and the basis of linearly polarized modes ($LP_{\eta\mu}$ modes, where η is the azimuthal number and μ is the radial number) is used in simulations [51–53].

The maximum number of LP modes (herein, we refer to this simply as modes) able to propagate in a fiber with given parameters is limited and can be estimated for SI and GI fibers, as presented in Equations (1) and (2), respectively,

$$M = \frac{V^2}{2}, \tag{1}$$

$$M = V^2\left[\frac{\alpha}{2(\alpha+2)}\right], \tag{2}$$

$$V = ak_0NA, \tag{3}$$

where V is the normalized frequency, a is the core radius, k_0 is the free-space wave number ($2\pi/\lambda$), NA is the numerical aperture of the fiber, and α is the profile parameter [51]. Depending on the launching conditions, all modes can be equally excited in the extreme case (overfilled launch), or the power carried by each mode can vary from mode to mode, providing a particular MPD [54].

If the fiber modes are excited by a coherent light source, they interfere at the fiber end-face, forming a complex interference pattern with multiple interference peaks, which is called a speckle pattern (SP) (Figure 2). The characteristics of SPs are strongly related to the fiber parameters, including the number of excited modes and the refractive index profile. Figure 2b demonstrates the SPs recorded at the end-face of MMFs with a multimode launch provided by a mode scrambler. Generally, the higher number of excited modes provides a higher number of speckles of smaller size, while a lower number of excited modes leads to a lower number of speckles with larger size (see Figure 2b, GI cases, where the different number of excited modes is provided by different wavelengths of 0.63, 1.3, and 1.55 μm). For an SI fiber, the "boundaries" of the SP are sharp (Figure 2b, SI case) unlike the case of the GI fiber, where intensity decays while approaching the core–cladding boundary (Figure 2b, GI cases). This is according to the profile of the core refractive index. Figure 2c shows interference patterns formed in the case of an underfilled launch realized by a connection of coaxially aligned SMFs and MMFs. The presence of ring-shaped structures for the 1 m SI MMF indicates the excitation of circularly symmetric $PL_{0\mu}$ modes, which is a typical case for interferometric SMS sensors (Figure 2c, 1 m SI case). With increasing fiber length, power is redistributed to other $PL_{\eta\mu}$ modes due to mode coupling, so the intensity profile at the MMF end-face obtains the form of an SP rather than a ring-shaped structure (Figure 2, 20 m SI case). In the case of a GI MMF, the recorded intensity profile remains almost unchanged with length due to very low mode coupling in the undisturbed fiber (Figure 2c, 1 m and 20 m GI cases). The SP is stable in the absence of any EFPs, while it transforms under disturbances applied to an MMF due to different propagation constants β of fiber modes. The type of fiber interferometer utilizing the interference of propagating modes in the MMF is known as an intermodal fiber interferometer (IFI) (Figure 1a).

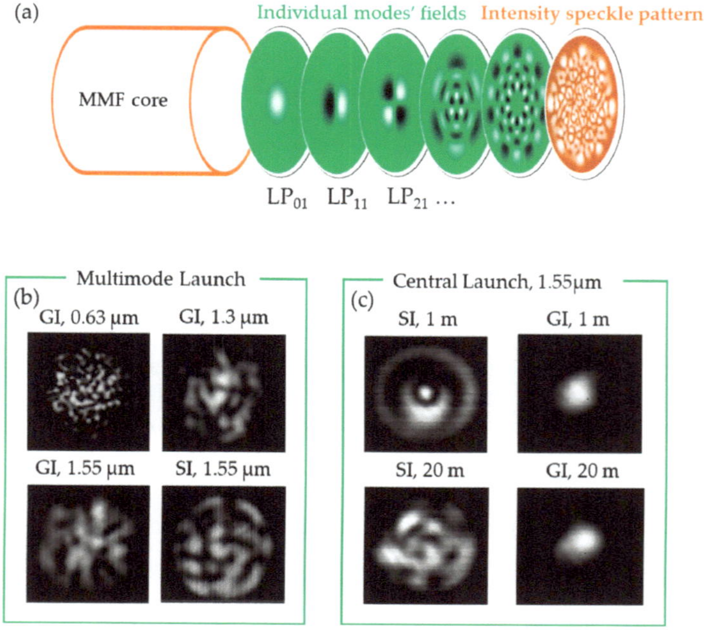

Figure 2. (**a**) Illustration of a speckle pattern formation at the end face of a GI MMF, (**b**) examples of near-field speckle patterns recorded at the end of 50 μm GI and SI MMFs at different wavelengths (multimode launch using mode scrambler), and (**c**) near-field patterns formed at the output of 50 μm GI and SI MMFs of 1 and 20 m length with underfilled launch (by centrally aligned SMF).

Due to the interferometric nature of the SP, the light intensity at a certain point (r,ϕ) of the fiber output facet can be described as [53]

$$I(r,\varphi,M,L) = \sum_{k=1}^{m} (A_k)^2 (E_k(r,\varphi))^2 + \sum_{i=2}^{m} \sum_{k=1}^{i-1} A_k A_i E_k(r,\varphi) E_i(r,\varphi) \cos[\Delta\beta_{ik} L + \Delta\phi_{ik}] \quad (4)$$

where the first term is a constant component and the second term is an interferometric component representing the sum of interferences of all pairs of excited modes, m is the number of excited modes ($2 \leq m \leq M$), L is the fiber length, A_i is the amplitude of the i-th mode, $E_i(r,\phi)$ is the mode function of the i-th mode (normalized mode field distribution), and $\Delta\beta_{ik} = \beta_i - \beta_k$ is the difference in propagation constants β of modes i and k. The cosine argument also contains a phase shift difference $\Delta\varphi_{ik} = \varphi_i - \varphi_k$, where $\varphi_{i(k)}$ is the $i(k)$-th mode's phase shift, which is usually considered random (uniformly distributed in the range of $[-\pi; \pi]$). This is due to the phases of the initial mode at the fiber input and accumulated random phase shifts obtained at some fiber inhomogeneities, bends, twists, etc. If the fiber remains static, $\varphi_{i(k)}$ remains random but static. It is also assumed in Equation (4) that the fiber is lossless and the light source is monochromatic and highly coherent; the polarization of modes is not taken into account (i.e., we use a scalar IFI model). A consideration of polarization would increase the complexity of analysis; however, the practical effect is mainly a reduction in the SP contrast, as orthogonally polarized modes do not interfere; in practice, a polarizer is often used at the output of an MMF to maximize the SP contrast [55]. The IFI signal is considered to be recorded in a near-field. The surface area of a PD is considered to be smaller or similar to the typical size of the speckles to provide maximum signal contrast and dynamic range.

If the fiber experiences external perturbation, the arguments of the cosines in Equation (4) change, leading to an SP transformation. In simulations, EFPs are often considered as changes in fiber length: $L = L_0 + \delta L$, where L_0 is the initial fiber length and δL is the fiber length alteration (deterministic value) caused by EFPs. Also, it is usually considered in simulations that the IFI signal is caused only by modal interferences but not mode coupling, mode parameter transformations, and other effects. Otherwise, the complexity of the model and simulations rises drastically. This explains the lack of theoretical models in papers related to specklegram systems, for example, with an exposed core, where contact of the core with the surrounding media can change not only the modes' phases but also local propagation constants, mode coupling coefficients, and field distributions [56].

It should be mentioned that in several works, the actual mode functions and mode amplitudes are replaced by a random value of the product A_iE_i at a PD coordinate (r,ϕ) for simplicity [53]. This is substantiated by the fact that the local mode intensities are very different at the particular point of fiber cross-section even for a known MPD. The latter is due to the mode field complexity, which rises with the order of modes.

An example of an IFI signal I versus fiber elongation δL is presented in Figure 3. Here, a large δL was used to show the general dependency $I(\delta L)$. It is seen that the signal response to fiber elongation is quite complex and highly nonlinear. The complexity of the signal is conditioned by the number of pairs of interfering modes with their propagation constant differences $\Delta\beta_{ik}$, phase shifts $\Delta\varphi_{ik}$, and local products of $A_iE_i(r,\phi)$ at the PD position in the fiber cross-section. If only two modes are excited, the signal has a sine shape, while the complexity of the signal increases with the number of excited modes: each pair of interfering modes contributes with a sine-shaped signal with its period (provided by $\Delta\beta_{ik}$) and phase. Moreover, fiber rearrangement, bending, or environment condition changes will result in changes in L_0 and $\Delta\varphi_{ik}$ and, therefore, in different dependences $I(\delta L)$ in Figure 3. The IFI response can be very different at different coordinates of the SP (Figure 3a,b) that is used for signal processing such as averaging over the fiber cross-section or correlation processing (Section 4).

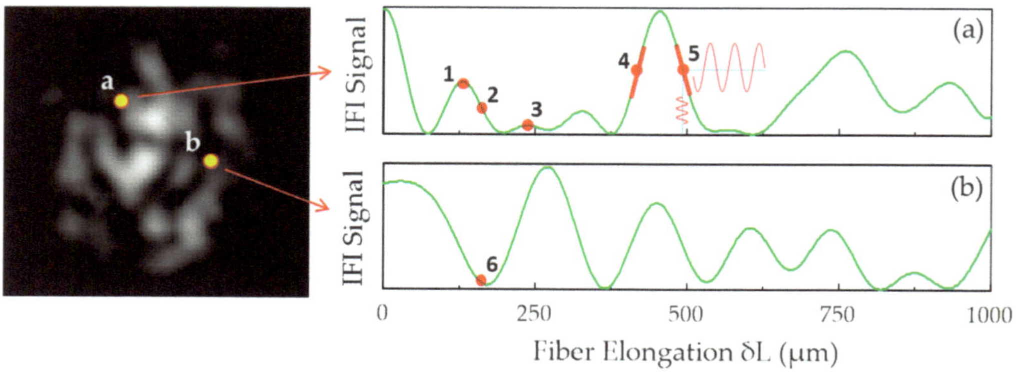

Figure 3. Illustration of IFI signal response to fiber elongation. (**a**,**b**) show that the response can be very different at different coordinates of the speckle pattern. Red points show examples of operating points for small EFPs. It is seen that points 2, 4, and 5 provide a quasi-linear response, while the response at operating points 1 and 3 will be highly nonlinear; the sensitivity at point 3 is close to zero (fading effect). Even in quasi-linear zones, the sensitivity and the dynamic range (points 2 and 5) as well as the phase of the signal (points 4 and 5) can differ. While the response is quasi-linear at one coordinate (graph (**a**), point 2), at the same time, it can be nonlinear at another coordinate (graph (**b**), point 6).

It is seen that the entire signal has certain quasi-linear zones (for example, points 2, 4, and 5 in Figure 3). On the contrary, at points 1 and 3, the response would be nonlinear. Moreover, the sensitivity at point 3 would be close to zero (fading effect). It is also seen that while the response at one coordinate of the SP is quasi-linear, at the same time, it can be nonlinear at another coordinate (points 2 and 6 in Figure 3). Even in quasi-linear zones, the sensitivity and the dynamic range (points 2 and 5) as well as the phase of the signal (points 4 and 5) can differ.

Thus, in the case of weak EFPs, a quasi-linear response can be achieved by choosing a correct operating point. However, in practice, it is very problematic to keep the operating point fixed: slowly changing environmental conditions (e.g., temperature), fiber rearrangements, and laser frequency instabilities lead to operating point drift or jumps that change sensitivity (fading effect) and a transfer function. Thus, the fundamental problems of IFIs and FSSs are signal fading, nonlinearities, and quasi-random responses in general. It should be mentioned that in several works, two mechanisms of the mode's phase difference modulation have been suggested: the first is related to the modes' optical path changes described above, and the second is related to fiber bending, twisting, and other perturbations causing a mode coupling [57–59]. In the latter case, the modulation of the phase difference of the modes occurs due to mode coupling but not due to simply δL. This complicates the analysis and makes fading behavior even more unpredictable [58]. To overcome the problems described above, various signal-processing solutions have been developed and investigated. They include averaging, correlation analysis, machine learning, and other techniques that are reviewed in Section 4.

It is seen from Equation (4) that the IFI properties (range of sensitivities, complexity of the signal, spectral components) depend on a set of parameters: mode amplitudes $A_{i(k)}$ and number of excited modes m (or MPD), a coordinate of the PD at the fiber cross-section, propagation constant differences $\Delta\beta_{ik}$, and mode field distributions $E_{i(k)}(r,\phi)$. The last two parameters ($\Delta\beta_{ik}$ and $E_{i(k)}(r,\phi)$) significantly depend on the refractive index profile of the MMF. In the case of an SI fiber, fiber modes are well approximated by Bessel modes [51,52,54], while modes of graded-index fibers are usually approximated by Laguerre–Gaussian (LG) modes [60]. For LG modes with a low radial number μ, the mode fields are "concentrated" closer to the core center, while they tend to occupy the entire physical space of the core cross-section with the rise in μ. This is not the case for Bessel modes where mode fields occupy approximately the entire cross-section of the core (see, for example, Figure 2c). This affects the IFI properties depending on the PD position in the fiber cross-section: the averaged IFI sensitivity decreases with the displacement of the PD from the core center towards the core–cladding boundary in the case of the GI fiber [55,61].

An even more important parameter is the difference in the modes' propagation constant $\Delta\beta$. For SI fibers, propagation constant β can be calculated by solving the characteristic equation for the tangential field components at the core–cladding boundary [51,52]. In the SI case, modes have different β and disordered spacing $\Delta\beta$ between "neighbouring" modes (i.e., modes with adjacent values of β) (Figure 4a). (We note that graphs in Figure 4 are plotted using the modes' refractive index n for convenience, according to relation $\beta = 2\pi n/\lambda$; the vertical axis in Figure 4a takes into account the presence of two degenerated modes for modes with non-zero azimuthal number η.) Therefore, the interferometric signal formed by the interference of different neighboring modes (and, therefore, by other pairs of modes) will demonstrate different sensitivity, period, and beat length $\Lambda_{ik} = 2\pi/\Delta\beta_{ik}$. Note that some of the modes have very close propagation constants (Figure 4a) and, therefore, quite a long beat length (several centimeters). Thus, the resulting IFI signal $I(\delta L)$ becomes quite complex with many spectral components.

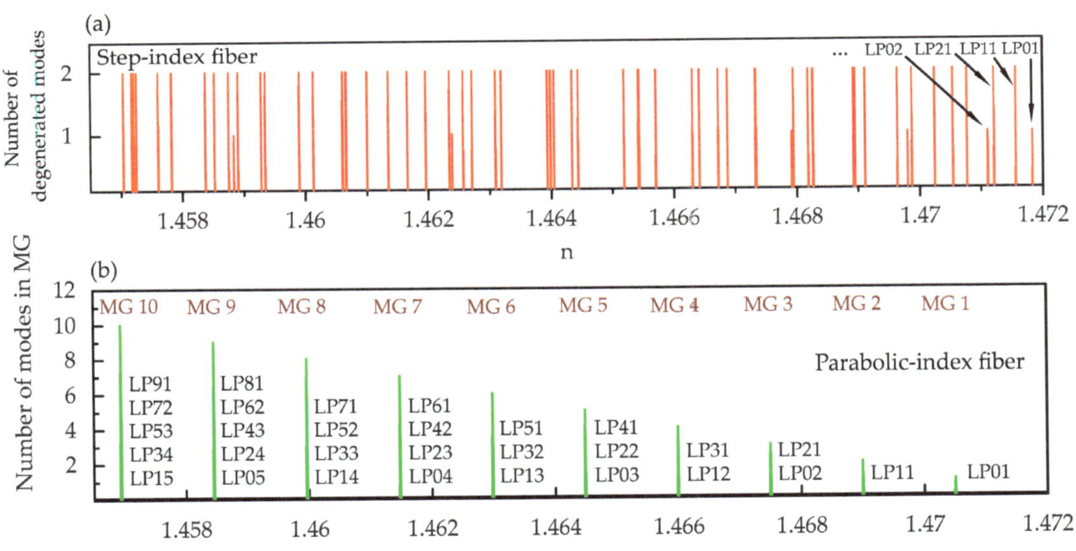

Figure 4. Refractive indices for modes of (**a**) SI and (**b**) GI ($\alpha = 2$) fiber, calculated for the fiber with parameters $a = 25$ μm, $\lambda = 1.55$ μm, $n_0 = 1.472$, and NA = 0.21. The corresponding β can be found according to the formula $\beta = 2\pi n/\lambda$. For an SI fiber, each mode has its β; the $\Delta\beta$ between adjacent modes depends on the mode numbers. For a GI fiber ($\alpha = 2$), modes are combined into mode groups (MGs); all adjacent MGs have the same $\Delta\beta$.

A completely different case takes place for the GI fiber. Here, particular sets of modes are combined into mode groups (MGs): each MG contains modes with (approximately) equal propagation constants [51,54]. Using the Wentzel–Kramers–Brillouin (WKB) approximation, the analytical expression for each MG's propagation constant can be derived [51,54,62]:

$$\beta_m = \frac{2\pi}{\lambda} n_0 \left[1 - 2\Delta \left(\frac{m_g}{M_g} \right)^{\frac{2\alpha}{\alpha+2}} \right]^{\frac{1}{2}}, \tag{5}$$

where m_g is the MG number taking integer values from 1 to M_g (index g emphasizes the modes' groups instead of separate modes), Δ is the refractive index contrast, n_0 is the core refractive index at the fiber axis, α is the profile parameter, and M_g is the maximum number of MGs able to propagate:

$$M_g = a\sqrt{\Delta}\frac{2\pi}{\lambda}n_0\sqrt{\frac{\alpha}{\alpha+2}} \tag{6}$$

The number of modes belonging to a given MG is defined by the number of combinations (μ; η), which can be found for a given m_g according to the formula $m_g = 2\mu + \eta - 1$. Taking into account two degenerated modes for modes with non-zero azimuthal number η, it can be seen that the number of modes belonging to an MG is equal to an MG number. The consideration of two polarization states for each mode will further double the number of modes in each MG.

An interesting case is a parabolic profile of refractive index ($\alpha = 2$) where propagation constants of MGs are equidistant (Figure 4b). This means that the interferometric signal formed by the interference of any neighboring MGs has the same period and beat length according to the same propagation constant difference of adjacent MGs $\Delta\beta_p$ (index p emphasizes the parabolic profile of refractive index). Therefore, the IFI signal $I(\delta L)$ is expected to be less complex with significantly fewer spectral components compared to the step-index case. The spectral components are equidistant as well; if all modes are excited, the first spectral component corresponds to the interference of the adjacent MGs (provided by $\Delta\beta_p$), while the last spectral component corresponds to the interference of the fundamental mode and the modes belonging to the highest-order MG (provided by the maximum $\Delta\beta_{pmax} = \Delta\beta_p(M_g - 1)$). As the propagation constants within the MGs are approximately equal, the interferometric signal formed by modes belonging to the same MG is insensitive to EFPs ($\Delta\beta_{ik} \approx 0$ in Equation (4)). Thus, in the case of the GI profile, the number of modes m in Equation (4) can represent the number of mode groups m_g, and the mode amplitudes and field distributions are those of mode groups as well.

To complete the topic of IFI signal formation, the effect of laser frequency modulation (LFM) must be discussed. Apart from the EFPs, the LFM also generates the IFI signal as $\Delta\beta_{ik}$ in Equation (4) is sensitive to optical frequency shifts $\delta\nu$. In fiber communications, this effect has been studied within the topic of modal noise [63]. To consider the LFM as a source of the IFI signal, the product $\Delta\beta_{ik}L_0$ can be decomposed into Taylor series in the vicinity of the optical frequency mean value ν_0 [53]:

$$\Delta\beta_{ki}L_0 = (\beta_k(\nu_0) - \beta_i(\nu_0))L_0 + \left(\frac{d\beta_k(\nu_0)}{d\nu} - \frac{d\beta_i(\nu_0)}{d\nu}\right)\delta\nu \cdot L_0 = (\beta_k(\nu_0) - \beta_i(\nu_0))L_0 + \left(\frac{1}{V_k} - \frac{1}{V_i}\right)\bigg|_{\nu_0}\delta\nu \cdot L_0 \tag{7}$$

where $V_{i(k)}$ is the group velocity of the $i(k)$-th mode (a linear approximation was applied when decomposing into a Taylor series). It is seen that the first term in Equation (7) is fixed, while the second term is proportional to frequency shift $\delta\nu$ and fiber length L_0. Thus, the IFI sensitivity to laser frequency modulation rises with fiber length. This is a principal difference from the case of EFPs, where the IFI sensitivity depends on the fiber section subjected to the EFP but not on the entire fiber length. Also, the IFI sensitivity to LFM is radically stronger for the SI fiber compared to the GI fiber with $\alpha \approx 2$, where the mode propagation times are equalized [53,64]. Even though the LFM caused by fluctuations in laser frequency can be considered a source of noise for IFIs [65], intentional frequency modulation can be useful. It has been proposed as an averaging method of IFI signals [55] and for advanced IFI signal processing providing signal linearity and fading mitigation [66]. The LFM principle is also used in interferometric SMS sensor schemes, where periodically updating wavelength scans are used to extract the signal [42]. It should be mentioned that the mechanisms of the IFI with LFM have been utilized in other specklegram-related applications such as MMF-based spectrometers [67].

As discussed above, if no processing algorithms are used, the IFI demonstrates a quasi-random transfer function and sensitivity, the response can be highly nonlinear, and the useful signal experiences fading. This complicates both the effective sensing and the determination of the IFI properties, which depend on fiber parameters, the laser wavelength, and the MPD. To enable a convenient analysis of IFI properties, the apparatus of so-called averaged amplitude characteristics (ACs) has been developed [53]. It is based on ensemble averaging of the IFI response to EFPs and plotting the graph of the averaged standard deviation of the signal versus EFP amplitude, for example, fiber elongation δL (Figure 5a). Unaveraged ACs are quasi-random and strongly nonlinear (gray curves in Figure 5a), while the averaged AC is stable and reproducible (red dashed curve in Figure 5a). It contains a rising part and a saturation part. The saturation occurs when the EFP amplitude exceeds the maximum beat length of adjacent modes (modes with minimum $\Delta\beta$). The AC method allows for effective analysis (theoretically and experimentally) of the IFI properties depending on fiber parameters, wavelength, and MPD [35,55,61]. Figure 5b shows an example of averaged ACs for two wavelengths of 0.63 and 1.3 μm and two MPDs for the 0.63 μm wavelength. The graphs allow us to compare averaged sensitivities and to estimate a range of δL when the quasi-linear response can be achieved. In the given example, it is seen that the average sensitivity in the rising part of the curves is significantly stronger for the multimode case than for the few-mode case at 630 nm, while the multimode case at 630 nm is more sensitive to EFPs than the multimode case at 1300 nm due to a higher number of modes able to propagate at 630 nm than at 1300 nm. It should be mentioned that the quasi-linear part of the dependences ends at significantly lower EFP amplitudes for the cases with stronger sensitivities (630 nm, multimode case) than for the cases with lower sensitivities (1300 nm multimode, and 630 nm few-mode cases).

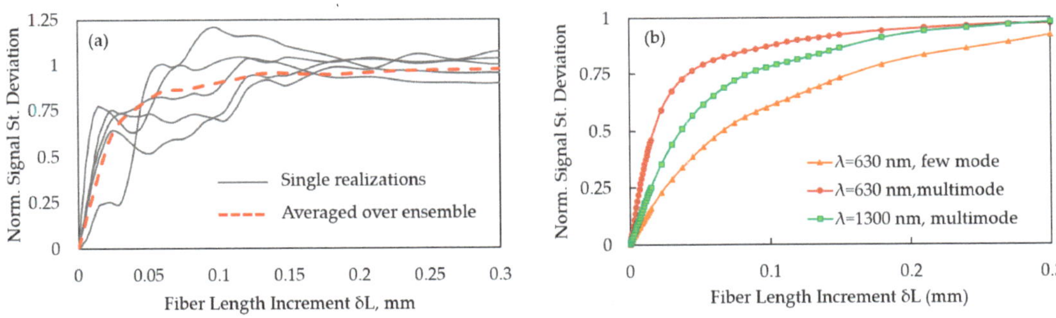

Figure 5. Example of the amplitude characteristics experimentally obtained in [55,61]: (**a**) several unaveraged ACs and the AC obtained by ensemble averaging, and (**b**) the ACs spatially averaged over fiber cross-section for two wavelengths (630 and 1300 nm) and two cases of launching conditions for 630 nm.

Further, various averaging methods have been experimentally analyzed using the AC method [55,61]. They include ensemble averaging, averaging "over long realization", laser frequency averaging, spatial averaging over the fiber cross-section, and their combinations. While the first two methods are useful for the assessment of IFI properties, the spatial and frequency averaging methods have also been used for real-time sensing applications. To complement the ACs for the EFPs of strong amplitudes, the averaged spectral characteristics (SC) have been introduced as well; thus, the IFI can be characterized by the pair of AC and SC for the full range of EFP amplitudes [53].

It should be noted that the IFI signal registration can be performed in the near-field and far-field zones. Practically, the selection of one of these two methods to use depends on the equipment and convenience: near-field registration allows us to use a camera matrix and to analyze the entire SP, but it requires the use of imaging optics (e.g., microscope objective), while registration in the far-field using one or multiple photodetectors is simpler, but it does not allow us to analyze the entire speckle pattern, rather only parts of it. For theoretical investigations of IFIs, in particular, for the analysis of IFI characteristics depending on fiber parameters (refractive index profile, MPD, etc.), the use of the near-field is more straightforward. The far-field distribution of modes can be calculated by the Fourier transform of the near-field distribution, and the results can be different depending on the refractive index profile of the MMF [54]. It has been demonstrated that the modes' field shapes are the same for the near- and far-field for a GI profile, while they significantly differ in the case of an SI fiber [54]. The same behavior can be expected for speckle patterns.

In this section, we considered the basic physics of IFIs, and we discussed signal characteristics and limitations and the relations between fiber parameters and IFI properties. We note that the theoretical background was limited by a weakly guiding approximation and two cases of refractive index profiles (step-index and graded-index), while a great variety of specialty optical fibers can be applied for specklegram sensing. The consideration of this variety of fibers, however, would significantly complicate the theory and is beyond the scope of the format of this article. In the following sections, signal processing approaches and the FSS applications are reviewed in more detail.

4. Signal Processing and Data Analysis

In Section 3, we mentioned two main problems of FSSs: the quasi-random and nonlinear transfer function (changing under drifting environmental conditions) and, as a result, signal fading. The first problem does not allow precise measurement of EFPs; rather, the sensor can detect some events disturbing the fiber, for example, heartbeats in medical applications [68,69] or intrusion into a secured territory in perimeter surveillance applications [70]. If the interferometric signal is detected using one PD (or camera pixel), the second problem, signal fading, does not allow reliable measurement of the perturbation amplitude. Moreover, the perturbation can be completely missed, since the sensitivity can drop down to almost zero. Therefore, the main purpose of signal processing algorithms and approaches is to overcome the above problems. Researchers have been trying to reach this goal either via a spatial domain (analyzing different parts of the speckle pattern, i.e., performing multichannel signal detection and processing, or analyzing the speckle pattern as a whole), or spectral domain, using a scanning light source or broadband source with an OSA (or spectrometer). Below, various signal processing techniques are categorized and reviewed.

4.1. Intensity Approach

In the simplest case, intensity-based FSSs can be realized by receiving an intensity-modulated signal after spatial filtering of the SP using one PD. As it was mentioned above, the signal will experience nonlinearity and fading under changing environmental conditions. As a solution, multichannel signal detection provides a mitigation of fading that allows the realization of reliable threshold sensors (e.g., heartbeat monitoring, perimeter surveillance). This approach is well investigated [36,59,71]. For instance, in works [36,59], the authors compared algorithms based on calculation moduli of the signals from several photodetectors; the work [71] analyzes the influence of the number of photodetectors on the effectiveness of fading mitigation. The authors of these papers assume that the useful signal has a high-frequency spectrum compared to slowly changing (quasi-static) environmental conditions. Therefore, the signal was subjected to high-pass filtering before applying the algorithms. It should be noted that amplitude single-channel and multichannel detection (spatial filtration) can also be realized by using one or several single-mode fibers serving as spatial filters at the output of the MMF [72]. In this case, the electronic part of the scheme can be a good distance away.

A more advanced approach consists of the use of a CCD or camera matrix instead of a set of PDs [37]. This allows camera pixels to be used as separate PDs that significantly increases the number of spatial channels and, therefore, can improve the effectiveness of data processing. One camera matrix can even be used to record SPs from several MMFs simultaneously [73]. However, the use of cameras can significantly limit the bandwidth of the sensor, which can be inappropriate in some applications. Generally, the same algorithms investigated for the case of multiple photodetectors can be applied with the use of cameras. For example, in works [37,74], the sum of absolute values of signal changes is analyzed; in work [55], the effectiveness of the averaging of signal amplitudes for fading mitigation (and, therefore, for the stability of the response) is experimentally investigated for different wavelengths (that define the maximum number of propagating modes) and MPDs. It should be mentioned that the use of the absolute value of the alternating signal (e.g., sine shape) leads to the doubling of signal spectrum width [37]. In works [69,70,75], authors applied a time differentiation of the camera pixel's signal. The work [76] demonstrates the results of the application of various image processing algorithms (both intensity-based and correlation-based) for speckle pattern processing. Authors assumed that the type of EFPs, microbends, causes not only DPM but also strong mode coupling between core modes and further between core and cladding modes. The latter means that the condition of constant total intensity at the fiber output is no more valid that was considered in the investigated processing approaches. It should be noted that authors compare the processing algorithms in terms of linearity, sensitivity, and dynamic range, while the effectiveness of fading mitigation is out of the scope of this work. In [77], the low frame rate and response time of cameras and CCD arrays usually considered as disadvantages, were proposed as a way to measure the amplitude of high-frequency EFPs by calculating the contrast of the SP: the higher the amplitude of the EFP, the lower the contrast of the SP due to averaging of the SPs over the response time of the camera matrix (a similar principle is also applied for SP elimination, i.e., minimization of the SP contrast towards zero, in MPD and encircled flux measurements).

4.2. Correlation Approach

In general, the correlation approach is based on the comparison of an SP recorded by a camera matrix at a given moment in time with a certain reference SP recorded in the absence of EFPs. In this case, the problem of signal fading is effectively solved, since the entire SP is analyzed. Several research groups have developed the SP correlation approach for FSSs. For example, work [78] demonstrates the correlation approach for the case of fiber axial strain. The authors confirmed the stability of the sensor response for different initial conditions of the SP, which indicates the robustness to signal fading. The authors also showed the influence of the radius of the analyzed part of the SP on the correlation curve, and the influence of the camera saturation on measurement errors; they demonstrated a method of increasing the sensor's dynamic range by periodically updating the reference image. In work [79], the authors considered microbend perturbations of an MMF and proposed the normalized inner product coefficient for analysis of EFPs; the "updating reference" technique was also applied for the extension of sensor's dynamic range. In work [80], the authors investigated the effectiveness of the correlation method when just a few modes were excited in an MMF; they proposed the utilization of diffusion transparency to improve the quality and stability of the measured signals (Figure 6a). In [76], the authors investigated various image processing correlation techniques for FSSs. As mentioned in the previous subsection, the authors assumed the possibility of changes in the total output power due to the possible coupling of core and cladding modes under fiber microbendings.

Various image processing techniques utilizing the correlation approach are proposed and investigated in other works as well [81,82]. For example, in paper [82], the authors propose the application of morphological image processing for strain sensing: they converted SP images into binary form and compared them with a set of reference images recorded in advance during a controlled ramp of strains applied to the fiber. This approach demonstrated an enhanced dynamic range towards stronger amplitudes of perturbations. We suppose, however, that this approach requires stable environmental conditions (temperature); otherwise, the procedure of periodical recalibration is necessary.

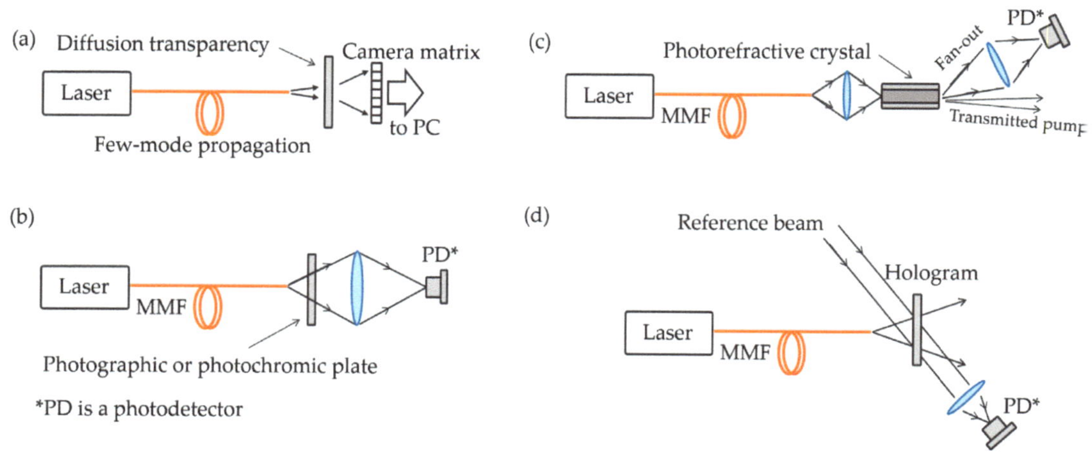

Figure 6. Variety of modified schemes for interferometric signal detection used in correlation approach of signal processing: (**a**) using a diffusion transparency in case of a few-mode propagation regime, (**b**) using a photographic or photochromic plate and a photodetector, (**c**) using a photorefractive crystal and a photodetector, and (**d**) holographic scheme with a reference laser beam.

The correlation approach can be realized not only by camera image processing, but also using a single PD. In early works [83,84], authors proposed an FSS with a photographic plate that served as an amplitude-transmission spatial filter (Figure 6b). The filter was recorded onto a photographic plate using the reference SP (in the absence of EFPs). The light that exited from the MMF and transmitted through the filter was focused into a PD. After the filter recording, the SP and the negative image in the filter were spatially matched if the fiber was undisturbed. Therefore, the PD signal was minimal. The alternation of the SP from its reference position under acoustic pressure applied to the MMF generated a PD signal proportional (under a condition of weak enough perturbations) to the acoustic pressure amplitude. This approach was attractive since it required only one PD; however, its practical use was limited since the reference SP must be unchanged and must exactly match the spatial amplitude filter, which is problematic in real conditions in the presence of temperature drift and possible fiber rearrangements. The spatial amplitude filter approach was further developed to allow its adaptive self-adjustment under slowly changing environmental conditions (and therefore under slowly transforming SP). In work [85], the authors proposed the adaptive correlation filter based on the photorefractive fanning effect (Figure 6c), while in [86], the author proposed the use of photochromic glasses for the same purpose (Figure 6b).

Finally, the holographic correlation approach has been proposed and investigated in [27]. This method provides a significant increase in the sensor's sensitivity. Notably, it exhibits potential for multiplexing capabilities through the use of reference laser beams with different angular separations or wavelengths [28]. However, since it utilizes principles of holography, it requires a reference laser beam that increases the complexity of the sensor (Figure 6d). In addition, the problem of reference SP drift under environmental conditions remains a reality.

4.3. Machine Learning Approach

The machine learning (ML) approach in FSS data processing and analysis is the newest among others. An impressive number of publications dedicated to fiber-specklegram-related applications have been published in the past few years: FSSs; MMF-based image transmission, recognition, and reconstruction; and MMF spectrometers [44,87]. Various network types, such as VGGNet, AlexNet, ANN, ResNet, and others, have been trained, and their effectiveness and reliability have been investigated for these applications. The feature of the ML approach to FSSs is that it does not involve any theoretical basis on MMF light propagation and SP formation; it operates simply with images, being trained to recognize various scenarios of fiber perturbation from the SP evolution. FSSs studied with the use of ML include bending and curvature [88–90], tactile [91], biomedical [92], and intrusion detection [93] sensors. In several works, the investigation of the possibility of EFP localization along the MMF is presented [94–96]. In general, the ML techniques applied to FSSs have demonstrated impressive results of measurement accuracy and have opened new sensing capabilities; however, this approach also has significant practical limitations. For reliable measurements, the system must be trained for a given sample of MMF, keeping it in a particular, unchanged position, and in perfectly stabilized environmental conditions. Any fiber parameter changes (length, core diameter, refractive index profile), rearrangement, and temperature drift drastically reduce system performance. To the best of our knowledge, the ML-based FSSs resistive to environmental condition changes have still not been demonstrated. A detailed review of recent achievements of the ML application to FSSs is presented in [87].

4.4. Spectral Domain Approach

The spectral domain approach utilizes the phenomenon of the dependency of the modes' propagation constant on laser frequency (wavelength) (Section 3). As a result of different group velocities of modes, the SP is sensitive to laser frequency changes. This effect is widely used in SMS-type sensors [41]. As discussed in Section 2, the basic scheme of the SMS sensor contains a short piece of MMF sandwiched between two SMFs. Few modes in the MMF are excited by a scanning light source via the input SMF. The interferometric signal spatially filtered by the output SMF is recorded by a photodetector. As a result, a spectral response (which is a dependency of the interferometric signal versus laser frequency) of the SMS structure can be formed and updated in real-time (Figure 7). The same effect can be obtained using a broadband light source and an optical spectrum analyzer (or spectrometer). EFPs, applied to the MMF section, cause spectrum transformations. Due to a small number of excited modes and the short length of the MMF, the transmission spectrum represents a small number of interferometric maxima and minima, while spectrum transformations obtain a form of movements of the spectrum towards higher or lower wavelengths proportionally to EFPs (the shape of the spectrum remains almost unchanged). Thus, tracking a wavelength position of a spectral maximum or minimum (red arrows in Figure 7) opens a possibility for a simple measurement technique. In this case, the problem of signal fading is not explicitly present, which is a strong advantage of this sensor configuration. Also, a good linearity of the sensor response can be achieved. It should be noted that the spectral-domain SMS sensor can also be considered a set of FSSs receiving a signal via one spatial channel (output SMF fiber) and operating at different wavelengths within the range of scanning wavelength (the number of FSSs is equal to the scanning range divided by the OSA resolution or by the sampling rate of the photodetector). Interferometric time signals formed by fiber length modulation at two wavelengths λ_1 and λ_2 are schematically illustrated by black curves in Figure 7. Spectral-domain SMS sensors have a great number of variations based on fiber types, interrogation techniques, and combinations with other sensing techniques, applications, etc., that are comprehensively reviewed in several articles [42–47]. It should be noted that a potential problem of the peak-tracking technique can occur in the case of a highly multimode propagation regime and long lengths of MMFs increasing the sensor's sensitivity. Here, spectrum transformation ceases to represent simply shifts towards higher and lower wavelengths, but obtains more complicated behavior, changing the shape of the spectrum. Therefore, peaks that are supposed to be tracked can appear and disappear, and their tracking becomes impossible.

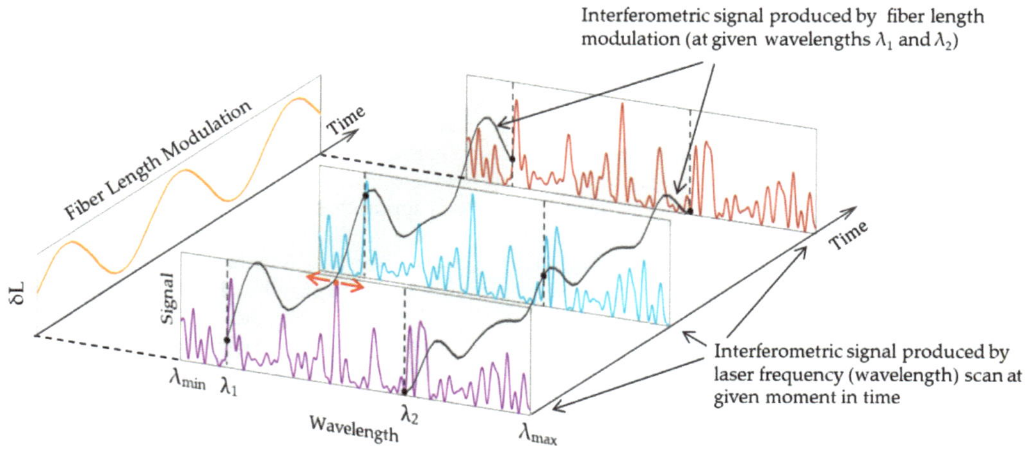

Figure 7. Schematic illustration of spectral-domain intermodal fiber interferometer signal. The signal is formed by continuously updated wavelength scans. Measurements can be performed by the tracking of some interferometric peak shifting under EFPs (red arrows). The scheme can also be considered a set of interferometers operating at different wavelengths (black curves at λ_1 and λ_2).

Another promising approach that can be used instead of spectral peak tracking is the analysis of the fast Fourier transform (FFT) of the SMS transmission spectrum. The FFT spectrum of continuously updating wavelength scans demonstrates several spectral peaks (harmonics); each harmonic corresponds to a particular pair of interfering modes. Thus, by analyzing the phase on the frequency of the particular harmonic, one can measure EFPs [97–101]. This approach provides a linear relation between the phase and EFPs, which is a strong advantage. The sensor's dynamic range can be extended towards strong perturbations by implementing a phase unwrapping technique. Phase demodulation for SMS structures with a length of the MMF section up to 50 m (GI MMF, few-mode regime) has been demonstrated [102]. However, obtaining reliable results, as well as in the previous technique of peak tracking, becomes more complicated with the increasing number of excited modes, since the spectral components corresponding to different pairs of interfering modes can be very close to each other and even overlap, making the phase tracking of harmonics unreliable.

One more approach proposed for a spectral domain FSS consists of the use of the auto-correlation function of the spectral interferometric signal [66]. The correlation coefficient is calculated between the reference wavelength scan, which is recorded under the absence of EFPs, and the signal scan, updating in real-time and transforming under EFPs. This approach can be considered an alternative to the correlation processing of the specklegram image described in Section 4.2; however, it allows for obtaining a linear sensor response by applying a certain wavelength shift to the reference scan, thus shifting the operating point to a linear part of the correlation function. This approach confirmed its reliability even with the highly multimode regime of the GI MMF.

5. Sensing Applications

Optical fiber specklegram sensors have proven their capability in measuring a diverse array of parameters, including micro-bending, strain, vibration, temperature, and refractive index. These sensors find applications in various sectors such as healthcare, structural health monitoring, and security or surveillance.

The sensing parameters can be categorized into two main groups, i. physical and ii. chemical parameters, as summarized in Table 1. Within the physical parameters group, we can include strain, pressure, vibration, sound, bending, and temperature. On the other hand, the second group primarily focuses on liquid or gas measurements, where light must physically interact with the ambient environment, encompassing refractive index and relative humidity.

Table 1. Categorization of the different basic parameters, which can be measured using optical fiber sensors.

Group A—Physical	Group B—Chemical
Strain	Refractive index
Temperature	Relative Humidity
Pressure	Gas sensing
Curvature	pH
Vibrations	

Regarding the materials used in developing these sensors, the main types are silica and polymer fibers. It is noteworthy that depending on the material, certain parameters may transition from one group to another. For example, many polymers exhibit hydrophilic characteristics, enabling a mechanism where the polymer absorbs water content, causing the fiber to swell. This swelling can lead to local expansion of the fiber dimensions. In cases where only the cladding of the fiber absorbs water, pressure is applied to the core. Consequently, fluctuations in relative humidity can be directly detected through strain or pressure sensitivity.

Figure 8 illustrates these two main categories of sensing parameters, physical and chemical, and also shows an overlapping area where some chemical parameters can be measured indirectly through physical parameters. For example, the relative humidity concentration can be measured indirectly when a hydrophilic material absorbs the water content and swells. As a result, a strain is applied to the fiber core.

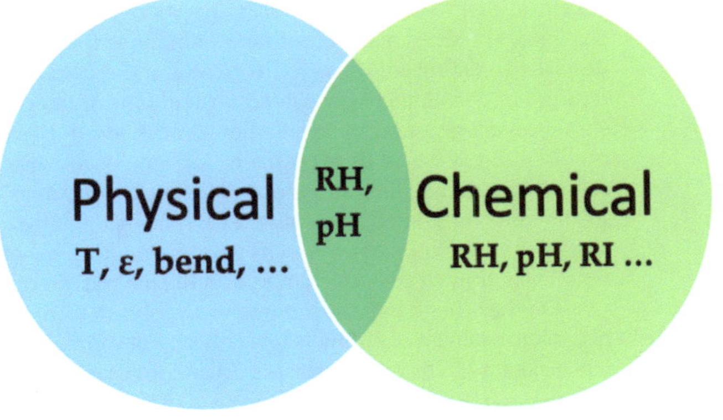

Figure 8. Schematic diagram of the categorization of the different sensing parameters that can be measured using optical fiber sensors.

In FSSs designed for physical parameters (group A), the sensing mechanism and sensitivity are conditioned only by properties of the fiber (possible number of propagating modes, refractive index profile, fiber material, elasticity, Young's modulus, strain-optic coefficient, thermal expansion, and thermo-optic coefficients). As a result, the sensing procedure and analysis are more straightforward. On the other hand, for the measurement of parameters of group B, the fiber needs to be specifically designed to make light in the fiber core interact with the ambient environment. Such specialty fiber solutions include designing fibers with an exposed core or fiber tapers, designing them to be side-polished or coreless, among others. In most cases, the second group of sensors has been realized in the form of spectral-domain SMS sensor [42]. However, sensor designs utilizing a static wavelength and analyzing the entire speckle pattern have been investigated as well [103].

As mentioned in Section 2, the very early FSS works were dedicated to acoustic and hydroacoustic sensing [31–33,83]. Further, the scope of potential applications expanded towards vibration measurement [37] and perimeter surveillance [104]. These applications as well as new directions have also been investigated in the following years: vibration [105], intrusion detection [70], security of museum collections [72], temperature sensors [106–108], strain [109,110], force sensors [111,112], tactile sensors for human–system interaction applications [111,113], monitoring of gas leakage along the pipelines [114], current sensors [115], displacement sensors [116–119], and bending sensors [120].

From the 2010s, medical applications of FSSs, especially the topic of vital sign monitoring, became quite popular in the research community: human movement detection, and heartbeat rate and breath rate measurements [68,69,121,122]. Indeed, these measurands produce event-like (pulsed) signals that can be easily detected by FFSs with simple signal processing. Recently, the topic has been extended to other applications such as pulse wave velocity monitoring [123], blood pressure monitoring [124], and blood glucose sensing [92].

In recent years, significant attention has been paid to specklegram chemical- and biosensors involving various specialty fibers or standard fiber pre-processing, allowing light interaction with surrounding media. The applications of chemical sensors include salinity sensors [125,126], refractive index sensors [103,127–132], biosensors [133], and even DNA-related sensors [134].

Finally, we would like to mention some other fiber-speckle-related applications whose principle of operation involved intermodal interference effects: fiber spectrometers [135–137], optical tweezers [138,139], and estimation of MMF bandwidth [140,141].

6. Challenges and Future Directions

Fiber specklegram sensing has emerged as a promising low-cost technique that demonstrated a potential for application in various fields, from physical (strain, vibration, temperature, etc.) to chemical applications. As demonstrated in previous sections, the main challenges of FSSs are the quasi-random nonlinear transfer function and signal fading. We described existing solutions for overcoming these main problems. The problem of signal fading has been effectively solved for highly multimode regimes utilizing multichannel signal processing and by correlation processing of SPs or in the spectral domain. The problem of the stable linear transfer function seems more complicated, and it has been solved with several limitations usually related to the dynamic range of the linear response. At the same time, spectral-domain SMS sensors demonstrate good linearity. However, the length of a sensitive element can be a restriction for some applications in the form of limited sensitivity, limited dynamic range, and the inability to distribute sensing.

On the other hand, for some recent applications of FSSs, especially utilizing specialty fibers, the problem of fading is less critical. For example, for refractive index measurement, it is necessary to record a reference SP and then perform contact with the measured liquid. This requires quite a short time, during which the environmental conditions are expected to be stable enough and will not provoke noticeable additional transformations of the SP. Thus, a correlation approach can be effectively applied in this case. One of the improvements for FSS performance seems to be a selection of fibers with low-temperature cross-sensitivity: temperature-insensitive fibers could be a solution for strain-related applications with enhanced stability [142].

One of the significant challenges and a prospective research direction is the problem of spatially resolvable distributed measurements. The topic of external impact localization has been addressed in several works. Some early solutions dedicated to intrusion detection applications consisted of simultaneously using several separate FSSs covering different zones, so the perturbation location can be found by analyzing all FSSs together [143]. Another approach for the same application also utilized multiple FSSs and involved the mechanical isolation of specific parts of MMFs [104]. The patent [144] describes a distributed sensor where the EFP localization is realized using time-multiplexing based on measurements of the time of flight of light impulses (that should significantly increase the sensor's cost), and the specklegram technique is mentioned as one of the possible sensing principles.

Recent works involving machine learning approaches have also addressed the EFP localization problem [94–96] with promising results for identifying the EFP's location for up to 10 fiber sections. However, as discussed in Section 4.3, the main limitation of the practical application of the machine learning method is the requirement of perfect stability of the setup. Therefore, more research should be focused on obtaining reliable results under changing environmental conditions, fiber rearrangements, laser frequency drift, and other factors leading to SP transformations.

Another concept proposed as a localization problem solution suggests the use of changing the number of propagating modes along the GI MMF. It was found that the sensitivity of the FSS depends on the number of interfering modes at the location of perturbation, while the SP can be formed by a significantly higher number of modes [34,145]. Therefore, if the experimental setup provides an increasing number of excited modes along the MMF using mode controllers or the mode coupling effect, the sensitivity of the output SP will depend on the number of excited modes at the perturbation location. The proof of concept has been realized with three MMF segments and several FSS schemes, including two counter-propagating light beams (bi-directional scheme) [146,147] or two co-propagating beams at two different wavelengths [148]. In both cases, two SPs were simultaneously analyzed and by comparing the amplitude of the recorded SP signals, the place of perturbation can be found. Air gaps between MMFs [147,148], as well as long-period gratings [149], have been tested as mode controllers. Moreover, by applying spatial averaging or a correlation approach for signal amplitude determination, the system becomes robust to changing environmental conditions, i.e., becomes fading-free. However, this approach has two main factors that would limit the number of resolvable fiber segments: a limited number of mode groups in the GI MMF and difficulties with the selective modal launch and their transformations; in practice, modes are excited not stepwise, but as a certain MPD containing a set of modes with different amplitudes.

In addition to the prospects described above, the development of the FSS topic could evolve towards the application of various types of specialty multimode fibers that can exhibit new sensing features [56,103,150]. Also, combinations of different signal processing methods (e.g., spectral domain approach and multichannel signal processing with analysis of the entire SP) could provide new measurement capabilities such as enhanced linearity, EFP localization, and multiparameter sensing. The latter can also be achieved by combining FSSs with other sensing techniques, e.g., fiber Bragg gratings (this is already widely implemented for short-length spectral-domain SMS sensors [42,44]). Advanced data processing methodologies, including machine learning and AI, are crucial for improving measurement accuracy and real-time decision making. The development of algorithms resistant to changing mechanical conditions of the MMF and changing environmental conditions would radically improve the efficiency of the machine learning and AI approaches.

7. Conclusions

While fiber specklegram sensing has achieved significant milestones, acknowledging current challenges and charting future directions are imperative for its sustained success. This field holds significant potential, promising substantial contributions to industrial applications due to its simplicity and low cost. Consequently, it remains an exciting area for continuous research and innovation. As technology evolves, the integration of fiber specklegram sensing into various applications could soon become commonplace, ushering in advancements across diverse fields, from healthcare to infrastructure monitoring.

In conclusion, the trajectory of fiber specklegram sensing is poised for transformative enhancements, driven by a synergy of cutting-edge technologies. The incorporation of machine learning and artificial intelligence emerges as a pivotal force, refining speckle pattern analysis, improving measurement accuracy, and expediting decision-making processes. Moreover, the integration with other sensing elements holds the promise of fostering a comprehensive understanding of systems, expanding the application scope in complex environments. As advancements in miniaturization techniques unfold, the prospect of more compact and portable fiber specklegram sensing devices takes center stage, unlocking transformative applications, especially in medical procedures and on-site diagnostics. The embracement of robustness, adaptability, and technological innovation is certain to propel fiber speckle sensing to new frontiers of capability and utility.

Author Contributions: Conceptualization, A.T. and I.C.; writing—original draft preparation, A.T. and I.C.; writing—review and editing, A.T., I.C., A.S. and P.M.; funding acquisition, A.T. All authors have read and agreed to the published version of the manuscript.

Funding: This research was funded by the Research and Innovation Foundation (RIF) of Cyprus under the project POLYMEROSCOPY(CULTURE/AWARD-YR/0421B/0004).

Data Availability Statement: Not applicable.

Conflicts of Interest: Author Antreas Theodosiou was employed by the company Photonics Research Center (PRC), Lumoscribe Ltd. The remaining authors declare that the research was conducted in the absence of any commercial or financial relationships that could be construed as a potential conflict of interest.

References

1. Theodosiou, A.; Kalli, K. Recent trends and advances of fibre Bragg grating sensors in CYTOP polymer optical fibres. *Opt. Fiber Technol.* **2020**, *54*, 102079. [CrossRef]
2. Mizuno, Y.; Theodosiou, A.; Kalli, K.; Liehr, S.; Lee, H.; Nakamura, K. Distributed polymer optical fiber sensors: A review and outlook. *Photonics Res.* **2021**, *9*, 1719–1733. [CrossRef]
3. Sabri, N.; Aljunid, S.A.; Salim, M.S.; Ahmad, R.B.; Kamaruddin, R. Toward Optical Sensors: Review and Applications. *J. Phys. Conf. Ser.* **2013**, *423*, 012064. [CrossRef]
4. Poeggel, S.; Tosi, D.; Duraibabu, D.; Leen, G.; McGrath, D.; Lewis, E. Optical Fibre Pressure Sensors in Medical Applications. *Sensors* **2015**, *15*, 17115–17148. [CrossRef] [PubMed]
5. Leal-Junior, A.G.; Diaz, C.A.; Avellar, L.M.; Pontes, M.J.; Marques, C.; Frizera, A. Polymer Optical Fiber Sensors in Healthcare Applications: A Comprehensive Review. *Sensors* **2019**, *19*, 3156. [CrossRef] [PubMed]
6. Xiao, G.; Guo, H.; Mrad, N.; Rocha, B.; Sun, Z. Towards the simultaneous monitoring of load and damage in aircraft structures using fiber Bragg grating sensors. In Proceedings of the OFS2012 22nd International Conference on Optical Fiber Sensors, Beijing, China, 14–19 October 2012; Volume 8421, p. 8421BD-8421BD–4. [CrossRef]
7. Theodosiou, A.; Komodromos, M.; Kalli, K. Carbon cantilever beam health inspection using a polymer fiber bragg grating array. *J. Light. Technol.* **2017**, *36*, 986–992. [CrossRef]
8. André, P.; Varum, H.; Antunes, P.; Ferreira, L.; Sousa, M. Monitoring of the concrete curing process using plastic optical fibers. *Measurement* **2012**, *45*, 556–560. [CrossRef]
9. Theodosiou, A.; Savva, P.; Mendoza, E.; Petrou, M.F.; Kalli, K. In-Situ Relative Humidity Sensing for Ultra-High-Performance Concrete Using Polymer Fiber Bragg Gratings. *IEEE Sens. J.* **2021**, *21*, 16086–16092. [CrossRef]
10. Leal-Junior, A.G.; Theodosiou, A.; Diaz, C.R.; Marques, C.; Pontes, M.J.; Kalli, K.; Frizera, A. Simultaneous measurement of axial strain, bending and torsion with a single fiber bragg grating in cytop fiber. *J. Light. Technol.* **2019**, *37*, 971–980. [CrossRef]
11. Leal-junior, A.G. Polymer Optical Fiber Sensors for Healthcare Devices: From Material Analysis to Practical Applications. Ph.D. Thesis, Digital Repository of the Universidade Federal do Espírito Santo, Vitória, Brasil, 2018; p. 201.
12. Domingues, M.F.; Tavares, C.; Leitão, C.; Frizera-Neto, A.; Alberto, N.; Marques, C.; Radwan, A.; Rodriguez, J.; Postolache, O.; Rocon, E.; et al. Insole optical fiber Bragg grating sensors network for dynamic vertical force monitoring. *J. Biomed. Opt.* **2017**, *22*, 091507. [CrossRef]
13. Chiavaioli, F.; Baldini, F.; Tombelli, S.; Trono, C.; Giannetti, A. Biosensing with optical fiber gratings. *Nanophotonics* **2017**, *6*, 663–679. [CrossRef]
14. Pospori, A.; Webb, D.J. Stress Sensitivity Analysis of Optical Fiber Bragg Grating-Based Fabry–Pérot Interferometric Sensors. *J. Light. Technol.* **2017**, *35*, 2654–2659. [CrossRef]
15. Teng, C.; Min, R.; Zheng, J.; Deng, S.; Li, M.; Hou, L.; Yuan, L. Intensity-Modulated Polymer Optical Fiber-Based Refractive Index Sensor: A Review. *Sensors* **2021**, *22*, 81. [CrossRef] [PubMed]
16. Lee, H.; Hayashi, N.; Mizuno, Y.; Nakamura, K. Slope-Assisted Brillouin Optical Correlation-Domain Reflectometry Using Polymer Optical Fibers with High Propagation Loss. *J. Light. Technol.* **2017**, *35*, 2306–2310. [CrossRef]
17. Theodosiou, A.; Kalli, K. All-in-Fiber Cladding Interferometric and Bragg Grating Components Made via Plane-by-Plane Femtosecond Laser Inscription. *J. Light. Technol.* **2019**, *37*, 4864–4871. [CrossRef]
18. Efendioglu, H.S. A Review of Fiber-Optic Modal Modulated Sensors: Specklegram and Modal Power Distribution Sensing. *IEEE Sens. J.* **2017**, *17*, 2055–2064. [CrossRef]

19. Puttnam, B.J.; Rademacher, G.; Luís, R.S. Space-division multiplexing for optical fiber communications. *Optica* **2021**, *8*, 1186–1203. [CrossRef]
20. Ukil, A.; Braendle, H.; Krippner, P. Distributed Temperature Sensing: Review of Technology and Applications. *IEEE Sens. J.* **2012**, *12*, 885–892. [CrossRef]
21. Ding, Z.; Wang, C.; Liu, K.; Jiang, J.; Yang, D.; Pan, G.; Pu, Z.; Liu, T. Distributed Optical Fiber Sensors Based on Optical Frequency Domain Reflectometry: A review. *Sensors* **2018**, *18*, 1072. [CrossRef]
22. Leal-Junior, A.G.; Theodosiou, A.; Min, R.; Casas, J.; Diaz, C.R.; Dos Santos, W.M.; Pontes, M.J.; Siqueira, A.A.G.; Marques, C.; Kalli, K.; et al. Quasi-Distributed Torque and Displacement Sensing on a Series Elastic Actuator's Spring Using FBG Arrays Inscribed in CYTOP Fibers. *IEEE Sens. J.* **2019**, *19*, 4054–4061. [CrossRef]
23. Sanders, W.C. *Basic Principles of Nanotechnology*; CRC Press: Boca Raton, FL, USA, 2018. [CrossRef]
24. Theodosiou, A.; Komodromos, M.; Kalli, K. Accurate and Fast Demodulation Algorithm for Multipeak FBG Reflection Spectra Using a Combination of Cross Correlation and Hilbert Transformation. *J. Light. Technol.* **2017**, *35*, 3956–3962. [CrossRef]
25. Leal-Junior, A.; Frizera-Neto, A.; Marques, C.; Pontes, M.J. Measurement of Temperature and Relative Humidity with Polymer Optical Fiber Sensors Based on the Induced Stress-Optic Effect. *Sensors* **2018**, *18*, 916. [CrossRef]
26. Bilro, L.; Alberto, N.; Pinto, J.L.; Nogueira, R. Optical Sensors Based on Plastic Fibers. *Sensors* **2012**, *12*, 12184–12207. [CrossRef]
27. Wu, S.; Yin, S.; Yu, F.T.S. Sensing with fiber specklegrams. *Appl. Opt.* **1991**, *30*, 4468–4470. [CrossRef]
28. Wu, S.; Yin, S.; Rajan, S.; Yu, F.T.S. Multichannel sensing with fiber specklegrams. *Appl. Opt.* **1992**, *31*, 5975–5983. [CrossRef]
29. Takahara, H. Visibility of speckle patterns: Effect of the optical guide length in coherent light. *Appl. Opt.* **1976**, *15*, 609–610. [CrossRef]
30. Crosignani, B.; Daino, B.; Di Porto, P. Speckle-pattern visibility of light transmitted through a multimode optical fiber. *J. Opt. Soc. Am.* **1976**, *66*, 1312–1313. [CrossRef]
31. Kingsley, S.; Davies, D. Multimode optical-fibre phase modulators and discriminators: I-theory. *Electron. Lett.* **1978**, *14*, 322–324. [CrossRef]
32. Layton, M.R.; Bucaro, J.A. Optical fiber acoustic sensor utilizing mode-mode interference. *Appl. Opt.* **1979**, *18*, 666–670. [CrossRef]
33. Bucaro, J.A.; Dardy, H.D.; Carome, E.F. Optical fiber acoustic sensor. *Appl. Opt.* **1977**, *16*, 1761–1762. [CrossRef]
34. Kotov, O.; Chapalo, I.; Medvedev, A. *Mode-Mode Interference Sensor with Increasing Number of Modes along the Multimode Optical Fiber*; SPIE Optical Engineering + Applications: San Diego, CA, USA, 2014; p. 92030B. [CrossRef]
35. Chapalo, I.; Theodosiou, A.; Kalli, K.; Kotov, O. Multimode Fiber Interferometer Based on Graded-Index Polymer CYTOP Fiber. *J. Light. Technol.* **2020**, *38*, 1439–1445. [CrossRef]
36. Kotov, O.I.; Kosareva, L.I.; Liokumovich, L.B.; Markov, S.I.; Medvedev, A.V.; Nikolaev, V.M. Multichannel signal detection in a multimode optical-fiber interferometer: Ways to reduce the effect of amplitude fading. *Tech. Phys. Lett.* **2000**, *26*, 844–848. [CrossRef]
37. Spillman, W.B.; Kline, B.R.; Maurice, L.B.; Fuhr, P.L. Statistical-mode sensor for fiber optic vibration sensing uses. *Appl. Opt.* **1989**, *28*, 3166–3176. [CrossRef]
38. Petrov, A.; Bisyarin, M.; Kotov, O.I. Broadband intermodal fiber interferometer for sensor application: Fundamentals and simulator. *Appl. Opt.* **2022**, *61*, 6544–6552. [CrossRef]
39. Petrov, A.V.; E Chapalo, I.; I Kotov, O. Theoretical analysis of averaging methods for intermodal fiber interferometer. *J. Phys. Conf. Ser.* **2019**, *1326*, 012023. [CrossRef]
40. Petrov, A.; Chapalo, I.; Kotov, O. *Intermodal Fiber Interferometer with Scanning Laser and Correlation Signal Processing: An Experimental Study*; Springer: Cham, Switzerland, 2021; pp. 307–316.
41. Mehta, A.; Mohammed, W.; Johnson, E.G. Multimode interference-based fiber-optic displacement sensor. *IEEE Photonics Technol. Lett.* **2003**, *15*, 1129–1131. [CrossRef]
42. Wu, Q.; Qu, Y.; Liu, J.; Yuan, J.; Wan, S.-P.; Wu, T.; He, X.-D.; Liub, B.; Liuc, D.; Ma, Y.; et al. Singlemode-Multimode-Singlemode Fiber Structures for Sensing Applications—A Review. *IEEE Sens. J.* **2021**, *21*, 12734–12751. [CrossRef]
43. Wang, K.; Dong, X.; Kohler, M.H.; Kienle, P.; Bian, Q.; Jakobi, M.; Koch, A.W. Advances in Optical Fiber Sensors Based on Multimode Interference (MMI): A Review. *IEEE Sens. J.* **2021**, *21*, 132–142. [CrossRef]
44. Wang, K.; Mizuno, Y.; Dong, X.; Kurz, W.; Koehler, M.; Kienle, P.; Lee, H.; Jakobi, M.; Koch, A.W. Multimode optical fiber sensors: From conventional to machine learning-assisted. *Meas. Sci. Technol.* **2024**, *35*, 022002. [CrossRef]
45. Wang, P.; Zhao, H.; Wang, X.; Farrell, G.; Brambilla, G. A Review of Multimode Interference in Tapered Optical Fibers and Related Applications. *Sensors* **2018**, *18*, 858. [CrossRef] [PubMed]
46. Guzmán-Sepúlveda, J.R.; Guzmán-Cabrera, R.; Castillo-Guzmán, A.A. Optical Sensing Using Fiber-Optic Multimode Interference Devices: A Review of Nonconventional Sensing Schemes. *Sensors* **2021**, *21*, 1862. [CrossRef] [PubMed]
47. Frazão, O.; Silva, S.O.; Viegas, J.; Ferreira, L.A.; Araújo, F.M.; Santos, J.L. Optical fiber refractometry based on multimode interference. *Appl. Opt.* **2011**, *50*, E184–E188. [CrossRef]
48. Wang, P.; Brambilla, G.; Ding, M.; Semenova, Y.; Wu, Q.; Farrell, G. Investigation of single-mode–multimode–single-mode and single-mode–tapered-multimode–single-mode fiber structures and their application for refractive index sensing. *J. Opt. Soc. Am. B* **2011**, *28*, 1180–1186. [CrossRef]
49. Kumar, A.; Varshney, R.K.; Antony, C.S.; Sharma, P. Transmission characteristics of SMS fiber optic sensor structures. *Opt. Commun.* **2003**, *219*, 215–219. [CrossRef]

50. Donlagic, D.; Culshaw, B. Microbend sensor structure for use in distributed and quasi-distributed sensor systems based on selective launching and filtering of the modes in graded index multimode fiber. *J. Light. Technol.* **1999**, *17*, 1856–1868. [CrossRef]
51. Gowar, J. *Optical Communication Systems*; Prentice/Hall International: London, UK, 1984.
52. Gloge, D. Weakly Guiding Fibers. *Appl. Opt.* **1971**, *10*, 2252–2258. [CrossRef]
53. Kotov, O.I.; Bisyarin, M.A.; Chapalo, I.E.; Petrov, A.V. Simulation of a multimode fiber interferometer using averaged characteristics approach. *J. Opt. Soc. Am. B* **2018**, *35*, 1990–1999. [CrossRef]
54. Hallam, A. *Mode Control in Multimode Optical Fiber and Its Applications*; Aston University: Birmingham, UK, 2007.
55. Chapalo, I.; Petrov, A.; Bozhko, D.; Bisyarin, M.A.; Kotov, O.I. Averaging Methods for a Multimode Fiber Interferometer: Experimental and Interpretation. *J. Light. Technol.* **2020**, *38*, 5809–5816. [CrossRef]
56. Cabral, T.D.; Fujiwara, E.; Warren-Smith, S.C.; Ebendorff-Heidepriem, H.; Cordeiro, C.M.B. Multimode exposed core fiber specklegram sensor. *Opt. Lett.* **2020**, *45*, 3212–3215. [CrossRef]
57. Taylor, H. Bending effects in optical fibers. *J. Light. Technol.* **1984**, *2*, 617–628. [CrossRef]
58. Kosareva, L.I.; Kotov, O.I.; Liokumovich, L.B.; Markov, S.I.; Medvedev, A.V.; Nikolaev, V.M. Two mechanisms of phase modulation in multimode fiber-optic interferometers. *Tech. Phys. Lett.* **2000**, *26*, 70–74. [CrossRef]
59. Kotov, O.I.; Liokumovich, G.B.; Markov, S.I. *Registration of Influence on Optical Fiber by Mode-Mode Interference*; Lasers for Measurements and Information Transfer: St. Petersburg, Russia, 2004; pp. 91–102.
60. Berdagué, S.; Facq, P. Mode division multiplexing in optical fibers. *Appl. Opt.* **1982**, *21*, 1950–1955. [CrossRef] [PubMed]
61. Chapalo, I.; Petrov, A.; Bozhko, D.; Bisyarin, M.; Kotov, O. Methods of signal averaging for a multimode fiber interferometer: An experimental study. In *Optical Sensors 2019*; SPIE Optics + Optoelectronics: Prague, Czech Republic, 2019; p. 110282Q. [CrossRef]
62. Marcuse, D. *Light Transmission Optics*; Van Nostrand Reinhold Company: New York, NY, USA, 1982.
63. Rawson, E.G.; Goodman, J.W.; Norton, R.E. Frequency dependence of modal noise in multimode optical fibers. *J. Opt. Soc. Am.* **1980**, *70*, 968–976. [CrossRef]
64. Kotov, O.I.; Chapalo, I.E.; Petrov, A.V. Signals of an intermodal fiber interferometer induced by laser frequency modulation. *Tech. Phys. Lett.* **2016**, *42*, 11–14. [CrossRef]
65. Kotov, O.; Chapalo, I. *Signal-to-Noise Ratio for Mode-Mode Fiber Interferometer*; SPIE Optical Metrology: Munich, Germany, 2017; p. 1032945. [CrossRef]
66. Petrov, A.; Chapalo, I.E.; Bisyarin, M.; Kotov, O.I. Intermodal fiber interferometer with frequency scanning laser for sensor application. *Appl. Opt.* **2020**, *59*, 10422–10431. [CrossRef] [PubMed]
67. Redding, B.; Popoff, S.M.; Cao, H. All-fiber spectrometer based on speckle pattern reconstruction. *Opt. Express* **2013**, *21*, 6584–6600. [CrossRef] [PubMed]
68. Lomer, M.; Rodriguez-Cobo, L.; Revilla, P.; Herrero, G.; Madruga, F.; Lopez-Higuera, J.M. Speckle POF Sensor for Detecting Vital Signs of Patients. In Proceedings of the OFS2014 23rd International Conference on Optical Fiber Sensors, Santander, Spain, 2–6 June 2014; p. 91572I. [CrossRef]
69. Rodríguez-Cuevas, A.; Peña, E.R.; Rodríguez-Cobo, L.; Lomer, M.; López-Higuera, J.M. Low-cost fiber specklegram sensor for noncontact continuous patient monitoring. *J. Biomed. Opt.* **2017**, *22*, 037001. [CrossRef] [PubMed]
70. Dhall, A.; Chhabra, J.; Aulakh, N. Intrusion detection system based on speckle pattern analysis. *Exp. Tech.* **2005**, *29*, 25–31. [CrossRef]
71. Kotov, O.I.; Kosareva, L.I.; Liokumovich, L.B.; Markov, S.I.; Medvedev, A.V.; Nikolaev, V.M. Multichannel signal detection in a multimode optical-fiber interferometer: Reducing the effect of spurious signals. *Tech. Phys. Lett.* **2000**, *26*, 991–995. [CrossRef]
72. Szustakowski, M.; Ciurapiński, W.; Życzkowski, M.; Wróbel, J.; Dulski, R.; Markowska, P. Security–orientated Plastic Optical Fiber sensor in Modalmetric Configuration. *Procedia Eng.* **2012**, *47*, 916–923. [CrossRef]
73. Rodriguez-Cobo, L.; Lomer, M.; Lopez-Higuera, J.-M. Fiber Specklegram-Multiplexed Sensor. *J. Light. Technol.* **2015**, *33*, 2591–2597. [CrossRef]
74. Pan, K.; Uang, C.-M.; Cheng, F.; Yu, F.T.S. Multimode fiber sensing by using mean-absolute speckle-intensity variation. *Appl. Opt.* **1994**, *33*, 2095–2098. [CrossRef] [PubMed]
75. Lujo, I.; Klokoc, P.; Komljenovic, T.; Sipus, Z. Measuring structural vibrations with a multimode fiber optical sensor. In Proceedings of the 2007 19th International Conference on Applied Electromagnetics and Communications (ICECom), Dubrovnik, Croatia, 24–26 September 2007; pp. 1–4. [CrossRef]
76. Fujiwara, E.; Ri, Y.; Wu, Y.T.; Fujimoto, H.; Suzuki, C.K. Evaluation of image matching techniques for optical fiber specklegram sensor analysis. *Appl. Opt.* **2018**, *57*, 9845–9854. [CrossRef] [PubMed]
77. Yu, F.T.S.; Zhang, J.; Pan, K.; Zhao, D.; Ruffin, P.B. Fiber vibration sensor that uses the speckle contrast ratio. *Opt. Eng.* **1995**, *34*, 236–239. [CrossRef]
78. Kulchin, Y.N.; Vitrik, O.B.; Lantsov, A.D. Correlation method for processing speckles of signals from single-fibre multimode interferometers by using charge-coupled devices. *Quantum Electron.* **2006**, *36*, 339–342. [CrossRef]
79. Yu, F.T.S.; Wen, M.; Yin, S.; Uang, C.-M. Submicrometer displacement sensing using inner-product multimode fiber speckle fields. *Appl. Opt.* **1993**, *32*, 4685–4689. [CrossRef]
80. Kulchin, Y.N.; Vitrik, O.B.; Lantsov, A.D. Correlation processing of signals of a single-fibre multimode interferometer upon excitation of few modes during measurements of deformation effects. *Quantum Electron.* **2008**, *38*, 56–58. [CrossRef]

81. Fujiwara, E.; dos Santos, M.F.M.; Suzuki, C.K. Optical fiber specklegram sensor analysis by speckle pattern division. *Appl. Opt.* **2017**, *56*, 1585–1590. [CrossRef]
82. Rodriguez-Cobo, L.; Lomer, M.; Cobo, A.; Lopez-Higuera, J.M. Optical fiber strain sensor with extended dynamic range based on specklegrams. *Sens. Actuators A Phys.* **2013**, *203*, 341–345. [CrossRef]
83. Kulchin, Y.N.; Obukh, V.F. Spatial filtering of radiation from a multimode waveguide in determination of hydroacoustic pressures. *Sov. J. Quantum Electron.* **1986**, *16*, 424–426. [CrossRef]
84. A Bykovskiĭ, Y.; Vitrik, O.B.; Kulchin, Y.N. Amplitude spatial filtering in the processing of signals from a single-fiber multimode interferometer. *Sov. J. Quantum Electron.* **1990**, *20*, 1288–1290. [CrossRef]
85. Kamshilin, A.A.; Jaaskelainen, T.; Kulchin, Y.N. Adaptive correlation filter for stabilization of interference-fiber-optic sensors. *Appl. Phys. Lett.* **1998**, *73*, 705–707. [CrossRef]
86. Potton, R.J. Adaptive spatial filtering using photochromic glass. *Meas. Sci. Technol.* **1999**, *10*, 1315–1318. [CrossRef]
87. Newaz, A.; Faruque, O.; Al Mahmud, R.; Sagor, R.H.; Khan, M.Z.M. Machine-Learning-Enabled Multimode Fiber Specklegram Sensors: A Review. *IEEE Sens. J.* **2023**, *23*, 20937–20950. [CrossRef]
88. Liu, Y.; Li, G.; Qin, Q.; Tan, Z.; Wang, M.; Yan, F. Bending recognition based on the analysis of fiber specklegrams using deep learning. *Opt. Laser Technol.* **2020**, *131*, 106424. [CrossRef]
89. Li, G.; Liu, Y.; Qin, Q.; Zou, X.; Wang, M.; Ren, W. Feature Extraction Enabled Deep Learning From Specklegram for Optical Fiber Curvature Sensing. *IEEE Sens. J.* **2022**, *22*, 15974–15984. [CrossRef]
90. Lu, S.; Tan, Z.; Li, G.; Jingya, Y. A Sensitized Plastic Fiber Sensor for Multi-Point Bending Measurement Based on Deep Learning. *IEEE Photonics J.* **2021**, *13*, 8600107. [CrossRef]
91. Ding, Z.; Zhang, Z. 2D tactile sensor based on multimode interference and deep learning. *Opt. Laser Technol.* **2020**, *136*, 106760. [CrossRef]
92. Pal, D.; Agadarov, S.; Beiderman, Y.; Beiderman, Y.; Kumar, A.; Zalevsky, Z. Non-invasive blood glucose sensing by machine learning of optic fiber-based speckle pattern variation. *J. Biomed. Opt.* **2022**, *27*, 097001. [CrossRef]
93. Chen, P.; You, C.; Ding, P. Event classification using improved salp swarm algorithm based probabilistic neural network in fiber-optic perimeter intrusion detection system. *Opt. Fiber Technol.* **2020**, *56*, 102182. [CrossRef]
94. Cuevas, A.R.; Fontana, M.; Rodriguez-Cobo, L.; Lomer, M.; Lopez-Higuera, J.M. Machine Learning for Turning Optical Fiber Specklegram Sensor into a Spatially-Resolved Sensing System. Proof of Concept. *J. Light. Technol.* **2018**, *36*, 3733–3738. [CrossRef]
95. Gao, H.; Hu, H. Spatially-resolved bending recognition based on a learning-empowered fiber specklegram sensor. *Opt. Express* **2023**, *31*, 7671–7683. [CrossRef] [PubMed]
96. Wei, M.; Tang, G.; Liu, J.; Zhu, L.; Liu, J.; Huang, C.; Zhang, J.; Shen, L.; Yu, S. Neural Network Based Perturbation-Location Fiber Specklegram Sensing System Towards Applications With Limited Number of Training Samples. *J. Light. Technol.* **2021**, *39*, 6315–6326. [CrossRef]
97. Markvart, A.A.; Liokumovich, L.B.; Ushakov, N.A. Simultaneous measurement of strain and bend with a fiber optic SMS structure. In Proceedings of the Optical Fiber Sensors Conference 2020 Special Edition, Washington, DC, USA, 8–12 June 2020; p. W4.57. [CrossRef]
98. Markvart, A.A.; Liokumovich, L.B.; Ushakov, N.A. Fiber Optic SMS Sensor for Simultaneous Measurement of Strain and Curvature. *Tech. Phys. Lett.* **2022**, *48*, 34. [CrossRef]
99. Galarza, M.; Perez-Herrera, R.A.; Leandro, D.; Judez, A.; López-Amo, M. Spatial-frequency multiplexing of high-sensitivity liquid level sensors based on multimode interference micro-fibers. *Sens. Actuators A Phys.* **2020**, *307*, 111985. [CrossRef]
100. Rota-Rodrigo, S.; Lopez-Aldaba, A.; Perez-Herrera, R.A.; Bautista, M.d.C.L.; Esteban, O.; Lopez-Amo, M. Simultaneous Measurement of Humidity and Vibration Based on a Microwire Sensor System Using Fast Fourier Transform Technique. *J. Light. Technol.* **2016**, *34*, 4525–4530. [CrossRef]
101. Cardona-Maya, Y.; Del Villar, I.; Socorro, A.B.; Corres, J.M.; Matias, I.R.; Botero-Cadavid, J.F. Wavelength and Phase Detection Based SMS Fiber Sensors Optimized With Etching and Nanodeposition. *J. Light. Technol.* **2017**, *35*, 3743–3749. [CrossRef]
102. Petrov, A.; Kotov, O.; Golovchenko, A. Measurement of External Impact by an Intermodal Fiber Interferometer with Spectral Interrogation and Fourier Transform of Output Signals. In Proceedings of the 2023 International Conference on Electrical Engineering and Photonics (EExPolytech), St. Petersbourg, Russia, 20–21 October 2023; pp. 406–409. [CrossRef]
103. Theodosiou, A. Adaptive Refractive Index Measurements via Polymer Optical Fiber Speckle Pattern Analysis. *IEEE Sens. J.* **2024**, *24*, 287–291. [CrossRef]
104. Leung, C.-Y.; Chang, I.-F.; Hsu, S. Fiberoptic Line-Sensing System for Perimeter Protection Against Intrusion. In *Optical Fiber Sensors*; Optica Publishing Group: Tokyo, Japan, 1986; p. 54. [CrossRef]
105. Leng, J.; Asundi, A. NDE of smart structures using multimode fiber optic vibration sensor. *NDT E Int.* **2002**, *35*, 45–51. [CrossRef]
106. Wang, J.-J.; Yan, S.-C.; Xu, F. Speckle-based fiber sensor for temperature measurement. In Proceedings of the 2017 16th International Conference on Optical Communications and Networks (ICOCN), Wuzhen, China, 7–10 August 2017; pp. 1–3. [CrossRef]
107. Wang, J.-J.; Yan, S.-C.; Ruan, Y.-P.; Xu, F.; Lu, Y.-Q. Fiber-Optic Point-Based Sensor Using Specklegram Measurement. *Sensors* **2017**, *17*, 2429. [CrossRef] [PubMed]
108. Musin, F.; Mégret, P.; Wuilpart, M. Fiber-Optic Surface Temperature Sensor Based on Modal Interference. *Sensors* **2016**, *16*, 1189. [CrossRef] [PubMed]

109. Regez, B.; Sayeh, M.; Mahajan, A.; Figueroa, F. A novel fiber optics based method to measure very low strains in large scale infrastructures. *Measurement* **2009**, *42*, 183–188. [CrossRef]
110. Fujiwara, E.; da Silva, L.E.; Marques, T.H.R.; Cordeiro, C.M.B. Polymer optical fiber specklegram strain sensor with extended dynamic range. *Opt. Eng.* **2018**, *57*, 116107. [CrossRef]
111. Fujiwara, E.; Wu, Y.T.; dos Santos, M.F.M.; Schenkel, E.A.; Suzuki, C.K. Development of a tactile sensor based on optical fiber specklegram analysis and sensor data fusion technique. *Sens. Actuators A Phys.* **2017**, *263*, 677–686. [CrossRef]
112. Garcia-Valenzuela, A.; Tabib-Azar, M. Fiber-optic force and displacement sensor based on speckle detection with 0.1 nN and 0.1 Å resolution. *Sens. Actuators A Phys.* **1993**, *36*, 199–208. [CrossRef]
113. Wang, X.; Song, L.; Wang, X.; Lu, S.; Li, J.; Zhang, P.; Fang, F. An Ultrasensitive Fiber-End Tactile Sensor With Large Sensing Angle Based on Specklegram Analysis. *IEEE Sens. J.* **2023**, *23*, 30394–30402. [CrossRef]
114. Kinet, D.; Wuilpart, M.; Reginster, M.; Caucheteur, C.; Musin, F. Cost-effective optical fiber gas leakage detector around buried pipelines. In *Optical Sensing and Detection V*; SPIE Photonics Europe: Strasbourg, France, 2018; p. 1068018. [CrossRef]
115. Ramprasad, B.; Bai, T.R. Speckle-based fibre-optic current sensor. *Opt. Laser Technol.* **1984**, *16*, 156–159. [CrossRef]
116. Wang, B.; Fu, J.Y.; Liu, Y.; Guo, R.; Yu, F.T.S. Displacement Sensing with Hetero-Core Fiber Specklegram. In Proceedings of the Optical Science and Technology, the SPIE 49th Annual Meeting, Denver, CO, USA, 2–6 August 2004; p. 164. [CrossRef]
117. Liu, Y.; Qin, Q.; Liu, H.-H.; Tan, Z.-W.; Wang, M.-G. Investigation of an image processing method of step-index multimode fiber specklegram and its application on lateral displacement sensing. *Opt. Fiber Technol.* **2018**, *46*, 48–53. [CrossRef]
118. Chen, W.; Feng, F.; Chen, D.; Lin, W.; Chen, S.-C. Precision non-contact displacement sensor based on the near-field characteristics of fiber specklegrams. *Sens. Actuators A Phys.* **2019**, *296*, 1–6. [CrossRef]
119. Osório, J.H.; Cabral, T.D.; Fujiwara, E.; Franco, M.A.; Amrani, F.; Delahaye, F.; Gérôme, F.; Benabid, F.; Cordeiro, C.M. Displacement sensor based on a large-core hollow fiber and specklegram analysis. *Opt. Fiber Technol.* **2023**, *78*, 103335. [CrossRef]
120. Wang, X.; Yang, Y.; Li, S.; Wang, X.; Zhang, P.; Lu, S.; Yu, D.; Zheng, Y.; Song, L.; Fang, F. A reflective multimode fiber vector bending sensor based on specklegram. *Opt. Laser Technol.* **2024**, *170*, 110235. [CrossRef]
121. Warren-Smith, S.C.; Kilpatrick, A.D.; Wisal, K.; Nguyen, L.V. Multimode optical fiber specklegram smart bed sensor array. *J. Biomed. Opt.* **2022**, *27*, 067002. [CrossRef] [PubMed]
122. Bennett, A.; Beiderman, Y.; Agdarov, S.; Beiderman, Y.; Hendel, R.; Straussman, B.; Zalevsky, Z. Monitoring of vital bio-signs by analysis of speckle patterns in a fabric-integrated multimode optical fiber sensor. *Opt. Express* **2020**, *28*, 20830–20844. [CrossRef] [PubMed]
123. Markvart, A.; Kulik, D.; Petrov, A.; Liokumovich, L.; Ushakov, N. Pulse Wave Measurement Using Fiber-Optic Intermodal Interferometric Sensor. In *International Youth Conference on Electronics, Telecommunications and Information Technologies: Proceedings of the YETI 2021, St. Petersburg, Russia*; Springer: Berlin/Heidelberg, Germany, 2022; pp. 429–436.
124. Pang, Y.-N.; Liu, B.; Liu, J.; Wan, S.-P.; Wu, T.; Yuan, J.; Xin, X.; He, X.-D.; Wu, Q. Singlemode-Multimode-Singlemode Optical Fiber Sensor for Accurate Blood Pressure Monitoring. *J. Light. Technol.* **2022**, *40*, 4443–4450. [CrossRef]
125. Zhao, F.; Lin, W.; Guo, P.; Hu, J.; Liu, S.; Yu, F.; Zuo, G.; Wang, G.; Liu, H.; Chen, J.; et al. Demodulation of DBR Fiber Laser Sensors With Speckle Patterns. *IEEE Sens. J.* **2023**, *23*, 26022–26030. [CrossRef]
126. Lin, W.; Liu, B.; Liu, H.; Yang, C.; Zhang, H. Fibre-optic salinity sensor based on multimode fibre specklegram analysis. *Meas. Sci. Technol.* **2021**, *32*, 115110. [CrossRef]
127. Fujiwara, E.; da Silva, L.E.; Cabral, T.D.; de Freitas, H.E.; Wu, Y.T.; Cordeiro, C.M.d.B. Optical Fiber Specklegram Chemical Sensor Based on a Concatenated Multimode Fiber Structure. *J. Light. Technol.* **2019**, *37*, 5041–5047. [CrossRef]
128. Mu, G.; Liu, Y.; Qin, Q.; Tan, Z.; Li, G.; Wang, M.; Yan, F. Refractive Index Sensing Based on the Analysis of D-Shaped Multimode Fiber Specklegrams. *IEEE Photonics Technol. Lett.* **2020**, *32*, 485–488. [CrossRef]
129. Arı, F.; Şerbetçi, H.; Navruz, I. Tapered fiber optic refractive index sensor using speckle pattern imaging. *Opt. Fiber Technol.* **2023**, *79*, 103366. [CrossRef]
130. Al Zain, M.; Karimi-Alavijeh, H.; Moallem, P.; Khorsandi, A.; Ahmadi, K. A High-Sensitive Fiber Specklegram Refractive Index Sensor With Microfiber Adjustable Sensing Area. *IEEE Sens. J.* **2023**, *23*, 15570–15577. [CrossRef]
131. Cai, L.; Wang, M.; Zhao, Y. Investigation on refractive index sensing characteristics based on multimode fiber specklegram. *Meas. Sci. Technol.* **2023**, *34*, 015125. [CrossRef]
132. Guo, P.; Peng, X.; Li, J.; Zhou, Z.; Hu, J.; Xue, C.; Hu, J.; Li, J.; Dang, H.; Chen, J.; et al. Refractive Index Detection of Liquid Analyte in Broad Range Using Multimode Fiber Speckle Sensor. In Proceedings of the 2022 Asia Communications and Photonics Conference (ACP), Shenzhen, China, 5–8 November 2022; pp. 1981–1983. [CrossRef]
133. Liu, Y.; Lin, W.; Zhao, F.; Liu, Y.; Sun, J.; Hu, J.; Li, J.; Chen, J.; Zhang, X.; Vai, M.I.; et al. A Multimode Microfiber Specklegram Biosensor for Measurement of CEACAM5 through AI Diagnosis. *Biosensors* **2024**, *14*, 57. [CrossRef] [PubMed]
134. Feng, F.; Chen, W.; Chen, D.; Lin, W.; Chen, S.-C. In-situ ultrasensitive label-free DNA hybridization detection using optical fiber specklegram. *Sens. Actuators B Chem.* **2018**, *272*, 160–165. [CrossRef]
135. Inalegwu, O.C.; Ii, R.E.G.; Huang, J. A Machine Learning Specklegram Wavemeter (MaSWave) Based on a Short Section of Multimode Fiber as the Dispersive Element. *Sensors* **2023**, *23*, 4574. [CrossRef]
136. Tan, H.; Li, B.; Crozier, K.B. Optical fiber speckle spectrometer based on reversed-lens smartphone microscope. *Sci. Rep.* **2023**, *13*, 12958. [CrossRef] [PubMed]
137. Redding, B.; Alam, M.; Seifert, M.; Cao, H. High-resolution and broadband all-fiber spectrometers. *Optica* **2014**, *1*, 175. [CrossRef]

138. Jamali, R.; Nazari, F.; Ghaffari, A.; Velu, S.K.P.; Moradi, A.-R. Speckle tweezers for manipulation of high and low refractive index micro-particles and nano-particle loaded vesicles. *Nanophotonics* **2021**, *10*, 2915–2928. [CrossRef]
139. Volpe, G.; Kurz, L.; Callegari, A.; Volpe, G.; Gigan, S. Speckle optical tweezers: Micromanipulation with random light fields. *Opt. Express* **2014**, *22*, 18159–18167. [CrossRef]
140. Moslehi, B.; Goodman, J.W.; Rawson, E.G. Bandwidth estimation for multimode optical fibers using the frequency correlation function of speckle patterns. *Appl. Opt.* **1983**, *22*, 995–999. [CrossRef]
141. Ahn, T.-J.; Moon, S.; Kim, S.; Oh, K.; Kim, D.Y.; Kobelke, J.; Schuster, K.; Kirchhof, J. Frequency-domain intermodal interferometer for the bandwidth measurement of a multimode fiber. *Appl. Opt.* **2006**, *45*, 8238–8243. [CrossRef] [PubMed]
142. Feng, Z.; Sakr, H.; Hayes, J.R.; Fokoua, E.R.N.; Ding, M.; Poletti, F.; Richardson, D.J.; Slavík, R. Hollow-core fiber with stable propagation delay between −150 °C and +60 °C. *Opt. Lett.* **2023**, *48*, 763–766. [CrossRef] [PubMed]
143. Kumar, S.H.; Kumar, C.J.; Somnath, B.; Kumar, G.P. Multi-Fiber Optic 2d-Array Device for Sensing and Localizing Environment Perturbation Using Speckle Image Processing. US6590194B2, 27 March 2001.
144. Thompson, D.; Joshua, G.; Keenan, R. Fiber-Optic Mat Sensor. US7532781B2, 19/772007, 12 May 2009.
145. Kotov, O.I.; Chapalo, I.E.; Medvedev, A.V. Dependence of the signal of a multimode fiber-optic interferometer on the mode power distribution. *Tech. Phys. Lett.* **2014**, *40*, 509–512. [CrossRef]
146. Kotov, O.I.; Chapalo, I.E. External impact localization in a distributed intermodal fiber interferometer. *Tech. Phys. Lett.* **2015**, *41*, 1139–1142. [CrossRef]
147. Kotov, O.; Chapalo, I. *Mode-Mode Fiber Interferometer with Impact Localization Ability*; SPIE Photonics Europe: Brussels, Belgium, 2016; p. 98992J. [CrossRef]
148. Chapalo, I.; Kotov, O.; Petrov, A. Dual-wavelength one-directional multimode fiber interferometer with impact localization ability. In *Optical Sensing and Detection V*; SPIE Photonics Europe: Strassburg, France, 2018; p. 106801V. [CrossRef]
149. Chapalo, I.; Theodosiou, A.; Kalli, K.; Kotov, O. Multimode fiber interferometer with embedded long period grating. In *Seventh European Workshop on Optical Fibre Sensors; In Proceedings of the Seventh European Workshop on Optical Fibre Sensors*; Limassol, Cyprus, 1–4 October 2019, p. 111990Q. [CrossRef]
150. Chapalo, I.; Chah, K.; Gusarov, A.; Ioannou, A.; Pospori, A.; Nan, Y.-G.; Kalli, K.; Mégret, P. Gamma-radiation enhancement of sensing properties of FBGs in a few-mode polymer CYTOP fiber. *Opt. Lett.* **2023**, *48*, 1248–1251. [CrossRef]

photonics

Review

Fluorometric Mercury (II) Detection Using Heteroatom-Doped Carbon and Graphene Quantum Dots

Mosayeb Chaghazardi [1], Soheila Kashanian [1,2,*], Maryam Nazari [1], Kobra Omidfar [3,4], Yvonne Joseph [5,6] and Parvaneh Rahimi [5,6,*]

[1] Faculty of Chemistry, Razi University, Kermanshah 67149, Iran; m.chaghazardi@razi.ac.ir (M.C.); nazari.maryam@razi.ac.ir (M.N.)
[2] Department of Nanobiotechnology, Faculty of Innovative Science and Technology, Razi University, Kermanshah 67149, Iran
[3] Biosensor Research Center, Endocrinology and Metabolism Molecular—Cellular Sciences Institute, Tehran University of Medical Sciences, Tehran 1416753955, Iran; omidfar@tums.ac.ir
[4] Endocrinology and Metabolism Research Center, Endocrinology and Metabolism Clinical Sciences Institute, Tehran University of Medical Sciences, Tehran 1416753955, Iran
[5] Institute of Nanoscale and Biobased Materials, Faculty of Materials Science and Technology, Technische Universität Bergakademie Freiberg, 09599 Freiberg, Germany; yvonne.joseph@esm.tu-freiberg.de
[6] Freiberg Center for Water Research, Technische Universität Bergakademie Freiberg, 09599 Freiberg, Germany
* Correspondence: kashanian_s@yahoo.com (S.K.); parvaneh.rahimi@esm.tu-freiberg.de (P.R.)

Abstract: Mercury ion (Hg^{2+}) is one of the most toxic pollutants that can exist throughout the environment and be diffused into water, soil, air, and eventually the food chain. Even a very low level of Hg^{2+} diffused in living organisms can hurt their DNA and cause the permanent damage of the central nervous system and a variety of consequential disorders. Hence, the development of a sensitive and specific method for the detection of Hg^{2+} at trace ranges is extremely important as well as challenging. Fluorometric detection assays based on graphene quantum dots (GQDs) and carbon quantum dots (CQDs) offer considerable potential for the determination and monitoring of heavy metals due to their fascinating properties. Although the quantum yield of GQDs and CQDs is sufficient for their use as fluorescent probes, doping with heteroatoms can significantly improve their optical properties and selectivity toward specific analytes. This review explores the primary advances of CQDs and GQDs in their great electronic, optical, and physical properties, their synthetic methods, and their use in Hg^{2+} fluorimetry detection.

Keywords: mercury; carbon quantum dots; graphene quantum dots; fluorescent sensors

Citation: Chaghazardi, M.; Kashanian, S.; Nazari, M.; Omidfar, K.; Joseph, Y.; Rahimi, P. Fluorometric Mercury (II) Detection Using Heteroatom-Doped Carbon and Graphene Quantum Dots. *Photonics* **2024**, *11*, 841. https://doi.org/10.3390/photonics11090841

Received: 16 August 2024
Revised: 29 August 2024
Accepted: 3 September 2024
Published: 5 September 2024

1. Introduction

Mercury ion (Hg^{2+}) is known as one of the most toxic heavy metals, with features such as bioaccumulation and strong toxicity. It has been proven that Hg^{2+} can simply diffuse through the tissues and skin, thereby impairing central mitosis, DNA, and the nervous system. So even at very low concentrations, it can cause the permanent damage of the central nervous system and critical problems for human health [1]. Actually, mercury toxicity is attributed to cellular dysfunction in organisms because of its strong affinity and conjugation with the thiol groups (-SH) and seleno groups (-SeH) of enzymes and proteins [2]. Thus, the major building blocks of enzymes and proteins are made due to forming strong mercury complexes with amino acids, and serious neurotoxicity and hepatotoxicity occur [3]. For example, Hg^{2+} can be converted by bacteria in aquatic sediments into methylmercury, a highly toxic form of mercury, and can consequently accumulate in the human body through the food chain, causing many brain and neurological diseases by blocking related proteins. According to the United States Environment Protection Agency (EPA) and the World Health Organization (WHO), the maximum permissible concentrations of Hg^{2+} in drinking water are 2.0 and 6.0 μg/L, respectively. Therefore, designing a sensitive, accurate, and selective analytical technique for Hg^{2+} sensing in water samples is highly important and challenging for water quality considerations. There are numerous analytical procedures for the mercury recognition in practice, including auger-electron spectroscopy [4], inductively coupled plasma-mass spectrometry (ICP-MS) [5,6], cold-vapor atomic absorption spectrometry (CV-AAS) [7,8], cold-vapor atomic fluorescence spectroscopy (CV-AFS) [9,10], etc. However, some disadvantages have been reported for these techniques, including being time-consuming, requiring sophisticated instruments, and a need for a large volume of samples. Simple and eco-friendly procedures have been developed for the sensitive and specific sensing of Hg^{2+}, and they have become greatly desirable to overcome these limitations [11]. In Table S1, the different analytical characteristics of these detection techniques for the quantification of Hg^{2+} are compared.

Fluorescence analysis is a very promising method due to its rapid process, operational simplicity, and high sensitivity. In this regard, numerous fluorescent probes have been fabricated for Hg^{2+} detection, including metal nanoclusters [12–14], fluorescent dyes [15,16], semiconductor quantum dots (QDs) [17–19], and carbon-based quantum dots [20]. Among them, carbon-based quantum dots, especially graphene quantum dots (GQDs) and carbon quantum dots (CQDs), have received special attention owing to their easy synthesis from natural resources, non-toxicity, water solubility, biocompatibility, and excellent optical properties. Moreover, the optical properties of CQDs and GQDs can be tailored by doping with different metals or non-metals. However, due to the toxicity of metal ions and the resulting inefficient and non-uniform doping due to their larger radii than those of carbon atoms, their use is restricted [21]. In comparison, doping CQDs and GQDs with non-metallic heteroatoms such as phosphorus (P), nitrogen (N), sulfur (S), and boron (B) has proven to be an effective means of altering their optical properties, improving their fluorescent quantum yields (QY), and making them more practical tools for a broader range of applications [1,22].

In this review, the synthesis of heteroatom-doped CQDs and GQDs, their QYs, their fluorescence characteristics, and their application as optical probes in the fluorescence sensing of Hg^{2+} have been discussed.

2. CQDs and GQDs as Promising Fluorescent Probes

Common fluorescent dyes have inflexible optical properties, low photo stability, and low QY. The alternative way to overcome these limitations is the use of QDs, because they have good photostability with size-tunable optical properties and high quantum efficiency [23]. Conventional metallic and semiconductor QDs are environmentally questionable and seriously toxic, even in relatively low concentrations [11]. Therefore, the demand for non-metallic QDs with sufficient photo stability, easy and inexpensive production, low toxicity, and high fluorescence QY has become a vital challenge. Carbon-based QDs such as CQDs and GQDs meet all these requirements and have become an interesting class of advanced fluorescent probes as they are mainly composed of carbon, a generally non-toxic element, which is likely to offer a notable advantage for their application, particularly in the biological field.

CQDs were discovered accidentally during the purification of single-walled carbon nanotubes (SWNTs) in 2004 and were subsequently named in 2006. CQDs with a size of 1–10 nm, consisting of a core of sp^2 graphitic carbon and/or a surface of amorphous carbon shell (sp^2/sp^3 carbon) coated with oxygen-rich functional groups, are a great alternative to conventional fluorescent materials, as their functional groups give them high water solubility, high biological activity, and facilitate conjugation with various inorganic and organic substances [24,25]. The optical properties can also vary considerably depending on the choice of synthesis method, passivation, doping, and size of the CQDs.

GQDs composed of sp^2-hybridized carbon are QDs with oxygen-containing functional groups at their edges and photoluminescent (PL) properties. Both their height and lateral dimensions are less than 100 nm. They can be manufactured using "top-down" or "bottom-up" procedures. GQDs exhibit unique properties, including variable bandgaps, the presence of hydroxyl and carboxyl functional groups on their edges and consequently, an ease of functionalization, water solubility, high biocompatibility, a large surface area, chemical inertness, quantum confinement, electrochemical, optical, and electrochemiluminescent characteristics, and are energy size-dependent [26]. Therefore, they have attracted a great deal of research interest, not only in the fluorescence detection of chemical and biological analyte, but also in other application areas like light emitting diodes, photocatalysis, bioimaging, and photovoltaic devices.

3. Optical Features of Carbon and Graphene Quantum Dots

CQDs normally demonstrate strong absorption in the ultraviolet (UV) spectral area with a visible stretching absorption tail. Furthermore, a shoulder or a weak peak is sometimes seen at a wavelength that is rather longer than the UV absorption peak. The UV absorption at a shorter wavelength, 200–350 nm, is linked to the π-π^* electronic transitions of C=C, which are found in the core of CQDs nanoparticles. UV absorption with a longer wavelength, which normally manifests as a shoulder or a weaker peak of 300–400 nm, is linked to n-π^* transitions of C=O on the surface of CQDs and hybridization with heteroatoms like N, S, P, etc. [27]. Moreover, it is possible to tune the absorption properties of CQDs via a modification process or surface passivation. More prominently, the CQDs have uniform fluorescence properties with a wide emission peak range from the ultraviolet to the near-infrared. Their fluorescence properties can be tuned purposefully by monitoring their size, shape, and heteroatom doping, as well as by altering their surfaces and edges due to the remarkable edge effect, quantum confinement, and surface effect [28]. For example, it has been exhibited that the blue fluorescent emission with higher QY can be achieved via N/S/P- functionalization at the edges of the CQDs [29].

The GQDs optical characteristics rely on the presence of surface functional groups, doping, and flaws [30]. GQDs demonstrate two different peaks at the UV-visible region. We can observe a strong peak at ~230 nm and a weak edge at ~300 nm, which originate from aromatic C=C bonds related to π-π^* transition and aromatic C=O bonds related to n-π^*transition, respectively [30]. Just like CQDs, the absorption spectrum of GQDs strongly depends on the size, synthesis techniques, functionalization, and doping. By reducing the size of GQDs, the absorption peak will be shifted to blue [31]. Remarkably, the absorption spectrum of the GQDs will be altered via various synthesis methods. For instance, hydrothermally synthesized GQDs with a diameter of 9.6 nm show a UV absorption (320 nm) peak similar to solvothermally obtained GQDs with a diameter of 5 nm [32].

Moreover, most GQD materials exhibit blue emissions with low photoluminescence QY, significantly restricting their applications. Additionally, their production processes are often time-consuming and complex. To address this issue, GQD geometry is altered through functionalization at various edge locations, leading to significant red shifts and enhanced electronic localization. This advancement promotes their use as novel two-dimensional nanomaterials for optical tuning [33]. In this context, Tang et al. [34] were able to realize a tunable emission wavelength from 505 nm (cyan) to 560 nm (yellow) via the co-doping of GQDs with N and S (N, S-GQDs) and through further adjusting the concentration of the N, S-GQDs solution (Figure 1). They produced N, S-GQDs using catechol and o-phenylendiamine as the raw substances in a dimethyl sulfoxide solution and under UV irradiation conditions for 2 h. The synthesized N,S-GQDs possessed good crystallinity and uniform size and showed a clear red shift with an increasing concentration of N,S-GQDs.

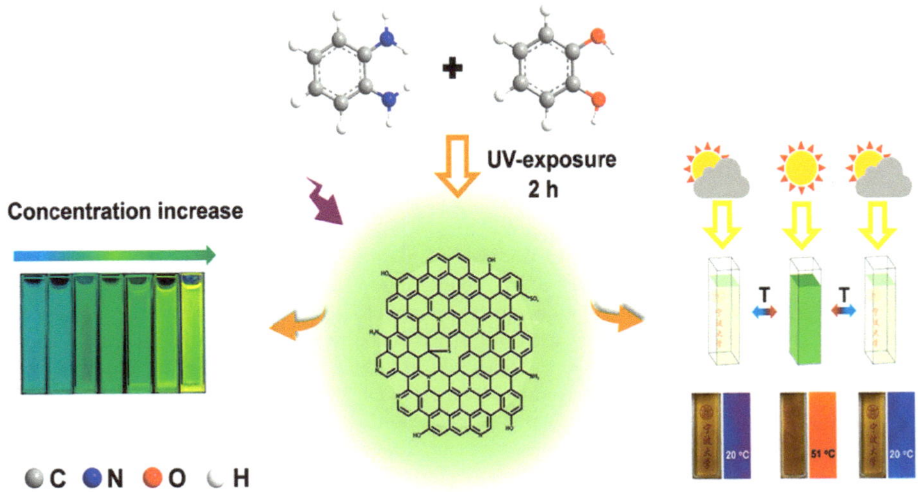

Figure 1. Schematic representation of the synthesis of N,S-GQDs via UV irradiation, illustrating the concentration-dependent properties (**left**) and the photothermal conversion process (**right**) [34]. Reprinted with permission from Elsevier.

4. Synthesis Methods of Carbon and Graphene Quantum Dots

In recent years, many synthesis strategies (Figure 2) based on both top-down (destroying or dispersing macromolecules into small CQDs via physical or chemical methods) and bottom-up (chemical polymerization and carbonization of small molecules to CQDs) methods for the production of CQDs have been developed and reviewed [28,35–38]. It is therefore not our intention to provide a detailed account of these matters, as more comprehensive information can be found in the relevant reviews.

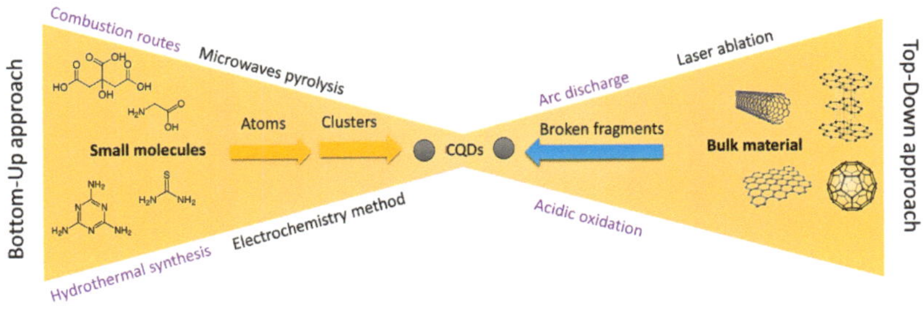

Figure 2. Bottom-up and top-down strategies for synthesis of CQDs.

Arc discharge, laser ablation, and electrochemical oxidation processes are employed as top-down approaches that use graphite dust or SWCNTs and multi-walled carbon nanotubes (MWCNTs) as starting materials. These procedures usually require post-processing and a strong acidic or alkaline solution to produce CQDs, but the QY is still not high enough (lower than 10%) and the size of the particles cannot be controlled [20,24]. Hydrothermal/solvothermal, microwave pyrolysis, and ultrasonic synthesis processes are used as bottom-up approaches, in which macromolecular precursors with surface passivation agents, biomass, and materials with carbon sources such as glucose, fructose, chitosan, and chicken eggs are used for CQDs synthesis. Compared to the top-down methods, they do not require harsh reaction conditions and materials. Additionally, the QY is higher, and the size of the produced particles is more controllable and homogeneous. The hydrothermal method is the most preferred of the aforementioned bottom-up techniques due to its numerous advantages. These include the fact that the synthesis process is both cost-effective and rapid, that environmentally friendly components are utilized (such as orange, coconut and banana peel, rice husk and papaya, among others), and that the particles produced are smaller and more homogeneous with a higher QY [39,40]. For example, Zhu and coworkers could produce the CQDs using citric acid and ethylenediamine with the highest QY of 80 % and average size of 2.81 nm using the hydrothermal method [41]. In another study, Jing et al. produced CQDs with a size of ~2.4 nm and a QY of 76.9% from carbon-derived biomass, including hydrochar and carbonized biomass, through mild oxidation [42].

Similar to the CQDs, the GQDs were synthesized using either top-down or bottom-up approaches (Figure 3), which are reviewed in several papers [43–45]. Chemical and electrochemical oxidation, electron beam lithography, and microwave-assisted hydrothermal, solvothermal, and microfluidization are the top-down methods that utilize graphite [46], graphene and its derivatives (like graphene oxide and graphene nanoribbons) [47–49], carbon nanofibers [50], and carbon black [51] as carbon source materials for GQDs production. However, the application of these top-down processes is limited by some drawbacks, such as the need for special equipment, difficult reaction conditions, and the formation of poorly crystalline and nonhomogeneous GQDs with relatively low QYs. Bottom-up approaches such as microwave hydrothermal, pyrolysis, and oxidation were used to prepare GQDs based on the condensation of smaller units of benzene derivatives [52] as well as biomass waste [53]. Using bottom-up techniques, it is possible to produce homogeneous GQDs with desired sizes, shapes, and properties [43,54]. Nevertheless, a disadvantage of the bottom-up route is the purification of the synthesized GQDs from the starting materials and by-products.

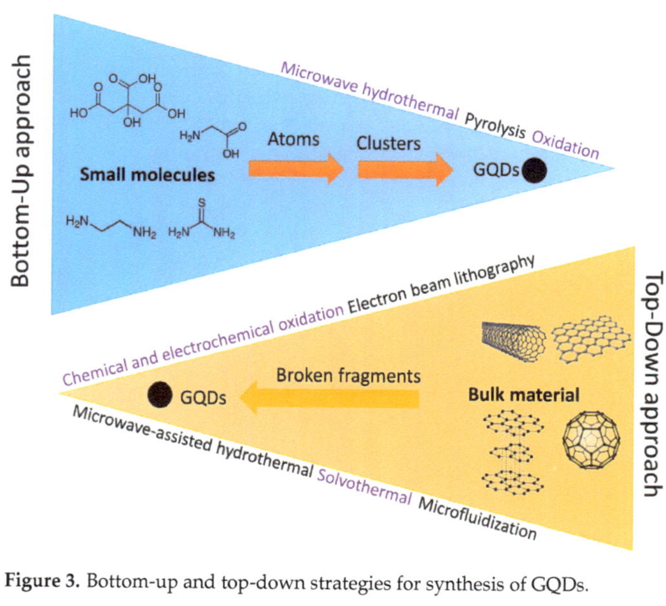

Figure 3. Bottom-up and top-down strategies for synthesis of GQDs.

5. Doping Carbon and Graphene Quantum Dots with Heteroatoms

Despite the development of numerous straightforward and efficacious synthesis techniques for the fabrication of CQDs and GQDs, the large-scale synthesis of reproducible and controllable CQDs and GQDs with a considerably higher QY comparable to that of semiconductor QDs remains a significant challenge. Consequently, ongoing efforts are directed towards overcoming these limitations through the further surface functionalization, passivation, and doping of CQDs and GQDs. Heteroatom doping is regarded as a promising strategy to modulate the bandgap and electron density as well as to improve the QY of CQDs and GQDs due to their inherent scalability, simplicity, and low cost [44]. Among the various heteroatom dopants, N and S are the most commonly used elements for CQDs doping, because they have similar electronegativity and a comparable atomic size, as well as strong valence bonds [55]. It has been theoretically and experimentally investigated that N- and S-doping introduces additional electron pairs and consequently more new defects on the surface of CQDs/GQDs and induces the formation of more sp^2-hybridized sites, leading to an enhancement of fluorescence. Moreover, the band gap between the lowest unoccupied molecular orbitals (LUMO) and the highest occupied molecular orbitals (HOMO) is reduced, which has a positive effect on fluorescence enhancement [55–58].

In the case of N-doping, due to the similar size and structure of N and C, doping amplifies the surface defects of CQDs/GQDs and introduces more reactive groups (such as N=C, CN, NH_2), which leads to an improvement in the electrical and optical properties. Therefore, it has been adopted as the most commonly used doping-strategy for improving the fluorescence properties of CQDs/GQDs. As an example, Cheng et al. produced nitrogen-doped CQDs (N-CQDs) using urea and citric acid hydrothermal treatment (Figure 4). The obtained N-CQDs show a high relative QY of 82.4% in a wide-range pH from 4 to 9, good fluorescence stability, and strong blue luminescence [59]. In another study, Luo et al. [60], using the hydrothermal technique and 2,4,6-triaminopyrimidine as the precursor, synthesized the CQDs containing 41% nitrogen with a bright blue fluorescence and a QY of 74.9%. Guo et al. used a one-step solvothermal molecule fusion method to introduce amine functional groups onto atomic-Fe-rich CQDs as photocatalysts. They feature an impressive Fe-loading capability and well-defined Fe-N_4 active sites that causes a synergistic effect between their electron-donating amine functional groups and Fe-N_4 active centers, enabling an outstanding CO_2-to-CO conversion performance and inhibiting a photocatalytic reduction inCO_2 to CO [61].

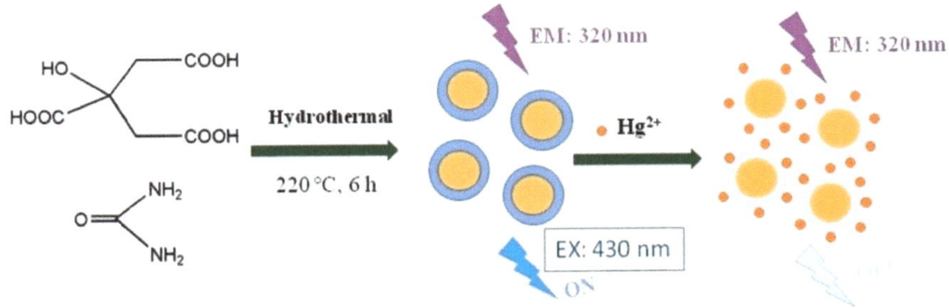

Figure 4. Nitrogen-doped carbon quantum dots as "turn-off" fluorescent probes for highly selective and sensitive detection of mercury (II) ions [59]. Reprinted with permission from Wiley.

In contrast, S-doping is more difficult due to the larger size of the S atom compared to the carbon (C) atom and the small difference in electronegativity of both atoms, so that a significant charge transfer seems almost impossible between C-S [62]. This is despite the fact that there are several reports on the synthesis of S-doped CQDs/GQDs, where S-doping has improved the electronic and optical properties as well as the surface chemical reactivities of CQDs/GQDs. This improvement can be attributed to the electron-rich nature of the S atom, which increases the electron-donating ability of the CQDs/GQDs [63–66]. Thiophene structures can create GQDs containing S atoms with good chemical stabilities and optical properties, and ultrahigh QY [2].

Shen et al. reported oxygen-doped CQDs (O-CQDs) that demonstrated amazing selectivity and outstanding performance as electrocatalysts for H_2O_2 production [67]. Experimental results reveal that the introduction of a series of functionalized GQDs during the synthesis of near-atom layer 2H-MoS_2 nanosheets remediated with GQDs plays a crucial role in this process [68].

6. Using Carbon and Graphene Quantum Dots for Quantitative Analysis

The unique optical properties of CQDs make them excellent candidates for quantitative analysis. Typically, CQD-based fluorometric sensing can be structured in both turn-on and turn-off modes. In the turn-off mode, the interaction between CQDs and specific analytes leads to quenching, resulting in a reduction in the emitted fluorescence. Conversely, in the turn-on mode, this interaction enhances the fluorescence emission of the CQDs [25]. The turn-off and turn-on modes provide versatile approaches for designing fluorometric sensing systems based on CQDs, allowing for the detection and quantification of various analytes like metal ions [69–71], drugs [72,73], explosives [74,75], pesticides [76–78], small organic molecules [2,79,80], proteins [81,82], nucleic acids [83], and bacteria [44,84,85] through distinct fluorescence responses. Indeed, the turn-off sensing process has encountered challenges due to background fluorescence interference, making it less practical in sensing applications. However, to address this limitation, a recent introduction of an on-off-on switching strategy shows significant promise. This innovative approach holds great potential to create highly sensitive sensing platforms capable of detecting trace amounts of desired analytes [86,87].

The selection of appropriate precursors for CQDs preparation allows the creation of a finely tuned probe with high selectivity, sensitivity, and linearity between CQDs fluorescence intensity and analyte concentration. In the CQD-based fluorescence sensing of metal ions, the mechanism often involves electron transfer from chelation between the metal ions and the surface functional groups of the CQDs. This interaction leads to changes in fluorescence intensity, which is the basis for detecting and quantifying the presence of specific metal ions in a sample [88]. To date, various fluorometric sensors have been successfully developed for Hg^{2+} detection based on simple CQDs, heteroatom-doped CQDs, and modified CQDs, as summarized in Table 1 [1,3].

Table 1. Developed fluorometric sensors based on CQDs for Hg^{2+} detection.

CQDs Precursor	Synthesis Method	QY	LOD	Linear Range	Reference
Citric acid, tartaric acid, ethanediamine	Solvothermal	-	83.5 nM	0–18 μM	[1]
Citric acid, melamine	Solid thermal method	-	0.44 μM	2–14 μM	[3]
Citric acid, ethylenediamine, $Mg(OH)_2$	Hydrothermal	-	0.02 μM	0.05–5 μM	[24]
Citric acid, glutathione or thiourea	Microwave hydrothermal	26%	5.4 μM	5–50 μM	[25]
Citric acid, urea	Hydrothermal	-	1.3 nM	0.005–250 μM	[59]
Citric acid, L-cysteine,	Hydrothermal	-	4.2 pM	0.01–0.75 nM	[88]
Citric acid, triethylenetetramine, (TETA)	Thermal	54%	0.2 nM	1–20 nM	[89]
Citric acid, ethylene diamine	Hydrothermal	-	-	-	[90]
Citric acid, 2,2-dimethyl-1,3-propanediamine	Microwave-assisted synthesis	51.20%	7.63 nM	0–4.2 μM	[91]
Citric acid, urea	Hydrothermal	-	8.7 μM	10–70 μM	[92]
Citric acid, glycine	Hydrothermal	-	38 ppb	0.12–2 ppm	[93]
Citric acid, spermine	Hydrothermal	-	2.2 nM	0.01–1.0 μM	[94]
Citric acid, aminopropyltriethoxysilane (APTEOS)	Hydrothermal		0.015 μM	0.02–5.0 μM	[95]
Citric acid, chitosan, thiourea	Hydrothermal	33.00%	4 nM	5–160 nM	[96]
2,4,6-Triaminopyrimidine	Hydrothermal metho.	-	11.4 nM	-	[60]
Xylose	Solvothermal	-	10 nM	50–800 nM	[97]
Glutathione	Solvothermal	41.90%	0.5 μM	0.5–15.0 μM	[98]
Ortho-phenylenediamine (OPDA)	Solvothermal	-	60 nM	30–60 μM	[99]
Methyl orange	-	29.4	237 nM	-	[100]

Table 1. *Cont.*

CQDs Precursor	Synthesis Method	QY	LOD	Linear Range	Reference
Malic acid, urea	Microwave-assisted hydrothermal synthesis	-	0.90 μM	0–40 μM	[101]
Trisodium citrate dihydrate, DL-thioctic acid,	Hydrothermal	-	33.3 nM	0.05–5.8 μM	[102]
Aconitic acid, oligomeric polyethyleneimine	Thermal	44.20%	84 nM	0–800 μM	[103]
Polyamidoamine (PAMAM), and (3-aminopropyl)triethoxysilane (APTES)	Hydrothermal	-	87 fM	0.2 nM–10 μM	[104]
Diaminomaleonitrile (DAMN), thymine-1-acetic acid	Microwave-assisted hydrothermal synthesis	-	0.15 nM	1.0–500 nM	[105]
Glucose, $HAuCl_4$, reduced glutathione	Microwaving	-	8.7 nM	50–1000 nM	[2]
Glucose, boric acid, thiourea, phosphoric	Hydrothermal	-	16.5 μM	25.0 μM–1500.0 mM	[106]
Trisodium citrate dihydrate, melamine	Microwave-assisted hydrothermal	-	42 nM	0–6 μM	[107]
Methyl glycine diacetic acid trisodium salt (MGDA), m-phenylenediamine (MPD)	Hydrothermal	63.8	0.9 μM	0–100 μM	[68]
D-Glucose, aspartic acid, and branched polyethyleneimine	Hydrothermal	-	10 nM	20–800 nM	[108]
Honey	Hydrothermal	-	1.02 nM	0–10 nM	[11]
Tamarindus indica leaves	-	-	6 nM	0–0.1 μM	[20]
Hongcaitai	Hydrothermal	-	0.06 μM	0.2–15 μM	[72]
Black wolfberry	Hydrothermal	-	0.12 nM	0–300 μM	[109]
Highland barley, Ethylenediamine	Hydrothermal	14.40%	0.48 μM	10–160 μM	[110]
Peach palm (Bactris gasipaes) peels	Microwave assisted	25.4	0.19 μM	-	[111]

As illustrated in Table 1, citric acid is recognized as a highly prevalent precursor in conjunction with other suitable precursors in the synthesis of CQDs through a bottom-up approach. The majority of the employed precursors in combination with citric acid for the synthesis of CQDs are utilized for the purpose of doping N and S heteroatoms in the structure of CQDs. Additionally, complex compounds comprising heavy molecules [108] or bulk structures [11] have been documented in the preparation of CQDs via top-down approaches. As evidenced by the data presented in Table 1, a variety of strategies have been employed in the synthesis or modification of CQDs with the objective of achieving optimal sensor performance. These strategies include the use of novel materials, modifications, redesigned sensors, and other innovative techniques, with the aim of developing a highly effective procedure for the detection of Hg, either alone or in conjunction with other analytes.

The highly sensitive and selective fluorescence quenching of N-, S-, doped CQDs, and N, S- co-doped CQDs by Hg^{2+} can be attributed to two primary mechanisms. First, Hg^{2+}, due to its larger ionic radius and stronger chemical interaction with functional groups of doped CQDs, easily forms stable, non-fluorescent complexes. Second, the quenching may result from the aggregation of doped/co-doped CQDs, where Hg^{2+} simultaneously binds to multiple nitrogen and oxygen atoms in the doped/co-doped CQDs. This complexation induces aggregation, leading to changes in the electronic structure of doped/co-doped CQDs and ultimately quenching their fluorescence.

Han et al. employed a strategy to design a highly sensitive sensor for Hg^{2+} detection by functionalizing CQDs with sulfhydryl groups (HS-CQDs), which have a high affinity for the target metal ions, mediated by Ag^{+}. The sensing mechanism (Figure 5) involved quenching the fluorescence of HS-CQDs through the induced agglomeration caused by adding Ag^{+}, and then restoring the fluorescence by adding Hg^{2+}. The higher affinity between -SH and Hg^{2+} results in the substitution of Hg^{2+} for Ag^{+} in the CQDs/Ag agglomerate and a subsequent amalgamation. This design provided the sensor with a linear detection range of 0.01 to 0.75 nM and an ultra-low detection limit of 4.2 pM for Hg^{2+} [88].

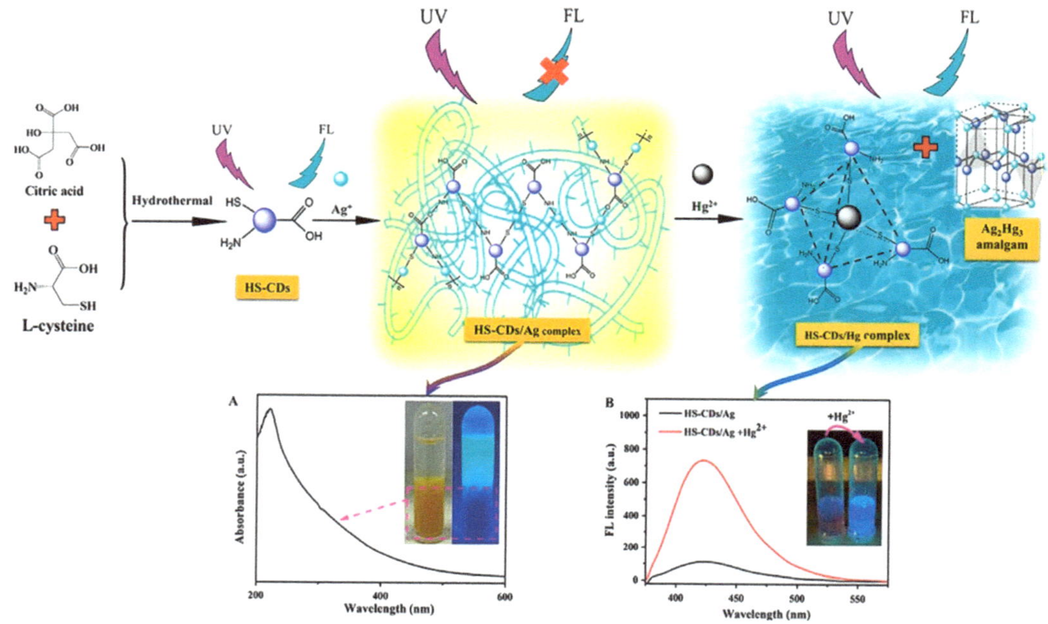

Figure 5. Fluorescent sensing method to determine Hg^{2+} in aqueous solutions using sulfhydryl functionalized CQDs (HS-CQDs) mediated by Ag^{+}. (**A**) Absorption spectrum of the HS-CDs/Ag precipitate; the insets are corresponding photos of HS-CQDs/Ag under visible light (**left**) and UV light at 365 nm (**right**). (**B**) FL spectra of HS-CDs/Ag and HS-CQDs/Ag + Hg^{2+}; the insets are corresponding photos of the HS-CQDs/Ag sensor (**left**) and the HS-CQDs/Ag + Hg^{2+} (**right**) under UV light at 365 nm [88]. Reprinted with permission from Elsevier.

To design a more sensitive sensor for Hg, a dual-emission fluorescence strategy was developed by modifying CQDs with Europium (III, Eu^{3+}) [89]. This involved combining blue-emissive CQDs (at 443 nm) with red-emissive europium complexes (at 617 nm) to create the new dual-emission fluorescence approach. The addition of Hg^{2+} to the designed sensor resulted in a notable enhancement in dual-emission fluorescence intensity. As the concentration of Hg^{2+} increased, the sensing platform exhibited a discernible change in color, from a deep blue to a light blue, then to a lavender hue, and finally to a deep purple. The sensor displayed remarkable sensitivity and selectivity towards Hg^{2+}, with a linear concentration range of 1 to 20 nM and a LOD of 0.2 nM. This procedure was effectively applied for the ultrasensitive and visual sensing of Hg^{2+} in milk and drinking water samples with acceptable recovery from 97.6% to 105.4%.

Guo et al. introduced another dual-emission ratiometric fluorescent system for Hg^{2+} detection by embedding CQDs into europium metal–organic frameworks (Eu-MOF) using a straightforward in situ hydrothermal method [109]. The proposed fluorescence composite material, which exhibited good stability in an aqueous solution while possessing the excellent luminescence properties of Eu^{3+} and CQDs, was successfully employed to detect Hg^{2+} in environmental water samples. In this system, the fluorescence of the encapsulated CQDs within the Eu-MOF was significantly quenched by Hg^{2+}, while the fluorescence emission of the Eu-MOFs remained unaffected. This dual-functionality, providing both a recognition part (CQDs) and a reference part (Eu-MOF), offers a distinct advantage for constructing a ratiometric fluorescence sensor for Hg^{2+} detection in aqueous solutions. The developed sensor demonstrated an excellent response to Hg^{2+} over a wide linear range (0–300 μM) with a lower detection limit of 0.12 nM, surpassing the performance of a previously reported ratiometric fluorescence system [89].

In a further strategy for the rapid and ultrasensitive co-analysis of Hg^{2+} and Pb^{2+} based on photoluminescence and magnetic properties, Ahmadian-Fard-Fini and colleagues proposed combining CQDs, nanofibers, and magnetic nanoparticles [90]. The CQDs were initially prepared via a hydrothermal method, utilizing citric acid (from lemon extract) and ethylene diamine as precursors. The magnetic nanoparticles were then synthesized through ball milling, while the cellulose acetate nanofibers were produced through electrospinning. To construct the magnetic nanofiber fluorescent sensor, the magnetic nanoparticles were incorporated into the polymeric cellulose nanofibers. Subsequently, the CQDs were coated onto the fibers through a sonochemical-assisted hydrothermal method. Following the exposure of the designed sensor to Hg^{2+} and Pb^{2+}, a reduction in the photoluminescent intensity (blue color) of the CQDs was observed. This reduction occurs because electrons from the excited CQDs are transferred to the d-orbitals of Pb^{2+} and Hg^{2+}, leading to the formation of a complex.

In another study, Muthurasu and Ganesh reported a ratiometric sensing strategy for Hg using a hybrid system in which N-CQDs serve as the donor and Rhodamine B as the acceptor [92]. The intermolecular interaction between N-CQDs and Rhodamine B enables effective fluorescence resonance energy transfer (FRET), leading to a tunable shift in the fluorescence emission color of N-CQDs from blue to red. Interestingly, varying the concentration of Rhodamine B resulted in different emission colors, and the concentration that emitted violet fluorescence was selected for the sensing experiments. The addition of varying concentrations of Hg^{2+} ions effectively quenched the photoluminescence properties of N-CQDs due to the complexation of Hg^{2+} with the carboxylic acid and amine functional groups present in N-CQDs. In contrast, no interaction was detected for RhB with Hg^{2+} ions.

In a recent study, Lu et al. developed a dual-mode sensing platform for the colorimetric and turn-on fluorometric detection of Hg^{2+}. This platform was based on the modification of N-CQDs with a sulfhydryl group (–SH) and their subsequent combination with gold nanoparticles (AuNPs). Following the integration of SH-modified N-CQDs, a reduction in fluorescence intensity was observed, which was attributed to the fluorescence quenching effect of AuNPs. But, upon the addition of Hg^{2+}, the fluorescence intensity of the SH-modified N-CQDs was gradually restored, accompanied by a visible color change, transitioning from red to blue-violet, and eventually to blue. The sensing mechanism could be explained as follows: Initially, the fluorescence of SH-modified N-CQDs was quenched due to their strong interaction with AuNPs. When Hg^{2+} was introduced, citrate ions on the surface of the AuNPs reduce Hg^{2+} to Hg^0. This reduction leads to the formation of a gold–mercury alloy through Hg–Au metallophilic interactions on the AuNP surface. As Hg^0 occupies the surface, the modified N-CQDs were released, resulting in the restoration of their fluorescence. Furthermore, to emphasize the significance of the modification of N-CQDs with the SH group in the sensing performance of the proposed sensor, they presented a sensor based on unmodified N-CQDs with AuNPs. It was observed that under visible light, SH-modified N-CQDs did not affect the color of AuNPs. However, unmodified N-CQDs were found to cause the color of AuNPs to change from red to bluish violet, which interfered with the determination of Hg^{2+} through the use of a color reaction. This finding confirmed the crucial role of the SH functional group on CQDs in enhancing their optical properties and contributing to the development of a sensitive and selective Hg sensor [98].

Building on the observed increase in QYs and fluorescence properties after doping and functionalizing CQDs, an attempt was made to dope CQDs with S using thioctic acid, chosen for its strong chelating properties with metal ions [102]. The sensing mechanism relied on the quenching of S-doped CQDs fluorescence upon exposure to Hg^{2+} ions, attributed to electron transfer between Hg^{2+} and the S-doped CQDs, leading to complex formation (turn-off). The fluorescence was then restored (turn-on) by adding thiophanate methyl, which forms thiophanate methyl-Hg complexes due to the strong affinity between Hg^{2+} and the two mercapto groups in thiophanate methyl. Based on this mechanism, the proposed sensor demonstrated dual functionality, enabling the detection of both Hg^{2+} ions and thiophanate methyl, a systemic fungicide. Under optimal conditions, the fluorescence sensor system achieved limits of detection of 7.6 nM for thiophanate methyl and 33.3 nM for Hg^{2+}. Consequently, the developed sensor can be utilized as a multifunctional tool for monitoring trace contaminants in the environment and ensuring the quality control of food safety products.

Ge et al. developed a reusable sensor for detecting Hg and Cu using functionalized CQDs synthesized with aconitic acid for core carbon nanoparticle formation and oligomeric polyethyleneimine as a functionalization agent [103]. The fluorescence of the prepared CQDs was effectively quenched upon the addition of Cu^{2+} or Hg^{2+} across a broad linear concentration range, achieving low detection limits of 70 nM for Cu^{2+} and 84 nM for Hg^{2+}. Additionally, the sensor could be regenerated by introducing aspartic acid or L-cysteine which serve as metal ion chelators, forming complexes with the metal ions. It is important to note that the proposed "on–off–on" sensing system can be utilized not only for the detection of Hg^{2+} and Cu^{2+} but also for aspartic acid and L-cysteine.

GQDs have gained prominence as other carbon-based fluorescence materials due to their exceptional optical properties, high QY, and unique characteristics. Their excellent photoluminescence, high stability, and tunable properties make them promising candidates for various applications, particularly in sensing and biosensing. GQDs present analogous advantages to CQDs and have gained attention as valuable substitutes in fluorescence-based sensor technologies. Due to various interactions such as electrostatic interaction or electron transfer as well as π-π^* conjugation between GQDs and the target analyte, the fluorescence intensity of the GQDs is modulated either by quenching (switch-off mode) or through enhancement mechanisms (switch-on mode) and used as a sensor response [46,112,113]. Similar to CQDs, fluorescence-based sensors based on GQDs using the "on-off-on" fluorescence switching mode have been proposed for the ultrasensitive and selective detection of various analytes, especially for metal ions, as well as for simultaneous multi-analytes detection [114–116]. Various developed fluorometric sensors and biosensors based on GQDs for the detection of Hg^{2+} and their sensing performance are compared and showed in Table 2. As indicated in Table 2, the studies conducted in relation to the fabrication of GQDs are based on bottom-up [117] and top-down methods [118]. In the bottom-up approach, citric acid and glucose are the most frequently used carbon sources for the synthesis of GQDs. In contrast, the top-down procedure has predominantly employed graphene oxide (GO) for the production of GQDs.

Table 2. Developed fluorometric sensors based on GQDs for Hg^{2+} detection.

GQDs Precursor	Synthesis Method	QY	LOD	Linear Range	Reference
Glucose, urea, and ammonia sulfate	Infrared (IR)-assisted pyrolysis	-	10 ppb	10 ppb–10 ppm	[117]
Citric acid and Thiourea	Hydrothermal	41.90%	0.14 nM	0.1–15 µM	[119]
Citric acid and ethylene diamine	Hydrothermal	-	0.45 nM	100–1000 nM	[120]
Citric acid and Gly	Hydrothermal	35.50%	8.3 nM	0–3.0 µM	[121]
Graphene oxide	Electrochemical	-	2.5 µM	2.5–800 µM	[118]
Graphene oxide and 5,10,15,20-tetrakis(1-methyl-4-pyridinio)porphyrin tetra(p-toluenesulfonate) (TMPyP)	Two-step hydrothermal method	-	0.32 nM	2–200 nM	[122]
Graphene oxide, urea, citric acid	Solvotermal	-	-	-	[123]

In a pioneering study, Gue and colleagues reported the development of a Hg^{2+} fluorescence sensor utilizing S and N co-doped GQDs. This sensor was synthesized through a rapid IR approach, whereby citric acid and ammonium sulfate underwent thermal pyrolysis under IR irradiation [117]. In order to evaluate the effectiveness of S,N co-doping in GQDs in comparison to N-doped GQDs, the sensing performance of both types was examined across a Hg^{2+} concentration range of 10 ppb to 10 ppm. The findings demonstrated that the sensitivity of the co-doped GQDs was 4.23 times greater than that of the N-doped GQDs. This augmented sensitivity could be ascribed to the coordination of sulfur with the phenolic groups at the edges of the S,N co-doped GQDs, which induced a reduction or interruption of photon injection pathways, resulting in notable fluorescence quenching.

In another study, S,N co-doped GQDs were synthesized using a one-pot hydrothermal method to develop a Hg^{2+} sensor [115]. The doping with N was found to enhance the QY, while the introduction of S improved the selectivity for Hg^{2+} detection through strong coordination interactions. Moreover, S,N co-doped GQDs were coated onto cellulose filter paper strips, which were then utilized as a simple sensor for the rapid and in situ screening of Hg^{2+} in wastewater, yielding satisfactory results.

To enhance the QY and sensitivity of GQDs towards Hg^{2+}, Zhu et al. proposed a two-step strategy for the synthesis and subsequent functionalization of GQDs [121]. In this context, GQDs were first synthesized by pyrolyzing citric acid and then functionalized with glycine under alkaline conditions. This process involved a carbamate reaction between the GQDs and glycine, resulting in the successful covalent attachment of glycine to the GQDs, forming Gly-GQDs. Furthermore, the prepared Gly-GQDs revealed an enhanced sensitive and selective detection ability for Hg^{2+} in aqueous solutions and a LOD of 8.3 nM was obtained.

As previously mentioned, while bottom-up techniques are generally preferred over top-down methods, there are a few reports on the fabrication of GQDs using top-down approaches, primarily involving graphene oxide (GO) but employing various preparation techniques, as detailed in Table 2. For example, Liu et al. successfully produced N-doped GQDs in a single step using an electrochemical method in the presence of ammonium hydroxide. They subsequently designed a sensor for Hg^{2+} detection, which exhibited excellent sensing performance across a wide concentration range of Hg^{2+} [118].

Peng et al. introduced an innovative sensing strategy for the detection of trace levels of Hg^{2+} [122]. It was found that the incorporation rate of porphyrin and small Mn^{2+} ions was markedly increased in the presence of trace N-doped GQDs (NGQD) and nanomolar concentrations of Hg^{2+} (Figure 6). The proposed synergistic mechanism posits that the larger Hg^{2+} ions deform the porphyrin nucleus, thereby rendering it more susceptible to the smaller Mn^{2+} ions delivered by the NGQDs. The formation of the metalloporphyrin resulted in a red-shift in the absorption and fluorescence quenching of the porphyrin, while the fluorescence of the NGQDs was gradually enhanced due to the inner filter effect of the porphyrin on the NGQDs. This unique phenomenon was used to develop sensitive and selective ratiometric fluorescence and colorimetric methods for the determination of trace Hg^{2+}.

A FRET sensing system has been recently introduced, employing a combination of GQDs as the donor and carbon nanodots (CDs) as the acceptor for the quantification of arsenic (As^{5+}) and Hg^{2+}. The system takes advantage of the excellent spectral overlap between the emission profile of the GQDs and the absorption profile of the CDs, making them an ideal FRET pair. Upon the addition of As^{5+} and Hg^{2+}, the FRET signal was significantly quenched due to the strong affinity of these metal ions to the carboxylic groups on both GQDs and CDs. Furthermore, the fluorescence lifetime of the FRET system decreased from 2.93 ns to 0.44 ns and 1.97 ns in the presence of As^{5+} and Hg^{2+}, respectively, confirming the effectiveness of this sensing system [123].

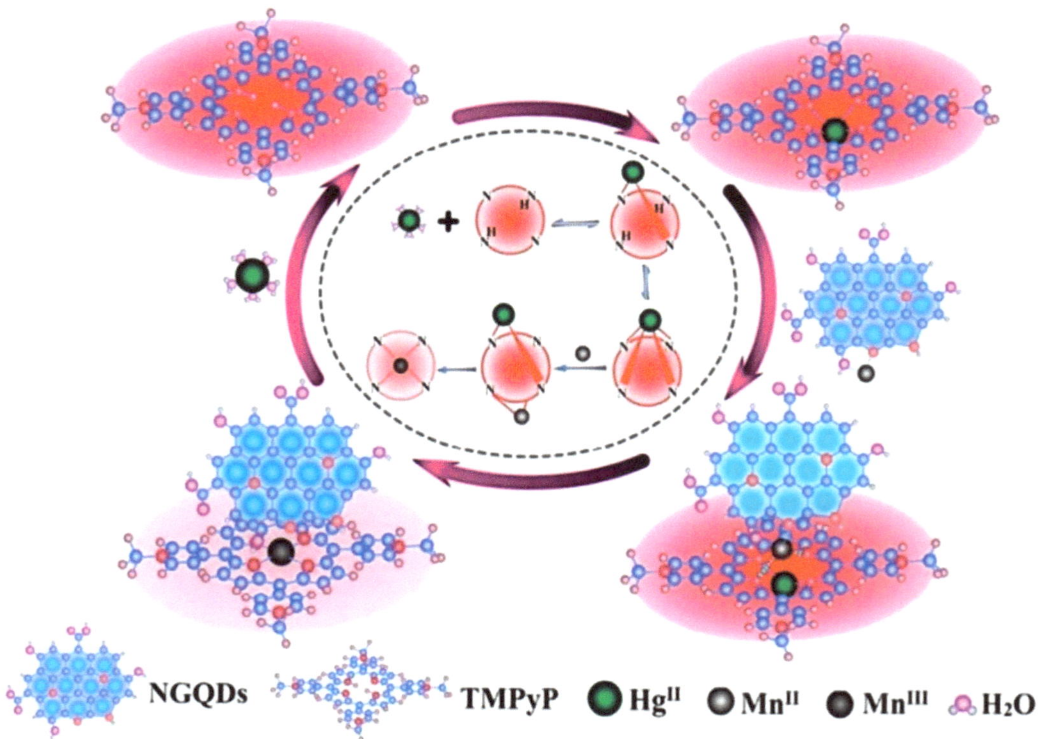

Figure 6. The scheme illustration of the synergistic effect of Hg^{2+} and NGQDs in accelerating the coordination rate of Mn^{2+} and 5,10,15,20-tetrakis(1-methyl-4-pyridinio) porphyrin tetra(p-toluenesulfonate) (TMPyP) [122]. Reprinted with permission from ACS.

7. Conclusions

Hg^{2+} ranks among the most toxic cations in the environment, posing a significant risk to human health. Therefore, developing a sensitive and specific method for detecting Hg^{2+} at trace levels is both critically important and challenging. Among the various proposed techniques, fluorometric sensors are notable for their simplicity, rapid sample preparation, and quick response times. In particular, fluorometric sensing assays utilizing GQDs and CQDs demonstrate considerable potential for determining and monitoring heavy metals, thanks to their unique optical, chemical, and physical properties. These nanomaterials can be synthesized using various top-down and bottom-up methods. However, studies indicate that bottom-up methods for producing CQDs and GQDs are preferred over top-down approaches, as they avoid harsh reaction conditions and yield particles with higher QY and more uniform sizes. Furthermore, the optical properties of CQDs and GQDs can be tailored through doping and functionalization, which subsequently enhances their sensing performance. In this context, doping with heteroatoms such as N and S, as well as co-doping with both N and S, has been predominantly utilized in Hg^{2+} sensors to enhance the electronic and optical properties of CQDs and GQDs, as discussed in this review. In addition, a range of strategies, including the use of new materials, modifications, redesigned sensors, and other innovative techniques, have been reviewed and debated, all of which aimed at developing highly effective methods for detecting Hg^{2+}, both individually and in combination with other analytes.

Supplementary Materials: The following supporting information can be downloaded at https://www.mdpi.com/article/10.3390/photonics11090841/s1, Table S1: Comparing the analytical characteristics of different detection techniques for quantifying Hg^{2+}.

Author Contributions: M.C.: Writing—original draft, investigation, visualization, conceptualization. S.K.: Investigation, writing—review and editing, supervision, project administration. M.N.: Writing—review and editing, supervision, project administration, conceptualization. K.O.: Writing—review and editing, visualization. Y.J.: Writing—review and editing, visualization. P.R.: Writing—review and editing, conceptualization, visualization. All authors have read and agreed to the published version of the manuscript.

Funding: This research received no external funding.

Institutional Review Board Statement: Not applicable.

Data Availability Statement: Data are contained within the article.

Conflicts of Interest: The authors declare no conflicts of interest.

References

1. Huang, H.; Weng, Y.; Zheng, L.; Yao, B.; Weng, W.; Lin, X. Nitrogen-doped carbon quantum dots as fluorescent probe for "off-on" detection of mercury ions, L-cysteine and iodide ions. *J. Colloid Interface Sci.* **2017**, *506*, 373–378. [CrossRef] [PubMed]
2. Wang, G.; Guo, Q.; Chen, D.; Liu, Z.; Zheng, X.; Xu, A.; Yang, S.; Ding, G. Facile and Highly Effective Synthesis of Controllable Lattice Sulfur-Doped Graphene Quantum Dots via Hydrothermal Treatment of Durian. *ACS Appl. Mater. Interfaces* **2018**, *10*, 5750–5759. [CrossRef]
3. Pajewska-Szmyt, M.; Buszewski, B.; Gadzała-Kopciuch, R. Carbon dots as rapid assays for detection of mercury (II) ions based on turn-off mode and breast milk. *Spectrochim. Acta Part A Mol. Biomol. Spectrosc.* **2020**, *236*, 118320. [CrossRef] [PubMed]
4. Ghaedi, M.; Reza Fathi, M.; Shokrollahi, A.; Shajarat, F. Highly selective and sensitive preconcentration of mercury ion and determination by cold vapor atomic absorption spectroscopy. *Anal. Lett.* **2006**, *39*, 1171–1185. [CrossRef]
5. Lee, S.H.; Suh, J.K. Determination of mercury in tuna fish tissue using isotope dilution-inductively coupled plasma mass spectrometry. *Microchem. J.* **2005**, *80*, 233–236.
6. Krata, A.; Vassileva, E.; Bulska, E. Reference measurements for total mercury and methyl mercury content in marine biota samples using direct or species-specific isotope dilution inductively coupled plasma mass spectrometry. *Talanta* **2016**, *160*, 562–569. [CrossRef]
7. KopyŚĆ, E.; PyrzyŃslka, K.; Garbos, S.; Buska, E. Determination of mercury by cold-vapor atomic absorption spectrometry with preconcentration on a gold-trap. *Anal. Sci.* **2000**, *16*, 1309–1312. [CrossRef]
8. Pourreza, N.; Ghanemi, K. Solid phase extraction of cadmium on 2-mercaptobenzothiazole loaded on sulfur powder in the medium of ionic liquid 1-butyl-3-methylimidazolium hexafluorophosphate and cold vapor generation–atomic absorption spectrometric determination. *J. Hazard. Mater.* **2010**, *178*, 566–571. [CrossRef]
9. Erxleben, H.; Ruzicka, J. Atomic Absorption Spectroscopy for Mercury, Automated by Sequential Injection and Miniaturized in Lab-on-Valve System. *Anal. Chem.* **2005**, *77*, 5124–5128. [CrossRef]
10. Pfeil, D.L.; Bruce, M.L. Automated determination of mercury cold vapor atomic fluorescence with gold amalgamation. *Am. Lab.* **2001**, *33*, 26–33.
11. Srinivasan, K.; Subramanian, K.; Murugan, K.; Dinakaran, K. Sensitive fluorescence detection of mercury (II) in aqueous solution by the fluorescence quenching effect of MoS_2 with DNA functionalized carbon dots. *Analyst* **2016**, *141*, 6344–6352. [CrossRef] [PubMed]
12. Li, S.; Cao, W.; Kumar, A.; Jin, S.; Zhao, Y.; Zhang, C.; Zou, G.; Wang, P.C.; Li, F.; Liang, X.-J. Highly sensitive simultaneous detection of mercury and copper ions by ultrasmall fluorescent DNA–Ag nanoclusters. *New J. Chem.* **2014**, *38*, 1546–1550. [CrossRef]
13. Nain, A.; Tseng, Y.-T.; Lin, Y.-S.; Wei, S.-C.; Mandal, R.P.; Unnikrishnan, B.; Huang, C.-C.; Tseng, F.-G.; Chang, H.-T. Tuning the photoluminescence of metal nanoclusters for selective detection of multiple heavy metal ions. *Sens. Actuators B Chem.* **2020**, *321*, 128539. [CrossRef]
14. Panthi, G.; Park, M. Synthesis of metal nanoclusters and their application in Hg^{2+} ions detection: A review. *J. Hazard. Mater.* **2022**, *424*, 127565. [CrossRef] [PubMed]
15. Su, M.; Liu, C.; Zhang, Y.; Rong, X.; Wang, X.; Li, X.; Wang, K.; Zhu, H.; Zhu, B. Rational design of a water-soluble TICT-AIEE-active fluorescent probe for mercury ion detection. *Anal. Chim. Acta* **2022**, *1230*, 340337. [CrossRef]
16. Yoon, S.; Miller, E.W.; He, Q.; Do, P.H.; Chang, C.J. A bright and specific fluorescent sensor for mercury in water, cells, and tissue. *Angew. Chem.* **2007**, *119*, 6778–6781. [CrossRef]
17. Ribeiro, D.S.M.; Castro, R.C.; Páscoa, R.N.M.J.; Soares, J.X.; Rodrigues, S.S.M.; Santos, J.L.M. Tuning CdTe quantum dots reactivity for multipoint detection of mercury (II), silver (I) and copper (II). *J. Lumin.* **2019**, *207*, 386–396. [CrossRef]

18. Labeb, M.; Sakr, A.-H.; Soliman, M.; Abdel-Fattah, T.M.; Ebrahim, S. Effect of capping agent on selectivity and sensitivity of CdTe quantum dots optical sensor for detection of mercury ions. *Opt. Mater.* **2018**, *79*, 331–335. [CrossRef]
19. Ke, J.; Li, X.; Zhao, Q.; Hou, Y.; Chen, J. Ultrasensitive quantum dot fluorescence quenching assay for selective detection of mercury ions in drinking water. *Sci. Rep.* **2014**, *4*, 5624. [CrossRef]
20. Bano, D.; Kumar, V.; Singh, V.K.; Hasan, S.H. Green synthesis of fluorescent carbon quantum dots for the detection of mercury (II) and glutathione. *New J. Chem.* **2018**, *42*, 5814–5821. [CrossRef]
21. Kou, X.; Jiang, S.; Park, S.-J.; Meng, L.-Y. A review: Recent advances in preparations and applications of heteroatom-doped carbon quantum dots. *Dalton Trans.* **2020**, *49*, 6915–6938. [CrossRef] [PubMed]
22. Azami, M.; Wei, J.; Valizadehderakhshan, M.; Jayapalan, A.; Ayodele, O.O.; Nowlin, K. Effect of Doping Heteroatoms on the Optical Behaviors and Radical Scavenging Properties of Carbon Nanodots. *J. Phys. Chem. C* **2023**, *127*, 7360–7370. [CrossRef]
23. Resch-Genger, U.; Grabolle, M.; Cavaliere-Jaricot, S.; Nitschke, R.; Nann, T. Quantum dots versus organic dyes as fluorescent labels. *Nat. Methods* **2008**, *5*, 763–775. [CrossRef] [PubMed]
24. Liu, T.; Li, N.; Dong, J.X.; Luo, H.Q.; Li, N.B. Fluorescence detection of mercury ions and cysteine based on magnesium and nitrogen co-doped carbon quantum dots and IMPLICATION logic gate operation. *Sens. Actuators B Chem.* **2016**, *231*, 147–153. [CrossRef]
25. Pajewska-Szmyt, M.; Buszewski, B.; Gadzała-Kopciuch, R. Sulphur and nitrogen doped carbon dots synthesis by microwave assisted method as quantitative analytical nano-tool for mercury ion sensing. *Mater. Chem. Phys.* **2020**, *242*, 122484. [CrossRef]
26. Tabish, T.A.; Hayat, H.; Abbas, A.; Narayan, R.J. Graphene Quantum Dots-Based Electrochemical Biosensing Platform for Early Detection of Acute Myocardial Infarction. *Biosensors* **2022**, *12*, 77. [CrossRef]
27. Javed, N.; O'Carroll, D.M. Carbon dots and stability of their optical properties. *Part. Part. Syst. Charact.* **2021**, *38*, 200027_. [CrossRef]
28. Wang, X.; Feng, Y.; Dong, P.; Huang, J. A mini review on carbon quantum dots: Preparation, properties, and electrocatalytic application. *Front. Chem.* **2019**, *7*, 671. [CrossRef]
29. Jagannathan, M.; Dhinasekaran, D.; Soundharraj, P.; Rajendran, S.; Vo, D.-V.N.; Prakasarao, A.; Ganesan, S. Green synthesis of white light emitting carbon quantum dots: Fabrication of white fluorescent film and optical sensor applications. *J. Hazard. Mater.* **2021**, *416*, 125091. [CrossRef]
30. Kim, S.; Hwang, S.W.; Kim, M.-K.; Shin, D.Y.; Shin, D.H.; Kim, C.O.; Yang, S.B.; Park, J.H.; Hwang, E.; Choi, S.-H. Anomalous behaviors of visible luminescence from graphene quantum dots: Interplay between size and shape. *ACS Nano* **2012**, *6*, 8203–8208. [CrossRef]
31. Ahirwar, S.; Mallick, S.; Bahadur, D. Electrochemical method to prepare graphene quantum dots and graphene oxide quantum dots. *ACS Omega* **2017**, *2*, 8343–8353. [CrossRef] [PubMed]
32. Kumar, P.; Dhand, C.; Dwivedi, N.; Singh, S.; Khan, R.; Verma, S.; Singh, A.; Gupta, M.K.; Kumar, S.; Kumar, R. Graphene quantum dots: A contemporary perspective on scope, opportunities, and sustainability. *Renew. Sustain. Energy Rev.* **2022**, *157*, 111993. [CrossRef]
33. Sheely, A.; Gifford, B.; Tretiak, S.; Bishop, A. Tunable Optical Features of Graphene Quantum Dots from Edge Functionalization. *J. Phys. Chem. C* **2021**, *125*, 9244–9252. [CrossRef]
34. Tang, S.; Chen, D.; Wang, C.; Yang, Y.; Li, X.; Li, T.; Zhang, X. Preparation of concentration dependent nitrogen sulfur Co-doped graphene quantum dots by UV irradiation and reversible optical switch of hydroxypropyl methyl cellulose. *Dye. Pigment.* **2022**, *198*, 110022. [CrossRef]
35. Alas, M.O.; Alkas, F.B.; Aktas Sukuroglu, A.; Genc Alturk, R.; Battal, D. Fluorescent carbon dots are the new quantum dots: An overview of their potential in emerging technologies and nanosafety. *J. Mater. Sci.* **2020**, *55*, 15074–15105. [CrossRef]
36. Vibhute, A.; Patil, T.; Gambhir, R.; Tiwari, A.P. Fluorescent carbon quantum dots: Synthesis methods, functionalization and biomedical applications. *Appl. Surf. Sci. Adv.* **2022**, *11*, 100311. [CrossRef]
37. Gulati, S.; Baul, A.; Amar, A.; Wadhwa, R.; Kumar, S.; Varma, R.S. Eco-Friendly and Sustainable Pathways to Photoluminescent Carbon Quantum Dots (CQDs). *Nanomaterials* **2023**, *13*, 554. [CrossRef]
38. Xu, A.; Wang, G.; Li, Y.; Dong, H.; Yang, S.; He, P.; Ding, G. Carbon-based quantum dots with solid-state photoluminescent: Mechanism, implementation, and application. *Small* **2020**, *16*, 2004621. [CrossRef]
39. Gaurav, A.; Jain, A.; Tripathi, S.K. Review on Fluorescent Carbon/Graphene Quantum Dots: Promising Material for Energy Storage and Next-Generation Light-Emitting Diodes. *Materials* **2022**, *15*, 7888. [CrossRef]
40. Magesh, V.; Sundramoorthy, A.K.; Ganapathy, D. Recent advances on synthesis and potential applications of carbon quantum dots. *Front. Mater.* **2022**, *9*, 906838. [CrossRef]
41. Zhu, S.; Meng, Q.; Wang, L.; Zhang, J.; Song, Y.; Jin, H.; Zhang, K.; Sun, H.; Wang, H.; Yang, B. Highly photoluminescent carbon dots for multicolor patterning, sensors, and bioimaging. *Angew. Chem. Int. Ed.* **2013**, *52*, 3953–3957. [CrossRef] [PubMed]
42. Jing, S.; Zhao, Y.; Sun, R.-C.; Zhong, L.; Peng, X. Facile and high-yield synthesis of carbon quantum dots from biomass-derived carbons at mild condition. *ACS Sustain. Chem. Eng.* **2019**, *7*, 7833–7843. [CrossRef]
43. Kalluri, A.; Debnath, D.; Dharmadhikari, B.; Patra, P. Chapter Twelve—Graphene Quantum Dots: Synthesis and Applications. In *Methods in Enzymology*; Kumar, C.V., Ed.; Academic Press: Cambridge, MA, USA, 2018; pp. 335–354.
44. Yuan, F.; Li, S.; Fan, Z.; Meng, X.; Fan, L.; Yang, S. Shining carbon dots: Synthesis and biomedical and optoelectronic applications. *Nano Today* **2016**, *11*, 565–586. [CrossRef]

45. Fernando, K.A.S.; Sahu, S.; Liu, Y.; Lewis, W.K.; Guliants, E.A.; Jafariyan, A.; Wang, P.; Bunker, C.E.; Sun, Y.-P. Carbon quantum dots and applications in photocatalytic energy conversion. *ACS Appl. Mater. Interfaces* **2015**, *7*, 8363–8376. [CrossRef] [PubMed]
46. Luo, Z.; Qi, G.; Chen, K.; Zou, M.; Yuwen, L.; Zhang, X.; Huang, W.; Wang, L. Microwave-assisted preparation of white fluorescent graphene quantum dots as a novel phosphor for enhanced white-light-emitting diodes. *Adv. Funct. Mater.* **2016**, *26*, 2739–2744. [CrossRef]
47. Zhuo, S.; Shao, M.; Lee, S.-T. Upconversion and downconversion fluorescent graphene quantum dots: Ultrasonic preparation and photocatalysis. *ACS Nano* **2012**, *6*, 1059–1064. [CrossRef]
48. Li, L.L.; Ji, J.; Fei, R.; Wang, C.Z.; Lu, Q.; Zhang, J.R.; Jiang, L.P.; Zhu, J.J. A facile microwave avenue to electrochemiluminescent two-color graphene quantum dots. *Adv. Funct. Mater.* **2012**, *22*, 2971–2979. [CrossRef]
49. Rajender, G.; Giri, P.K. Formation mechanism of graphene quantum dots and their edge state conversion probed by photoluminescence and Raman spectroscopy. *J. Mater. Chem. C* **2016**, *4*, 10852–10865. [CrossRef]
50. Peng, J.; Gao, W.; Gupta, B.K.; Liu, Z.; Romero-Aburto, R.; Ge, L.; Song, L.; Alemany, L.B.; Zhan, X.; Gao, G.; et al. Graphene Quantum Dots Derived from Carbon Fibers. *Nano Lett.* **2012**, *12*, 844–849. [CrossRef]
51. Dong, Y.; Chen, C.; Zheng, X.; Gao, L.; Cui, Z.; Yang, H.; Guo, C.; Chi, Y.; Li, C.M. One-step and high yield simultaneous preparation of single- and multi-layer graphene quantum dots from CX-72 carbon black. *J. Mater. Chem.* **2012**, *22*, 8764–8766. [CrossRef]
52. Jin, H.; Huang, H.; He, Y.; Feng, X.; Wang, S.; Dai, L.; Wang, J. Graphene Quantum Dots Supported by Graphene Nanoribbons with Ultrahigh Electrocatalytic Performance for Oxygen Reduction. *J. Am. Chem. Soc.* **2015**, *137*, 7588–7591. [CrossRef] [PubMed]
53. Abbas, A.; Tabish, T.A.; Bull, S.J.; Lim, T.M.; Phan, A.N. High yield synthesis of graphene quantum dots from biomass waste as a highly selective probe for Fe^{3+} sensing. *Sci. Rep.* **2020**, *10*, 21262. [CrossRef] [PubMed]
54. Ghaffarkhah, A.; Hosseini, E.; Kamkar, M.; Sehat, A.A.; Dordanihaghighi, S.; Allahbakhsh, A.; van der Kuur, C.; Arjmand, M. Synthesis, applications, and prospects of graphene quantum dots: A comprehensive review. *Small* **2022**, *18*, 2102683. [CrossRef] [PubMed]
55. Somaraj, G.; Mathew, S.; Abraham, T.; Ambady, K.G.; Mohan, C.; Mathew, B. Nitrogen and Sulfur Co-Doped Carbon Quantum Dots for Sensing Applications: A Review. *ChemistrySelect* **2022**, *7*, e202200473. [CrossRef]
56. Liu, Y.; Jiang, L.; Li, B.; Fan, X.; Wang, W.; Liu, P.; Xu, S.; Luo, X. Nitrogen doped carbon dots: Mechanism investigation and their application for label free CA125 analysis. *J. Mater. Chem. B* **2019**, *7*, 3053–3058. [CrossRef]
57. Nguyen, K.G.; Baragau, I.-A.; Gromicova, R.; Nicolaev, A.; Thomson, S.A.J.; Rennie, A.; Power, N.P.; Sajjad, M.T.; Kellici, S. Investigating the effect of N-doping on carbon quantum dots structure, optical properties and metal ion screening. *Sci. Rep.* **2022**, *12*, 13806. [CrossRef]
58. Yang, S.; Sun, J.; Li, X.; Zhou, W.; Wang, Z.; He, P.; Ding, G.; Xie, X.; Kang, Z.; Jiang, M. Large-scale fabrication of heavy doped carbon quantum dots with tunable-photoluminescence and sensitive fluorescence detection. *J. Mater. Chem. A* **2014**, *2*, 8660–8667. [CrossRef]
59. Cheng, Z.; Du, F.; Sun, L.; Jiang, L.; Ruan, G.; Li, J. Nitrogen-Doped Carbon Quantum Dots as a "Turn-Off" Fluorescent Probes for Highly Selective and Sensitive Detection of Mercury (II) Ions. *ChemistrySelect* **2019**, *4*, 2122–2128. [CrossRef]
60. Luo, L.; Wang, P.; Wang, Y.; Wang, F. pH assisted selective detection of Hg (II) and Ag (I) based on nitrogen-rich carbon dots. *Sens. Actuators B Chem.* **2018**, *273*, 1640–1647. [CrossRef]
61. Guo, H.; Raj, J.; Wang, Z.; Zhang, T.; Wang, K.; Lin, L.; Hou, W.; Zhang, J.; Wu, M.; Wu, J. Synergistic Effects of Amine Functional Groups and Enriched-Atomic-Iron Sites in Carbon Dots for Industrial-Current–Density CO_2 Electroreduction. *Small* **2024**, *20*, 2311132. [CrossRef]
62. Gao, H.; Liu, Z.; Song, L.; Guo, W.; Gao, W.; Ci, L.; Rao, A.; Quan, W.; Vajtai, R.; Ajayan, P.M. Synthesis of S-doped graphene by liquid precursor. *Nanotechnology* **2012**, *23*, 275605. [CrossRef] [PubMed]
63. Ge, J.; Jia, Q.; Liu, W.; Guo, L.; Liu, Q.; Lan, M.; Zhang, H.; Meng, X.; Wang, P. Red-emissive carbon dots for fluorescent, photoacoustic, and thermal theranostics in living mice. *Adv. Mater.* **2015**, *27*, 4169–4177. [CrossRef]
64. Wu, F.; Yang, M.; Zhang, H.; Zhu, S.; Zhu, X.; Wang, K. Facile synthesis of sulfur-doped carbon quantum dots from vitamin B1 for highly selective detection of Fe^{3+} ion. *Opt. Mater.* **2018**, *77*, 258–263. [CrossRef]
65. Li, S.; Li, Y.; Cao, J.; Zhu, J.; Fan, L.; Li, X. Sulfur-Doped Graphene Quantum Dots as a Novel Fluorescent Probe for Highly Selective and Sensitive Detection of Fe^{3+}. *Anal. Chem.* **2014**, *86*, 10201–10207. [CrossRef] [PubMed]
66. Park, C.H.; Yang, H.; Lee, J.; Cho, H.-H.; Kim, D.; Lee, D.C.; Kim, B.J. Multicolor Emitting Block Copolymer-Integrated Graphene Quantum Dots for Colorimetric, Simultaneous Sensing of Temperature, pH, and Metal Ions. *Chem. Mater.* **2015**, *27*, 5288–5294. [CrossRef]
67. Shen, X.; Wang, Z.; Guo, H.; Lei, Z.; Liu, Z.; Wang, L. Solvent engineering of oxygen-enriched carbon dots for efficient electrochemical hydrogen peroxide production. *Small* **2023**, *19*, 2303156. [CrossRef]
68. Hu, B.; Huang, K.; Tang, B.; Lei, Z.; Wang, Z.; Guo, H.; Lian, C.; Liu, Z.; Wang, L. Graphene quantum dot-mediated atom-layer semiconductor electrocatalyst for hydrogen evolution. *Nano-Micro Lett.* **2023**, *15*, 217. [CrossRef] [PubMed]
69. Yarur, F.; Macairan, J.-R.; Naccache, R. Ratiometric detection of heavy metal ions using fluorescent carbon dots. *Environ. Sci. Nano* **2019**, *6*, 1121–1130. [CrossRef]
70. Li, C.; Liu, W.; Ren, Y.; Sun, X.; Pan, W.; Wang, J. The selectivity of the carboxylate groups terminated carbon dots switched by buffer solutions for the detection of multi-metal ions. *Sens. Actuators B Chem.* **2017**, *240*, 941–948. [CrossRef]

71. Bhamore, J.R.; Park, T.J.; Kailasa, S.K. Glutathione-capped Syzygium cumini carbon dot-amalgamated agarose hydrogel film for naked-eye detection of heavy metal ions. *J. Anal. Sci. Technol.* **2020**, *11*, 13. [CrossRef]
72. Amjadi, M.; Jalili, R. A molecularly imprinted dual-emission carbon dot-quantum dot mesoporous hybrid for ratiometric determination of anti-inflammatory drug celecoxib. *Spectrochim. Acta Part A Mol. Biomol. Spectrosc.* **2018**, *191*, 345–351. [CrossRef]
73. Shahshahanipour, M.; Rezaei, B.; Ensafi, A.A.; Etemadifar, Z. An ancient plant for the synthesis of a novel carbon dot and its applications as an antibacterial agent and probe for sensing of an anti-cancer drug. *Mater. Sci. Eng. C* **2019**, *98*, 826–833. [CrossRef] [PubMed]
74. Campos, B.B.; Contreras-Cáceres, R.; Bandosz, T.J.; Jiménez-Jiménez, J.; Rodríguez-Castellón, E.; da Silva, J.C.E.; Algarra, M. Carbon dots as fluorescent sensor for detection of explosive nitrocompounds. *Carbon* **2016**, *106*, 171–178. [CrossRef]
75. Babar, D.G.; Garje, S.S. Nitrogen and phosphorus co-doped carbon dots for selective detection of nitro explosives. *ACS Omega* **2020**, *5*, 2710–2717. [CrossRef]
76. Li, J.; Du, B.; Li, Y.; Wang, Y.; Wu, D.; Wei, Q. A turn-on fluorescent sensor for highly sensitive mercury (II) detection based on a carbon dot-labeled oligodeoxyribonucleotide and MnO_2 nanosheets. *New J. Chem.* **2018**, *42*, 1228–1234. [CrossRef]
77. Mandal, P.; Sahoo, D.; Sarkar, P.; Chakraborty, K.; Das, S. Fluorescence turn-on and turn-off sensing of pesticides by carbon dot-based sensor. *New J. Chem.* **2019**, *43*, 12137–12151. [CrossRef]
78. Ashrafi Tafreshi, F.; Fatahi, Z.; Ghasemi, S.F.; Taherian, A.; Esfandiari, N. Ultrasensitive fluorescent detection of pesticides in real sample by using green carbon dots. *PLoS ONE* **2020**, *15*, e0230646. [CrossRef] [PubMed]
79. Moonrinta, S.; Kwon, B.; In, I.; Kladsomboon, S.; Sajomsang, W.; Paoprasert, P. Highly biocompatible yogurt-derived carbon dots as multipurpose sensors for detection of formic acid vapor and metal ions. *Opt. Mater.* **2018**, *81*, 93–101. [CrossRef]
80. Gu, J.; Hu, D.; Wang, W.; Zhang, Q.; Meng, Z.; Jia, X.; Xi, K. Carbon dot cluster as an efficient "off–on" fluorescent probe to detect Au (III) and glutathione. *Biosens. Bioelectron.* **2015**, *68*, 27–33. [CrossRef]
81. Maiti, S.; Das, K.; Das, P.K. Label-free fluorimetric detection of histone using quaternized carbon dot–DNA nanobiohybrid. *Chem. Commun.* **2013**, *49*, 8851–8853. [CrossRef]
82. Freire, R.; Le, N.D.; Jiang, Z.; Kim, C.S.; Rotello, V.M.; Fechine, P. NH_2-rich Carbon Quantum Dots: A protein-responsive probe for detection and identification. *Sens. Actuators B Chem.* **2018**, *255*, 2725–2732. [CrossRef]
83. Godavarthi, S.; Kumar, K.M.; Vélez, E.V.; Hernandez-Eligio, A.; Mahendhiran, M.; Hernandez-Como, N.; Aleman, L.; Gomez, L.M. Nitrogen doped carbon dots derived from Sargassum fluitans as fluorophore for DNA detection. *J. Photochem. Photobiol. B Biol.* **2017**, *172*, 36–41. [CrossRef] [PubMed]
84. Bhattacharya, S.; Nandi, S.; Jelinek, R. Carbon-dot–hydrogel for enzyme-mediated bacterial detection. *RSC Adv.* **2017**, *7*, 588–594. [CrossRef]
85. Ahmadian-Fard-Fini, S.; Ghanbari, D.; Salavati-Niasari, M. Photoluminescence carbon dot as a sensor for detecting of Pseudomonas aeruginosa bacteria: Hydrothermal synthesis of magnetic hollow $NiFe_2O_4$-carbon dots nanocomposite material. *Compos. Part B Eng.* **2019**, *161*, 564–577. [CrossRef]
86. Cao, J.; Zhang, H.; Nian, Q.; Xu, Q. Electrospun chitosan/polyethylene oxide nanofibers mat loaded with copper (II) as a new sensor for colorimetric detection of tetracycline. *Int. J. Biol. Macromol.* **2022**, *212*, 527–535. [CrossRef]
87. Msto, R.K.; Othman, H.O.; Al-Hashimi, B.R.; Salahuddin Ali, D.; Hassan, D.H.; Hassan, A.Q.; Smaoui, S. Fluorescence Turns on-off-on Sensing of Ferric Ion and L-Ascorbic Acid by Carbon Quantum Dots. *J. Food Qual.* **2023**, *2023*, 5555608. [CrossRef]
88. Han, Y.; Shi, L.; Luo, X.; Chen, X.; Yang, W.; Tang, W.; Wang, J.; Yue, T.; Li, Z. A signal-on fluorescent sensor for ultra-trace detection of Hg^{2+} via Ag^{+} mediated sulfhydryl functionalized carbon dots. *Carbon* **2019**, *149*, 355–363. [CrossRef]
89. Gan, Z.; Hu, X.; Huang, X.; Li, Z.; Zou, X.; Shi, J.; Zhang, W.; Li, Y.; Xu, Y. A dual-emission fluorescence sensor for ultrasensitive sensing mercury in milk based on carbon quantum dots modified with europium (III) complexes. *Sens. Actuators B Chem.* **2021**, *328*, 128997. [CrossRef]
90. Ahmadian-Fard-Fini, S.; Ghanbari, D.; Amiri, O.; Salavati-Niasari, M. Electro-spinning of cellulose acetate nanofibers/Fe/carbon dot as photoluminescence sensor for mercury (II) and lead (II) ions. *Carbohydr. Polym.* **2020**, *229*, 115428. [CrossRef]
91. Ghanem, A.; Al-Marjeh, R.A.-Q.B.; Atassi, Y. Novel nitrogen-doped carbon dots prepared under microwave-irradiation for highly sensitive detection of mercury ions. *Heliyon* **2020**, *6*, e03750. [CrossRef]
92. Muthurasu, A.; Ganesh, V. Tuning optical properties of nitrogen-doped carbon dots through fluorescence resonance energy transfer using Rhodamine B for the ratiometric sensing of mercury ions. *Anal. Methods* **2021**, *13*, 1857–1865. [CrossRef]
93. Yahyazadeh, E.; Shemirani, F. Easily synthesized carbon dots for determination of mercury (II) in water samples. *Heliyon* **2019**, *5*, e01596. [PubMed]
94. He, J.H.; Cheng, Y.Y.; Yang, T.; Zou, H.Y.; Huang, C.Z. Functional preserving carbon dots-based fluorescent probe for mercury (II) ions sensing in herbal medicines via coordination and electron transfer. *Anal. Acta* **2018**, *1035*, 203–210. [CrossRef] [PubMed]
95. Liao, S.; Zhang, L.; Li, S.; Yue, S.; Wang, G. Selective detection of mercury ions via single and dual signals by silicon-doped carbon quantum dots. *New J. Chem.* **2023**, *47*, 14242–14248. [CrossRef]
96. Chaghaghazardi, M.; Kashanian, S.; Nazari, M.; Omidfar, K.; Joseph, Y.; Rahimi, P. Nitrogen and sulfur co-doped carbon quantum dots fluorescence quenching assay for detection of mercury (II). *Spectrochim. Acta Part A Mol. Biomol. Spectrosc.* **2023**, *293*, 122448. [CrossRef]
97. Omer, K.M.; Aziz, K.H.H.; Mohammed, S.J. Improvement of selectivity via the surface modification of carbon nanodots towards the quantitative detection of mercury ions. *New J. Chem.* **2019**, *43*, 12979–12986. [CrossRef]

98. Lu, C.; Ding, H.; Wang, Y.; Xiong, C.; Wang, X. Colorimetric and turn-on fluorescence determination of mercury (II) by using carbon dots and gold nanoparticles. *Nanotechnology* **2021**, *32*, 155501. [CrossRef]
99. Li, B.; Ma, H.; Zhang, B.; Qian, J.; Cao, T.; Feng, H.; Li, W.; Dong, Y.; Qin, W. Dually emitting carbon dots as fluorescent probes for ratiometric fluorescent sensing of pH values, mercury (II), chloride and Cr (VI) via different mechanisms. *Microchim. Acta* **2019**, *186*, 341. [CrossRef]
100. Ren, H.; Labidi, A.; Sun, J.; Allam, A.A.; Ajarem, J.S.; Abukhadra, M.R.; Wang, C. Facile synthesis of nitrogen, sulfur co-doped carbon quantum dots for selective detection of mercury (II). *Environ. Chem. Lett.* **2024**, *22*, 35–41. [CrossRef]
101. Hao, X.; Dai, S.; Wang, J.; Fang, Z. Synthesis of blue fluorescent carbon dots and their application in detecting mercury and iodine based on "off–on" mode. *Luminescence* **2021**, *36*, 721–732. [CrossRef]
102. Wang, S.; Chen, H.; Xie, H.; Wei, L.; Xu, L.; Zhang, L.; Lan, W.; Zhou, C.; She, Y.; Fu, H. A novel thioctic acid-carbon dots fluorescence sensor for the detection of Hg^{2+} and thiophanate methyl via S-Hg affinity. *Food Chem.* **2021**, *346*, 128923. [CrossRef] [PubMed]
103. Ge, L.; Hu, G.; Zhao, F.; Wang, X.; Ma, Z.; Liu, R. Carbon dots prepared by thermal reactions and selective detections of copper and mercury ions in visible spectrum. *Appl. Phys. A* **2021**, *127*, 388. [CrossRef]
104. Tang, W.; Wang, Y.; Wang, P.; Di, J.; Yang, J.; Wu, Y. Synthesis of strongly fluorescent carbon quantum dots modified with polyamidoamine and a triethoxysilane as quenchable fluorescent probes for mercury (II). *Microchim. Acta* **2016**, *183*, 2571–2578. [CrossRef]
105. Achadu, O.J.; Revaprasadu, N. Microwave-assisted synthesis of thymine-functionalized graphitic carbon nitride quantum dots as a fluorescent nanoprobe for mercury (II). *Microchim. Acta* **2018**, *185*, 461. [CrossRef] [PubMed]
106. Karami, C.; Taher, M.A.; Shahlaei, M. A simple method for determination of mercury (II) ions by PNBS-doped carbon dots as a fluorescent probe. *J. Mater. Sci. Mater. Electron.* **2020**, *31*, 5975–5983. [CrossRef]
107. Lan, M.; Zhang, J.; Chui, Y.-S.; Wang, P.; Chen, X.; Lee, C.-S.; Kwong, H.-L.; Zhang, W. Carbon nanoparticle-based ratiometric fluorescent sensor for detecting mercury ions in aqueous media and living cells. *ACS Appl. Mater. Interfaces* **2014**, *6*, 21270–21278. [CrossRef]
108. Aziz, K.H.H.; Omer, K.M.; Hamarawf, R.F. Lowering the detection limit towards nanomolar mercury ion detection via surface modification of N-doped carbon quantum dots. *New J. Chem.* **2019**, *43*, 8677–8683. [CrossRef]
109. Guo, H.; Wang, X.; Wu, N.; Xu, M.; Wang, M.; Zhang, L.; Yang, W. In-situ synthesis of carbon dots-embedded europium metal-organic frameworks for ratiometric fluorescence detection of Hg^{2+} in aqueous environment. *Anal. Acta* **2021**, *1141*, 13–20. [CrossRef]
110. Xie, Y.; Cheng, D.; Liu, X.; Han, A. Green hydrothermal synthesis of N-doped carbon dots from biomass highland barley for the detection of Hg^{2+}. *Sensors* **2019**, *19*, 3169. [CrossRef]
111. Gómez Pineros, B.S.; Granados-Oliveros, G. L-cysteine modified N-doped carbon quantum dots derived from peach palm (Bactris gasipaes) peels for detection of mercury ions. *J. Mol. Struct.* **2024**, *1317*, 138990. [CrossRef]
112. Ananthanarayanan, A.; Wang, X.; Routh, P.; Sana, B.; Lim, S.; Kim, D.H.; Lim, K.H.; Li, J.; Chen, P. Facile synthesis of graphene quantum dots from 3D graphene and their application for Fe^{3+} sensing. *Adv. Funct. Mater.* **2014**, *24*, 3021–3026. [CrossRef]
113. Gao, B.; Chen, D.; Gu, B.; Wang, T.; Wang, Z.; Yang, Y.; Guo, Q.; Wang, G. Facile and highly effective synthesis of nitrogen-doped graphene quantum dots as a fluorescent sensing probe for Cu^{2+} detection. *Curr. Appl. Phys.* **2020**, *20*, 538–544. [CrossRef]
114. Kadian, S.; Sethi, S.K.; Manik, G. Recent advancements in synthesis and property control of graphene quantum dots for biomedical and optoelectronic applications. *Mater. Chem. Front.* **2021**, *5*, 627–658. [CrossRef]
115. Kongsanan, N.; Pimsin, N.; Keawprom, C.; Sricharoen, P.; Areerob, Y.; Nuengmatcha, P.; Oh, W.-C.; Chanthai, S.; Limchoowong, N. A fluorescence switching sensor for sensitive and selective detections of cyanide and ferricyanide using mercuric cation-graphene quantum dots. *ACS Omega* **2021**, *6*, 14379–14393. [CrossRef]
116. Li, J.; Wang, Z.; Yang, J.; Xia, X.; Yi, R.; Jiang, J.; Liu, W.; Chen, J.; Chen, L.; Xu, J. "On-off-on" fluorescence switch of graphene quantum dots: A cationic control strategy. *Appl. Surf. Sci.* **2021**, *546*, 149110. [CrossRef]
117. Gu, S.; Hsieh, C.-T.; Tsai, Y.-Y.; Ashraf Gandomi, Y.; Yeom, S.; Kihm, K.D.; Fu, C.-C.; Juang, R.-S. Sulfur and nitrogen co-doped graphene quantum dots as a fluorescent quenching probe for highly sensitive detection toward mercury Ions. *ACS Appl. Nano Mater.* **2019**, *2*, 790–798. [CrossRef]
118. Liu, Y.; Tang, X.; Deng, M.; Cao, Y.; Li, Y.; Zheng, H.; Li, F.; Yan, F.; Lan, T.; Shi, L. Nitrogen doped graphene quantum dots as a fluorescent probe for mercury (II) ions. *Microchim. Acta* **2019**, *186*, 140. [CrossRef]
119. Anh, N.T.N.; Chowdhury, A.D.; Doong, R.-A. Highly sensitive and selective detection of mercury ions using N, S-codoped graphene quantum dots and its paper strip based sensing application in wastewater. *Sens. Actuators B Chem.* **2017**, *252*, 1169–1178. [CrossRef]
120. Yang, Y.; Xiao, X.; Xing, X.; Wang, Z.; Zou, T.; Wang, Z.; Zhao, R.; Wang, Y. One-pot synthesis of N-doped graphene quantum dots as highly sensitive fluorescent sensor for detection of mercury ions water solutions. *Mater. Res. Express* **2019**, *6*, 095615. [CrossRef]
121. Zhu, Q.; Mao, H.; Li, J.; Hua, J.; Wang, J.; Yang, R.; Li, Z. A glycine-functionalized graphene quantum dots synthesized by a facile post-modification strategy for a sensitive and selective fluorescence sensor of mercury ions. *Spectrochim. Acta Part A Mol. Biomol. Spectrosc.* **2021**, *247*, 119090. [CrossRef]

122. Peng, D.; Zhang, L.; Liang, R.-P.; Qiu, J.-D. Rapid detection of mercury ions based on nitrogen-doped graphene quantum dots accelerating formation of manganese porphyrin. *ACS Sens.* **2018**, *3*, 1040–1047. [CrossRef]
123. Chini, M.K.; Kumar, V.; Javed, A.; Satapathi, S. Graphene quantum dots and carbon nano dots for the FRET based detection of heavy metal ions. *Nano-Struct. Nano-Objects* **2019**, *19*, 100347. [CrossRef]

MDPI AG
Grosspeteranlage 5
4052 Basel
Switzerland
Tel.: +41 61 683 77 34

Photonics Editorial Office
E-mail: photonics@mdpi.com
www.mdpi.com/journal/photonics

www.ingramcontent.com/pod-product-compliance
Lightning Source LLC
Chambersburg PA
CBHW041431210926
51850CB00013B/3398
* 9 7 8 3 7 2 5 8 4 1 0 5 9 *